INTERPRETATION OF MS–MS MASS SPECTRA OF DRUGS AND PESTICIDES

INTERPRETATION OF MS–MS MASS SPECTRA OF DRUGS AND PESTICIDES

WILFRIED M. A. NIESSEN
hyphen MassSpec, The Netherlands

RICARDO A. CORREA C.
Trans-Laboratory, Brussels, Belgium

WILEY

This edition first published 2017

Registered Office
John Wiley & Sons, Inc., 111 River Street, Hoboken, NJ 07030, USA

Editorial Office
111 River Street, Hoboken, NJ 07030, USA

For details of our global editorial offices, customer services, and more information about Wiley products visit us at www.wiley.com.

Wiley also publishes its books in a variety of electronic formats and by print-on-demand. Some content that appears in standard print versions of this book may not be available in other formats.

Library of Congress Cataloging-in-Publication Data:

Names: Niessen, W. M. A. (Wilfried M. A.), 1956- author. | Correa C., Ricardo A., 1961- author.
Title: Interpretation of MS-MS mass spectra of drugs and pesticides / Wilfried M.A. Niessen, Ricardo A. Correa C.
Other titles: Wiley-Interscience series on mass spectrometry.
Description: Hoboken, New Jersey : John Wiley & Sons, 2016. | Series: Wiley series on mass spectrometry | Includes bibliographical references and index.
Identifiers: LCCN 2016031593 (print) | LCCN 2016046526 (ebook) | ISBN 9781118500187 (cloth) | ISBN 9781119294245 (pdf) | ISBN 9781119294252 (epub)
Subjects: LCSH: Tandem mass spectrometry. | Liquid chromatography. | Drugs–Analysis. | Pesticides–Analysis.
Classification: LCC QD96.M3 N525 2016 (print) | LCC QD96.M3 (ebook) | DDC 543/.65–dc23
LC record available at https://lccn.loc.gov/2016031593

Cover Design by Wiley.
Cover image: Curtsey of the authors

Set in 10/12pt, TimesLTStd by SPi Global, Chennai, India.

Printed in the United States of America

10 9 8 7 6 5 4 3 2 1

CONTENTS

PREFACE

In the 1980s, tandem mass spectrometry was introduced for the structural elucidation of even-electron ions (protonated or deprotonated molecules) generated by soft ionization techniques such as fast-atom bombardment, thermospray, and electrospray. When compared to the fragmentation of odd-electron ions generated by electron ionization, scientists were well aware of the fact that different rules apply to the fragmentation of even-electron ions. Surprisingly, no major fundamental research was carried out on trying to understand and describe these differences. More effort was placed on the development of improved instrumentation and advanced applications for the emerging technologies. This particular effort paid off, as exemplified by tandem mass spectrometry which, often in combination with gas or liquid chromatography, has been a major contributor to the progress of many scientific disciplines, for example, pharmaceutical, biochemical, and environmental sciences; food safety; sports doping analysis; clinical diagnostics; forensics; and toxicology.

This work is an attempt to add to the understanding of the fragmentation of even-electron ions. This has been done by studying the fragmentation of a wide variety of compounds, with a special focus on chemical structure similarities, that is, from the same class. The basic data set used comprises a number of mass spectral libraries developed for general unknown screening in toxicology. In this respect, we need to thank Dr Wolfgang Weinmann (originally at the Institute of Legal Medicine, University of Freiburg, Germany, and currently at the Institute of Forensic Medicine, University of Bern, Switzerland) for providing public access to his toxicology library and the library of designer drugs via the Internet (http://www.chemicalsoft.de/index.html); Dr Pierre Marquet (of the Faculty of Medicine, Department of Pharmacology, Toxicology, and Pharmacovigilance at the University Hospital of Limoges, France) for providing his mass spectral library of negative-ion mass spectra; and Dr Bernhard Wüst of Agilent Technologies for his help with using the Agilent Broecker, Herre & Pragst PCDL for forensic toxicology. The information from these libraries and other data sets is complemented by data from the scientific literature.

The origins of this book can be found in two publications describing the fragmentation of toxicologically relevant drugs in both positive-ion tandem mass spectrometry (Niessen, 2011) and negative-ion tandem mass spectrometry (Niessen, 2012). Soon after, the authors decided to develop the project further by extending the number of compounds covered and the detail of the information provided. The fragmentation of some 1300 compounds and the product-ion mass spectra of even more are studied and interpreted in this book.

This volume consists of five chapters. Chapters 3 and 4 are the main chapters, where proposed fragmentation rules for the "Fragmentation of Even-Electron Ions" (Chapter 3) are derived from the behavior of the "Fragmentation of Drugs and Pesticides" (Chapter 4) pertaining to many different classes of compounds. Chapter 1, "Introduction to LC–MS–MS Technology", provides a concise introduction to mass spectrometry technology. Chapter 2, "Interpretation of Mass Spectra" gives the basic concepts and definitions related to the information that can be extracted from mass spectra. Finally, Chapter 5, "Identification Strategies" gives an overview of the different classes of unknowns and identification strategies that exist as well as how they relate to multiple areas of application.

Last but not least, special thanks go to our families, and the many people who have inspired us to continue working on this project. We hope that you, as our reader, find this material useful and inspirational to further extend our understanding of the fragmentation of even-electron ions in tandem mass spectrometry.

WILFRIED M. A. NIESSEN
hyphen MassSpec
Herenweg 95, 2361 EK Warmond, The Netherlands
mail@hyphenms.nl; www.hyphenms.nl

RICARDO A. CORREA C.
Trans-Laboratory
Rue François Stroobant 41, 1050 Brussels, Belgium
ricardo.correa@translaboratory.com;
www.translaboratory.com

REFERENCES

Niessen WMA. 2011. Fragmentation of toxicologically relevant drugs in positive-ion liquid chromatography–tandem mass spectrometry. Mass Spectrom Rev, 30: 626–663.

Niessen WMA. 2012. Fragmentation of toxicologically relevant drugs in negative-ion liquid chromatography–tandem mass spectrometry. Mass Spectrom Rev, 31: 626–665.

ABBREVIATIONS

AC	alternating current potential
ADC	analog-to-digital converter
APCI	atmospheric-pressure chemical ionization
API	atmospheric-pressure ionization
APPI	atmospheric-pressure photoionization
BPC	base-peak chromatogram
CE	charge exchange
CECI	charge-exchange chemical ionization
CI	chemical ionization
CID	collision-induced dissociation
CIS	coordination electrospray ionization
CLND	chemiluminescence nitrogen detector
CNS	central nervous system
CPA	chlorinated phenoxy acid
CRF	charge-remote fragmentation
CRM	charge-residue model
DC	direct current potential
DDA	data-dependent acquisition
DESI	desorption electrospray ionization
DIA	data-independent acquisition
EA	electron affinity
ECD	electron-capture dissociation
ECNI	electron-capture negative ionization
EE^+ and EE^-	even-electron ion
EI	electron ionization
EM	electron multiplier
EPA	environmental protection agency
EPI	enhanced product-ion
ESI	electrospray ionization
ETD	electron-transfer dissociation
EU	European Union
FAB	fast-atom bombardment ionization
FAIMS	high-field asymmetric waveform ion mobility spectrometry
FDI	field-desorption ionization
FT-ICR	Fourier-transform ion cyclotron resonance
FWHM	full peak width at half maximum height
GC	gas chromatography
GC–MS	gas chromatography–mass spectrometry
G_R	reagent gas
H/D exchange	hydrogen/deuterium exchange
HCD	higher-energy collision-induced dissociation
HFBPC	(*S*)-(−)-*N*-(heptafluorobutanoyl)prolyl chloride
HILIC	hydrophilic interaction (liquid) chromatography
HIV	human immunodeficiency virus
HRAM-MS	high-resolution accurate-mass mass spectrometry
IE	ionization energy
IEM	ion-evaporation model
IMS	ion-mobility spectrometry
IMS–MS	ion-mobility spectrometry–mass spectrometry
IRMPD	infrared multiphoton photodissociation

IUPAC	International Union for Pure and Applied Chemistry
LC	liquid chromatography
LC–MS	liquid chromatography–mass spectrometry
LIFDI	liquid injection field desorption ionization
LINAC	linear-acceleration high-pressure collision cell
LIT	linear ion trap
LOD	limit of detection
LOQ	limit of quantification
m/z	mass-to-charge ratio
MALDI	matrix-assisted laser desorption ionization
MAOI	monoamine oxidase inhibitor
MCP	microchannel plate
MDA	3,4-methylenedioxy-amphetamine
MDEA	3,4-methylenedioxy-ethylamphetamine
MDF	mass-defect filtering
MDMA	3,4-methylenedioxy-methamphetamine
Met-ID	metabolite identification
MS	mass spectrometry
MS–MS; MS^n	tandem mass spectrometry
NCE	new chemical entity
NICI	negative-ion chemical ionization
NLA	neutral-loss analysis
NMR	nuclear magnetic resonance spectroscopy
NPLC	normal-phase liquid chromatography
nNRTI	non-nucleoside reverse transcriptase inhibitors
NRTI	nucleoside reverse transcriptase inhibitors
NSAIDs	non-steroidal anti-inflammatory drugs
$OE^{+\bullet}$ and $OE^{-\bullet}$	odd-electron ion
PA	proton affinity
PDA	photodiode array spectrometry
PDI	^{252}Cf plasma desorption ionization
PIA	precursor-ion analysis
PICI	positive-ion chemical ionization
Q–LIT	quadrupole–linear ion-trap hybrid
Q–TOF	quadrupole–time-of-flight hybrid
QuEChERS	quick, easy, cheap, effective, rugged, and safe
RDA	retro-Diels–Alder
RDBE	ring double-bond equivalent
RF	radiofrequency alternating current potential
RPLC	reversed-phase liquid chromatography
RSD	relative standard deviation
S/N	signal-to-noise ratio
SIL-IS	stable-isotope-labeled internal standard
SNRI	selective non-serotonin reuptake inhibitor
SPE	solid-phase extraction
SQ	single-quadrupole
SRM	selected-reaction monitoring
SSRI	selective serotonin reuptake inhibitor
STA	systematic toxicological analysis
SWATH	sequential windowed acquisition of all theoretical fragment ion mass spectra
TCA	tricyclic antidepressant
TDC	time-to-digital converter
TeCA	tetracyclic antidepressant
TIC	total-ion chromatogram
TOF	time-of-flight
TQ	tandem-quadrupole
TSI	thermospray ionization
UHPLC	ultra-high-performance liquid chromatography
UV	ultraviolet spectroscopy
WADA	world anti-doping agency
XIC	extracted-ion chromatogram

1

INTRODUCTION TO LC–MS TECHNOLOGY

Interpretation of MS–MS Mass Spectra of Drugs and Pesticides, First Edition. Wilfried M. A. Niessen and Ricardo A. Correa C.

1.1 INTRODUCTION

In order to separate and quantify ions using mass spectrometry (MS), one must first generate and then send them to the mass analyzer, which is no easy task by any means. This process takes place in the ion source, where the introduced neutral atoms or molecules (the sample) are rendered ionized and in the gas phase. From there, they are sent into the mass analyzer and separated according to their *m/z* (mass-to-charge ratio (Section 2.2), where *m* is the mass number of an ion and *z* is the number of elementary charges regardless of sign). The order in which ionization and vaporization happen depends on the chosen technique, but ultimately the ions will have to find themselves under vacuum so that the mean free path between them is long enough to avoid random collisions, for example, fragment–fragment reactions. This is essential for the tenet of unimolecular reactions in MS to hold, whereby all the ions seen in the mass spectrum arise from the initially ionized sample in question. The ions generated can be odd-electron ions ($OE^{+\bullet}$ or $OE^{-\bullet}$) or even-electron ions (EE^{+} or EE^{-}). Providing the *m/z* for all ions and especially for the ions related to the intact molecule, for example, molecule ion or (de)protonated molecule, is the main reason of MS success as an analytical technique. In general, one can say that there are two main types of ionization techniques: hard and soft ionization techniques. In the former case, the molecular ion undergoes significant fragmentation (even with no molecular ion detection), whereas in the latter case ions do not undergo extensive (or any) fragmentation and an ion related to the intact molecule is readily detected.

In practice, chemical analysis begins with two critical steps that determine the ultimate quality of the experiments: sample collection and preparation, which should always strive at getting the highest purity specimen possible. The ion source contribution to the overall instrumental sensitivity arises from the two main events taking place within: sample ionization and ion transmission to the mass analyzer. Ionization efficiency is defined as the ratio of the number of ions generated to the number of molecules consumed in the ion source of a mass spectrometer: the method for determining the number of molecules consumed has to be clearly stated. The transmission efficiency is defined as the ratio of the number of ions leaving a region of a mass spectrometer to the number of ions entering that region. Since the performance of a source is tightly related to its actual components and their operating principles, sensitivity optimization depends on the kind and model of instrument used.

Sample introduction to the source is done by several methods: the most common being directly via a direct vapor inlet, or a direct insertion or exposure probe; indirectly via hyphenated techniques such as gas chromatography–mass spectrometry (GC–MS) and liquid chromatography–mass spectrometry (LC–MS), or surface-related desorption techniques such as thermally or laser-assisted techniques. Hyphenated techniques refer to the coupling of two (or more) separate analytical techniques by means of an appropriate hardware interface. In such cases, the instruments used in the hyphenated techniques work together in an automated manner as a single integrated unit (Hirschfeld, 1980). Particularly interesting is the coupling of powerful separation techniques, for example, GC, LC, thin-layer chromatography, electrophoresis, with spectrometry-related methods, for example, MS, infrared, ultraviolet–visible, atomic absorption, fluorescence, light scattering, Raman, nuclear magnetic resonance, for the analysis and characterization of all kinds of known matter.

1.2 ANALYTE IONIZATION: ION SOURCES

1.2.1 Electron Ionization

Electron ionization (EI) is a hard ionization technique and one of the oldest ionization methods in existence, yet still one the most widely used (Märk & Dunn, 1985). Vaporization of sample molecules must take place before their ionization, and therefore this limits the scope of the technique to volatile and thermostable compounds. EI furnishes ions by extracting one (or more) electron (e^-) out of the neutral sample molecule (M), according to Eq. 1.1. This process is carried out with high-energy electrons produced by means of thermionic emission from a heated (tungsten or rhenium) filament inside the source. Typically, the electrons are accelerated with a potential difference of 70 V. The energetic electrons interact with the analyte molecules, transfer part of their energy to the molecules, and render them ionic. The result is the production of a radical cation $M^{+\bullet}$ (molecular ion) and two electrons: the electron ejected from the neutral molecule and the ionizing electron after transferring part of its energy to M.

$$M + e^- \rightarrow M^{+\bullet} + 2e^- \tag{1.1}$$

The fate of the radical cation ($M^{+\bullet}$) produced depends on its internal energy at the moment of formation, which is determined by the kind and number of chemical bonds present in the sample molecule. It is $M^{+\bullet}$ and its fragmentation products (when present) that constitute the EI mass spectrum of the sample, and in principle for a given set of experimental conditions, each individual compound analyzed gives a unique mass spectrum (except for enantiomers).

1.2.1.1 Ionization Using Electrons The general operating components of an EI source are illustrated in Figure 1.1. These are contained within a heated (to avoid condensation of sample and ions) metal housing called the source block. EI uses thermionic emission as the main working principle for the production of high-energy (usually 70 eV, $1\ eV = 1.602177 \times 10^{-19}$ J) electrons under vacuum (0.1–1 Pa; 10^{-3}–10^{-2} mbar) in order to disrupt the nonbonding and bonding electrons of molecules.

An appropriately housed (coiled) tungsten or rhenium filament (cathode) is heated by passing a current through it (2–5 A). Once it reaches a certain temperature, the thermal energy of the electrons (greater than the work function of the metal) at the metal surface is sufficient to allow them to leave the metal thereby creating a flow of electrons. This is the thermionic emission of electrons from the filament. Concurrently, a negative potential (−70 V) is applied to the filament (e^- energy), and the electrons are thus accelerated

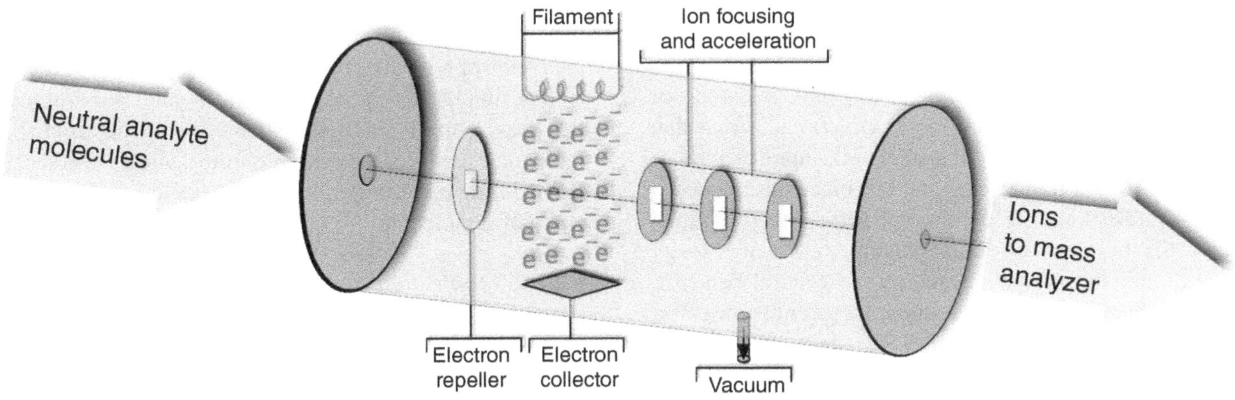

FIGURE 1.1 Schematic diagram of an electron ionization (EI) source.

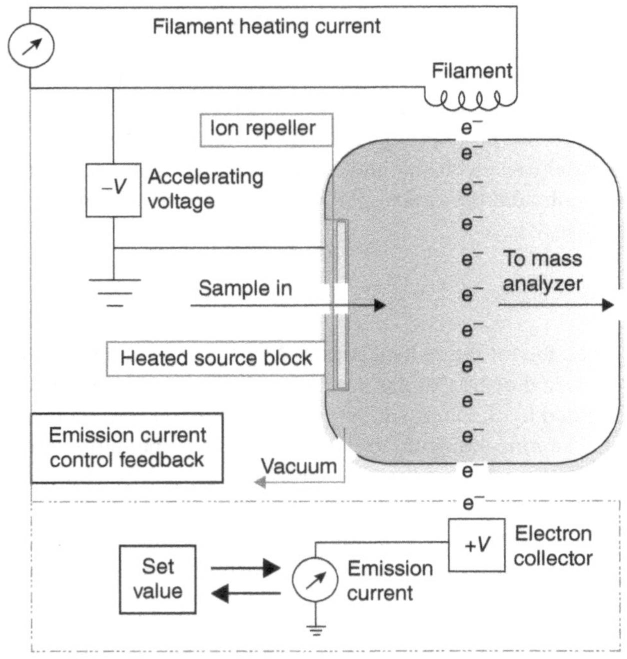

FIGURE 1.2 Scheme for the generation of ionizing electrons in an EI source.

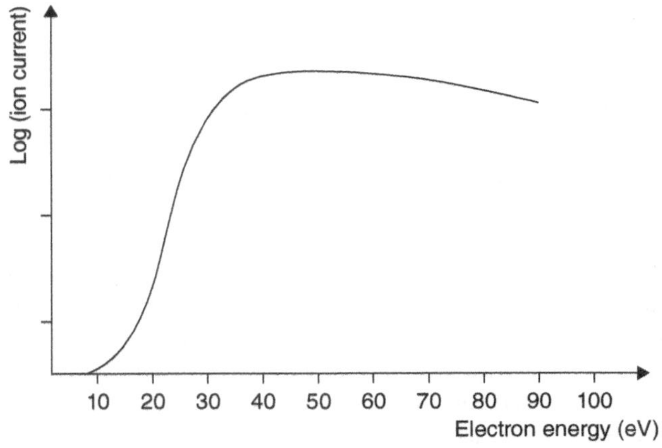

FIGURE 1.3 Relationship between ion current and electron energy.

and travel across from the surface of the metal filament to within the volume of the ion source. These electrons are attracted (by a positive voltage) to the e$^-$ collector (anode) located opposite and on-axis to the filament. This filament current (emission current) is measured and kept constant (150 μA) via a feedback mechanism with the heating current driven through the filament. This ensures constant ionization conditions (the number of electrons emitted by the filament is constant). Effectively, this setup places a shower of electrons that analyte molecules must cross as they are transmitted from the inlet (sample in) to the outlet (to mass analyzer) of the EI source (Figure 1.2). Often, by using a magnet, the flight path of the electrons is made helical; since the electrons must travel a longer path, their interaction with analyte molecules is enhanced.

Fortunately, the value of 70 eV has been used for the electron energy (and to less extent 150 μA for the emission current) throughout the years, and this has allowed for the creation of searchable EI mass spectral libraries that are of critical importance to the analytical applications of MS. By controlling the energy of the electrons, one can achieve different ionizing conditions for a given sample. The plot of the ion current versus the electron energy for most atoms and molecules shows the general behavior illustrated in Figure 1.3. A rise in the ion current is observed once the analyte ionization energy (IE, minimum energy required to eject an e$^-$ out of a neutral atom or molecule in its ground state) is reached. As the electron energy increases (≈20 eV), so does the ion current, mostly due to the formation of molecular ions. Further increase in energy (>30 eV) promotes fragmentation until a plateau is reached (around 70 eV); higher electronic energies actually cause a decrease in the ion current (Hübschmann, 2015). Operating the source at 70 eV for the electron energy, that is, at the plateau in Figure 1.3, ensures stable performance of the EI source. The EI efficiency is evaluated by the ratio of the number of ions formed to the number of electrons used in an ionization process.

Considering that helium has the highest ionization energy of any element (24.6 eV), along with the fact that the IE for most organic compounds lies between 5 and 12 eV, electrons with 70 eV will have more energy than the IE required to ionize incoming neutral species (Montalti et al., 2006). In chemistry, eV (non-SI unit) is expressed in molar terms and thus 70 eV = 6,754 kJ mol^{-1}. The amount of excess energy transferred from the electron to the molecule, typically a few eV (≈5 eV), and the structure of the molecule will determine the degree of fragmentation. The general trend of atomic IE is the same as the one for electronegativity, for example, F > Cl > Br > I. For molecules, nonbonding (nb) electrons are easier to ionize than bonding electrons, for example, IE of F-nb > N-nb > O-nb > S-nb. The greater the s character of a covalent bond, the more the electronegative it is; thus, the IE of a sigma sp bond (alkynes) > sp^2 sigma bond (alkenes) > sp^3 sigma bond (alkanes) > nb electrons. Special molecular features, for example, conjugation, which can help stabilize the resulting radical cation, greatly influence the IE value of a molecule.

1.2.1.2 Ionization and Fragmentation As the sample is introduced into the source (perpendicular to the electron axis), electrons and neutral molecules interact. When the rapprochement of sample molecules and electrons is within the ionization cross-sectional area (area the electron must cross to lead to an effective ionization) of the analyte molecule and the energy transferred is at least equal to the ionization energy, the loss of one (or more) electron is observed, along with the eventual fragmentation of the molecular ion thus produced. In the vacuum of the EI source, a random collision between an e^- and a sample molecule is extremely unlikely. Furthermore, the electrostatic repulsion of valence electrons makes it even more improbable. It is the electric field of the fast-moving charge (e^-) that causes a distortion in the orbits of the valence electrons. This interaction leads to a kinetic energy transfer from the e^- to the analyte cloud of electrons. If enough energy is transferred (IE) during this process, a valence electron is ejected from the analyte molecule, thereby forming an $M^{+\bullet}$. It is worthwhile noting that the de Broglie wavelength (λ) of the ionizing electrons must be of the same order as the bond length of the sample molecule, otherwise the energy transfer from the electrons to the analyte molecule will not happen effectively, for example, a 70 eV electron has a λ of 150 pm, an sp^2 hybridized C—C double bond has a bond length of ≈130 pm (Allen et al., 2006).

Approximately speaking, molecules have a diameter ranging from 0.1 nm for the smallest molecule (H_2), through macromolecules and supramolecular assemblies with diameters between 10 and 90 nm, for example, polymers, ATP synthase, to viruses and complex biological structures with >100 nm in diameter, for example, influenza virus, phages, chromosomes (Goodsell, 2009). Considering that the reaction in Eq. 1.1 is happening between two classical particles, an e^- with an energy of 70 eV travels approximately at a speed of 5000 km s^{-1} (0.017*c*, where *c* is the speed of light), which means that for a molecule like sucrose (nominal mass of 342 Da) with a 1 nm molecular diameter (Ramm et al., 1985), the electron will pass by the molecule in 2×10^{-16} s. In this timescale, the interaction between the electron and the molecule occurs much faster than that of an sp^3 O—H bond stretching vibration (10^{-14} s). As this electronic transition happens before any change occurs in the position of the nuclei involved (Franck–Condon principle), it can happen vertically from the electronic ground state of M to a (meta)stable excited electronic state of $M^{+\bullet}$ (or higher energy states) as illustrated in Figure 1.4. Taking a homodiatomic molecule as an example (Demtröder, 2010), its electronic ground state can be represented as shown in Figure 1.4a: the potential energy well is defined by the bond dissociation energy and the bond length. When the high-energy electrons match an electronic transition *i* (Figure 1.4b), the energy transfer leads to a stable excited electronic state (molecular ion), plus an e^- ejected off from the neutral sample. It is important to notice that electronic states higher than the ground state have potential energy wells with shallower minima and longer internuclear separations. Therefore, the bond is both weaker and elongated as a result of the ionization process (Figure 1.4b). Equally, if the energy of the electrons matches an electronic transition like *j* in Figure 1.4b, the formation of the radical cation will lead to an unstable excited state and fragmentation ensues.

What happens to the newly formed ions depends on their total energy and the ease with which they dissipate the excess energy among their other modes of motion, namely translational, vibrational, and rotational. Generally, the ions can be stable and last long enough to be detected, they can rapidly decompose producing fragment ions, or they can be

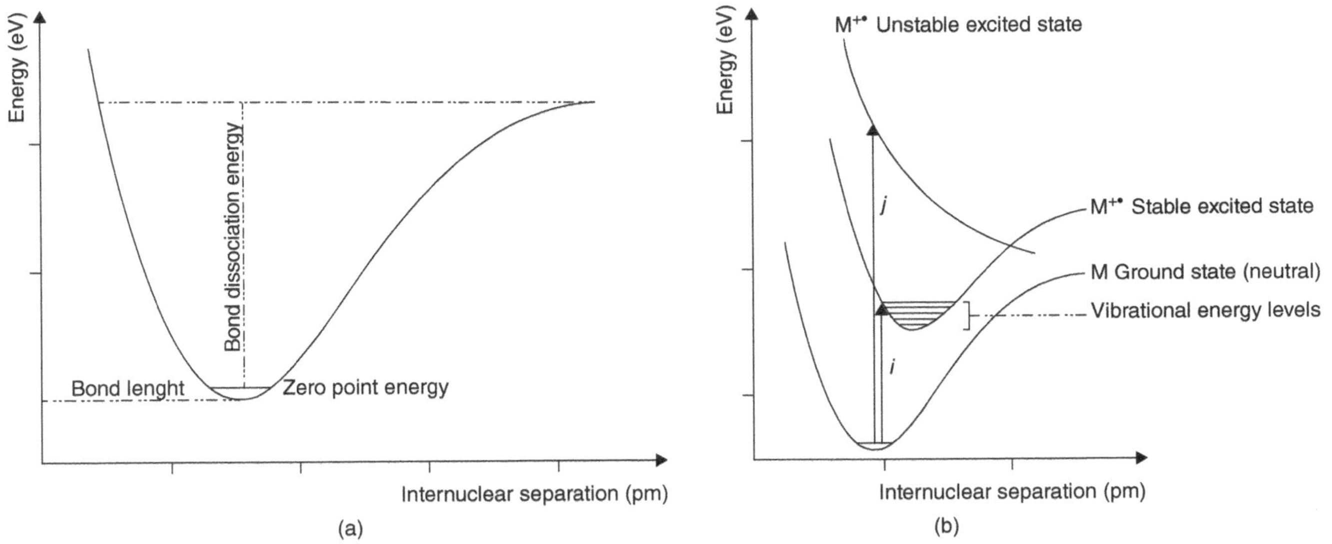

FIGURE 1.4 Ground electronic state of a neutral homodiatomic molecule (a). Vertical transitions depicting the ionization process in an EI source (b).

metastable and decompose in their flight to the detector. It is a process that is tightly related to the exact chemical structure of a molecule (Blanksby & Ellison, 2003).

1.2.1.3 Ion Transmission Ion transmission refers to the process of moving ions from one section to another within the mass spectrometer, for example, from the source through the analyzer and furthest to the detector. This process is not always necessarily accompanied by an *m/z* separation. In fact, in an EI source when transferring the ions produced into the analyzer, the goal is to do so with highest efficiency and lowest *m/z* spreading. Two complementary and simultaneous devices are applied (Figure 1.5). First, as the ions are being produced, a potential difference of the same sign is applied to the ion repeller, which is a plate placed before and perpendicular to the electron flux. This ion repeller pushes the ions toward the mass analyzer.

Second, three parallel (exact design changes depending on manufacturer) electrostatic lenses of equal sign are placed opposite and on-axis to the ion repeller, between the e^- flux and the mass analyzer. A potential difference of opposite sign to the ion repeller is applied in order to extract the ions out of the source, followed by a lower potential difference in order

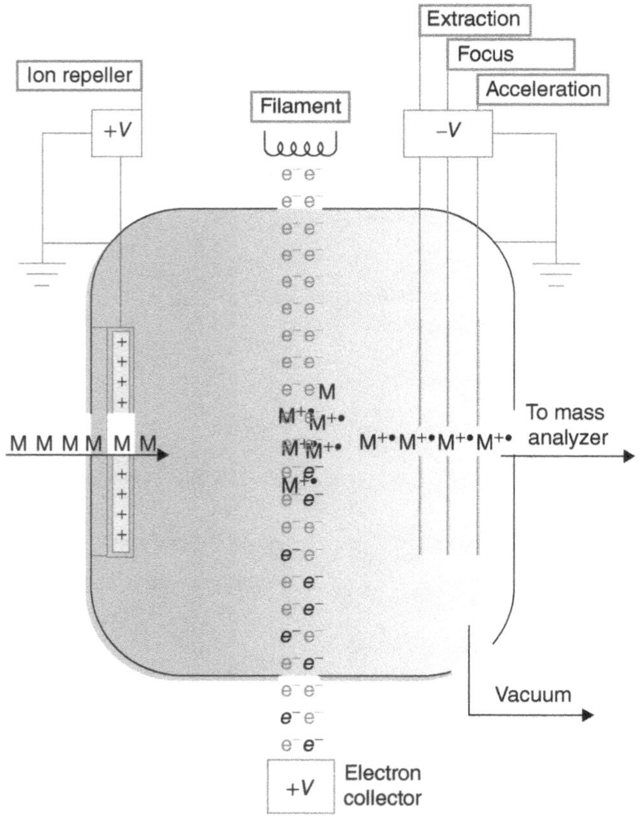

FIGURE 1.5 Devices for ion transmission from the EI ion source to the mass analyzer.

to focus the ions to finally reaccelerate them as they are sent into the mass analyzer, where separation according to their *m/z* takes place. Typical fragmentation characteristics under EI conditions are briefly discussed in Section 3.3.

1.2.1.4 Analytical Applications of Electron Ionization EI is probably the most widely applied ionization technique in MS. It is extensively used in GC–MS, where it provides good sensitivity for most compounds and structure-informative fragmentation in highly reproducible mass spectra. Besides, after basic tuning of the ion source, which can be performed automatically under software control, there are essentially no experimental parameters to set or optimize. In terms of qualitative analysis, interpretation of the EI mass spectra can be performed based on a solid understanding of the fragmentation behavior of $M^{+\bullet}$ (Section 3.3) (McLafferty & Tureček, 1993; Smith, 2004). In addition, elaborate and searchable mass spectral libraries have been compiled to assist in the identification of compounds (Atwater et al., 1985; Stein & Scott, 1994; Ausloos et al., 1999; Koo et al., 2013). The results of these library searching routines can be quite powerful. If a mass spectrum of the unknown compound is present in the library, expert comparison of library and experimental mass spectra can lead to compound identification. If the compound is not present in the library, the computer library search often provides insight about the presence of substructures or other structural features of the unknown compound, which facilitates further spectrum interpretation. Although many researchers take the result of the library search for granted, a thorough and critical evaluation of the agreement between experimental and library spectrum is recommended. In addition, GC–MS with EI is also frequently used in quantitative analysis using either extracted-ion chromatograms (Section 1.3.1.1) or selected-ion monitoring (Section 1.5.2) before peak area determination. More recently, gas chromatography tandem mass spectrometry (GC–MS–MS) in selected-reaction monitoring (SRM) (Section 1.5.2) mode has become the method of choice in routine quantitative analysis of compounds present at very low levels in complex biological matrices.

As EI is limited to the analysis of volatile and thermostable analytes, analyte derivatization strategies have been developed to enhance the volatility and stability of more polar analytes. Derivatization obviously changes the fragmentation behavior of the analyte because the fragmentation may be directed from a different site in the molecule (Zaikin & Halket, 2009; Sparkman et al., 2011). Silylation and oximation reactions are most frequently carried out. Characteristic fragment ions derived from the derivatizing agent are readily seen, thereby improving analysis selectivity. For instance, the trimethylsilyl ether derivative ($(CH_3)_3SiOR$) of hydroxy group (OH) containing molecules show the trimethylsilyl group ion with *m/z* 73 ($[(CH_3)_3Si]^+$) and an ion with *m/z* 75 corresponding to protonated dimethylsilanone

($[(CH_3)_2SiO+H]^+$). When the target compound has several trimethylsilyl ether moieties, the formation of the pentamethyldisiloxane cation ($[(CH_3)_2SiOSi(CH_3)_3]^+$) with *m/z* 147 is observed (a commonly seen ion from GC column bleeding). These ions may undergo ion–neutral reactions with analyte molecules (M), one of these reactions is the adduct formation of an ion with *m/z* (M+73) (Carles et al., 2007).

After seeing the power of EI in GC–MS, the implementation of EI in LC–MS has been pursued as well. However, given the gas load of the mostly aqueous mobile-phase vapor admitted into the ion source and the MS vacuum system in LC–MS, it is more complicated to achieve the high-vacuum ion source conditions required for successful EI. The most successful approaches to EI in LC–MS (which were also commercialized) were the moving-belt interface (Arpino, 1989) and the particle-beam interface (Creaser & Stygall, 1993), both quite complex instrumental solutions. Unfortunately, these solutions did not provide the reliability, user-friendliness, and sensitivity required. More recently, the so-called direct-EI interface has been described, which provides nebulization of the effluent of a nano-LC column (flow rates < 100 nL min^{-1}), directly into the EI source (Cappiello et al., 2011).

1.2.2 Chemical Ionization

Chemical ionization (CI) is a soft ionization technique used to study chemical structure and reactivity. A CI source uses a reagent gas (G_R) inside a modified EI source to create conditions of high source pressure, such that G_R ions–molecule and molecule–e^- reactions can occur in high yield (Harrison, 1992; Munson, 2000). In fact, most instruments are equipped with a source that can be switched between EI and CI conditions. As seen so far, an EI source is an environment where neutral molecules (or atoms) and radicals, radical cations, cations, and electrons coexist. Intuitively, the presence of electrons in the source begs the question of whether or not positive ions are the only ions present in the source. As expected, negative-ion formation is an inherent process in EI and formation of radical anions is also observed (Bowie, 1984).

Thus, there can be a simultaneous presence of positive and negative ions inside an EI/CI source. Their transmission and detection are a matter of choice and depend on the voltage polarities chosen to carry out the experiments, for example, when analyzing negative ions except for the e^- collector voltage in Figure 1.5, all other voltages must be switched in polarity. CI creates conditions that favor the production of EE^+ and EE^-, and as a result, CI can be carried out in two different modes: positive mode as in positive-ion chemical ionization (PICI) and negative mode as in negative-ion chemical ionization (NICI) and electron-capture negative ionization (ECNI). Both modes can use the same source and often but not necessarily use the same G_R. Nevertheless, the function of the G_R serves a different purpose on each mode, and experimental conditions must be optimized for each type of analyte in relation to the mode of CI chosen. Ionization in CI happens without the transfer of large excess of energy from a G_R (and ions thereof) or from a secondary e^-; thus, the initially generated ions do not undergo extensive fragmentation. CI is a technique that offers both high sensitivity and selectivity. Nevertheless, it is not suitable to all kinds of molecules as the analytes must be volatile and thermostable and must present special structural features in order to be responsive to the technique.

1.2.2.1 Electron Ionization of the Reagent Gas, G_R For particles of similar shape and at a given temperature, the mean free path between them is inversely proportional to the pressure. Usually, in EI the mean free path is ≥1 m, and caution must be taken as mean free paths of ≤0.5 m lead to ion–ion reactions, generating an atypical mass spectrum. As the G_R flows into the CI source, it establishes conditions of high pressure (1–100 Pa; 10^{-2}–1 mbar; while the pressure in the vacuum manifold is ≤10^{-3} Pa; 10^{-5} mbar) and its ionization by primary 70 eV electrons readily yields molecular ions ($G_R^{+\bullet}$). In many CI sources, higher electron energies (up to 400 eV) are applied in order to ensure that the electrons penetrate well the high-pressure environment of the ion source. Ensuing fragmentation of $G_R^{+\bullet}$ occurs by forming cations (G_{EE}^+), other radical cations ($G_{OE}^{+\bullet}$), neutral species (R, $R^\bullet$), and secondary electrons (e^-) (Eqs 1.2 and 1.3).

$$G_R + e^- \rightarrow G_R^{+\bullet} + 2e^- \quad (1.2)$$

$$G_R^{+\bullet} \rightarrow G_{EE}^+ + R^\bullet \quad (1.3a)$$

$$G_R^{+\bullet} \rightarrow G_{OE}^{+\bullet} + R \quad (1.3b)$$

Given a controlled flow of G_R into the source, it is the most abundant species and reacts (ion–molecule reactions) with the newly formed $G_R^{+\bullet}$, G_{EE}^+, $G_{OE}^{+\bullet}$ yielding reactive electrophilic cations that can undergo further reactions with analytes of interest. While EI is a unimolecular process, in CI bimolecular and even termolecular reactions generate a steady-state plasma inside the source as shown in Figure 1.6; methane is used as an example to illustrate the reactions observed.

When the sample is introduced into the source, it encounters a plasma of both positive and negative (low-energy electrons) reactive species. The most common reactions taking place involve proton transfer, electron capture, or adduct formation between the analyte of interest and charged species of the reactants. In this technique, the presence of the (de)protonated molecule is characteristic, which serves as a complementary tool to other types of MS methods. The ions generated in PICI, NICI, and ECNI happen via different mechanisms; nevertheless, all three can happen concurrently.

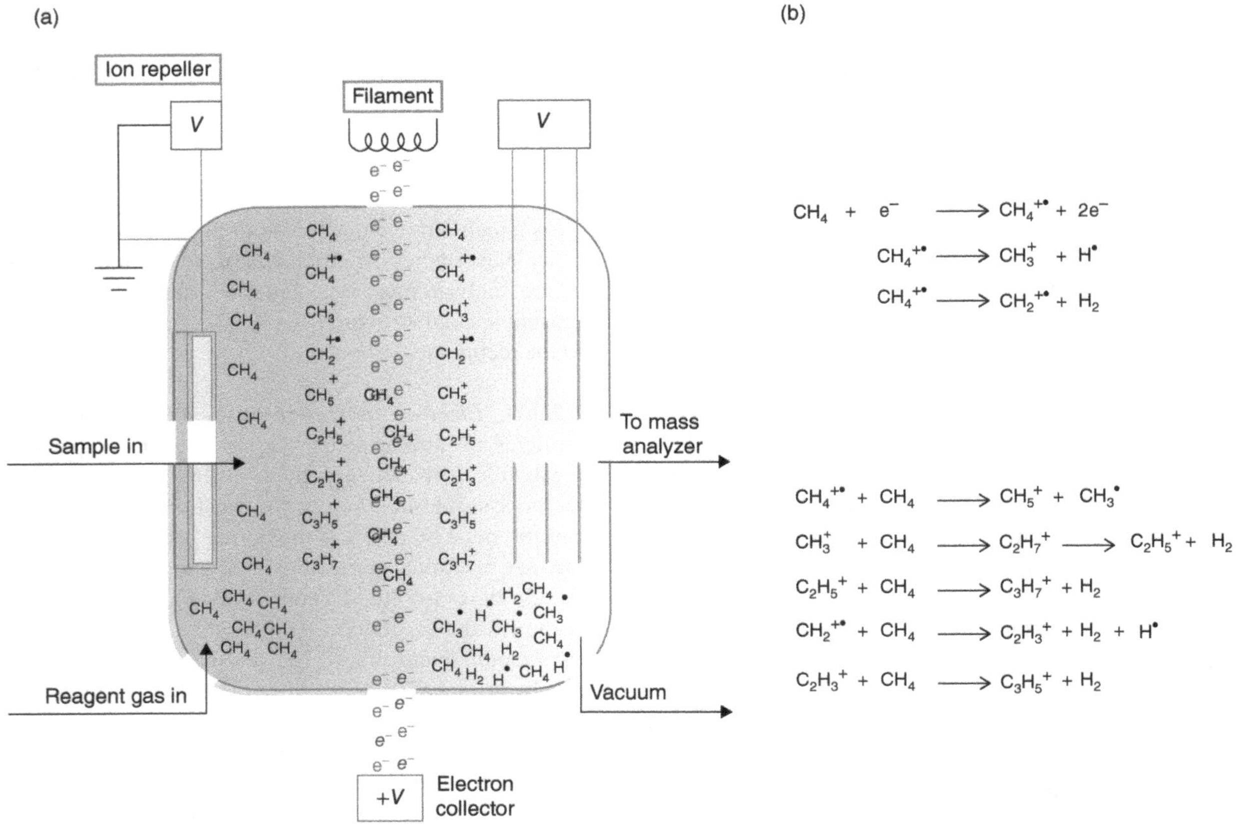

FIGURE 1.6 Ongoing processes inside a chemical ionization (CI) source during reagent gas ionization (methane) in a CI experiment (a). Main chemical reactions involved in the ionization of methane reagent gas in an EI source during a CI experiment (b).

1.2.2.2 Positive-Ion Chemical Ionization The main pathways that explain the experimental observations regarding ion formation in PICI between analyte molecules and G_R plasma are as follows: (i) proton transfer, (ii) electrophilic addition, (iii) anion abstraction, and (iv) charge exchange (CE).

Proton transfer Proton transfer is the most commonly observed reaction and serves as the basis for PICI measurements. These Brønsted–Lowry acid–base reactions afford protonated analyte molecules as long as their gas-phase basicity is greater than that of the reactive species present in the source. However, hydride (H^-) abstraction from the analyte molecules can also occur. The former case yields a cation $[M+H]^+$ with *m/z* (M+1), where M is the (monoisotopic) mass of the analyte molecule (Eq. 1.4a), whereas the latter case yields a cation $[M–H]^+$ with *m/z* (M−1) (Eq. 1.4b).

$$[G_R{+}H]^+ + M \rightarrow [M{+}H]^+ + G_R \qquad (1.4a)$$

$$G_{EE}{}^+ + M \rightarrow [M{-}H]^+ + G_RH \qquad (1.4b)$$

In addition to methane (Section 1.2.2.1), several other gases including propane, butane, isobutane, and ammonia can form cations that serve as G_R in Brønsted–Lowry acid–base reactions. If the reaction is exothermic, these cations will readily transfer protons to analyte molecules (M) forming $[M+H]^+$ cations. The exothermicity of the reaction is determined by the proton affinity (PA) difference between the reacting species (Table 1.1). In general, the more exothermic the reaction is, the more fragmentation is observed (more energy transferred to analyte molecule).

Careful choice of acid–base pairs allows control of the extent of the ionization and fragmentation process, thus either inducing or eliminating ionization and/or fragmentation. Eq. 1.5 shows the protonation and hydride abstraction reactions of an analyte molecule (M) when using methane as G_R.

$$[CH_5]^+ + M \rightarrow [M{+}H]^+ + CH_4 \qquad (1.5a)$$

$$[C_2H_5]^+ + M \rightarrow [M{-}H]^+ + CH_3CH_3 \qquad (1.5b)$$

TABLE 1.1 Proton affinities of compounds commonly used in GC–MS and LC–MS.

Compound	PA ($kJ\ mol^{-1}$)	Compound	PA ($kJ\ mol^{-1}$)
Methane (CH_4)	552	Methyl acetate (CH_3COOCH_3)	828
Ethyne ($HC{\equiv}CH$)	641	Ethenone ($H_2C{=}C{=}O$)	830
Ethene ($H_2C{=}CH_2$)	680	Diethyl ether ($C_2H_5OC_2H_5$)	838
Water (H_2O)	697	Ammonia (NH_3)	854
Hydrogen sulfide (H_2S)	712	Aniline ($C_6H_5NH_2$)	877
Formaldehyde ($H_2C{=}O$)	718	Methylamine (CH_3NH_2)	896
Propene ($CH_3CH{=}CH_2$)	752	Alanine ((CH_3CHNH_2)COOH)	899
Benzene (C_6H_6)	759	Ethyl amine ($CH_3CH_2NH_2$)	908
Methanol (CH_3OH)	761	Dimethylamine ($(CH_3)_2NH$)	923
Ethanol (C_2H_5OH)	788	Pyridine (C_5H_5N)	924
Acetonitrile ($CH_3C{\equiv}N$)	788	Dimethyl aniline ($C_6H_5N(CH_3)_2$)	935
Toluene ($C_6H_5CH_3$)	794	Trimethylamine ($(CH_3)_3N$)	942
Ethyl formate ($HCOOC_2H_5$)	808	Piperidine ($C_5H_{11}N$)	947
iso-Butene ($(CH_3)_2C{=}CH_2$)	820	Quinoline (C_9H_7N)	948
Acetone (CH_3COCH_3)	823	Triethylamine ($(C_2H_5)_3N$)	972

Source: Adapted from Lias, 1984 and Hunter, 1998. Reproduced with permission of the American Institute of Physics.

The methanium ion ($[CH_5]^+$) with *m/z* 17 is a good example of a G_R ionic species reacting in both protonation (Eq. 1.5a) and hydride abstraction reactions with analyte molecules (Eq. 1.5c).

$$[CH_5]^+ + M \rightarrow [M{-}H]^+ + CH_4 + H_2 \quad (1.5c)$$

Electrophilic addition Electrophilic addition (adduct formation, e.g., alkylation) is another type of acid–base reaction that occurs when analyte molecules have Lewis base character, for example, presence of heteroatoms with nonbonding electrons or π-electrons, allowing their reaction with electrophiles (even-electron cations, $G_{EE}{}^+$) present in the G_R plasma (Eq. 1.6).

$$G_{EE}{}^+ + M \rightarrow [M{+}G_{EE}]^+ \quad (1.6)$$

Some examples of adduct formation when using methane as G_R are shown in Eq. 1.7. Knowing the mass of the alkylating cation allows one to find the molecular mass of the target compound. For methane, these ions are found with *m/z* (M+15), (M+29), and (M+41).

$$[CH_3]^+ + M \rightarrow [M{+}CH_3]^+ \quad m/z\ (M{+}15) \quad (1.7a)$$

$$[C_2H_5]^+ + M \rightarrow [M{+}C_2H_5]^+ \quad m/z\ (M{+}29) \quad (1.7b)$$

$$[C_3H_5]^+ + M \rightarrow [M{+}C_3H_5]^+ \quad m/z\ (M{+}41) \quad (1.7c)$$

Conditions within the source can be changed in order to promote or inhibit a given type of acid–base reaction from happening. This can be achieved by establishing physical conditions, for example, e^- energy and G_R pressure, in the source that will favor the formation of the G_R ions needed for either proton transfer or adduct formation. Table 1.2 shows the most common CI reagent gases used in MS, along with the adducts formed from analyte molecules–G_R plasma reactions.

Anion abstraction Anion abstraction happens when $G_{EE}{}^+$ ions react with sample molecules to form an analyte-derived cation and a neutral species as shown in Eq. 1.8. Proton abstraction is a good example (exothermic reaction with the nitrosonium cation (NO^+) for most alkanes) leading to $[M{-}H]^-$ ions with *m/z* (M−1). Alcohols (1° and 2°), aldehydes, and ketones undergo this kind of reaction. Tertiary alcohols undergo abstraction of hydroxy group (OH) leading to a stable tertiary carbocation $[M{-}OH]^+$ with *m/z* (M−17).

$$G_{EE}{}^+ + M \rightarrow [M{-}A]^+ + G_{EE}A \quad (1.8)$$

Hydride abstraction from alkanes when using cations such as $[C_2H_5]^+$ (Eq. 1.5b) and $[CF_3]^+$ is a good example as well; group electronegativity is useful in this respect (Wells, 1968). There is no reagent gas system exclusively developed for this mode of CI; the nitrosyl radical ($^{\bullet}NO$) or a mixture of nitrogen/nitrous oxide (N_2/NO_2) are reagent gases used to produce NO^+, which acts as hydrogen abstractor, and can also participate in adduct formation and charge-transfer reactions.

Charge exchange (CE) CE is the outcome of the interaction between a $G_R{}^{+\bullet}$ and a neutral analyte molecule. Ionization takes place when there is a transfer of charge to the analyte molecule producing an $M^{+\bullet}$ and a neutral G_R. The reaction

TABLE 1.2 Common reagent gases used in positive-ion CI and adducts formed thereof.

Reagent Gas (G_R)	G_{EE}^+ Plasma Ions	Adducts Formed	m/z
Methane (CH_4)	$[CH_3]^+$	$[M+CH_3]^+$	M+15
	$[CH_5]^+$	$[M+H]^+$	M+1
		$[M-H]^+$	M−1
	$[C_2H_3]^+$	$[M-H]^+$	M−1
	$[CH_2CH_3]^+$	$[M+C_2H_5]^+$	M+29
	$[CH_2CHCH_2]^+$	$[M+C_3H_5]^+$	M+41
Isobutane $((CH_3)_2CHCH_3)$	$[(CH_3)_3C]^+$	$[M+(CH_3)_3C]^+$	M+57
	$[CH_3CHCH_3]^+$	$[M+H]^+$	M+1
		$[M+C_3H_7]^+$	M+43
	$[C_3H_3]^+$	$[M+C_3H_3]^+$	M+39
Ammonia (NH_3)	$[NH_4]^+$	$[M+H]^+$	M+1
		$[M+NH_4]^+$	M+18
	$[NH_4+NH_3]^+$	$[M+[NH_4+NH_3]]^+$	M+35

is observed when the recombination energy (exothermicity of the reaction $G_R^{+\bullet} + e^- \rightarrow G_R$) of G_R is higher than the IE of M (Eq. 1.9). The degree of fragmentation of $M^{+\bullet}$ depends on the exothermicity of the reaction. However, the molecular ions produced are usually of low internal energy. The presence of protonating species must be kept at a minimum in order to avoid formation of G_RH. Pure compounds are usually used as G_R for charge-exchange chemical ionization (CECI), nonetheless, mixtures with an inert buffer gas such as N_2 find application. Despite the fact that alkanes, for example, CH_4, and aromatic compounds, for example, benzene, chlorobenzene, can be used as G_R for CECI, aprotic solvents are preferred: rare gases, for example, Ne, Ar, Xe, methanedithione (S=C=S), sulfanylidenemethanone (S=C=O), nitrosyl ($^\bullet$NO).

$$G_R^{+\bullet} + M \rightarrow M^{+\bullet} + G_R \qquad (1.9)$$

In addition to its routine application as an analytical tool, CI has also been used in mechanistic studies, such as the study of gas-phase ion–molecule reactions (organic chemistry in the high-vacuum gas phase), *regio-* and *stereo-* selectivity questions, conformational analysis, and the measurement of relative reaction rate constants.

1.2.2.3 Negative-Ion Chemical Ionization The study of reactions between negative ions of G_R and neutral sample molecules has not been carried out as thoroughly as it has been done for their positive counterparts. This mode of ionization happens in two different methods: NICI and ECNI. In the former case, it is the result from reactions of G_R anions present in the source and neutral analyte molecules (M). This occurs readily when stable anions of the G_R can be formed. ECNI, in contrast, is the process by which thermal electrons present in the source (e^-) react with neutral analyte molecules generating radical anions ($OE^{-\bullet}$) and anions (EE^-).

The main reactions in NICI can be grouped as (i) proton transfer, (ii) nucleophilic addition, (iii) nucleophilic displacement, and (iv) CE.

Proton transfer Proton transfer occurs when an anion (G_R^-) derived from a G_R or a G_R mixture reacts with a neutral analyte molecule containing a removable proton. This happens when the PA (or gas-phase basicity) of G_R^- is greater than the PA of the conjugate base of the analyte ($[M-H]^-$), according to Eq. 1.10.

$$G_R^- + M \rightarrow [M-H]^- + G_RH \qquad (1.10)$$

Molecules with acidic H-atoms (removable) such as carboxylic acids and phenols are common examples of functional groups undergoing proton-transfer reactions. Therefore, the PA of typical anions can be used to predict the outcome of NICI proton-transfer reactions. Some examples of G_R^- are as follows: Cl^-, $[CN]^-$, $[O_2]^{-\bullet}$, F^-, $[CH_2CN]^-$, $[CH_3O]^-$, $O^{-\bullet}$, $[OH]^-$, H^-, $[NH_2]^-$, and $[C_5F_5]^-$ (Table 1.3).

There exist many gas mixtures to generate the anions of interest, for example, the use of fluorocarbons (trifluoromethane, CHF_3) and chlorofluorocarbons (CF_2Cl_2) to generate F^- and Cl^-, respectively, and the use of ammonia (NH_3) to generate $[NH_2]^-$ (Dougherty, 1981). Most of these anionic reactive species themselves are produced by associative electron-capture reactions, for example, formation of $[O_2]^{-\bullet}$. The reaction between methoxide ion ($[CH_3O]^-$, PA $\approx 1580\,kJ\,mol^{-1}$) and cyclopentadiene producing the cyclopentadiene anion ($[C_5H_5]^-$) (ΔPA $\approx -100\,kJ\,mol^{-1}$) serves as an example (Eq. 1.11).

$$H_3C\!-\!O\!-\!N{\overset{O}{\parallel}} \ + \ e^- \ \longrightarrow \ H_3C\!-\!O^- \ + \ ^\bullet N{=}O$$

$$(1.11a)$$

(1.11b)

Methyl nitrite (CH_3ONO) undergoes dissociative electron capture to produce the reactive species of interest CH_3O^- (Eq. 1.11a), which deprotonates cyclopentadiene producing the $[C_5H_5]^-$ (Eq. 1.11b). Superoxide ($O_2^{-\bullet}$, PA ≈ 1465 kJ mol^{-1}), formed by electron capture of nitrous oxide (NO_2) or a molecular oxygen/argon gas mixture, can behave as a basic species and deprotonates acidic compounds such as 4-nitrophenol producing the corresponding phenoxide ion (PA_{calc} ≈ 1350 kJ mol^{-1}) (Chandra & Uchimaru, 2002) and hydroperoxyl radical ($HOO^\bullet$), as illustrated in Eq. 1.12.

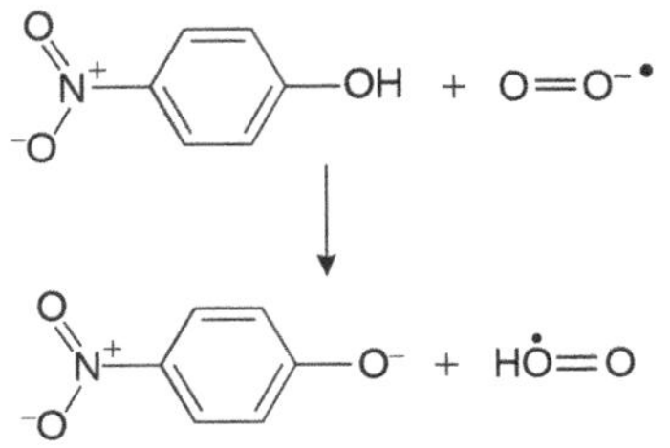

(1.12)

Hydroxide ions (HO^-, PA ≈ 1635 kJ mol^{-1}) are frequently used for their ability to produce NICI mass spectra of a diversity of functional groups: alcohols, ethers, neutral lipids, and hydrocarbons.

TABLE 1.3 Anions used for neutral analyte negative ionization in GC–MS and LC–MS.

Anion	PA (kJ mol^{-1})
NH_2^- (amide)	1689
H^- (hydride)	1676
OH^- (hydroxide)	1636
$O^{-\bullet}$ (atomic oxygen radical anion)	1595
CH_3O^- (methoxide)	1583
$(CH_3)_2CHO^-$ (isopropoxide)	1565
$^-CH_2CN$ (cyanomethide)	1556
F^- (fluoride)	1554
$C_5H_5^-$ (cyclopentadiene anion)	1480
$O_2^{-\bullet}$ (molecular oxygen radical anion)	1465
CN^- (cyanide)	1462
Cl^- (chloride)	1395
$HCOO^-$ (formate)	1444*
CH_3COO^- (acetate)	1458*
CF_3COO^- (trifluoroacetate)	1350*

Source: Bruno & Svoronos, 2010; *Harrison, 1992. Reproduced with permission of American Chemical Society.

Nucleophilic addition Nucleophilic addition can occur when anions do not have very high proton affinities (e.g., $O_2^{-\bullet}$, $[CN]^-$ (PA ≈ 1460 kJ mol^{-1}), Cl^- (PA ≈ 1395 kJ mol^{-1}). Instead of undergoing acid–base reactions leading to deprotonated products, they form adducts by nucleophilic addition to analyte molecules (Eq. 1.11a).

$$G_R^- + M \rightarrow [M{+}G_R]^- \quad (1.13)$$

Examples of this reaction are hydrogen-bonded adducts formed by chloride ions (Cl^-) with analyte molecules containing functional groups with electrophilic H-atom, such as carboxylic acids, amides, aromatic amines, phenols, and organophosphorus pesticides. This leads to the production of $[M{+}Cl]^-$ ions with *m/z* (M+35) and *m/z* (M+37) in a ≈3:1 ratio of relative intensities. For instance, 4-nitrophenol reacts with Cl^- as shown in Eq. 1.14.

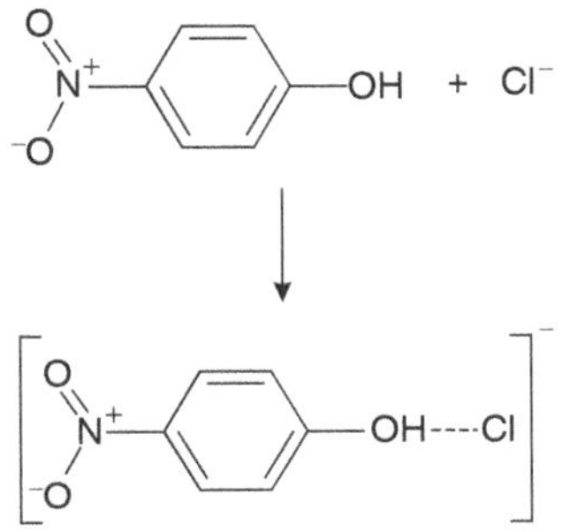

(1.14)

Nucleophilic addition is also observed with $O_2^{-\bullet}$ and compounds of low acidity such as aliphatic compounds forming the corresponding $[M{+}O_2]^{-\bullet}$ radical ion. Alcohols also undergo nucleophilic addition adduct formation. For instance, it was found that 11 different anionic species form adducts with neutral oligosaccharides (Jiang & Cole, 2005).

Nucleophilic displacement Nucleophilic displacement is a substitution reaction where an electrophilic center of an analyte molecule undergoes nucleophilic attack (e.g., S_N2). The leaving group thus produced can be a neutral radical or a new anionic species as illustrated in Eq. 1.15.

$$G_R^{-\bullet} + M \rightarrow [MG_R{-}H]^- + H^\bullet \quad (1.15a)$$

$$G_R^- + MA \rightarrow MG_R + A^- \quad (1.15b)$$

Many strongly basic anions such as atomic oxygen radical anion ($O^{-\bullet}$, PA ≈ 1595 kJ mol^{-1}) and HO^- usually react in proton-transfer reactions. Nonetheless, with certain analytes, they participate in gas-phase nucleophilic reactions. Both of these ions can be produced by using N_2O as G_R (e.g., N_2O, N_2O/CH_4). Examples of this mechanism are the gas-phase reactions of $O^{-\bullet}$ with phthalic acid alkyl esters (Stemmeler et al., 1994; Lépine et al., 1999) and the analysis of steroids with HO^- where both proton abstraction and nucleophilic displacement are observed (Roy et al., 1979).

Charge exchange (CE) CE occurs when a G_R (Lewis base) with lower electron affinity (EA) than that of the neutral analyte (Lewis acid) is allowed to react in the CI ion source and an electron transfer is effected as shown in Eq. 1.16. The degree of fragmentation depends on the exothermicity of the reaction. An important characteristic of this type of reaction is the possibility of obtaining single peak mass spectra, consisting of the anionized analyte molecule.

$$G_R^{-\bullet} + M \rightarrow M^{-\bullet} + G_R \tag{1.16}$$

As an example, the analysis of dibenzothiophene using $[O_2]^{-\bullet}$ as G_R delivered $M^{-\bullet}$, while the G_R was oxidized to molecular oxygen (O_2) (Hunt et al., 1976). Care must be taken to avoid the presence of competing species that would react with $M^{-\bullet}$, thereby lowering the sensitivity of the analysis. For instance, the presence of fluorine radicals ($F^{\bullet}$) would lead to the formation of fluoride ions (F^-) and neutral analyte M.

Despite the successes of NICI as an analytical tool, the most common technique used for the generation of negative ions is ECNI. Strictly speaking, these electron–molecule reactions are not chemical ionization processes. If at a given temperature there is an equilibrium between the generation and recombination of electrons, the electrons are said to be in thermal equilibrium. Thermal electrons have a kinetic energy $\leq$2 eV. Under these conditions, they can be captured by electronegative atoms present in analyte molecules, thereby forming radical anions ($OE^{-\bullet}$). The thermionic emission of electrons from heated filaments is the usual way of producing high-energy primary electrons in EI. The main source of secondary (thermal) electrons is the deceleration of primary electrons by collisional energy transfer with gases inside the source, such as G_R ionization as shown in Eq. 1.17.

$$2G_R + e^-_{70\,eV} \rightarrow G_R^* + G_R^{+\bullet} + 2e^-_{2\,eV} \tag{1.17}$$

Polyatomic gases are more efficient collisional energy sinks than diatomic and monoatomic gases, and therefore their rate of e^- thermalization is higher (e.g. $NH_3 > CO_2 > i\text{-}C_4H_{10} > CH_4 > N_2 > Ar$). After the reaction of the secondary electrons with the analyte molecules, the presence of a G_R (or a buffer gas) is essential for collisional stabilization of the newly formed excited radical anion $OE^{-\bullet}$. Otherwise, e^- detachment can happen and no analyte anion is observed.

Neutral analyte molecules undergo EC to form radical anions ($OE^{-\bullet}$). The ease, with which this process happens, depends on the EA of the neutral analyte and its ability to dissipate the excess internal energy after its formation (Eq. 1.18).

$$M + e^- \rightarrow M^{-\bullet} \tag{1.18}$$

Since charge density leads to instability, for example, HO^- is less stable than H_2O, charge dissipation must be effective. Therefore, analyte molecules must have electronic features that promote electron capture. Factors that contribute most prominently in the stabilization of a negative charge are as follows: orbital hybridization of the atom bearing the charge, for example, for carbanions the stability follows $sp > sp^2 > sp^3$, the presence of geminal or vicinal electronegative elements ($F > O > Cl > N > Br > I > S > C > P$) and/or electron-withdrawing functional groups or substituents ($—CF_3 > —CCl_3 > —CH_3$; $—CN \approx —CCH > —CHCH_2 \approx —C_6H_5$; $—OH > —NO_2 > —NH_2$), charge delocalization by resonance or aromaticity, and molecular polarizability whereby small atoms and molecules dissipate a charge less effectively than large ones, for example, the I-atom is more polarizable than an F-atom, thus I^- is a much better leaving group than F^- in substitution reactions. Usually, the most electronegative element present in the molecule determines its EA. For this reason, molecules with electronegative elements or groups, for example, nitro (NO_2), acyl (RCO), and cyano (CN), are attractive targets of ECNI. The main processes that explain the formation of negative species in ECNI are as follows: (i) associative electron capture, (ii) dissociative electron capture, and (iii) ion-pair formation reactions (Hiraoka, 2003; Stemmeler & Hites, 1988).

Associative electron capture Associative electron capture as shown in Eq. 1.18 gives the molecular radical anion $M^{-\bullet}$ after reaction of M with a low energy e^- (<2 eV). The molecular anion is formed without great excess energy, and additional collisional stabilization with (buffer) gases present in the source explains the high relative intensity of $M^{-\bullet}$ observed.

Dissociative electron capture Dissociative electron capture happens when electrons inside the ion source with a kinetic energy of up to ≈15 eV react with analyte molecules containing electronegative atoms or substituent groups that can form good leaving groups, for example, halogens, benzyl ($C_6H_5CH_2^-$), and methoxy (CH_3O^-), according to Eq. 1.19. The formation of a stable anion $[M–X]^-$ or X^- is the basis for this sensitive and selective type or CI analysis.

$$MX + e^- \rightarrow [M–X]^- + X^{\bullet} \tag{1.19a}$$

$$MX + e^- \rightarrow M^{\bullet} + X^- \tag{1.19b}$$

As expected, all these reactions are exothermic, and the outcome depends on the difference between the bond energy of the X group in the analyte and the EA of the analyte [M–X] and X fragments.

Ion-pair formation Ion-pair formation happens with electrons of ≈10–15 eV. The initially formed $OE^{-\bullet}$ has enough internal energy to dissociate into positive and negative ions

(Eq. 1.20). This process is not very common and does not find widespread use as an analytical method.

$$MX + e^- \rightarrow [M–X]^- + X^+ + e^- \quad (1.20a)$$

$$MX + e^- \rightarrow [M–X]^+ + X^- + e^- \quad (1.20b)$$

Attention must be given when choosing the buffer gases in such a way that they do not form stable negative ions or reactive species, in order to avoid competition reactions or reactions with neutral or charged analyte molecules, which inevitably lower the sensitivity of the analysis. Equally important is keeping matrix effects and impurities to a minimum. In addition, the vacuum pump speed must also be adequate to fulfill the pressure requirements of CI experiments.

1.2.2.4 Analytical Applications of Chemical Ionization

CI is not applied in combination with GC–MS as widely as is EI. In terms of analytical applications, the various modes of performing CI have different application areas. PICI is mainly used to determine or confirm the mass of the intact analyte molecule, for example, in cases where $M^{+\bullet}$ is not observed or is present with a very low relative intensity under EI conditions. In this context, PICI may become more important in GC–MS in the future, given the increasing use of SRM in tandem-quadrupole (TQ) instruments. The introduction of atmospheric-pressure chemical ionization (Section 1.2.5) for GC–MS is also highly interesting (van Bavel et al., 2015; Li et al., 2015). Different CI reactions can be achieved under those conditions, which are largely dependent on the reagent gas used and the instrumental parameters for attaining the sought-after results.

GC–MS with ECNI has found a wide range of applications in targeted quantitative analysis, for instance in forensic toxicology and pharmacology for the analysis of polar compounds. For such applications, pentafluoropropyl or pentafluorobenzyl ester derivatives are produced. As such, GC–ECNI-MS is routinely applied in forensic toxicology to determine illicit drugs, for instance for the presence of tetrahydrocannabinol (THC) in hair (Foltz, 1992; Moore et al., 2006). Enantioselective analysis of amphetamines has been reported after derivatization with (*S*)-(−)-*N*-(heptafluorobutanoyl)prolyl chloride (HFBPC) (Lim et al., 1993). HFBPC and its related compounds are very efficient chiral derivatizing reagent of amino groups (Leis & Windischhofer, 2012). GC–ECNI-MS also plays an important role in the analysis of environmental pollutants such as polybrominated compounds of both synthetic (polybrominated diphenyl ethers as fire retardants) and natural (polybrominated hexahydroxanthene derivatives) origins. In such cases, bromide ions (Br^-) are produced during dissociative ECNI (Eq. 1.19b). The high selectivity of the analysis lies in the production of ions with *m/z* 79 and 81 ($^{79}Br^-$ and $^{81}Br^-$ with ≈1:1 relative intensity) (Rosenfelder & Vetter, 2009).

Another possibility of dissociative electron capture leads to retention of charge by the analyte molecule, to effectively produce $[M–H]^-$ of the underivatized analyte, in combination with the production of a neutral radical ($X^\bullet$) leaving group (Eq. 1.19a). This behavior is applied in the GC–ECNI-MS analysis of fatty acids (RCOOH) such as arachidonic acid analogs after derivatization to their pentafluorobenzyl esters. In this case, the dissociative ECNI process leads to an ion corresponding to the deprotonated acid with *m/z* (M−1) and pentafluorobenzyl radical, as shown in Eq. 1.21 (Hadley et al., 1988).

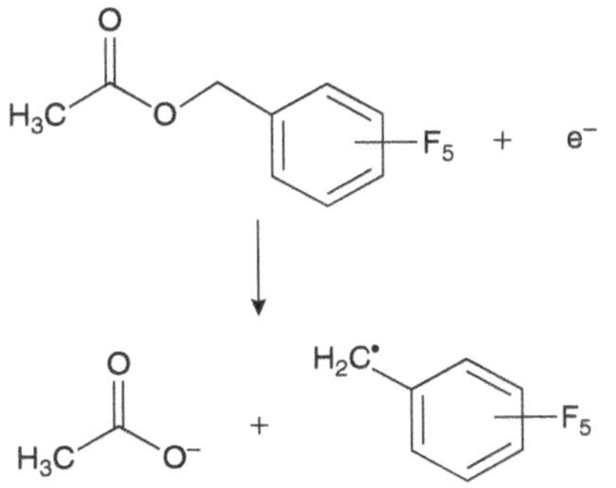

(1.21)

When comparing modes of ionization in CI, sensitivity is a parameter often employed to quantitatively gauge them. Inherently, neither NICI nor PICI is a more sensitive technique than the other. What determines the sensitivity is the number of extractable and detectable analyte ions present in the source at any time. For that reason and when possible, the relative second-order reaction rates in ECNI versus proton transfer and adduct formation in PICI are used to determine the sensitivity of a particular method. Generally speaking, electron-capture rate constants can be up to 1000 times larger or smaller than proton transfer, for example, methanol gas-phase H/D-exchange rate constant is $\approx 10^{-11}\ cm^3\ molecule^{-1}\ s^{-1}$, Green & Lebrilla, 1997). Therefore, CI experiments must be carefully planned to use G_R-analyte partners that will offer optimum sensitivity and selectivity.

1.2.3 Atmospheric-Pressure Ionization

GC enjoys the advantage of being able to deliver the analyte molecules inside the source in the gas phase, and that makes it suitable when using an EI source. Notwithstanding the technological challenges, precedents exist shortly after its development of GC coupling to MS (Holmes & Morrell, 1957). LC coupling to MS presents a greater challenge: analytes elute out of the LC column dissolved in liquid solvents of varying volumes and polarities (volatilities). The *conditio sine qua non* for MS is to have ions under vacuum and in the gas phase. Therefore, in order to couple LC to MS, devising a way to desolvate sample molecules, ionize, and transmit them to the high-vacuum environment of the mass analyzer was indispensable. Atmospheric-pressure ionization (API) sources were developed to achieve that task, and three kinds of API are routinely used: electrospray ionization (ESI), atmospheric-pressure chemical ionization (APCI), and atmospheric-pressure photoionization (APPI). API techniques provide soft-ionization processes where the post-ionization energy of analyte molecules is not large enough to cause extensive fragmentation (if any), with an ion related to the intact molecule (as a cationized or anionized molecule) usually present. Equally important, API techniques offer an alternative ionization way apt for polar, low volatility (high molecular mass), and thermolabile compounds. Figure 1.7 is an approximate chart showing the molecular mass and polarity ranges of application for the most common ionization techniques in MS.

The three techniques accomplish the same task in different but related ways, the main difference being the process of analyte ionization itself. Desolvation and ion transmission share the same electromechanical principles in all three techniques: sample nebulization in an atmospheric-pressure chamber, inert gasses and thermal energy for desolvation, and reduced pressure. The source is also designed to keep neutral molecules from reaching the detector (lower background noise).

Since the analyte is dissolved in the mobile phase, one must make sure that prior to mass analysis the removal of unwanted material is as complete as possible, for example, remnants of solvents, buffers, and additives used to guarantee the ionization of neutral compounds while avoiding signal suppression by interfering chemicals. Therefore, the use of volatile solvents and additives is indicated. In this respect, gradient elution must be carefully planned not to adversely affect the mass spectrum. A flow reduction of the eluting mobile phase leads to more efficient analyte desolvation and analyte ionization. Several techniques exist to reduce the flow rate to the ESI source such as pre-source flow split (for concentrated samples as well) or the use of $nL\,min^{-1}$ flow rates with LC columns of 10–100 μm internal diameter (Chervet et al., 1996).

In an API source, the coupling to an LC system column effluent or any other liquid flow is done via the sample inlet, where the liquid is nebulized into a fine aerosol of small droplets. The nebulization process in ESI (Section 1.2.4.1) differs from the one used for APCI and APPI (Section 1.2.5.1). In the course of droplet solvent evaporation mediated by heated desolvation gas, for example, nitrogen (N_2), analyte ionization is achieved by different processes in ESI (Section 1.2.4.2), APCI (Section 1.2.5.2), and APPI

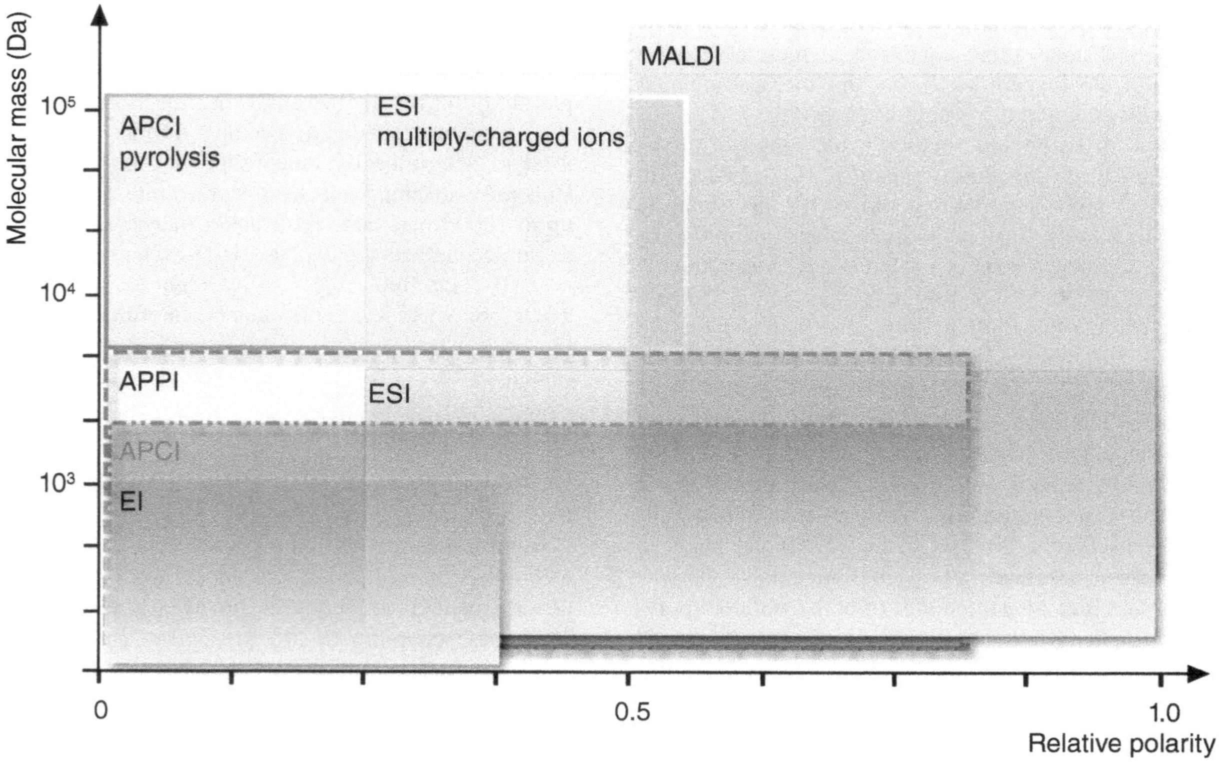

FIGURE 1.7 Approximate range of molecular mass and polarity for the most common ionization sources in MS.

(Section 1.2.5.3). The resulting mixture (containing analyte ions) is passed through the ion-sampling orifice into the first vacuum chamber of the differentially pumped interface between the API source and the mass analyzer. Before analyte ion transmission to a high-vacuum region, where mass analysis is performed, two or three stages of vacuum pumping are applied to remove as much mobile-phase vapors and N_2 as possible. A schematic diagram of an API source with an ESI inlet is shown in Figure 1.8. The API source parts are discussed in more detail, and inlet components are discussed separately for ESI (Section 1.2.4.1), APCI, and APPI (Section 1.2.5.1). Liquid nebulization at the sample inlet (electrospray needle or heated nebulizer) results in a fine aerosol of very small droplets. These droplets are then stripped of their solvent. In APCI and APPI, this is done within a heated nebulizer (Section 1.2.5.1), whereas in ESI, this is done in the API source. In order to achieve electrospray nebulization in ESI, a voltage difference (1–5 kV) is established between the electrospray needle and the ion-sampling orifice (Section 1.2.4.1). At higher flow rates (>10 μL min^{-1}), N_2 gas is used to assist and support the nebulization process (pneumatically assisted ESI). This voltage difference can be applied in two different ways: in some ESI sources, the ESI needle is grounded and the voltage is applied to the ion-sampling orifice region, whereas in ESI sources from most instrument manufacturers, the voltage is applied to the ESI needle. The ions generated are transmitted through the vacuum interface toward the mass analyzer by means of voltages applied at different points in the source: ion-sampling orifice, skimmer, and RF-only multipole ion guide.

When flow rates are in excess of 1 μL min^{-1}, solvent evaporation in ESI must be seconded by the application of heat. Typically, this is done using N_2 as heated desolvation gas. Depending on the instrument design, heat exchange between the plume and the heated-nitrogen flow is implemented in different ways, that is, concurrent flow, counter-current flow, and orthogonal flow (or just off-axis) to the direction of liquid introduction. The solvent evaporation and ionization process, discussed in more detail in Section 1.2.4.2, results in a mixture of analyte ions, N_2 gas, solvent vapors, and neutral analyte molecules in the API source. A small part of this mixture is sampled into the first vacuum chamber aided by a voltage applied to the ion-sampling orifice. The remainder of the mixture leaves the API source region via an exhaust connected to a fume hood at atmospheric pressure.

An important practical problem in operating an API source with a wide variety of samples is the contamination of the ion-sampling orifice area by nonvolatile materials present in the liquid flow. This may seriously compromise the performance of the ESI source. The most important design feature to reduce ion-sampling orifice contamination is the orthogonal sample introduction (Hiraoka et al., 1995). In addition, several different designs are available making use of counter-current dry N_2 desolvation gas flow (Bruins et al., 1987). This is done in order to push unwanted materials away from the area of the ion-sampling orifice (Cole, 2010).

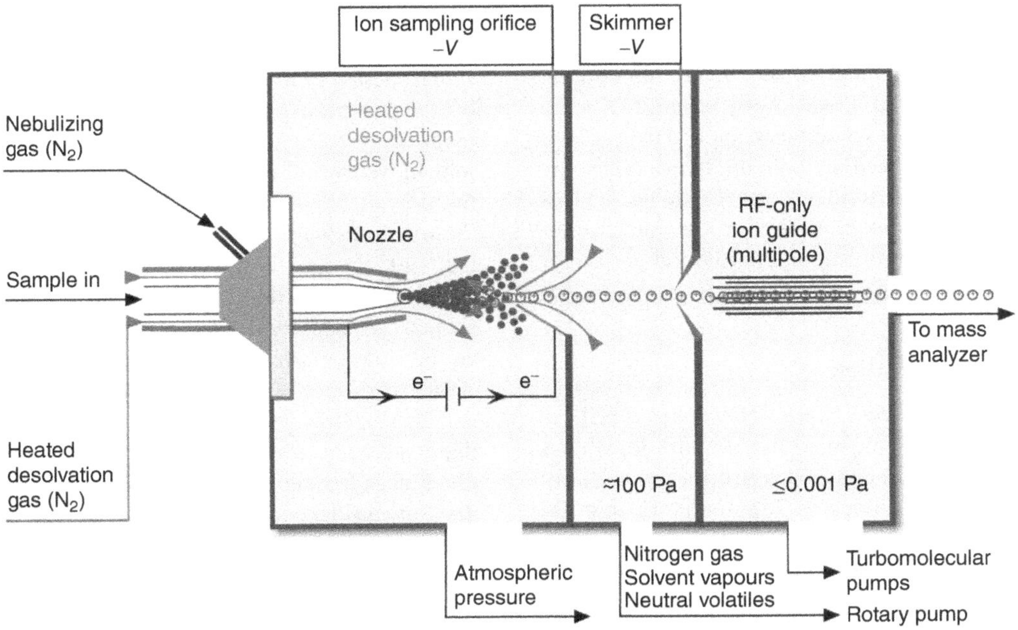

FIGURE 1.8 Schematic representation of an atmospheric-pressure ionization (API) source with an electrospray ionization (ESI) inlet.

The geometry of the ion-sampling orifice depends on the instrument at hand. In one of the original ESI source designs by Fenn (Whitehouse et al., 1985) and still in use today, a glass capillary with metallized inlet and outlet ends is used as ion-sampling orifice. It allows having different voltages at the inlet and outlet ends of the capillary, electrically decouples the API source region from the vacuum interface, and thereby enables the application of the high voltage to the ion-sampling orifice region rather than to the ESI needle. This also facilitates the coupling of capillary electrophoresis to MS. Other ion-sampling orifice designs include a heated stainless steel capillary and an orifice in a flat plate or a cone.

At the low-pressure (typically 100 Pa; 1 mbar) side of the ion-sampling orifice, expansion of the gas mixture occurs. Since the analyte ions usually have a higher mass than the N_2 and solvent molecules, they are preferentially found in the core of the expansion. Then, the core of the expansion is sampled by a skimmer into the second vacuum stage. Electrostatic or quadrupole lenses are applied between ion-sampling orifice and skimmer, that is, at the high-pressure side of the skimmer, in some API source designs to achieve focusing of the ions in order to enhance ion transmission in this region of the API source. The ions present at the low-pressure side of the skimmer are transmitted and focused through the vacuum chamber by means of RF-only multipole(s), which may be a quadrupole, a hexapole, or an octapole.

Source contribution to overall experimental performance will depend on the ionization and ion transmission efficiencies. Significant developments are made in trying to improve ion transmission from the vacuum interface to the mass analyzer. In some instrument designs, the ion optics described earlier (and even the skimmer) have recently been replaced with two-stage ion funnels, which consist of a series of parallel electrode plates orthogonally placed to the direction of ion transmission (Giles et al., 2004; Kelly et al., 2010). RF voltages and/or DC transient voltages are applied to guide the ions through them while speeding them up. In principle, any type of mass analyzer can be used, nevertheless some devices require additional instrumentation, for example, ion acceleration for time-of-flight instrument and pulsed ion-gating for ion-trap instruments.

1.2.4 Electrospray Ionization

1.2.4.1 Electrospray Nebulization Nowadays, ESI is the most widely used technique in MS for the ionization of liquid samples. Even though ESI has been successfully used as an ionization technique in MS for more than 30 years (Yamashita & Fenn, 1984a,1984b; Whitehouse et al., 1985), electrospray-related phenomena had already been recorded over 400 years ago by Gilbert (1600) who noted: "in the presence of a charged piece of amber, a drop of water deformed into a cone." Another more colorful example is the observation by Nollet (1754), who was the first to perform electrostatic spraying. While experimenting with human blood and electricity, he concluded that "a person, electrified by connection to a high-voltage generator, would not bleed normally if he were to cut himself; blood would spray from the wound". Electrospraying is a technique that exploits the electrohydrodynamic behavior of a liquid meniscus at the tip of a conductive hollow emitter, for example, a metallic or a contact electrode silica capillary, and under the influence of electrical shear stress (directly proportional to the voltage applied). This is done to electrostatically charge an electrically conductive liquid flowing through the emitter in order to atomize it, generating a spray containing charged self-dispersive microdroplets in a very fine aerosol. Electrospraying finds application in a wide variety of disciplines: aerosol sciences, coating processes[1], electronics, energy generation, food technology, fuel delivery, mass spectrometry, medical sciences, meteorology, mining, nuclear fission, just to mention a few. In ESI-MS, electrospraying or electrospray nebulization is achieved by placing the ESI emitter or ESI needle (electrode) under high voltage (1–5 kV) relative to the ion-sampling orifice (counter-electrode) (Figure 1.9a). In this way, a special electrolytic cell, where part of the ion transport is done in the gas phase, establishes a liquid–gas redox reaction (anode–cathode, respectively), with the flow of electrons as indicated (Kebarle & Verkerk, 2009). Accumulation of charge is effected at each electrode with the tip of the emitter been positively charged (in positive-ion mode). Heated nitrogen gas is supplied to the aerosol to assist in solvent evaporation and declustering of ion-solvent and ion-additive clusters that might have formed in the flowing liquid. The arrangement illustrated in Figures 1.9a is for the study of positive ions, but the sign is a matter of choice and negative-ion ESI is widely performed as well.

The action of the electric field in the solution makes (part of) the present negative ions undergo electrophoretic movement away from the counter-electrode, while the positive ions move toward it. The removal of negative charges from the flowing liquid causes a build-up of positive charge on the meniscus surface at the emitter tip (Figure 1.9b) (Bruins, 1998). Because of the potential difference applied, the liquid at the tip of the emitter is elongated into an elliptically shaped meniscus, where for every point of the surface there is equilibrium between the two main forces acting upon it: the cohesive surface tension that tends to hold the liquid back, and the electrostatic attraction from the counter-electrode that tends to draw the liquid out of the emitter. When the applied electric field is strong enough (GV m^{-1}) to overcome the (solvent-dependent) meniscus surface tension, the elliptically shaped meniscus suddenly elongates into a regular axisymmetric cone shape named Taylor cone (Taylor, 1964).

[1] Dole ran into electrospray during a visit to a car manufacturer where he saw car painting done with electrospraying (Dole et al., 1968; Mack et al., 1970).

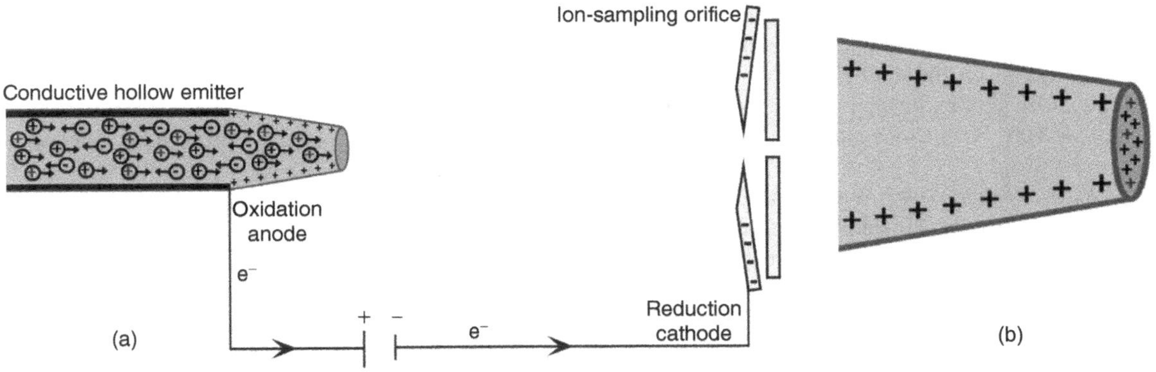

FIGURE 1.9 (a) Electrospray setup in a positive-ion ESI-MS experiment. (b) Positively charged meniscus formation at the electrospray emitter tip.

Depending on the experimental conditions, the Taylor cone can be linear, concave, or convex (Wilm, 2011). The meniscus adopts this shape because a cone can hold more charge than a sphere. The cone surface area depends on the flow rate of the liquid passing through the emitter. For a given experimental setup, there is a minimum flow rate below which formation of Taylor cone does not occur. Under controlled flow rate conditions, the threshold electric potential (V) needed for the formation of Taylor cone is directly proportional to $\sqrt{\gamma R}\ln(4d/R)$ (γ, the meniscus surface tension, R, the emitter tip inner radius, d, the distance between the emitter tip and the counter-electrode in Figure 1.9a) (Smith, 1986). The applied voltage accelerates the charges at the meniscus surface toward the apex of Taylor cone where the electric field is highest. When this electric field is strong enough, that is, the applied electric potential exceeds a certain threshold, the apex of Taylor cone becomes unstable and droplets of controllable and narrow size distribution (approximately monodisperse) begin to leave the cone apex forming a fine jet of charged droplets. Due to the electrostatic repulsion among the newly formed charged droplets and the V applied at the ion-sampling orifice, this jet is then radially dispersed into the so-called (spray) plume, containing a fine aerosol of charged droplets (Figure 1.10).

The liquid flow rate is an important experimental parameter in ESI-MS. Conventional ESI sources (e.g., Whitehouse et al., 1985) are limited to flow rates up to $10\,\mu L\,min^{-1}$. Since ESI-MS was specifically developed as an ionization technique for LC–MS, the possibility to operate the source at higher flow rates was thoroughly investigated. A spraying process by means of a surrounding high-speed N_2 gas flow (nebulizer gas) (Bruins et al., 1987) was introduced as ionspray. This term has become a registered trademark, and the more general term of pneumatically assisted ESI was adopted. As pneumatically assisted ESI is the most widely applied ionization method, in most cases it is simply called ESI. Other alternative spraying modifications that have not found wide (commercial) application include sonic spray (no

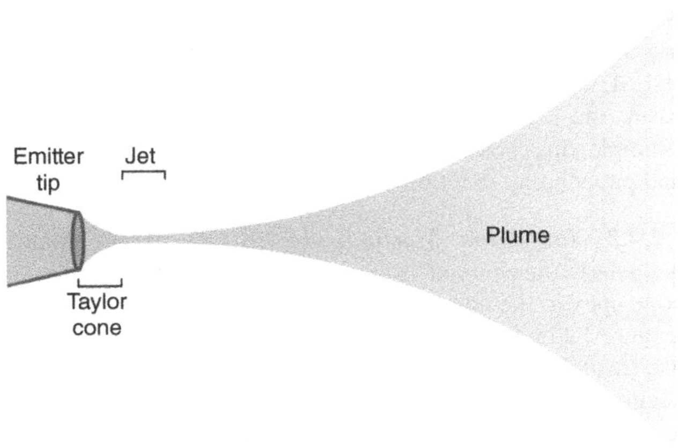

FIGURE 1.10 Taylor cone formation and charged aerosol generation in an ESI experiment.

use of electric field) (Hirabayashi et al., 1994, 1995), electrosonic spray (Schmid et al., 2011), and ultrasonic nebulizer ES (Banks Jr., et al., 1994).

An important development in the field was the introduction of nano-electrospray ionization (nano-ESI), which is extensively used today for applications where sample availability is limited, especially in the field of proteomics. Nano-ESI is a more efficient way of generating gas-phase ions than conventional and pneumatically assisted ESI. This is achieved by the use of low flow rates ($<100\,nL\,min^{-1}$), which requires narrower ESI needle capillaries, or capillaries with a narrower tip diameter (1–5 μm internal diameter). Borosilicate glass capillary emitters with electrically conductive coatings such as a sputtered gold film are used as nano-ESI needles. The low flow rate reduces the energy needed for droplet liquid evaporation, and no pneumatic assistance is needed. Therefore, the distance between the emitter tip and the counter-electrode can be shorter. As a result, a lower spraying voltage (0.5–1.5 kV) can be applied, permitting the use of very polar solvents. Once the voltage is applied the test solution flows by capillarity, refilling the emitter tip as droplets leave the Taylor cone apex. The flow rate and the droplet size depend on the diameter of the emitter tip. The aerosol thus formed contains nanodroplets (<200 nm in diameter), that is, 100–1000 times smaller in volume than conventional microdroplets (1–2 µm in diameter) seen in ESI experiments (Wilm & Mann, 1994; Wilm & Mann, 1996). In effect, more efficient analyte ionization is achieved and the loss of generated ions at the ion-sampling orifice is greatly reduced. The smaller droplets with higher surface-to-volume ratios lead to less discrimination effects where hydrophobic analytes are favored over hydrophilic analytes.

Variations in the way analyte ions are produced have also given way to newer ESI techniques such as fused droplet and extractive ESI. A more recent development is desorption techniques combined with electrospraying techniques. Sample molecules desorbed from a solid or liquid surface are made to ionize with ESI, for example, desorption ESI (DESI) and electrospray laser ionization (ELDI). In some cases, the use of an emitter can be avoided altogether, for example, direct electrospray probe (DEP), probe ESI (PESI), and paper spray (Section1.2.6).

1.2.4.2 Ionization Mechanisms in ESI What exactly happens to the charged droplets in their flight from the Taylor cone apex to the counter-electrode is still a matter of debate (Figure 1.11). Due to desolvation of the droplets with heated gas (usually N_2), they shrink in size as solvent evaporation occurs. Provided that charges are not shed from the droplet as the solvent evaporates, there is a decrease in droplet radius with a concomitant increase in charge density. This causes the uniformly distributed like charges (Q) on the droplet surface to be closer together, resulting in an increase in electrostatic repulsion. A stability limit (Rayleigh limit) is reached when the charge of the droplet (Q_R) and its radius (R) satisfy Rayleigh equation (Eq. 1.22), where ε_o is the vacuum permittivity (Rayleigh, 1882):

$$Q_R = 8\pi(\varepsilon_o \gamma R^3)^{1/2} \tag{1.22}$$

At $Q > Q_R$, the electrostatic repulsion is greater than the surface tension of the droplet. Then, the droplet breaks up in turn jetting out several smaller charged droplets, which have a much higher charge-to-mass ratio (Grimm & Beauchamp, 2002). This phenomenon is known as coulombic fission (explosion) and it is partly responsible for the formation of the spray plume. Evaporation of solvent continues and ultimately from ionized analyte–solvent nanodroplets free gas-phase ions are produced and accelerated toward the ion-sampling orifice counter-electrode.

Even though the exact mechanism of ESI is a complex physicochemical process where many variables play a role, for example, analyte solubility and surface activity, solvent polarity, and surface tension, there are two generally accepted models to explain the ionization process: the charge residue model (CRM; Dole et al., 1968; Mack et al., 1970) and the ion evaporation model (IEM; Iribarne & Thomson, 1976).

In the CRM after enough solvent evaporation-coulombic fission events, minute charged droplets (≈1 nm radius), each containing one analyte ion undergoes solvent evaporation until a solvent-free (gas-phase) analyte ion is obtained, thus retaining the droplet residual charge. Refinement of the model and experimental evidence have been put forward supporting the CRM model (Schmelzeisen-Redeker et al., 1989), for instance ionization of large macromolecules like proteins involves CRM (Winger et al., 1993).

In contrast, in the IEM, the liquid-phase analyte ions evaporate out of the droplets. As they leave the droplet, they remove a charge or several ones from it. After enough solvent evaporation-coulombic fission events, when a droplet has reached a radius of ≈10 nm, the electric field on the surface of the droplet is strong enough to make solvated charged analyte ions leave the charged droplets. Further desolvation results in a gas-phase ion. Experimental evidence exists supporting the IEM for small inorganic and organic ions (Kebarle & Verkerk, 2009). A modification of the IEM stipulates that the charge resides on the surface of the droplet, and the analyte

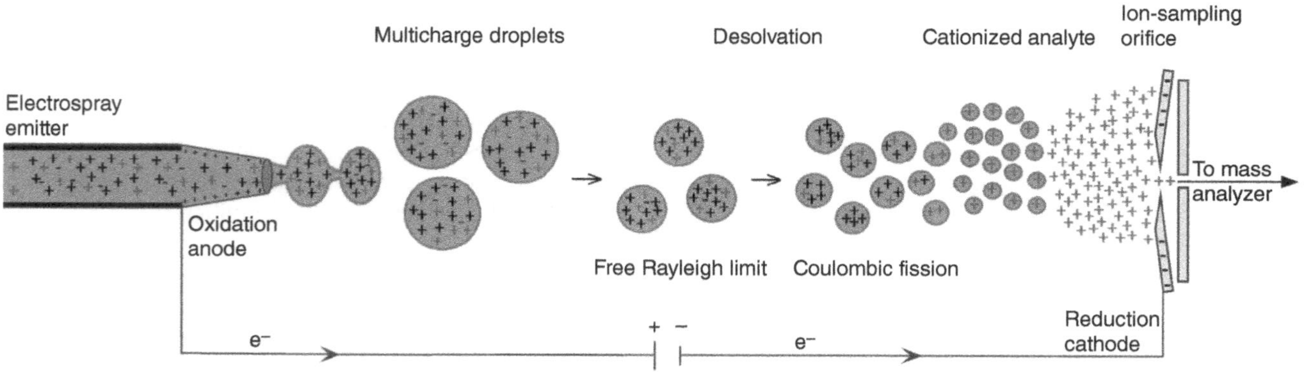

FIGURE 1.11 Analyte ion formation illustrating evaporation-coulombic fission events in positive-ion ESI-MS.

remains neutral inside it while in constant Brownian motion (Wong et al., 1988; Fenn, 1993; Nguyen & Fenn, 2007). After enough solvent evaporation-coulombic fission events, the surface charge density increases, for example, protons (H^+) in positive ESI. Because of the steric proximity of the surface protons to one or several electron-rich functional groups of the analyte, the analyte gets protonated. In this way, an analyte ion is produced, which finds itself in close proximity to the surface of the droplet. The newly formed ions or ionic complexes of analyte molecules then escape the droplet, carrying with them one or more charges. The ease with which an analyte ion leaves the droplet is reflected in the signal intensities of the mass spectrum. In general, the smaller the droplet from where the charged analyte escapes is, the more charges it will have. Many factors intervene such as shape, size, and fugacity of the analytes, ionic strength, and viscosity of the solvent. In fact, if other ionic species have more affinity for the droplet surface than the analyte molecules, ionization suppression is readily observed. The production of multiply-charged ions in ESI has been of tremendous importance for the analysis of macromolecules of both natural and synthetic origins (Fenn et al., 1989).

Usually in ESI, the type of reaction charging the analytes is a Brønsted–Lowry acid–base reaction. In positive ESI, ionization involves a protonation reaction resulting in $[M+H]^+$ with m/z (M+1), for example, an amine to an ammonium ion with m/z (M+1) (Eq. 1.23a). Multiple protonation reaction can occur as well, leading to $[M+nH]^{n+}$ with m/z $((M+n)/n)$ (Eq. 1.23b). However, Lewis acid–base reactions also occur leading to adduct formation such as $[M+Na]^+$ with m/z (M+23), (Eq. 1.23c). Doubly-charged ions can also be seen in positive ESI, for example, $[M+H+Na]^{2+}$ with m/z ((M+24)/2) (Section 2.2.8).

$$RNH_2 + H^+ \rightarrow [RNH_3]^+ \quad m/z\ (M+1) \tag{1.23a}$$

$$R(NH_2)_n + nH^+ \rightarrow [R(NH_3)_n]^{n+} \quad m/z\ (M+n)/n \tag{1.23b}$$

$$C_n(H_2O)_n + Na^+ \rightarrow [C_n(H_2O)_n{+}Na]^+ \quad m/z\ (M+23) \tag{1.23c}$$

In the negative mode, deprotonation reactions are common, leading to deprotonated molecules $[M-H]^-$ with m/z (M−1), for example, a carboxylic acid to a carboxylate ion (Eqs 1.24a and 1.24b). In addition, adduct formation occurs with anions such as chloride (Cl^-), acetate (CH_3COO^-), or formate ($HCOO^-$) leading to ions with m/z $(M+X)$, for example, where $X = 35$ for Cl^-, $X = 59$ for CH_3COO^-, or $X = 45$ for $HCOO^-$ (Eq. 1.24c). Production of multiply-charged (multiple deprotonation) negative ions also occur, for example, $[M-2H]^{2-}$ with m/z ((M−2)/2).

$$RCOOH + B \rightarrow RCOO^- + [BH]^+ \quad m/z\ (M-1) \tag{1.24a}$$

$$RCOOH + B^- \rightarrow RCOO^- + BH \quad m/z\ (M-1) \tag{1.24b}$$

$$C_n(H_2O)_n + Cl^- \rightarrow [C_n(H_2O)_n{+}Cl]^- \quad m/z\ (M+35) \tag{1.24c}$$

The contribution of ESI to the overall sensitivity of the analysis is a direct result of the efficiency of the two main events carried out in the source: ionization and ion transmission (Page et al., 2007).

The ionization efficiency in ESI measures the production of gas-phase ions from analyte molecules present in the solution. This is a multifaceted problem that depends on interface design (sample introduction), analyte properties, solvent composition, and flow rate. Analyte properties and solvent composition must be carefully evaluated in order to promote ion formation of the desired charge, while having the environment needed to perform the ESI. The flow is a twofold limiting factor: there is a minimum flow required for a stable LC performance and ESI operation, and ionization efficiency increases as the flow rate through the emitter decreases. This seems contradictory because for a given set of conditions, the total Taylor cone-jet ESI current is proportional to the square root of the flow rate (Fernández de la Mora & Loscertales, 1994). The explanation lies in the fact that as the flow rate decreases the size of the droplet in the Taylor cone diminishes and less evaporation-coulombic fission events are needed before having free gas-phase analyte ions. Also, as the droplet size decreases, there is an increase in charge density (provided charges are not shed while solvent evaporation happens), thus there is more charge available per analyte molecule.

Another factor that affects overall sensitivity is the ion transmission efficiency between the atmospheric-pressure regions of the source where ionization is done and the mass analyzer high vacuum. The ion-sampling orifice functions as a conductance limit allowing only a small fraction of analyte ions to go through. As the plume is formed, dispersion of the sample is done over an area much larger that the ion-sampling orifice. Therefore, sampling of ions is done only over a small fraction (10^{-3}–10^{-5}) of all the ions actually produced. Reducing the distance between the emitter and the ion-sampling orifice does increase the overall efficiency. However, as this distance gets smaller, droplet solvent evaporation time is reduced and solvent stripping might not be that effective. This hampers gas-phase analyte ion production, and the ion transmission efficiency is reduced. Alternatively, in order to sample a larger area of the plume spray, the size of the ion-sampling orifice can be increased. Unfortunately, this causes strain on the vacuum needs as well as turbulent flows through the ion-sampling orifice. Care must be taken to prevent that overall ion transmission efficiency is not lowered. Several examples of systems developed to improve ion transmission exemplify well the scope of the problem (Ibrahim et al., 2006).

Careful observation must be done to realize the impact that each source parameter has upon the strength of the signal observed (sensitivity) in the mass spectrum. Four key parameters are as follows: LC effluent and emitter flow rates, source temperature, desolvation gas flow, and distance between the emitter tip and the ion-sampling orifice (Figure 1.8).

1.2.4.3 Analytical Applications of ESI The introduction of ESI and especially the observation of multiply-charged ions for proteins and other biomacromolecules (Fenn et al., 1989; Covey et al., 1988; Mann et al., 1989) have significantly transformed the analytical application of MS. Although some ionization techniques introduced in the 1980s, for example, fast-atom bombardment and thermospray ionization, enabled the MS analysis of increasingly polar molecules and also allowed the MS analysis of larger biomolecules, it was the introduction of both ESI and matrix-assisted laser desorption ionization (MALDI) (Tanaka et al., 1988; Karas & Hillenkamp, 1988) that really opened MS analysis for biochemical and biotechnological applications.

With respect to ESI, the ability to generate multiply-charged ions for biomacromolecules with high efficiency enabled the accurate determination of the molecular weight of proteins, oligonucleotides, and other biomacromolecules. Compared to the established method at the time, that is, two-dimensional (2D) gel electrophoresis, ESI-MS provides an easier, faster, and more accurate molecular weight determination (Section 2.8) (Smith et al., 1990; Smith et al., 1991). In fact, the introduction of ESI-MS can be considered one of the most important enabling technologies for current proteomic workflows and research.

In addition, the ESI technology has been developed into a highly robust, sensitive, and more user-friendly interface for LC–MS (compared to previous interface designs). In this respect, ESI-MS is applied not only for the LC–MS analysis of biological and synthetic macromolecules but also in applications concerning small-molecule analysis. Over 90% of all LC–MS methods comprising many areas of application are performed using ESI as the ionization technique. This is also reflected in the data collected in this book: ESI was used as an ionization technique for more than 95% of the drugs and pesticides whose fragmentation characteristics are discussed in Chapter 4. Several alternative ionization techniques have been introduced and developed over the past 20 years, yet none of them challenges the position of ESI as the leading ionization and interface strategy in LC–MS.

1.2.5 Atmospheric-Pressure Chemical Ionization and Photoionization

APCI and APPI are ionization techniques that can be considered as alternatives to ESI in LC–MS. In contrast to ESI, where solvent evaporation and ion formation are tightly coupled, the two processes happen separately in APCI. This enables the use of low-polarity solvents that would not favor analyte ionization in ESI because of insufficient conductivity. Another important difference with ESI is the higher flow rates used ($\approx 1\,\mathrm{mL\,min^{-1}}$). In general, APCI and APPI are less affected by chemical interferences and enjoy high ionization efficiency. APCI has also been used for the analysis of flames and for environmental air pollution control. Commercial systems based on APCI have long been available in combination with both ion-mobility spectrometry (IMS) (Karasek, 1974) and mass spectrometry (Horning et al., 1973; Carroll et al., 1974; Carroll et al., 1981). As primary source of ionizing electrons, either a radioactive ^{63}Ni foil (Carroll et al., 1974; Carroll et al., 1981) or a corona discharge needle was applied (Shahin, 1966; Carroll et al., 1975). The actual commercial breakthrough of APCI for LC–MS can be attributed to the high-speed quantitative analysis of the drug phenylbutazone and three of its metabolites in plasma and urine using a prototype heated pneumatic nebulizer (Covey et al., 1986) and the (almost) simultaneous introduction of ionspray ionization mass spectrometry (Bruins et al., 1987). As a result, a commercially available API tandem mass spectrometry system was successfully introduced for LC–MS. The present discussion focuses on instrumentation, ionization, and application of APCI and APPI for LC–MS.

1.2.5.1 Instrumentation: The Heated Nebulizer The hardware of APCI for LC–MS is almost identical to the hardware used in ESI, that is, the same atmospheric-pressure ion source and atmospheric-pressure-to-vacuum interface are used (Section 1.2.3). However, there are two important changes to the hardware: (1) the ESI inlet device is replaced by a heated nebulizer, and (2) a corona discharge needle is installed in between the heated nebulizer and the ion-sampling orifice, off-axis and perpendicular to the incoming nebulized solvent flow (Figure 1.12).

The APCI heated nebulizer is a concentric pneumatic nebulizer attached to a heated quartz or stainless steel tube (300–600 °C). A high-flow stream of nitrogen (N_2) is used as nebulizer gas. The liquid flow (typically $1\,\mathrm{mL\,min^{-1}}$), for example, the LC column effluent, is nebulized into a fine aerosol of small droplets, which is passed through the heated vaporization region where droplet desolvation by evaporation occurs. The droplet evaporation leads to a soft desolvation of analyte molecules from the liquid stream. In this way, analyte molecules are transferred from the liquid phase to the gas phase and made amenable to gas-phase ionization at atmospheric pressure. Optimization of the temperature regimes in the heated nebulizer, leading to a first high-temperature zone (up to 800 °C) and a second low-temperature zone, has been described. This reduces thermal degradation and memory effects in the heated nebulizer and results in an overall improvement in performance (Covey et al., 2001).

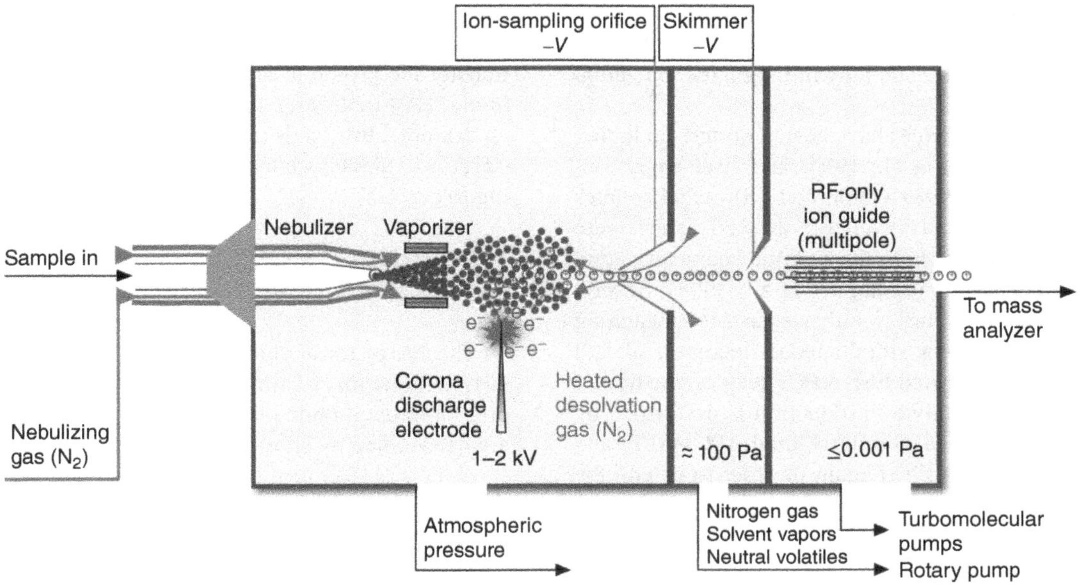

FIGURE 1.12 Schematic representation of an atmospheric-pressure ionization (API) source with an atmospheric-pressure chemical ionization (APCI) inlet.

Corona discharge is a pulsed atmospheric phenomenon (Saint Elmo's fire) that has been known for a long time. It resembles lightning in terms of an electrical discharge, in that it is an arc of very short duration and high power density. This produces visible plasma between a grounded object and a strong electric field in the atmosphere, such as those seen in thunderstorms created by volcanic eruptions. Thus, the presence of ionizable gases in the surroundings is a prerequisite. These arcs can happen between two electrodes, and furthermore they can be controlled in time and space. They are called transient discharges when they are in the nanosecond timescale (sparks are microsecond). In an APCI source, one uses an asymmetric electrode pair, where an electrode (corona discharge anode) having a strong curvature (e.g., sharp needle) is placed between the vaporizer and the ion-sampling orifice (cathode) (Figure 1.12). The corona discharge electrode is applied a voltage of 1–2 kV, thereby creating a high field region, which generates a transient discharge that spreads out toward the ion-sampling orifice. Notwithstanding the fact that it is a complex physicochemical process, corona discharge generators are technologically very reliable (van Veldhuizen & Rutgers, 2001).

The corona discharge acts as a source of primary electrons, and depending on the voltage applied to the needle (anode), the electrons produced will acquire several electronvolts of energy that are indispensable for the ionization of the reagent gas present in the APCI source. A corona discharge can be operated in either positive or negative mode, changing on whether the needle electrode is connected to the positive or the negative pole of the power supply, respectively.

In APPI-MS, the hardware is almost identical to that of APCI. However, no corona discharge is needed. The heated nebulizer is combined with a krypton (Kr) discharge lamp that replaces the corona discharge electrode. This lamp can

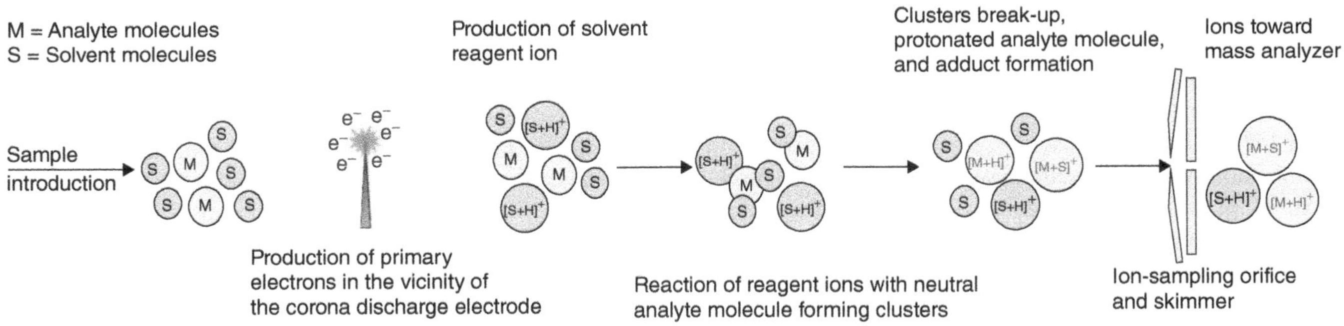

FIGURE 1.13 Schematic representation of the solute ionization process in positive-ion APCI-MS.

produce photons of 10.03 and 10.64 eV in a 4:1 ratio (Short et al., 2007), which interact with the vapors of the nebulized eluting solvent from the LC, thereby initiating the ionization process.

Several related API devices have been reported, including sonic spray (Hirabayashi et al., 1994) and laser spray ionization (Hiraoka et al., 1998). Combined ESI–APCI sources have become commercially available as well. They were especially developed for high-throughput characterization of combinatorial libraries (Gallagher et al., 2003). Several designs are available, featuring either scan-wise switching between ESI and APCI or simultaneous operation of ESI and APCI. In these combined ESI–APCI devices, the heated nebulizer is not used; analyte introduction is performed by pneumatically assisted ESI. Similarly, dual APCI/APPI and ESI/APPI have been described again in order to extend the applicability range of LC–MS in screening of combinatorial libraries and other studies in early drug discovery (Cai et al., 2005).

1.2.5.2 Ionization Mechanisms in APCI Like in conventional CI (Section 1.2.2), the ionization process in APCI is initiated by electron ionization, in this case by electrons from the corona discharge. The sequence of events is assumed to start with the ionization of the nitrogen bath gas (Huertas & Fontan, 1975; Carroll et al., 1981), according to Eq. 1.25.

$$N_2 + e^- \rightarrow [N_2]^{+\bullet} + 2e^- \quad (1.25)$$

In the presence of traces of water, the nitrogen molecular ion ($[N_2]^{+\bullet}$) enters a series of ion–molecule reactions, resulting in protonated water clusters, according to Eq. 1.26.

$$[N_2]^{+\bullet} + H_2O \rightarrow [H_2O]^{+\bullet} + N_2 \quad (1.26a)$$

$$[H_2O]^{+\bullet} + H_2O \rightarrow [H_3O]^+ + HO^\bullet \quad (1.26b)$$

$$[H_3O]^+ + nH_2O \rightarrow [H_3O(H_2O)_n]^+ \quad (1.26c)$$

The charge-exchange reaction (Eq. 1.26a) is likely to occur because the ionization energy of water (≈12.6 eV) is lower than that of nitrogen (≈15.6 eV). When an APCI-MS system is run with pure water as mobile phase, a series of protonated water clusters ($[H_3O(H_2O)_n]^+$) or solvated protons ($[(H_2O)_n+H]^+$) can be observed, with the ion with $n = 4$ is especially abundant due to the magic numbers determining the stability of such clusters (Tsuchiya et al., 1989).

In LC–MS applications, APCI is mostly performed in combination with reversed-phase LC (RPLC), featuring mixtures of water, an organic solvent, mostly acetonitrile (CH_3CN) or methanol (CH_3OH), and an ammonium-based buffer as the most frequently used mobile-phase constituents. This determines the gas/vapor mixture present in the APCI source. As a result, proton-transfer ion–molecule reactions can take place between the protonated water clusters and the mobile-phase constituents. The protonated water clusters transfer the proton to any species in the gas mixture with a higher PA (Table 1.1). In a CH_3CN–water mobile phase, as an example, this leads to the formation of mixed protonated CH_3CN–water clusters (with $m = 1–3$, and $p = 0–1$), according to Eq. 1.27.

$$[H_3O(H_2O)_n]^+ + CH_3CN \rightarrow [(CH_3CN)_m(H_2O)_p{+}H]^+ \quad (1.27)$$

The PA of these cluster ions resembles more the PA of CH_3CN than that of water. If an ammonium-based buffer is present in the mobile phase, the final reagent gas conditions are determined by ammonium-related ions (Eq. 1.28).

$$[H_3O(H_2O)_n]^+ + CH_3CN + [NH_4]^+ \rightarrow [(CH_3CN)_m{+}NH_4]^+ \quad (1.28a)$$

$$[H_3O(H_2O)_n]^+ + CH_3CN + [NH_4]^+ \rightarrow [(CH_3CN)_m(H_2O)_p{+}NH_4]^+ \quad (1.28b)$$

The size of any of these ion clusters depends on the experimental conditions ($m = 1–2$ is a common feature); the numbers given reflect the situation commonly observed in APCI for LC–MS. The ionization process can be described as a solvent-mediated CI process. It may be concluded that the PA of the APCI reagent gas in LC–MS is determined by the PA of the mobile-phase constituent with the highest PA.

Eventually, the protonated solvent cluster ions ($[S+H]^+$) enter into a proton-transfer ion–molecule reactions with the analyte molecules M, according to Eq. 1.29.

$$M + [S+H]^+ \rightarrow [M+H]^+ + S \quad (1.29)$$

This reaction proceeds to the protonated analyte molecule ($[M+H]^+$) as long as the PA of the analyte exceeds that of the protonated solvent cluster (Table 1.1). Initially, the analyte-related ions may be formed in clusters with water or/and solvent ions as well. Subsequent declustering is achieved by collision-induced processes in the vacuum interface.

The change of the APCI reagent gas as a function of the mobile-phase composition can be advantageous or disadvantageous. In principle, it introduces selectivity in the ionization process because analyte molecules with PAs lower than the PA of the cluster ions in the reagent gas are not ionized by proton-transfer CI. Thus, sample constituents with low PA are not ionized. On the other hand, in multiresidue analysis, the PAs of the target analytes may be greatly different. When the presence of an ammonium-based buffer is necessary in LC for those analytes in the mixture that show protolytic properties, the high PA of the reagent gas may exclude target analytes from being ionized. The ammonium-containing reagent gas predominantly ionizes analyte molecules containing N-atoms, whereas compounds without an N-atom are excluded (Table 1.1). This is an issue deserving attention, especially in untargeted screening for unknowns, for example, in toxicology, in environmental and food safety analysis, and in impurity profiling of drugs (Chapter 5).

In addition to proton-transfer CI, electrophilic addition or adduct formation can take place, which is frequently observed with ammonium-containing mobile phases (Eq. 1.30).

$$M + [S{+}NH_4]^+ \rightarrow [M{+}NH_4]^+ + S \quad (1.30)$$

The formation of an ammoniated analyte molecule, $[M{+}NH_4]^+$, can be observed as long as the PA of the analyte is within $\approx\pm 30\,kJ\,mol^{-1}$ of the PA of NH_3. As a result, adduct-ion formation broadens the applicability range of APCI-MS with a particular mobile-phase composition because analyte molecules with PA slightly lower than that of the protonated solvent cluster can be ionized. It must be pointed out that the above discussion is based on the thermodynamics of the proton-transfer reactions and electrophilic addition reactions. Several (mostly collision-induced) processes take place in between the actual ionization event and the moment the analyte ions enter the mass analyzer. As a result of declustering processes due to collisions in the API source or the vacuum interface, $[M{+}NH_4]^+$ may be dissociated and $[M{+}H]^+$ is observed instead.

Other reaction types, described for positive-ion CI, such as CECI (Section 1.2.2.2), do not readily occur under RPLC conditions. Due to the aqueous solvent conditions, solvent molecular ions that could enter into charge-exchange reactions are not available because they would have had already reacted with solvent constituents, especially water. Charge-exchange reactions may occur under normal-phase LC (NPLC) conditions, but this has not been thoroughly investigated.

In summary, protonated solvent clusters are generated by ion–molecule reactions initiated by the corona discharge. These cluster ions act as reagent gas ions in the solvent-mediated APCI. The reagent gas properties are determined by the mobile-phase constituent with the highest PA. Analyte ionization results from proton-transfer reactions or electrophilic addition reactions (Figure 1.13). The high-pressure conditions in the API source give way to a high collision frequency between the ions, neutrals, and reagents, explaining the high ionization efficiency in APCI. Since collisions are not very energetic, little or no fragmentation of the ionized analyte molecule is produced.

In negative-ion mode, a similar treatment holds. Reactions leading to analyte anion formation share mechanistic similarities with NICI and ECNI observed in conventional CI (Section 1.2.2.3). Proton transfer begins with an ionization reaction leading to superoxide ($[O_2]^{-\bullet}$) as an important initial ionic species. Ion–molecule reactions of $[O_2]^{-\bullet}$ with the mobile-phase constituents of RPLC predominantly lead to deprotonated solvent molecules (Eq. 1.31).

$$O_2 + e^- \rightarrow [O_2]^{-\bullet} \quad (1.31a)$$

$$S + [O_2]^{-\bullet} \rightarrow [S{-}H]^- + HOO^\bullet \quad (1.31b)$$

Alternatively, a direct reaction of a solvent molecule with a primary electron leads to the production of a base, which in turn deprotonates neutral analyte molecules (Eq. 1.32).

$$S + e^- \rightarrow [S{-}X]^\bullet + X^- \quad (1.32a)$$

$$M + X^- \rightarrow [M{-}H]^- + XH \quad (1.32b)$$

Although one could envisage the methoxide (CH_3O^-) or the cyanomethide anion ($^-CH_2CN$), the reagent gas in negative-ion conditions in practice is determined by the buffer constituents, that is, by deprotonated acids such as formate ($HCOO^-$) and acetate (CH_3COO^-) anions. Thus, proton-transfer reactions between the deprotonated solvent ($[S{-}H]^-$) and the analyte molecule M are observed, according to Eq. 1.33.

$$M + [S{-}H]^- \rightarrow [M{-}H]^- + S \quad (1.33)$$

The reaction is determined by the relative gas-phase acidities ΔH_{acid} of the analyte and the reagent gas molecules (Section 1.2.2.3). The proton-transfer or proton-abstraction reaction proceeds if the ΔH_{acid} of the solvent-related anion exceeds that of the analyte molecule. This limits the applicability of negative-ion APCI to compounds with acidic H-atoms, that is, carboxylic acids, phenols, and compounds exhibiting tautomeric equilibria (amide–iminol or alike). Again, adduct formation, for example, attachment of $HCOO^-$ or CH_3COO^-, can take place according to Eq. 1.34.

$$M + HCOO^- \rightarrow [M{+}HCOO]^- \quad (1.34)$$

For many polar molecules, the latter reaction is thermodynamically favorable, leading to the generation of $[M{+}HCOO]^-$ or $[M{+}CH_3COO]^-$ in mobile phases containing formic and acetic acid, respectively, as observed for corticosteroids (Section 4.6.6).

ECNI is an important ionization technique in GC–MS (Section 1.2.2). For APCI in LC–MS, the applicability of ECNI seems limited. It has been demonstrated that low-energy thermal electrons, generated in the initial step of the APCI process, can be captured by compounds with favorable electron affinities. Pentafluorobenzyl derivatives of steroids and prostaglandins could be detected with 25–100 times improved detection limits compared to conventional negative-ion APCI (e.g., Singh et al., 2000; Mesaros et al., 2010).

Charge exchange, although not very common, can happen according to Eq. 1.35.

$$X^{-\bullet} + M \rightarrow M^{-\bullet} + X \quad (1.35)$$

Superoxide ($[O_2]^{-\bullet}$) can be the radical anion initially formed. Nitroaromatics undergo charge exchange most likely with mobile-phase solvent-based radical anions, for example, methanol ($[CH_3OH]^{-\bullet}$). An example of this is the analysis of 1,3,5-trinitrotoluene (TNT), which shows both $M^{-\bullet}$ with *m/z* 227 and $[M-H]^-$ with *m/z* 226. In addition, two main fragment ions were observed with *m/z* 210 and 197 after the loss from $M^{+\bullet}$ of either the hydroxyl ($^\bullet$OH) or the nitrosyl ($^\bullet$NO) radical, respectively (Holmgren et al., 2005).

With some analytes, apparent fragmentation occurs under APCI-MS conditions. In most cases, these fragment ions are due to the ionization by APCI of thermal degradation products generated in the heated nebulizer. An interesting example of this process is the thermally induced reduction of an aromatic nitro group ($-NO_2$) into an amine group ($-NH_2$), observed in the positive-ion APCI-MS analysis of aromatic nitro compounds (Karancsi & Slégel, 1999). A fragment due to the loss of 30 Da was observed. H/D-exchange experiments showed that this loss is not due to the loss of $^\bullet$NO, but rather due to the indicated reduction.

1.2.5.3 Ionization Mechanisms in APPI APPI was introduced as a new ionization technique for LC–MS in 2000 by two groups simultaneously (Robb et al., 2000; Syage et al., 2000). APPI-MS was considered a highly promising alternative to ESI-MS and APCI-MS. It has been extensively reviewed (Raffaelli & Saba, 2003; Bos et al., 2006; Robb & Blades, 2008; Marchi et al., 2009). Two different APPI source designs have become commercially available.

The initial concept of APPI is that the absorption of a photon ($h\nu$) from the krypton (Kr) lamp results in an electronically excited molecules ([M]*) with sufficient energy for the ejection of an electron to happen, with the formation of the analyte molecular ion ($M^{+\bullet}$) (Eq. 1.36).

$$M + h\nu \rightarrow [M]^* \rightarrow M^{+\bullet} + e^- \quad (1.36)$$

Ionization happens if the photon energy is larger than the first ionization energy of the target compound ($h\nu > IE_M$), and a single photon ionization occurs, forming a molecular ion $M^{+\bullet}$ (or $M^{-\bullet}$). The reason for using a Kr discharge lamp is because the energy of the photons produced (10.03 eV) is greater than the ionization energies (IE) of most organic compounds (7–10 eV) and lower than the IE for the most commonly used LC solvents, for example, methanol (IE = 10.8 eV), water (IE = 12.6 eV), and acetonitrile (IE = 12.2 eV) (Robb et al., 2000). Since argon (Ar) discharge lamps can emit photons with an energy of 11.7 eV, it generates ≈100-fold more solvent ions than when using Kr lamps, and the abundance of $M^{+\bullet}$ is higher as well. In general, Kr lamps give a better signal-to-noise ratio for a low solvent flow rate and Ar lamps do so for high solvent flow rates (Marchi et al., 2009).

In case $IE_M > h\nu$, [M]* may undergo de-excitation mainly via photodissociation (Eq. 1.37a), photon emission (Eq. 1.37b), or collisional quenching with gases present in the source (Eq. 1.37c).

$$M^* \rightarrow A + B \quad (1.37a)$$

$$M^* \rightarrow M + h\nu \quad (1.37b)$$

$$M^* + G \rightarrow M + G^* \quad (1.37c)$$

This direct-APPI approach is primarily promoted by one of the two initial research groups (Syage et al., 2000; Hanold et al., 2004). However, $M^{+\bullet}$ shows a high tendency to react with other compounds in the API source. As a result, the direct-APPI process is not very efficient. Alternatively, an easily ionizable compound, that is, a so-called dopant D, can be added to the mobile phase or to the nebulizing gas to enhance the response. Toluene, chlorobenzene, anisole, or acetone is frequently used as dopant (Robb et al., 2000; Kauppila et al., 2004a). In the presence of a dopant, the APPI process occurs via a charge-exchange reaction between the dopant molecular ion ($D^{+\bullet}$) and the analyte molecule M (Eq. 1.38). This reaction proceeds only when the EA of the analyte is higher than the EA of the dopant.

$$D + h\nu \rightarrow D^{+\bullet} \quad (1.38a)$$

$$D^{+\bullet} + M \rightarrow D + M^{+\bullet} \qquad EA_M > EA_D \quad (1.38b)$$

Unfortunately, although in both direct-APPI and dopant-APPI an analyte $M^{+\bullet}$ would be expected, a protonated molecule $[M+H]^+$ is observed for many analytes. This is due to ionization of the mobile-phase constituents by APPI. In the direct-APPI approach, the protonated molecule is formed due to a reaction of the analyte $M^{+\bullet}$ with a solvent molecule S (Eq. 1.39a):

$$M^{+\bullet} + S \rightarrow [M+H]^+ + [S-H]^\bullet \quad (1.39a)$$

The formation of $[M+H]^+$ is especially important in protic solvents, that is, under RPLC conditions. Similar processes are applicable in negative-ion APPI (Kauppila et al., 2004b). If the PA of the analyte is higher than the PA of the deprotonated dopant ion, solvent molecules can serve as intermediates between the dopant ion and the analyte M, leading to the formation of $[M+H]^+$ when the $PA_M > PA_S$ (Eqs 1.39b and 1.39c). In dopant-APPI, the formation of $[M+H]^+$ is attributed to internal proton rearrangement in the solvated dopant ion clusters (Robb & Blades, 2005).

$$D^{+\bullet} + nS \rightarrow [S_n+H]^+ + [D-H]^\bullet \qquad PA_S > PA_{[D-H]^\bullet} \quad (1.39b)$$

$$[S_n+H]^+ + M \rightarrow [M+H]^+ + nS \qquad PA_M > PA_S \quad (1.39c)$$

The prospect of being able to generate $M^{+\bullet}$ of analyte molecules under LC–MS conditions is that via fragmentation by collision-induced dissociation (CID) in MS–MS product-ion mass spectra can be obtained that show great resemblance to EI mass spectra, and thus may be searched against the large mass spectral libraries available (Section 1.2.1.4).

1.2.5.4 Analytical Applications of APCI and APPI Both APCI and APPI can be used as an alternative to ESI in LC–MS, especially in the analysis of the less polar analytes (Figure 1.7). APCI can be effectively used in the analysis of a wide variety of drugs in biological matrices. Compared to ESI, the ionization process of APCI is less prone to ionization suppression by matrix effects (Matuszewski et al., 2003; Matuszewski, 2006). Therefore, if similar sensitivity can be reached in quantitative bioanalysis, APCI should often be preferred over ESI in method development. Despite this, many researchers continue working with ESI, probably because they need to invest some effort in understanding the specific practical features of APCI. Perhaps the instrument manufacturers should invest more effort in optimizing the performance of their APCI devices.

Despite its initial promise, APPI did not become a major ionization technique in LC–MS. From a recent review (Marchi et al., 2009), a clear view can be obtained on the most important application areas of APPI, that is, especially for the analysis of steroids and polycyclic aromatic hydrocarbons and synthetic organic chemicals.

Over the years, a large number of comparative studies of the performance of APPI relative to APCI and ESI has been reported, for example, in the LC–MS analysis of flavonoids in plant extracts (Rauha et al., 2001), anabolic steroids for sports doping analysis (Leinonen et al., 2002), dinitropyrene and aminonitropyrene in biological matrices (Straube et al., 2004), cyclosporin A in rat plasma (Wang et al., 2005), lipids (Cai & Syage, 2006), chiral pharmaceuticals by NPLC (Cai et al., 2007), estrogens in water (Lien et al., 2009), hexabromocyclododecane enantiomers in environmental samples (Ross & Wong, 2010), several drugs in municipal wastewater (Garcia-Ac et al., 2011), environmental contaminants in water (Wang & Gardinali, 2012), and ergot alkaloids from endophyte-infected sleepy grass (*Achnatherum robustum*) (Jarmusch et al., 2016).

In general, considerable attention was paid at optimizing solvent composition to obtain optimum performance. For the quantitative analysis of flavonoids, negative-ion ESI gave the best results (Rauha et al., 2001). In negative-ion mode, flavonoids show less fragmentation than in positive-ion mode. Therefore, the negative-ion mode is preferred for quantitative analysis, while the positive-ion mode is more favorable for confirmation of identity or identification purposes. For the analysis of anabolic steroids, positive-ion ESI was found best for the purpose. Although in-source fragment ions involving the loss of water were observed with all three ionization methods tested, they were far more abundant in APCI and APPI (Leinonen et al., 2002).

Dinitropyrenes and their metabolites aminonitropyrenes and diaminopyrenes may be used as biomarkers for monitoring human exposure to diesel engine emissions. Dinitropyrene itself is not effectively ionized by ESI. Best results were obtained with APPI, where $[M-30]^{+\bullet}$ for dinitropyrene and $[M+H-30]^{+}$ for aminonitropyrene were observed, with similar detection limits in RPLC and NPLC (Straube et al., 2004).

For cyclosporin A, comparable results were obtained with all three ionization techniques (Wang et al., 2005).

For free fatty acids and their esters, monoacylglycerols, diacylglycerols, and triacylglycerols, APPI is two to four times more sensitive than APCI and much more sensitive than ESI, unless mobile-phase additives such as ammonium formate or sodium acetate are used in combination with ESI (Cai & Syage, 2006). The ability to use APPI under NPLC conditions can be useful in chiral separations. This was tested for several chiral drugs. In comparison with APCI, APPI generated higher peak area as well as lower baseline noise, that is, 2–500 times better signal-to-noise ratio (S/N) (Cai et al., 2007).

In the analysis of estrogens, that is, estrone, 17β-estradiol, estriol, 17α-ethinylestradiol, 4-nonylphenol, 4-*tert*-octylphenol, and bisphenol A, in sewage treatment plant effluent and in river water, the best performance was achieved for the dansylated derivatives in positive-ion ESI. For native compounds, negative-ion ESI outperformed APCI, APPI, and a combined APCI/APPI, all operated in negative-ion mode (Lien et al., 2009). In another study on native estrogens and other steroids, that is, testosterone, equilenin, progesterone, equilin, 17β-estradiol, 17α-ethynylestradiol, estrone, androsterone, mestranol, and estriol (Wang & Gardinali, 2012), negative-ion APPI using toluene as dopant was found to provide a better performance than ESI and APCI.

Anion-attachment APPI with 1,4-dibromobutane as dopant generated $[M+Br]^{-}$ for hexabromocyclododecane enantiomers. Compared to APPI and ESI, anion-attachment APPI shows better S/N and reduced matrix effects (Ross & Wong, 2010).

In the analysis of cyclophosphamide, methotrexate, bezafibrate, enalapril, and orlistat in wastewater samples, ESI provided better S/N than APCI and APPI (Garcia-Ac et al., 2011).

In the analysis of ergot alkaloids in extracts of the grass *Achnatherum robustum* infected with the *Epichloë* fungus, comparable results were obtained with ESI and APPI (Jarmusch et al., 2016). From these comparative studies, it may be concluded that all three ionization methods, ESI, APPI, and APCI, have their merits with specific analytes or samples.

1.2.6 Other Ionization Techniques

Apart from EI, CI, ESI, APCI, and APPI, there is a wide variety of other ionization techniques that are used or have been used in MS. Some of these techniques are briefly discussed in this section.

1.2.6.1 Energy-Sudden Desorption Ionization Techniques In two interesting review papers, it was concluded that many soft ionization techniques have some common features (Arpino & Guiochon, 1982; Vestal, 1983): they all need some kind of matrix component and the ionization is effected by a short-duration energy input, thus the term "energy-sudden" ionization methods was coined (Vestal, 1983). The ionization techniques discussed include field ionization and field desorption ionization (FDI), ^{252}Cf plasma desorption ionization (PDI), fast-atom bombardment (FAB), laser-desorption ionization (LDI), and thermospray ionization (TSI). Ionization techniques such as ESI and MALDI, which had not been introduced at the time, readily fit this model. The matrix involved can be a specific compound mixed with the analyte to achieve analyte ionization, that is, nitrocellulose in PDI; glycerol in FAB; or sinapinic acid (SA), 2,5-dihydroxybenzoic acid (DHB), or α-cyano-4-hydroxycinnamic acid (CHCA) in MALDI, while in TSI and ESI the matrix is the liquid phase from which droplets are generated. Important common processes in the ionization mechanism of these methods are the formation of analyte ions in the sample matrix prior to evaporation or desorption combined with rapid evaporation prior to ionization. The latter is achieved by very rapid heating or by sputtering with high-energy photons or particles. The energy deposited on the sample surface, which can also come from a strong electric field as in ESI, can provide preformed ions in the condensed phase with sufficient kinetic energy to leave the matrix and/or can cause (gas-phase) ionization reactions to occur near the interface of the solid or liquid and the vacuum (the so-called selvedge).

In general terms, a simplified view on the ionization process involves the formation of primary matrix ions because (clusters of) matrix molecules (S) undergo acid–base reaction (Eq. 1.40a).

$$\mathrm{S + energy \rightarrow [S{+}H]^+ + [S{-}H]^-} \qquad (1.40a)$$

This is followed by secondary ion formation from matrix ion–analyte reactions. The ions produced are usually protonated (Eq. 1.40b) or deprotonated analyte molecules (Eq. 1.40c) although radical cations or anions can also be observed.

$$\mathrm{[S{+}H]^+ + M \rightarrow S + [M{+}H]^+} \qquad (1.40b)$$

$$\mathrm{[S{-}H]^- + M \rightarrow S + [M{-}H]^-} \qquad (1.40c)$$

In FDI, the sample is deposited on a thin FDI emitter (a few μm in diameter, activated to provide for carbon microneedles on the surface). The emitter is kept at a high potential (>5 kV) in the high-vacuum ion source and a current is passed through to achieve slow heating of the emitter. As a result, nonvolatile analytes can be desorbed and ionized by various mechanisms (Beckey, 1977; Schulten, 1982; Lattimer & Schulten, 1989). For nonpolar analytes, mainly $M^{+\bullet}$ (by electron tunneling from the sample molecules into the emitter) is observed, whereas for more polar analytes like glycosides, lipids and peptides $[M{+}H]^+$ and/or $[M{+}Na]^+$ are observed. Liquid injection field desorption ionization (LIFDI) is a more user-friendly alternative to FDI because it enables sample application to the emitter while keeping the system under vacuum (Linden, 2004).

In FAB and liquid secondary ion mass spectrometry (LSIMS), the analyte of interest is dissolved in an appropriate matrix solvent, such as glycerol, diethanolamine, or other polar solvents with low vapor pressure. The solution is applied as a thin film onto a metal target, which subsequently is exposed to a beam of high-energy particles, that is, Ar or Xe atoms or Cs^+ ions. Analyte ionization is achieved by three processes: (1) desorption of preformed ions by energy transfer upon particle impact, (2) desolvation of preformed ions in the splash droplets resulting from disruption of the liquid layer upon particle impact, and (3) gas-phase ion–molecule reactions in the selvedge (Barber et al., 1981; Bélanger & Paré, 1986; Fenselau & Cotter, 1987). In FAB mass spectra, mostly $[M{+}H]^+$, $[M{+}Na]^+$, and/or $[M{+}K]^+$ are detected in positive-ion mode and $[M{-}H]^-$ in negative-ion mode, often along with some fragment ions. Apart from the analyte-related ions, (abundant) background ions are observed, which are due to matrix cluster ions, for example, $[(\text{glycerol})_n{+}H]^+$ with $n = 1$–10. With the introduction of ESI and MALDI, FAB has become obsolete.

TSI was developed in the mid-1970s as an interface for LC–MS (Blakley et al., 1978; Blakley & Vestal, 1983; Arpino, 1990, 1992). During this development, the system evolved from highly complex hardware into an easy-to-use interface for LC–MS that has been successfully commercialized and applied in the 1980s and the early 1990s, until it started to lose territory in favor of ESI. Nowadays, TSI is obsolete as an LC–MS interface. The heated source block of a typical TSI-MS system contains a gas-tight cylindrical tube, which at one end has the vaporizer probe and at the other end has a connection to a rotary pump. It is equipped with a (liquid-nitrogen) cold trap to avoid pump-oil contamination by solvent vapors. A temperature sensor is placed downstream to monitor the temperature of the vapor jet. The ion source contains an off-axis ion-sampling cone that acts as the entrance slit to the mass analyzer and opposite to that of a repeller electrode. A filament behind an electron entrance slit and/or a discharge electrode may be positioned upstream (Blakley & Vestal, 1983; Vestal & Fergusson, 1985).

The TSI hardware provides several ionization modes. Apart from electron-ionization initiated processes, where electrons from the filament or discharge electrode are the primary source of ionization, two liquid-based ionization modes are available. With ionic analytes and preformed ions in solution, ionization can be achieved by ion evaporation processes (Section 1.2.4.2). With neutral analytes, TSI buffer ionization is predominant: ionization takes place by either gas-phase ion–molecule reactions or rapid proton-transfer reactions at the selvedge. In the latter case, the addition of ammonium acetate or any other volatile buffer to the LC effluent is obligatory (Blakley et al., 1980; Vestal, 1983; Arpino, 1990; Katta et al., 1991).

1.2.6.2 Matrix-Assisted Laser Desorption Ionization MALDI was introduced in 1988 by two research groups simultaneously (Tanaka et al., 1988; Karas & Hillenkamp, 1988). MALDI was developed as a solution for the ionization of nonvolatile and high-molecular mass analytes. In a typical MALDI experiment, a mixture of the sample solution and an appropriate matrix solution is deposited onto a metal target (sample holder). Upon drying, co-crystallization of matrix and analyte molecules takes place. However, analysis of (nonvolatile) liquid matrices containing the analyte of interest is also possible when compounds are capable of absorbing UV and/or IR radiation. The matrix serves several purposes, that is, it is the mechanical support for the analyte, it reduces intermolecular hydrogen bonding and thereby results in isolated analyte molecules, while it also serves as energy-transfer agent between the excitation source and the analyte in question. The crystals on the target under vacuum are then bombarded by laser pulses delivering photons with an energy that matches the maximum absorption frequency of the matrix, for example, with 337 nm from an N_2 laser for the matrices mentioned earlier. A two-step ionization process is assumed to take place, where first the laser energy is absorbed by the matrix molecules, which are then desorbed and ionized by protonation. In the hot plume generated in this ablation step, proton transfer between matrix ions and analyte molecules leads to protonated analytes (Knochenmuss, 2006; Karas & Krüger, 2003). Gas-phase analyte ions are generated in the selvedge, according to Eq. 1.40. The ions produced tend to be stable and undergo little or no fragmentation, thus the protonated analyte is an important feature of the mass spectrum (Figure 1.14). The ions generated can be mass analyzed, which is mostly done using a time-of-flight mass spectrometer (Section 1.3.4).

MALDI is an important ionization technique for peptides and proteins. Unlike in ESI, where ion envelopes of multiply-charged ions are generated (Sections 1.2.4.3 and 2.8), in MALDI-MS mostly singly-charged $[M+H]^+$ ions are generated, together with less abundant doubly-charged $[M+2H]^{2+}$ and proton-bound dimeric $[2M+H]^+$ ions.

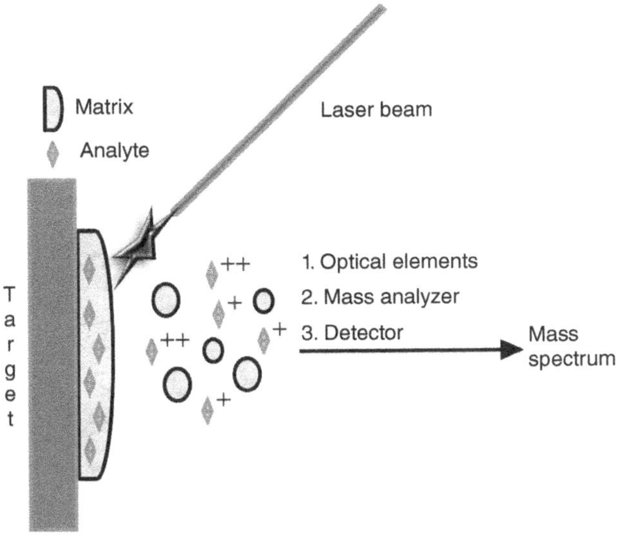

FIGURE 1.14 Schematic representation of analyte evaporation and ionization in a MALDI-MS experiment.

In addition to the analysis of peptides and proteins, MALDI is especially useful in the analysis of bio(macro) molecules such as oligosaccharides and glycans, oligonucleotides, and lipids, but it is also used for synthetic polymers, and in some cases even of small molecules such as drugs and antibiotics (Mann & Talbo, 1996; Karas, 1996). Moreover, MALDI-MS plays an important role in imaging mass spectrometry (Angel & Caprioli, 2013) and in the identification of bacteria and microbial fingerprinting (Clark et al., 2013), which are two emerging application areas of MS.

1.2.6.3 Atmospheric-Pressure Desorption Ionization Techniques In recent years, several atmospheric-pressure desorption ionization techniques have been introduced (Van Berkel et al., 2008; Huang et al., 2010). The first and most widely used technique is DESI. In DESI, the high-velocity spray of charged microdroplets from a (pneumatically assisted) electrospray needle is directed at a surface, which is mounted in front of the ion-sampling orifice of an API source. Surface constituents are released and ionized. These gas-phase ions are then introduced into the MS (Takáts et al., 2005). DESI-MS enables analyte ionization directly from solid surfaces. No extensive sample pretreatment or prior separation is performed. DESI-MS has been applied in the analysis of drugs in tablets, for example, for illicit drugs, or of natural products in plants. Chemical imaging of surfaces such as thin-layer chromatography plates and tissue sections can also be performed.

Among the other atmospheric-pressure desorption ionization techniques available are atmospheric-pressure matrix-assisted laser desorption ionization (AP-MALDI) (Creaser & Ratcliffe, 2006), direct analysis in real time (DART) (Cody et al., 2005), and atmospheric-pressure solids analysis probe (ASAP) (McEwen et al., 2005).

1.3 MASS SPECTROMETER BUILDING BLOCKS

1.3.1 Introduction

Mass spectrometry involves the generation of gas-phase ions from analyte molecules, the subsequent separation or mass analysis of these ions according to their *m/z*-values, and their detection. The instrument must be equipped with computing capabilities for instrument setup, data acquisition, and (advanced) data processing (Figure 1.15). Prior to analyte ionization, sample introduction must be performed. This may involve the introduction of individual samples by means of a direct insertion technique. However, sample introduction via hyphenated chromatographic techniques, that is, GC or LC, is performed more frequently. In such a hyphenated setup, the mass spectrometer can be used as a detector to provide mass spectrometric information on the analytes eluting after a chromatographic separation. GC–MS and LC–MS are very powerful and widely used analytical tools in many areas of chemistry, pharmacy, biology, and plenty other fields.

In analytical chemistry, six basic types of mass analyzers are used, two that provide unit-mass resolution, that is, the quadrupole and the ion-trap mass spectrometers, and four that provide high-resolution accurate-mass (HRAM) analysis, that is, time-of-flight (TOF), sector, orbitrap and Fourier-transform ion cyclotron resonance (FT-ICR) mass spectrometers. Except for sector instruments where other definition applies, the resolution of a high-resolution mass spectrometer is measured from the FWHM for a given *m/z* value (Section 2.6). The value for resolution is calculated from the ratio of *m/z* and FWHM. Instruments providing unit-mass resolution show FWHM of $\approx$0.7 for singly-charged ions over the entire applicable *m/z* range.

In addition to mass spectrometric resolution, the achievable mass accuracy is another important figure of merit (Section 2.6). When proper calibration of the *m/z* axis is performed, a unit-mass resolution instrument can provide a mass accuracy of $\pm$0.1 for singly-charged ions over the entire applicable *m/z* range. For high-resolution instruments, accurate-mass determination can be achieved, currently down to 1 ppm. For singly-charged ions with *m/z* <1000, this means that the error is in the third decimal place (e.g., >0.001). Internal mass calibration or frequent external calibration is required to routinely maintain the high mass accuracy of HRAM instruments.

An MS experiment generally requires high-vacuum conditions (pressure $\leq 10^{-3}$ Pa; $\leq 10^{-5}$ mbar) in both the mass analyzer and the ion detection system. Depending on the technique applied, analyte ionization may be performed either in high vacuum or at atmospheric pressure (Section 1.2). In the latter case, a vacuum interface is required to transfer ions from the API source into the high-vacuum mass analyzer region.

1.3.1.1 Basic Data Acquisition and Data Processing

In its basic operation in GC–MS or LC–MS, the mass spectrometer can be set to continuously acquire mass spectra between a low *m/z* and a high *m/z* within a preset time period ($\leq$1 s). This is the full-scan mode. However, since several types of mass spectrometers do not actually scan, the more general term full-spectrum mode is preferred over the term full-scan mode. The ionization technique applied and the resolution of the mass spectrometer utilized determine the information content of the mass spectrum. Initially, the mass spectra are acquired in continuous or profile mode, for example, with $\approx$10 data points per *m/z* value for a unit-mass resolution instrument, whereas far more data points per *m/z* are required in HRAM-MS to provide the appropriate resolution and mass accuracy. The data system digitally stores the information in

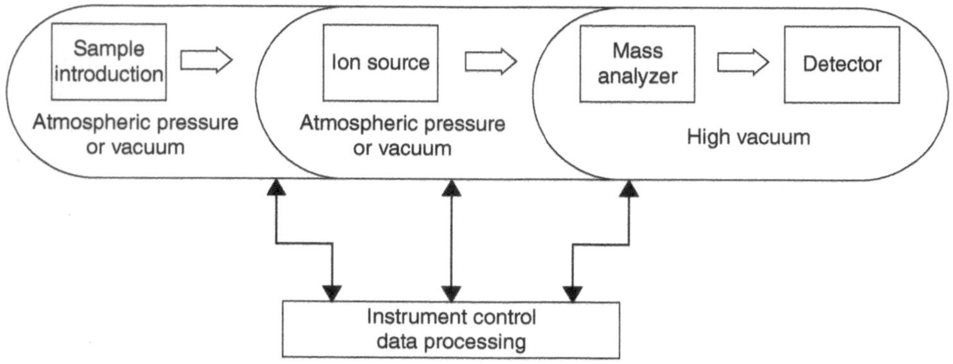

FIGURE 1.15 Flow chart of a typical MS experimental setup.

either the profile mode, that is, continuum spectra, possibly after data reduction like in apodization, in order to reduce the size of the data file (e.g., Scigelova et al., 2011), or the centroid mode, where only a weighted average of the mass peak is saved (Urban et al., 2014) (Figure 1.16). Centroiding procedures greatly reduce the data file size but may also reduce the information content of the initial raw data acquired. Post-acquisition data processing tools may require either profile or centroid data. Data acquisition modes for MS are further discussed in Section 1.5.

In the full-spectrum mode, a three-dimensional (3D) data array is acquired with time, *m/z*, and ion intensity (often expressed in counts) as the three axes. The data acquired can be visualized in different ways. The total-ion chromatogram (TIC) is a plot of the sum of the ion counts of the individual mass spectra as a function of time (or spectrum number). In a base-peak chromatogram (BPC), the ion count recorded for the most abundant ion in each spectrum, that is, irrespective of the *m/z* of that ion, is plotted as a function of time. BPCs are especially useful for peak searching in chromatograms with relatively high chemical background, such as in LC–MS. At any given time in the TIC or BPC, a mass spectrum can be obtained, which represents a slice of the data array of the ion counts as a function of *m/z*.

It is often useful to generate summed, averaged, and/or background subtracted mass spectra. The mass spectrum may be computer searched against a mass spectral library to assist in compound identification. The information from the TIC (Figure 1.17a) and the mass spectra can be combined in either a 3D representation, with time, *m/z*, and ion intensity/counts as the three axes *x*, *y*, and z, respectively (Figure 1.17b and c), or a contour plot, which basically is a 2D representation of the time against *m/z* where the ions detected are seen as spots; colors may be used to represent relative intensity/counts (Figure 1.17d). In an extracted-ion chromatogram (XIC), the counts for an ion with a particular *m/z* are plotted as a function of time. By default, a selection width or selection window of $m/z \pm 0.5$ is used to generate the XIC. However, with HRMS, a narrower selection window, for example, $m/z \pm 0.01$, may be used to achieve XIC with a greatly improved S/N.

Quadrupole, ion-trap, and sector mass analyzers can also acquire data in the selected-ion mode. In that case, the mass analyzer is programmed to select a particular *m/z* for transmission to the detector during a preset period (the so-called dwell time, typically 5–200 ms) and to subsequently jump to other preselected *m/z* values; after monitoring all the preselected *m/z* values, the same function is repeated, for example, during (part of) the chromatographic run-time. Compared to the full-spectrum mode, the selected-ion mode provides a longer measurement time for the selected ion (or ions), which results in enhanced S/N. The data can be displayed in terms of XICs. This acquisition mode is especially applied in targeted quantitative analysis. With instruments not capable of a selected-ion mode, that is, TOF and orbitrap mass analyzers, improved S/N and targeted quantitative analysis can be achieved post-acquisition in narrow-window XICs, as discussed earlier.

1.3.1.2 Ion Detection The ion detection device must be capable of converting the tiny electric current of the incoming ions into a measurable and usable signal. The actual detector employed depends on the type of mass analyzer. In general, ion detection systems must be backed by sufficiently fast electronics, including analog-to-digital converters (ADCs), to enable the high-speed data acquisition required in MS (de Hoffmann & Stroobant, 2007). Broadly speaking, ions are made to collide onto a special surface called the conversion dynode, made of a low IE material such as Pb, Be–Cu, which upon impact by the fast incoming ions emits electrons and possibly other secondary particles (Figure 1.18). These electrons are then converted into a usable current by a signal amplification system (e.g., an electron multiplier (EM)). The EM may be a device of either the continuous dynode type or the discrete dynode type (Allen, 1947). In an EM, the secondary particles from the conversion dynode hit the first dynode (EM entrance) and cause the emission of secondary electrons, which in turn do the same as they are directed toward the subsequent dynodes. Finally, the resulting current is detected over a collector plate and amplified by an electrometer. This repeated emission of secondary electrons creates a cascade effect with a typical current gain for an EM with an order of magnitude of 10^6–10^7. A conversion dynode, held at a high electric potential (5–20 kV), is positioned in front of the multiplier to increase the signal intensity of ions, especially in the high-mass region, as well as to enable the detection of negative ions. The EM is used for ion detection in quadrupole, ion-trap, and sector instruments.

In some instruments, a photomultiplier is used instead of an EM. In this system, a conversion dynode is used to generate electrons from the incoming ions by secondary emission. These secondary electrons in turn are directed toward a phosphorescent screen, which upon electronic excitation emits photons. These photons are sent to a photomultiplier, where typically signal amplification with an order of magnitude of 10^5 is achieved.

With TOF instruments, where the ion beam shows more spatial spreading, microchannel plate (MCP) detectors are applied. An MCP is an array of parallel miniature electron multipliers (Wiza, 1979). In order to generate a mass spectrum from the ion arrival events in TOF instruments, either a time-to-digital converter (TDC) or an ADC has to be used. TDCs provide excellent time resolution and low random noise, but do not discriminate in the intensity of the pulse. Therefore, high ion densities may lead to saturation effects, thus greatly limiting the dynamic range. In an ADC, the integrated circuit chip receives a time-dependent signal and typically generates a 10-bit digital output: both

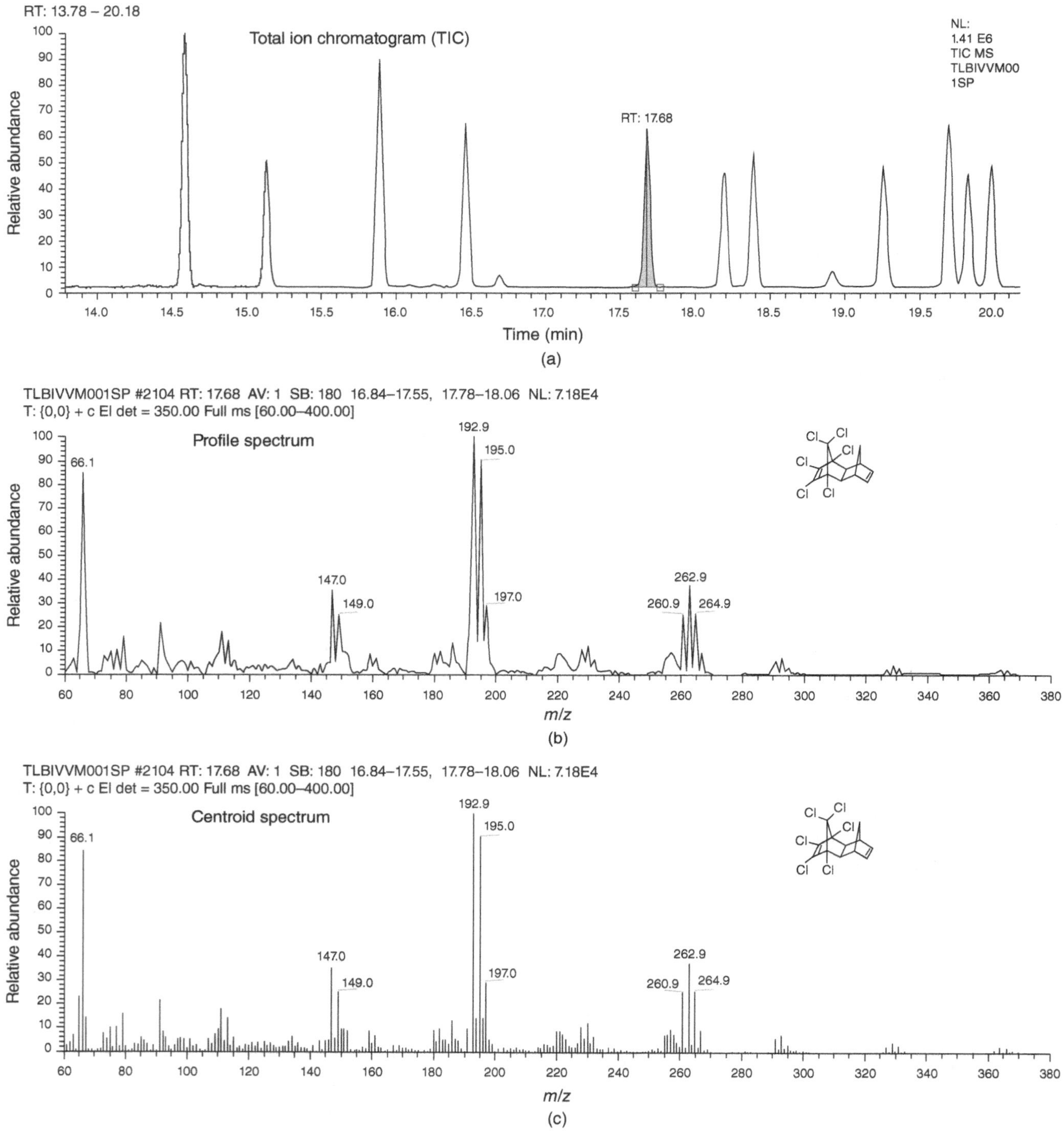

FIGURE 1.16 GC–MS analysis of a mixture of organochlorine insecticides. (a) Total-ion chromatogram (TIC), (b) profile or continuum mass spectrum for isodrin, and (c) centroid mass spectrum for isodrin.

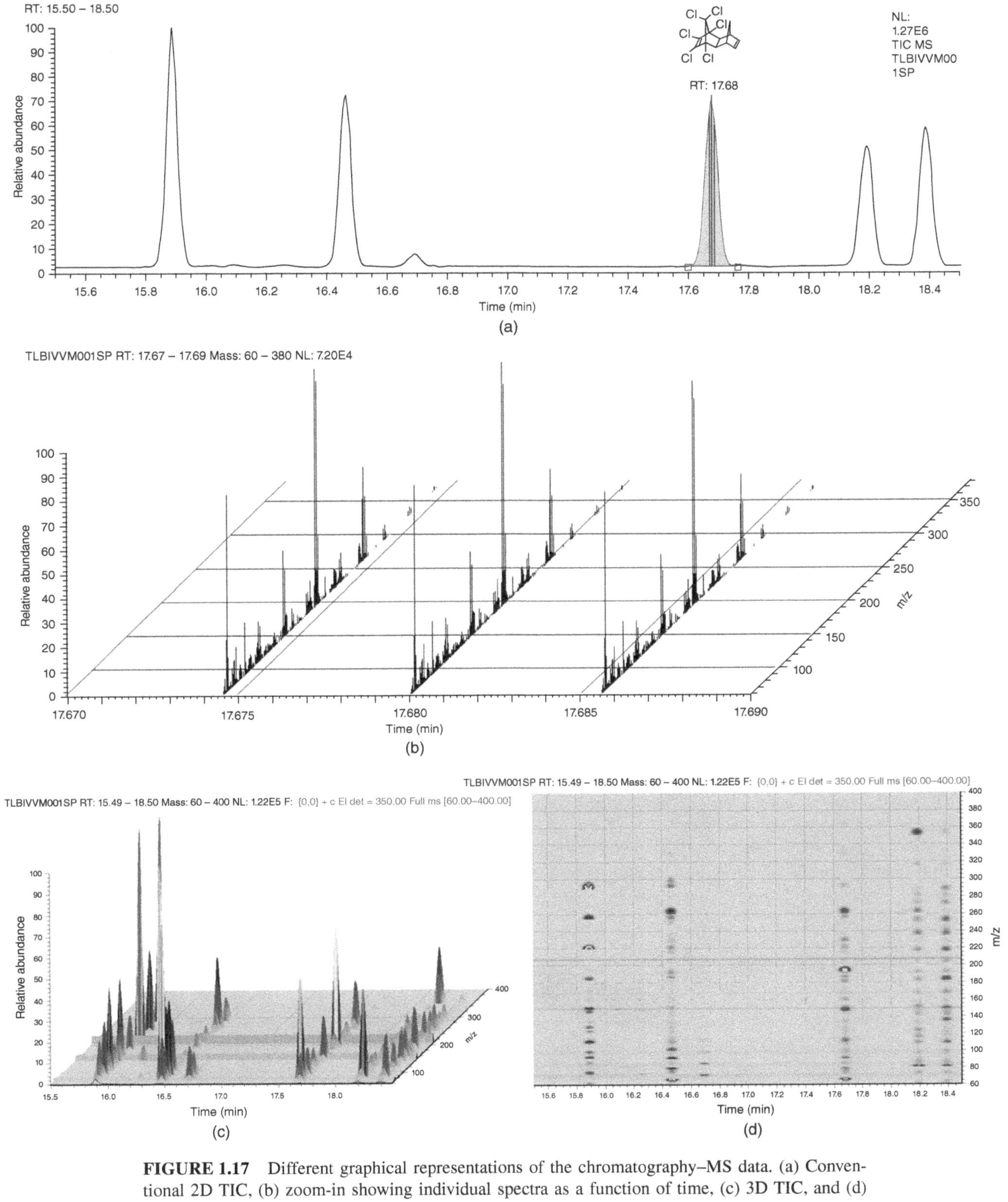

FIGURE 1.17 Different graphical representations of the chromatography–MS data. (a) Conventional 2D TIC, (b) zoom-in showing individual spectra as a function of time, (c) 3D TIC, and (d) contour plot of the 3D TIC shown in (c).

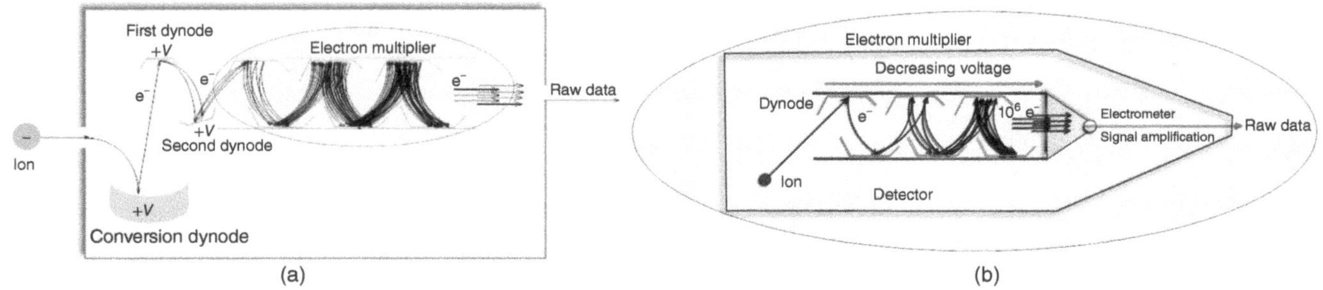

FIGURE 1.18 Schematic diagram of an electron multiplier for ion detection.

arrival time and the number of colliding ions are recorded. Currently applied ADCs can provide 1–4 GHz time resolution and discriminate 1024 different ion intensity levels. In FT-ICR-MS and orbitrap-MS systems, ion detection is based on the detection of high-frequency image currents of the coherently moving ions (Marshall et al., 1998). The signals of all ions with different *m/z* values are detected simultaneously. The time-domain signal is Fourier-transformed to the frequency-domain signal, which can be converted into mass spectra.

1.3.2 Quadrupole Mass Analyzer

The majority of mass spectrometers in use in laboratories around the world are based on quadrupole mass analyzer technology. A quadrupole mass analyzer consists of two pairs of rods of hyperbolic or circular cross section that are accurately positioned parallel to each other and in a radial array. Generally, stainless-steel or metal-coated ceramic rods are employed. Each pair of rods is charged by either a positive or a negative direct-current (DC) potential with a superimposed alternating-current (AC) radiofrequency potential (RF, MHz) as shown in Figure 1.19. The latter successively reinforces and overwhelms the DC field.

Ions coming from the source are introduced into the quadrupole field by means of a low accelerating potential. Due to the applied oscillating fields, the ions are sequentially attracted and repelled by the rod pairs, and they oscillate in the *yz*- and *xz*-planes as they traverse the quadrupole filter. The theoretical description of the trajectory the ions follow in the quadrupole electric field involves a large number of physical parameters. Solutions to these Mathieu differential equations can be simplified by defining the a and q terms, where a is proportional to the DC and q is proportional to the RF (the mass m is in the denominator). This allows the construction of a stability diagram (a–q diagram), which is useful in understanding the features of quadrupole mass analysis (Figure 1.20). A limited number of combinations of a and q, that is, of DC and RF, leads to stable trajectories for the ions in the hyperbolic space defined by the quadrupole field between the rods, allowing them to travel the length of the analyzer and reach the detector.

The quadrupole mass filter is operated with a fixed ratio of DC and RF; the ratio determines the resolution of the device. Now assume that the device is operated at unit-mass resolution. At a given DC/RF combination within the stable region of the stability diagram, the ions with only one *m/z* (actually $m/z \pm 0.35$ as FWHM at unit-mass resolution is 0.7) show a stable trajectory toward the end of the rods and are thus transmitted to the detector, while ions with unstable trajectories do not pass the mass filter because the amplitude of their oscillation becomes infinite. Consequently, they are

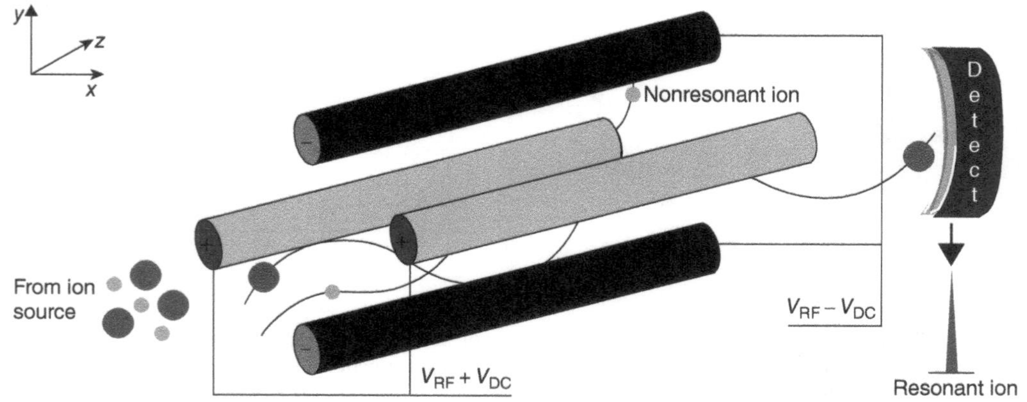

FIGURE 1.19 Schematic diagram of a quadrupole mass analyzer.

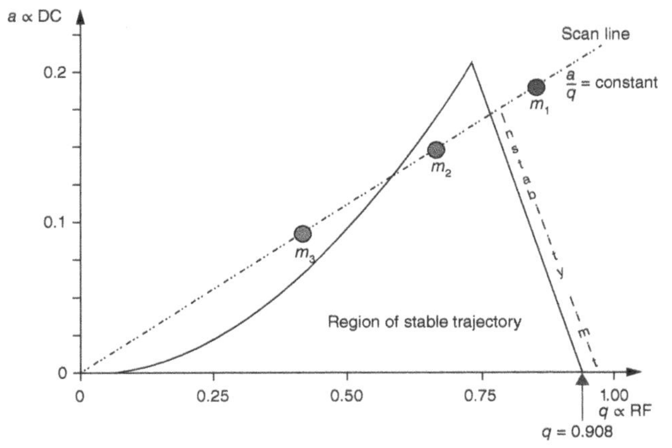

FIGURE 1.20 Stability diagram for a quadrupole mass filter.

discharged against the rods and/or lost in the vacuum system. Thus, the quadrupole mass analyzer can be considered as a variable band-pass filter (Miller & Denton, 1986). By ramping DC and RF voltages at a fixed ratio, that is, moving along the scan line in the stability diagram, ions of increasing *m/z* values are transmitted one after another to the ion detector, as they pass the instability limit ($q = 0.908$) (Figure 1.20).

Since the ramping of voltages can be done quite fast in modern electronics, scan speeds as high as 10,000 *m/z* s^{-1} can be achieved. In principle, the resolution of the quadrupole mass analyzer depends on the ratio of DC and RF, that is, the slope of the scan line in the stability diagram. However, operation at higher than unit-mass resolution generally greatly compromises the ion transmission and thereby the sensitivity (Tyler et al., 1996). Therefore, the quadrupole is generally operated at unit-mass resolution. Recently, it has been demonstrated that by improving the quadrupole design, the stability of the RF power supply and the temperature control enhanced mass resolution (FWHM of down to 0.2 instead of usual 0.7) on a quadrupole mass analyzer can be achieved without dramatic losses in signal intensity (Yang et al., 2002). This feature has not found wide application. Thus, in full-spectrum mode, the quadrupole mass analyzer provides at least unit-mass resolution and nominal monoisotopic mass determination and can be operated with great ease and versatility. It provides fast spectrum acquisition at limited costs, clearly justifying its popularity.

Apart from full-spectrum mode, featuring ramping of DC and RF at a fixed ratio, the quadrupole mass analyzer can be operated in three additional modes. It can be applied in selected-ion monitoring (SIM) mode, dwelling on selected *m/z* values for 10–200 ms, and capable of rapidly switching (within ≤5 ms) between different *m/z* values. In SIM mode, significantly improved S/N can be achieved, making the SIM mode of a quadrupole ideal for routine targeted quantitative analysis. Another important mode of operation is the RF-only mode. In this mode, the quadrupole can be used as an ion transport and focusing device. As such, RF-only quadrupoles have been used in vacuum interfaces of API-MS systems and as collision cells and/or ion-transport devices in MS–MS instruments (Section 1.4.2). Finally, it has been demonstrated that a quadrupole mass analyzer can be applied as linear ion trap. In this mode of operation, the quadrupole provides similar features as a conventional 3D ion trap (Section 1.4.3) (Hager, 2002; Schwartz et al., 2002; Douglas et al., 2005).

1.3.3 Ion-Trap Mass Analyzer

The introduction of the 3D ion trap has been an important development in quadrupole technology (March, 1997; Jonscher & Yates, 1997). Ion-trap MS is based on the same principles of quadrupole technology developed by Paul (Paul & Steinwedel, 1953; Paul & Raether, 1955; Paul, 1990). Two major advances, made by the group of Stafford, revived the interest in ion-trap technology: the development of the *m/z*-selective instability mode of operation and the use of He as damping gas (Stafford et al., 1984; Louris et al., 1987; Stafford, 2002). Additional improvements followed soon after.

A 3D quadrupole ion trap consists of a ring electrode with a hyperbolic geometry to which an RF voltage is applied and two end-cap electrodes resembling inverted hyperbolic saucers (Figure 1.21). The ring electrode is positioned symmetrically in between the two end-cap electrodes. The electrodes are electrically isolated by means of nonconducting

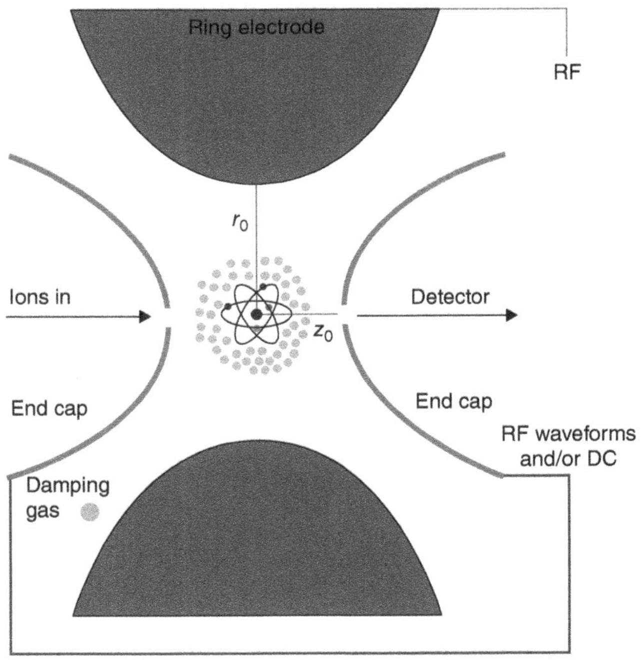

FIGURE 1.21 Schematic diagram of an ion-trap mass analyzer.

spacers. The internal volume of a typical 3D ion trap is ≈1 cm^3. Both end-cap electrodes contain holes: one of them for the introduction of ions from an external ion source into the trap and the other one for ion ejection from the trap toward an external EM. A He bath gas (≈0.2 Pa; ≈0.002 mbar) is used to stabilize the ion trajectories in the trap by acting as energy sink to help keep the ions in tight small orbits in the center of the trap. The overall effect is a dramatic improvement in resolution and reproducibility.

The basic mass analysis process consists of two steps, performed consecutively in time. First, a pulse of ions is injected and stored in the trap by the application of an appropriate low-RF voltage to the ring electrode. As a result, all ions above a low-*m/z* cutoff are stored in the trap. The ion injection pulse has a variable duration, depending on the ion current, because too high a number of ions in the ion trap, that is, in excess of $\approx 10^5$ ions, adversely influences mass resolution and accuracy due to space-charge effects. Automatic gain or ion charge control (software controlled) has been developed, which dynamically adjusts the duration of the ion injection pulse from the external ion source (March, 1997). Filling the ion trap by means of the ion injection pulse results in an accumulation of ions, which in practice leads to an enhanced full-spectrum sensitivity when compared to the linear quadrupole mass analyzer.

Once the ions are trapped, an *m/z*-selective instability scan is performed. The (fundamental) RF voltage applied to the ring electrode is ramped to consecutively eject from the trap and toward the external detector first the low-*m/z* ions, and likewise all other ions in order of increasing *m/z* values (resonant ion ejection at $q = 0.908$) (Jonscher & Yates, 1997). The resonant ion ejection may be supported by additional waveforms applied to the end-cap electrodes. Alternatively, a (resonant) supplementary RF can be applied to the end-cap electrodes to cause the ions to gain energy: the amplitude of the ion trajectory expands and the ions approach the end-cap electrodes until they are ejected from the trap at values of q lower than 0.908. The q value, where ions are ejected under these conditions, depends on the frequency of the supplementary RF potential. This operational mode is needed to achieve the ejection of ions with high-*m/z* values ($m/z \gtrsim 600$), but it is also important in the selection of precursor ions in an MS^n experiment (Section 1.4.3).

The achievable resolution depends on the scan speed. Ion ejection and subsequent detection can be achieved with unit-mass resolution or at enhanced resolution by slowing down the scan rate of the RF voltage on the ring electrode. In this respect, improvements have been made over time. Older ion traps provide peak widths (FWHM) of 0.2 at a scan speed of ≈300 $m/z\,s^{-1}$, unit-mass resolution (FWHM ≈ 0.7) at 5500 $m/z\,s^{-1}$, and degraded resolution (FWHM of 3.0 at 55,000 $m/z\,s^{-1}$), whereas more recently introduced systems provide better resolution at higher scan speeds, for example, FWHM of 0.1 at 4600 $m/z\,s^{-1}$ and of 0.58 at 52,000 $m/z\,s^{-1}$. An FWHM of 0.1 enables almost baseline resolution for a quadruply-charged ion (e.g., $[M+4H]^{4+}$) of a peptide.

More recently, 2D or linear ion traps (LITs) have been introduced as an alternative to 3D ion traps (Hager, 2002; Schwartz et al., 2002; Douglas et al., 2005). Similar to a quadrupole, the ions are confined radially by a 2D RF field. Ion ejection can be done either radially, as is done in a stand-alone LIT, or axially, as is done in quadrupole–linear ion-trap (Q–LIT) hybrid instruments (Section 1.4.4) and in LITs applied in orbitrap or FT-ICR hybrid instruments (Section 1.4.6). Because an LIT is less prone to space-charging effects, a higher number of ions can be accumulated and enhanced sensitivity (up to 60-fold) can be achieved. LITs are extensively used in hybrid MS–MS technology, but stand-alone versions of LITs have been introduced as well, thus competing against the 3D ion traps.

1.3.4 Time-of-Flight Mass Analyzer

A basic TOF mass spectrometer consists of a pulsed ion source, an accelerating grid, a field-free flight tube, and a detector (Guilhaus et al., 2000; Lacorte & Fernandez-Alba, 2006). In TOF-MS, owing to the applied accelerating potential, all ions begin their flight toward the detector at the same time and with the same initial kinetic energy. Because of their higher velocity, low-*m/z* ions will arrive to the detector before higher-*m/z* ions. Therefore, the time it takes an ion to reach the detector is related to its *m/z* (Figure 1.22).

Thus, if a particular ion with a given *m/z* is accelerated by a potential V, the flight time t to reach the detector placed at a distance d can be calculated from Eq. 1.41.

$$t = d\sqrt{\frac{m}{2zeV}} \tag{1.41}$$

Pulsing of the ion introduction into the flight tube is required to avoid the simultaneous arrival of ions of different

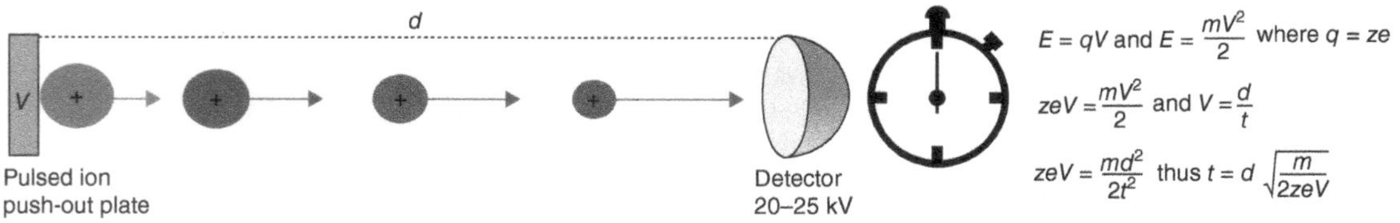

FIGURE 1.22 Principle of time-of-flight mass spectrometry.

m/z to the detector. The introduction of MALDI as a powerful ionization technique in the MS analysis of large biomolecules led to a revaluation of TOF-MS (Karas & Hillenkamp, 1988; Karas, 1996). The pulse rate of the laser used in MALDI (typically <1 kHz) makes the use of a TOF-MS instrument highly attractive. In addition, the *m/z* range of a TOF-MS instrument is unlimited in principle. Much higher pulse frequencies are applied (20–50 kHz) in combinations of TOF-MS with continuous ion sources (e.g., in ESI). As the data system cannot process such a high acquisition rate, mass spectra from multiple ionization events or pulses are accumulated. This leads to enhanced spectrum quality by averaging random noise. Acquisition rates as high as 100 spectra s^{-1} have been reported for LC–MS applications, although lower acquisition rates (1–10 spectra s^{-1}) are used more frequently.

The initial ion kinetic energy spread of the ions arising from their generation in the ion source is the most important limiting factor determining the achievable resolution in TOF-MS. With the progress in fast electronics, the speed of detection and acquisition electronics is no longer a limiting factor nowadays. In MALDI-TOF-MS, delayed extraction has been applied to reduce the ion kinetic energy spread of the ions (Vestal et al., 1995), whereas in ESI-TOF-MS, orthogonal acceleration has been a powerful tool (Guilhaus et al., 2000). Even more important in reducing the deteriorating effect of the ion kinetic energy spread on the resolution is the use of a reflectron (Doroshenko & Cotter, 1989; Guilhaus et al., 2000). A reflectron consists of a series of equally spaced grid or ring electrodes connected to a resistive network (Figure 1.23). It creates a homogeneous or curved retarding field that acts like an ion mirror. If two ions with equal *m/z* but slightly different kinetic energy enter the reflectron, the ion with the higher kinetic energy penetrates deeper into the field and thus has a slightly longer flight path than the ion with the somewhat lower kinetic energy. In effect, the two ions reach the detector more synchronously. It is important to keep in mind that while the arrival of the ions to the detector is nearly synchronous, the kinetic energy of the ions is the same as it was before they entered. Added value of the reflectron is that it effectively approximately doubles the flight distance d.

Significant progress has been made in TOF instrumentation, mainly directed at enhancing resolution and improving sensitivity and dynamic range. With a reflectron TOF in combination with orthogonal acceleration (with an ESI source), a mass resolution in excess of 15,000 (FWHM) can be readily achieved, enabling accurate-mass determination (<3 ppm) (Xie et al., 2012). Currently, commercial TOF-MS systems are available with a mass resolution in excess of 70,000 (FWHM).

1.3.5 Orbitrap Mass Analyzer

An orbitrap consists of three electrodes: a spindle-like central electrode (A) and two cup-shaped outer electrodes (B) facing each other (Figure 1.24). Ions are injected tangentially into the electric field present in the volume between the central and the outer electrodes. A radial electric field resulting from the voltage applied between the central and outer electrodes leads to circular movements of the ions around the central electrode. The electrostatic attraction of the ions to the inner electrode is balanced by their centrifugal forces. In addition, the axial electric field caused by the shape of the electrodes initiates harmonic axial oscillations of the ions along the central electrode. Thus, ions circle around the central electrode in rings, as they move back and forth along the axis of the central electrode. This oscillation is proportional to $(m/z)^{-1/2}$ and, furthermore, independent of the ion velocity. By sensing the ion oscillation frequency in a similar manner as done in FT-ICR-MS, the orbitrap can be used as a mass analyzer. The image current resulting from these axial oscillations is measured using the outer electrodes. The digitized time-domain image current is Fourier-transformed into the frequency domain. The *m/z* value of an ion is related to the frequency ω of the axial oscillations as $\omega = (k \times z/m)^{1/2}$. Thus, the frequency-domain spectrum can be converted into a mass spectrum (Hu et al., 2005; Makarov et al., 2006; Zubarev & Makarov, 2013).

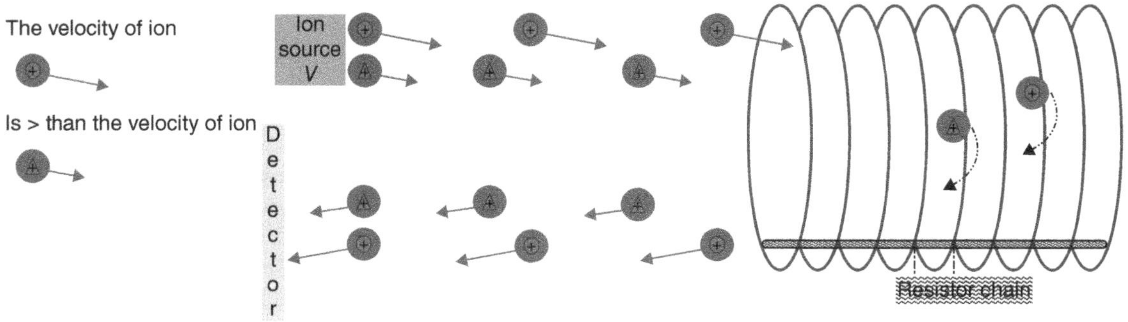

FIGURE 1.23 Operation principle of kinetic-energy focusing in a reflectron time-of-flight mass spectrometer.

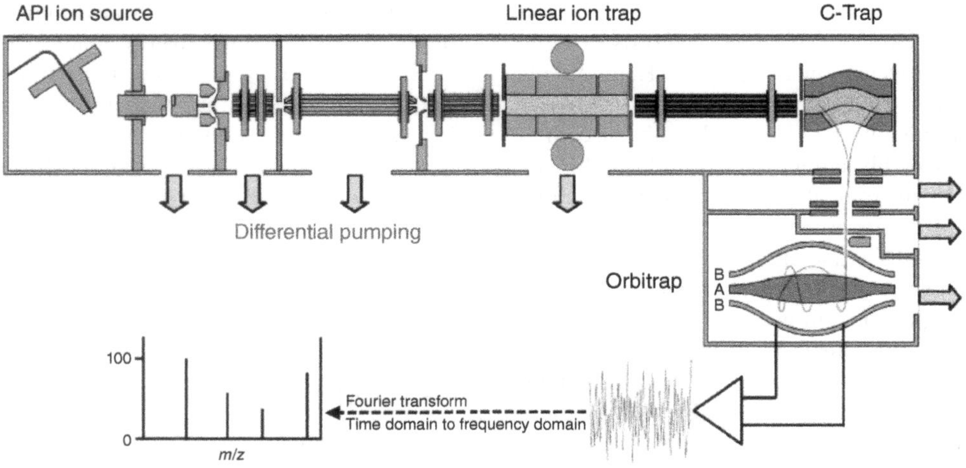

FIGURE 1.24 Schematic diagram on a linear-ion-trap–orbitrap hybrid instrument. (Source: Reprinted and adapted from (Makarov et al., 2006) with permission, ©2006, American Chemical Society.)

An important practical aspect is the adequate delivery of ions into the orbitrap. To this end, a curved high-pressure RF-only quadrupole, the so-called C-trap, is applied in combination with two flat electrodes. The C-trap is filled with N_2 bath gas (≈10^{-2} Pa; ≈1 mbar) for collisional damping. After filling the C-trap with ions, the ion package is compressed by applying 200 V to the flat electrodes. Rapidly ramping the RF voltage at the C-trap rods (within 200 ns) in combination with high voltages on the flat electrodes accelerates a concise ion package via a dual electrostatic deflector and through three stages of differential pumping into the orbitrap (2×10^{-5} Pa; 2×10^{-10} mbar) (Makarov et al., 2006). A high-field compact orbitrap was introduced in 2011, providing enhanced performance characteristics (Michalski et al., 2011; Zubarev & Makarov, 2013).

An orbitrap mass spectrometer allows ultra-high-resolution measurements (in excess of 10^5, FWHM). The achievable mass resolution depends on the spectrum acquisition time. The initially introduced orbitrap systems needed ≈1.6 s to acquire a mass spectrum with a resolution of 100,000 (at *m/z* 200, FWHM). The more recently introduced high-field orbitrap instruments provide mass spectra with a resolution of 17,500 (FWHM) with 64 ms and a resolution of 140,000 with 512 ms spectrum acquisition time. At a resolution in excess of 100,000, a mass accuracy within 1 ppm can be achieved. Although a stand-alone version of the orbitrap has been produced (Geiger et al., 2010), far better analytical capabilities can be achieved with hybrid systems (Section 1.4.6).

1.3.6 Other Mass Analyzers

In addition to the four types of mass analyzers just discussed, brief attention should be paid to two other types of mass analyzers: the sector instrument and the FT-ICR-MS instrument. Perhaps any discussion on mass analyzers should start with sector instruments because historically they are at the basis of all MS developments. However, since sector instrument are no longer used in combination with ESI and APCI, their principles and operation are not very relevant to the topic of this book.

A basic example of a sector instrument applies only a magnetic sector. Ions with mass m and z elementary charges e are accelerated with a voltage V into a magnetic field B, where they follow a path with a radius of curvature r. From the fundamental equation $m/z = B^2r^2e/2V$, describing the relationship between m/z and the experimental parameters, one can derive equations to describe how the separation of ions with different m/z can be achieved in three different ways. By variation of the radius of curvature, ions with different m/z are separated in space; an array of detectors is needed to detect the ions and acquire the mass spectrum. In a more practical approach, ions of different m/z can be detected one after another by means of a single-point detector (e.g., an EM) at a fixed position by variation of either B or V (de Hoffmann & Stroobant, 2007). The performance of the sector instrument in terms of mass resolution can be greatly improved by combining the magnetic sector with an electrostatic analyzer, resulting in a double-focusing instrument, which provides HRAM determination. Several geometries of double-focusing instruments have been described. Since the mid-1990s, the sector instrument has been replaced as the instrument of choice for HRMS and HRAM-MS by alternatives that are easier to operate and less expensive, for example, instruments based on time-of-flight or orbitrap technology.

In an FT-ICR-MS instrument, the ions are trapped in a strong magnetic field B (up to 15 T). An ICR cell consists of two opposite trapping plates, two opposite excitation plates, and two opposite receiver plates. Both cylindrical and cubic ICR cells have been produced. In the magnetic field, ions with *m/z* describe cyclotron motions with a radius *r* perpendicular to the magnetic field lines. The ions moving in this way induce an image current at the receiver plate with a cyclotron frequency of $\omega_c = 2\pi f = v/r = Bez/m$, where f is the frequency in hertz. The cyclotron frequency is thus inversely proportional to the *m/z* value. In a typical FT-ICR-MS experiment, the ions, trapped in their cyclotron motion in the cell, are excited by means of an RF pulse on the excitation plates. This increases the radius of the cyclotron motion and, more importantly, the ions with the same *m/z* values start to move in phase. This coherent movement of the ions generates an image current at the receiver plates, which decays in time because of disturbance of the coherency of the ion movement in time. The time-domain signal from the receiver plates contains the frequency information of all the ions present in the cell. By Fourier transformation, the time-domain signal can be transformed into a frequency-domain signal and subsequently converted into a mass spectrum (Marshall et al., 1998; Scigelova et al., 2011). Similar to the orbitrap mass spectrometer, the resolution in FT-ICR-MS increases with measurement time; longer measurement times require extreme high vacuum in the ICR cell ($\approx 10^{-4}$ Pa; $\approx 10^{-6}$ mbar). FT-ICR-MS instruments can provide (*m/z*-dependent) ultra-high resolution, typically in excess of 10^5 (FWHM). A high-resolution spectrum can be achieved in a shorter time in instruments with a higher magnetic field strength. Commercial FT-ICR-MS systems are available that provide a resolution of ≈650,000 (at *m/z* 400, FWHM) with 1 spectrum s^{-1}.

1.4 TANDEM MASS SPECTROMETRY

1.4.1 Introduction

MS–MS involves the combination of two mass analyzers in series with a reaction chamber in between, enabling first-stage mass analysis in MS_1 and second-stage mass analysis in MS_2. A basic MS–MS experiment (product-ion analysis) consists of three steps. In the first step, a precursor ion with a particular *m/z* is selected from the population of ions generated in the ion source. In the second step, the precursor ion is fragmented, in most cases using CID. In the third step, the product ions formed in the CID of the precursor ion are mass analyzed.

The two mass analyzers can be combined either in space or in time (Johnson et al., 1990). There are tandem-in-space mass spectrometers, where the two mass analyzers are of the same type, for example, a TQ instrument, and tandem-in-space mass spectrometers, where the two mass analyzers are different, the so-called hybrid instruments, for example, a quadrupole–time-of-flight (Q–TOF) hybrid instrument with a quadrupole mass analyzer for MS_1 and a time-of-flight mass analyzer for MS_2. In tandem-in-space instruments, the three steps of the MS–MS process (precursor-ion selection, CID, mass analysis of product ions) are performed in spatially separated devices. The ion-trap instrument is an example of a tandem-in-time mass spectrometer, where the three steps are performed one after another in the same device. The reaction chamber in the tandem-in-space instrument is a collision cell, whereas in the tandem-in-time instrument CID is performed in the same region as the mass analysis, although dual-cell ion-trap instrument have been developed as well (Olsen et al., 2009).

In an MS–MS instrument, the *m/z* values of ions are measured before and after the collision cell. In practice, as most MS–MS instruments have only one ion detector, two separate experiments have to be performed, one involving the acquisition of the MS spectrum (without collision energy and collision gas) and another involving the acquisition of the product-ion MS–MS spectrum (with collision energy and gas). In most cases, the reaction in the collision cell leads to a change in *m/z*. For positive ions, the precursor or parent ion m_p^+ is converted into the product or daughter ion m_d^+ by the loss of a neutral fragment m_n. Thus, fragmentation of the precursor ion occurs. Although the neutral fragment m_n itself is not detected, its mass can be deduced from the *m/z* difference of m_p^+ and m_d^+. In principle, the fragmentation observed for a particular ionized molecule is compound specific. Even the protonated or deprotonated molecules of the same compound often show different fragmentation behavior.

1.4.1.1 Ion Dissociation Techniques

In most analytical applications of MS–MS, the fragmentation reactions are induced by ion collisions with a neutral target gas, that is, helium (He), nitrogen (N_2), or argon (Ar), which is present in the collision cell at a pressure of typically 10^{-2}–0.1 Pa (10^{-4}–10^{-3} mbar). In the region between MS_1 and the collision cell (tandem-in-space instrument), the precursor ion selected in MS_1 is accelerated by a potential difference (*V*), which is generally called the collision energy. As a result, the ions have increased translation energy. Part of this translation energy is converted into vibrational energy of the ions upon the collisions. Thus, the ions are collisionally activated, which means that they are at higher vibrational states (larger amplitudes in the stretching and bending vibrational modes of the molecule). These excited ions can then fragment in a compound-specific way. Depending on the available internal energy and the structural features of the precursor ion, several competitive unimolecular fragmentation pathways may be available, thus leading to different fragment ions. At low collision energy, only the weakest bonds in the ion can be cleaved. By increasing the collision energy, an increasing number of pathways may be available. At very high collision energy, extensive C—C bond cleavages occur, resulting in uncontrolled fragmentation. In addition, scattering of ions reduces the transmission of ions under these conditions. The process in the collision cell is generally called CID. One should be aware of the fact that CID is a two-step process: the actual collision is an ultrafast event ($\approx 10^{-15}$ s, i.e., the Franck–Condon approximation applies), followed by the unimolecular decomposition of the excited ions in competing reaction pathways, happening 10^{-10}–10^{-5} s after ion excitation. In between the two steps of the CID process, energy redistribution within the ion may take place.

CID can be performed in two different energy regimes (Sleno & Volmer, 2004). In most instruments, low-energy CID is performed with (laboratory) collision energies between 10 and 100 eV. One should discriminate between collision-cell CID and ion-trap CID. In collision-cell CID, applicable to TQ and Q–TOF instruments, multiple ion collisions are performed with N_2 or Ar. Typical residence time of the ions in a collision cell is 10–20 μs. In ion-trap CID, collisions are performed with a smaller target (He instead of Ar) and ion excitation is achieved by means of an *m*/*z*-dependent RF waveform pulse; the interaction time is in the ms range in ion-trap CID (Jonscher & Yates, 1997). At the first stage of fragmentation, ion-trap CID is generally softer than collision-cell CID, that is, less fragment ions are formed. In sector and TOF–TOF instruments, high-energy CID can be performed, involving single keV collisions with He as target gas. In high-energy collisions, more informative but often more complex product-ion MS–MS spectra may be obtained because a wider range of fragmentation pathways is opened. The mass spectral data discussed in Chapter 4 were acquired using low-energy CID in either collision-cell CID or ion-trap CID. In the discussion on fragmentation rules in Chapter 3, some data from high-energy CID are included as well.

It should be mentioned that fragmentation can also be induced in a process called in-source or up-front CID. By increasing the voltage applied to the ion-sampling orifice (or the voltage difference between orifice and skimmer), the ions experience more energetic collisions with neutrals in the high-pressure region. The resulting gain in internal energy in the ion can result in fragment ions, which can be mass analyzed. Because all ions present in the source can be fragmented in this way, the method only leads to useful, interpretable data if either pure compounds or well-separated analytes in not-too-complex matrices are analyzed. In-source CID initially found application in general unknown screening in toxicology (Marquet & Lachâtre, 1999; Marquet et al., 2000; Weinmann et al., 1999), but has since been replaced by MS–MS-based procedures (Section 5.5).

Next to CID, several other ion activation methods have been used, mostly in specific applications and/or instruments (Sleno & Volmer, 2004; Laskin & Futrell, 2005). As performing CID in the ultra-high-vacuum ICR cell of an FT-ICR-MS may compromise its performance, especially in terms of resolution and sensitivity, several alternative methods have been developed to induce fragmentation in FT-ICR-MS instruments, for example, infrared multiphoton photodissociation (IRMPD), sustained off-resonance irradiation (SORI), and black-body infrared radiative dissociation (BIRD) (Sleno & Volmer, 2004; Laskin & Futrell, 2005). Other ion-activation methods such as surface-induced dissociation and laser photodissociation have been mostly used by a limited number of research groups. Currently, the most widely applied alternative ion-activation methods are electron-capture dissociation (ECD) and electron-transfer dissociation (ETD) (Kim & Pandey, 2012; Zhurov et al., 2013), which are especially important in the fragmentation of multiply-charged ions of peptide, proteins, glycopeptides, and phosphorylated peptides. ECD can be applied in FT-ICR-MS, whereas ETD can be implemented on other types of mass analyzers such as ion-trap, Q–TOF, and orbitrap instruments.

1.4.1.2 Product-Ion Analysis The most straightforward operational mode of the tandem mass spectrometer is the product-ion analysis mode, where the precursor ion m_p^+ is selected in MS_1, while the product (or daughter) ions m_d^+ are mass analyzed in MS_2 and then detected (Yost & Enke, 1978; Busch et al., 1988). The resulting mass spectrum, called the product-ion mass spectrum, provides structural information on the precursor ion and thus on the compound analyzed. The selection of precursor ions in most MS–MS instruments is done either with unit-mass resolution or with somewhat degraded resolution to transmit the complete isotope pattern of the precursor ion. In the latter case, the presence of particular elements in the precursor ion, for example, Cl, Br, and/or S, can be followed in the product ions.

The interpretation of the product-ion mass spectra is the topic of this book, and especially using protonated molecules $[M+H]^+$ or deprotonated molecules $[M-H]^-$ as precursor ions. In order to learn about the fragmentation reactions and to derive general fragmentation rules (Chapter 3), the fragmentation in MS–MS of a large number of compounds from a wide variety of compound classes (mainly drugs and pesticides) was studied. These compounds were analyzed either in positive-ion mode as $[M+H]^+$ or in negative-ion mode as $[M-H]^-$, or in some cases as both. If the fragmentation pathways of a number of compounds from the same structural class are compared, often both class-specific (or group-specific) fragmentation and compound-specific fragmentation can be recognized. The class- or group-specific fragmentation is observed for all members of that particular group or class and may involve the observation of particular product ions and/or of particular neutral losses. Ample examples of such features are discussed in Chapter 4. Compound-specific fragmentation is only observed for a particular compound, and thus involves structural features by which that particular compound differs from other members of the group or class.

1.4.1.3 Development of MS–MS Instruments Historically, the starting point of MS–MS was in the 1940s with the observation and subsequent explanation of metastable ions in magnetic sector instruments (Hipple et al., 1946; Busch et al., 1988; de Hoffmann & Stroobant, 2007). Metastable ions are ions with sufficient internal energy to fragment, but which survive long enough to be extracted from the ion source before they fragment. When fragmentation occurs, they do so in the mass analyzer region before reaching the detector. The charged fragments of metastable ions that dissociate in the reaction region of the instrument may be detected. It was discovered that the abundance of the metastable ions can be increased by energetic collisions of the ions with a neutral bath gas in a collision cell. This observation led to the development of MS–MS instruments, initially based on sector instruments, where high-energy CID (keV collisions) is applied (Section 3.4).

The extensive use of MS–MS in analytical applications can be attributed to the introduction by Yost and Enke of the so-called triple-quadrupole instruments in 1978, featuring two analytical quadrupole mass analyzers and a collision cell in an RF-only quadrupole, which works as an ion guide that does not provide *m/z* separation (Section 1.3.2) (Yost & Enke, 1978). As in many modern instruments, the RF-only quadrupole collision cell has been replaced by other types of RF-only multipole collision cells, for example, hexapoles, octapoles, ion tunnels, and travelling-wave stacked-ring ion tunnels; nowadays, the term tandem quadrupole (TQ) is preferred over triple-quadrupole.

Subsequently, the possibility of multistage MS–MS (MS^n) in ion-trap instruments was introduced by the group of Stafford in 1987 (Louris et al., 1987). With the increased analytical application of MS–MS instruments, which in part is the result of the introduction of new soft ionization techniques (e.g., FAB, TSI, and ESI), several hybrid MS–MS instruments were introduced, also providing additional analytically interesting features, such as HRAM determination. Thus, Q–TOF (Morris et al., 1996), hybrid Q–LIT instruments (Hager, 2002), hybrid LIT–orbitrap instruments (Hu et al., 2005; Makarov et al., 2006), and hybrid quadrupole–orbitrap instruments (Michalski et al., 2011) were introduced. These instruments are more cost effective and easier to operate than sector instruments.

MS–MS has become an indispensable tool in fundamental studies on ion generation, ion–molecule reactions, unimolecular fragmentation reactions, and the identity of ions. It plays an important role both in qualitative analytical applications of MS involving the online coupling of MS to GC or LC, for example, identification of drug metabolites (Section 5.7.1), and in quantitative analytical applications based on selected-reaction monitoring (Section 1.5.2).

1.4.2 Tandem Quadrupole Instruments

The most widely used MS–MS configuration is the TQ instrument. Initially, triple-quadrupole instruments were used where mass analysis is performed in the first and third quadrupoles and CID in the second quadrupole (collision cell), that is, in a Q–q_{coll}–Q configuration (Yost & Enke, 1978). The gas-filled collision cell, operated in RF-only mode to transmit all ions without mass analysis, provides refocusing of ions scattered by the collisions and thereby somewhat reduces the transmission losses. Alternative RF-only collision cells have been developed in order to further reduce such transmission losses (e.g., RF-only hexapoles or octapoles). In a linear-acceleration high-pressure collision cell (LINAC), an axial voltage and tilted rods are used to reduce the residence time of the ions in the collision cell and to reduce crosstalk (Mansoori et al., 1998). A stacked-ring collision cell, featuring an axial travelling-wave or transient DC voltage, has been reported to reduce the transit times in the collision cell as well (Giles et al., 2004). Compared to the sector instruments used for MS–MS prior to 1978, the TQ instrument yielded significantly better product-ion resolution and greatly facilitated the acquisition of product-ion mass spectra.

The extent of fragmentation in collision-cell CID, as applied in TQ (and also in Q–TOF instruments), is determined by the gas pressure in the collision cell, the *m/z* of the precursor ion, and can be readily controlled by the collision energy, that is, the potential applied to the ions upon entering the collision cell. Optimization of the relevant instrumental parameters is needed for a particular application. For structure elucidation, it can be useful to acquire product-ion mass spectra at several collision energies. A plot of the relative abundance of the various product ions observed as a function of the collision energy is referred to as a breakdown curve; it plays an important role in optimizing collision energy conditions for optimizing SRM transitions. The optimum collision energy for a particular SRM transition, where one aims at preferably one intense product ion, is often not the optimum collision energy for structure elucidation, where one aims at a range of structure-informative fragment ions. One should be aware that setting a particular value for the collision energy in a particular instrument does not reflect the actual internal energy gained by the precursor ion. Setting the same collision energy for the fragmentation of the precursor ion of a particular compound in instruments from different manufacturers may yield widely different product-ion mass spectra, especially in terms of relative abundance of the product ions. For doubly-charged peptide ions, that is, $[M+2H]^{2+}$, a linear relationship is observed between the optimum collision energy and its *m/z*-value, where the slope of the curve is instrument-dependent (Haller et al., 1996; Holman et al., 2012). Some instruments allow ramping the collision energy during the acquisition of the product-ion mass spectrum: higher collision energies are applied when scanning for low *m/z* and lower collision energies for the high *m/z*.

In the past, TQ instruments were frequently used in confirmation of identity and structure elucidation of unknowns. This role has diminished, on the one hand, due to its limitations in terms of resolution, mass accuracy, and full-spectrum sensitivity and, on the other hand, due to the introduction of other, more powerful alternatives. However, at present, the TQ instrument is still first choice in routine-targeted quantitative analysis where it is operated in SRM mode (Section 1.5.2) (van Dongen & Niessen, 2012).

1.4.3 Ion-Trap Instruments

In a typical ion-trap MS^2 experiment, one starts with a population of ions, generated in the ion source (e.g., by ESI) from which the precursor ion is selected, excited, and fragmented. This results in a new population of (product or daughter) ions. The latter population either can be scanned out to be detected or can serve in a new series of subsequent steps of the process: selection of a product ion as precursor ion in a new MS^n experiment, which is to be excited and fragmented, and leads to a new population of (granddaughter) ions. Note that the term MS–MS is applied to denote collision-cell CID, as achieved in TQ and Q–TOF instruments, whereas MS^n is applied in the case of ion-trap CID in ion-trap instruments.

The selection of a precursor ion from a population of ions can be done in several ways. A combination of DC and RF potentials can be applied, which drive the precursor ion to the apex of the stability diagram (Figure 1.20), where the motion of the ions with *m/z* values other than the *m/z* value to be selected becomes unstable. Alternatively, the amplitude of the fundamental RF voltage can be scanned in a reverse-then-forward manner, while applying a resonant supplementary voltage. In this way, ions with *m/z* values higher than the *m/z* of the precursor ion are ejected first, followed by ejection of ions with smaller *m/z* values.

In the next step of the process, the fundamental RF voltage applied to the ring electrode is lowered and a resonant RF potential is applied to the end-cap electrodes with such an amplitude that ion trajectory is enlarged but not to such an extent as to eject the ion out of the trap. This ion excitation results in more energetic collisions with the He bath gas and may lead to CID of the precursor ions. With a sufficiently high number of collisions, the ions gain sufficient internal energy to cause fragmentation of the precursor ions into product ions, generally with relatively high efficiency. Depending on the fundamental RF voltage, product ions can be trapped. Often, product ions with *m/z* values below 25–33% of the *m/z* of the precursor ion cannot be trapped and are ejected and lost without detection. This may result in limitations if CID of the precursor ion forms predominantly low-*m/z* product ions, as is the case for tramadol (Hakala et al., 2006). Different from collision-cell CID, the voltage (amplitude) of the *m/z* selective RF potential applied as collision energy results in an on/off situation. When the voltage is too low, no fragmentation is observed. If by increasing the voltage the onset of fragmentation is reached, further increase in the voltage does not change the appearance of the product-ion spectrum anymore.

Either the population of product or daughter ions produced by CID can be ejected and detected to obtain the MS^2 spectrum, or one of the product ions may be selected to serve as precursor ion in subsequent sequence of excitation and fragmentation. The resulting population of granddaughter ions (secondary product ions) can be scanned out and detected to obtain the MS^3 spectrum or again one of the product ions may be selected to serve as precursor ion in subsequent sequence of excitation and fragmentation. Most ion-trap MS^n instruments allow for $n = 10$, although due to sensitivity limitations in most cases a maximum of only six stages can be achieved. A nice example of six stages of ion-trap MS^n has been demonstrated in the negative-ion ESI-MS analysis of glycosylated saponins (Wolfender et al., 1998). Step-wise fragmentation involving subsequent losses of sugar monomers in subsequent MS^n steps is achieved. Step-wise fragmentation in MS^n has been obtained for other compound classes as well, for example, chlorpromazine (Section 4.2.1), citalopram (Section 4.2.3.3), simvastatin (Section 4.4.3.2), dicloxacillin (Section 4.8.3.1), and chlorotriazines (Section 4.11.1.1). The ion-trap CID provides fragmentation with effectively lower energy involved. As such, it facilitates the acquisition of a wealth of structural information, for example, by step-wise fragmentation and the generation of fragmentation trees (Kind & Fiehn, 2010; van der Hooft et al., 2011). A fragmentation tree is generated by further fragmenting selected fragment ions of a particular stage of MS^n into a next stage, that is, MS^{n+1}. Another interesting feature of ion-trap CID is the ability to fragment sodiated molecules $[M+Na]^+$, which is often not possible in collision-cell CID. This difference may be due to the longer residence time of ions in an ion-trap (ms) compared to their residence time in an RF-only collision cell (10–20 μs). Fragmentation of $[M+Na]^+$ may result in different information, compared to the fragmentation of $[M+H]^+$. This has been nicely demonstrated in the MS analysis of the iridoid glycoside globularin (Section 3.8) (Es-Safi et al., 2007) and by the ability to obtain linkage information in the fragmentation of $[M+Na]^+$ of disaccharides (Asam & Glish, 1997). Ion-trap MS^n of $[M+Na]^+$ is also important in the structure elucidation of compound classes such as avermectins (Section 4.10.1.2) and polyether ionophores (Section 4.10.2.3).

As discussed earlier (Section 1.3.3), 2D or linear ion traps (LITs) have been introduced as an alternative to 3D ion traps. In principle, they provide the same features in MS^n but with improved full-spectrum sensitivity (Douglas et al., 2005). LIT-MS^n systems as stand-alone devices have been widely used in combination with LC–MS in a wide range of application areas. In addition, LITs have been employed as part of hybrid systems, such as the second stage of mass analysis in a Q–LIT hybrid system (Hager, 2002), as well as a first-stage mass analyzer (eventually also providing CID) in combination with an orbitrap and an FT-ICR. In the latter type of hybrid instruments, the ion-trap MS^n features in MS_1 are combined with HRMS and accurate-mass determination in MS_2 (Hu et al., 2005; Makarov et al., 2006; Zubarev & Makarov, 2013) (Section 1.4.6). Another interesting development is the introduction of dual-pressure LIT instruments (Olsen et al., 2009). The first high-pressure (0.5 Pa, 5×10^{-3} mbar) ion trap provides storage of ions, selection of precursor ions, their excitation, and fragmentation. Threefold reduction of ion fragmentation time (10 ms) can be achieved. The fragment ions are then transferred to the second reduced-pressure (0.035 Pa; 3.5×10^{-4} mbar) ion trap to enable very high scan speeds.

1.4.4 Quadrupole–Linear Ion-Trap Hybrid Instruments

The hybrid Q–LIT instrument, introduced in 2002, has the general layout of a TQ instrument, but MS_2 can be operated (under software control) as either a normal linear quadrupole or a linear ion trap (Hager, 2002). For full-spectrum product-ion spectrum acquisition, the Q–LIT instrument shows several advantages. Precursor-ion fragmentation can be performed using collision-cell CID (in the LINAC, Section 1.4.2). The resulting product ions are accumulated in the LIT, providing enhanced product-ion (EPI) spectra with improved full-spectrum sensitivity. If necessary, additional MS^3 experiments can be done by selection of one of the product ions in the LIT as a precursor ion for subsequent excitation, fragmentation (ion-trap CID), and detection. Since its introduction in 2002, hardware, electronics, and software control of the Q–LIT instruments have been further optimized to provide a wide variety of MS–MS operating modes and to allow very rapid switching between various MS and MS–MS experiments (Hopfgartner et al., 2004). Because MS_2 can be rapidly switched between linear quadrupole and LIT mode of operation, combination of SRM and full-spectrum product-ion analysis can be achieved (Section 1.5.4).

1.4.5 Quadrupole–Time-of-Flight Hybrid Instruments

Similar to the Q–LIT hybrid, discussed in Section 1.4.4, the Q–TOF hybrid mass spectrometer can be considered as a modified TQ instrument, where the MS_2 quadrupole has been replaced by an orthogonal-acceleration reflectron-TOF mass analyzer (Morris et al., 1996). The first commercially available Q–TOF instrument was especially developed to facilitate peptide sequencing analysis, but the instrument found much wider application, especially in small-molecule structure elucidation studies. Thus, in MS mode, the quadrupole (MS_1) is operated in RF-only mode and the RF-only collision cell with low collision energy, whereas in MS–MS mode the quadrupole MS_1 performs the selection of the precursor ion with unit-mass resolution, and fragmentation of the precursor ion is achieved by collision cell CID. In both modes, the ions are orthogonally accelerated into the flight tube and HRAM analysis is performed in the reflectron-TOF analyzer (MS_2). Q–TOF instruments are now available from several instrument manufacturers (Xie et al., 2012). They are widely used in structure elucidation, metabolite identification, and sequencing of peptides. Because collision-cell CID is applied, the fragmentation characteristics are the same as in TQs. In structure elucidation, a significant advantage of Q–TOF is the ability to perform accurate-mass determination (<3 ppm) for both precursor and product ions. Principles and applications of Q–TOF hybrid instruments have been reviewed (Chernushevich et al., 2001; Campbell & Le Blanc, 2012).

In the context of this book, the use of Q–TOF instruments for structure elucidation is important, given the fact that the most important source of (HRAM) mass spectral data for Chapter 4 was a mass spectral library generated on a Q–TOF instrument.

1.4.6 Orbitrap Hybrid Instruments for MS–MS and MS^n

The initial instrumental setup of a commercial orbitrap mass spectrometer consisted of a hybrid LIT–orbitrap configuration, featuring an LIT to control the number of ions transferred to the orbitrap and to perform MS^n, a C-trap to direct the package of ions into the orbitrap, and the orbitrap itself (Figure 1.24) (Hu et al., 2005; Makarov et al., 2006). In this instrument, the orbitrap is used to perform HRAM analysis. Any precursor-ion selection and/or precursor-ion fragmentation is performed prior to delivering the ions to the orbitrap. Initially, precursor-ion selection and fragmentation, in order to generate MS^n product-ion mass spectra, are performed in the LIT. As the LIT in the commercial instrument is equipped with two separate off-axis detectors, simultaneous acquisition of precursor-ion HRAM mass spectrum and (several) unit-mass resolution product-ion MS^n mass spectra can be achieved. If both precursor-ion and product-ion mass spectra are acquired using the orbitrap, high resolution (≈100,000, FWHM) is used for the precursor ions, whereas medium resolution (≈15,000–30,000, FWHM) is generally sufficient for the HRAM analysis of the product ions of a well-characterized precursor ion.

Subsequently, it was demonstrated that CID could be achieved in the gas-filled quadrupole C-trap (Olsen et al., 2007). The fragmentation behavior in the C-trap is more similar to collision-cell CID. In commercial instruments, this was implemented by means of the installation of a separate RF-only collision octapole device (higher-energy collision-induced dissociation, HCD) to optimize the use of this feature. In the resulting system, fragmentation can be achieved in the LIT, featuring ion-trap MS^n, as well as in the HCD cell, featuring collision-cell CID. Additional hardware to perform ETD has been made available as well. The introduction of the HCD cell enabled the development of the stand-alone orbitrap, providing all-ion fragmentation without prior precursor-ion selection (Geiger et al., 2010). In addition, a quadrupole–orbitrap hybrid system was developed (Michalski et al., 2011). The configuration of this system comprises a quadrupole mass analyzer, a C-trap, an HCD cell, and an orbitrap mass analyzer. In MS mode, ions are passed through the quadrupole, operated in RF-only mode, to the C-trap, from where they are transferred into the orbitrap mass analyzer for HRAM analysis. In MS–MS mode, a precursor ion is selected in the quadrupole mass analyzer, transferred through the C-trap into the HCD cell for fragmentation. The fragment ions are transported back to the C-trap and then transferred into the orbitrap for HRAM product-ion analysis. Compared to the initially introduced LIT-orbitrap hybrid, the quadrupole-orbitrap hybrid is much faster, for example, with full-spectrum analysis with 140,000 resolution (FWHM, at *m/z* 200) within 1 s, and up to 12 full-spectrum product-ion mass spectra at 17,500 resolution (FWHM) within 1 s.

Most recently, more advanced orbitrap-based systems have been introduced, featuring both a quadrupole mass analyzer, a HCD cell, and a LIT system, next to the C-trap and orbitrap (Senko et al., 2013). In MS^n with a commercial version of this system, the precursor-ion selection is performed in the quadrupole mass analyzer. The ions are transferred through the C-trap to the high-pressure cell of a dual-cell LIT for excitation and fragmentation, eventually performing multistage MS^n. For multiply-charged peptide ion fragmentation, ETD can also be performed in the high-pressure LIT either as a separate experiment or in combination with HCD fragmentation. The fragment ions are transported back to the C-trap and from there transferred into the orbitrap for HRAM product-ion analysis.

Any orbitrap hybrid system for MS–MS or MS^n provides HRAM analysis at present with 140,000 or 280,000 resolution (FWHM for *m/z* 200) and with a mass accuracy of typically $\leq$1 ppm. As orbitrap technology is relatively young, further developments and improvements are to be expected in the future. Orbitrap-based instruments play an important role in metabolite identification, in peptide sequencing, and in solving other structure elucidation problems. They have been used in elucidating the identity of fragment ions of drugs such as furosemide (Section 4.1.4.3) and stanozolol (Section 4.6.4).

1.4.7 Other Instruments for MS–MS and MS^n

In the previous sections, the most widely used system configurations for MS–MS and/or MS^n in combination with ESI have been discussed. Two other tandem mass spectrometry configurations should be briefly mentioned: ion-trap–time-of-flight (IT–TOF) hybrid instruments and MS–MS on FT-ICR-MS instruments.

In addition to the ion-trap-based hybrid systems discussed so far, that is, Q–LIT and LIT–orbitrap instruments, another ion-trap-based hybrid system became commercially available, with an ion-trap as MS_1 and a TOF as MS_2, that is, the IT–TOF hybrid system. IT–TOF systems have been pioneered by the group of Lubman (Michael et al., 1992, 1993). It has subsequently become commercially available for both MALDI and ESI applications (Liu, 2012). Unlike CID in other ion-trap devices where He is used to stabilize the ion trajectories and as collision gas, in the IT–TOF instrument pulses of Ar are used to prevent precursor ions to be lost from the trap and to perform MS^n. It readily provides high-resolution MS and MS^n data, and currently it finds extensive use. However, the mass resolution of the commercial instrument (up to $\approx$15,000, FWHM) is not as good as in Q–TOF hybrid instruments (up to $\approx$60,000, FWHM). The fragmentation of the anabolic steroid norethisterone (Section 4.6.4) has been studied using an IT–TOF hybrid instrument.

An FT-ICR-MS instrument shows attractive features for use in MS–MS or MS^n, that is, its ultra-high resolution and accurate-mass capabilities, as well as the possibility of selectively trapping targeted ions in the ICR cell, while unwanted ions can be eliminated by the application of RF pulses. Thus, the MS^n procedures in an FT-ICR-MS instrument greatly resemble those in an ion-trap instrument. However, the extreme low pressures ($\leq 10^{-7}$ Pa; $\leq 10^{-9}$ mbar) required in the FT-ICR cell exclude the use of CID in the FT-ICR cell (Marshall et al., 1998). This problem can be solved in several ways, that is, the FT-ICR cell serves as MS_2 either in a hybrid system while fragmentation is performed in MS_1 or in a collision cell in between MS_1 and MS_2. To this end, LIT–FT-ICR hybrid systems have been introduced enabling MS^n of a selected precursor ion prior to transferring the product ions to the FT-ICR cell (Wu et al., 2004). Alternatively, Q–FT-ICR hybrid systems can be used (Patrie et al., 2004; Syka et al., 2004). Both types of instrument have become commercially available. FT-ICR-MS instruments have been used to elucidate the fragmentation of drugs such as galantamine (Section 4.5.4), anabolic steroids (Section 4.6.4), and polyether ionophores (Section 4.10.2.3). Due to the success of orbitrap technology, the product line of LIT–FT-ICR has recently been discontinued.

Another solution to the problem with CID in FT-ICR-MS is the use of alternative ion-activation methods to induce fragmentation in the FT-ICR. To this end, ion-activation methods such as infrared multiphoton photodissociation (IRMPD) and sustained off-resonance irradiation (SORI) have been introduced (Sleno & Volmer, 2004; Laskin & Futrell, 2005). More recently, ECD has been used as a powerful ion dissociation tool in FR-ICR-MS, which is applicable to multiply-charged ions of peptides and proteins (Kim & Pandey, 2012; Zhurov et al., 2013).

Another instrumental development that may have large impact on the way structure elucidation by MS–MS is performed is the commercial introduction of hybrid MS–MS systems featuring IMS. IMS is a powerful analytical tool routinely applied for field detection of explosives, drugs, and chemical weapons (e.g., at airports and in field forensics). In its simplest form, a drift-tube ion-mobility spectrometer measures how fast a given ion moves in a uniform electric field through a given atmosphere, for example, a (counter-current) buffer gas (He, N_2, Ar). Thus, an ion-mobility system separates ions by shape and charge. The hyphenation of IMS and MS combines a separation technique based of the analysis of molecular conformation and shape as performed in IMS, with the analysis of *m/z* and the gathered information on molecular structural features as performed in MS. IMS–MS has been pioneered by the groups of Bowers (Wyttenbach et al., 1996) and Clemmer (Clemmer & Jarrold, 1997; Srebalus et al., 1999). In IMS–MS, IMS provides a rapid gas-phase separation step prior to MS analysis, enabling the identification of ions with different drift times, thus with different collisional cross sections. There are several ways to implement IMS in IMS–MS (Kanu et al., 2008; Wyttenbach et al., 2014).

The groups of Bowers and Clemmer use the type of drift tubes also applied in stand-alone IMS. The conventional drift-tube approach is the oldest method to perform ion mobility in combination with MS, whereas drift-tube IMS–MS systems have become commercially available only very recently. The first commercial implementation of IMS–MS was based on the use of traveling-wave stacked-ring ion guides, which were initially developed to replace RF-only hexapole ion guides in vacuum interfaces for API or as collision cells (Giles et al, 2004; Pringle et al., 2007). The collision-cell region of a hybrid Q–IMS–TOF instrument features three traveling-wave stacked-ring ion guides, of which the middle one is used as a ion-mobility drift tube, operated at pressures up to 100 Pa (1 mbar) and with up to 200 mL min^{-1} of N_2 gas, whereas the other two may be used as a collision cell, operated at 1 Pa (10^{-2} mbar) when applicable (Pringle et al., 2007). A third way to perform IMS–MS is high-field asymmetric waveform ion mobility spectrometry (FAIMS), which involves the gas-phase mobility separation of ions in an electric field at atmospheric pressure. The FAIMS device is positioned in the API source in between the ESI needle and the vacuum interface (Kolakowski & Mester, 2007; Tsai et al., 2012). Commercially available FAIMS devices are primarily applied to improve sensitivity and to reduce matrix effects in quantitative analysis using LC–ESI-MS (Tsai et al., 2012; Xia et al., 2008). An IMS–MS system with separation in the ion tunnel area of atmospheric-pressure-to-vacuum interfaces is also under development. An application of IMS–MS in structure elucidation of hydroxylated metabolites is discussed in Section 5.7.1.

1.4.8 MS–MS and MS^n in the Analysis of Drugs and Pesticides

This book focuses on the interpretation of product-ion MS–MS and MS^n mass spectra of drugs, pesticides, and related compounds. The basic data sets comprised mass spectral libraries acquired using TQ and Q–LIT hybrid instruments and especially Q–TOF hybrid instruments. However, the interpretation of the product-ion mass spectra in these libraries were complemented with extensive literature data, which were acquired not only with the three aforementioned instrument types, but also with ion-trap MS^n, IT–TOF MS^n, LTQ–orbitrap, and various FT-ICR-MS systems. Obviously, an important issue in the reliable interpretation of product-ion mass spectra is the availability of accurate-mass data. If in some instances the interpretation was not supported by accurate-mass data, it is mentioned in the text. No attempt was made to explicitly show the differences on the information content of the product-ion mass spectra between collision-cell CID and ion-trap CID. Given the instruments used to acquire the product-ion mass spectra in the mass spectral libraries, most of the data refers to collision-cell CID. However, when comparison between the two CID modes could be made, in most cases the same information content was found, except for ions with low *m/z* (Section 1.4.2).

1.5 DATA ACQUISITION

1.5.1 Introduction

The two general modes of mass spectrometric data acquisition, that is, the full-spectrum mode and the selected-ion mode, have already been introduced in Section 1.3.1.1. The full-spectrum mode in single-MS and the full-spectrum product-ion analysis mode in MS–MS and MS^n (Section 1.4.1.2) have been explained in sufficient detail. In this section, additional and more advanced modes of data acquisition are briefly discussed. A more extensive discussion on the application of these modes follows in Chapter 5.

For a proper understanding of the possibilities and limitations of data acquisition in MS, one should be aware of the fact that, when equipped with one ion-detection system, a mass spectrometer can perform only one experiment at a time. Several experiments in different acquisition modes can be performed in series. Functions may be defined to perform a series of such experiments repeatedly. Given the speed of current data acquisition and processing systems, decisions for a next step in a series of experiments may be based on the data acquired in the previous experiment, that is, data-dependent acquisition (DDA). The time required for individual MS experiments very much depends on the type of instrument used (and its model). Due to the huge progress in electronics, modern instruments can perform much faster than older ones.

1.5.2 Selected-Ion and Selected-Reaction Monitoring

The selected-ion acquisition mode, that is, selected-ion monitoring (SIM) in single-MS instruments and selected-reaction monitoring (SRM) in MS–MS instruments, is a powerful tool especially with beam instruments, that is, quadrupole and sector instruments, to improve S/N in targeted analysis by elongating the dwell time at a particular *m/z*. Ion-trapping devices, both ion-traps and FT-ICR-MS systems, can perform a selected-ion acquisition mode as well, but the gain in S/N will be generally less than that in beam instruments under similar conditions as no significant gain in dwell time is achieved.

In TQ instruments, SRM is a powerful tool to greatly enhance selectivity, and thereby achieve excellent lower limits of quantification in targeted quantitative analysis. In the SRM mode, both stages of mass analysis perform the selection of ions with a particular *m/z* value, that is, in MS_1 a precursor ion, mostly $[M+H]^+$ or $[M-H]^-$ of the target analyte is selected, subjected to dissociation in the collision cell, while in MS_2 a preferably structure-specific product ion of the selected precursor is selected and detected (Figure 1.25). The SRM mode makes the TQ instrument as the instrument of choice in routine quantitative analysis of target compounds in complex (biological) matrices (van Dongen & Niessen, 2012). Recently, it has also been implemented in quantitative analytical strategies using GC–MS as well (e.g., Cherta et al., 2013; Pang et al., 2015). In many instances, SRM is referred to as MRM, multiple-reaction monitoring, to indicate that multiple product ions of one precursor ion are monitored, even if only one product ion is monitored. Given the fact that SRM stands for selected-reaction monitoring (and not single-reaction monitoring), the term MRM is neither useful nor needed and it is therefore deprecated.

An SRM transition is a combination of a precursor ion *m/z*, a product ion *m/z*, and all MS parameters, for example, collision energy, required to measure this transition with the best sensitivity in a particular TQ instrument (and eventually the chromatographic retention time of the compound with which the SRM transition is defined). Important practical aspects, applications, as well as advantages and limitations of SRM are discussed in more detail in Section 5.3.

1.5.3 Structure-Specific Screening: Precursor-Ion and Neutral-Loss Analysis

Apart from full-spectrum product-ion analysis mode and the SRM mode, a TQ instrument has two additional modes of operation that can be useful as structure-specific screening procedures, for example, the precursor-ion analysis (PIA) and the neutral-loss analysis (NLA) modes (Hunt et al., 1983; Johnson & Yost, 1985).

In the PIA mode, MS_1 is operated in full-spectrum mode, whereas MS_2 is operated in selected-ion mode to monitor a structure-specific product ion. In PIA mode, a signal is detected if an ion transmitted in MS_1 upon CID generates the common product ion selected in MS_2. In the resulting mass spectrum, the peaks are labeled with their precursor-ion *m/z* value. An early example of PIA involved the screening for phthalate plasticizers in environmental samples by means of the common fragment ion with *m/z* 149 due to protonated phthalic anhydride (Hunt et al., 1983). The PIA mode can be used to determine from which precursor ion(s) a particular product ion originates. It may thus help answering the question whether a particular product ion is formed from a particular precursor ion in a one-step dissociation reaction or whether an intermediate fragmentation step is involved. As such, the PIA mode enables more detailed studies on fragmentation pathways, as for instance demonstrated for morphine and related opiates (Section 4.7.5) (Bijlsma et al., 2011).

In the NLA mode, both MS_1 and MS_2 are operated in scanning mode, but with a fixed *m/z* offset corresponding to a

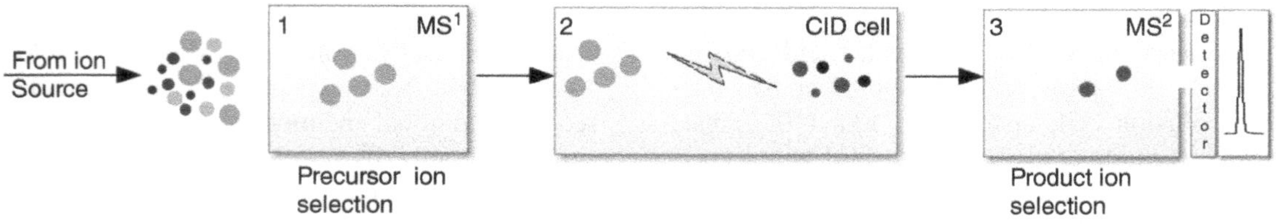

FIGURE 1.25 Schematic representation of selected-reaction monitoring (SRM) in a tandem-quadrupole instrument.

structure-specific neutral loss in the fragmentation reaction. In the NLA mode, a signal is detected if an ion transmitted in MS_1 upon CID loses a neutral molecule with a mass matching the fixed *m/z* difference. An early example of NLA is the monitoring of CO_2 losses from deprotonated aromatic carboxylic acids (Hunt et al., 1983).

The PIA and NLA modes of acquisition in TQ instruments have been successfully applied for structure-specific screening, that is, to search for specific compound classes in complex matrices. This can be demonstrated by the screening for anthocyanins in black raspberries, red raspberries, highbush blueberries, and grapes (Tian et al., 2005) and for two classes of designer drugs in urine (Montesano et al., 2013). The PIA and NLA modes are also frequently applied in drug metabolism studies (Kostiainen et al., 2003), especially using the constant neutral losses of 80 and 176 Da, characteristic for phase II sulfate and glucuronate conjugation, and for glutathione and cyanide-trapped reactive drug metabolites (Jian et al., 2012) (Section 5.7.1).

1.5.4 Data-Dependent Acquisition

The data-dependent acquisition (DDA) mode (also called information dependent acquisition) is another powerful mode of MS data acquisition (Stahl et al., 1996; Wenner & Lynn, 2004; Ma & Chowdhury, 2013). In DDA, the instrument performs a rule-based automatic switching between a survey and a dependent mode. In the most widely applied DDA mode, the instrument switches between full-spectrum MS mode and a full-spectrum product-ion (MS–MS) analysis mode. The switching is controlled by the intensity of a possible precursor ion observed and eventually by additional criteria such as isotope pattern, charge state, or specific *m/z* values on an inclusion or an exclusion list. In this way, highly efficient data acquisition is possible: MS and MS–MS data of unknown compounds in a mixture are acquired simultaneously in one chromatographic run. The DDA mode is widely applied in metabolite identification strategies (Section 5.7.1), in proteomics (Stahl et al., 1996; Wenner & Lynn, 2004), in general unknown screening in (clinical) toxicology (Oberacher & Arnhard, 2015), among other areas. In the MS analysis of complex LC chromatograms of digested proteomes, multiple precursor ions, for example, the 4–10 most abundant precursor ions in a survey MS spectrum, may be used to perform product-ion analysis in the dependent scan, that is, one survey MS spectrum followed by 4–10 product-ion MS–MS spectra before returning to the survey MS mode (ddTopN with $N = 4–10$). Switching between a targeted SRM method as the survey mode and product-ion analysis as the dependent mode has also been demonstrated, for example, in metabolite identification (Li et al., 2005; Soglia et al., 2004). As an example, general unknown screening in toxicology has been described involving data-dependent switching between scheduled-SRM, using ≈1250 SRM transitions, and full-spectrum product-ion analysis (Sections 5.3.3.4 and 5.5) (Dresen et al., 2009; Dresen et al., 2010).

1.5.5 Data-Independent Acquisition

An alternative to DDA is data-independent acquisition (DIA), where scan-wise switching between MS and MS–MS is performed to obtain fragments for all precursor ions present. This is especially useful in combination with HRAM instruments. DIA modes have been described in several ways, for example, as MS^E for Q–TOF instrument (Plumb et al., 2006) and as such widely applied, as MS^M for LIT–orbitrap instruments (Cho et al., 2012), and SWATH for another type of Q–TOF instrument (Arnhard et al., 2015). Examples of these data-acquisition modes are given in Section 5.5.

1.6 SELECTED LITERATURE ON MASS SPECTROMETRY

In this section, useful references are given to general books on mass spectrometry, which are considered relevant to the current discussion.

General mass spectrometry

- Busch KL, Glish GL, McLuckey SA. 1988. Mass spectrometry–mass spectrometry. Techniques and applications of tandem mass spectrometry. VCH Publishers, Inc, New York, NY. ISBN: 978-0-895-73275-0.
- Harrison AG. 1992. Chemical ionization mass spectrometry, 2nd ed. CRC Press, Boca Raton, FL. ISBN: 978-0-849-34254-7.
- Chapman JR. 1993. Practical organic mass spectrometry, 2nd ed. John Wiley & Sons, Ltd., Chichester, UK. ISBN: 978-0-471-92753-8.
- Sparkman OD. 2006. Mass spectrometry desk reference, 2nd ed. ISBN: 978-0-966-08139-8.
- de Hoffmann E, Stroobant V. 2007. Mass spectrometry. Principles and applications, 3rd ed. John Wiley & Sons, Ltd., Chichester, UK. ISBN: 978-0-470-03310-4.
- Watson JT, Sparkman OD. 2007. Introduction to mass spectrometry, 4th ed. Wiley Interscience, Hoboken, NJ. ISBN: 978-0-470-51634-8.
- Boyd RK, Basic C, Bethem RA. 2008. Trace quantitative analysis by mass spectrometry. Wiley Interscience, Hoboken, NJ. ISBN: 978-0-470-05771-1.
- Gross JH. 2011. Mass spectrometry, 2nd ed. Springer-Verlag Berlin-Heidelberg. ISBN: 978-3-642-10711-5.
- Carey FA, Sundberg RJ. 2007. Advanced organic chemistry, 5th ed. Springer New York, NY. ISBN: 978-0-387-44897 (part A) and ISBN: 978-0-387-68350 (part B).

Fragmentation in mass spectrometry

- McLafferty FW, Tureček F. 1993. Interpretation of mass spectra, 4th ed. University Science Books, Mill Valley, CA. ISBN: 978-0-935702-25-3.
- Kinter M, Sherman NE. 2000. Protein sequencing and identification using tandem mass spectrometry. Wiley Interscience, NY. ISBN: 978-0-47132-249-8.
- Smith RM. 2004. Understanding mass spectra. A basic approach, 2nd ed. Wiley Interscience, Hoboken, NJ. ISBN: 978-0-471-42949-4.
- Ham BM. 2008. Even electron mass spectrometry with biomolecule applications. Wiley Interscience, Hoboken, NJ. ISBN: 978-0-470-11802-3.

Liquid chromatography–mass spectrometry

- Cole RB (Ed.). 2010. Electrospray and MALDI mass spectrometry, fundamentals, instrumentation & applications, 2nd ed. Wiley Interscience, Hoboken, NJ. ISBN: 978-0-471-74107-7.
- Lee MS. 2002. LC/MS applications in drug development, 2nd ed. Wiley Interscience, Hoboken, NJ. ISBN: 978-0-471-40520-7.
- Ardrey RE. 2003. Liquid chromatography-mass spectrometry: An introduction. John Wiley & Sons, Ltd., Chichester, UK. ISBN: 978-0-471-49799-8.
- Niessen WMA. 2006. Liquid chromatography-mass spectrometry, 3rd ed. CRC Press, Boca Raton, FL. ISBN: 978-0-824-74082-5.
- Ferrer I, Thurman EM. (Ed.). 2009. Liquid chromatography time of flight mass spectrometry. Wiley Interscience, Hoboken, NJ. ISBN: 978-0-470-13797-0.
- Li W, Zhang J, Tse FLS. (Ed.). 2013. Handbook of LC-MS bioanalysis. Wiley Interscience, Hoboken, NJ. ISBN: 978-1-118-15924-8.

REFERENCES

Allen FH, Watson DG, Brammer L, Orpen AG, Taylor R. 2006. Typical interatomic distances: Organic compounds. Int Tables Crystallogr C 9.5: 790–811.

Allen JS. 1947. An improved electron multiplier particle counter. Rev Sci Instrum 18: 739–749.

Angel PM, Caprioli RM. 2013. Matrix-assisted laser desorption ionization imaging mass spectrometry: In situ molecular mapping. Biochemistry 52: 3818–3828.

Arnhard K, Gottschall A, Pitterl F, Oberacher H. 2015. Applying 'Sequential Windowed Acquisition of All Theoretical Fragment Ion Mass Spectra' (SWATH) for systematic toxicological analysis with liquid chromatography-high-resolution tandem mass spectrometry. Anal Bioanal Chem 407: 405–414.

Arpino PJ, Guiochon G. 1982. Optimization of the instrumental parameters of a combined LC–MS, coupled by an interface for DLI. III. Why the solvent should not be removed in LC–MS interfacing methods. J Chromatogr 251: 153–164.

Arpino P. 1989. Combined liquid chromatography mass spectrometry. Part I. Coupling by means of a moving belt interface. Mass Spectrom Rev 8: 35–55.

Arpino PJ. 1990 Combined liquid chromatography mass spectrometry. Part II. Techniques and mechanisms of thermospray. Mass Spectrom Rev 9: 631–669.

Arpino PJ. 1992. Combined liquid chromatography mass spectrometry. Part III. Applications of thermospray. Mass Spectrom Rev 11: 3–40.

Asam MR, Glish GL. 1997. Tandem mass spectrometry of alkali cationized polysaccharides in a quadrupole ion trap. J Am Soc Mass Spectrom 8: 987–995.

Atwater BL, Stauffer DB, McLafferty FW, Peterson DW. 1985. Reliability ranking and scaling improvements to the probability based matching system for unknown mass spectra. Anal Chem 57: 899–903.

Ausloos P, Clifton CL, Lias SG, Mikaya AI, Stein SE, Tchekhovskoi DV, Sparkma OD, Zaikin V, Zhu D. 1999. The critical evaluation of a comprehensive mass spectral library. J Am Soc Mass Spectrom 10: 287–299.

Banks JF Jr., Shen S, Whitehouse CM, Fenn JB. 1994. Ultrasonically assisted electrospray ionization for LC/MS determination of nucleosides from a transfer RNA digest. Anal Chem 66: 406–414.

Barber M, Bordoli RS, Sedgwick RD, Tyler AN, Whalley ET. 1981. Fast atom bombardment mass spectrometry of bradykinin and related oligopeptides. Biomed Mass Spectrom 8: 337–342.

Beckey HD. 1977. Principles of field ionization and field desorption mass spectrometry. Pergamon, Oxford, UK. ISBN 978-0-080-20612-3.

Bélanger J, Paré JRJ. 1986. Fast atom bombardment mass spectrometry in the pharmaceutical analysis of drugs. J Pharm Biomed Anal 4: 415–441.

Bijlsma L, Sancho JV, Hernández F, Niessen WMA. 2011. Fragmentation pathways of drugs of abuse and their metabolites based on QTOF MS/MS and MSE accurate-mass spectra. J Mass Spectrom 46: 865–875.

Blakley CR, McAdams MJ, Vestal ML. 1978. Crossed-beam liquid chromatograph–mass spectrometer combination, J Chromatogr 158: 261–276.

Blakley CR, Carmody JJ, Vestal ML. 1980. A new soft ionization technique for mass spectrometry of complex molecules. J Am Chem Soc 102: 5931–5933.

Blakley CR, Vestal ML. 1983. Thermospray interface for liquid chromatography/mass spectrometry. Anal Chem 55: 750–754.

Blanksby SJ, Ellison GB. 2003. Bond dissociation energies of organic molecules. Acc Chem Res 36: 255–263.

Bos SJ, van Leeuwen SM, Karst U. 2006. From fundamentals to applications: Recent developments in atmospheric pressure photoionization mass spectrometry. Anal Bioanal Chem 384: 85–99.

Bowie JH. 1984. The formation and fragmentation of negative ions derived from organic molecules. Mass Spectrom Rev 3: 161–207.

Bruins AP, Covey T, Henion JD. 1987. Ion spray interface for combined liquid chromatography/atmospheric pressure ionization mass spectrometry. Anal Chem 59: 2642–2646.

Bruins AP. 1998. Mechanistic aspects of electrospray ionisation. J Chromatogr A 794: 345–357.

Bruno TJ, Svoronos DN. 2010. CRC handbook of basic tables for chemical analysis. 3rd ed. CRC Press. Boca Raton, FL. ISBN: 978-1-420-08042-1.

Busch KL, Glish GL, McLuckey SA. 1988. Mass spectrometry–mass spectrometry. Techniques and applications of tandem mass spectrometry. VCH Publishers, Inc, New York, NY. ISBN 978-0-895-73275-0.

Cai SS, Syage JA. 2006. Comparison of atmospheric pressure photoionization, atmospheric pressure chemical ionization, and electrospray ionization mass spectrometry for analysis of lipids. Anal Chem 78: 1191–1199.

Cai SS, Hanold KA, Syage JA. 2007. Comparison of atmospheric pressure photoionization and atmospheric pressure chemical ionization for normal-phase LC/MS chiral analysis of pharmaceuticals. Anal Chem 79: 2491–2498.

Cai Y, Kingery D, McConnell O, Bach AC, II. 2005. Advantages of atmospheric pressure photoionization mass spectrometry in support of drug discovery. Rapid Commun Mass Spectrom 19: 1717–1724.

Campbell JL, Le Blanc JCY. 2012. Using high-resolution quadrupole TOF technology in DMPK analyses. Bioanalysis 4, 487–500.

Cappiello A, Famiglini G, Palma P, Pierini E, Termopoli V, Trufelli H. 2011. Direct-EI in LC-MS: Towards a universal detector for small-molecule applications. Mass Spectrom Rev 30: 1242–1255.

Carles S, Le Garrec JL, Mitchel JBA. 2007. Electron and ion reactions with hexamethyldisoloxane and pentamethyldisiloxane. J Chem Phys 127: 144308.

Carroll DI, Dzidic I, Stillwell RN, Horning MG, Horning EC. 1974. Subpicogram detection system for gas-phase analysis based upon atmospheric-pressure ionization (API) mass spectrometry. Anal Chem 46: 706–710.

Carroll DI, Dzidic I, Stillwell RN, Haegele KD, Horning EC. 1975. Atmospheric-pressure ionization mass spectrometry: Corona discharge ion source for use in liquid chromatograph–mass spectrometer–computer analytical system. Anal Chem 47: 2369–2373.

Carroll DI, Dzidic I, Horning EC, Stillwell RN. 1981. Atmospheric-pressure ionization mass spectrometry. Appl Spectrosc Rev 17: 337–406.

Chandra AK, Uchimaru T. 2002. The O—H bond dissociation energies of substituted phenols and proton affinities of substituted phenoxide ions: A DFT study. Int J Mol Sci 3: 407–422.

Chernushevich IV, Loboda AV, Thomson BA. 2001. An introduction to quadrupole–time-of-flight mass spectrometry. J Mass Spectrom 36: 849–865.

Cherta L, Portolés T, Beltran J, Pitarch E, Mol JG, Hernández F. 2013. Application of gas chromatography-(triple quadrupole) mass spectrometry with atmospheric pressure chemical ionization for the determination of multiclass pesticides in fruits and vegetables. J Chromatogr A 1314: 224–240.

Chervet JP, Ursem M, Salzman, JP. 1996. Instrumental requirements for nanoscale liquid chromatography. Anal Chem 68: 1507–1512.

Cho R, Huang Y, Schwartz JC, Chen Y, Carlson TJ, Ma J. 2012. MS^M, an efficient workflow for metabolite identification using hybrid linear ion trap orbitrap mass spectrometer. J Am Soc Mass Spectrom 23: 880–888.

Clark AE, Kaleta EJ, Arora A, Wolk DM. 2013. Matrix-assisted laser desorption ionization-time of flight mass spectrometry: A fundamental shift in the routine practice of clinical microbiology. Clin Microbiol Rev 26: 547–603.

Clemmer DE, Jarrold MF. 1997. Ion mobility measurements and their applications to clusters and biomolecules. J Mass Spectrom 32: 577–592.

Cody RB, Laramée JA, Durst HD. 2005. Versatile new ion source for the analysis of materials in open air under ambient conditions. Anal Chem 77: 2297–2302.

Cole RB. 2010. Electrospray and MALDI mass spectrometry, 2nd ed. John Wiley & Sons, Hoboken, NJ. ISBN 978-0-471-74107-7.

Covey TR, Lee ED, Henion JD. 1986. High-speed liquid chromatography tandem mass spectrometry for the determination of drugs in biological samples. Anal Chem 58: 2453–2480.

Covey TR, Bonner RF, Shushan BI, Henion JD. 1988. The determination of protein, oligonucleotide and peptide molecular weights by ion-spray mass spectrometry. Rapid Commun Mass Spectrom 2: 249–256.

Covey TR, Jong R, Javahari H, Liu C, Thomson C, LeBlanc Y. 2001. Design optimization of APCI instrumentation, Proceedings of the 49th ASMS Conference on Mass Spectrometry and Allied Topics, May 27–32, 2001, Chicago, IL.

Creaser CS, Stygall JW. 1993. Particle beam liquid chromatography – mass spectrometry: Instrumentation and aplications. A review. Analyst 118: 1467–1480.

Creaser CS, Ratcliffe L. 2006. Atmospheric pressure matrix-assisted laser desorption/ionization mass spectrometry: A review. Curr Anal Chem 2: 9–15.

de Hoffmann E, Stroobant V. 2007. Mass spectrometry. Principles and applications, 3rd ed. John Wiley & Sons, Ltd., Chichester, UK. ISBN 978-0-470-03310-4.

Demtröder W. 2010. Atoms, molecules and photons, 2nd ed. Springer, Heidelberg. ISBN 978-3-642-10297-4.

Dole M, Mack LL, Hines RL, Mobley RC, Ferguson LD, Alice MB. 1968. Molecular beams of macroions. J Chem Phys 49: 2240–2249.

Doroshenko VM, Cotter RJ. 1989. Ideal velocity focusing in a reflectron time-of-flight mass spectrometer. J Am Soc Mass Spectrom 10: 992–999.

Dougherty RC. 1981. Negative chemical ionization mass spectrometry. Anal Chem 53: 625A-636A.

Douglas DJ, Frank AJ, Mao D. 2005. Linear ion traps in mass spectrometry. Mass Spectrom Rev 24: 1–29.

Dresen S, Gergoc M, Politi, L, Halter C, Weinmann W. 2009. ESI-MS–MS library of 1,253 compounds for application in forensic and clinical toxicology. Anal Bioanal Chem 395: 2521–2526.

Dresen S, Ferreirós N, Gnann H, Zimmermann R, Weinmann W. 2010. Detection and identification of 700 drugs by multi-target screening with a 3200 QTRAP® LC–MS–MS system and library searching. Anal Bioanal Chem 396: 2425–2434.

Es-Safi NE, Kerhoas L, Ducrot PH. 2007. Fragmentation study of iridoid glucosides through positive and negative electrospray ionization, collision-induced dissociation and tandem mass spectrometry. Rapid Commun Mass Spectrom 21: 1165–1275.

Fenn JB, Mann M, Meng CK, Wong SF, Whitehouse CM. 1989. Electrospray ionization for mass spectrometry of large biomolecules. Science 246: 64–71.

Fenn JB. 1993. Ion formation from charged droplets: Roles of geometry, energy, and time. J Am Soc Mass Spectrom 4: 524–535.

Fenselau C, Cotter RJ. 1987. Chemical aspects of fast atom bombardment. Chem Rev 87: 501–512.

Fernández de la Mora J, Loscertales IG. 1994. The current emitted by highly conducting Taylor cones. J Fluid Mech 260: 155–184.

Foltz RL. 1992. Recent applications of mass spectrometry in forensic toxicology. Int J Mass Spectrom Ion Processes 118/119: 237–263.

Gallagher RT, Balogh MP, Davey P, Jackson MR, Southern LJ. 2003. Combined electrospray ionization–atmospheric pressure chemical source for use in high-throughput LC-MS applications. Anal Chem 75: 973–977.

Garcia-Ac A, Segura PA, Viglino L, Gagnon C, Sauvé S. 2011. Comparison of APPI, APCI and ESI for the LC-MS/MS analysis of bezafibrate, cyclophosphamide, enalapril, methotrexate and orlistat in municipal wastewater. J Mass Spectrom 46: 383–390.

Geiger T, Cox J, Mann M. 2010. Proteomics on an orbitrap benchtop mass spectrometer using all-ion fragmentation. Mol Cell Proteomics 9: 2252–2261.

Gilbert W. 1600. De Magnete. Peter Short, London. Translation by Mottelay PF. 1958. Dover Publications Inc., New York. ISBN: 978-0-486-26761-X (unabridged and unaltered republication of the John Wiley & Sons, New York, 1893 edition).

Giles K, Pringle SD, Worthington KR, Little D, Wildgoose JL, Bateman RH. 2004. Applications of a travelling wave-based radio-frequency-only stacked ring ion guide. Rapid Commun Mass Spectrom 18: 2401–2414.

Goodsell DS. 2009. The machinery of life, 2nd ed. Copernicus Books, Springer, New York. ISBN 978-0-387-84924-9.

Green KM, Lebrilla CB. 1997. Ion-molecule reactions as probes of gas-phase structures of peptides and proteins. Mass Spectrom Rev 16: 53–71.

Grimm RL, Beauchamp, JL. 2002. Evaporation and discharge dynamics of highly charged droplets of heptane, octane, and *p*-xylene generated by electrospray ionization. Anal Chem 74: 6291–6297.

Guilhaus M, Selby D, Mlynski V. 2000. Orthogonal acceleration time-of-flight mass spectrometry. Mass Spectrom Rev 19: 65–107.

Hadley JS, Fradin A, Murphy RC. 1988. Electron capture negative ion chemical ionization analysis of arachidonic acid. Biomed Environ Mass Spectrom 15: 175–178.

Hager JW. 2002. A new linear ion trap mass spectrometer. Rapid Commun Mass Spectrom 16: 512–526.

Hakala KS, Kostiainen R, Ketola RA. 2006. Feasibility of different mass spectrometric techniques and programs for automated metabolite profiling of tramadol in human urine. Rapid Commun Mass Spectrom 20: 2081–2090.

Haller I, Mirza UA, Chait BT. 1996. Collision induced decomposition of peptides. Choice of collision parameters. J Am Soc Mass Spectrom 7: 677–681.

Hanold KA, Fischer SM, Cormia PH, Miller CE, Syage JA. 2004. Atmospheric pressure photoionization. 1. General properties for LC/MS. Anal Chem 76: 2842–2851.

Harrison AG. 1992. Chemical ionization mass spectrometry, 2nd ed. CRC Press, Boca Raton, FL. ISBN-978-0-849-34254-7.

Hipple JA, Fox RE, Condon EU. 1946. Metastable ions formed by electron impact in hydrocarbon gases. Phys Rev 69: 347–356.

Hirabayashi A, Sakairi M, Koizumi H. 1994. Sonic spray ionization method for atmospheric pressure ionization mass spectrometry. Anal Chem 66: 4557–4559.

Hirabayashi A, Sakairi M, Koizumi H. 1995. Sonic spray mass spectrometry. Anal Chem 67: 2878–2882.

Hiraoka K, Fukasawa H, Matsushita F, Aizawa K. 1995. High-flow liquid chromatography/mass spectrometry interface using a parallel ion spray. Rapid Commun Mass Spectrom 9: 1349–1355.

Hiraoka K, Saito S, Katsuragawa J, Kudaka I. 1998. A new liquid chromatography mass spectrometry interface: Laser spray, Rapid Commun Mass Spectrom 12: 1170–1174.

Hiraoka K. 2003. Gas-phase ion/molecule reactions. in: fundamentals of mass spectrometry: 109–144. Springer Science+Business Media, New York. ISBN 978-1-461-47233-9.

Hirschfeld T. 1980. The Hy-phen-ated methods. Anal Chem 52: 297A-312A.

Holman SW, Sims PF, Eyers CE. 2012. The use of selected reaction monitoring in quantitative proteomics. Bioanalysis 4: 1763–1786.

Holmes J, Morrell F. 1957. Oscillographic mass spectrometric monitoring of gas chromatography. Appl Spectrosc 11: 86–87.

Holmgren E, Carlsson H, Goede P, Crescenzi C. 2005. Determination and characterization of organic explosives using porous graphitic carbon and liquid chromatography-atmospheric pressure chemical ionization mass spectrometry. J Chromatogr A 1099: 127–135.

Hopfgartner G, Varesio E, Tschäppät V, Grivet C, Bourgogne E, Leuthold LA. 2004. Triple quadrupole linear ion trap mass spectrometer for the analysis of small molecules and macromolecules. J Mass Spectrom 39: 845–855.

Horning EC, Horning MG, Carroll DI, Dzidic I, Stillwell RN. 1973. New picogram detection system based on a mass spectrometer with an external ionization source at atmospheric pressure. Anal Chem 45: 936–943.

Hu Q, Noll RJ, Li H, Makarov A, Hardman M, Cooks GR. 2005. The orbitrap: A new mass spectrometer. J Mass Spectrom 40: 430–443.

Huang MZ, Yuan CH, Cheng SC, Cho YT, Shiea J. 2010. Ambient ionization mass spectrometry. Annu Rev Anal Chem 3: 43–65.

Hübschmann HJ. 2015. Handbook of GC-MS, 3rd ed. Wiley-VCH Verlag GmbH & Co. ISBN 978-3-527-33474-2.

Huertas ML, Fontan J. 1975. Evolution times of tropospheric positive ions. Atmos Environ 9: 1018–1026.

Hunt DF, Stafford GC, Crow FW, Russell JW. 1976. Pulsed positive negative ion chemical ionization mass spectrometry. Anal Chem 48: 2098–2104.

Hunt DF, Shabanowitz J, Harvey TM, Coates ML. 1983. Analysis of organics in the environment by functional group using a triple quadrupole mass spectrometer. J Chromatogr 271: 93–105.

Hunter EPL, Lias SG. 1998. Evaluated gas phase basicities and proton affinities of molecules: An update. J Phys Chem Ref Data 27: 413–656.

Ibrahim Y, Tang K, Tolmachev AV, Shvartsburg AA, Smith RD. 2006. Improving mass spectrometer sensitivity using a high-pressure electrodynamic ion funnel interface. J Am Soc Mass Spectrom 17: 1299–1305.

Iribarne, JV, Thomson BA. 1976. Evaporation of small ions from charged droplets. J Chem Phys 64: 2287–2294.

Jarmusch AK, Musso AM, Shymanovich T, Jarmusch SA, Weavil MJ, Lovin ME, Ehrmann BM, Saari S, Nichols DE, Faeth SH, Cech NB. 2016. Comparison of electrospray ionization and atmospheric pressure photoionization liquid chromatography mass spectrometry methods for analysis of ergot alkaloids from endophyte-infected sleepygrass (*Achnatherum robustum*). J Pharm Biomed Anal 117: 11–17.

Jian W, Liu HF, Zhao W, Jones E, Zhu M. 2012. Simultaneous screening of glutathione and cyanide adducts using precursor ion and neutral loss scans-dependent product ion spectral acquisition and data mining tools. J Am Soc Mass Spectrom 23: 964–976.

Jiang Y, Cole RB. 2005. Oligosaccharide analysis using anion attachment in negative mode electrospray mass spectrometry. J Am Soc Mass Spectrom 16: 60–70.

Johnson JV, Yost RA. 1985. Tandem mass spectrometry for trace analysis. Anal Chem 57: 758A-768A.

Johnson JV, Yost RA, Kelley PE, Bradford DC. 1990. Tandem-in-space and tandem-in-time mass spectrometry: Triple quadrupoles and quadrupole ion traps. Anal Chem 62: 2162–2172.

Jonscher KR, Yates JR, III, 1997. The quadrupole ion trap mass spectrometer – a small solution to a big challenge. Anal Biochem 244: 1–15.

Kanu AB, Dwivedi P, Tam M, Matz L, Hill HH, Jr., 2008, Ion mobility-mass spectrometry. J Mass Spectrom 43: 1–22.

Karancsi T, Slégel P. 1999. Reliable molecular mass determination of aromatic nitro compounds: Elimination of gas-phase reduction occurring during atmospheric pressure chemical ionization. J Mass Spectrom 34: 975–977.

Karas M, Hillenkamp F. 1988. Laser desorption ionization of proteins with molecular masses exceeding 10,000 Daltons. Anal Chem 60: 2299–2301.

Karas M. 1996. Matrix-assisted laser desorption ionization mass spectrometry: A progress report. Biochem Soc Trans 24: 897–900.

Karas M, Krüger R. 2003. Ion formation in MALDI: The cluster ionisation mechanism. Chem Rev 103: 427–439.

Karasek FW. 1974. Plasma chromatography. Anal Chem 46: 710A-720A.

Katta V, Rockwood AL, Vestal ML. 1991. Field limit for ion evaporation from charged thermospray droplets. Int J Mass Spectrom Ion Processes 103: 129–148.

Kauppila TJ, Kostiainen R, Bruins AP. 2004a. Anisole, a new dopant for atmospheric pressure photoionization mass spectrometry of low proton affinity, low ionization energy compounds. Rapid Commun Mass Spectrom 18: 808–815.

Kauppila TJ, Kotiaho T, Kostiainen R, Bruins AP. 2004b. Negative-ion atmospheric pressure photoionization mass spectrometry. J Am Soc Mass Spectrom 15: 203–211.

Kebarle P, Verkerk UH. 2009. Electrospray: From ions in solution to ions in the gas phase, what we know now. Mass Spectrom Rev 28: 898–917.

Kelly RT, Tolmachev AV, Page, JS, Tang K, Smith RD. 2010. The ion funnel: Theory, implementations, and applications. Mass Spectrom Rev 29: 294–312.

Kim M-S, Pandey A. 2012. Electron transfer dissociation mass spectrometry in proteomics. Proteomics 12: 530–542.

Kind T, Fiehn O. 2010. Advances in structure elucidation of small molecules using mass spectrometry. Bioanal Rev 2: 23–60.

Knochenmuss R. 2006. Ion formation mechanisms in UV-MALDI. Analyst 131: 966–986.

Kolakowski BM, Mester Z. 2007. Review of applications of high-field asymmetric waveform ion mobility spectrometry (FAIMS) and differential mobility spectrometry (DMS). Analyst 132: 842–864.

Koo I, Kim S, Zhang X. 2013. Comparative analysis of mass spectral matching-based compound identification in gas chromatography–mass spectrometry. J Chromatogr A 1298: 132–138.

Kostiainen R, Kotiaho T, Kuuranne T, Auriola S. 2003. Liquid chromatography/atmospheric pressure ionization-mass spectrometry in drug metabolism studies. J Mass Spectrom 38: 357–372.

Lacorte S, Fernandez-Alba AR. 2006. Time of flight mass spectrometry applied to the liquid chromatographic analysis of pesticides in water and food. Mass Spectrom Rev 25: 866–880.

Laskin J, Futrell JH. 2005. Activation of large ions in FT-ICR mass spectrometry. Mass Spectrom Rev 24: 135–167.

Lattimer RP, Schulten H-R. 1989. Field ionization and field desorption mass spectrometry: Past, present, and future. Anal Chem 61: 1201A-1215A.

Leinonen A, Kuuranne T, Kostiainen R. 2002. LC–MS in anabolic steroid analysis - optimization and comparison of three ionization techniques: Electrospray ionization, atmospheric pressure chemical ionization and atmospheric pressure photoionization. J Mass Spectrom 37: 693–698.

Leis HJ, Windischhofer W. 2012. (*S*)-(−)-*N*-(Pentafluorobenzylcarbamoyl)prolyl chloride: A chiral derivatisation reagent designed for gas chromatography/negative ion chemical ionisation mass spectrometry of amino compounds. Rapid Commun Mass Spectrom 26: 592–598.

Lépine F, Boismenu D, Milot S, Mamer O. 1999. Collision of molecular anions of benzenedicarboxylic esters with oxygen in a triple quadrupole mass spectrometer. J Am Soc Mass Spectrom 10: 1248–1252.

Li AC, Alton D, Bryant MS, Shou WZ. 2005. Simultaneously quantifying parent drugs and screening for metabolites in plasma pharmacokinetic samples using selected reaction monitoring information-dependent acquisition on a QTrap instrument. Rapid Commun Mass Spectrom 19: 1943–1950.

Li DX, Gan L, Bronja A, Schmitz OJ. 2015. Gas chromatography coupled to atmospheric pressure ionization mass spectrometry (GC-API-MS): Review. Anal Chim Acta 891: 43–61.

Lias SG, Liebman JF, and Levin RD. 1984. Evaluated gas phase basicities and proton affinities of molecules; heats of formation of protonated molecules. J Phys Chem Ref Data: 13: 695–808.

Lien GW, Chen CY, Wang GS. 2009. Comparison of electrospray ionization, atmospheric pressure chemical ionization and atmospheric pressure photoionization for determining estrogenic chemicals in water by liquid chromatography tandem mass spectrometry with chemical derivatizations. J Chromatogr A 1216: 956–966.

Lim KH, Su Z, Foltz RL. 1993. Stereoselective disposition: Enantioselective quantitation of 3,4-(methylenedioxy)methamphetamine and three of its metabolites by gas chromatography/electron capture negative ion chemical ionization mass spectrometry. Biol Mass Spectrom 22: 403–411.

Linden HB. 2004. Liquid injection field desorption ionization: A new tool for soft ionization of samples including air-sensitive catalysts and non-polar hydrocarbons. Eur J Mass Spectrom 10: 459–468.

Liu ZY. 2012. An introduction to hybrid ion trap/time-of-flight mass spectrometry coupled with liquid chromatography applied to drug metabolism studies. J Mass Spectrom 47: 1627–1642.

Louris JN, Cooks RG, Syka JEP, Kelley PE, Stafford GC, Jr., Todd JFJ. 1987. Instrumentation, applications, and energy deposition in quadrupole ion-trap tandem mass spectrometry. Anal Chem 59: 1677–1685.

Ma S, Chowdhury SK. 2013. Data acquisition and data mining techniques for metabolite identification using LC coupled to high-resolution MS. Bioanalysis 5: 1285–1297.

Mack LL, Kralik P, Rheude A, Dole M. 1970. Molecular beams of macroions. II. J Chem Phys 52: 4977-4986.

Makarov A, Denisov E, Kholomeev A, Balschun W, Lange O, Strupat K, Horning S. 2006. Performance evaluation of a hybrid linear ion trap/orbitrap mass spectrometer. Anal Chem 78: 2113–2120.

Mann M, Meng CK, Fenn JB. 1989. Interpreting mass spectra of multiply charged ions. Anal Chem 61: 1702–1708.

Mann M, Talbo G. 1996. Developments in matrix-assisted laser desorption/ionization peptide mass spectrometry. Curr Opin Biotechnol 7: 11–19.

Mansoori BA, Dyer EW, Lock CM, Bateman K, Boyd RK, Thomson BA. 1998. Analytical performance of a high-pressure radiofrequency-only quadrupole collision cell with an axial field applied by using conical rods. J Am Soc Mass Spectrom 9: 775–788.

March RE. 1997. An introduction to quadrupole ion trap MS. J Mass Spectrom 32: 351–369.

Marchi I, Rudaz S, Veuthey J-L. 2009. Atmospheric pressure photoionization for coupling liquid-chromatography to mass spectrometry: A review. Talanta 78: 1–18.

Märk TD, Dunn GH. 1985. Electron impact ionisation. Springer Verlag. Wien. ISBN 978-3-709-14030-7.

Marquet P, Lachâtre G. 1999. Liquid chromatography–mass spectrometry: Potential in forensic and clinical toxicology. J Chromatogr B 733: 93–118.

Marquet P, Venisse N, Lacassie É, Lachâtre G. 2000. In-source CID mass spectral libraries for the "general unknown" screening of drugs and toxicants. Analusis 28: 925–934.

Marshall A, Hendrickson CL, Jackson GS. 1998. Fourier transform ion cyclotron resonance mass spectrometry: A primer. Mass Spectrom Rev 17: 1–35.

Matuszewski BK, Constanzer ML, Chavez-Eng CM. 2003. Strategies for the assessment of matrix effect in quantitative bioanalytical methods based on HPLC-MS/MS. Anal Chem 75: 3019–3030.

Matuszewski BK. 2006. Standard line slopes as a measure of a relative matrix effect in quantitative HPLC-MS bioanalysis. J Chromatogr B 830: 293–300.

McEwen CN, McKay RG, Larsen BS. 2005 Analysis of solids, liquids, and biological tissues using solids probe introduction at atmospheric pressure on commercial LC/MS instruments. Anal Chem 77: 7826–7831.

McLafferty FW, Tureček F. 1993. Interpretation of mass spectra, 4th ed. University Science Books, Mill Valley, CA. ISBN 978-0-935702-25-3.

Mesaros C, Lee SH, Blair IA. 2010. Analysis of epoxyeicosatrienoic acids by chiral liquid chromatography/electron capture atmospheric pressure chemical ionization mass spectrometry using [^{13}C]-analog internal standards. Rapid Commun Mass Spectrom 24: 3237–3247.

Michael SM, Chien BM, Lubman DM. 1992. An ion trap storage/time-of-flight mass spectrometer. Rev Sci Instrum 63: 4277–4284.

Michael SM, Chien BM, Lubman DM. 1993. Detection of electrospray ionization using a quadrupole ion trap storage/reflectron time-of-flight mass spectrometer. Anal Chem 65: 2614–2620.

Michalski A, Damoc E, Hauschild JP, Lange O, Wieghaus A, Makarov A, Nagaraj N, Cox J, Mann M, Horning S. 2011. Mass spectrometry-based proteomics using Q Exactive, a high-performance benchtop quadrupole orbitrap mass spectrometer. Mol Cell Proteomics 10, M111.011015, 1–12.

Miller PE, Denton MB. 1986. The quadrupole mass filter: Basic operating concepts. J Chem Educ 63: 617–622.

Montalti M, Credi A, Prodi L, Gandolfi MT. 2006. Handbook of photochemistry, 3rd ed. Taylor & Francis Group, LLC, Boca Raton, FL. ISBN 978-0-824-72377-5.

Montesano C, Sergi M, Moro M, Napoletano S, Romolo FS, Del Carlo M, Compagnone D, Curini R. 2013. Screening of methylenedioxyamphetamine- and piperazine-derived designer drugs in urine by LC-MS/MS using neutral loss and precursor ion scan. J Mass Spectrom 48: 49–59.

Moore C, Rana S, Coulter C, Feyerherm F, Prest H. 2006. Application of two-dimensional gas chromatography with electron capture chemical ionization mass spectrometry to the detection of 11-nor-Δ^9-tetrahydrocannabinol-9-carboxylic acid (THC-COOH) in hair. J Anal Toxicol 30: 171–177.

Morris HR, Paxton T, Dell A, Langhorne J, Berg M, Bordoli RS, Hoyes J, Bateman RH. 1996. High-sensitivity collisionally-activated decomposition tandem mass spectrometry on a novel quadrupole–orthogonal acceleration time-of-flight mass spectrometer. Rapid Commun Mass Spectrom 10: 889–896.

Munson MSB. 2000. Development of chemical ionization mass spectrometry. Int J Mass Spectrom 200: 243–251.

Nguyen S, Fenn JB. 2007. Gas-phase ions of solute species from charged droplets of solutions. Proc Natl Acad Sci U S A 104: 1111–1117.

Nollet JA. 1754. Recherches sur les causes particulières des phénomènes électriques, Nouvelle Ed. H.L. Guerin & L.F. Delatour, Paris. (Modern re-edition ISBN: 978-1-276-08135-9).

Oberacher H, Arnhard K. 2015. Compound identification in forensic toxicological analysis with untargeted LC-MS-based techniques. Bioanalysis 7: 2825–2840.

Olsen JV, Macek B, Lange O, Makarov A, Horning S, Mann M. 2007. Higher-energy C-trap dissociation for peptide modification analysis. Nat Methods 4: 709–712.

Olsen JV, Schwartz JC, Griep-Raming J, Nielsen ML, Damoc E, Denisov E, Lange O, Remes P, Taylor D, Splendore M, Wouters ER, Senko M, Makarov A, Mann M, Horning S. 2009. A dual pressure linear ion trap orbitrap instrument with very high sequencing speed. Mol Cell Proteomics 8: 2759–2769.

Page JS, Kelly RT, Tang K, Smith RD. 2007. Ionization and transmission efficiency in an electrospray ionization-mass spectrometry interface. J Am Soc Mass Spectrom 18: 1582–1590.

Pang GF, Fan CL, Cao YZ, Yan F, Li Y, Kang J, Chen H, Chang QY. 2015. High throughput analytical techniques for the determination and confirmation of residues of 653 multiclass pesticides and chemical pollutants in tea by GC/MS, GC/MS/MS, and LC/MS/MS: Collaborative study, first action 2014.09. J AOAC Int 98: 1428–1454.

Patrie SM, Charlebois JP, Whipple D, Kelleher NL, Hendrickson CL, Quinn JP, Marshall AG, Mukhopadhyay B. 2004. Construction of a hybrid quadrupole–Fourier transform ion cyclotron resonance mass spectrometer for versatile MS–MS above 10 kDa. J Am Soc Mass Spectrom 15: 1099–1108.

Paul W, Steinwedel H. 1953. Ein neues Massenspektrometer ohne Magnetfeld. Z Naturforsh A 8: 448–450.

Paul W, Raether M. 1955. Das elektrische Massenfiler. Z Phys 140: 262–271.

Paul W. 1990. Electromagnetic traps for charged and neutral particles (Nobel lecture). Angew Chem Int Ed Engl 29: 739–748.

Plumb RS, Johnson KA, Rainville P, Smith BW, Wilson ID, Castro-Perez JM, Nicholson JK. 2006. UPLC/MS^E; A new approach for generating molecular fragment information for biomarker structure elucidation. Rapid Commun Mass Spectrom 20: 1989–1994.

Pringle SD, Giles K, Wildgoose JL, Williams JP, Slade SE, Thalassinos K, Bateman RH, Bowers MT, Scrivens JH. 2007. An investigation of the mobility separation of some peptide and protein ions using a new hybrid quadrupole/travelling wave IMS/oa-ToF instrument. Int J Mass Spectrom 261: 1–12.

Raffaelli A, Saba A. 2003. Atmospheric pressure photoionization mass spectrometry. Mass Spectrom Rev 22: 318–331.

Rauha J-P, Vuorela H, Kostiainen R. 2001. Effect of eluent on the ionization efficiency of flavonoids by ion spray, atmospheric pressure chemical ionization, and atmospheric pressure photoionization mass spectrometry. J Mass Spectrom 36: 1269–1280.

Rayleigh JWS. 1882. On the equilibrium of liquid conducting masses charged with electricity. Philos Mag series 5. 14: 184–186.

Ramm LE, Whitlow MB, Mayer MM. 1985. The relationship between channel size and the number of C9 molecules in the C5b-9 complex. J Immunol 134: 2594–2599.

Robb DB, Covey TR, Bruins AP. 2000. Atmospheric pressure photoionization: An ionization method for liquid chromatography–mass spectrometry. Anal Chem 72: 3653–3659.

Robb DB, Blades MW. 2005. Effects of solvent flow, dopant flow, and lamp current on dopant-assisted APPI for LC–MS. Ionization via proton transfer. J Am Soc Mass Spectrom 16: 1275–1290.

Robb DB, Blades MW. 2008. State-of-the-art in atmospheric pressure photoionization for LC/MS. Anal Chim Acta 627: 34–49.

Rosenfelder N, Vetter W. 2009. Gas chromatography coupled to electron capture negative on mass spectrometry with nitrogen as the reagent gas – an alternative method for the determination of polybrominated compounds. Rapid Commun Mass Spectrom 23: 3807–3812.

Ross MS, Wong CS. 2010. Comparison of electrospray ionization, atmospheric pressure photoionization, and anion attachment atmospheric pressure photoionization for the analysis of hexabromocyclododecane enantiomers in environmental samples. J Chromatogr A 1217: 7855–7863.

Roy TA, Field FH, Lin YY, Smith LL. 1979. Hydroxyl ion negative chemical ionization mass spectra of steroids. Anal Chem 51: 272–278.

Schmelzeisen-Redeker G, Bütfering L, Röllgen FW. 1989. Desolvation of ions and molecules in thermospray mass spectrometry. Int J Mass Spectrom Ion Processes 90: 139–150.

Schmid S, Jecklin MC, Zenobi R. 2011. Electrosonic spray ionization–an ideal interface for high-flow liquid chromatography applications. J Chromatogr A 1218: 3704–3710.

Schulten HR. 1982. Off-line combination of liquid chromatography and field desorption mass spectrometry: Principles and environmental, medical and pharmaceutical applications. J Chromatogr 251: 105–128.

Schwartz JC, Senko MW, Syka JEP. 2002. A two-dimensional quadrupole ion trap mass spectrometer. J Am Soc Mass Spectrom 13: 659–669.

Scigelova M, Hornshaw M, Giannokopulos A, Makarov A. 2011. Fourier transform mass spectrometry. Mol Cell Proteomics 10: M111.009431.1-19.

Senko MW, Remes PM, Canterbury JD, Mathur R, Song Q, Eliuk SM, Mullen C, Earley L, Hardman M, Blethrow JD, Bui H, Specht A, Lange O, Denisov E, Makarov A, Horning S, Zabrouskov V. 2013. Novel parallelized quadrupole/linear ion trap/orbitrap tribrid mass spectrometer improving proteome coverage and Peptide identification rates. Anal Chem 85: 11710–11714.

Shahin MM. 1966. Mass-spectrometric studies of corona discharges in air at atmospheric pressures. J Chem Phys 45: 2600–2605.

Short LC, Cai SS, Syage JA. 2007. APPI-MS: Effects of mobile phases and VUV lamps on the detection of PAH compounds. J Am Soc Mass Spectrom 18: 589–599.

Singh G, Gutierrez A, Xu K, Blair AI. 2000. Liquid chromatography/electron capture atmospheric pressure chemical ionization/mass spectrometry: Analysis of pentafluorobenzyl derivatives of biomolecules and drugs in the attomole range. Anal Chem 72: 3007–3013.

Sleno L, Volmer DA. 2004. Ion activation methods for tandem mass spectrometry. J Mass Spectrom 39: 1091–1112.

Smith DPH. 1986. The electrohydrodynamic atomization of liquids. IEEE Trans Ind Appl 22: 527–535.

Smith RD, Loo JA, Edmonds CG, Barinaga CJ, Udseth HR. 1990. New developments in biochemical mass spectrometry: Electrospray ionization. Anal Chem 62: 882–899.

Smith RD, Loo JA, Ogorzalek Loo RR, Busman M, Udseth HR. 1991. Principles and practice of electrospray ionization-mass spectrometry for large polypeptides and proteins. Mass Spectrom Rev 10: 359–452.

Smith RM. 2004. Understanding mass spectra. A basic approach, 2nd ed. Wiley Interscience, Hoboken, NJ. ISBN 978-0-471-42949-4.

Soglia JR, Harriman SP, Zhao S, Barberia J, Cole MJ, Boyd JG, Contillo LG. 2004. The development of a higher throughput reactive intermediate screening assay incorporating micro-bore liquid chromatography-micro-electrospray ionization-tandem mass spectrometry and glutathione ethyl ester as an in vitro conjugating agent. J Pharm Biomed Anal 36: 105–116.

Sparkman OD, Penton ZE, Kitson FG. 2011. Gas chromatography and mass spectrometry. A practical guide, 2nd ed. Academic Press, Burlington, MA. ISBN 978-0-123-73628-4.

Srebalus CA, Li J, Marshall WS, Clemmer DE. 1999. Gas-phase separations of electrosprayed peptide libraries. Anal Chem 71: 3918–3927.

Stafford GC Jr., Kelley PE, Syka, JEP, Reynolds WE, Todd JFJ. 1984. Recent improvements in and analytical applications of advanced ion trap technology. Int J Mass Spectrom Ion Processes 60: 85–98.

Stafford GC Jr., 2002. Ion trap mass spectrometry: A personal perspective. J Am Soc Mass Spectrom 13 (2002) 589–596.

Stahl DC, Swiderek KM, Davis MT, Lee TD. 1996. Data-controlled automation of liquid chromatography/tandem mass spectrometry analysis of peptide mixtures. J Am Soc Mass Spectrom 7: 532–540.

Stein SE, Scott DR. 1994. Optimization and testing of mass spectral library search algorithms for compound identification. J Am Soc Mass Spectrom 5: 859–866.

Stemmeler EA, Diener JL, Swift A. 1994. Gas-phase reactions of $O_2^{-\bullet}$ with alkyl and aryl esters of benzenedicarboxylic acids. J Am Soc Mass Spectrom 5: 990–1000.

Stemmeler EA, Hites RA. 1988. The fragmentation of negative ions generated by electron capture negative ion mass spectrometry: A review with new data. Biomed Environ Mass Spectrom 17: 311–328.

Straube EA, Dekant W, Völkel W. 2004. Comparison of electrospray ionization, atmospheric pressure chemical ionization, and atmospheric pressure photoionization for the analysis of dinitropyrene and aminonitropyrene LC-MS/MS. J Am Soc Mass Spectrom 15: 1853–1862.

Syage JA, Evans MD, Hanold KA. 2000. Photoionization mass spectrometry. Am Lab 32: 24–29.

Syka JE, Marto JA, Bai DL, Horning S, Senko MW, Schwartz JC, Ueberheide B, Garcia B, Busby S, Muratore T, Shabanowitz J, Hunt DF. 2004. Novel linear quadrupole ion trap/FT mass spectrometer: Performance characterization and use in the comparative analysis of histone H3 post-translational modifications. J Proteome Res 3: 621–626.

Takáts Z, Wiseman JM, Cooks RG. 2005. Ambient mass spectrometry using desorption electrospray ionization (DESI): Instrumentation, mechanisms and applications in forensics, chemistry and biology. J Mass Spectrom 40: 1261–1275.

Tanaka K, Waki H, Ido Y, Akita S, Yoshida Y, Yoshida T. 1988. Protein and polymer analyses up to *m/z* 100 000 by laser ionization time-of-flight mass spectrometry. Rapid Commun Mass Spectrom 2: 151–153.

Taylor G. 1964. Disintegration of water drops in an electric field. Proc R Soc London Ser A 280: 383–397.

Tian Q, Giusti MM, Stoner GD, Schwartz SJ. 2005. Screening for anthocyanins using high-performance liquid chromatography coupled to electrospray ionization tandem mass spectrometry with precursor-ion analysis, product-ion analysis, common-neutral-loss analysis, and selected reaction monitoring. J Chromatogr A 1091: 72–82.

Tsai CW, Yost RA, Garrett TJ. 2012. High-field asymmetric waveform ion mobility spectrometry with solvent vapor addition: A potential greener bioanalytical technique. Bioanalysis 4: 1363–1375.

Tsuchiya M, Aoki E, Kuwabara H. 1989. Clusters of water under atmospheric pressure studied by field ionization and liquid ionization mass spectrometry. Int J Mass Spectrom Ion Processes 90: 55–70.

Tyler AN, Clayton E, Green BN. 1996. Exact mass measurement of polar organic molecules at low resolution using electrospray ionization and a quadrupole mass spectrometer. Anal Chem 68: 3561–3569.

Urban J, Afseth NK, Štys D. 2014. Fundamental definitions and confusions in mass spectrometry about mass assignment, centroiding and resolution. Trends Anal Chem 52: 126–136.

van Bavel B, Geng D, Cherta L, Nácher-Mestre J, Portolés T, Ábalos M, Sauló J, Abad E, Dunstan J, Jones R, Kotz A, Winterhalter H, Malisch R, Traag W, Hagberg J, Ericson Jogsten I, Beltran J, Hernández F. 2015. Atmospheric-pressure chemical ionization tandem mass spectrometry (APGC/MS/MS) an alternative to high-resolution mass spectrometry (HRGC/HRMS) for the determination of dioxins. Anal Chem 87: 9047–9053.

Van Berkel GJ, Pasilis SP, Ovchinnikova O. 2008. Established and emerging atmospheric pressure surface sampling/ionization techniques for mass spectrometry. J Mass Spectrom 43: 1161–1180.

van der Hooft JJ, Vervoort J, Bino RJ, Beekwilder J, de Vos RC. 2011. Polyphenol identification based on systematic and robust high-resolution accurate mass spectrometry fragmentation. Anal Chem 83: 409–416.

van Dongen WD, Niessen WMA. 2012. LC-MS systems for quantitative bioanalysis. Bioanalysis 4: 2391–2399.

van Veldhuizen EM, Rutgers WR. 2001. Corona discharges: Fundamentals and diagnostics. Proceedings of Frontiers in Low Temperature Plasma Diagnostics IV, Rolduc, The Netherlands, March 2001: 40–49.

Vestal ML. 1983. Ionization techniques for nonvolatile molecules. Mass Spectrom Rev 2: 447–480.

Vestal ML, Fergusson GJ. 1985. Thermospray liquid chromatograph/mass spectrometer interface with direct electrical heating of the capillary. Anal Chem 57: 2373–2378.

Vestal ML, Juhasz P, Martin SA. 1995. Delayed extraction matrix-assisted laser desorption time-of-flight mass spectrometry. Rapid Commun Mass Spectrom 9: 1044–1050.

Wang C, Gardinali PR. 2012. Comparison of multiple API techniques for the simultaneous detection of microconstituents in water by on-line SPE-LC-MS/MS. J Mass Spectrom 47: 1255–1268.

Wang G, Hsieh Y, Korfmacher WA. 2005. Comparison of atmospheric pressure chemical ionization, electrospray ionization, and atmospheric pressure photoionization for the determination of cyclosporin A in rat plasma. Anal Chem 77: 541–548.

Weinmann W, Wiedemann A, Eppinger B, Renz M, Svoboda M. 1999. Screening for drugs in serum by electrospray ionization/collision-induced dissociation and library searching. J Am Soc Mass Spectrom 10: 1028–1037.

Well PR. 1968. Group electronegativities. Prog Phys Org Chem 6: 111–145.

Wenner BR, Lynn BC. 2004. Factors that affect ion trap data-dependent MS/MS in proteomics. J Am Soc Mass Spectrom 15: 150–1507.

Whitehouse CM, Dreyer RN, Yamashita M, Fenn JB. 1985. Electrospray interface for liquid chromatographs and mass spectrometers. Anal Chem 57: 675–679.

Wilm M, Mann M. 1994. Electrospray and Taylor-cone theory, Dole's beam of macromolecules. Int J Mass Spectrom Ion Processes 136: 167–180.

Wilm M, Mann M. 1996. Analytical properties of the nanoelectrospray ion source. Anal Chem 68: 1–8.

Wilm M. 2011. Principles of electrospray ionization. Mol Cell Proteomics 10(7): M111.009407-1-8.

Winger BE, Light-Wahl KJ, Ogorzalek Loo RR, Udseth HR, Smith RD. 1993. Observation and implications of high mass-to-charge ratio ions from electrospray ionization mass spectrometry. J Am Soc Mass Spectrom 4: 536–545.

Wiza JL. 1979. Microchannel plate detectors. Nucl Instrum Methods 162: 587–601.

Wolfender J-L, Rodriguez S, Hostettmann K. 1998. Liquid chromatography coupled to mass spectrometry and nuclear magnetic resonance spectroscopy for the screening of plant constituents. J Chromatogr A 794: 299–316.

Wong SF, Meng CK, Fenn JB. 1988. Multiple charging in electrospray ionization of poly(ethylene glycols). J Phys Chem 92: 546–550.

Wu SL, Jardine I, Hancock WS, Karger BL. 2004. A new and sensitive on-line liquid chromatography/mass spectrometric approach for top-down protein analysis: The comprehensive analysis of human growth hormone in an *E. coli* lysate using a hybrid linear ion trap/Fourier transform ion cyclotron resonance mass spectrometer. Rapid Commun Mass Spectrom 18: 2201–2207.

Wyttenbach T, von Helden G, Bowers MT. 1996. Gas-phase conformation of biological molecules: Bradykinin. J Am Chem Soc 118: 8355–8364.

Wyttenbach T, Pierson NA, Clemmer DE, Bowers MT. 2014. Ion mobility analysis of molecular dynamics. Annu Rev Phys Chem 65: 175–196.

Xia YQ, Wu ST, Jemal M. 2008, LC-FAIMS-MS/MS for quantification of a peptide in plasma and evaluation of FAIMS global selectivity from plasma components. Anal Chem 80: 7137–7143.

Xie C, Zhong D, Yu K, Chen X. 2012. Recent advances in metabolite identification and quantitative bioanalysis by LC-Q-TOF MS. Bioanalysis 4: 937–959.

Yamashita M, Fenn FB. 1984a. Electrospray ion source. Another variation on the free-jet theme. J Phys Chem 88: 4451–4459.

Yamashita M, Fenn FB. 1984b. Negative ion production with the electrospray ion source. J Phys Chem 88: 4671–4675.

Yang L, Amad M, Winnik WM, Schoen AE, Schweingruber H, Mylchreest I, Rudewicz PJ. 2002. Investigation of an enhanced resolution triple quadrupole mass spectrometer for high-throughput liquid chromatography/tandem mass spectrometry assays. Rapid Commun Mass Spectrom 16: 2060–2066.

Yost RA, Enke CG. 1978. Selected ion fragmentation with a tandem quadrupole mass spectrometer. J Am Chem Soc 100: 2274–2275.

Zaikin V, Halket J. 2009. A handbook of derivatives for mass spectrometry. IM Publications LLP, Charlton, Chichester. ISBN 978-1-901-01909-4.

Zhurov KO, Fornelli L, Wodrich MD, Laskay ÜA, Tsybin YO. 2013. Principles of electron capture and transfer dissociation mass spectrometry applied to peptide and protein structure analysis. Chem Soc Rev 42: 5014–5030.

Zubarev RA, Makarov A. 2013. Orbitrap mass spectrometry. Anal Chem 85: 5288–5296.

2

INTERPRETATION OF MASS SPECTRA

2.1 MASS SPECTROMETRY: A NUCLEAR AFFAIR

It is well established that the atomic number (Z) is determined by the sum of the positively charged particles (protons, p^+) contained in the atomic nucleus. In fact, Z uniquely characterizes every atom of the periodic table, and thus determines the place an element holds in Mendeleev's tabular arrangement of the elements. As a consequence of charge neutrality, an uncharged atom must have as many negative charges as there are p^+. Therefore, the periodic table also gives us the number of negatively charged particles (electrons, e^-) for all elements as well as their electronic distribution. It is these electrons that largely determine the chemical properties of the elements of the periodic table. Apart from p^+, the atomic nucleus also consists of neutral particles (neutrons, n^0), the sum of these n^0 (neutron number, N) and the number of protons Z constitute the nucleons, which determine the atomic mass number (A) or nucleon number of any given atomic species or nuclide ($A = Z + N$)[1].

In mass spectrometry (MS), the obvious concern is the actual masses of the atoms, molecules, or fragment thereof. Due to the small masses of nucleons (rest mass of an n^0 is $1.674927471 \times 10^{-27}$ kg and of a p^+ is $1.672621898 \times 10^{-27}$ kg), a unit has been defined to easily deal with atomic masses: the unified atomic mass unit (u) or Dalton (Da) (non-SI unit). The IUPAC definition (IUPAC, 2014) states that it is one-twelfth of the mass of a carbon-12 atom (^{12}C) in its nuclear and electronic ground state and is used to express masses of atomic particles. Since one mole (mol, SI unit[2]) of ^{12}C has been given an exact mass of 0.012 kg, Avogadro constant (N_A, $6.022140857 \times 10^{23}$ mol^{-1}) dictates that 1 u or 1 Da is approximately $1.660539040 \times 10^{-27}$ kg. Given the fact that electrons with a rest mass of $5.48579909070 \times 10^{-4}$ Da are approximately less than 0.055% as massive as n^0 (1.00866491588 Da) or p^+ (1.007276466879 Da), for

[1] All constants were taken from NIST (http://physics.nist.gov/cuu/Constants/index.html) (not subject to copyright protection within the United States and is considered to be in the public domain pursuant to title 15 United States Code Section 105, see disclaimer in references). Definitions were taken from Murray et al. (2013).

[2] SI definition of a mole = the amount of substance of a system that contains as many elementary entities as there are atoms in 0.012 kg of carbon-12.

Interpretation of MS–MS Mass Spectra of Drugs and Pesticides, First Edition. Wilfried M. A. Niessen and Ricardo A. Correa C.

practical purposes they can be neglected when accounting for the total mass of an atom. It is of use to realize that the actual mass of any given nuclide is within 0.1 Da of its atomic mass number *A*. However, in order to avoid systematic errors, caution must be exercised with this approximation when dealing with high accuracy measurements in MS.

Except for the hydrogen atom, all other nuclides contain neutrons. An increment of one nuclear p^+ leads to a new atom, and by the necessity of neutrality at least one more e^- must also be added. In fact, as one goes from one element to another, the number of n^0 added varies and per nuclide it is either equal or higher than the number of p^+. Nucleons are held together inside the atomic nucleus by the strong nuclear force. This fundamental force of nature acts at extremely short distances, and it is such that it overcomes the electrostatic coulombic repulsion of all p^+ present in the nucleus, which has a diameter in the range of approximately $1–10 \times 10^{-15}$ m. With such a density (approximately 2.3×10^{17} kg m^{-3} for a ^{12}C atomic nucleus) and the knowledge of Einstein's mass–energy equivalence, it is not difficult to infer that nuclear reactions involve the colossal amounts of energy observed in either fission or fusion processes (Fernandez, 2006).

As the number of nucleons increases, the stability of the atomic nucleus breaks down due to an excited nuclear state (there is no stable nuclide beyond bismuth with $Z=83$). In such a state, the repulsive electromagnetic force (a fundamental force of nature) overcomes the cohesive strong nuclear force and spontaneously the atom releases its excess energy, leading to more stable products that can include other elements (nuclear transmutation). This natural nuclear disintegration process called radioactivity happens via (mostly) three types of mechanisms referred to as α-, β-, and γ-radioactive decay. α-Decay is observed in heavy atoms and involves the production of new nuclides and/or the release of an α-particle (helium nuclei ^4He, two p^+ and two n^0). γ-Decay involves the release of electromagnetic radiation (high-energy photons) by internal pair production (creation of a particle and its antiparticle, e.g., an e^- plus a positron) or by internal conversion where the transfer of energy from the nucleus to orbital electrons leads to the ejection of an e^-. γ-Decay happens without a change in the number of p^+ or n^0, thus no change of atomic species occurs (strictly speaking it is not a decay). This phenomenon is commonly seen as the release of residual excess energy left over from α and β radioactive decay processes (there are no pure γ sources). In contrast, β-decay happens by transforming n^0 into p^+ and vice versa, in both cases leading to a more stable nucleus. This phenomenon is mediated by another fundamental force of nature: the weak nuclear force (acting at distances of 10^{-18} m). This process is accompanied by the emission of a β-particle (an e^-/electron antineutrino pair or a positron/electron neutrino pair in β^-- and β^+-decay, respectively). Despite the fact that there is n^0 and p^+ interconversion, this process does not entail a change in the nucleon number of the nuclides involved. Additionally, e^- capture exists where a K or L shell e^- is captured in the nucleus, and a proton is converted into an n^0 (plus an electron neutrino) with the concomitant emission of either an X-ray photon or an e^- (Auger effect). β^--decay is well illustrated by radioactive carbon-14 (^{14}C), which is the basis of the radiocarbon dating technique for some materials up to about 75,000 years old (Currie et al., 2006; Povinee et al., 2009).

The natural formation of ^{14}C begins through a process of spallation, where high-energy p^+ present in cosmic rays collide with atoms and produce n^0 which in turn react with stable nitrogen-14 atoms (^{14}N, seven p^+ and seven n^0) present in the atmosphere (upper troposphere and lower stratosphere), producing the nuclide ^{14}C (six p^+ and eight n^0) plus a p^+. Naturally radioactive ^{14}C undergoes β^--decay, and one of its n^0 is converted into a p^+ producing stable ^{14}N (seven p^+ and seven n^0) while ejecting a β-particle (an e^-). This process happens with a radioactive half-life[3] of 5,730 years. The ^{14}C formed rapidly oxidizes forming $^{14}CO_2$, which via the carbon cycle enters the atmospheric, terrestrial, and oceanic carbon reservoirs (Figure 2.1). Using MS techniques to measure the ratio of ^{12}C to ^{14}C, one can determine the age of objects that incorporated ^{14}C into their structure such as once living fossils.

2.2 ISOMERS, ISOTONES, ISOBARS, ISOTOPES

Not all atoms do exist exclusively as one of a kind. In addition to isomers, the number of nucleons reveals three different arrangements: isotones, isobars, and isotopes. These differences are important because they have a direct impact in the observable physicochemical properties of the elements.

Isomers are nuclides that have the same number of p^+ *Z* and the same number of n^0 *N*, but have different energy states and can undergo different radioactive decays. In MS, isomers are defined for molecules and fragments thereof.

Isomers (in MS) are compounds with the same elemental composition (same molecular formula) but with different atom connectivity as in constitutional (structural) isomers, or the same atom connectivity but with different spatial atomic arrangement as in stereoisomers. This also applies to compounds where the isotopic composition varies, as defined later.

Isotopomers (isotopic isomers) are isomers that have the same number of each isotopic atom but differ in their positions. They can be constitutional isomers as with CH_2DCH_2OH and CH_3CHDOH, or stereoisomers as with the enantiomers (*R*)- and (*S*)-CH_3CHDOH or the geometric isomers (*Z*)- and (*E*)-$CH_3CH{=}CHD$].

[3]Radioactive half-life ($t_{1/2}$): For a single radioactive decay process, the time required for the activity to decrease to half its value by that process.

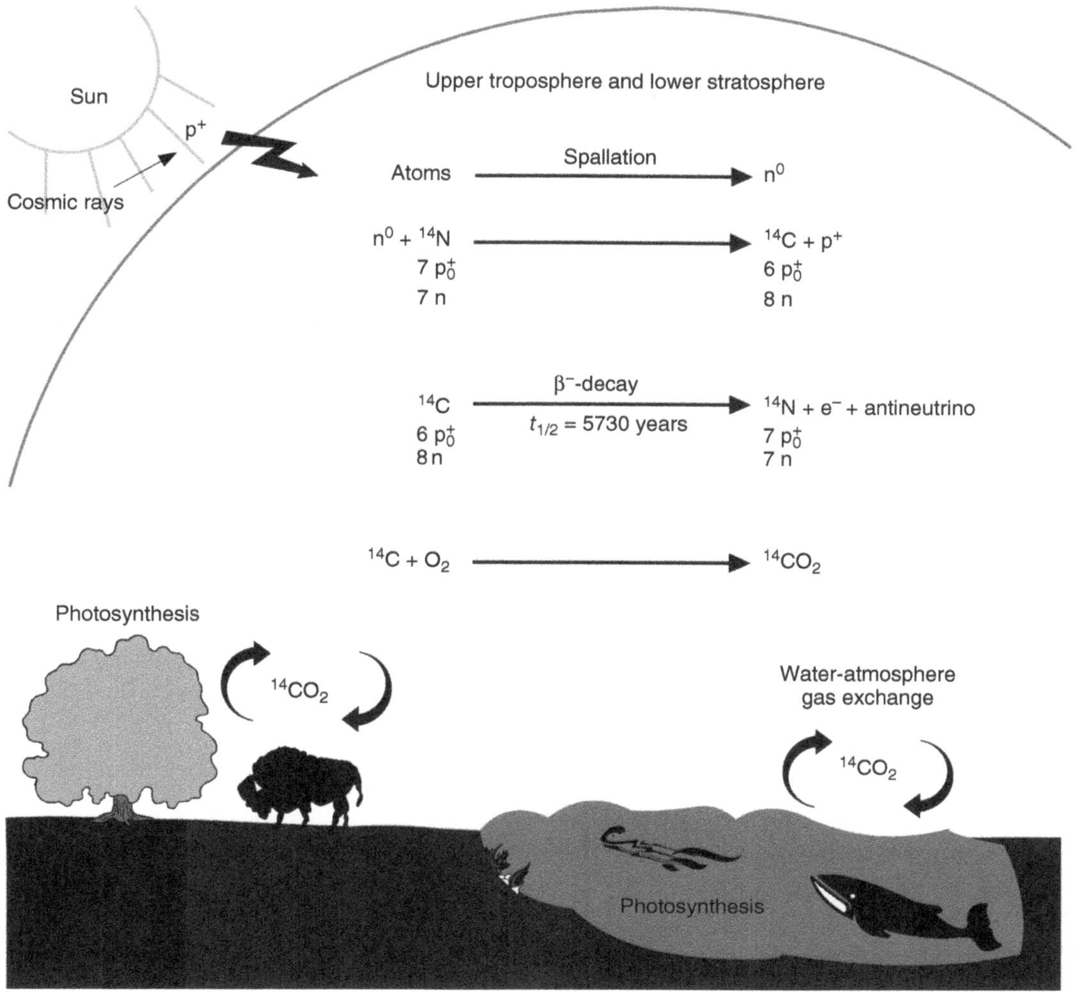

FIGURE 2.1 Formation of ^{14}C and its incorporation into the carbon cycle.

Isotopologues are molecular entities that differ only in isotopic composition (number of isotopic substitutions) as with CH_4, CH_3D, CH_2D_2, and CHD_3.

Isotones are nuclides that have the same number of n^0 N but different number of p^+ Z. There are several examples of observable naturally occurring stable isotones: ^{11}B and ^{12}C; ^{13}C and ^{14}N; ^{30}Si, ^{31}P, and ^{32}S; ^{36}S, ^{37}Cl, ^{38}Ar, ^{39}K, and ^{40}Ca; among others. Since they are not the same atoms (different Z), their physicochemical properties vary accordingly.

Isobars, as the name suggests, are nuclides with the same total number of nucleons and, therefore, the same atomic mass A. As in the case of isotones, their chemical properties will differ since the atomic number Z is different, while sharing many of their physical properties. Many examples of observable naturally occurring stable isobars are also found: ^{36}S and ^{36}Ar; ^{40}Ar, ^{40}K, and ^{40}Ca; ^{46}Ar and ^{46}Ti; ^{50}Ti, ^{50}V, and ^{50}Cr; ^{64}Ni and ^{64}Zn; ^{76}Ge and ^{76}Se; ^{96}Zr, ^{96}Mo, and ^{96}Ru; ^{130}Te, ^{130}Xe, and ^{130}Ba; ^{204}Hg and ^{204}Pb; among others. In MS, isobars are defined for molecules and ions.

Isobars (in MS) are atomic or molecular species (or ions) with the same nominal mass (Da), but different exact masses (Da). A well-known trio of isobars is CO, N_2, and C_2H_4, all with nominal mass 28 Da but with exact (monoisotopic) masses of 27.994914 Da, 28.006148 Da, and 28.031300 Da, respectively.

Isotopes hold a special place not only in MS but also, in general terms, in chemistry and physics. The reason for this lies in the fact that isotopes have the same atomic number Z, and therefore are different versions of the same atom. This of course can only happen when the number of n^0 changes for a given atom of the periodic table. Since their atomic number Z is the same, so is their electronic configuration and thus they share the same chemical properties. In contrast, by virtue of their different number of nucleons (atomic mass number A), their physical properties do change. The origin of isotopes is mostly the result of cataclysmic stellar events, such as cosmological (hydrogen, helium, and lithium formed soon after the big bang), stellar (interior of stars), explosive (supernovae), and cosmogenic, for example, lithium and beryllium formed by spallation of interstellar gas atoms, or nucleosyntheses. Other isotopes result from natural radioactive decay, as well as from anthropogenic sources as in artificial nuclear fusion and fission. Naturally occurring stable isotopes do exist for many of the elements of the periodic table (Earnshaw & Greenwood, 1997). The fact they exist with a given natural abundance is the basis for determining the standard atomic weight of the elements found in most periodic tables. These values are calculated by taking into consideration the percentage abundance (relative abundance) of each isotope as well as its atomic mass. All the isotopic contributions are added up to give the average weight for any given element. This explains why neither hydrogen has an exact standard atomic weight of 1 nor carbon has an exact standard atomic weight of 12, as well as to why many atoms do not have standard atomic weights close to whole number values (given that a nucleon has a mass of approximately 1 Da), for example, magnesium, chlorine, nickel, copper, rubidium, hafnium, mercury, and so on. Since MS is all about measuring the nuclear mass, it takes full advantage of this natural phenomenon to get a plethora of information extremely valuable when doing MS analysis and interpretation of results.

Organic MS deals with ionized organic molecular entities or fragments derived thereof, whose mass properties are the result of the collective sum of the individual atomic masses that make them up. The study and characterization of these gaseous ions (with or without fragmentation) constitute the hallmark of MS. In this process, a beam of ions is separated according to the mass-to-charge ratio (*m/z*) of the ionic species contained within it. This ratio denotes a dimensionless quantity formed by dividing the mass number *A* of an ion by its charge number (independent of its sign), for example, for H^+, *m/z* 1. The end result of the technique, the mass spectrum, is a plot of the relative intensities of the ions forming a beam or other collection as a function of their *m/z* values (independent variable) (Figure 2.2).

Relative intensity in turn is defined as the ratio of intensity of a resolved peak to the intensity of the resolved peak that has the greatest intensity (base peak). This is generally measured as the normalized ratio of the heights of the respective peaks in the mass spectrum to the height of the base peak taken as 100. The intact initial ion produced by the removal (radical cation $M^{+\bullet}$) or addition (radical anion $M^{-\bullet}$) of one electron to the analyte and from where all other fragment ions derive is called molecular ion ($M^{+\bullet}$ or $M^{-\bullet}$). It is clearly seen that a cluster of peaks is present around M and the main fragment ion peaks of the mass spectrum. This cluster is a group of peaks representing ions of the same elemental composition but different isotopic compositions. Table 2.1 lists the atoms of the periodic table most commonly found in MS of organic compounds, including their standard atomic weights, as well as their representative isotopic composition (abundance).

It is worthwhile noting that the lightest isotope (except for lithium and iron) of each atom in Table 2.1 is the most abundant one, explaining the observation of isotope clusters in the mass spectrum with peaks for instance at *m/z* (M+1) and (M+2) due to sample ions containing isotopes other than that of the most abundant ones.

2.3 MASSES IN MS

As a consequence of the existence of isotopes, several different concepts of mass are routinely used in MS.

The **average mass** (Da) is the mass of an ion or molecule weighted for its isotopic composition. Using the standard atomic weight values listed in Table 2.1, the average mass of bromomethane (CH_3Br) is $12.0106 + 3 \times 1.007975 + 79.904 = 94.938525$ Da.

The **nominal mass** (Da) is the mass of any ion or molecule calculated using the isotope mass of the most abundant isotope of each element rounded to the nearest integer value (atomic mass number *A* multiplied by 1 Da) and multiplied

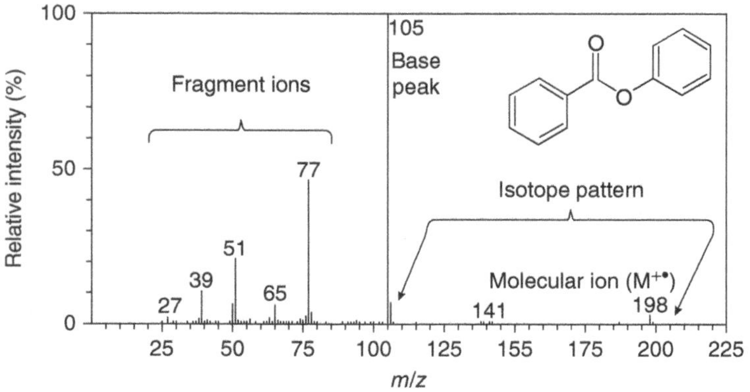

FIGURE 2.2 The electron ionization (EI) mass spectrum of phenyl benzoate.

TABLE 2.1 Standard atomic weights and isotopic abundances for atoms commonly present in bioorganic MS (NIST*).

Atomic Number (*Z*)	Symbol	Name	Standard Atomic Weight	Isotope Atomic Mass Number (*A*) – Symbol	Representative Isotopic Composition	Isotope Atomic Mass (Da)
1	H	Hydrogen	1.007975	1 – ^{1}H	0.999885	1.00782503223
				2 – ^{2}H or D	0.000115	2.01410177812
3	Li	Lithium	6.9675	6 – ^{6}Li	0.0759	6.0151228874
				7 – ^{7}Li	0.9241	7.0160034366
6	C	Carbon	12.0106	12 – ^{12}C	0.9893	12 (exactly)
				13 – ^{13}C	0.0107	13.00335483507
7	N	Nitrogen	14.006855	14 – ^{14}N	0.99636	14.00307400443
				15 – ^{15}N	0.00364	15.00010889888
8	O	Oxygen	15.9994	16 – ^{16}O	0.99757	15.99491461957
				17 – ^{17}O	0.00038	16.99913175650
				18 – ^{18}O	0.00205	17.99915961286
9	F	Fluorine	18.998403163	19 – ^{19}F	1	18.99840316273
11	Na	Sodium	22.98976928	23 – ^{23}Na	1	22.9897692820
14	Si	Silicon	28.085	28 – ^{28}Si	0.92223	27.97692653465
				29 – ^{29}Si	0.04685	28.97649466490
				30 – ^{30}Si	0.03092	29.973770136
15	P	Phosphorus	30.973761998	31 – ^{31}P	1	30.97376199842
16	S	Sulfur	32.0675	32 – ^{32}S	0.9499	31.9720711744
				33 – ^{33}S	0.0075	32.9714589098
				34 – ^{34}S	0.0425	33.967867004
				36 – ^{36}S	0.0001	35.96708071
17	Cl	Chlorine	35.4515	35 – ^{35}Cl	0.7576	34.968852682
				37 – ^{37}Cl	0.2424	36.965902602
19	K	Potassium	39.0983	39 – ^{39}K	0.932581	38.9637064864
				40 – ^{40}K	0.000117	39.963998166
				41 – ^{41}K	0.067302	40.9618252579
26	Fe	Iron	55.845	54 – ^{54}Fe	0.05845	53.93960899
				56 – ^{56}Fe	0.91754	55.93493633
				57 – ^{57}Fe	0.02119	56.93539284
				58 – ^{58}Fe	0.00282	57.93327443
35	Br	Bromine	79.904	79 – ^{79}Br	0.5069	78.9183376
				81 – ^{81}Br	0.4931	80.9162897
53	I	Iodine	126.90447	127 – ^{127}I	1	126.9044719

*NIST (http://www.nist.gov/pml/data/comp.cfm). Relative atomic masses (the ratio of the average mass of the atom to the unified atomic mass unit) applicable to elements in any normal sample with a high level of confidence. A normal sample is any reasonably possible source of the element or its compounds in commerce for industry and science and has not been subject to significant modification of isotopic composition within a geologically brief period. Information developed by employees of the National Institute of Standards and Technology (NIST), an agency of the Federal Government, is not subject to copyright protection within the United States and is considered to be in the public domain pursuant to title 15 United States Code Section 105 (see disclaimer in references).

by the number of atoms of each element. Using the values in Table 2.1, the nominal mass of bromomethane ($^{12}CH_3{}^{79}Br$) is $12+3\times1+79=94$ Da.

The **monoisotopic mass** (Da) is the exact mass of an ion or molecule calculated using the mass of the most abundant isotope of each element. Using the values in Table 2.1, the monoisotopic mass for bromomethane ($^{12}CH_3{}^{79}Br$) is $12+3\times1.007825+78.918338=93.941813$ Da.

The **exact mass** (Da) is the calculated mass of an ion or molecule with specified isotopic composition. As calculated earlier, the exact mass of $^{12}CH_3{}^{79}Br$ is clearly different from the mass of $^{13}CD_3{}^{81}Br$, which equals 99.961950 Da.

The **accurate mass** (Da) is the experimentally determined mass of an ion of known charge. This measurement, if made very precisely, can be used to determine elemental composition (Section 2.7).

A glance at Table 2.1 puts in evidence how the atomic mass (except for ^{12}C) is always different from the atomic mass number *A* for any of the nuclides listed. However, the atomic mass of a nuclide is approximately equal to its atomic mass number *A* expressed in Da. In fact, when the mass of an atomic nucleus is calculated from the rest masses of its components ($p^0 + n^0$), the value obtained is always greater than the experimentally found value. The origin of this difference lies in the binding energy of the nucleus. When the nucleons formed the nucleus, a part of their mass was converted into energy (mass of an n^0 is 939.565413 MeV and of a p^0 is 938.272081 MeV). This energy released is the nuclear binding energy and gives the magnitude of the cohesive forces (strong and weak nuclear forces) that keep the nucleus intact while overcoming the electrostatic repulsion of the nuclear p^+. This mass difference is called mass defect.

The **mass defect** is the mass difference found in all nuclides when the sum of its constituting nucleons is compared to the actual experimentally found values.

In MS, the concept of mass defect exists but it has been defined for atoms, molecules, or ions.

The **mass defect (in MS, Da)** is the difference between the monoisotopic mass of an atom, molecule, or ion and its nominal mass. Dependent on the elemental composition, the mass defect can be a positive value, as is the case for ^{1}H, ^{7}Li, and ^{14}N, or a negative value, as is the case for ^{19}F, ^{32}S, and ^{35}Cl. Hydrogen with a mass defect of approximately 0.007825 Da will contribute to a +1 Da mass defect for every 128 hydrogen atoms present in a molecule.

2.4 ISOTOPES AND STRUCTURE ELUCIDATION

It follows that it is statistically possible for any set of a given molecule or ion to exist in all possible isotopic elemental compositions. Looking closer at CH_3Br, we have that $^{12}CH_3{}^{79}Br$, $^{13}CH_3{}^{79}Br$, $^{12}CD_3{}^{79}Br$, $^{12}CH_3{}^{81}Br$, $^{12}CD_3{}^{81}Br$, $^{13}CD_3{}^{79}Br$, $^{13}CH_3{}^{81}Br$, and $^{13}CD_3{}^{81}Br$ could all be present in the set (note that all molecules can also exist as isotopologues with H_2D and HD_2). Therefore, the mass spectrum should show peaks reflecting the different atomic masses of the isotopes present. Taking the atomic mass of each isotope rounded to the nearest integer, these peaks should correspond to *m/z* 94, 95, 97, 96, 99, 98, 97, and 100, respectively, that is, with *m/z* (M), (M+1), (M+3), (M+2), (M+5), (M+4), (M+3), and (M+6). Whether one observes one or more of these possible ions depends on the natural

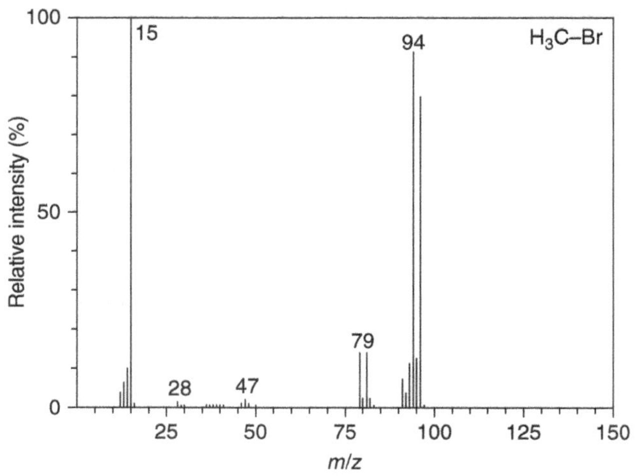

FIGURE 2.3 The EI mass spectrum of CH_3Br showing isotopic peaks due to $^{12}CH_3{}^+$ (*m/z* 15):$^{13}CH_3{}^+$ (*m/z* 16) and $CH_3{}^{79}Br^{+\bullet}$ (*m/z* 94):$CH_3{}^{81}Br^{+\bullet}$ (*m/z* 96) with a relative intensity of 1:0.011 and 1:0.97, respectively.

isotopic abundance of the atoms involved, on the quality of the instrument, and on the origin of the sample. The values in Table 2.1 (isotopic abundance) indicate that for $^{13}CH_3{}^{79}Br^{+\bullet}$, one should see a peak with *m/z* (M+1) with approximately 1.1% of the relative intensity of M, and a peak with *m/z* (M+2) due to $^{12}CH_3{}^{81}Br^{+\bullet}$ is expected with approximately 97.0% of the relative intensity of M (Figure 2.3). Care must be taken after realizing that the relative intensity of a peak might be due to contributions from several different ions, for example, the peak at *m/z* 95 has contributions from ions other than $^{13}CH_3{}^{79}Br^{+\bullet}$ because its intensity is larger than 1.1% of the peak at *m/z* 94. In contrast, the peak at *m/z* 16 corresponds to $^{13}CH_3{}^+$ as it is approximately 1.1% as intense as $^{12}CH_3{}^+$ at *m/z* 15; by the same token, the peaks at *m/z* 79 and 81 are indicative of the presence of bromine ion ($^{79}Br^+$ and $^{81}Br^+$).

When similar cases are analyzed, some generalizations can be made. Regarding their isotopes, in MS the elements can be classed as monoisotopic (X), isotopes with a mass of X+1 and isotopes with mass X+2. Taking the percentage abundance of an isotope relative to its most abundant isotope (100%) gives the factor (percentage relative abundance) by which an isotope will contribute to the relative intensity of a peak in a mass spectrum containing such isotope (Table 2.2). The isotope pattern (defined as the set of peaks related to ions with the same chemical formula but containing different isotopes that has a particular pattern associated with the relative abundance of the isotopes) is of key importance when determining the kind and number of atoms that make up a particular ion.

Since this is a statistical event, the more atoms of a given element are present in an ion, the more likely it will be to

TABLE 2.2 Elements commonly encountered in bioorganic MS and tabulated by type and percentage (%) relative abundance.

Element	Element Type	X		X+1		X+2	
		Atomic Mass (A)	Relative Abundance (%)	Atomic Mass (A)	Relative Abundance (%)	Atomic Mass (A)	Relative Abundance (%)
H	X	1	100	2	0.0115		
C	X+1	12	100	13	1.08		
N	X+1	14	100	15	0.365		
O	X+2	16	100	17	0.038	18	0.205
F	X	19	100				
Na	X	23	100				
Si	X+2	28	100	29	5.08	30	3.35
P	X	31	100				
S	X+2	32	100	33	0.79	34	4.47
Cl	X+2	35	100			37	32.0
K	X+2	39	100	40	0.0125	41	7.22
Br	X+2	79	100			81	97.3
I	X	127	100				

observe its isotopic clusters. For instance, hexadeuterobenzene (C_6D_6) has a peak at m/z 85 (M+1) with 6.6% relative intensity of that of M (at m/z 84), which corresponds to six times 1.1%, that is the relative abundance contribution per carbon atom (Figure 2.4a). The fact that this enhancing effect leads to patterns dependent on the structure of the ions explains why in the mass spectrum of (bromomethyl)sulfur pentafluoride (CH_2BrSF_5) in Figure 2.4b, the peak with m/z 222 (M+2) has a higher relative intensity than M with m/z 220 ($^{12}CH_2{}^{79}Br^{32}SF5$): the combined percentage relative abundance contributions of $CH_2{}^{81}BrSF_5$ and $CH_2Br^{34}SF_5$ (approximately 101.8% the relative intensity of M).

Tables of isotope patterns and percentage relative abundances, design of polynomials, and algorithm analysis are very useful for the interpretation of mass spectra and structure elucidation of ionic species. An example of such a table lists the percentage relative abundances of peaks corresponding to X+1 and X+2 isotope containing peaks for a given number of carbons, as well as the percentage relative abundance factor per atom (other than carbon) that must be taken into account when analyzing isotope patterns (McLafferty & Tureček, 1993).

In addition, ^{37}Cl and ^{81}Br are diagnostic of their presence (or absence) in an ion as a result of their high percentage relative abundance. In fact, one can tell how many chlorine or bromine atoms make part of an ion from the analysis of their isotope patterns (Figure 2.5 and Table 2.3). Polynomials and programs for obtaining computer-generated isotope patterns have been developed that include other atoms in addition to halogens (Gross, 2004; Stoll et al., 2006).

With the advent of electrospray ionization (ESI) in conjunction with liquid chromatography (LC) and of matrix-assisted laser desorption ionization (MALDI), the analysis of high molecular mass species, for example, biological polymers, has been made possible. The analysis of isotope patterns in these molecules is much more complex due to the importance of X+n isotope peaks. For instance, low abundance isotopic species such as deuterium (X+1 element) cannot be neglected, as they clearly contribute to the percentage relative intensity of isotopic peaks in large molecules. Consequently, there is a great deal of interest in theoretical models that help explain experimental findings (Li et al., 2008; Li et al., 2010). In order to include this kind of information into any MS analysis, one must first experimentally acquire the pertinent data, which in turn depends on the instrumentation available. Nevertheless, important concepts have been agreed upon for assessing quantitatively what the quality is for both the instrument used and the data obtained.

2.5 NITROGEN RULE, RING DOUBLE-BOND EQUIVALENT, AND HYDROGEN RULE

Nitrogen also offers valuable information regarding structure elucidation from mass spectral data. The nitrogen rule is borne out of an arithmetical observation when looking at the most common atoms found in organic molecules. Usually, elements with an odd nominal mass have odd valences, for example, H, F, Na, P, Cl, K, Br, I, and elements with an even nominal mass have even valences, for example, C, O, Si, S. However, in organic compounds, the N atom has an even nominal mass but an odd valence of three (odd–even parity, Moriwaka & Newbold, 2003). This exception is the basis of the nitrogen rule. When these elements form covalent bonds (σ- and π-bonds), nitrogen with an even nominal mass forms an odd number of bonds, whereas the other elements that have either an even or an odd nominal mass form an even

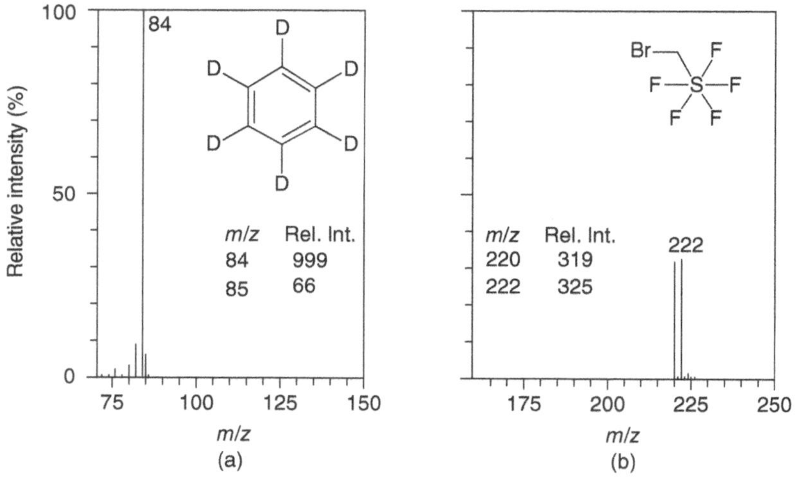

FIGURE 2.4 Combined effect of isotopes on the relative intensity of isotope containing ions.

or an odd number of bonds, respectively. The strength of the nitrogen rule lies in its usefulness to determine whether an ion (or molecule) has an even or an odd electron number, or whether an ion (or molecule) has an even or an odd number of N-atoms in its structure.

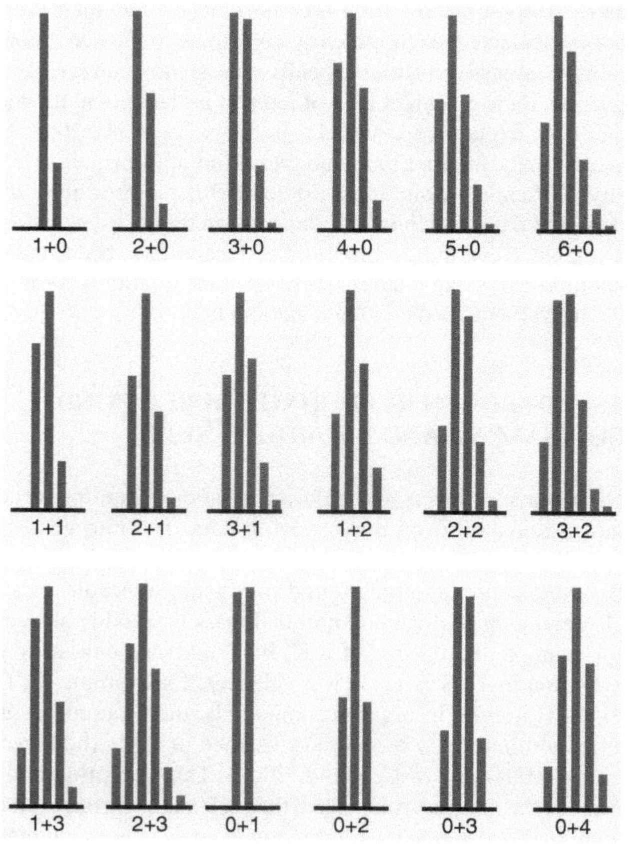

FIGURE 2.5 Expected isotopic patterns for ions containing Cl + Br atoms.

TABLE 2.3 Expected signal isotopic ratios for ions containing Cl + Br atoms as shown in Figure 2.5.

Cl + Br	X	X+2	X+4	X+6	X+8	X+10
Cl	100	32.5				
Cl_2	100	65.0	10.6			
Cl_3	100	97.5	31.7	3.4		
Cl_4	76.9	100	48.7	0.5	0.9	
Cl_5	61.5	100	65.0	21.1	3.4	0.2
Cl_6	51.2	100	81.2	35.2	8.5	1.1
ClBr	76.6	100	24.4			
Cl_2Br	61.4	100	45.6	6.6		
Cl_3Br	51.2	100	65.0	17.6	1.7	
$ClBr_2$	43.8	100	69.9	13.7		
Cl_2Br_2	38.3	100	89.7	31.9	3.9	
Cl_3Br_2	31.3	92.0	100	49.9	11.6	1.0
$ClBr_3$	26.1	85.1	100	48.9	8.0	
Cl_2Br_3	20.4	73.3	100	63.8	18.7	2.0
Br	100	98.0				
Br_2	51.0	100	49.0			
Br_3	34.0	100	98.0	32.0		
Br_4	17.4	68.0	100	65.3	16.0	

For singly-charged ions, the nitrogen rule can be summarized as follows (Table 2.4): Odd-electron ions ($OE^{+\bullet}$) with an odd number of nitrogen atoms have an odd nominal *m/z*. $OE^{+\bullet}$ ions with an even (or zero) number of nitrogen atoms have an even nominal *m/z*. Even-electron ions (EE^{+}) with an odd number of nitrogen atoms have an even nominal *m/z*. EE^{+} with an even (or zero) number of nitrogen atoms have an odd nominal *m/z*. Once the *m/z* of an ion is measured, the breaking, for example, by deprotonation, or forming, for example, by protonation, of a single covalent bond leads to a reversal of the rule. Caution must be exercised when applying this rule to other accurate-mass measurements; it becomes unreliable with masses higher than 500 Da, giving wrong

TABLE 2.4 Summary of the nitrogen rule for $OE^{+\bullet}/OE^{-\bullet}$ and EE^{+}/EE^{-} ions or molecules of a given nominal mass.

Ion or Molecule	Odd N-Atoms	Even N-Atoms (or None)
$OE^{+\bullet}/OE^{-\bullet}$	Odd mass	Even mass
EE^{+}/EE^{-}	Even mass	Odd mass

assignments for the number of even and uneven N-atoms present in a molecule (Kind & Fiehn, 2007).

Ring double-bond equivalent (RDBE) or index of hydrogen deficiency is another concept used in structure elucidation in order to determine the degree of unsaturation of a molecule. Deviation from an acyclic alkane molecular formula (C_nH_{2n+2}) happens when there is a presence of either π-bonds or rings (cyclic compounds). In practice, determining the number of H-atoms (2 × RBDE) needed to saturate π-bonds and to make rings into acyclic structures amounts to determining the RDBE, for example, benzene and allene, RDBE is 4 and 2, respectively. For molecules or ions containing carbon, hydrogen, halogens, nitrogen, oxygen, phosphorus, silicon, and sulfur, Eq. 2.1 is used to calculate the RDBE.

$$\mathrm{RDBE} = \mathrm{C} + \mathrm{Si} - \frac{1}{2}(\mathrm{H} + \mathrm{F} + \mathrm{Cl} + \mathrm{Br} + \mathrm{I}) + \frac{1}{2}(\mathrm{N} + \mathrm{P}) + 1 \quad (2.1a)$$

$$\mathrm{RDBE} = (\mathrm{C} + \mathrm{Si} + 1) - \left[\frac{(\mathrm{H}^{\dagger} - \mathrm{N}^{\ddagger})}{2}\right] \quad (2.1b)$$

where $H^{\dagger}$ = hydrogen + halogen atoms (univalent) and $N^{\ddagger}$ = nitrogen + phosphorus atoms (tervalent). Note that neither divalent atoms, for example, O- and S-atoms, nor different valence states of tervalent elements, for example, N- (V), P- (V), S- (IV, VI), contribute to the degree of unsaturation, for example, $R_3P{=}O$ compounds and S=O and SO_2 moieties. The RDBE of a molecule or an $OE^{+\bullet}/OE^{-\bullet}$ is an integer number, whereas the RDBE of an EE^{+}/EE^{-} is a half integer. It is 0.5 lower than that of the corresponding molecule for a positive ion (EE^{+}) and 0.5 higher than that of the corresponding molecule for a negative ion (EE^{-}). In this way, the RDBE allows discrimination between $OE^{+\bullet}/EE^{+}$ and $OE^{-\bullet}/EE^{-}$ ion fragments, which is important in mass spectral interpretation and structure elucidation. The number of different heteronuclei present in the molecular structure will cause large deviations, which along with the rising complexity of the calculation do not lead to a single solution but rather a range of RDBE results (Kind & Fiehn, 2006).

Hydrogen rule (H-rule) refers to the maximum number of monovalent H-atoms and halogens for a given number or C- and N-atoms in a molecular formula, and it is equal to C + N/2 + 1. This concept is useful for eliminating improbable structures, for example, a molecule with four C-atoms and one N-atom cannot have more than nine H-atoms. It is worthwhile noting that for a molecule containing C, N, H, O, S, and halogens, the RDBE is equal to half the difference between the maximum number of monovalent H-atoms and halogens (H-rule) minus $H^{\dagger}$, thus RDBE = $(2C + N + 2 - H^{\dagger})/2 = C + N/2 + 1 - H^{\dagger}/2$, as stated earlier (Eq. 2.1b).

These concepts can be used in conjunction with isotopic relative signal intensities to elucidate molecular formulae from mass spectra. Consider a molecule with a given mass spectrum, begin finding the molecular mass (M), and the nitrogen rule comes into play to assess the minimum number of N-atoms present (if any). When M is not the base peak of the spectrum, the signal must be normalized to find out the relative signal intensities of the M+1 and M+2 peaks. M+1 sheds information on the (maximum) number of carbon atoms present by using the percentage relative abundance of ^{13}C (Table 2.2). The relative signal intensity of M+2 provides information about other possible atoms present in the molecule, for example, Si, S, Cl, and Br, some of which will find confirmation in the relative signal intensity of M+1. The atoms found from the relative signal intensities of M+1 and M+2 can be subtracted from M, the remaining mass should be due to O-, N-, F-, I-, and H-atoms. The H-rule then helps in narrowing down the choices. Several possibilities must be considered and carefully chose the one (or several) that does not violate the nitrogen and hydrogen rules, and more importantly that is consistent with the *m/z* values and the relative signal intensities observed in the spectrum. The calculation of elemental composition is a constant challenge, and high-resolution accurate-mass MS (HRAM-MS) is of critical importance as it helps greatly in limiting the number of possible formulae for a given molecular mass. Better methods have been developed for the structure elucidation using elemental composition calculations (Kind & Fiehn, 2007, 2010).

2.6 RESOLVING POWER, RESOLUTION, ACCURACY

Even though there is no general agreement about the use of the terms resolving power and resolution, the concepts and definitions used in this work are of actual use by the IUPAC (Todd, 1995; Murray et al., 2013). Resolution and resolving power are very closely related concepts, and their difference lies mainly in the method used to find their respective numerical values. In both cases, it is a peak width (or the distance between two peaks) at a given peak height that determines their values. Generally speaking, resolving power and resolution give a measure of the ability of a mass spectrometer to separate two different ions of close *m/z* values with masses m_1 and m_2 (Figure 2.6).

The **resolving power** is defined as the ability of an instrument or measurement procedure to distinguish between

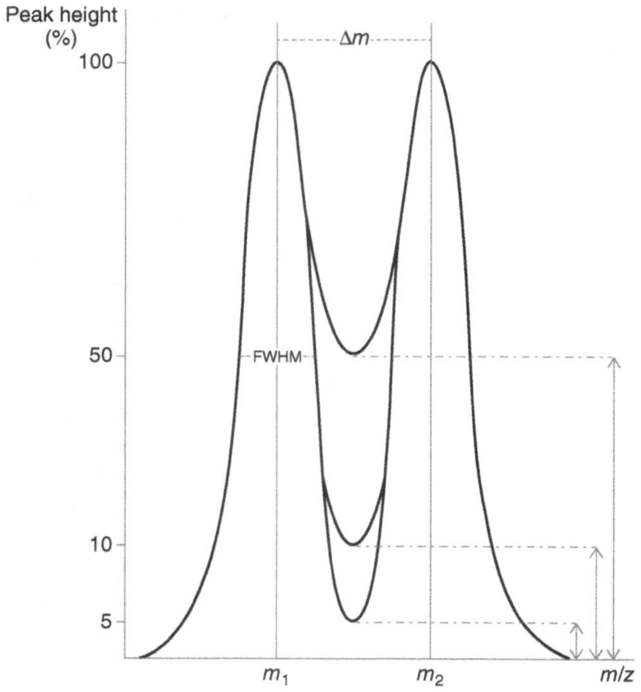

FIGURE 2.6 Parameters for the characterization of resolving power and resolution.

two peaks with *m/z* values differing by a small increment and expressed as the peak width in mass units. It may be characterized by giving the peak width measured in mass units, or expressed as a function of mass for at least two points on the peak, specifically for 50% and for 5% of the maximum peak height.

The definition of **resolution** in MS expresses this value as $m/\Delta m$, where m is the mass of the ion of interest and Δm is the peak width at a given height, or the spacing between two equal intensity peaks with a valley between them at no more than a defined percentage of their height.

The **resolution 10% valley definition** is used when Δm is characterized by two peaks of equal height resolved to at least 10% of their height.

The **resolution peak width definition** is used for a single peak made up of a singly-charged ion with a given *m/z* in a mass spectrum, the resolution may be expressed as $m/\Delta m$, where Δm is the width of the peak at a height that is a specified fraction of the maximum peak height. It is recommended that one of three values 50%, 5%, or 0.5% be used. A common standard is the definition of resolution based on Δm being the full width of the peak at half its maximum height (FWHM).

The reader must be aware that resolving power and resolution are not equivalent terms and should not be used indiscriminately. There is current use of these two terms defined in the opposite manner as described here (Sparkman, 2006; Benyon, 1978). Thus, although formally they are not synonyms, their use as such is widely spread in the literature. Therefore, it is important to know the definitions used before comparing values. Approximately speaking, the value obtained from the 10% valley definition of resolution is technically equivalent to the value obtained from the 5% peak width definition of resolution (with a system linear in that range), and half the value when comparing to the FWHM peak width definition of resolution. With sector instruments, the 10% valley definition of resolution is commonly used, whereas for Fourier-transform ion cyclotron resonance (FT-ICR), quadrupole and ion traps, orbitrap, and time-of-flight (TOF) mass analyzers, it is the FWHM peak width definition of resolution that is mostly used (Russell & Edmondson, 1997; Price, 1991). Despite this lack of uniformity, it is their physical relationship and implications that matter most.

Resolution gives the quotient of the *m/z* value observed in a mass spectrum, divided by the smallest difference (Δm) in *m/z* values for two ions that can be separated. In contrast, resolving power is a measure of the ability of a mass spectrometer to provide a specified value of mass resolution. These concepts are of utmost importance when dealing with isobars, where the resolution will determine whether or not a mass spectrometer will be able to separate them. As mentioned earlier (Section 2.2), CO, N_2, and C_2H_4 all have a nominal mass of 28 Da but monoisotopic masses of 27.994914, 28.006148, and 28.031300 Da, respectively. Figure 2.7 shows the effect of resolution upon the separation of these isobars (Muenster & Taylor, 2009).

Using the FWHM definition of resolution, one would need a resolution of ($\Delta m_{N_2-CO} = 0.011234$ Da) of 2,500 for the peaks of CO and N_2 to be resolved to 50% of their heights, as shown for the peaks under the area of the unresolved peaks at a resolution of 2,300 (Figure 2.7a). Similarly, one would need a resolution of 5,000 (or 2,500 when using the 10% valley definition) to obtain a separation of at least 10% of the peaks heights (Figure 2.7b). It is this kind of performance that has come to be known as HRAM-MS, which finds multiple uses especially when in need of accurate-mass measurements for elemental analysis (where isotopic information is a key factor) or high selectivity for the identification and characterization of compounds.

The **accuracy (of a measuring instrument)** is the ability of a measuring instrument to give responses close to a true value.

The **accuracy (of a measurement)** is defined as the closeness of the agreement between the result of a measurement and a true value of the measurand.

Accuracy in MS is used to assess in Da, the closeness of a measurement (accurate mass) to its actual value (exact mass). The absolute error is numerically equal to the difference between the exact and the accurate masses. This value

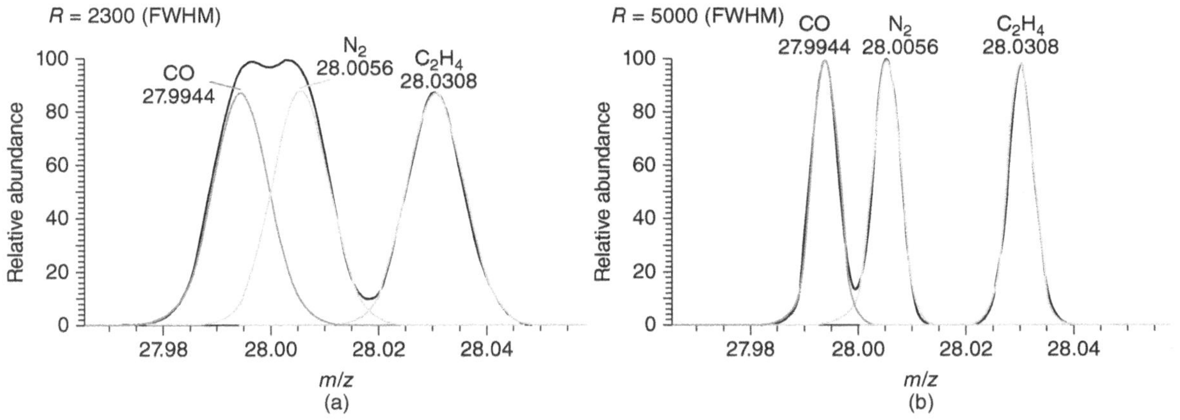

FIGURE 2.7 Effect of resolution on the identification of 28 Da isobars. (Source: Muenster & Taylor, 2009. Reproduced with permission of Thermo Fisher Scientific Inc.)

is also expressed as a relative error in parts per million (ppm = [(accurate mass − exact mass)/exact mass] × 10^6), which implies that 1 ppm represents an error of 0.0001% of the mass of the ion being evaluated. Better resolution means a smaller peak width, thus enabling better mass accuracy. This is illustrated for the triply-charged ion of a peptide with *m/z* 962.9 in Figure 2.8. Changing resolution from 5,000 (FWHM 0.6) via 15,000 (FWHM 0.2) to 25,000 (FWHM 0.1) greatly improves the quality of the spectrum by achieving complete separation of the isotope envelope. It is important to stress that the ion charge state (single or multiple) is significant because the mass of the electron (0.000548 Da) matters in high-resolution accurate-mass measurements using MS.

One should never lose sight of the fact that before embarking in MS measurements, it is imperative to have a well-calibrated instrument. This *m/z* scale calibration (external or internal) is achieved by evaluating the accuracy for ions of known exact masses (Brenton & Godfrey, 2010). It is commonly accepted that HRAM-MS begins when the accuracy is equal to 0.005 mDa for every Da of mass present in an ion which corresponds to a relative error of 5 ppm. Given the linear relationship between resolution and mass, a plot of the resolution (R) versus the mass (m) gives a slope equal to the inverse of the accuracy ($1/\Delta m$). This tells us that for a given mass m_1, the ratio of the accuracy of such mass (Δm_1) to any other accuracy Δm_2 is the factor by which the resolution of m_1 will change. Equally, the plot of the resolution (R) versus the inverse of the accuracy ($1/\Delta m$) gives a slope equal to m, which tells us that for any given mass m_1 with a given accuracy (Δm_1), the ratio of any other mass m_2 to such mass m_1 is the factor by which the resolution of m will change for an accuracy of Δm_1. Thus, any resolution for a given mass (m_1) and accuracy (Δm_1) multiplied by ($m_2 \Delta m_1 / m_1 \Delta m_2$) will give the resolution for mass m_2 with an accuracy of Δm_2 (Eq. 2.2).

$$R_{m_2,\Delta m_2} = R_{m_1,\Delta m1} \left(\frac{m_2 \Delta m_1}{m_1 \Delta m_2} \right) \tag{2.2}$$

This shows that as the mass increases and the mass accuracy improves, so must do the resolution. For instance, a measurement of a mass of 100 Da (m_1) with a 5 ppm relative error ($\Delta m_1 = 0.5$ mDa) would require a resolution of 200,000; whereas a mass of 1000 Da (m_2) would require a resolution 2,000,000 to achieve the same accuracy ($\Delta m_1 = \Delta m_2$). Different kinds of instruments deal with this problem in various manners, including those with a constant resolution over the mass range, as well as others with variable resolution as the mass changes.

Lack of resolution leads to coalesced peaks with no discernible isotope patterns, resulting in broad signals representative of average masses rather than accurate masses of the species making part of the isotope pattern. In fact, changes in the expected isotope patterns should be a warning of other underlying phenomena such as different ionization mechanisms at play, isotopic enrichment, and the presence of compounds with similar masses (isobars). As the accuracy of the measurement increases, the number of possible molecular structures for a given elemental composition (exact mass) is greatly reduced (Figure 2.9).

However, even if MS can tell which and how many atoms are present in an ion, it does not give outright how those atoms are linked to one another. Consequently, user information coupled to computer assisted elemental composition determination of fragment ions are indispensable elements in many fields of chemistry, as exemplified by metabolomics (Kumar et al., 1992; Rojas-Chertó et al., 2011). An accuracy of 1 ppm relative error does not imply automatically one single molecular formula for a given set of atoms (Kind & Fiehn, 2006). High mass accuracy is of prime importance,

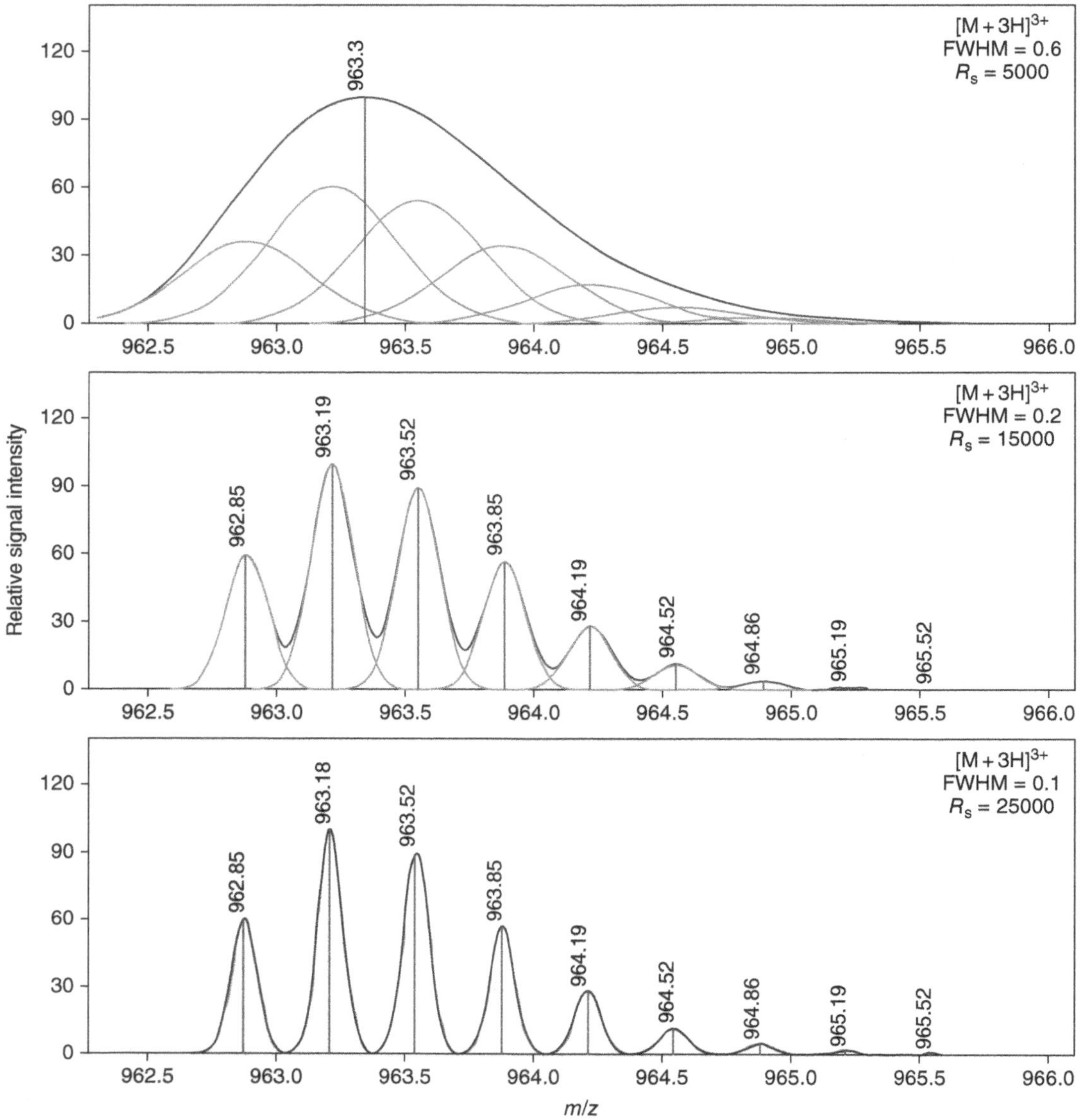

FIGURE 2.8 Relationship between peak width and resolution. The isotope patterns are simulated for the $[M+3H]^{3+}$ of the 1–24 fragment of follicle-stimulating hormone (MDYYRKYAAI FLVTLSVFLH VLHS) with *m/z* 962.9.

and combined with isotopic information are the main parameters when elucidating structures from HRAM-MS spectral data.

2.7 CALCULATING ELEMENTAL COMPOSITION FROM ACCURATE *m/z*

There are several software tools available to estimate the elemental composition of an ion from its accurate mass. The more advanced tools implement important rules from organic chemistry, such the nitrogen rule, RDBE, and the hydrogen rule. The choice of elements to be considered in performing these calculations depends on possible prior knowledge on the elements that might be present in the ion, for example, because of knowledge of the synthetic protocol. The number of elements can be limited to an expert inspection of the isotope patterns, as the presence of elements such as Si, S, Cl, and Br can be readily observed (Table 2.2 and Figure 2.5). The number of possible elemental compositions for a given *m/z* value depends on the number of elements considered and the accuracy with which the *m/z* could be measured: the higher the accuracy, the lower the number of possible elemental compositions (Figure 2.9) (Kind & Fiehn, 2006).

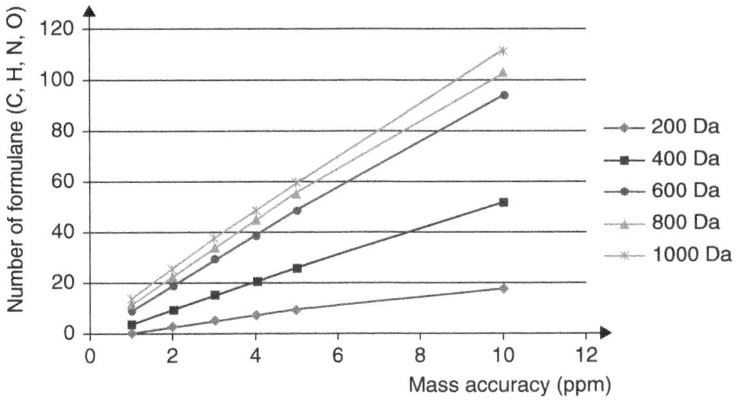

FIGURE 2.9 Variation of the number of possible elemental formulae as the mass accuracy changes.

Some more advanced software tools also take advantage of the information available in an accurately measured isotope pattern (Kind & Fiehn, 2007; Pelander et al., 2009; Pluskal et al., 2012). After calculating possible elemental compositions for a particular *m/z*, the theoretically predicted isotope patterns for all these elemental compositions are also calculated and compared to the experimentally found patterns. The elemental compositions providing a better fit to the measured isotope patterns are more likely to be correct (Pelander et al., 2009). Ultra-HRAM-MS instruments featuring a resolution in excess of ≈60,000 (FWHM) have additional possibilities when deriving elemental composition from spectra, as they can even separate the contributions of different atoms to the M+1 and M+2 isotope peaks (Nakabayashi et al., 2013).

In presenting the results of the elemental composition calculations, the software tools generally also provide the RDBE for the results. This is useful to help decide whether $OE^{+\bullet}$ or EE^+ (or $OE^{-\bullet}$ and EE^-) are observed: $OE^{+\bullet}$ show integer numbers for RDBE, while for EE^+ the RDBE calculations result in half integer numbers. Thus, if the calculations are being performed for a precursor ion in an ESI mass spectrum, generally speaking one only has to consider the EE^+ solutions. However, exceptions to this behavior are known (Section 3.9), as seen for carotenoids where both $M^{+\bullet}$ and $[M+H]^+$ may be generated in positive-ion ESI (Bijttebier et al., 2013). Since the carotenoids do not contain N-atoms, discrimination between $OE^{+\bullet}$ and EE^+ can be readily made by applying the nitrogen rule.

2.8 PROTONATED AND DEPROTONATED MOLECULES AND ADDUCT IONS

Special attention must be paid to the ions generated under ESI and atmospheric-pressure chemical ionization (APCI) conditions. Currently, these are the most widely applied ionization techniques in LC–MS. For the analysis of polar compounds, either reversed-phase LC (RPLC) or hydrophilic interaction LC (HILIC) is used for their separation, featuring mobile phases composed of water, an organic solvent (usually acetonitrile or methanol), and a buffer, an acid or a base. In LC–MS, volatile mobile-phase additives must be used because nonvolatile additives lead to ion source contamination, ionization suppression, and excessive adduct ion formation. Under appropriate solvent conditions, most analytes generate EE^+, that is, protonated molecules $[M+H]^+$ with *m/z* (M+1) in positive-ion mode, and also EE^-, that is, deprotonated molecules $[M-H]^-$ with *m/z* (M−1) in negative-ion mode. The *m/z* of these ions can be readily used to determine the molecular mass of the analyte molecule. Whether or not useful data can be gathered for a particular analyte molecule, either in positive-ion or negative-ion mode or in both modes, depends to a large extent on the functional groups present in the analyte molecule, that is, gas-phase basicity (accept a H^+) and/or gas-phase acidity (donate a H^+) (Section 1.2.3).

In addition to $[M+H]^+$ and/or $[M-H]^-$, other analyte-related ions may be observed, especially adduct ions. Depending on the mobile-phase composition, adduct ions such as $[M+NH_4]^+$ with *m/z* (M+18), $[M+CH_3OH+H]^+$ with *m/z* (M+33), and/or $[M+CH_3CN+H]^+$ with *m/z* (M+42) may be observed in positive-ion mode, while adduct ions such as $[M+HCOO]^-$ with *m/z* (M+45), $[M+CH_3COO]^-$ with *m/z* (M+59), and/or $[M+CF_3COO]^-$ with *m/z* (M+113) may be observed in negative-ion mode. In addition to the adduct ions that depend on the mobile-phase compositions, adduct ions such as $[M+Na]^+$ with *m/z* (M+23), $[M+K]^+$ with *m/z* (M+39), or $[M+Cl]^-$ with *m/z* (M+35) and (M+37) (with a 3:1 relative signal intensity ratio) may be observed due to the ubiquitous presence of Na^+, K^+, and Cl^- in the solvents used (typically at 10^{-5}–10^{-4} mol L^{-1}) and/or sample analyzed. Examples of multiple adduct ions are given for nucleoside antiviral agents (Kamel et al., 1999). Compounds with several O-atoms in neighboring positions seem to be especially susceptible to the formation of adduct ions with Na^+

and/or K^+, for example, oligosaccharides (Zaia, 2004), avermectins (Section 4.10.1.2), and polyether ionophores (Section 4.10.2.3). In some cases, $[M+H]^+$ may be absent and only $[M+Na]^+$ and/or $[M+K]^+$ is observed. Owing to the well-known *m/z* differences, the occurrence of adduct ions can be of help when confirming the analyte molecular mass derived from the mass spectrum. However, in quantitative analysis, the generation of adduct ions is generally not wanted, as they spread out the signal over several different ion species and relative signal intensity ratios among the various adduct ions can be nonreproducible, compromising the reliability of the analyte quantification.

In analyte molecules with one or more acidic functional groups, which are nevertheless analyzed in positive-ion mode, in addition to or instead of $[M+Na]^+$ and/or $[M+K]^+$, other ions may be observed. This behavior is illustrated in Figure 2.10, that is, the ESI mass spectra of adenosine triphosphate (ATP). The protonated molecule is the ion with *m/z* 508. Due to the *m/z* difference of 22, the ion with *m/z* 530 may be interpreted as $[M+Na]^+$, but this leads to problems in interpreting the identity of the ion with *m/z* 552, showing another *m/z* difference of 22. Therefore, the ions with *m/z* 530 and 552 are best interpreted as the result of a solution-phase H^+/Na^+ exchange by the acidic functional groups. Thus, the ion with *m/z* 530 is $[M+H]^+$ of the monosodium salt, $[(M-H+Na)+H]^+$, and the ion with *m/z* 552 is $[M+H]^+$ of the disodium salt, $[(M-2H+2Na)+H]^+$.

The occurrence of H^+/Na^+ exchange can be demonstrated using the negative-ion ESI mass spectrum shown in Figure 2.10. The ion with *m/z* 506 is the deprotonated molecule $[M-H]^-$. The ions with *m/z* 528 and 550 show an *m/z* difference of 22 and 2×22 relative to $[M-H]^-$, respectively, which suggests the presence of Na^+. Thus, the ion with *m/z* 528 can be interpreted as the deprotonated monosodium salt, that is, $[(M-H+Na)-H]^-$, and the ion with *m/z* 550 as the deprotonated disodium salt, that is, $[(M-2H+2Na)-H]^-$. This type of H^+/Na^+ exchange behavior is frequently observed in compounds with multiple acidic functional groups such as peptides with Asp and/or Glu residues and oligonucleotides.

Another type of ions frequently observed under ESI conditions are adduct-bound dimers, of which the proton-bound dimer, that is, $[2M+H]^+$ or $[2M-H]^-$, is frequently the most abundant. However, one must be aware of the fact that other adduct-bound dimers may be observed in addition to or instead of the ones just mentioned, for example, $[2M+NH_4]^+$ or $[2M+Na]^+$ (Kamel et al., 1999). Since these ions are produced only at high analyte concentrations, this leads to problems when constructing calibration curves over large concentration ranges. In some cases, when interpreting the mass spectrum of an unknown compound, the ions due to $[M^1+H]^+$ and $[2M^1+H]^+$ may be confused with $[M^2+2H]^{2+}$ and $[M^2+H]^+$, respectively, where $M^2 = 2 \times M^1$. Zooming in on the isotope envelopes of the ions can reveal whether they are singly-charged or doubly-charged ions.

For compounds with higher mass, that is, typically above 500 Da, multiply-charged ions may be generated under ESI conditions. This requires that multiple sterically unhindered protonatable or deprotonatable sites be present in the molecule. Multiply-charged ions are readily observed for peptides and proteins, oligosaccharides, oligonucleotides, as well as in aminoglycoside antibiotics (Section 4.8.5) and disulfate steroid conjugates (Section 4.6.8). In principle, multiply-charged ions may be observed in both positive-ion and negative-ion modes, that is, $[M+nH]^{n+}$ and $[M-nH]^{n-}$. Again, Na^+ or K^+ adduct formation or (multiple) H^+/Na^+ exchange may occur, which may lead to ions such as $[M+nH+Na]^{(n+1)+}$ or $[(M-mH+mNa)+nH]^{n+}$, respectively. The relative abundance of the multiply-charged ion species may depend on analyte concentration, and in most cases it depends on experimental conditions, such as solvent composition and ESI ion source voltages.

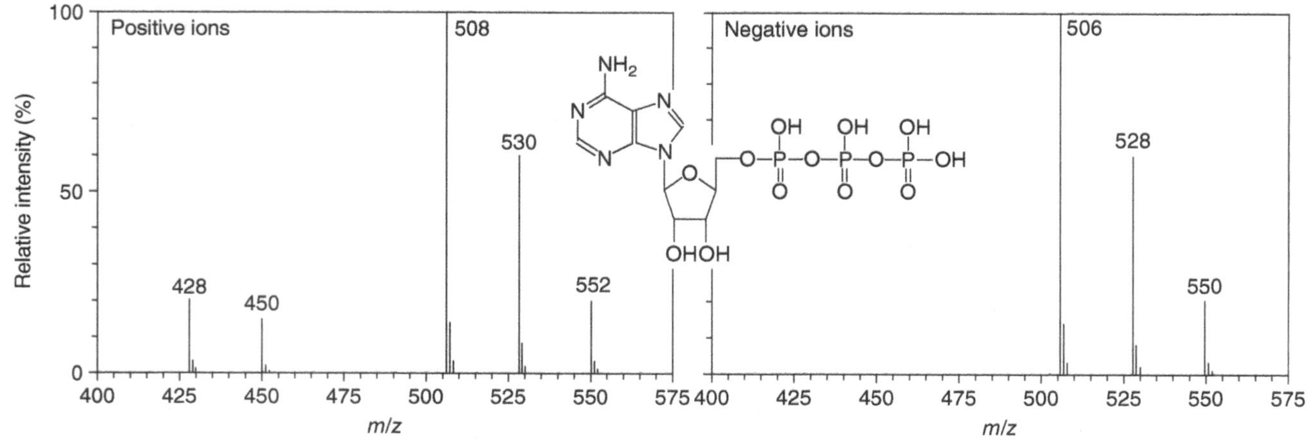

FIGURE 2.10 H^+/Na^+ exchange by the acid functional groups of adenosine triphosphate in positive-ion and negative-ion ESI mass spectra.

With larger biomolecules, for example, proteins, the multiple charging in ESI leads to ion envelopes of multiply-charged ions (Mann et al., 1989; Covey et al., 1988). When using MS instruments with a limited mass range, this allows accurate molecular-weight determination of the protein. Note that for large molecules, typically in excess of 10,000 Da, molecular weight (average mass) rather than monoisotopic mass is determined by MS. This is due to the fact that current MS instrumentation lacks the resolution to fully resolve the isotope patterns of the multiply-charged ion envelopes. In addition, for a large molecule, the isotope peak corresponding to the monoisotopic mass has very low abundance, for example, for horse myoglobin ($C_{769}H_{1212}N_{210}O_{218}S_2$, average mass 16951.602 Da), the relative abundance of the monoisotopic mass (16940.966 Da) is only 0.03% of that of the most abundant isotope (16950.992 Da).

Assuming protonation in the positive-ion mode, for each peak in the ion envelope of a protein with mass M and n charges, the m/z equals $(M+n)/n$, where n is the number of charges. For two adjacent peaks with $(m/z)_1$ and $(m/z)_2$, where $(m/z)_1 < (m/z)_2$, it holds that $n_1 = n_2 + 1$. Therefore, the charge for the ion leading to one of the peaks can be calculated with the following equation: $n_2 = ((m/z)_1 - 1)/((m/z)_2 - (m/z)_1)$. After rounding off to an integer number, the molecular weight of the protein can be calculated from the equation $M = n_2 \times ((m/z)_2 - 1)$. Since this calculation can be made for each pair of adjacent ions in the ion envelope, accurate molecular-weight determination can be achieved. For this purpose, averaging algorithms have been developed, which apply the calculations outlined earlier (Mann et al., 1989; Covey et al., 1988). More advanced software tools, based on maximum entropy algorithms, have been developed for this purpose (Ferrige et al., 1991; Reinhold & Reinhold, 1992; Ferrige et al., 1992). In positive-ion mode, the multiply-charged ions arise from protonation at the N-terminal of the proteins and possibly on each of its basic amino acid (Arg, His, Leu) (Smith et al., 1990). In negative-ion mode, similar series of multiply-charged ions can be observed for proteins (Loo et al., 1992) as well as for oligonucleotides (Covey et al., 1988; Potier et al., 1994; Lin et al., 2007).

One should be aware of the fact that the distance between the isotope peaks of an ion on the m/z scale is $1/n$ (the reciprocal of the charge). Thus, for a singly-charged ion, the isotope peaks are 1 unit apart, for doubly-charged ions 0.5 units, and for triply-charged ions 0.33 units. Using a mass analyzer with unit-mass resolution, that is, a quadrupole mass analyzer (Section 1.3.2), the isotope peaks of the multiply-charged ions are not resolved. As discussed (Section 1.3.3), the ion-trap mass analyzer provides enhanced resolution by slow scanning. In this way, resolution of isotope peaks is possible up to quadruply-charged ions. High-resolution instruments allow the resolution of higher charge states, of course depending on the resolution available. In the interpretation of mass spectra of unknown compounds, the ability to determine the charge state of an ion can be very important, for example, in discriminating between adduct-bound dimers and doubly-charged ions (see earlier).

REFERENCES

Benyon JH. 1978. Recommendation for symbolism and nomenclature for mass spectroscopy. Pure Appl Chem 50: 65–73.

Bijttebier SK, D'Hondt E, Hermans N, Apers S, Voorspoels S. 2013. Unravelling ionization and fragmentation pathways of carotenoids using orbitrap technology: A first step towards identification of unknowns. J Mass Spectrom 48: 740–754.

Brenton AG, Godfrey RA. 2010. Accurate mass measurement: Terminology and treatment of data. J Am Soc Mass Spectrom 21: 1821–1835.

Covey TR, Bonner RF, Shushan BI, Henion JD. 1988. The determination of protein, oligonucleotide and peptide molecular weights by ion-spray mass spectrometry. Rapid Commun Mass Spectrom 2: 249–256.

Currie K, Brailsford G, Nichol S, Gomez A, Riedel K, Sparks R, Lassey K. 2006. $^{14}CO_2$ in the southern hemisphere atmosphere—the rise and the fall. Chem N Z April 2006: 20–22.

Earnshaw A, Greenwood NN. 1997. Chemistry of the elements. 2nd ed. Ch1: 1–19. Butterworth-Heinemann Elsevier. Oxford. ISBN. 978-0-750-63365-9.

Fernandez B. 2006. De l'atome au noyau. Ellipses Éditions Marketing. Paris. ISBN 978-2-729-82784-6.

Ferrige AG, Seddon MJ, Jarvis S. 1991. Maximum entropy deconvolution in electrospray mass spectrometry. Rapid Commun Mass Spectrom 5: 374–377.

Ferrige AG, Seddon MJ, Green BN, Jarvis SA, Skilling J. 1992. Disentangling electrospray spectra with maximum entropy. Rapid Commun Mass Spectrom 6: 707–711.

Gross RA. 2004. A mass spectral chlorine rule for use in structure determinations in sophomore organic chemistry. J Chem Educ 81: 1161–1168.

IUPAC (International Union of Pure and Applied Chemistry). 2014. Compendium of Chemical Terminology Gold Book, Version 2.3.3. 2014-02-24. World Wide Web Gold Book PDF.

Kamel AM, Brown PR, Munson B. 1999. Effects of mobile-phase additives, solution pH, ionization constant, and analyte concentration on the sensitivities and electrospray ionization mass spectra of nucleoside antiviral agents. Anal Chem 71: 5481–5492.

Kind T, Fiehn O. 2006. Metabolomic database annotations via query of elemental compositions: Mass accuracy is insufficient even at less than 1 ppm. BMC Bioinf 7: 234.

Kind T, Fiehn O. 2007. Seven Golden Rules for heuristic filtering of molecular formulas obtained by accurate mass spectrometry. BMC Bioinf 8: 105.

Kind T, Fiehn O. 2010. Advances in structure elucidation of small molecules using mass spectrometry. Bioanal Rev 2: 23–60.

Kumar K, Menon AG, Sastry PS. 1992. Computer-assisted determination of elemental composition of fragments in mass spectra. Rapid Commun Mass Spectrom 6: 585–591.

Li L, Kresh JS, Karabacak NM, Cobb JS, Agar JN, Hong P. 2008. A hierarchical algorithm for calculating the isotopic fine structures of molecules. J Am Soc Mass Spectrom 19: 1867–1874.

Li L, Karabacak NM, Cobb JS, Hong P, Agar JN. 2010. Memory-efficient calculation of the isotopic mass states of a molecule. Rapid Commun Mass Spectrom 24: 2689–2696.

Lin ZJ, Li W, Dai G. 2007. Application of LC–MS for quantitative analysis and metabolite identification of therapeutic oligonucleotides. J Pharm Biomed Anal 44: 330–341.

Loo JA, Loo RR, Light KJ, Edmonds CG, Smith RD. 1992. Multiply charged negative ions by electrospray ionization of polypeptides and proteins. Anal Chem 64: 81–88.

McLafferty FW, Tureček F. 1993. Interpretation of mass spectra. 4th ed. University Science Books, Mill Valley, CA. ISBN 978-0-935-70225-5.

Mann M, Meng CK, Fenn JB. 1989. Interpreting mass spectra of multiply charged ions. Anal Chem 61: 1702–1708.

Mohr JP, Newell DB, Taylor BN. 2015. CODATA recommended values of the fundamental physical constants: 2014. National Institute of Standards and Technology. http://arxiv.org/pdf/1507.07956.

Moriwaka T, Newbold BT. 2003. Analogous odd-even parities in mathematics and chemistry. Chemistry 12: 445–450.

Muenster H, Taylor L. 2009. Mass resolution and resolving power. Thermo Fisher Scientific publication XX30175_E 02/09C.

Murray KK, Boyd RK, Eberlin MN, Langley GJ, Li L, Naito Y. 2013. Definitions of terms relating to mass spectrometry (IUPAC Recommendations 2013). Pure Appl Chem 85, 1515–1609.

Nakabayashi R, Sawada Y, Yamada Y, Suzuki M, Hirai MY, Sakurai T, Saito K. 2013. Combination of liquid chromatography-Fourier transform ion cyclotron resonance-mass spectrometry with ^{13}C-labeling for chemical assignment of sulfur-containing metabolites in onion bulbs. Anal Chem 85: 1310–1315.

The NIST Reference on Constants, Units, and Uncertainty. 2014. CODATA (Mohr et al. 2015) (http://physics.nist.gov/cuu/Constants/index.html). Atomic Weights and Isotopic Compositions with Relative Atomic Masses as of 20.04.2016 (http://www.nist.gov/pml/data/comp.cfm).

NIST Information/Data/Publication Disclaimer Information /data/publications ("Information" collectively) developed by employees of the National Institute of Standards and Technology (NIST), an agency of the Federal Government, is not subject to copyright protection within the United States and is considered to be in the public domain pursuant to title 15 United States Code Section 105. Your use of NIST developed Information is subject to the terms and conditions of this Disclaimer.

NIST Information is provided as a public service and is expressly provided "AS IS." NIST MAKES NO WARRANTY OF ANY KIND, EXPRESS, IMPLIED OR STATUTORY, INCLUDING, WITHOUT LIMITATION, THE IMPLIED WARRANTY OF MERCHANTABILITY, FITNESS FOR A PARTICULAR PURPOSE, NON-INFRINGEMENT AND DATA ACCURACY. NIST does not warrant or make any representations regarding the use of the Information or the results thereof, including but not limited to the correctness, accuracy, reliability or usefulness of the Information. NIST SHALL NOT BE LIABLE AND YOU HEREBY RELEASE NIST FROM LIABILITY FOR ANY INDIRECT, CONSEQUENTIAL, SPECIAL, OR INCIDENTAL DAMAGES (INCLUDING DAMAGES FOR LOSS OF BUSINESS PROFITS, BUSINESS INTERRUPTION, LOSS OF BUSINESS INFORMATION, AND THE LIKE), WHETHER ARISING IN TORT, CONTRACT, OR OTHERWISE, ARISING FROM OR RELATING TO THE INFORMATION (OR THE USE OF THIS INFORMATION), EVEN IF NIST HAS BEEN ADVISED OF THE POSSIBILITY OF SUCH DAMAGES.

Pelander A, Tyrkkö E, Ojanperä I. 2009. In silico methods for predicting metabolism and mass fragmentation applied to quetiapine in liquid chromatography/time-of-flight mass spectrometry urine drug screening. Rapid Commun Mass Spectrom 23: 506–514.

Pluskal T, Uehara T, Yanagida M. 2012. Highly accurate chemical formula prediction tool utilizing high-resolution mass spectra, MS/MS fragmentation, heuristic rules, and isotope pattern matching. Anal Chem 84: 4396–4403.

Potier N, Van Dorsselaer A, Cordier Y, Roch O, Bischoff R. 1994. Negative electrospray ionization mass spectrometry of synthetic and chemically modified oligonucleotides. Nucleic Acids Res 22: 3895–3903.

Povinee PP, Liherland AE, von Reden KF. 2009. Developments in radiocarbon technologies: From the libby counter to compound-specific AMS analyses. Radiocarbon 51: 45–78.

Price P. 1991. Standard definitions of terms relating to mass spectrometry. J Am Soc Mass Spectrom 2: 336–348.

Reinhold BB, Reinhold VN. 1992. Electrospray ionization mass spectrometry: Deconvolution by an entropy-based algorithm. J Am Soc Mass Spectrom 3: 207–215.

Rojas-Chertó M, Kasper PT, Willinghagen EL, Vreeken R, Hankemeier T, Reijmers T. 2011. Elemental composition determination based on MS^n. Bioinformatics 27: 2376–2383.

Russell DH, Edmondson RD. 1997. High-resolution mass spectrometry and accurate mass measurements with emphasis on the characterization of peptides and proteins by matrix-assisted laser desorption/ionisation time-of-flight mass spectrometry. J Mass Spectrom 32: 263–276.

Smith RD, Loo JA, Edmonds CG, Barinaga CJ, Udseth HR. 1990. New developments in biochemical mass spectrometry: Electrospray ionization. Anal Chem 62: 882–899.

Sparkman OD. 2006. Mass spectroscopy desk reference. 2nd ed. Global View Publishing. ISBN-13 978-0-966-08139-8.

Stoll N, Schmidt E, Thurow K. 2006. Isotope pattern evaluation for the reduction of elemental compositions assigned to high-resolution mass spectral data from electrospray ionization Fourier transform ion cyclotron resonance mass spectrometry. J Am Soc Mass Spectrom 17: 1692–1699.

Todd JFJ. 1995. Recommendations for nomenclature and symbolism for mass spectroscopy. Int J Mass Spectrom Ion Processes 142: 209–240.

Zaia J. 2004. Mass spectrometry of oligosaccharides. Mass Spectrom Rev 23: 161–227.

3

FRAGMENTATION OF EVEN-ELECTRON IONS

Interpretation of MS–MS Mass Spectra of Drugs and Pesticides, First Edition. Wilfried M. A. Niessen and Ricardo A. Correa C.

3.1 INTRODUCTION

In this chapter, important fragmentation reactions of even-electron ions (EE^+/EE^-) are discussed and to some extent compared to the fragmentation of odd-electron ions ($OE^{+\bullet}/OE^{-\bullet}$). It was decided to have this chapter preceding the data collection on fragmentation of drugs, pesticides, and related compounds in Chapter 4. To some degree, this chapter can be considered as a summary of the information gathered there. Frequent reference is made to specific examples of fragmentation behavior in Chapter 4. In fact, it is from studying the fragmentation of drugs and pesticides from literature sources and from their own experience that the authors have tried to develop rules for the fragmentation of EE^+/EE^-. This experience originates from 30 years of research and application in the fields of liquid chromatography–mass spectrometry (LC–MS) and liquid chromatography–tandem mass spectrometry (LC–MS–MS). In the last 20 years, the analyte ionization in LC–MS has been mostly performed with electrospray ionization (ESI; Section 1.2.4) or atmospheric-pressure chemical ionization (APCI; Section 1.2.5). These developing rules have been tested against the fragmentation of other compounds and compound classes. This chapter as well as the data collection in Chapter 4 can be considered as another step in a long-term project directed at understanding the fragmentation of EE^+/EE^- (Niessen, 1998, 2000, 2005, 2011, 2012).

Understanding fragmentation reactions is very much connected to interpretation of product-ion mass spectra. As our mass spectral data collection originates mainly from collision-induced dissociation (CID) in tandem quadrupole (TQ) and quadrupole–time-of-flight (Q–TOF) instruments, emphasis is put on fragmentation under collision-cell CID conditions (Section 1.4.1.1). Data obtained with ion-trap CID in three-dimensional or linear ion traps, in some cases combined with accurate-mass information from orbitrap mass analyzers, have been added whenever available. In some cases, the fragmentation behavior has been compared between the two different conditions, that is, collision-cell CID and ion-trap CID, for example, for furosemide (Section 4.1.4.3), citalopram (Section 4.2.3.3), benzodiazepines (Section 4.2.4), sulfonamides (Section 4.8.1), β-lactam antibiotics (Section 4.8.3.1), avermectins (Section 4.10.1.2), chlorotriazine herbicides (Section 4.11.1.1), and the dinitrophenol herbicides dinoseb and dinoterb (Section 4.11.9.8).

The most powerful tool in the interpretation of product-ion mass spectra is the determination of accurate mass with high-resolution accurate-mass MS (HRAM-MS), currently obtained mostly with Q–TOF (Section 1.4.5) and orbitrap (Section 1.4.6) instruments. HRAM-MS enables the accurate-mass determination of both precursor and product ions within 5 ppm, which for ions of up to *m/z* 1000 correspond to an absolute mass accuracy within 5 mDa. The accurate mass can be used to calculate possible molecular formulae for an ion, as discussed in Section 2.7. It is important to stress that the possible molecular formulae for product ions are restricted by the molecular formula of the precursor ion. In principle, product ions cannot contain elements that are not already present in the precursor ion and their molecular formulae cannot contain a higher number of each element than already present in the precursor ion.

Discrimination between $OE^{+\bullet}$ and EE^+ is of utmost importance in spectrum interpretation. Knowing the molecular formula of an ion enables the use of the nitrogen rule (Section 2.5), which should only be applied to singly-charged ions. The nitrogen rule states that a molecule, a molecular ion $M^{+\bullet}$, or any other $OE^{+\bullet}$ with an odd nominal mass or *m/z* should contain an odd number of N-atoms, whereas an EE^+ with an odd nominal *m/z* should contain even number of N-atoms.

Knowledge of the molecular formula also enables determination of the degree of unsaturation, that is, the number of double bonds and rings, present in the ion structure. The parameter used is the ring-double-bond equivalent (RDBE) (Section 2.5). The calculation of the RDBE results in an integer number for a molecule or an $OE^{+\bullet}/OE^{-\bullet}$ and a number ending in +0.5 for an EE^+ and −0.5 for an EE^-. Available software tools to calculate molecular formulae from accurate mass make use of the nitrogen rule and calculate the RDBE for the ions (Section 2.7) (Kind & Fiehn, 2007).

With the recent advent of HRAM-MS instruments, the interpretation of product-ion mass spectra had to be corrected in some cases because accurate-mass determination indicated a different molecular formula for a particular fragment ion than initially proposed. This issue is for instance discussed for β-lactam antibiotics in negative-ion mode (Section 4.8.3.1), trimethoprim (Section 4.8.9), and fluazifop-butyl (Section 4.11.7). HRAM-MS also plays a pivotal role in the recognition and confirmation of skeletal rearrangements (Section 3.5.8).

Depending on the problem at hand, there is a variety of additional tools that can be applied to assist in structure elucidation and/or interpretation of product-ion mass spectra, for example, studies involving stable-isotope labeling or hydrogen/deuterium (H/D) exchange, multistage MS^n in ion-trap instruments, advanced data-acquisition strategies such as precursor-ion and neutral-loss analysis, and software tools to predict fragmentation behavior.

Site-specific covalent isotopic labeling with stable isotopes such as D (^2H) and/or ^{13}C is a powerful tool for the elucidation of fragmentation mechanisms and/or confirmation of H-rearrangements, such as observed in the McLafferty rearrangement and in many EE^+ fragmentation reactions (McLafferty, 1959; Kingston et al., 1974). A nice example of this is the structure elucidation described for the fragment ion with *m/z* 109, observed in the product-ion mass spectrum of protonated testosterone (Section 4.6.2) (Williams et al., 1999). Other examples involve for instance the fragmentation of amphetamine (Section 4.7.2) and of Δ^9-tetrahydrocannabinol (Section 4.7.3).

A dynamic alternative to covalent isotopic labeling is the use of H/D-exchange experiments. In an H/D-exchange experiment, the *m/z* shift is measured between a compound and its H/D-exchange product. The *m/z* shift results from the exchange of any proton by deuterium at the hydroxy (–OH), carboxylic acid (–COOH), and amino (–NH_2) functional groups in a compound. As an example, an H/D-exchange experiment with ethanolamine ($HOCH_2CH_2NH_2$) would result in an *m/z* shift of 4: the exchange of three protons (to $DOCH_2CH_2ND_2$) and the formation of $[M+D]^+$ rather than $[M+H]^+$ (Liu & Hop, 2005; Shah et al., 2013). In an LC–MS experiment, H/D exchange can be achieved during the ESI process by replacing water in the mobile phase by heavy water (D_2O). This and various other ways to perform the H/D-exchange experiments have been discussed (Shah et al., 2013). The application of H/D-exchange experiments is discussed for instance in the fragmentation of simvastatin (Section 4.4.3.2), stanozolol (Section 4.6.4), oxytetracycline (Section 4.8.6), and glyphosate (Section 4.11.4).

Data-acquisition strategies in MS–MS instruments also provide powerful tools to get more insight into fragmentation reactions. Multistage MS^n in an ion-trap instrument enables step-by-step fragmentation (Section 1.4.3). In this way, fragmentation trees can be built showing the relationships between product ions and precursor ions, which can be very helpful in structure elucidation (Wolfender et al., 1998; Kind & Fiehn, 2010; van der Hooft et al., 2011; Kasper et al., 2012; Schwarzenberg et al., 2013). Examples of multistage MS^n for the interpretation of product-ion mass spectra are discussed for benzodiazepines (Section 4.2.4), simvastatin (Section 4.4.3.2), and chlorotriazine herbicides (Section 4.11.1.1). A somewhat similar approach is the so-called pseudo-MS^3, which can be performed on instruments featuring collision-cell CID. In pseudo-MS^3, a particular fragment ion generated by in-source CID is selected as precursor ion in MS_1, subsequently fragmented in the collision cell, and mass analyzed in MS_2. In this way, the fragmentation of particular fragment ions can be studied. This procedure has for instance been applied in the structure elucidation of ceramide lipids (Hsu et al., 2002), in understanding a skeletal rearrangement observed in the fragmentation of a farnesyl transferase inhibitor (Qin, 2001) (Section 3.5.8), and in discriminating between isobaric or isomeric flavonoid glycosides (Wojakowska et al., 2013; Abrankó & Szilvássy, 2015). The precursor-ion analysis (PIA) and neutral-loss analysis (NLA) modes (Section 1.5.3) in a TQ instrument can also be powerful tools in structure elucidation. In PIA, the precursor ions of a particular product ion can be established. This has been applied for instance in the elucidation of the fragmentation patterns of 6-hydroxychlorzoxazone (Anari et al., 2003) and of morphine and related opiates (Bijlsma et al., 2011).

Finally, software packages have been introduced in recent years that perform *a priori* prediction (with variable success) of the fragment ions one may observe for a given structure. Two widely used commercially available packages are Mass Frontier (www.highchem.com) and Mass Fragmenter (http://www.acdlabs.com/products/adh/ms/ms_frag/). Other software tools assist in the interpretation of MS–MS fragmentation of a given structure (or its related substances) based on accurate-mass data from Q–TOF instruments (Pelander et al., 2009; Hill & Mortishire-Smith, 2005; Pluskal et al., 2012; Gerlich & Neumann, 2013). Although complete prediction of the fragment ions for a particular structure is still far away, significant progress can be expected in this area in years to come.

3.2 ANALYTE IONIZATION REVISITED

As discussed in Section 1.2, a wide variety of analyte ionization techniques is available in MS. Any ionization technique should lead to gas-phase analyte ions, either in high vacuum or transferable to high vacuum, in order to enable their subsequent mass analysis. The data discussed in Chapter 4 have been obtained mainly by LC–MS–MS using

ESI or APCI. In the context of this book, only a limited number of ionization techniques is relevant. The ionization techniques can be classified in different ways.

Ionization techniques can be classified based on the physical state of the analyte molecule in either gas-phase or condensed-phase techniques. The condensed-phase techniques can be further classified into liquid-phase and solid-phase or surface-ionization techniques. Electron ionization (EI), chemical ionization (CI), and APCI are typical examples of gas-phase ionization techniques, ESI of a liquid-phase ionization technique, and matrix-assisted laser desorption/ionization (MALDI) and desorption electrospray ionization (DESI) are examples of surface-ionization techniques (Section 1.2). To achieve gas-phase ionization, either a gas-phase sample or sample evaporation prior to ionization is required. In this respect, gas chromatography–mass spectrometry (GC–MS) is a powerful tool, as it not only achieves separation of the sample into individual components but also delivers the individual components in the gas phase. In liquid-phase ionization, as applied in LC–MS, the sample solution, for example, the LC mobile phase, is nebulized into small droplets, from which gas-phase analyte ions are generated, for example, by ESI. Surface-ionization techniques are frequently considered to be the so-called energy-sudden techniques (Section 1.2.6.1) (Arpino & Guiochon, 1982; Vestal, 1983). Intense localized energy is applied to the sample, for example, by means of a laser pulse as in MALDI, to simultaneously ionize and transfer the ion from the solid phase to the gas phase. Instrumental and mechanistic topics related to analyte ionization are discussed in more detail in Section 1.2.

From a fragmentation point of view, it is useful to distinguish between hard- and soft-ionization techniques. In a hard-ionization technique, a substantial amount of energy is transferred to the molecule upon generation of the ion. The most frequently applied hard-ionization technique is EI, where during ionization a few electronvolts (≈2–5 eV, a non-SI unit) of energy is transferred to the molecular ion, $M^{+\bullet}$, resulting in an ion in an excited state with high internal energy (1 eV is the amount of energy acquired by an electron that passed across a potential difference of 1 V; $1\,eV = 1.6 \times 10^{-19}\,J$). This excess of internal energy leads to a rapid in-source fragmentation of $M^{+\bullet}$. The resulting mixture of $M^{+\bullet}$ and its fragment ions is then mass analyzed. In a soft-ionization technique, on the other hand, hardly any energy is transferred to the ion and no in-source fragmentation occurs. The type of ions generated in a soft-ionization technique is often different: instead of $M^{+\bullet}$ an EE^+/EE^- is generated, often being a protonated molecule $[M+H]^+$ or a deprotonated molecule $[M-H]^-$. If fragmentation is required, it has to be induced by increasing the internal energy of the ion, for instance by collisions with gas (CID) in the collision cell of the MS–MS instrument (Section 1.4.1). Thus, a further classification of the ionization techniques involves the type of ions generated, that is, $OE^{+\bullet}/OE^{-\bullet}$ or EE^+/EE^-.

3.3 FRAGMENTATION OF ODD-ELECTRON IONS

In EI, analyte ionization results from the interaction of gas-phase molecules with 70 eV electrons in the EI ion source under high-vacuum. Because it provides good sensitivity and robustness, 70 eV is used as a standard value for the electron energy (Section 1.2.1.1). The interaction between the electrons and the analyte molecule M results in the loss of an electron, thus resulting in the formation of the molecular ion $M^{+\bullet}$. After ionization, $M^{+\bullet}$ may typically carry an excess of 2–5 eV in internal energy. This may lead to fast in-source fragmentation reactions, where from a singly-charged ion at least two products are generated, that is, an ion with a lower *m/z* and a neutral molecule or a radical. Since only ions are observed in the MS, the mass of the neutral molecule or radical may be inferred from the *m/z* difference between the $M^{+\bullet}$ and its fragment ion.

The fragmentation of $M^{+\bullet}$ has been investigated systematically and in considerable detail (McLafferty & Tureček, 1993; Smith, 2004). In the context of this chapter, it is useful to discuss five important fragmentation reactions in EI: (1) the σ-homolytic cleavage, (2) the α-homolytic cleavage, (3) the inductive heterolytic cleavages, (4) the McLafferty rearrangement, and (5) the retro-Diels–Alder (RDA) fragmentation. When drawing fragmentation mechanisms, either a curved single-headed arrow (movement of one electron) or a curved double-headed arrow (movement of two electrons) is used.

In the $M^{+\bullet}$ of a molecule without double bonds or heteroatoms, that is, an (cyclo) alkane (C_nH_{2n+2}), the charge and radical can be on any of the C-atoms. Fragmentation results in a homolytic σ-C—C bond cleavage, with preference to cleave at 3° and 4° carbon atoms (branching sites) to generate 2° and 3° carbocations, respectively. The loss of the largest alkyl radical is favored. This reflects the ability of the fragment ions to stabilize the positive charge and the degree of stabilization of the excited $M^{+\bullet}$ (loss of internal energy). Thus, the so-called σ-cleavage leads to the formation of an EE^+ carbocation and an alkyl radical (Eq. 3.1).

$$R^1CH_2CHR^2{+\bullet}CH_2R^3 \rightarrow R^1CH_2\overset{+}{C}HR^2 + {}^{\bullet}CH_2R^3 \tag{3.1}$$

These σ-cleavages are especially important in petrochemistry. In the presence of a heteroatom with low ionization energy, for example, Si, S, or P, a σ-cleavage may occur and lead to an EE^+ with the charge on the heteroatom. Although thermodynamically less favored, the EE^+ may be accompanied

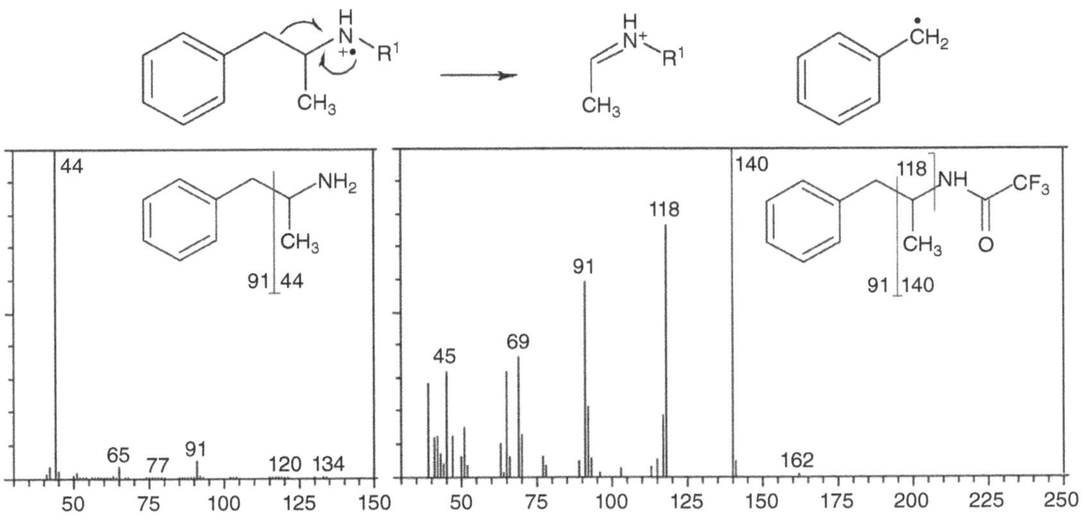

FIGURE 3.1 Mechanism of the α-cleavage in $OE^{+\bullet}$, illustrated for amphetamine ($M^{+\bullet}$ with *m/z* 135) and its trifluoroacetyl derivative ($M^{+\bullet}$ with *m/z* 231).

by an $OE^{+\bullet}$ (and a neutral molecule) due to an intramolecular shift of a hydrogen radical ($H^{\bullet}$) in $M^{+\bullet}$ (Eq. 3.2).

$$R^1CH_2CHR^2{+}{\bullet}CH_2R^3 \rightarrow R^1CH\overset{+\bullet}{=}CHR^2 + CH_3R^3 \quad (3.2)$$

For instance, in the EI mass spectrum of 4-methyldecane ($M^{+\bullet}$ with *m/z* 156), a series of EE^+ are observed. However, the ion with *m/z* 113 ($[C_8H_{17}]^+$), which results from the loss of a propyl radical (${}^{\bullet}C_3H_7$), is accompanied by an ion with *m/z* 112 ($[C_8H_{16}]^{+\bullet}$) due to the loss of propane (C_3H_8).

In a molecule with a heteroatom, for example, N- or O-atoms, or a double bond, the charge and radical are initially localized on the heteroatom or the double bond. In this case, the radical-initiated homolytic cleavage of an α,β-C—C bond is more likely to occur, leading to an EE^+ fragment and the loss of a radical, as shown for an ether (Eq. 3.3a), a secondary amine (Eq. 3.3b), a ketone (Eq. 3.3c), and an alkene (Eq. 3.3d).

$$R^1CH_2\overset{+\bullet}{O}CH_2CH_2R^2 \rightarrow R^1CH_2\overset{+}{O}{=}CH_2 + {}^{\bullet}CH_2R^2 \quad (3.3a)$$

$$R^1CH_2\overset{+\bullet}{N}HCH_2CH_2R^2 \rightarrow R^1CH_2\overset{+}{N}H{=}CH_2 + {}^{\bullet}CH_2R^2 \quad (3.3b)$$

$$R^1\overset{+\bullet}{C}OCH_2R^2 \rightarrow R^1\overset{+}{C}{=}O + {}^{\bullet}CH_2R^2 \quad (3.3c)$$

$$R^1CH_2CH\overset{+\bullet}{=}CHR^2 \rightarrow {}^{\bullet}R^1 + CH_2{=}CH\overset{+}{C}HR^2 \quad (3.3d)$$

Note that the ion produced in the homolytic cleavage of a ketone (Eq. 3.3c) can be written as either an acylium ion, $[R^1C{=}O]^+$, with the charge on the C-atom or an oxonium ion, $[R^1C{\equiv}O]^+$, with the charge on the O-atom, and thus it is resonance stabilized.

This type of fragmentation is often referred to as α-cleavage. It is the most important fragmentation reaction of $OE^{+\bullet}$ in EI (Smith, 2004). In Figure 3.1, the α-cleavage is illustrated for amphetamine ($M^{+\bullet}$ with *m/z* 135) and its trifluoroacetyl derivative ($M^{+\bullet}$ with *m/z* 231). Derivatization is performed to avoid the formation of the ethaniminium ion ($[CH_3CH{=}NH_2]^+$) with *m/z* 44 as the α-cleavage fragment because this ion is isobaric with $[CO_2]^{+\bullet}$, which may arise from a vacuum leak. For the trifluoroacetyl derivative, an *N*-trifluoroacetylethaniminium ion ($[CH_3CH{=}NHCOCF_3]^+$) with *m/z* 140 is formed instead. In both cases, the benzyl radical ($C_6H_5CH_2{}^{\bullet}$) is lost as a neutral.

Alternatively, a charge-initiated heterolytic bond cleavage of the C—heteroatom bond occurs, again leading to an EE^+ fragment and the loss of a radical containing the heteroatom. The heterolytic or inductive cleavage is observed especially in compounds where the heteroatom is a halogen, an O-atom, or an S-atom. The reaction is shown for a chloroalkane and a ketone in Eq. 3.4. Heterolytic cleavages (i-cleavages) of $M^{+\bullet}$ are generally less important than homolytic cleavages (σ- and α-cleavages).

$$R^1CH_2Cl^{+\bullet} \rightarrow R^1\overset{+}{C}H_2 + {}^{\bullet}Cl \quad (3.4a)$$

$$R^1CH_2\overset{+\bullet}{C}OR^2 \rightarrow R^1\overset{+}{C}H_2 + R^2\overset{\bullet}{C}{=}O \quad (3.4b)$$

The homolytic and heterolytic cleavages are in fact competing reactions. In the homolytic cleavage, the charge is

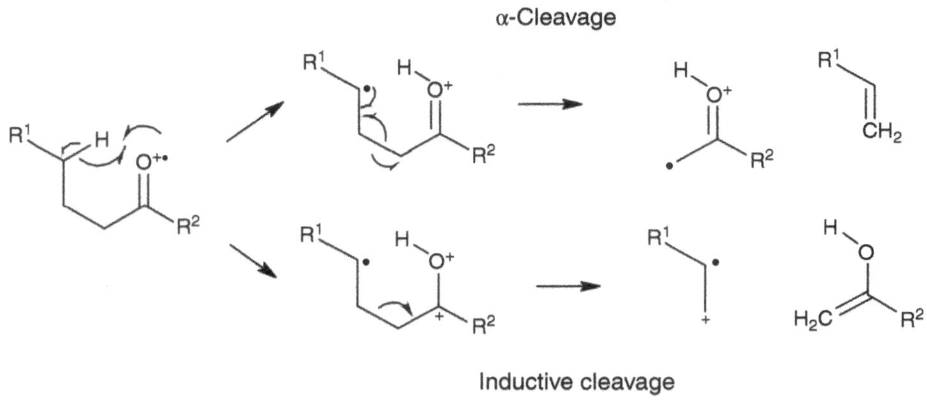

FIGURE 3.2 Mechanism of the McLafferty rearrangement. For the trifluoroacetyl derivative of amphetamine, the McLafferty rearrangement results in an ion with *m/z* 118 (Figure 3.1).

retained on the original site, whereas charge migration occurs in the heterolytic or inductive cleavage. The competition for charge retention or charge migration follows Stevenson's rule, which states that the unpaired electron is preferentially retained by the fragment with the higher ionization energy. Thus, the formation of a fragment ion with the lower ionization energy is more likely.

The three fragmentation reactions discussed thus far involve the loss of a radical and the formation of an EE^+ fragment. In the case of the McLafferty rearrangement and the RDA fragmentation, an $OE^{+\bullet}$ fragment ion is produced and a neutral molecule is lost. These fragmentation reactions involve the cleavage of two σ-bonds.

The McLafferty rearrangement may occur for instance in ketones with a γ-H-atom and the ability to form a six-membered ring-like transition state with the carbonyl-O-atom (Figure 3.2) (McLafferty, 1959; Kingston et al., 1974). The fragmentation involves two steps: a radical-initiated H-rearrangement, followed by either a homolytic (Eq. 3.5a) or a heterolytic (Eq. 3.5b) cleavage (determined by Stevenson's rule).

$$R^1CH_2CH_2CH_2\overset{+\bullet}{C}OR^2 \rightarrow R^1CH{=}CH_2 + {}^{\bullet}CH_2\overset{+}{C}OHR^2 \leftrightarrow CH_2{=}CR^2{-}\overset{+\bullet}{O}H \quad (3.5a)$$

$$R^1CH_2CH_2CH_2\overset{+\bullet}{C}OR^2 \rightarrow R^1CH\overset{+\bullet}{=}CH_2 + CH_2{=}CR^2OH \quad (3.5b)$$

The McLafferty rearrangement forms characteristic $OE^{+\bullet}$ for aldehydes, ketones, esters, acids, amides, carbonates, alkenes, alkynes, and phenylalkanes (among others). A McLafferty rearrangement is observed for the trifluoroacetyl derivative of amphetamine in Figure 3.1. In this case, the $OE^{+\bullet}$ fragment resulting from the heterolytic cleavage, that is, the ion with *m/z* 118 ($[C_6H_5CH{-}CHCH_3]^{+\bullet}$), is more abundant than the ion resulting from the homolytic cleavage, that is, the ion with *m/z* 113 ($[NH_2COCF_3]^{+\bullet}$).

The RDA fragmentation occurs in a six-membered ring with a double bond and results in the formation of a (substituted) diene and a (substituted) ethene (Tureček & Hanuš, 1984). The RDA fragmentation involves an α-cleavage followed by either another *α*-cleavage or an i-cleavage. Depending on the ionization energy of the two reaction products, the charge and the radical reside at one of the two products (Stevenson's rule), that is, resulting in $M^{+\bullet}$ while the other product is lost as a neutral molecule. An example of the RDA fragmentation and the competition between the two cleavage reactions (*α*- or i-cleavage) is shown in Figure 3.3 for 4-methyl-cyclohexene ($M^{+\bullet}$ with *m/z* 96) and 4-phenyl-cyclohexene ($M^{+\bullet}$ with *m/z* 158). In 4-methyl-cylcohexene, the ion with *m/z* 54 ($[C_4H_6]^{+\bullet}$; $M^{+\bullet}$ of 1,3-butadiene) results from an α-cleavage with charge retention because the ionization energy of propene (9.7 eV) is higher than that of butadiene (9.1 eV). In 4-phenyl-cyclohexene, on the other hand, the ion with *m/z* 104 ($[C_8H_8]^{+\bullet}$; $M^{+\bullet}$ of styrene) results from an i-cleavage with charge migration because the ionization energy of the styrene (8.4 eV) is lower than that of butadiene (9.1 eV). The RDA fragmentation clearly shows that this fragmentation in MS can be considered as a retrosynthesis: the fragments from the RDA fragmentation are the starting reagents of a Diels–Alder cycloaddition reaction.

A characteristic feature of the fragmentation reactions of $OE^{+\bullet}$ discussed in this section is that when both fragment ions are actually observed, the sum of their *m/z* values equals the *m/z* value of the $M^{+\bullet}$, that is, *m/z* (M).

3.4 HIGH-ENERGY COLLISIONS OF PROTONATED MOLECULES

From the discussion in the previous section, it may be concluded that radical-initiated fragmentation of a molecular ion

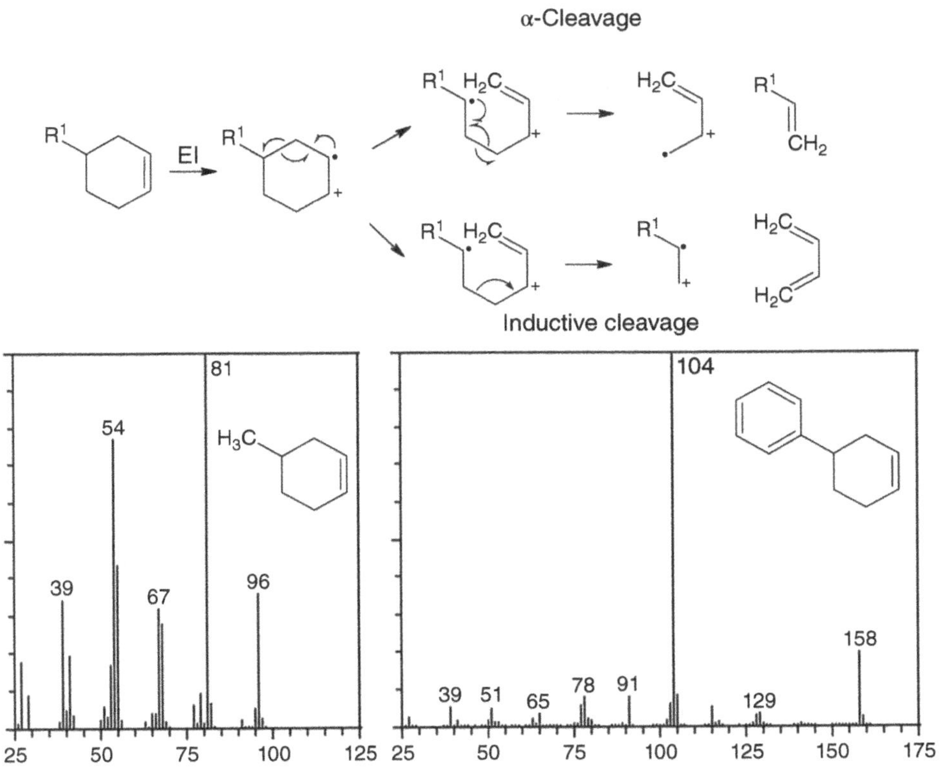

FIGURE 3.3 Mechanism of the retro-Diels–Alder fragmentation for $OE^{+\bullet}$, illustrated for 4-methyl-cyclohexene ($M^{+\bullet}$ with m/z 96) and 4-phenyl-cyclohexene ($M^{+\bullet}$ with m/z 158).

$M^{+\bullet}$ is very important in EI. In soft-ionization techniques like ESI, however, primarily EE^+/EE^- are generated, that is, $[M+H]^+$ in positive-ion mode and $[M-H]^-$ in negative-ion mode. As no odd electron (radical) is present in an EE^+, the fragmentation of an EE^+ is significantly different from that of an $OE^{+\bullet}$. Several systematic studies on fragmentation of EE^+ have been reported, mostly in relation to proton-transfer CI (Sigsby et al., 1979; McLafferty, 1980; Wagner et al., 1980; Kingston et al., 1983; Zollinger & Seibl, 1985).

3.4.1 General Aspects

The first general rule involved in the fragmentation of EE^+ is the even-electron rule, stating that upon fragmentation an EE^+ does not usually lose a radical to form an $OE^{+\bullet}$ fragment (Karni & Mandelbaum, 1980; McLafferty, 1980). As a consequence, the loss of a neutral molecule from an EE^+ resulting in an EE^+ fragment is more likely to occur than the loss of a radical resulting in an $OE^{+\bullet}$ fragment. Violation of the even-electron rule is discussed in Section 3.5.7. For $EE^+ \rightarrow EE^+$ fragmentation reactions, a curved double-headed arrow (movement of two electrons) should be drawn; the use of two curved single-headed arrows (movement of one electron each) instead implies a violation of the even-electron rule.

Unimolecular fragmentation of CI-generated EE^+ has been reviewed (McLafferty, 1980). Four types of fragmentation reactions are discussed, that is, (1) an inductive cleavage involving the cleavage of a σ-bond with charge migration, (2) an acyclic EE^+ cleavage of a σ-bond with charge retention, (3) EE^+ fragmentation yielding $OE^{+\bullet}$ fragment ions and the loss of a radical (discussed in more detail in Section 3.5.7), and (4) a cyclic EE^+ cleavage of two σ-bonds with charge retention, for example, an RDA fragmentation (Section 3.5.6).

The first fragmentation reaction is the inductive cleavage. In $[M+H]^+$, the proton is generally located on a carbon-bound heteroatom, for example, an N- or O-atom. The inductive cleavage involves a cleavage of the heteroatom–C bond with charge migration to the C-atom containing fragment. In a protonated ether, $[R^1OHR^2]^+$, the inductive cleavage involves a cleavage of the O—C bond. Thus, an alkyl cation $[R^1]^+$ or $[R^2]^+$ is produced with the loss of either alcohol, R^2OH, or R^1OH, respectively (Eq. 3.6).

$$[R^1OHR^2]^+ \rightarrow [R^1]^+ + R^2OH \qquad (3.6)$$

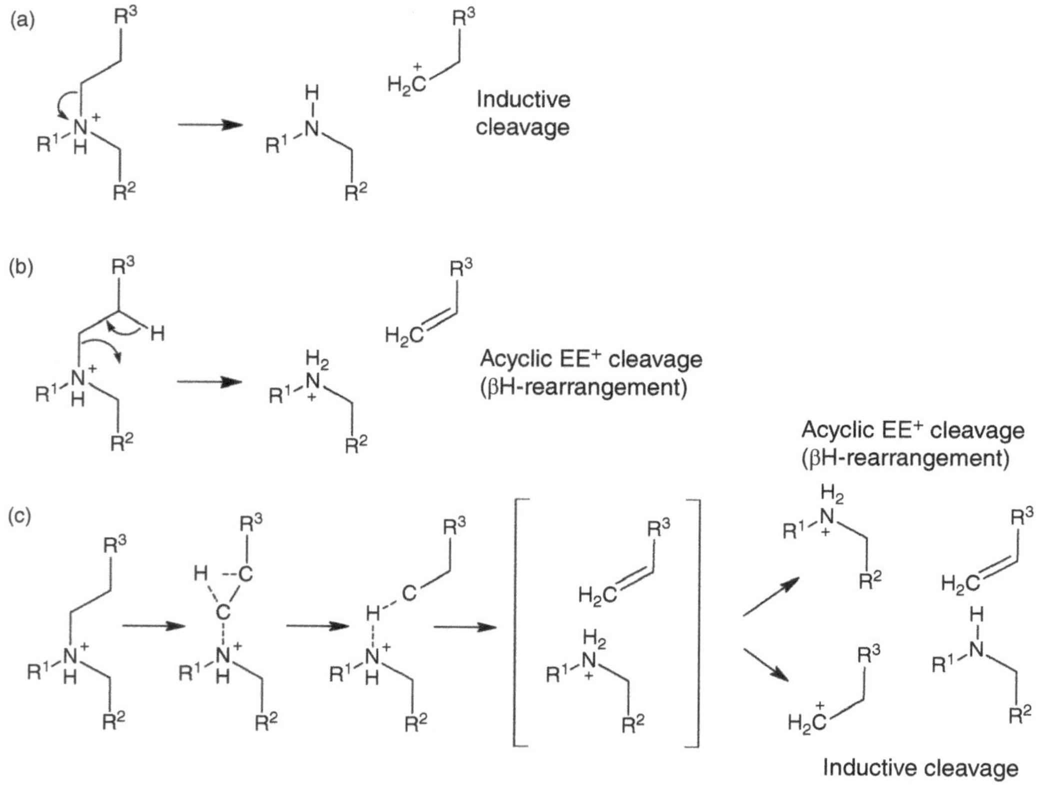

FIGURE 3.4 Proposed mechanisms for EE^+ fragmentation of a tertiary amine, with (a) the inductive cleavage, (b) the concerted four-center fragmentation mechanism for the acyclic EE^+ cleavage involving βH-rearrangement, and (c) the fragmentation mechanism of the acyclic EE^+ cleavage via an ion–molecule complex.

The inductive cleavage is depicted for a protonated tertiary alkyl amine in Figure 3.4a. Field's rule predicts that the loss of a neutral molecule with a low proton affinity is more likely (Field, 1972). Thus, charge retention on the product with the higher proton affinity is more favorable. Field's rule can be considered as the even-electron counterpart of Stevenson's rule in EI (Section 3.3). The abundance ratio of the butyl ion $[C_4H_9]^+$ with *m/z* 57 relative to that of $[M+H]^+$ for various *n*-butyl compounds was found to be correlated with the proton affinity of the leaving molecule, for example, 0.04 for NH_3 with 854 kJ mol^{-1}, through 20 for H_2O with 697 kJ mol^{-1} and up to 240 for HI with 611 kJ mol^{-1} (Audier et al., 1978). Obviously, the stability of the product ion is also an important factor. For various butylamine isomers, the abundance ratio of $[C_4H_9]^+$ relative to that of $[M+H]^+$ was found to be far higher for *tert*-butylamine than for *sec*- or *n*-butylamine (Audier et al., 1978). The elimination of butene (C_4H_8) resulting in ammonium ion ($[NH_4]^+$) with *m/z* 18 is also observed. Unfortunately, the situation is not always that straightforward. In molecules with several functional groups, steric factors and intramolecular H-bonding may influence the fragmentation. After the loss of a neutral molecule, a cyclized product ion may enhance the inductive cleavage (McLafferty, 1980). Isomerization can occur to form a more stable product ion. This is probably the case in the loss of formaldehyde ($H_2C{=}O$) from *n*-, *iso*-, and *sec*-butyl(methylidene)oxonium ions ($[C_4H_9O{=}CH_2]^+$), where similar relative abundances were observed irrespective of the butyl isomer (Bowen et al., 1978).

The acyclic EE^+ cleavage with charge retention is a competing process with the inductive cleavage. Mechanistically, this cleavage is often described as a βH-rearrangement, that is, a 1,2-elimination, via a four-center mechanism, for example, in secondary fragmentation of EE^+ fragment ions produced from α-cleavages of $OE^{+\bullet}$ (McLafferty, 1959). Labeling studies indicate that often not only a 1,2-elimination but also a 1,3-elimination is an important reaction. However, there has been considerable discussion on the actual fragmentation mechanism. The βH-rearrangement in a four-center mechanism (McLafferty, 1959) implies a concerted fragmentation mechanism: two bonds are broken and two new bonds are formed simultaneously, as depicted for a protonated tertiary amine in Figure 3.4b. In a protonated ether, $[R^1OHR^2]^+$, the acyclic EE^+ cleavage would yield a protonated alcohol $[R^1OH_2]^+$ or $[R^2OH_2]^+$ with the loss of an alkene $[(R^2–H)]$ or $[(R^1–H)]$, respectively (Eq. 3.7).

$$[R^1OHCH_2CH_2R^2]^+ \rightarrow [R^1OH_2]^+ + CH_2{=}CHR^2 \quad (3.7)$$

Alternatively, the fragmentation may be described in terms of a non-concerted mechanism via the formation of an ion–molecule or an ion–dipole complex (Bowen et al., 1978; Morton, 1980; Julian et al., 2007). Thus, the fragmentation proceeds through an initial rearrangement into a proton-bound or ion–dipole complex of the two product molecules. Dissociation of the proton-bound or ion–dipole complex occurs with charge retention on either of the two fragments, determined by their proton affinity according to Field's rule (Figure 3.4c).

Irrespective of the fragmentation mechanism, the competition between the inductive cleavage and acyclic EE^+ cleavage (βH-rearrangement) may result in two (complementary) fragment ions (Figure 3.4). A characteristic feature of these two fragment ions is that their *m/z* values sum up to *m/z* ([M+H]+1). In this way, the complementary ions may be easily recognized in the mass spectrum (Section 3.5.2).

3.4.2 Selected Examples

Four major primary fragmentation routes are observed in the fragmentation of alkylamines under high-energy CID conditions (Sigsby et al., 1979; Reiner et al., 1989). The four routes involve (1) the loss of an amine to form an alkyl cation, which implies an inductive cleavage, (2) the loss of an alkene to form a protonated amine, which implies an acyclic EE^+ cleavage (βH-rearrangement), (3) the elimination of alkane to produce an iminium ion, and (4) the loss of alkyl radicals. The competition reactions between the first two routes are discussed in terms of complementary ions (Section 3.5.2). Fragmentation of butylamines ($[M+H]^+$ with *m/z* 74) leads to complementary EE^+ with *m/z* 57 ($[C_4H_9]^+$) and $[NH_4]^+$ with *m/z* 18 (Audier et al., 1978). For the various isomeric pentylamines ($[M+H]^+$ with *m/z* 88), reactions (1) and (2) are most prominent resulting in the complementary ions $[C_5H_{11}]^+$ with *m/z* 71 and $[NH_4]^+$ (Reiner et al., 1989). For the C_5-dialkyl- and trialkylamines (Figure 3.5 or Table 3.1), alkene elimination is generally most predominant, with the loss of the largest alkene usually leading to the most abundant fragment ion. Also, alkane losses lead to ions with considerable abundance under the high-energy CID conditions used in the experiment. There is substantial evidence that the alkane loss does not occur via a concerted mechanism, but rather as a stepwise process. In fact, a stepwise fragmentation has been suggested via an initial homolytic cleavage, the formation of a complex of an alkyl radical and a radical cation, followed by a hydrogen or alkyl radical abstraction by the alkyl radical (Reiner et al., 1989). However, in such a mechanism, the products of the initial homolytic cleavage violate the even-electron rule (Karni & Mandelbaum, 1980; McLafferty, 1980). Along with alkane and alkene losses, alkyl cations and losses of a methyl radical ($^\bullet CH_3$) are often observed (Reiner et al., 1989). In another study, also involving high-energy CID, the loss of both ethane (C_2H_6) and ethene (C_2H_4) is observed for diethylamine, but the most abundant fragment ions were found to be the methaniminium ion ($[H_2C{=}NH_2]^+$) with *m/z* 30, the ethyl cation ($[C_2H_5]^+$) with *m/z* 29, and $[NH_4]^+$ with *m/z* 18. Interestingly, next to the loss of C_2H_6, losses of methane (CH_4) and propene ($CH_3CH{=}CH_2$) are observed, which can also be rationalized in terms of a four-center mechanism. For protonated *n*-propylamine ($[C_3H_7NH_2+H]^+$), the most abundant fragment ions result from the loss of ammonia (NH_3) and $CH_3CH{=}CH_2$. The loss of C_2H_6 to an ion with *m/z* 30 ($[H_2C{=}NH_2]^+$) involves a cleavage of the protonated amine α,β-C–C bond (Wagner et al., 1980); this shows resemblance to the α-cleavage in EI (Eq. 3.3b in Section 3.3).

Deuterium ([D]) labeling studies have been reported for the loss of $CH_3CH{=}CH_2$ from phenyl propyl ether ($M^{+\bullet}$ with *m/z* 136; $[M+H]^+$ with *m/z* 137) (Benoit & Harrison, 1976). For $[M+H]^+$, a non-random non-site-specific H-rearrangement occurs, that is, with αH (≈30%), βH (≈20%), and γH (≈50%). This indicates that for the phenyl propyl ether, a H-rearrangement occurs to the ether O-atom rather than to the phenyl ring. In the same way, it has been shown that in $M^{+\bullet}$ of phenyl propyl ether, a H-rearrangement occurs to the ether O-atom rather than the expected (site-specific) McLafferty H-rearrangement from the βH of the propyl group to the 2-position of the aromatic ring. These findings suggest that not only the four-membered cyclic transition state is of importance in the fragmentation but also the formation of three-, five-, and six-membered cyclic intermediates could be feasible (Benoit & Harrison, 1976). Other studies also show a low site specificity for the H-rearrangement in $[M+H]^+$, which is in striking contrast to the (often) high site specificity observed in $M^{+\bullet}$, for example, in the McLafferty rearrangement (Djerassi & Fenselau, 1965; Kingston et al., 1983). This suggests that mechanistically the fragmentation of $[M+H]^+$ is more complex. Therefore, the term H-rearrangement rather than βH-rearrangement is preferred here.

The fragmentation of $[M+H]^+$ of amines, alcohols, ethers, sulfides, halides, ketones, aldehydes, carboxylic acids, and nitriles has been compared with that of their respective $M^{+\bullet}$ (Wagner et al., 1980). Fragmentation of protonated acetone ($[M+H]^+$ with *m/z* 59) results in a fragment ion with *m/z* 43 due to the loss of CH_4. [D]-labeling studies indicate that two mechanisms are involved (in a 2:1 ratio), that is, (1) the formation of an acylium ion ($[CH_3C{\equiv}O]^+$) and (2) the formation of protonated ethenone ($[H_2C{=}C{=}O+H]^+$). The labeling studies also show that the O—H and the S—H bonds, obtained by protonation, are particularly strong. The fragmentation of

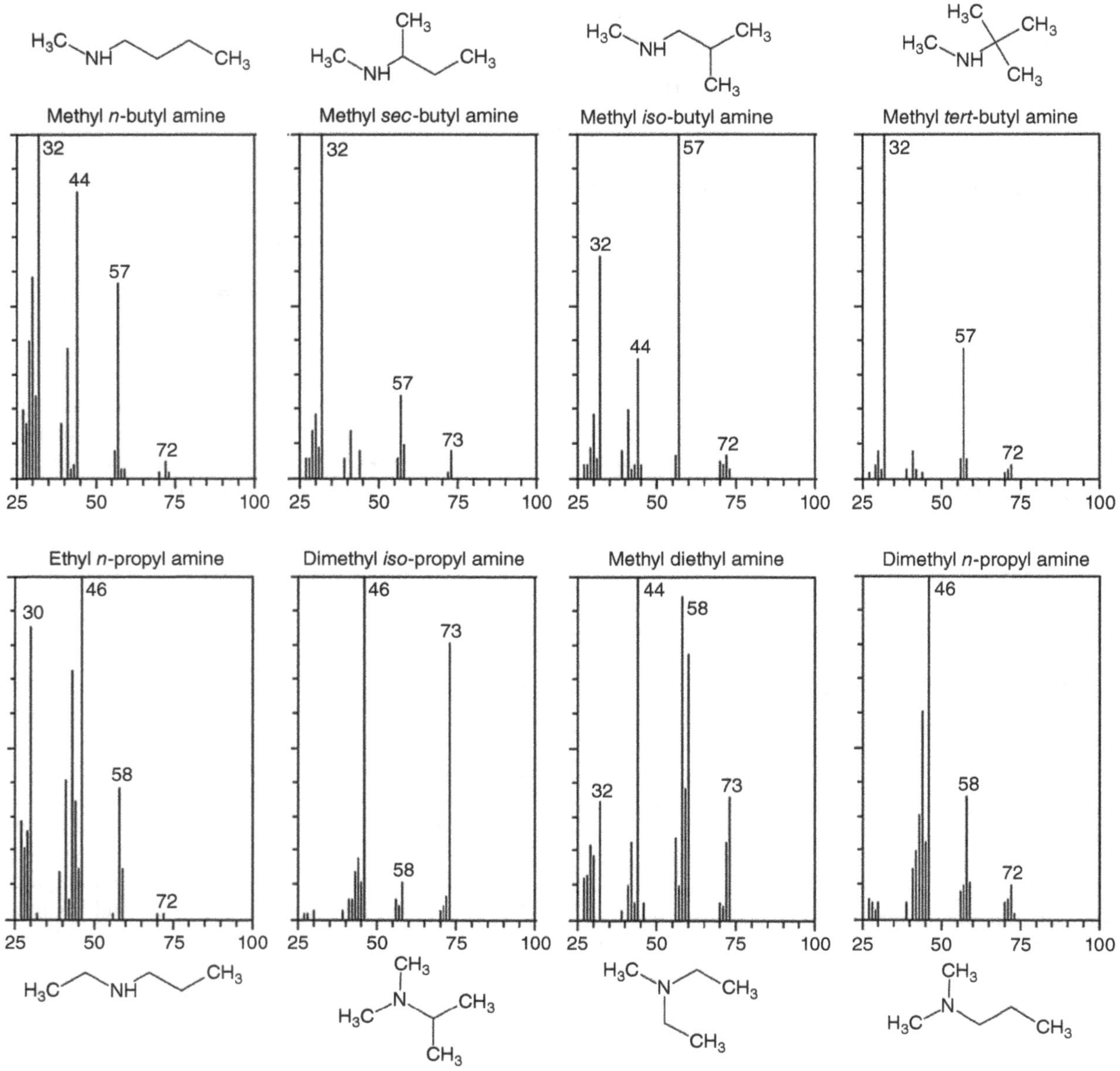

FIGURE 3.5 Product-ion mass spectra for eight protonated C_5-dialkyl- and trialkylamines. (Source: Reiner et al., 1989. Reproduced with permission of Canadian Science Publishing.)

3-pentanone results in protonated prop-1-en-1-one with *m/z* 57 ($[CH_3CH{=}C{=}O{+}H]^+$) due to C_2H_6 elimination. The two most abundant fragment ions of protonated diethyl ether ($[(C_2H_5)_2O{+}H]^+$; $[M{+}H]^+$ with *m/z* 75) and diethyl thioether ($[(C_2H_5)_2S{+}H]^+$; $[M{+}H]^+$ with *m/z* 91) are the protonated ethanol ($[C_2H_5OH{+}H]^+$) with *m/z* 47 and protonated ethanethiol ($[C_2H_5SH{+}H]^+$) with *m/z* 63, both due to the loss of C_2H_4, and $[C_2H_5]^+$ with *m/z* 29. The elimination of C_2H_4 is suggested to occur via a four-center transition state. For both compounds, substantial loss of C_2H_6 is also observed. Protonated diethyl thioether also shows loss of an ethyl radical ($^{\bullet}C_2H_5$) to an ion with *m/z* 62 ($[C_2H_5SH]^{+\bullet}$). The major fragment ion of protonated ethyl bromide ($[M{+}H]^+$ with *m/z* 109) and ethyl iodide ($[M{+}H]^+$ with *m/z* 157) is $[C_2H_5]^+$ with *m/z* 29 due to loss of HBr and HI, respectively. The fragment ions from an $EE^+ \rightarrow OE^{+\bullet}$ reaction involving the loss of $^{\bullet}C_2H_5$ to ions with *m/z* 80 ($[HBr]^{+\bullet}$) or *m/z* 128 ($[HI]^{+\bullet}$) are also observed (Wagner et al., 1980).

To achieve a better insight into the mechanistic aspects of EE^+ fragmentation, a careful evaluation of literature data and additional experiments has been performed to investigate the occurrence of McLafferty rearrangements

TABLE 3.1 Fragmentation (expressed as percentage relative abundance, %RA) of protonated C_5-dialkyl- and trialkylamines ($[C_5H_{13}N+H]^+$; $[M+H]^+$ with *m/z* 88).

			Loss of Alkene			Loss of Alkane		
	$-CH_3^\bullet$	$-CH_5N$	$-C_2H_4$	$-C_3H_6$	$-C_4H_8$	$-C_2H_6$	$-C_3H_8$	$-C_4H_{10}$
m/z	73	57	60	46	32	58	44	30
Me-*n*Bu-NH	*	57			100	*	84	59
Me-*s*Bu-NH	*	24			100	10	*	19
Me-*i*Bu-NH	*	100			63		35	19
Me-*t*Bu-NH		38			100	*	*	*
Et-*n*Pr-NH				100	*	39	35	86
DiMe-*i*Pr-N	81	*		100		11	18	*
Me-diEt-N	36	10	78	*	35	95	100	19
diMe-*n*Pr-N	*	10		100		36	61	*
ID	$C_4H_{11}N^{+\bullet}$	$C_4H_9^+$	$C_3H_{10}N^+$	$C_2H_8N^+$	CH_6N^+	$C_3H_8N^+$	$C_2H_6N^+$	CH_4N^+

*%RA < 10.

in EE^+ (Zollinger & Seibl, 1985). A McLafferty type γH-rearrangement has been suggested for the alkene loss observed in alkyl iminium ions, produced by α-cleavage of $M^{+\bullet}$ in tertiary amines (Figure 3.6) (Djerassi & Fenselau, 1965). However, this has been questioned by others (Veith, 1983). Studies involving a variety of alkene losses in EE^+ indicate that a McLafferty-type rearrangement can only be proven unambiguously in a limited number of cases, for example, for 2-hexyloctanoic acid. In general, it seems to be only a minor fragmentation pathway (Zollinger & Seibl, 1985). As an explanation, it has been suggested that protonation of a functional group diminishes the number of electrons available for further bond formation, and thus reduces the ability to function as a proton acceptor, as is required in a McLafferty rearrangement (Field, 1972). It has been suggested that two lone electron pair-bearing heteroatoms should be present for the H-rearrangement to take place in an EE^+, as in esters for instance, but this is probably an incomplete explanation because it would rule out a H-rearrangement in amines, where the N-atom bears only one lone electron pair. No conclusive evidence is available on the specificity of the H-rearrangement in such cases (Benoit & Harrison, 1978; Sigsby et al., 1979; Kingston et al., 1983).

The data discussed in this section for $[M+H]^+$ generated by proton-transfer ion–molecule reactions in CI were obtained using mostly high-energy CID in sector instruments. Their applicability in describing and understanding the fragmentation reactions in low-energy collision-cell CID and ion-trap CID may be somewhat limited. It is for instance well established that direct bond cleavages tend to prevail over the rearrangement reactions at higher-energy collisions (McLafferty et al., 1973). In fact, some fragmentation reactions are hardly observed in low-energy CID, for example, losses of an alkane from a secondary or a tertiary amine; the loss of an alkene tends to be favored.

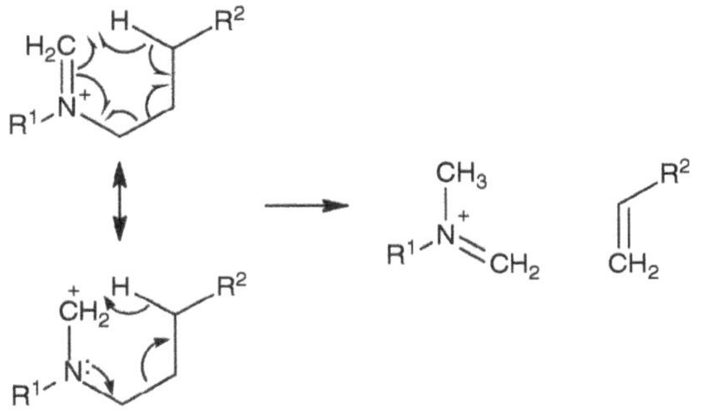

FIGURE 3.6 Two fragmentation mechanisms for an alkene loss from alkyl iminium ions via a McLafferty-type γH-rearrangement. As the top mechanism involving homolytic cleavages (Djerassi & Fenselau, 1965) is a violation of the even-electron rule, an alternative mechanism involving heterolytic cleavages is shown at the bottom.

3.5 FRAGMENTATION OF PROTONATED MOLECULES

In the previous section, fragmentation of $[M+H]^+$ generated by proton-transfer chemical ionization was discussed. In relation to ESI-MS–MS, few systematic studies have been performed on the fragmentation of $[M+H]^+$ (Levsen et al., 2007; Holčapek et al., 2010; Niessen, 2010; Weissberg & Dagan, 2011; Niessen, 2011; Weissberg et al., 2016). Monographs or comprehensive reviews are hardly available, except for the fragmentation of biomolecules (Ham, 2008). In the limited number of mechanistic studies reported, it is often found that, in contrast to EI fragmentation, even a relatively simple neutral loss involves multiple fragmentation pathways. Fragmentation of (mainly) ESI-generated EE^+, that is, $[M+H]^+$, and fragments thereof are discussed in this section. The fragmentation of EE^+ usually complies with the

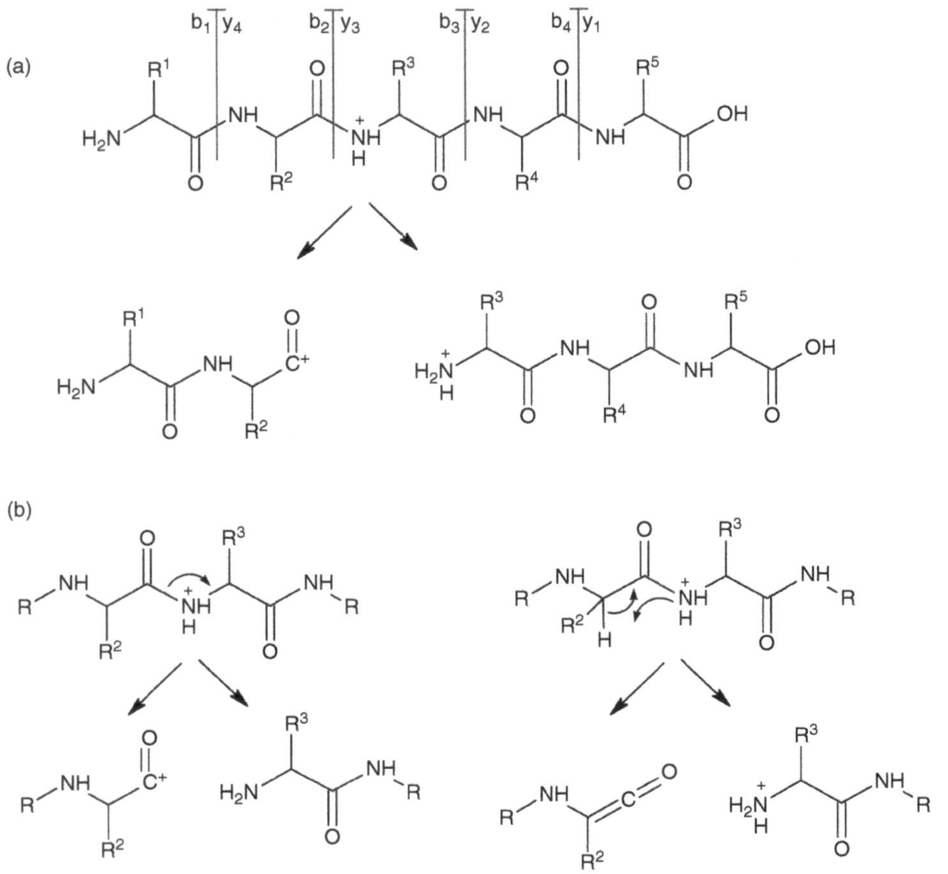

FIGURE 3.7 Fragmentation of $[M+H]^+$ of a peptide, with (a) cleavage of the peptide bond to generate either an acylium ion or a (shorter) protonated peptide, and (b) simplified mechanism, assuming an inductive cleavage to generate an acylium ion and an acyclic EE^+ cleavage (H-rearrangement) to generate the (shorter) protonated peptide.

even-electron rule (Karni & Mandelbaum, 1980). Violation of the even-electron rule is observed in specific cases, as discussed further in Section 3.5.7. Frequent reference is made to the detailed information on the fragmentation of drugs, pesticides, and related small molecules presented in the data collection of Chapter 4.

3.5.1 Singly-Charged Peptides

Probably, the most extensively studied fragmentation of EE^+ is that of protonated peptides (Hunt et al., 1986; Johnson et al., 1988; Biemann, 1988; Papayannopoulos, 1995; Paizs & Suhai, 2005; Mouls et al., 2007). The discussion in this section is restricted to the fragmentation of singly-charged peptides.

Protonation of a peptide occurs at the N-terminal amine group and/or at the basic amino acids Lys, His, and Arg (Smith et al., 1990; Smith et al., 1991). However, in order to describe the fragmentation from a charge-localized, charge-directed perspective, it is easier to consider the proton migrating along the peptide backbone prior to the fragmentation (mobile proton hypothesis, see references in Lee et al., 1998; Boyd & Somogyi, 2010). In low-energy CID, the primary cleavage takes place at the peptide (amide) bond. Assuming that the charge can go on either of the two resulting fragments, either a protonated (shorter) peptide with the loss of a ketene or an acylium ion with the loss of a (shorter) peptide is observed (Figure 3.7a). The protonated (shorter) peptide results from a H-rearrangement with charge retention on the N-atom, whereas charge migration leads to the acylium ion. In fact, since the cleavage can in principle occur at any of the peptide bonds, a series of both protonated peptides and acylium ions will be observed. From a simplistic mechanistic point of view, the fragmentation involves an inductive cleavage resulting in the acylium ion and an (acyclic) backbone cleavage via a four-center H-rearrangement resulting in the protonated (shorter) peptide (Figure 3.7b) (Johnson et al., 1988). As a result, the fragmentation at each peptide

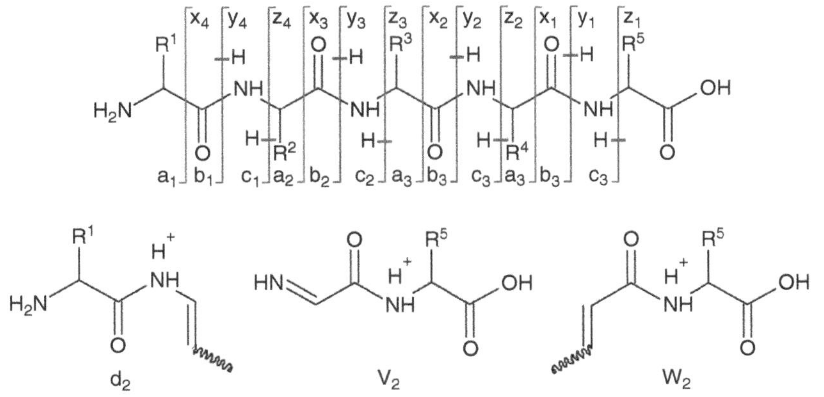

FIGURE 3.8 Nomenclature of peptide sequence ions. The N-terminal sequence ions a, b, and c and the C-terminal ions x, y, and z are due to backbone cleavages. Substituent-chain cleavage to ions d, v, and w only occurs in high-energy collisions and with specific amino acids, which helps to determine the identity of the peptide.

bond would lead to two complementary ions characterized by the sum of the *m/z* values of the two fragment ions being equal to *m/z* ([M+H]+1).

Based on the observed backbone cleavages in product-ion mass spectra of FAB-generated protonated peptides, nomenclature rules have been proposed to annotate the peptide fragment ions (Roepstorff & Fohlmann, 1984; Biemann, 1988). The general rules are outlined in Figure 3.8. The a-, b-, c-, and d-ions contain the N-terminal and are numbered from the N-terminal side of the peptide. The w-, x-, y-, and z-ions contain the C-terminal and are numbered from the C-terminal side of the peptide. Thus, for low-energy CID discussed earlier, the protonated (shorter) peptides are y-ions (or according to the initial nomenclature rule, y"-ions) and the acylium ions are b-ions. In low-energy peptide spectra, a-ions may also be observed. They may be due to secondary fragmentation involving the loss of carbon monoxide (CO) from the respective b-ions. Other ions defined by the nomenclature rules, such as the backbone cleavage ions, that is, the c-, x-, and z-ions, as well as the ions resulting from side-chain cleavages, that is, the d-, v-, and w-ions, are not observed under low-energy CID conditions. They may be observed under high-energy CID conditions. The occurrence of c- and $z^{\bullet}$-ions has been reported as the major fragment ions in electron-capture dissociation and electron-transfer dissociation (ECD and ETD, respectively) of multiply-charged peptides (Zhurov et al., 2013).

Although this description of the fragmentation of a protonated peptide may serve for simple cases, detailed mechanistic studies have shown that the actual fragmentation is more complex, as discussed in Section 3.5.3.

3.5.2 Protonated Small Molecules: Complementary Fragment Ions

In the previous sections, a general concept was introduced to rationalize the fragmentation of protonated molecules as a backbone cleavage via either an inductive cleavage or a four-center H-rearrangement (acyclic EE^+ cleavage). It involves cleavages of the C—heteroatom (N, O, S) bond, either by inductive cleavage (with charge migration to the α-C-atom) (Eq. 3.6) or involves a H-rearrangement (with charge retention on the heteroatom) (Figure 3.4 and Eq. 3.7). These two reactions may be considered as competing reactions. The relative abundance of the two fragments is primarily determined by Field's rule, indicating that charge retention on the fragment with the higher proton affinity is more favorable (Field, 1972). In addition, their relative abundance is also influenced by their relative stability. A characteristic feature of these two complementary fragments is that the sum of their *m/z* values equals *m/z* ([M+H]+1). The "+1" results from the H-rearrangement in the acyclic EE^+ cleavage; that particular H-atom is present in both fragment ions. The characteristic EE^+ fragmentation patterns for esters, secondary amines, ethers, and amides are presented in Figure 3.9. The proton affinities for a number of representative compounds are given in Table 3.2, enabling one to estimate the proton affinity of other small molecules. More extensive tables are available elsewhere (Lias et al., 1984; Lias et al., 1988; Hunter & Lias, 1998). Similar to peptide fragmentation, the mobile proton hypothesis seems to play an important role in small-molecule fragmentation (Wright et al., 2016).

As already discussed in Section 3.4, there is considerable discussion on the actual mechanisms for the fragmentation, for example, the site specificity of the H-rearrangement, whether the fragmentation occurs via a concerted mechanism, or should be rather described in terms of a non-concerted mechanism via the formation of a ion–molecule or ion–dipole complex (Figure 3.4) (Bowen et al., 1978; Morton, 1980; Julian et al., 2007). Despite the lack of mechanistic clarity, the fragmentation mechanisms outlined here are important in interpreting not only the fragmentation of peptides, but also that of EE^+ of many

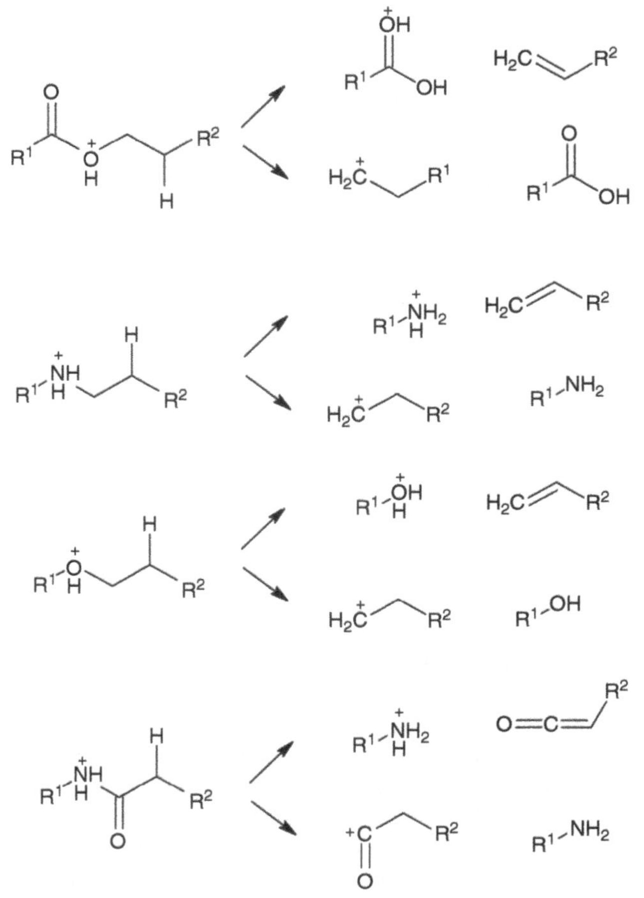

FIGURE 3.9 Characteristic EE^+ fragmentation of esters, secondary amines, ethers, and amides. For ethers and secondary amines, only the results from the cleavages of the C—O and C—N bonds, respectively, of the heteroatom—R^2-moiety are shown; the results from the cleavage in the heteroatom—R^1-moiety are not shown. For each compound class, the top reaction involves a four-center H-rearrangement (acyclic EE^+ cleavage) and the bottom reaction involves an inductive cleavage.

other compound classes. The fragmentation patterns of protonated timolol (Section 4.1.1) and metformin (Section 4.4.1.3) are illustrative in this respect. For both compounds, three pairs of complementary ions are observed.

The general fragmentation pattern for $[M+H]^+$ can be readily illustrated with the fragmentation of protonated bromhexine ($[M+H]^+$ with *m/z* 375). In this particular case, the precursor ion of the isotopologue with $^{79}Br^{81}Br$ was selected, thus $[M+H]^+$ with *m/z* 377 (Figure 3.10). The two major fragments, the ions with *m/z* 114 and 264, are due to the cleavage of the C—N bond between the substituted benzyl group and the tertiary amino group. The *m/z* values for the two fragments add up to 378, which is equal to *m/z* ([M+H]+1). Cleavage of the cyclohexyl C—N bond may explain the cyclohexyl carbocation ($[C_6H_{11}]^+$) with *m/z* 83, but because the complementary fragment ion with *m/z* 295 is missing, the fragment with *m/z* 83 is most likely due to a secondary fragmentation of the fragment ion with *m/z* 114 involving the loss of methylamine (CH_3NH_2). Because the proton affinity of the protonated 2,4-dibromo-6-[(methyl)amino)methyl]aniline fragment with *m/z* 295 is higher than that of the cyclohexene, its occurrence in the spectrum would certainly be expected. Other fragments in the bromhexine spectrum are discussed in Section 3.5.7.

In a conceptual way, the fragmentation of $[M+H]^+$ can be considered as the cleavage of a C—heteroatom bond resulting in two fragment molecules, for example, a primary amine and an alkene from the fragmentation of a secondary amine, which compete for the proton (H^+) charge. In the end, the charge (H^+) resides on the fragment molecule with highest proton affinity. If both fragment molecules have comparable proton affinities, both protonated fragment molecules can be observed. Thus, both fragment ions can be considered as protonated (fragment) molecules. The concept can be illustrated for protonated bromhexine, just discussed. The cleavage of the C—N bond between the substituted benzyl group and the tertiary amino group leads to 2,4-dibromo-6-methylidenecyclohexa-2,4-dien-1-imine and *N*-methylcyclohexanamine, both having comparable proton affinities, so both fragments are observed as ions, that is, the ions with *m/z* 264 (protonated 2,4-dibromo-6-methyl idenecyclohexa-2,4-dien-1-imine ($[C_7H_5{}^{79}Br^{81}BrN+H]^+$) and *m/z* 114 (protonated *N*-methylcyclohexanamine, $[C_6H_{11}NHCH_3+H]^+$) in Figure 3.10. The ion with *m/z* 264 is chemically more adequately described as the 2-amino-3,5-dibromobenzyl cation ($[^{79}Br^{81}BrNH_2C_6H_2CH_2]^+$). The cleavage of the cyclohexyl C—N bond would lead to *N*-(2-amino-3,5-dibromobenzyl)methylamine and cyclohexene. Because the proton affinity of cyclohexene is low compared to its counterpart and the protonated *N*-(2-amino-3,5-dibromobenzyl)methylamine ($[C_8H_{10}{}^{79}Br^{81}BrN_2+H]^+$) is not observed, rather the occurrence of $[C_6H_{11}]^+$ with *m/z* 83 is described as a secondary fragmentation of $[C_6H_{11}NHCH_3+H]^+$ with *m/z* 114 instead. In doing so, one would also expect to observe the protonated methylamine ($[CH_3NH_2+H]^+$) with *m/z* 32 (outside the *m/z* range measured). Finally, formation of *N*-desmethyl-bromhexine due to methyl C—N bond cleavage is not observed either, because the loss of methylene ($:CH_2$) does not occur in EE^+ fragmentation.

This fragmentation behavior can also be illustrated with the product-ion mass spectra of three esters (Figure 3.11). *n*-Butyl-4-hydroxybenzoate ($[M+H]^+$ with *m/z* 195) shows a fragment ion with *m/z* 139 due to the loss of butene (C_4H_8), but the complementary fragment with *m/z* 57 ($[C_4H_9]^+$) is not observed because the proton affinity of the protonated 4-hydroxybenzoic acid fragment outweighs that of

TABLE 3.2 Proton affinities (PA in kJ mol^{-1}) for a number of compounds.

Compound	PA (kJ mol^{-1})	Compound	PA (kJ mol^{-1})
Methane (CH_4)	552	Methyl acetate (CH_3COOCH_3)	828
Ethyne (HC≡CH)	641	Ethenone (H_2C=C=O)	830
Ethene (C_2H_4)	680	Diethyl ether ($C_2H_5OC_2H_5$)	838
Water (H_2O)	697	Ammonia (NH_3)	854
Hydrogen sulfide (H_2S)	712	Aniline ($C_6H_5NH_2$)	877
Formaldehyde (H_2C=O)	718	Methylamine (CH_3NH_2)	896
Propene (CH_3CH=CH_2)	752	Alanine ($CH_3CH(NH_2)COOH$)	899
Benzene (C_6H_6)	759	Ethyl amine ($CH_3CH_2NH_2$)	908
Methanol (CH_3OH)	761	Dimethylamine ($(CH_3)_2NH$)	923
Ethanol (C_2H_5OH)	788	Pyridine (C_5H_5N)	924
Acetonitrile (CH_3C≡N)	788	Dimethyl aniline ($C_6H_5N(CH_3)_2$)	935
Toluene ($C_6H_5CH_3$)	794	Trimethylamine ($(CH_3)_3N$)	942
Ethyl formate ($HCOOC_2H_5$)	808	Piperidine ($C_5H_{11}N$)	947
iso-Butene ($(CH_3)_2C$=CH_2)	820	Quinoline (C_9H_7N)	948
Acetone (CH_3COCH_3)	823	Triethylamine ($(C_2H_5)_3N$)	972

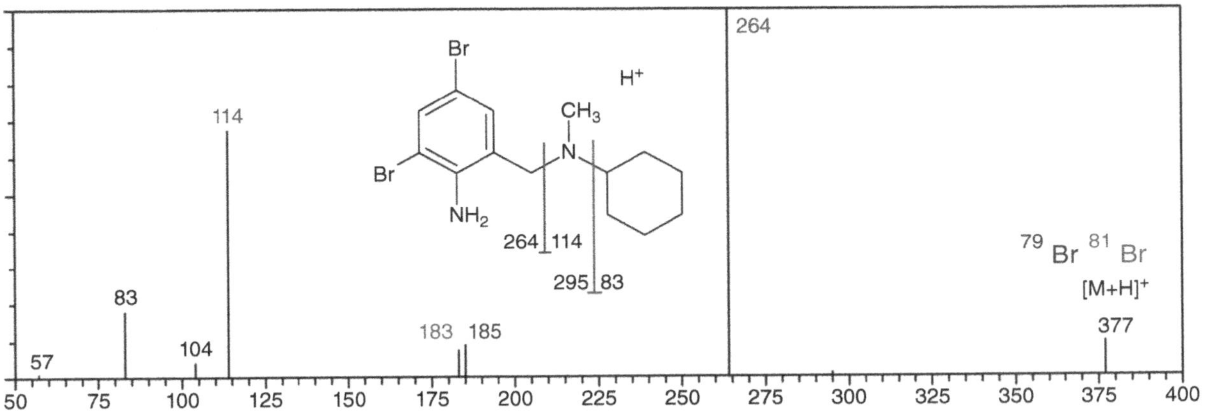

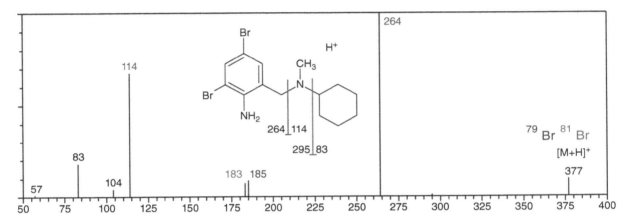

FIGURE 3.10 Product-ion mass spectrum of bromhexine ([M+H]$^+$ with *m/z* 375; the ion of the isotopologue with ^{79}Br,^{81}Br with *m/z* 377 was selected as precursor ion.

C_4H_8. The same argument holds for the loss of ethylene (C_2H_4) from meperidine ([M+H]$^+$ with *m/z* 248). In anisodamine ([M+H]$^+$ with *m/z* 306) (Chen et al., 2005), on the other hand, the cleavage of the ester C—O bond results in the loss of neutral tropic acid and charge retention on 6-hydroxytropine, that is, the ion with *m/z* 140 ([$C_8H_{14}NO$]$^+$). This is due to the high proton affinity of the tertiary amine group. Other aspects of these data are discussed in more detail in Sections 3.5.5 and 3.6.4.

Complementary fragments are also important in the fragmentation of, for instance, β-blockers (Section 4.1.1); phenothiazines (Section 4.2.1); antidiabetic drugs (Section 4.3.1); sulfonamide, β-lactam, and aminoglycoside antibiotics (Sections 4.8.1, 4.8.3, and 4.8.5); antimalarial drugs (Section 4.10.2.4); antiviral drugs (Section 4.10.3); organophosphorus pesticides (Section 4.11.4); sulfonylurea herbicides (Section 4.11.6); as well as many individual examples provided in the data collection in Chapter 4. The same fragmentation is also observed for peptides (Section 3.5.1) and for oligosaccharides, resulting in B- and Y-ions (Domon & Costello, 1988; Ham, 2008; Zaia, 2004).

At this stage of the discussion, it is appropriate to compare fragmentation in EI-MS and ESI-MS–MS. The amine fragmentation in the herbicide propazine (M$^{+\bullet}$ with *m/z* 229; [M+H]$^+$ with *m/z* 230; Section 4.11.1) can serve as an example (Figure 3.12). In EI-MS, a homolytic cleavage of M$^{+\bullet}$ leads to an α,β-C—C cleavage in the *iso*-propyl substituent, involving the loss of $^\bullet CH_3$ and resulting in an EE$^+$ fragment with *m/z* 214. In contrast, the primary cleavage in [M+H]$^+$ involves the cleavage of the *iso*-propyl C—N amine bond, along with a H-rearrangement to the amine N-atom, resulting in the corresponding protonated des-*iso*-propyl propazine. This loss of propene (CH_3CH=CH_2) leads to an EE$^+$ fragment with *m/z* 188. Some other fragment ions in the EI-MS spectrum attract attention as well. In fact, losses of CH_3CH=CH_2 are observed both from the M$^{+\bullet}$, possibly involving a McLafferty rearrangement with the N^3-atom of the triazine resulting in an OE$^{+\bullet}$ fragment with *m/z* 187,

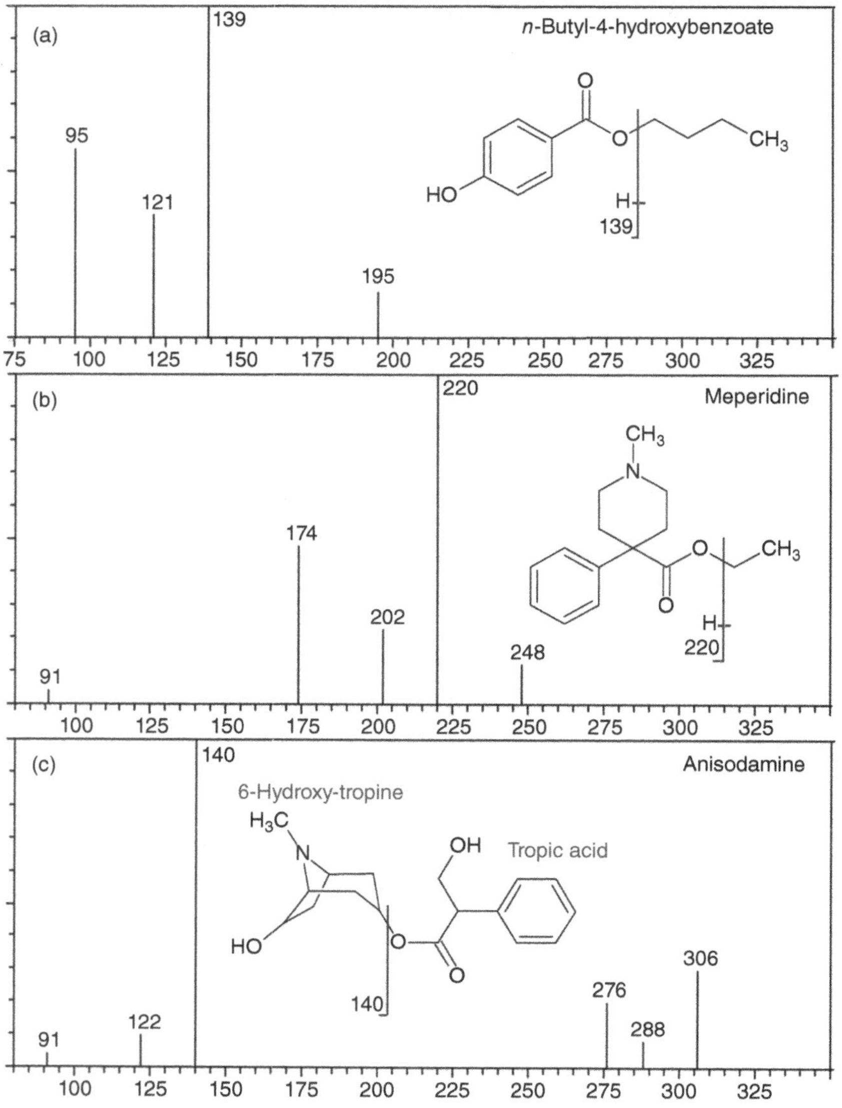

FIGURE 3.11 Product-ion mass spectra of (a) *n*-butyl 4-hydroxybenzoate, (b) meperidine, and (c) anisodamine. (Source: Data for anisodamine redrawn from (Chen et al., 2005) with permission from Elsevier.)

and in a secondary fragmentation from the EE^+ fragment with *m/z* 214, thus involving a H-rearrangement to the remaining amine N-atom resulting in an EE^+ with *m/z* 172 (Figure 3.12). Thus, the loss of the same group from either an $OE^{+\bullet}$ or an EE^+ is demonstrated in the same spectrum.

Several detailed studies have been published on the fragmentation mechanisms of some protonated molecules, using derivatization, isotope labeling, precursor-ion analysis experiments, accurate-mass determination, and other advanced MS methods. Some examples are the studies related to the fragmentation of testosterone (Section 4.6.3) (Williams et al., 1999), propranolol (Section 4.1.1) (Upthagrove et al., 1999), and polyamine spider toxins (Bigler & Hesse, 1995). In fact, one of these studies (Bigler & Hesse, 1995) indicated that the four-center mechanism, although useful in the rationalization of some fragmentation reactions, is actually not the correct mechanism for the fragmentation of *p*-hydroxycinnamoyl putrescine (*N*-(4-aminobutyl)-3-(4-hydroxyphenyl)prop-2-enamide). [D]-labeling experiments showed that the fragmentation in this case can be better explained by neighboring-group participation (Figure 3.13).

3.5.3 Fragmentation of Peptides Revisited

As indicated earlier (Section 3.5.1), mechanistic studies on the fragmentation of protonated peptides have shown that the

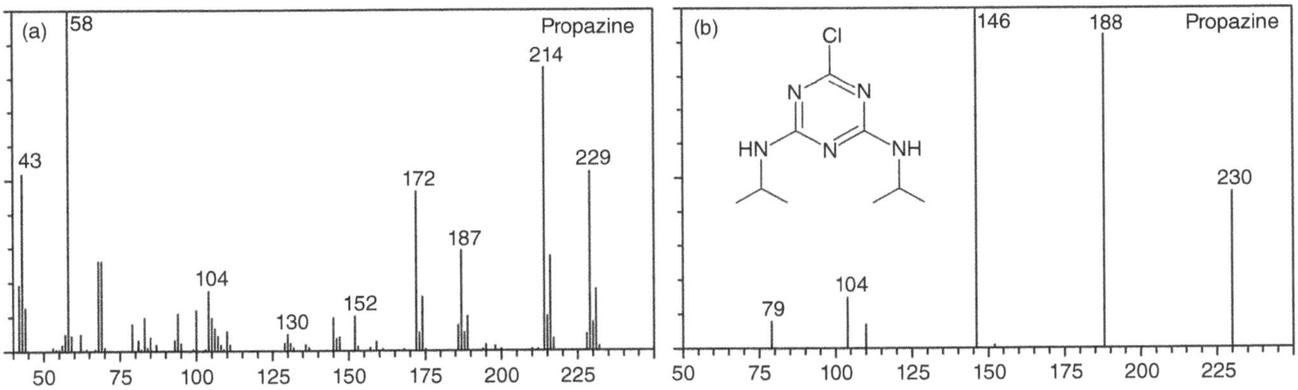

FIGURE 3.12 Mass spectra of the triazine herbicide propazine: (a) EI-MS spectrum, and (b) ESI-MS–MS spectrum.

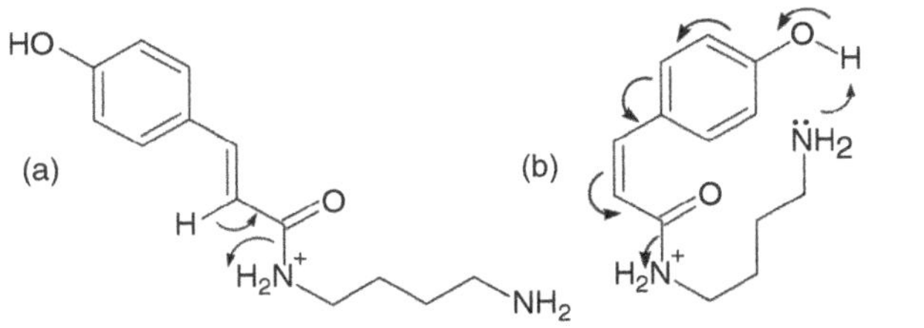

FIGURE 3.13 Fragmentation mechanisms for *p*-hydroxycinnamoyl putrescine (*N*-(4-aminobutyl)-3-(4-hydroxy phenyl)prop-2-enamide), with (a) general mechanism according to the four-center cleavage with H-rearrangement and (b) proposed mechanism involving neighboring-group participation. (Source: Bigler & Hesse, 1995. Reproduced with permission of Springer.)

actual fragmentation mechanism may be more intricate than discussed so far from the concept of a backbone cleavage via either a four-center H-rearrangement or an inductive cleavage to generate a protonated (shorter) peptide (charge retention and H-rearrangement) and an acylium ion (charge migration), respectively.

The fragmentation of a peptide at an amide bond starts with the migration of the proton to the site to be cleaved (mobile proton hypothesis, see references in Lee et al., (1998) and Boyd & Somogyi (2010)). Protonation of the amide N-atom weakens the amide bond and makes the carbonyl C-atom more electrophilic and thus a better target to a nucleophilic attack, for example, by the oxygen of the neighboring amino acid at the N-terminal side. This results in a protonated 5(4*H*)-oxazolone intermediate (Yalcin et al., 1995; Yalcin et al., 1996; Ambihapathy et al., 1997; Paizs & Suhai, 2005). From this intermediate, either the C-terminal fragment (a peptide) or the oxazolone intermediate is expelled as a neutral, thus resulting in the protonated oxazolone (b-ion) or the protonated peptide (y-ion), respectively (Figure 3.14). Field's rule applies in determining the relative abundance of the two fragments (Paizs & Suhai, 2005). The formation of the oxazolone intermediate also explains why the b_1-ion is frequently missing in product-ion mass spectra of peptides.

In addition to the fragmentation mechanism involving the oxazolone intermediate, a variety of other fragmentation pathways of peptides have been described, including the formation of diketopiperazine intermediates, fragmentation via amide–oxygen and aziridinone pathways, and fragmentation via a nucleophilic attack from the His, Gln, Asn, Lys, or Arg side chains (Paizs & Suhai, 2005).

Under ESI conditions, many peptides form doubly-charged or even multiply-charged ions. Protonation occurs at the N-terminal amine group and at the basic amino acids Lys, His, and Arg (Smith et al., 1990; Smith et al., 1991). Many peptides analyzed by LC–MS, for example, the peptides currently analyzed in proteomics strategies, result from tryptic digestion of proteins. Tryptic peptides are characterized by the presence of the basic amino acids Arg or Lys at the C-terminal. Thus, they typically generate doubly-charged ions, that is, protonation at the N-terminal and at the Arg or Lys of the C-terminal. The general description of the fragmentation pathway of these doubly-protonated peptides is very similar to that of singly-protonated peptides (Section 3.5.1). The proton from the N-terminal amine is mobile, whereas the other proton remains fixed at the C-terminal Arg or Lys. Following the fragmentation reactions outlined earlier and applying them to a doubly-protonated tryptic peptide, the reaction involving charge retention and H-rearrangement results in a doubly-protonated (shorter) peptide and a neutral oxazolone, whereas the charge migration reaction results in the singly-protonated oxazolone and a singly-protonated (shorter) peptide (Paizs & Suhai, 2005).

Whether singly-, doubly-, or even multiply-protonated peptides are fragmented, the cleavage reactions described typically result in two series of fragment ions, the so-called sequence ions, that is, the N-terminal b-ions and the C-terminal y-ions. From these two series, the amino-acid

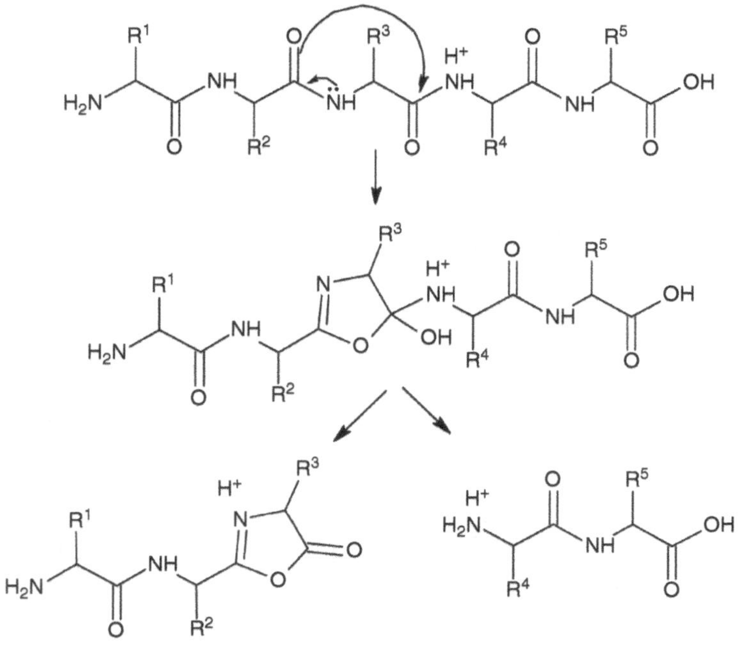

FIGURE 3.14 Fragmentation of $[M+H]^+$ of peptides via an oxazolone intermediate, generating a protonated oxazolone and a (shorter) protonated peptide. Illustration of mobile-proton hypothesis.

sequence of a peptide may be derived. Unfortunately, the interpretation of the product-ion mass spectrum may be complicated by losses of small neutrals from the sequence ions, for example, losses of ammonia (NH_3), water, carbon monoxide (CO), hydrogen sulfide (H_2S), and methanethiol (CH_3SH); by the occurrence of internal cleavage ions, such as b-ions resulting from secondary fragmentation of highly abundant y-ions, for example, protonated (shorter) peptides with Pro at their N-terminal side; and also by missing sequence ions, for example, when the cleavage occurs at the N-terminal side of Gly. Iminium ions ($[H_2N{=}CHR]^+$) with low *m/z* values, for example, *m/z* 86 for Leu and Ile and *m/z* 120 for Phe, often provide information on the amino-acid composition of the peptide. Interpretation strategies for product-ion mass spectra of peptides have been proposed (Hunt et al., 1986; Biemann, 1988; Papayannopoulos, 1995; Kinter & Sherman, 2000).

It should be mentioned that current high-throughput protein identification strategies in proteomics studies do not actually rely on the real interpretation of peptide product-ion mass spectra, but rather on bioinformatics tools and database searching strategies (Yates et al., 1995; Jensen et al., 1997; Yates, 1998; Zhang et al., 2013). The bioinformatics workflow relies on the availability of protein databases, eventually derived from DNA or genomic databases. Experimentally, a protein or a mixture of proteins, for example, a complete proteome, is digested by means of a protease like trypsin into a mixture of peptides. Given the known selectivity of the protease used, *in silico* digestion of all proteins in a relevant protein database is performed by the bioinformatics tool to predict the peptides. For each of these peptides, its mass or *m/z* value, amino-acid sequence, and protein ID are stored. Experimentally, MALDI-TOF-MS or LC–MS can be used to generate the so-called peptide mass fingerprint or the peptide map, which can be considered to be a list of *m/z* values of the tryptic peptides observed in the analysis of the digest. This list of experimental *m/z* values can be scored against the list of *in silico* generated *m/z* values. In most cases, the scoring algorithm converges to a particular protein from the database (or a list of proteins, when a mixture of proteins, for example, a complete proteome, was digested). Thus, a number of peptides in the peptide map correspond to partial sequences of a particular protein in the database. The software actually provides scoring values for the likeliness that a particular experimental peptide map is derived from a particular protein in the database. In this way, the peptide map may lead to protein identification. When HRAM-MS data are available, a surprisingly small number of peptides is needed to achieve protein identification. This workflow for protein identification can be extended and become even more powerful if additionally experimental MS–MS or MS^n data are available for the peptides in the peptide map. The information for each peptide in the *in silico* generated peptide list (mass or *m/z*, amino-acid sequence, and protein ID) can be readily complemented by predicted *m/z* values of the sequence ions of each of these peptides. Next, the *m/z* values for the fragment ions in the experimental product-ion mass spectra of all fragmented peptides can be cross-correlated with this information to achieve even more reliable protein identification (Yates et al., 1995; Jensen et al., 1997; Yates, 1998; Kinter & Sherman, 2000; Zhang et al., 2013). Such proteomics strategies and workflows have been implemented in various research fields, for example, directed toward the understanding of cell functions and cellular signaling cascades, the discovery of new drugs and (diagnostic) protein biomarkers, as well as the development and characterization of protein biopharmaceuticals (Liebler, 2002).

In the context of this chapter, it is interesting to note that careful interpretation of peptide product-ion mass spectra has shown that even for straightforward amino-acid sequencing of peptides, it can be made intricate by the complex chemistry of EE^+ fragmentation. It has been found that a nucleophilic attack within a b-ion may result in a cyclic peptide, which subsequently can reopen randomly, for example, at a different site. Upon further fragmentation of this "new" b-ion, sequence ions are generated that are no longer consistent with the initial amino-acid sequence of the peptide, but in fact lead to a scrambled amino-acid sequence and inconsistencies in data interpretation (Harrison et al., 2006; Jia et al., 2007; Bleiholder et al., 2008; Harrison, 2009).

3.5.4 Direct-Cleavage Reactions

The fragmentation of protonated molecules can often be rationalized as a backbone cleavage via either an acyclic EE^+ cleavage with a four-center H-rearrangement or an inductive cleavage, leading to two complementary ions, where the sum of their *m/z* is equal to *m/z* ([M+H]+1) (Eqs 3.6 and 3.7, and Figure 3.9). Numerous examples of this fragmentation behavior are discussed in Chapter 4. However, in a limited number of cases, a backbone cleavage leads to two ions with *m/z* values that add up to *m/z* ([M+H]−1). That means that both fragment ions are formed by an inductive cleavage and H-rearrangement does not occur. In order to distinguish this fragmentation behavior from the fragmentation reactions leading to the two complementary ions, it is called a direct-cleavage reaction. In some cases, the preference for a direct cleavage is mandated by structural features, for example, which prevent a H-rearrangement. The direct cleavage is observed for instance in protonated diaryl ketones ($[Ar^1COAr^2{+}H]^+$), for example, benziodarone (Section 4.1.7.2) and the naphthoylindole synthetic cannabinoids (JWH compounds, Section 4.7.7.1), according to Eq. 3.8:

$$[Ar^1COAr^2{+}H]^+ \rightarrow [Ar^1]^+ + Ar^2CHO \qquad (3.8a)$$

$$[Ar^1COAr^2{+}H]^+ \rightarrow Ar^1H + [Ar^2C{=}O]^+ \qquad (3.8b)$$

Direct-cleavage reactions are also observed in the fragmentation of phenylurea herbicides (Section 4.11.5.1), sulfonamides like sildenafil (Section 4.1.6.4), and sulfonamide antibiotics (Section 4.8.1). Backbone cleavages in deprotonated molecules also involve direct-cleavage reactions (Section 3.7.2).

3.5.5 Consecutive Small-Molecule Losses

Another useful approach is to describe the fragmentation of protonated molecules of many compound classes as a series of consecutive (and/or partly competing) small-molecule losses. This is certainly the case for ions with *m/z* close to that of $[M{+}H]^+$. For instance in the spectrum of anisodamine ($[M{+}H]^+$ with *m/z* 306) (Figure 3.11), losses of water and formaldehyde ($H_2C{=}O$) are observed, resulting in fragment ions with *m/z* 288 and 276, respectively. In some cases, a series of small molecules are lost in secondary fragmentation from an EE^+ fragment. Again, anisodamine can serve as an example (Figure 3.11). The fragment ions with *m/z* 122 and 91 result from consecutive losses of water and methylamine (CH_3NH_2) from the major fragment ion with *m/z* 140 ($[C_8H_{14}NO]^+$), discussed earlier (Section 3.5.2). Other examples are the fragment ions with low *m/z* in the product-ion mass spectra of dimethyl and diethyl esters of phosphates, phosphorothioates, and phosphorodithioates (Table 4.11.3). Approaching the interpretation of a product-ion mass spectrum in this way is obviously quite empirical and pragmatic, as no attention is paid to the underlying mechanistic issues of the fragmentation reactions.

For drugs, pesticides, and related compounds, numerous examples of (partial) interpretation of product-ion mass spectra based on consecutive small-molecule losses are discussed in Chapter 4. Some typical examples are the fragmentation of the high-ceiling loop diuretic drug torasemide (Section 4.1.4.3); the tetracyclic antidepressant mirtazapine (Section 4.2.3.2); the HMG-CoA reductase inhibitor simvastatin (Section 4.4.3.2); the anti-Alzheimer's drug galantamine (Section 4.5.4); the glucocorticosteroid beclomethasone (Section 4.6.6); the estrogen estrone (Section 4.6.7); the quinolone antibiotics nalidixic acid, oxolinic acid, piromidic acid, and pipemidic acid (Section 4.8.4); tetracycline antibiotics (Section 4.8.6); and the cyclohexanedione oxime herbicide alloxydim (Section 4.11.9.4).

The product-ion mass spectrum of the triazine herbicide propazine ($[M{+}H]^+$ with *m/z* 230) in Figure 3.12 can be interpreted from consecutive losses of propene ($CH_3CH{=}CH_2$) to a fragment ion with *m/z* 188; another $CH_3CH{=}CH_2$ to *m/z* 146; and either hydrogen chloride (HCl) to *m/z* 110, methanediimine (HN=C=NH) to *m/z* 104, or *N*-cyanomethanediimine (HN=C=NCN) to *m/z* 79. The fragmentation of the triazine herbicide atrazine is discussed in a similar way (Section 4.11.1.1 and Figure 4.11.1).

3.5.6 Other Fragmentation Reactions

In this section, several fragmentation reactions are discussed that show different patterns than what has been discussed so far. The fragmentation reactions are illustrated mainly with examples from the data collection in Chapter 4.

The RDA fragmentation, described for $OE^{+\bullet}$ in Section 3.3 and Figure 3.3, can also occur in EE^+, leading to two complementary fragment ions. It is one of four reaction types outlined in Section 3.4.1, that is, as a cyclic EE^+ cleavage with charge retention (McLafferty, 1980). The RDA fragmentation may take place in a six-membered homo- or heterocycle containing a double bond. An RDA fragmentation involves mainly the loss of ethene (C_2H_4) from a cyclohexene ring. Examples are discussed for the tetracyclic antidepressant maprotiline (Section 4.2.3.2), for the central nervous system stimulant methylphenidate (Section 4.2.8), and for the anti-Alzheimer's drug tacrine (Section 4.5.4).

The RDA fragmentation plays an important role in the structure elucidation of flavonoids, where it allows discrimination between isomeric forms with substitution on either of the two aromatic rings. Within the nomenclature system established for flavonoid fragmentation, outlined in Figure 3.15a (Ma et al., 1997), the RDA fragmentation in a flavonoid results in two characteristic (complementary) fragment ions, that is, the ^{1,3}A- and the ^{1,3}B-fragment (Figure 3.15b) (De Rijke et al., 2006; Justino et al., 2009), with a generally more abundant ^{1,3}A ion. For the isomeric 5,7-dimethoxycatechin and 3′,4′-dimethoxycatechin ($[M+H]^+$ with *m/z* 319), the characteristic RDA ^{1,3}A-fragment ions are found with *m/z* 167 and 139, respectively (Figure 3.15c) (Cren-Olivé et al., 2000).

In ethers and amines, apart from C—O or C—N bond cleavages via either a H-rearrangement or an inductive cleavage, one may observe an α,β-C—C bond cleavage, thus resulting in oxonium ($[RO{=}CH_2]^+$) or iminium ($[R_2N{=}CH_2]^+$) ions. The formation of iminium ions like the dimethyliminium ion ($[(CH_3)_2N{=}CH_2]^+$) with *m/z* 58 is frequently observed, for example, for phenothiazines with a 3-(*N*,*N*-dimethylamino)propyl substituent (Section 4.2.1), for tricyclic antidepressants (Section 4.2.3.1), and for histamine antagonists (Section 4.5.2). The cleavage of the α,β-C—C bond relative to the N-atom in a piperidine or the N^4-atom in a piperazine ring is also frequently observed, for example, for phenothiazines with a 3-(4-methylpiperazin-1-yl)propyl or a 3-[4-(2-hydroxyethyl)piperazin-1-yl]propyl substituent (Section 4.2.1), for urapidil (Section 4.1.6.2), and piperaquine (Section 4.10.2.4).

The fragmentation observed in a heterocyclic (aromatic) ring is often difficult to predict. Often, no ring fragmentation is observed and only losses of ring substituents

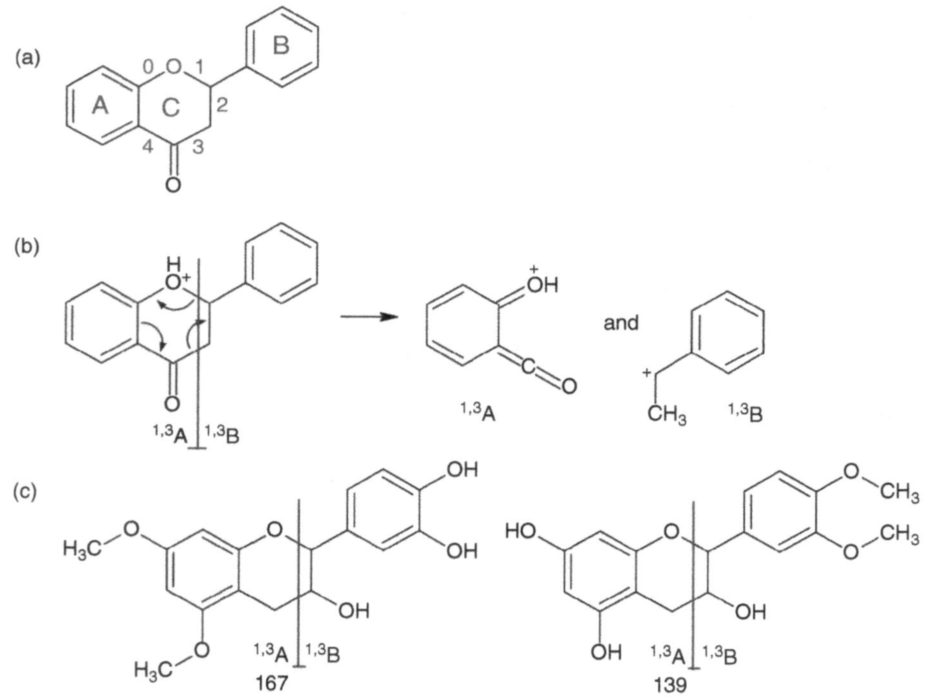

FIGURE 3.15 The retro-Diels–Alder (RDA) fragmentation illustrated for flavonoids, with (a) the nomenclature system for flavonoid fragment ions, (b) RDA fragmentation in flavonoids, resulting in two complementary fragment ions, ^{1,3}A and ^{1,3}B, and (c) RDA fragmentation illustrated for two isomeric dimethoxy-flavan-3-ols, that is, 5,7-dimethoxy-catechin and 3′,4′-dimethoxy-catechin ($[M+H]^+$ with *m/z* 319).

are found. Several examples are available of fragmentation in five-membered heterocyclic rings. For instance, in 2*H*-tetrazoles like sartans (Section 4.1.5), losses of N_2 are observed, resulting in a 3-substituted 1*H*-diazirine (R^1CHN_2), and of hydrogen azide (HN_3), resulting in a nitrile ($R^1C{\equiv}N$). For 1,2,4-oxadiazole, the loss of a 3-substituted 1*H*-diazirine (R^1CHN_2) results in an acylium ion $[R{-}C{=}O]^+$, as observed for butalamine (Section 4.1.6.3). For 5-methyl-1,2-oxazole, as in valdecoxib (Section 4.3.4), the loss of ethenone ($H_2C{=}C{=}O$) is observed. For compounds with a 1*H*-imidazole or a 1*H*-1,2,4-triazole ring, like for instance the imidazole and triazole fungicides (Sections 4.9.1 and 4.9.2), in most cases no fragmentation of the heterocyclic ring is observed, because cleavages at other sites of the molecule are more favorable. Ring fragmentation of the (six-membered) 1,3,5-triazine ring is discussed for triazine herbicides (Section 4.11.1).

3.5.7 Loss of Radicals from Even-Electron Ions

In the fragmentation of an EE^+, generally the even-electron rule applies (Karni & Mandelbaum, 1980; McLafferty, 1980). Thus, the fragmentation of an EE^+ results in an EE^+ fragment and a neutral molecule; the loss of a radical, resulting in an $OE^{+\bullet}$ fragment, is a violation of the even-electron rule. However, fragmentation of an EE^+ to an $OE^{+\bullet}$ fragment is one of the four types of reactions outlined in Section 3.4.1 (McLafferty, 1980). Thus, the loss of radicals from an EE^+ is observed, although less frequently than the loss of neutral molecules. In high-energy CID, radical losses are observed more frequently, for example, the loss of a methyl radical ($^\bullet CH_3$) from low molecular mass *N*-alkyl iminium ions ($[R_1R_2C{=}NR_3CH_3]^+$) (Bowen & Harrison, 1981), as well as from protonated acetone and acetonitrile. Although to a lesser extent, other low molecular mass compounds also show the same behavior, for example, ketones, ethers and amines (Wagner et al., 1980).

With low-energy collision-cell CID, the even-electron rule has been evaluated for 100 pesticides, amenable to both GC–MS and LC–MS (Thurman et al., 2007). Under EI-MS conditions, ≈35% of the fragment ions from the $M^{+\bullet}$ were $OE^{+\bullet}$, whereas under positive-ion ESI-MS conditions only 31 of the 432 (≈7%) fragment ions from the $[M+H]^+$ were found to be $OE^{+\bullet}$. The most important deviations from the even-electron rule involve the loss of relatively stable radicals like chlorine ($Cl^\bullet$), bromine ($Br^\bullet$), nitrosyl ($^\bullet NO$), and nitryl ($^\bullet NO_2$), mainly from aromatic ring systems, or the loss of $^\bullet CH_3$ or methoxy ($CH_3O^\bullet$) radicals from methoxy-substituted aromatic compounds (Thurman et al., 2007; Holčapek et al., 2010). The radical loss may not only occur directly from $[M+H]^+$, but also as secondary fragmentation from an EE^+ fragment. Some typical examples are discussed below.

The minor peaks in the spectrum of bromhexine (Figure 3.10) are readily interpreted as secondary fragmentation of the EE^+ fragment with *m/z* 264 involving the loss of $Br^\bullet$ resulting in two $OE^{+\bullet}$ fragments with *m/z* 185 and 183, corresponding to the loss of $^{79}Br^\bullet$ and $^{81}Br^\bullet$, respectively. Further loss of another $Br^\bullet$, which is likely to occur from an $OE^{+\bullet}$, results in the EE^+ fragment with *m/z* 104 ($[C_7H_6N]^+$) due to loss of $^{79}Br^\bullet$ and $^{81}Br^\bullet$ from the ions with *m/z* 183 and 185, respectively.

A careful evaluation, based on the large data collection in Chapter 4, shows that the loss of $Cl^\bullet$, $Br^\bullet$ or an iodine radical ($I^\bullet$) occurs frequently, whereas a fluorine radical ($F^\bullet$) loss has been observed in one case, that is, in the MS^4 product-ion mass spectrum of citalopram (Section 4.2.3.3). Generally, loss of hydrogen fluoride (HF) is observed instead. Losses of $Cl^\bullet$ and $Br^\bullet$ from aromatic systems have readily been observed for some phenothiazines (Section 4.2.1), benzodiazepines (Section 4.2.4), diphenhydramines and related compounds (Section 4.5.2) and for compounds like amiloride (Section 4.1.4.4), clozapine (Section 4.2.2.2), efavirenz (Section 4.10.2), and trichlocarban (Section 4.10.4). For some compounds, there seems to be a competition between the loss of $Cl^\bullet$ and HCl, $Br^\bullet$ and HBr, or $I^\bullet$ and HI. This is for instance observed in the fragmentation patterns of the psychotropic drug baclofen (Section 4.2.8), of the non-steroidal anti-inflammatory drug (NSAID) diclofenac (Section 4.3.3.1), of the histamine antagonists desloratadine and rupatadine (Section 4.5.2), and of X-ray contrast agents (Section 4.5.8). Interestingly, losses of HF and HCl, but no $Cl^\bullet$, have been observed for steroids and β-lactam antibiotics. In clenbuterol, the loss of $Cl^\bullet$ is followed by the loss of HCl (Section 4.5.1). The loss of a 2-chloroethyl radical ($ClCH_2CH_2^\bullet$) is observed for compounds like chlormequat (Section 4.11.3) and the alkylating agents carmustin and chlorambucil (Section 4.5.6.3).

Aromatic nitro compounds are mostly analyzed in negative-ion mode (Section 3.7.9). However, the data collection in Chapter 4 provides some examples of $^\bullet NO_2$ losses in positive-ion mode, for example, for nimodipine (Section 4.1.2), nitrazepam (Section 4.2.4), ranitidine (Section 4.4.2.2), and dimetridazole and related nitroimidazoles (Section 4.10.2.1). Under high-energy CID conditions, a wide range of fragmentation reactions has been observed for protonated aromatic nitro compounds, involving either the loss of neutral molecules such as water, nitroxyl (HNO) or nitrous acid (HNO_2) or the loss of radicals like hydroxyl ($^\bullet OH$), $^\bullet NO$, and $^\bullet NO_2$. Water loss may occur when a proton donor group is present in the β-position relative to the nitro group (*ortho*-effect) (Crombie & Harrison, 1988).

In the CI-MS spectrum of 4-nitro-1,3-diphenylurea, for example, an apparent fragment ion with *m/z* ($[M+H]-30$) is observed. High-resolution accurate-mass MS analysis as well as H/D-exchange experiments have demonstrated that this loss of 30 Da is not due to the loss of $^\bullet NO$, but rather due to a (thermally-induced) reduction of the nitro (—NO_2) group to an amino (—NH_2) group. Besides a fragment ion with *m/z* 139 due to protonated 4-nitroaniline, a fragment ion due to protonated 4-aminoaniline with *m/z* 109 is observed (Brophy et al., 1979). The same process has been studied for protonated trinitroaromatic compounds (Yinon & Laschever, 1981). Under APCI-MS conditions, frequently used for the analysis of nitroaromatic compounds, similar features are observed as in CI-MS conditions. This was demonstrated with H/D-exchange experiments in heavy water (D_2O) for a series of nitroaromatic compounds (Karancsi & Slégel, 1999).

The loss of $^{\bullet}CH_3$ is regularly observed under low-energy CID conditions in ESI-MS–MS. It is observed in the fragmentation of aromatic methoxy compounds, such as for methoxy-flavonoids like acacetin, chrysoeriol, and isorhamnetin (Ma et al., 1997; De Rijke et al., 2006), and also for drugs like ketoprofen and naproxen (Section 4.3.3.3), neostigmine (Section 4.5.4), mescaline (Section 4.7.6.5), trimethoprim (Section 4.8.9), and diaveridine (Section 4.10.2.2). The loss of an alkyl radical is also observed for steroids, for example, involving the loss of the C^{13}-substituent in gestrinone, THG and 17α-trenbolone (Section 4.6.4), as well as for some amphetamine-related compounds (Section 4.7.2). Trimethoprim (Section 4.8.9) and diaveridine (Section 4.10.2.2) show fragment ions due the loss of $CH_3O^{\bullet}$ or (less abundantly) $^{\bullet}CH_3$. The loss of $^{\bullet}CH_3$ is also observed for quaternary ammonium compounds with methyl substituents (Section 3.6.5).

The loss of a methanesulfonyl radical ($CH_3SO_2^{\bullet}$) from two different EE^+ fragments is observed for tiapride (Section 4.2.2). In some cases, cleavages at a sulfinyl or sulfonyl (R^1SOR^2 or $R^1SO_2R^2$) group results in radical losses and the formation of an $OE^{+\bullet}$ fragment. An example of this is discussed for the proton pump inhibitors, where the formation of radical cations is described, that is, ions with *m/z* 151 for omeprazole, with *m/z* 195 for rabeprazole, with *m/z* 205 for lansoprazole, and with *m/z* 153 for pantoprazole (Section 4.4.2.1). The loss of a 4-(*tert*-butyl)benzenesulfonyl radical ($(CH_3)_3C—C_6H_4SO_2^{\bullet}$) is observed for the vasodilator bosentan (Section 4.1.6.3). The resulting $OE^{+\bullet}$ fragment with *m/z* 311 shows the subsequent loss of 2-hydroxyphenoxyl radical ($HOC_6H_4O^{\bullet}$) in combination with a rearrangement of a CH_3 group leading to a transfer of a methylene group to the 2-pyrimidinyl ring.

The inability to achieve the loss of a neutral molecule from $[M+H]^+$ or an EE^+ fragment either via an inductive cleavage or via an acyclic EE^+ cleavage with H-rearrangement may lead to a radical loss. In some cases, the presence of special molecular features like aromaticity and/or conjugation gives rise to this behavior. This seems to play a role in the homolytic cleavage of the phenoxyphenyl group in the phenoxyphenylurea herbicides difenoxuron and chloroxuron (4.11.5.1), resulting in the $OE^{+\bullet}$ fragment with *m/z* 164 ($[C_9H_{12}N_2O]^{+\bullet}$). The loss of the propan-2-yl-1-carboxylic acid radical ($CH_3{}^{\bullet}CHCOOH$) from the protonated pirprofen may be explained along the same line (Section 4.3.3.3). The structure of the heterocyclic aromatic amine 3,8-dimethyl-3*H*-imidazo[4,5-*f*]quinoxaline-2-amine (MeIQx, $C_{11}H_{11}N_5$; $[M+H]^+$ with *m/z* 214) seems not to be very flexible due to the high degree of conjugation (RDBE = 9). The primary fragmentation involves the loss of a $^{\bullet}CH_3$; *D*-labeling studies showed that the N^3-methyl group is the one lost (Guy et al., 2000).

In some other examples, the formation of the $OE^{+\bullet}$ fragment instead of the EE^+ fragment is not readily understood. This can be the case for the loss of the *N*-substituents of the piperazine ring. In some cases, EE^+ fragments are observed. However, in most cases, $OE^{+\bullet}$ fragments instead of the expected EE^+ are observed. For piperaquine, the $OE^{+\bullet}$ fragment with *m/z* 245 due to the 4-(7-chloroquinolin-4-yl)-1,2,3,6-tetrahydropyrazin-1-yl-1-iminium radical cation ($[C_{13}H_{12}ClN_3]^{+\bullet}$) is observed (Section 4.10.2.4). For phenothiazines with a 3-(4-methylpiperazin-1-yl)propyl substituent (Section 4.2.1 and Table 4.2.1a), a class-specific $OE^{+\bullet}$ fragment is observed with *m/z* 98 due to the 4-methyl-1,2,3,6-tetrahydropyrazin-1-yl-1-iminium radical cation ($[C_5H_{10}N_2]^{+\bullet}$). For sildenafil, upon cleavage of the N—S bond in the 4-methylpiperazine-sulfonyl moiety, both the $OE^{+\bullet}$ and the EE^+ fragment are observed, that is, the 4-methylpiperazin-1-yl-1-ylium radical cation ($[C_5H_{12}N_2]^{+\bullet}$) with *m/z* 100 and the protonated 1-methyl-1,2,3,6-tetrahydropyrazine ($[C_5H_{11}N_2]^+$) with *m/z* 99, respectively (Section 4.1.6.4).

In summary, it can be stated that the loss of a radical from EE^+ is usually rare. It occurs when relatively stable radicals can be lost, for example, $Cl^{\bullet}$, $Br^{\bullet}$, $^{\bullet}CH_3$, and $^{\bullet}NO_2$, mainly from an aromatic or a highly conjugated system, which can readily stabilize the resulting radical. It may also occur when alternative EE^+ fragmentation routes do not seem readily available due to the high energy needed to produce EE^+ fragments by disrupting conjugation and/or aromaticity.

3.5.8 Skeletal Rearrangements in Protonated Molecules

So far, attention was paid to H-rearrangement in the fragmentation of protonated molecules. However, ever since methyl rearrangements have been reported in the methane–CI-MS spectra of alkenes and alkynes (Field, 1968), rearrangements of other groups rather than H have been observed and studied (Kingston et al., 1983). Under high-energy CID conditions, methyl rearrangement is involved in the loss of formaldehyde ($H_2C{=}O$) from protonated methoxymethyl acetate ($[CH_3COOCH_2OCH_3+H]^+$ with *m/z* 105) (Weeks & Field, 1970) and in the loss of carbon dioxide (CO_2) from protonated methyl benzoate ($[C_6H_5COOCH_3+H]^+$ with *m/z* 137) (Ichikawa & Harrison, 1978). Methyl rearrangements are

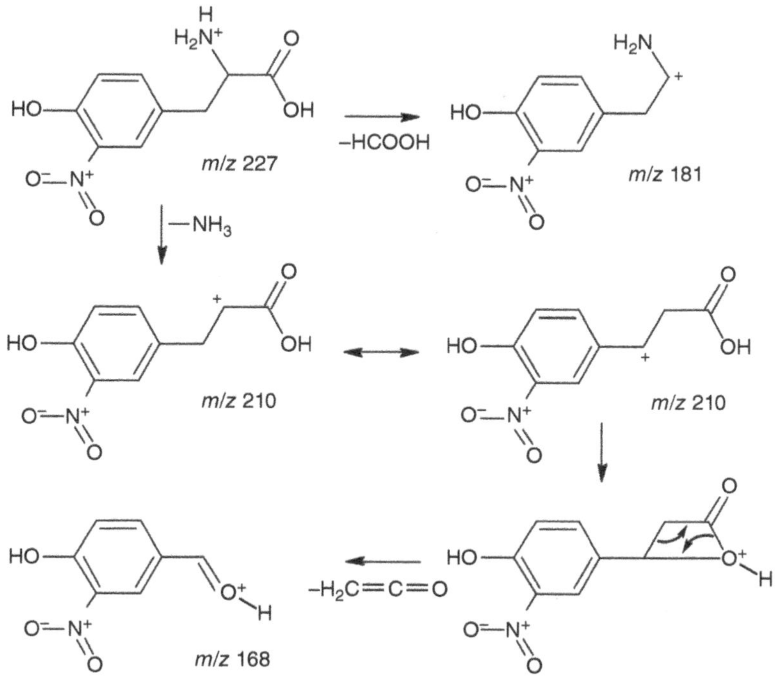

FIGURE 3.16 Fragmentation of 3-nitro-tyrosine.

also observed under ESI-MS–MS conditions, as for instance discussed for pirimicarb (Section 4.11.2.3) and for fluazifop-butyl ester (Section 4.11.7).

An O-atom rearrangement in conjunction with the loss of ethenone ($H_2C{=}C{=}O$) has been described for the amino acid Tyr and its 3-nitro analog (2-amino-3-(4-hydroxy-3-nitrophenyl)propanoic acid). After the loss of ammonia (NH_3), a H-rearrangement occurs from the 3- to the 2-position of the propanoic acid moiety to generate the more stable benzylic carbocation. Then, via a four-membered cyclic intermediate, ethenone ($H_2C{=}C{=}O$) is expelled to generate an oxonium ion, that is, protonated 4-hydroxy-3-nitrobenzaldehyde (Figure 3.16). The fragmentation mechanism was elucidated using [D]- and [^{13}C]-labeling experiments (Delatour et al., 2002). The same fragmentation is also observed for levodopa (Section 4.5.5) and melphalan (Section 4.5.6.3).

Under high-energy CID conditions, benzyl migration has been reported in the water loss from protonated dibenzyl ether (Kingston et al., 1981) and benzyl cyclohexyl ether (Diakiw et al., 1979). Phenyl migration has been observed in the CO_2 loss from protonated phenyl benzoate (Ichikawa & Harrison, 1978). Benzyl migration is also reported to occur under ESI-MS–MS conditions. Benzyl migration via a Claisen rearrangement is proposed to be involved in the loss of water or carbon monoxide (CO) from protonated benzyloxy indoles (Crotti et al., 2007). Rearrangement of a 4-cyanobenzyl group has been reported in relation to the loss of 4-methyl-1*H*-imidazole from a farnesyl transferase inhibitor (Qin, 2001). Rearrangement of a 4-fluorobenzyl group has been described in relation to the loss of 2,5-dimethylpiperazine from the p38α kinase inhibitor DMPIP (Falck et al., 2013). Such unusual fragmentation reactions require the use of accurate-mass determination for confirmation. Benzyl migration is also involved in the fragmentation of some imidazole fungicides, for example, econazole and isoconazole, where product ions consistent with protonated mono- or dichloro-1-benzyl-1*H*-imidazole, that is, ions with *m/z* 193 and 227, are observed (Section 4.9.1).

Remarkably, a rearrangement of a *tert*-butyl group seems to be involved in the loss of CO as a secondary fragmentation of the triazole fungicide triadimefon (Section 4.9.2).

A thiono–thiolo rearrangement is observed in the fragmentation of fungicides such as tolnaftate and tolciclate (Section 4.9.4), of thiocarbamate pesticides (Section 4.11.2.6), and of phosphorothioate diethyl esters, as discussed in detail for diazinon (Section 4.11.4). In the case of tolnaftate and tolciclate, compounds with a general structure $R^1NCH_3CSOR^2$, fragment ions with *m/z* 148 and 164, are observed, which are consistent with the iminium ions $[R^1CHON{=}CH_2]^+$ and $[R^1CHSN{=}CH_2]^+$, respectively (Section 4.9.4).

Another skeletal rearrangement, involving an internal glucose residue loss, has been reported for flavonoid-*O*-diglycosides. The isomeric *O*-diglycosides naringenin-7-*O*-neohesperidoside and naringenin-7-*O*-rutinoside ($[M+H]^+$ with *m/z* 581) show different spectra due to different fragmentation of the diglycosides. The neohesperidose (rhamnosyl-($\alpha 1 \rightarrow 2$)-glucose) analog shows a more abundant Y_0 fragment ion with *m/z* 273, consistent with the loss of the diglycoside. In the rutinose (rhamnosyl-($\alpha 1 \rightarrow 6$)-glucose), however, the Y_1-ion with *m/z* 435, consistent with the loss of the rhamnose unit, is more abundant. The latter compound also shows a far more abundant ion with *m/z* 419, indicated as a Y′-ion, which is actually due to the loss of the internal glucosyl residue, and thus should involve a rearrangement of the rhamnosyl unit to the aglycone (Ma et al., 2000; Ma et al., 2001). It should be mentioned that the internal glucose residue loss is not observed in the fragmentation of $[M+Na]^+$ or $[M-H]^-$.

3.6 CHARACTERISTIC POSITIVE-ION FRAGMENTATION OF FUNCTIONAL GROUPS

A particular functional group in a molecule may give rise to class-specific fragmentation. Whether such a class-specific fragmentation is actually observed depends on the other functional groups present in the molecule. With several functional groups in the molecule, their fragmentation pathways become competing processes. The result of these competitive pathways depends on a number of factors, including the internal energy available to the ion, and kinetic and thermodynamic parameters of the fragmentation reactions, for example, proton affinity, bond strengths, and stability of product ions. To some extent, the product-ion mass spectrum is a statistical reflection of these competitive processes. Prediction of relative abundance of fragment ions is often a delicate matter. With experience, it is possible to predict to some extent which fragment ions may be observed for a particular molecule. However, in some cases, a predicted fragmentation pathway does not lead to the observation of the expected fragment ion, where it is often difficult to explain which particular factors determine the outcome of the different competing pathways. In this section, the characteristic fragmentation of many functional groups in positive-ion mode is briefly summarized and illustrated with typical examples from the data collection of Chapter 4. In addition, data provided in systematic studies on the fragmentation of protonated molecules have been used (Levsen et al., 2007; Holčapek et al., 2010; Weissberg & Dagan, 2011). Emphasis is put on the most common fragmentation pathways for each functional group in $[M+H]^+$.

3.6.1 Cleavages of C—C Bonds

The cleavage of C—C bonds with the formation of secondary or tertiary carbocations ($[R^1R^2HC]^+$ or $[R^1R^2R^3C]^+$) is possible by fragmentation at a tertiary or a quaternary C-atom, respectively, according to Eq. 3.9.

$$[R^1R^2HCR^3+H]^+ \rightarrow [R^1R^2HC^+] + R^3H \qquad (3.9a)$$

$$[R^1R^2R^3CR^4+H]^+ \rightarrow [R^1R^2R^3C^+] + R^4H \qquad (3.9b)$$

At least one of the R-groups must have a substituent that can be protonated; alkanes are not protonated in ESI-MS. The fragmentation must have a charge-remote character, as the proton is not readily stabilized at a tertiary or quaternary C-atom. An example of this type of C—C bond cleavages is observed in the antihistamine pheniramine and related compounds (Section 4.5.2). A special case of this type of fragmentation is the loss of propene ($CH_3CH{=}CH_3$) or *iso*-butene ($(CH_3)_2C{=}CH_2$) from an aromatic ring with an *iso*-propyl or a *tert*-butyl substituent, according to Eq. 3.10.

$$[(CH_3)_3C{-}C_6H_4{-}NHR^1+H]^+ \rightarrow [C_6H_5{-}NHR^1+H]^+ + i\text{-}C_4H_8 \qquad (3.10)$$

The loss of $CH_3CH{=}CH_2$ is observed for instance in the phenylurea herbicide isoproturon (Section 4.11.5.1). In addition, C—C bond cleavages frequently occur in aliphatic fused-ring systems, such as steroids (Section 4.6).

A characteristic fragment ion, which is related to the presence of a benzyl group ($C_6H_5CH_2$—), often results from the cleavage of a C—C bond. The benzyl cation ($[C_6H_5CH_2]^+$) or seven-membered ring tropylium ion ($[C_7H_7]^+$), or related substituted ions, is an EE^+ fragment observed for both $OE^{+\bullet}$ and EE^+ generated by using EI and soft-ionization techniques such as CI and ESI (Eq. 3.11).

$$[C_6H_5CH_2R^1+H]^+ \rightarrow [C_6H_5CH_2]^+ \rightleftarrows [C_7H_7]^+ + R^1H \qquad (3.11)$$

The high stability of the benzyl cation ($[C_6H_5CH_2]^+$) can be explained by the five different resonance structures that can be written for it, where charge delocalization over the aromatic system stabilizes the carbocation. Skeletal rearrangement of the benzyl cation leads to the tropylium ion ($[C_7H_7]^+$), that is, cycloheptatrienylium ion. The tropylium ion in its planar conformation is an aromatic system where the charge is delocalized over the whole ring (seven resonance structures), also explaining the reason of its extra stability when compared to its acyclic analog (Kuck, 1990a; Kuck, 1990b).

This fragmentation is observed for amphetamines (Section 4.7.2), related structures such as the anorectic drug mefenorex (Section 4.4.4), and the antiparkinsonian selegiline (Section 4.5.5). Based on [D]-labeling studies, there is evidence that the tropylium ion is the actual form present rather than its isomeric benzyl cation (Section

4.7.2) (Bijlsma et al., 2011). Some other compounds for which a tropylium ion with *m/z* 91 is observed are the diuretic benzthiazide (Section 4.1.4.1), the antidepressant fluoxetine (Section 4.2.3.3), beclamide and primidone (Section 4.2.7), donepezil (Section 4.5.4), and penicillin G (Section 4.8.3.1). Some imidazole and triazole fungicides show benzyl cations with *m/z* 125 ($[ClC_6H_4CH_2]^+$), *m/z* 159 ($[Cl_2C_6H_3CH_2]^+$), or *m/z* 109 ($[FC_6H_4CH_2]^+$) (Sections 4.9.1 and 4.9.2). Benzyl cations with *m/z* 121 ($[CH_3O{-}C_6H_4CH_2]^+$) are observed for the histamine antagonists pyrilamine and thonzylamine (Section 4.5.2), with *m/z* 139 ($[(HO)_3C_6H_2CH_2]^+$) for benserazide (Section 4.5.5), with *m/z* 159 ($[F_3CC_6H_4CH_2]^+$) for dexfenfluramine (Section 4.4.4), and with *m/z* 169 ($[BrC_6H_4CH_2]^+$) for bromtripelennamine (Section 4.5.2).

3.6.2 Alcohols and Ethers

An aliphatic alcohol (R^1OH) shows the loss of water, thereby producing an alkyl cation according to Eq. 3.12.

$$[R^1OH{+}H]^+ \rightarrow [R^1]^+ + H_2O \qquad (3.12)$$

In an aromatic alcohol (Ar^1OH) like phenolic compounds, the loss of water is less common; such ions only occur as a minor fragment or else because other losses are even less likely. A benzyl alcohol (Ar^1CH_2OH) favors the loss of formaldehyde ($H_2C{=}O$) over the loss of water (Eq. 3.13).

$$[Ar^1CH_2OH{+}H]^+ \rightarrow [Ar^1H{+}H]^+ + H_2C{=}O \qquad (3.13)$$

Please note that both R^1 in Eq. 3.12 and Ar^1 in Eq. 3.13 should have a substituent that can be protonated; aliphatic and aromatic alcohols are not protonated in ESI-MS.

The loss of water in the MS–MS spectrum of anisodamine (Figure 3.11) results from the 6-hydroxy group of the tropine ring rather than from the 3-hydroxy group of tropic acid (2-phenyl-3-hydroxypropanoic acid). The loss of $H_2C{=}O$ is from the tropic acid 3-hydroxy carbon. Water loss is no assured indication for the presence of an alcohol group, as water loss may also occur from aldehydes, ketones, and carboxylic acids.

A dialkyl ether (R^1OR^2) shows cleavages of either of the C—O bonds to form a carbocation and/or a protonated alcohol, according to Eq. 3.14 (Figure 3.9). Note that either R^1 or R^2 (or both) should have a substituent that can be protonated; aliphatic ethers are not protonated in ESI-MS.

$$[R^1OR^2{+}H]^+ \rightarrow [R^1]^+ + R^2OH \qquad (3.14a)$$

$$[R^1OR^2{+}H]^+ \rightarrow [R^1OH{+}H]^+ + [R^2{-}H] \qquad (3.14b)$$

$$[R^1OR^2{+}H]^+ \rightarrow [R^2]^+ + R^1OH \qquad (3.14c)$$

$$[R^1OR^2{+}H]^+ \rightarrow [R^2OH{+}H]^+ + [R^1{-}H] \qquad (3.14d)$$

The relative abundance of the fragment ions depends on the proton affinity of the resulting neutral molecule (Field's rule, Section 3.4). If both ions are observed, that is, an alkyl cation and a protonated alcohol, these complementary ions show a sum of their *m/z* values equal to *m/z* ([M+H]+1). The fragmentation of the antihistamine clemastine ($[M{+}H]^+$ with *m/z* 344) is a good example. The cleavage of the C—O bond on the tertiary C-atom yields two complementary fragment ions, that is, the more abundant tertiary carbocation, that is, the α-(4-chlorophenyl)-α-(methyl)benzyl cation ($[(C_{14}H_{12}Cl]^+$) with *m/z* 215, and the protonated 2-(1-methylpyrrolidin-2-yl)ethanol with *m/z* 130 ($[C_7H_{15}NO{+}H]^+$) (Section 4.5.2). Cleavage of the C—O bond in the ethanol group would yield 1-(4-chlorophenyl)-1-phenylethanol, which is more difficult to protonate, and this is the reason why the pathway to the formation of the two more stable products is followed.

Ether fragmentation is an important route in oligosaccharides (Ham, 2008; Zaia, 2004) and glycosides (Cuyckens & Claeys, 2004), leading to B- and Y-ions or C- and Z-ions, depending on which glycosidic C—O bond is cleaved (Domon & Costello, 1988). Similar features in terms of glycosidic fragmentation are observed for aminoglycoside antibiotics (Section 4.8.5) and macrolide antibiotics such as erythromycin A (Section 4.8.8).

Aromatic methoxy compounds (Ar^1OCH_3) may show losses of a methyl radical ($^{\bullet}CH_3$), a methoxy radical ($CH_3O^{\bullet}$), as discussed in Section 3.5.6, as well as a loss of $H_2C{=}O$ (Eq. 3.15). Note that Ar^1 should have a substituent that can be protonated; otherwise, the compound is not observed in ESI-MS.

$$[Ar^1OCH_3{+}H]^+ \rightarrow [Ar^1OH]^{+\bullet} + {}^{\bullet}CH_3 \qquad (3.15a)$$

$$[Ar^1OCH_3{+}H]^+ \rightarrow [Ar^1H]^{+\bullet} + {}^{\bullet}OCH_3 \qquad (3.15b)$$

$$[Ar^1OCH_3{+}H]^+ \rightarrow [Ar^1H{+}H]^+ + H_2C{=}O \qquad (3.15c)$$

The discussion on the fragmentation of trimethoprim (Section 4.8.9) may serve as a good example for this. The loss of $^{\bullet}CH_3$ is also important in methoxy-substituted flavonoids (de Rijke et al., 2006).

Phenyl alkyl ethers (Ar^1OR^2) show a preferred fragmentation pathway yielding a protonated phenol with the loss of the alkyl group as an alkene (Eq. 3.16a). However, if the aliphatic group has significantly higher proton affinity, the phenol is lost as a neutral molecule and an aliphatic carbocation is formed (Eq. 3.16b). Note that Ar^1 or R^1 should have a substituent that can be protonated; otherwise, the phenyl alkyl ethers are not observed in ESI-MS.

$$[Ar^1OR^2{+}H]^+ \rightarrow [Ar^1OH{+}H]^+ + [R^2{-}H] \qquad (3.16a)$$

$$[Ar^1OR^2{+}H]^+ \rightarrow Ar^1OH + [R^2]^+ \qquad (3.16b)$$

Fragmentation according to Eq. 3.16b is for instance observed for tamsulosin, where the fragment ion with *m/z* 271 is produced rather than the fragment ion with *m/z* 139 (protonated 2-ethoxyphenol) (Section 4.1.6.1), or for paroxetine, where the fragment ion with *m/z* 192 is generated rather than the fragment ion with *m/z* 139 (protonated 3,4-(methylenedioxy)phenol) (Section 4.2.3.3). In imidazole fungicides such as isoconazole and econazole, the fragmentation of the aryl alkyl ether results in mono- and dichlorobenzyl cations with *m/z* 125 and 159, respectively, along with the loss of an alcohol (Section 4.9.1).

This type of behavior is relevant in the loss of the glycosidic group from an aromatic aglycone or in the loss of dehydroglucuronic acid (176 Da, $C_6H_8O_6$) from an aromatic glucuronic acid conjugate (Phase II metabolites of for instance estrogens, Section 4.6.8). In a glucuronic acid conjugate (or glycoside) attached to an aliphatic group, both C—O bonds can cleave, as demonstrated for instance in the fragmentation of ammoniated 3α-hydroxy-5α-estrane-17-one glucuronide, a metabolite of nandrolone (Kuuranne et al., 2003). In a glucuronic acid conjugate (or glycoside) attached to an aromatic group, only one C—O bond can be cleaved, resulting in a phenolic group and the dehydroglucuronic acid or dehydroglycoside; charge retention may occur on either fragment, as determined by the proton affinity (Field's rule).

In diaryl ethers (Ar^1OAr^2), for example, phenoxyphenyls, the C—O bond cleavage is often homolytic, resulting in an $OE^{+\bullet}$ fragment (Eq. 3.17), as is the case for the phenoxyphenylurea herbicides difenoxuron and chloroxuron (Section 4.11.5.2). Note that Ar^1 or Ar^2 should have a substituent that can be protonated; otherwise, the diphenyl ethers are not observed in ESI-MS.

$$[Ar^1OAr^2{+}H]^+ \rightarrow Ar^1O^\bullet + [Ar^2H]^{+\bullet} \quad (3.17)$$

In some cases, a β-cleavage occurs in an aryl alkyl ether (Ar^1OR^2), and to a lesser extent in a dialkyl ether (R^1OR^2), that is, the cleavage of the α,β-C—C bond relative to the O-atom, thus resulting in an oxonium ion, $[Ar^1O{=}CH_2]^+$ (Eq. 3.18).

$$[Ar^1OCH_2R^1{+}H]^+ \rightarrow [Ar^1O{=}CH_2]^+ + R_1H \quad (3.18)$$

This is observed for paroxetine, where the fragment ion with *m/z* 151 is an oxonium ion (Section 4.2.3.3) and for valaciclovir and valganciclovir (Section 4.10.3.4).

3.6.3 Aldehydes and Ketones

Simple aliphatic or aromatic aldehydes (R^1CHO, where R^1 can be either alkyl or aryl) can only be analyzed using ESI-MS or APCI-MS after analyte derivatization, for example, using 2,4-dinitrophenylhydrazine (Deng et al., 2012), which modifies the aldehyde group. If the aldehyde group is one of the functional groups in a more complex molecule, loss of water from the aldehyde may be observed. An aromatic aldehyde may show the loss of carbon monoxide (CO) (Levsen et al., 2007).

Ketones (R^1COR^2) may show the loss of water, as seen for cathinone and related compounds (Section 4.7.2.4). Water may also be lost from cyclic ketone groups, as seen for ketamine (Section 4.7.6.1). In addition, cleavages may occur on either of the α-C—C bonds to the carbonyl, resulting in an alkyl cation R^+ (or protonated alkene) and/or an acylium ion ($[RC{=}O]^+$) (Eq. 3.19). Note that an acylium ion, $[RC{=}O]^+$, with charge on the C-atom, and an oxonium ion, $[RC{\equiv}O]^+$, with charge on the O-atom, are resonance structures.

$$[R^1COR^2{+}H]^+ \rightarrow [R^1]^+ + R^2CHO \quad (3.19a)$$

$$[R^1COR^2{+}H]^+ \rightarrow R^1H + [R^2C{=}O]^+ \quad (3.19b)$$

$$[R^1COR^2{+}H]^+ \rightarrow [R^1C{=}O]^+ + R^2H \quad (3.19c)$$

$$[R^1COR^2{+}H]^+ \rightarrow R^1CHO + [R^2]^+ \quad (3.19d)$$

Note that a direct cleavage occurs in this case; the sum of the *m/z* values of $[R^1]^+$ and $[R^2C{=}O]^+$ or $[R^1C{=}O]^+$ and $[R^2]^+$ is equal to *m/z* ([M+H]−1) (Section 3.5.4).

In an aryl alkyl ketone (Ar^1COR^2), the major fragmentation pathway involves the loss of the aliphatic chain as an alkane, according to Eq. 3.19c. The loss of the aryl group is less likely. In the fragmentation of benziodarone ($[M{+}H]^+$ with *m/z* 519) (Section 4.1.7.2), a phenyl benzofuran ketone derivative (Ar^1COAr^2), fragment ions are observed from cleavages of both carbonyl substituents, with charge retention on either fragment, thus resulting in fragment ions with *m/z* 345 (Eq. 3.19a) and *m/z* 173 (Eq. 3.19b), when cleaving at the phenyl group, and with *m/z* 373 (Eq. 3.19c) and *m/z* 145 (Eq. 3.19d), when cleaving at the benzofuran group. The acylium ions with *m/z* 373 and 173 are more abundant than the aryl carbocations. Again, the sum of the *m/z* values is equal to *m/z* ([M+H]−1) (Section 3.5.4). This behavior is also observed in fenofibrate, where the chlorobenzoyl cation ($[ClC_6H_4{-}C{=}O]^+$) with *m/z* 139 is more abundant than the chlorophenylium ion ($[ClC_6H_4]^+$) with *m/z* 111 (Section 4.4.3.1). In cyclic ketones, the loss of CO may occur, most likely followed by ring closure. This CO loss is frequently observed in flavonoids (De Rijke et al., 2006).

3.6.4 Carboxylic Acids and Esters

Aliphatic and aromatic carboxylic acids (R^1COOH and Ar^1COOH) may show the loss of water, resulting in an alkyl or an aryl acylium ion ($[R^1C{=}O]^+$ or $[Ar^1C{=}O]^+$), or the loss of either formic acid (HCOOH) or carbon dioxide (CO_2). Note that R^1 or Ar^1 should have a substituent that can be protonated; otherwise, the carboxylic acids are not observed in positive-ion ESI-MS. The secondary fragmentation of the protonated acids resulting from the alkene loss in *n*-butyl-4-hydroxybenzoate and meperidine can be used to illustrate some features of the fragmentation of a protonated acid (Figure 3.11). For protonated 4-hydroxy-benzoic acid with *m/z* 139, the fragmentation involves the loss of water and CO_2, whereas for protonated 1-methyl-4-phenylpiperidine-4-carboxylic acid with *m/z* 220, the fragmentation involves the loss of water and HCOOH. The difference is due to the position of the carboxylic acid group on either the aromatic ring or the aliphatic group, respectively. This issue has been discussed in some detail for the secondary fragmentation of the dihydropyridine calcium channel blocker nimodipine (Section 4.1.2).

Thus, an aliphatic carboxylic acid may show the loss of HCOOH (Eq. 3.20).

$$[R^1COOH{+}H]^+ \rightarrow [R^1]^+ + HCOOH \qquad (3.20)$$

This is seen for captopril (Section 4.1.3.1), ketoprofen and naproxen (Section 4.3.3.3), levodopa (Section 4.5.5), mycophenolic acid (Section 4.5.7), and β-lactam antibiotics (Section 4.8.3). Aromatic carboxylic acids show the loss of CO_2 instead of HCOOH (Eq. 3.21).

$$[Ar^1COOH{+}H]^+ \rightarrow [Ar^1H{+}H]^+ + CO_2 \qquad (3.21)$$

Note that Ar^1 should have a substituent that can be protonated; otherwise, this fragmentation reaction does not occur. This is observed for (fluoro)quinolone antibiotics (Section 4.8.4) and in the secondary fragmentation of the herbicide imazapyr (Section 4.11.9.6). The loss of HCOOH instead of CO_2 in 5-aminosalicylic acid is probably due to an *ortho*-effect (Section 3.6.6). Surprisingly, a loss of HCOOH instead of CO_2 is also observed for THC-9-COOH (Section 4.7.3).

Methyl alkyl or aryl esters (R^1COOCH_3 or Ar^1COOCH_3) show the loss of methanol (CH_3OH), resulting in an acylium ion (Eq. 3.22).

$$[RCOOCH_3{+}H]^+ \rightarrow [RC{=}O]^+ + CH_3OH \qquad (R = \text{alkyl or aryl}) \qquad (3.22)$$

In esters with a larger alkyl group (R^1COOR^2 where R^1 can be either alkyl or aryl and R^2 is alkyl), the fragmentation pattern is determined by two favorable cleavage sites, that is, the carbonyl C—O bond to the ester moiety and the R^2 alkyl group C—O bond. The loss of the alcohol R^2OH results in the formation of a generally stable acylium ion $[R^1C{=}O]^+$ (Eq. 3.23a). Cleavage of the R^2 alkyl group C—O bond of the ester moiety yields two complementary ions, that is, a protonated acid (Eq. 3.23b) or an alkyl cation (Eq. 3.23c) (Figure 3.9). The relative abundance of the latter two ions is determined by the proton affinities of the products (Field's rule).

$$[R^1COOR^2{+}H]^+ \rightarrow [R^1C{=}O]^+ + R^2OH \qquad (R^1 = \text{alkyl or aryl}) \qquad (3.23a)$$

$$[R^1COOR^2{+}H]^+ \rightarrow [R^1COOH + H]^+ + [R^2{-}H] \qquad (R^1 = \text{alkyl or aryl}) \qquad (3.23b)$$

$$[R^1COOR^2{+}H]^+ \rightarrow R^1COOH + [R^2]^+ \qquad (R^1 = \text{alkyl or aryl}) \qquad (3.23c)$$

This is well demonstrated and discussed in Section 3.5.2 with the data in Figure 3.11. The dihydropyridine calcium channel blockers as a class make a good example of characteristic ester fragmentation with charge retention either on the acylium ion (Eq. 3.23a) or on the protonated acid (Eq. 3.23b) (Section 4.1.2). Some other examples of compounds that show this type of characteristic ester fragmentation are procaine (Section 4.2.5), oxybutynin (Section 4.5.3), cocaine and some of its related compounds (Section 4.7.4), dimethocaine (Section 4.7.6.2), and fluazifop-butyl (Section 4.11.7). In an ester with an aryl group (RCOOAr where R can be either alkyl or aryl), usually only the loss of the aryl alcohols (ArOH) with the formation of an acylium ion $[RC{=}O]^+$ is observed (Eq. 3.24).

$$[RCOOAr{+}H]^+ \rightarrow [RC{=}O]^+ + ArOH \qquad (R = \text{alkyl or aryl}) \qquad (3.24)$$

An important class of esters is the acylglycerols (Byrdwell, 2001; Griffiths, 2003; Cajka & Fiehn, 2014). Since mono-, di-, and triacylglycerols are not readily ionized to $[M{+}H]^+$ in ESI-MS, APCI-MS should be used instead. Alternatively, $[M{+}Li]^+$ or $[M{+}NH_4]^+$ may be generated by addition of Li^+ or $NH_4{}^+$ salts to the mobile phase (Section 3.8). Fragmentation of protonated triacylglycerols results in three major fragment ions consistent with the loss of any one of the three fatty acids from the glycerol backbone. Due to steric hindrance, the loss of the fatty acid in sn-2 position leads to the least abundant ion of the three. In addition, less abundant ions may be observed due to protonated fatty acids and/or the fatty acid acylium ions (with an *m/z* difference of 18). The fragmentation of related glycerophospholipids has been discussed in detail in a review publication (Hsu & Turk, 2009).

3.6.5 Amines and Quaternary Ammonium Compounds

Primary alkylamines (R^1NH_2) show the loss of ammonia (NH_3) and form an alkyl carbocation $[R^1]^+$. Because of its high proton affinity, NH_3 is not a good leaving group (Field's rule). In contrast, anilines (Ar^1NH_2) show the loss of hydrogen cyanide (HCN) rather than of NH_3. Methyl anilines (Ar^1NHCH_3) may show the loss of methyl radical ($^{\bullet}CH_3$) (Levsen et al., 2007).

Secondary and tertiary alkylamines (R^1R^2NH and $R^1R^2R^3N$) can show fragment ions due to cleavage of any of the C—N bonds with charge retention on either product, resulting in a protonated primary or secondary alkylamine, respectively, and/or an alkyl cation as complementary ion. For the cleavage of the C—N bond to the R^2 group, the reactions proceed according to Eq. 3.25 (Section 3.5.2 and Figure 3.9).

$$[R^1R^2NH{+}H]^+ \rightarrow [R^1NH_2{+}H]^+ + [R^2{-}H] \quad (3.25a)$$

$$[R^1R^2NH{+}H]^+ \rightarrow R^1NH_2 + [R^2]^+ \quad (3.25b)$$

$$[R^1R^2R^3N{+}H]^+ \rightarrow [R^1R^3NH{+}H]^+ + [R^2{-}H] \quad (3.25c)$$

$$[R^1R^2R^3N{+}H]^+ \rightarrow R^1R^3NH + [R^2]^+ \quad (3.25d)$$

The relative abundance of the fragment ions depends on the proton affinity (Field's rule) and stability of the products. This fragmentation is frequently observed in drugs, pesticides, and related compounds. It plays an important role in the fragmentation of compound classes such as β-blockers (Section 4.1.1), phenothiazines (Section 4.2.1), tricyclic antidepressants (Section 4.2.3.1), histamine antagonists (Section 4.5.2), and triazine herbicides (Section 4.11.1).

Diaryl amines (Ar^1NHAr^2) do not show cleavages of the C—N bonds. A reaction according to Eq. 3.25a would only occur if either Ar^1 or Ar^2 holds a substituent that readily leads to the necessary H-rearrangement to produce a protonated aniline analog ($[ArNH_2{+}H]^+$), whereas a reaction according to Eq. 3.25b would result in a phenylium-type ion ($[Ar]^+$) and the loss of an aniline derivative, which is not likely to occur because of Field's rule. This explains why no cleavages of the C—N bonds are observed in NSAIDs based on *N*-phenylanthranilic acid related structures (Section 4.3.3.1).

Another important fragmentation route, especially for tertiary alkylamines, is the cleavage of an α,β-C—C bond, resulting in stable (mono or dialkyl)iminium ions (Eq. 3.26) (Section 3.5.5).

$$[R^1R^2R^3N{+}H]^+ \rightarrow [R^1R^2N{=}CH_2]^+ + [(R^3{+}H){-}CH_2] \quad (3.26)$$

This is readily observed for substituent-chain fragmentation in phenothiazines (Section 4.2.1), tricyclic antidepressants (Section 4.2.3.1), and histamine antagonists (Section 4.5.2). In phenothiazines, cleavage of the propane-chain C^2—C^3 bond results in an abundant iminium ion, for example, the 1-methylidene-4-methylpiperazinium ion with *m/z* 113 ($[C_6H_{13}N_2]^+$) or the *N,N*-dimethylmethaniminium ion with *m/z* 58 ($[(CH_3)_2N{=}CH_2]^+$) (Section 4.2.1). Alternatively, iminium ions are generated by α,β-C—C bond cleavages relative to the N-atom of the phenothiazine ring (cleavage of the propane-chain C^1—C^2 bond) (Section 4.2.1).

The piperidine ring may show losses of NH_3 (loss of 17 Da), methylamine (CH_3NH_2, loss of 31 Da), and ethylamine ($CH_3CH_2NH_2$, loss of 45 Da). In contrast, a piperazine ring as well as a 4-methylidenepiperidine ring show losses of methanimine ($H_2C{=}NH$, loss of 29 Da) and ethanimine ($CH_3CH{=}NH$, loss of 43 Da), along with the loss of NH_3. If the N-atom lost is substituted, shifts of these losses are observed accordingly. In mirtazapine with an *N*-methyl substituent, losses of 31 Da (methylamine, CH_3NH_2), of 43 Da (*N*-methylmethanimine, $CH_3N{=}CH_2$), and of 57 Da (*N*-methylethanimine, $CH_3N{=}CHCH_3$) are observed (Section 4.2.3.2). Studies with *D*-labeling on the different C-atoms in the piperazine ring revealed that the NH_3 loss involves a βH-rearrangement, whereas ^{13}C-labeling demonstrated that the loss of 43 Da from the piperazine ring involves the loss of $CH_3CH{=}NH$ rather than the loss of $CH_3N{=}CH_2$ (Niessen & Honing, 2015). Fragmentation of the piperazine ring is relevant for terazosin (Section 4.1.6.1), dibenzazepines like clozapine (Section 4.2.2.2), some histamine antagonists such as cinnarizine and cetirizine (Section 4.5.2), and the quinolone antibiotic pipedemic acid (Section 4.8.4). An example of the loss of NH_3 and $H_2C{=}NH$ from a 4-methylidene-piperidine ring is observed for desloratadine (Section 4.5.2). Cleavage at the piperazine ring with charge retention on the ring is somewhat prone to the formation of $OE^{+\bullet}$ fragments, as discussed in Section 3.5.7.

Quaternary ammonium compounds are prone to showing losses of both molecules and radicals, according to Eq. 3.27.

$$[R^1N(CH_3)_3]^+ \rightarrow [R^1N(CH_3)_2]^{+\bullet} + {}^{\bullet}CH_3 \quad (3.27a)$$

$$[R^1N(CH_3)_3]^+ \rightarrow [R^1CH_3N{=}CH_2]^+ + CH_4 \quad (3.27b)$$

The loss of $^{\bullet}CH_3$ (Eq. 3.27a) is observed for instance in the acetylcholine esterase inhibitors neostigmine and pyridostigmine (Section 4.5.4). Both EE^+ and $OE^{+\bullet}$ fragments are observed for chlormequat (Section 4.11.3). Fascinating fragmentation has been reported for the quaternary ammonium herbicides paraquat and diquat (Section 4.11.3).

Losses of HCN or methanediimine (HN═C═NH) may be observed in the fragmentation of *N*-heterocyclic aromatic compounds (Eq. 3.28).

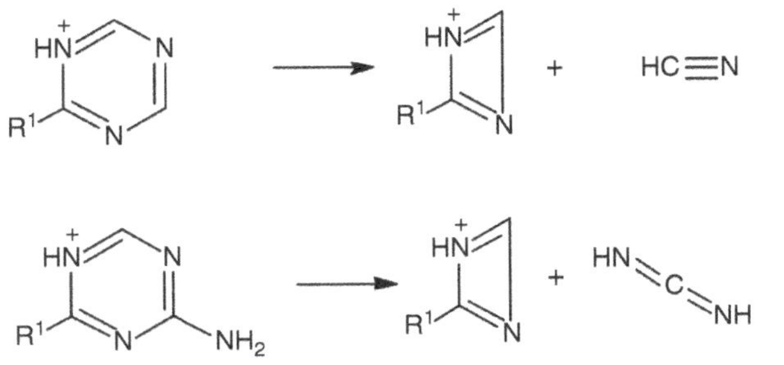

(3.28)

This is for instance observed for nucleobases (Section 4.5.6.1), carbendazim and related fungicides (Section 4.9.3), triazine herbicides (Section 4.11.1), and other triazine-containing compounds, such as the coccidiostats cyromazine and diaveridine (Section 4.10.2.2).

3.6.6 Amides, Sulfonyl Ureas, and Carbamates

The fragmentation of an amide (R^1CONHR^2, where R is alkyl or aryl) frequently results in two complementary fragment ions, that is, a protonated amine due to an acyclic EE^+ fragmentation with H-rearrangement and an acylium ion due to an inductive cleavage (Eq. 3.29).

$$[R^1CONHR^2{+}H]^+ \rightarrow [R^1C{=}O]^+ + R^2NH_2 \qquad (3.29a)$$

$$[R^1CONHR^2{+}H]^+ \rightarrow [(R^1{-}H){=}C{=}O] + [R^2NH_2{+}H]^+ \qquad (3.29b)$$

$$[R^1C{=}O]^+ \rightarrow [R^1]^+ + CO \quad (R^1 = \text{alkyl}) \qquad (3.29c)$$

The sum of the *m/z* values of the complementary fragment ions equals *m/z* ([M+H]+1). An aliphatic acylium ion may show a subsequent loss of carbon monoxide (CO) (Eq. 3.29c). This type of fragmentation is important not only in the fragmentation and amino-acid sequencing of peptides (Sections 3.5.1 and 3.5.3) but also for many drugs and pesticides. It is observed for the diuretic amiloride (Section 4.1.4.4), for acetaminophen (Section 4.3.1), for the antidiabetic drug nateglinide (Section 4.4.1.3), for lysergic acid diethylamide (Section 4.7.6.4), and for the antiretroviral protease inhibitors (Section 4.10.3.3).

Functional groups that are closely related to the amide (R^1CONHR^2) group, for example, urea ($R^1NHCONHR^2$), sulfonyl urea ($R^1NHCONHSO_2R^2$), and guanidine ($HN{=}CNHR^1NHR^2$) derivatives, show some similarities in their fragmentation characteristics. Fragmentation may occur on either σ-bond of the carbonyl, sulfone, or imine group with possible charge retention on either of the resulting products.

The fragmentation of phenylurea herbicides ($R^1NHCONR^2R^3$) is somewhat difficult to predict. For phenylurea with a 4-Cl, 4-Br, 4-trifluoromethyl (4-CF_3), or 4-*iso*-propyl ($(CH_3)_2CH$) substituted phenyl (Ar^1 in Eq. 3.30), two complementary ions are produced upon cleavage of one of the carbonyl C—N bonds according to Eq. 3.30a and Eq. 3.30b, leading to the ions with *m/z* 188 and 46, respectively, for diuron ($[M{+}H]^+$ with *m/z* 233) (Table 4.11.5). A direct cleavage takes place upon cleavage of the other carbonyl C—N bond according to Eqs 3.30c and 3.30d, leading to the ions with *m/z* 160 and 72, respectively, for diuron (Section 4.11.5.1). For phenoxyphenylurea, unsubstituted phenylurea, and urea herbicides with heterocyclic aryl substituents (Ar^2 in Eq. 3.30), fragment ions according to Eq. 3.30e rather than Eq. 3.30c are observed (Table 4.11.6).

$$[R^1NHCONR^2R^3{+}H]^+ \rightarrow [R^1NHC{=}O]^+ + R^2R^3NH \quad (R^1 = Ar^1 \text{ or } Ar^2) \qquad (3.30a)$$

$$[R^1NHCONR^2R^3{+}H]^+ \rightarrow R^1N{=}C{=}O + [R^2R^3NH{+}H]^+ \quad (R^1 = Ar^1 \text{ or } Ar^2) \qquad (3.30b)$$

$$[R^1NHCONR^2R^3{+}H]^+ \rightarrow [R^1NH]^+ + R^2R^3NCHO \quad (R^1 = Ar^1) \qquad (3.30c)$$

$$[R^1NHCONR^2R^3{+}H]^+ \rightarrow R^1NH_2 + [R^2R^3N{=}C{=}O]^+ \quad (R^1 = Ar^1 \text{ or } Ar^2) \qquad (3.30d)$$

$$[R^1NHCONHR^2{+}H]^+ \rightarrow [R^1NH_2{+}H]^+ + R^2N{=}C{=}O \quad (R^1 = Ar^2) \qquad (3.30e)$$

Sulfonamide antibiotics ($H_2NC_6H_4SO_2NHAr$, with Ar being mostly a (substituted) heterocyclic aryl group, Section 4.8.1) show characteristic class-specific and compound-specific fragmentation, involving cleavages of either σ-bond of the sulfone group, according to Eq. 3.31.

$$[H_2NC_6H_4SO_2NHAr{+}H]^+ \rightarrow [H_2NC_6H_4SO_2]^+ + ArNH_2 \qquad (3.31a)$$

$$[H_2NC_6H_4SO_2NHAr{+}H]^+ \rightarrow HN{=}C_6H_4{=}SO_2 + [ArNH_2{+}H]^+ \qquad (3.31b)$$

$$[H_2NC_6H_4SO_2NHAr{+}H]^+ \rightarrow [H_2NC_6H_4]^+ + ArNHSO_2H \qquad (3.31c)$$

$$[H_2NC_6H_4SO_2NHAr{+}H]^+ \rightarrow H_2NC_6H_5 + [ArNHSO_2]^+ \qquad (3.31d)$$

Cleavage of the sulfonamide S—N bond results in two complementary fragments, that is, the

4-aminobenzenesulfonyl cation with *m/z* 156 (Eq. 3.31a) and a protonated heterocyclic arylamine with *m/z* ([M+H]−155) (Eq. 3.31b). Cleavage of the S—C bond involves a direct cleavage, resulting in the 4-aminophenylium ion with *m/z* 92 (Eq. 3.31c) and an arylsulfamoyl cation with *m/z* ([M+H]−93) (Eq. 3.31d), (Figure 4.8.1).

Compounds with an *N*-carbamoyl sulfonamide moiety ($RNHCONHSO_2Ar$), such as the antidiabetic drugs carbutamide (Section 4.4.1.1) and glibenclamide (Section 4.4.1.2), and the sulfonylurea herbicides (Section 4.11.6, Tables 4.11.8 and 4.11.9), show cleavages of all σ-bonds in the *N*-carbamoyl sulfonamide moieties. The fragmentation shows cleavages characteristic to both phenylureas (Eq. 3.30) and sulfonamides (Eq. 3.31). For carbutamide (4-amino-*N*-(butylcarbamoyl)benzenesulfonamide, $[M+H]^+$ with *m/z* 272), this leads to (phenyl)urea-related fragment ions with *m/z* 173 according to Eq. 3.30e and *m/z* 74 according to Eq. 3.30b, and to sulfonamide-related fragment ions with *m/z* 92 according to Eq. 3.31c and *m/z* 156 according to Eq. 3.31a. Sulfonylurea herbicides with a 4,6-dimethoxypyrimidinyl group show urea-related fragment ions with *m/z* 154 according to Eq. 3.30c or with *m/z* 156 according to Eq. 3.30e, and with *m/z* 182 according to Eq. 3.30a; the complementary ion according to Eq. 3.30b may also be observed. A sulfonamide-related fragment ion according to Eq. 3.31a is also observed for some compounds. Similarly, sulfonylurea herbicides with a 4-methoxy-6-methyltriazinyl group show urea-related fragment ions with *m/z* 141 according to Eq. 3.30e, and with *m/z* 167 according to Eq. 3.30a, and a sulfonamide-related fragment ion according to Eq. 3.31a.

Similar fragmentation reactions are observed for the guanidine ($R^1NHC{=}NHNHR^2$) derivatives such as metformin (Section 4.4.1.3), proguanil (Section 4.10.2.4), moroxydine (Section 4.10.3.4), and chlorhexidine (Section 4.10.4).

In an interesting study on the EE^+ fragmentation of three acetamide-aminobenzenesulfonic acids, it was demonstrated that an *ortho*-effect, known from fragmentation in EI (McLafferty & Tureček, 1993; Smith, 2004), may also occur in the fragmentation of $[M+H]^+$. It was found that the product-ion mass spectrum of 2-acetamido-5-aminobenzenesulfonic acid with the acetamide group in the *ortho*-position relative to the sulfonic acid moiety differs from the spectra of 5-acetamido-2-aminobenzenesulfonic acid and 4-acetamido-2-aminobenzenesulfonic acid, which have the amino group in the *ortho*-position relative to the sulfonic acid moiety (Reddy et al., 2005). Due to an *ortho*-effect, the first compound shows a primary loss of ethenone ($H_2C{=}C{=}O$), whereas the other two compounds show a primary loss of water. The *ortho*-interaction is illustrated in Figure 3.17.

Carbamates ($R^1OCONHR^2$, where R can be either alkyl or aryl) generally fragment to produce two complementary ions, that is, a protonated alcohol $[R^1OH{+}H]^+$ (Eq. 3.32a) and the acylium ion $[R^2NHC{=}O]^+$ (Eq. 3.32b).

$$[R^1OCONHR^2{+}H]^+ \rightarrow [R^1OH{+}H]^+ + R^2N{=}C{=}O \quad (3.32a)$$

$$[R^1OCONHR^2{+}H]^+ \rightarrow R^1OH + [R^2N{=}C{=}O{+}H]^+ \quad (3.32b)$$

This fragmentation behavior is readily observed for *N*-methylcarbamate pesticides, where an abundant fragment ion is observed that corresponds to a characteristic loss of 57 Da (methyl isocyanate, $CH_3N{=}C{=}O$); the

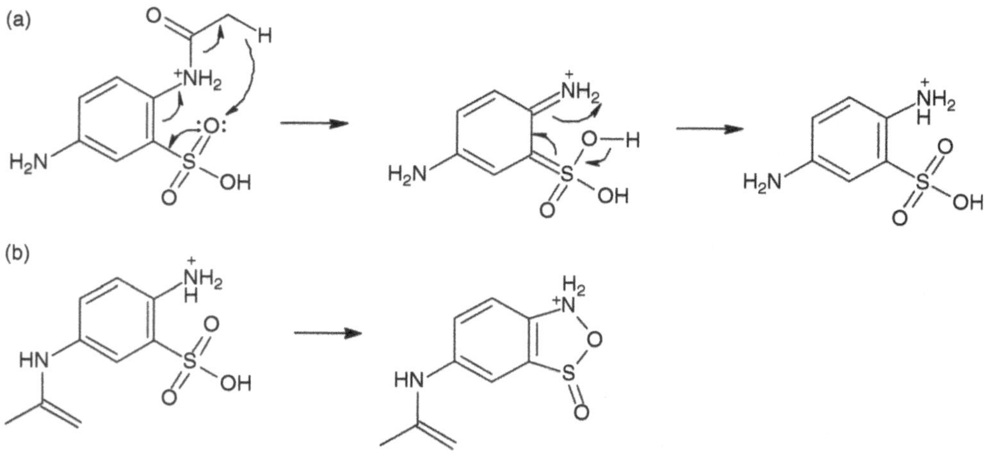

FIGURE 3.17 *ortho*-Effect in the fragmentation of (a) protonated 2-acetamido-5-aminobenzenesulfonic acid and (b) protonated 5-acetamido-2-aminobenzenesulfonic acid. (Source: Reddy et al., 2005. Reproduced with permission of Wiley.)

complementary ion with *m/z* 58 is generally far less abundant (Section 4.11.2.1). Similar fragmentation is observed for physostigmine (Section 4.5.4) and the fungicide carbendazim (Section 4.9.3).

3.6.7 Compounds Containing Phosphorus or Sulfur

The characteristic fragmentation of phosphate, phosphorothioate, and phosphorodithioate esters has been discussed in detail for organophosphorus pesticides in Section 4.11.4.

Thiols (R^1SH), thioethers (R^1SR^2, where R can be either alkyl or aryl), and thioesters (R^2CSOR^2) generally show fragmentation similar to their *O*-analogs, that is, alcohols, ethers, and esters, respectively. Thus, the loss of hydrogen sulfide (H_2S) from a thiol (cf. Eq. 3.12), a cleavage of either of the C—S bonds in a thioether (cf. Eq. 3.14), or the loss of an alcohol (R^2OH) from a thioester (cf. Eq. 3.23) may be observed. The fragmentation of cimetidine and famotidine (Section 4.4.2) serve as examples for the fragmentation of a thioether. Compared to ethers, a greater tendency toward a homolytic cleavage is observed for thioethers, as for instance demonstrated in the fragmentation of ranitidine and omeprazole (Section 4.4.2). The iminium 10-methylidenephenothiazinium ion with *m/z* 212 or its substituted analogs, frequently observed in the product-ion mass spectra of phenothiazine antipsychotics, may show the loss of an S-atom with ring closure to have a five-membered ring (Section 4.2.1).

A thiono–thiolo rearrangement is observed for thiocarbamate pesticides (Section 4.11.2.5), for fungicides such as tolnaftate and tolciclate (Section 4.9.4), as well as for phosphorothioate diethyl esters, as discussed in detail for diazinon (Section 4.11.4). In the case of diazinon, fragment ions with *m/z* 153 and 169 are observed, which are consistent with protonated 6-methyl-2-(*iso*-propyl)pyrimidin-4-ol ($[C_8H_{12}N_2O+H]^+$) and protonated 6-methyl-2-(*iso*-propyl) pyrimidine-4-thiol ($[C_8H_{12}N_2S+H]^+$), respectively (Section 4.11.4).

A sulfone ($R^1SO_2R^2$, where R can be either alkyl or aryl) may have a distinct influence on the fragmentation characteristics. To some extent, this is already discussed in Section 3.6.6 for sulfonamide antibiotics and sulfonylurea herbicides. In addition to the four fragment ions already discussed (Eq. 3.31), sulfonamide antibiotics show two more fragment ions related to the sulfonyl (SO_2) group: losses of sulfur monoxide (SO) and of hyposulfurous acid (H_2SO_2). The class-specific fragment ion with *m/z* 156 ($[H_2NC_6H_4SO_2]^+$) (Eq. 3.31a) shows a loss of SO to an ion with *m/z* 108, requiring a rearrangement in the SO_2-group (Eq. 3.33a) (Niessen, 1998; Klagkou et al., 2003; Wang et al., 2005).

$$[H_2NC_6H_4{-}SO_2]^+ \rightarrow [HN{=}C_6H_4{=}O{+}H]^+ + SO \qquad (3.33a)$$

$$[H_2NC_6H_4{-}SO_2{-}NHAr{+}H]^+ \rightarrow [HN{=}C_6H_4{=}NAr{+}H]^+ + H_2SO_2 \qquad (3.33b)$$

A similar loss is observed for the diuretic xipamide (Section 4.1.4.2), for the antidiabetic carbutamide (Section 4.4.1.1), and in the fragmentation pattern of sildenafil (Section 4.1.6.4). An internal loss of H_2SO_2 is observed from $[M+H]^+$ of sulfonamide antibiotics, thus requiring a rearrangement as well (Eq. 3.33b) (Section 4.8.1) (Niessen, 1998; Klagkou et al., 2003; Wang et al., 2005). Sulfur dioxide (SO_2) is readily lost from the sulfonyl-containing benzothiadiazine ring structures as observed for buthiazide (Section 4.1.4.1) or diazoxide (Section 4.1.6.3) and the cyclic sulfonamide substituent of sultiame (Section 4.2.7).

3.6.8 Miscellaneous Compound Classes

Characteristic fragmentation of nitroaryl compounds ($ArNO_2$) involves loss of a nitryl radical ($^{\bullet}NO_2$) (Eq. 3.34). Note that Ar should have a substituent that can be protonated; otherwise, the compounds are not observed in ESI-MS.

$$[ArNO_2{+}H]^+ \rightarrow [ArH]^{+\bullet} + {}^{\bullet}NO_2 \qquad (3.34)$$

Alkyl and aryl halides (R^1X and Ar^1X, where X is F, Cl, Br, or I) show the loss of hydrogen halide (HX) for all halides and/or the loss of a halide radical ($X^{\bullet}$) (Eq. 3.35). Examples of HX and/or $X^{\bullet}$ losses are discussed in Section 3.5.7.

$$[RX{+}H]^+ \rightarrow [R]^+ + HX \quad (R = \text{alkyl or aryl}, X = F, Cl, Br, \text{or } I) \qquad (3.35a)$$

$$[RX{+}H]^+ \rightarrow [RH]^{+\bullet} + X^{\bullet} \quad (R = \text{alkyl or aryl}, X = Cl, Br, \text{or } I) \qquad (3.35b)$$

The fragmentation of polycyclic structures with aliphatic rings as in steroids, with a combination of aromatic and aliphatic rings, or with conjugated polyaromatics is more difficult to summarize in fragmentation rules. Many class-specific fragment ions are observed for steroids (Section 4.6), but often it is unclear why certain fragment ions are observed with one class and not with another, even if they are structurally related. In some cases, class-specific fragmentation patterns can be recognized, as is the case for β-lactam antibiotics (Section 4.8.3). The fragmentation of tricyclic antidepressants (Section 4.2.3.1), benzodiazepines (Section 4.2.4), and cannabinoids (Section 4.7.3) can be interpreted, but it has not yet been possible to develop general fragmentation rules for them. Also, there can be considerable discrepancies on the fragmentation of complex ring structures, for example, as discussed for galantamine (Section 4.5.4).

3.7 FRAGMENTATION OF DEPROTONATED MOLECULES

The fragmentation of deprotonated molecules $[M-H]^-$ has not been as extensively studied as the protonated molecules $[M+H]^+$. The positive-ion mode is applied more often and for a wider range of compounds than the negative-ion mode. This is partly due to the specific analyte properties needed to generate $[M-H]^-$. In addition, in ESI-MS, the negative-ion mode is often somewhat less sensitive than the positive-ion mode, which may be partly due to the lower needle voltage required in order to avoid gas discharges in the ion source.

3.7.1 High-Energy CID with NICI-Generated Deprotonated Molecules

Under conventional CI conditions, negative ions can be generated in two ways (Section 1.2.2.3). Probably, the most widely applied method is electron-capture negative ionization (ECNI), resulting in either $M^{-\bullet}$ or, via a dissociative process, EE^-. Alternatively, proton-transfer reactions in negative-ion chemical ionization (NICI) with basic reagent ions such as hydroxide (HO^-), methoxide (CH_3O^-), and azanide (NH_2^-) can be used to generate $[M-H]^-$ (Bowie, 1990; Harrison, 1992).

Fragmentation of $[M-H]^-$ under high-energy CID collisions has been reviewed (Bowie, 1990). Four general fragmentation pathways are distinguished: (1) a homolytic cleavage, where a radical anion is generated by the loss of a radical, (2) reactions involving the formation of an anion–molecule complex, followed by loss of the molecule, which can take place via various mechanisms, (3) reactions involving the formation of an anion–molecule complex, followed by charge exchange and subsequent elimination of the resulting neutral, again via various mechanisms, and (4) various rearrangement reactions (Bowie, 1990). The fragmentation of various compound classes was discussed along these lines (Bowie, 1990). Partly based on this, typical functional group class-specific fragmentation reactions of $[M-H]^-$ for the data collection in Chapter 4 are summarized later.

3.7.2 General Aspects

In ESI–MS, the generation of $[M-H]^-$ is restricted to compounds with acidic protons, for example, carboxylic acids, sulfonic acids, sulfuric acids, phosphoric acids, and phenols, as well as to compounds in which acidic protons are available through keto–enol, amide–iminol, or a similar tautomerism. This is the case for esters, amides, and sulfonamides.

In positive-ion mode, the formation of two complementary ions due to cleavage of a C–heteroatom bond with charge retention on either resulting product was recognized as an important fragmentation pathway (Section 3.5.2). The complementary ions are characterized by the sum of the *m/z* values of the fragment ions equaling *m/z* ([M+H]+1). In addition, direct-cleavage reactions were discussed, where the *m/z* of the fragment ions sum up to *m/z* ([M+H]−1) (Section 3.5.4). In negative-ion mode, direct-cleavage reactions, where the sum of the *m/z* of the fragment ions equals *m/z* ([M−H]−1), are observed as well. An example of this is the cleavage of the C—C bond between the dihydropyridine ring and the substituted phenyl ring in nitrendipine and other dipines (Section 4.1.2). Other examples are given later.

For many compounds in the data collection of Chapter 4, the fragmentation of $[M-H]^-$ is most readily described as a series of (partly competing) consecutive small-molecule neutral losses. This type of fragmentation behavior is observed for deprotonated nitrazepam (Section 4.2.4), acifluorfen (Section 4.11.9.7), olsalazine (Section 4.3.2), and estrogens (Section 4.6.7 and Table 4.6.3), as well as for corticosteroids, using $[M+HCOO]^-$ or $[M+CH_3COO]^-$ as precursor ions (Section 4.6.6 and Table 4.6.2).

Another general observation from the data collection in Chapter 4 is the more frequent occurrence of radical losses from $[M-H]^-$. For dimethyl phosphorothioates and phosphorodithioates, $[M-CH_3]^{-\bullet}$ is observed rather than $[M-H]^-$ (Section 4.11.4). Losses of chlorine ($Cl^\bullet$), bromine ($Br^\bullet$), nitrosyl ($^\bullet NO$), and nitryl ($^\bullet NO_2$) radicals are frequently observed. Other examples of radical losses comprise the loss of a benzyl radical ($C_6H_5CH_2^\bullet$) from the diuretic benzthiazide, the loss of a 2,2,2-trifluoroethyl radical ($F_3CCH_2^\bullet$) from the diuretic polythiazide (Section 4.1.4.1), and the loss of propan-2-one-1-yl radical ($CH_3COCH_2^\bullet$) from deprotonated coumarin anticoagulants like warfarin (Section 4.5.9.1). Other examples of radical losses and radical anions appear in the following.

3.7.3 Alcohols and Ethers

Aliphatic alcohols do not readily form $[M-H]^-$ in ESI-MS, unless substituents are prone to negative-ion generation. An aliphatic alcohol (R^1OH) may show the loss of water, unless $[M-H]^-$ is an alkoxide ion ($[R^1O]^-$). The phenoxide ion ($[C_6H_5O]^-$) may show the loss of carbon monoxide (CO) to form an EE^- fragment with *m/z* 65 ($[C_5H_5]^-$). This CO loss is observed in the secondary fragmentation of salicylic acid, that is, after the loss of carbon dioxide (CO_2) (Section 4.3.2). Under high-energy CID, a H-rearrangement from phenoxide ion C^2 to C^1 leads to the loss of formyl radical ($^\bullet CHO$), resulting in an $OE^{-\bullet}$ with *m/z* 64 ($[C_5H_4]^{-\bullet}$).

Dialkyl ethers (R^1OR^2) do not readily form $[M-H]^-$ in ESI-MS, unless substituents are prone to negative-ion generation. Knowledge on the fragmentation of the ether bond in negative-ion mode comes from studying compounds that carry an additional functional group that can generate $[M-H]^-$, for example, acidic oligosaccharides and glycosides (see later). Compounds with a general structure $Ar^1OCHR^2CHR^3COOH$ (where R is H or an alkyl group) readily form $[M-H]^-$ and provide information on the fragmentation of phenyl alkyl ethers. Either an aroxide ion ($[Ar^1O]^-$) or a deprotonated alkenoic acid fragment ion $[R^2CH{=}CR^3COO]^-$ (or both) are observed, according to Eq. 3.36.

$$[Ar^1OCHR^2CHR^3COO]^- \rightarrow [Ar^1O]^- + R^2CH{=}CR^3COOH \quad \text{(with R = H or alkyl)} \tag{3.36a}$$

$$[Ar^1OCHR^2CHR^3COO]^- \rightarrow Ar^1OH + [R^2CH{=}CR^3COO]^- \quad \text{(with R = H or alkyl)} \tag{3.36b}$$

Given the weakness of the aryl alkyl ether bond and the stability of the aroxide ion, fragmentation according to Eq. 3.36a may be observed due to in-source CID in an ESI ion source, as is the case for 2,4-D and related phenoxy acid herbicides (Section 4.11.7 and Table 4.11.6), where oxiran-2-one ($C_2H_2O_2$) is lost as a neutral. This type of fragmentation is also found for fibrates like bezafibrate (Section 4.4.3.1).

Oligosaccharides with acidic sugars like sialic acid or sulfated sugar residues, glycosides with aglycones that readily provide deprotonated molecules, and glucuronic acid conjugates arising from Phase II drug metabolism behave like a dialkyl ether and show elimination of one of the sugar groups as a dehydrosugar (Zaia, 2004). Through a direct cleavage of one of the glycosidic C—O bonds, glucuronic acid conjugates show a characteristic loss of dehydroglucuronic acid ($C_6H_8O_6$) with *m/z* ($[M-H]-176$), also observed in the positive-ion mode (Section 3.6.4), and the dehydroglucuronate anion with *m/z* 175 ($[C_6H_7O_6]^-$. Secondary fragmentation of the ion with *m/z* 175 leads to the ions with *m/z* 113 ($[C_5H_5O_3]^-$) and 85 ($[C_4H_5O_2]^-$), which are due to the loss of water and CO_2 and of a subsequent loss of CO from the ion with *m/z* 113, respectively. This is observed for the glucuronic acid conjugate of oxazepam (Section 4.2.4), glucuromycophenol acid (Section 4.5.7), gemfibrozil glucuronide, and steroid glucuronides (Section 4.6.8). In the reported interpretation of the gemfibrozil glucuronide (Xia et al., 2003), the ions with *m/z* 113 and 85 are erroneously attributed to secondary fragmentation of the gemfibrozil aglycone. Careful inspection of the collision-cell MS^3 spectrum of the aglycone shows this is not the case. The loss of a hydroxy-sugar ($C_6H_{13}NO_3$) rather than a dehydrosugar ($C_6H_{11}NO_2$) is observed in deprotonated daunorubicin, where the aminosugar is attached to an aliphatic rather than to an aromatic ring (Section 4.5.6.2).

Interestingly, the data collection in Chapter 4 provides examples of a homolytic cleavage in a deprotonated phenyl alkyl ether (Ar^1OR^2) with the formation of a phenoxyl radical anion $[Ar^1O]^{-\bullet}$. This behavior is observed for some β-blockers (Figure 4.1.3). Similarly, a cleavage in deprotonated diphenyl ethers Ar^1OAr^2 may lead to radical anions $[Ar^1O]^{-\bullet}$ or $[Ar^2O]^{-\bullet}$, as observed for etiroxate (Section 4.4.3.1) and for benzoylphenylurea (Section 4.11.5.2).

3.7.4 Carboxylic Acid and Esters

Enolate anions resulting from the deprotonation of aldehydes and ketones are not generally observed under ESI-MS conditions. Carboxylate anions, that is, deprotonated carboxylic acids, are readily generated under ESI-MS conditions. The fragmentation of $[M-H]^-$ of carboxylic acids has been systematically investigated and rules have been developed to predict whether the loss of either carbon dioxide (CO_2) or formic acid (HCOOH) is observed (Bandu et al., 2004). CO_2 losses are observed from monocarboxylic acids in the following cases: (1) carboxylate groups in β-position to a conjugated system with at least three π-electron pairs, (2) carboxylate groups in γ-position to a conjugated system with at least three π-electron pairs, provided proton donors are not present for intramolecular hydrogen bonding, and (3) carboxylate groups at an sp^2-C-atom of a conjugated π-system with at least three π-electron pairs. In the latter case, the CO_2 loss is enhanced by the presence of electron-withdrawing substituents in the π-system. In dicarboxylic acids, CO_2 loss is observed from a carboxylate group at an sp^3-C-atom with the second carboxylate groups in α, β, or γ-position to it, provided proton donors are not present for intramolecular hydrogen bonding (Bandu et al., 2004). If no CO_2 loss is observed, the loss of HCOOH is favored. Thus, a CO_2 loss is observed for salicylic acid and for acetylsalicylic acid, although the latter case is preceded by the loss of ethenone ($H_2C{=}C{=}O$) from the ester moiety (Section 4.3.2). In 5-aminosalicylic acid, there is a competition between the loss of CO_2 at lower collision energy and of formyloxyl radical ($HCO_2^\bullet$) at higher collision energy (Section 4.3.2). Similarly, competition between CO_2 and HCOOH loss has been observed for tolfenamic acid (Section 4.3.3.1), carprofen, and flurbiprofen (Section 4.3.3.3). In aliphatic carboxylic acids, competition between the loss of HCOOH or CO_2 is observed.

Fragmentation of carboxylic acids is of special interest in the characterization of fatty acids and related substances (Griffiths, 2003; Murphy & Axelsen, 2011). In low-energy CID, the $[M-H]^-$ of saturated fatty acids show very little fragmentation. In fatty acids with one or two double bonds, cleavages of the β-bond to the double bond are observed, thus probably involving a charge-remote fragmentation (CRF) mechanism, that is, charge retention on a fragment bearing the carboxylate (—CO_2^-) group (Jensen et al., 1985; Kerwin et al., 1996). In polyunsaturated fatty acids like arachidonic acid, the fragmentation pattern is different: cleavages of the α-bond to the double bond are observed, with charge retention mostly on the polyunsaturated fragment (Kerwin

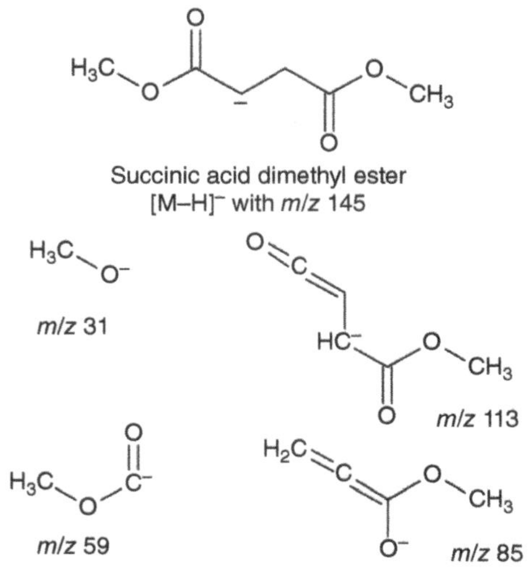

FIGURE 3.18 Fragmentation of deprotonated dimethyl succinate, resulting in two pairs of =C=O ions from direct cleavages.

et al., 1996). The fragmentation of $[M-H]^-$ of other classes of lipids under low-energy CID conditions has been reviewed as well (Murphy & Axelsen, 2011; Hsu & Turk, 2009). Under high-energy CID conditions, CRF in $[M-H]^-$ can be applied to determine double-bond positions in the molecule (Jensen et al., 1985).

The fragmentation of ester enolates has also been studied under high-energy CID conditions (Bowie, 1990). The fragmentation of the deprotonated dimethyl succinate ($[M-H]^-$ with *m/z* 145) has been described in terms of the formation of two possible anion–molecule complexes, which may result in two pairs of fragment ions: $[CH_3O]^-$ with *m/z* 31 and $[CH_3OCOCHCH{=}C{=}O]^-$ with *m/z* 113, and $[CH_3OCO]^-$ with *m/z* 59 and $[CH_2{=}C{=}COOCH_3]^-$ with *m/z* 85 (Figure 3.18). The sum of the *m/z* values of both pairs of fragment ions is equal to *m/z* ($[M-H]-1$). Interestingly, the loss of CO is observed as well (Bowie, 1990). The fragmentation pathways of $[M-H]^-$ of ethane-1,2-diyl diacetate (*m/z* 145) were studied in detail in order to achieve a better understanding of the fragmentation of deprotonated triacylglycerols (Stroobant et al., 1995). The direct cleavage in two ions with *m/z* 103 and 41 has been rationalized via an ion–molecule complex (Figure 3.19a). Some other fragment ions arise from a Claisen-type rearrangement derived from the initially formed diester enolate (Figure 3.19b).

Other examples of ester fragmentation show losses of alcohols, as is observed in dipines (Section 4.1.2), or losses of alkenes to produce an (aromatic) carboxylate group, as is observed for glafenine, for floctafenine (Section 4.3.3.2), and for dolasetron (Section 4.4.5). The simvastatin dihydroxy acid has a general structure of $R^1COOR^2R^3COOH$. The direct-cleavage fragmentation of the ester group results in fragment ions consistent with $[R^1COO]^-$ and $[[R^2\text{–}H]R^3COO]^-$ (Section 4.4.3.2).

3.7.5 Amines

Although examples of fragmentation of deprotonated amines have been discussed (Bowie, 1990), deprotonation of amines is not observed under ESI-MS conditions. The formation of $[M-H]^-$ in domperidone (Section 4.4.5) is most likely due to an amide–iminol tautomerism in the 1,3-dihydrobenzimidazol-2-one group. The compound shows cleavages of the amine C—N bonds, resulting in deprotonated 1,3-dihydrobenzimidazol-2-one with *m/z* 133 and its 5-chloro analog with *m/z* 167. In addition, cleavage at the piperidine ring with the loss of an alkene yields a fragment ion with *m/z* 250. In general, amines are best analyzed in positive-ion mode, where the fragmentation is readily understood (Section 3.6.5).

3.7.6 Amides and Ureas

Complex fragmentation patterns are observed for amides (R^1CONHR^2). Under high-energy CID conditions, the losses of hydrogen radical ($H^\bullet$), molecular hydrogen (H_2), water, R^2 as either an alkene or an alkyl radical, and R^1CHO are observed (Bowie, 1990).

To better understand amide fragmentation in drug-like molecules, available examples in the data collection of Chapter 4 have been studied. In fact, cleavages of all σ-bonds of the amide moiety have been observed (Eq. 3.37). The different possible fragmentation pathways of the amide function are illustrated for xipamide ($[M-H]^-$ with *m/z* 353) and lornoxicam ($[M-H]^-$ with *m/z* 370) in Figure 3.20, showing both the fragment ions observed and the neutral molecules lost for all three pathways.

$$[(R^1\text{–}H)CONHR^2]^- \rightarrow [(R^1\text{–}H)CONH_2]^- + [R^2\text{–}H] \tag{3.37a}$$

$$[(R^1\text{–}H)CONHR^2]^- \rightarrow [(R^1\text{–}H)+H]^- + R^2N{=}C{=}O \tag{3.37b}$$

$$[(R^1\text{–}H)CONHR^2]^- \rightarrow [(R^1\text{–}H){=}C{=}O] + [R^2NH]^- \tag{3.37c}$$

Fragmentation according to Eq. 3.37a is observed in the secondary fragmentation of xipamide (Section 4.1.4.2), leading to the 2-carbamoyl-5-chlorophenoxide ion ($[C_7H_5ClNO_2]^-$) with *m/z* 170 (Figure 3.20). Fragmentation according to Eq. 3.37b is observed in the secondary fragmentation of xipamide as well (Section 4.1.4.2), resulting in the 3-chlorophenoxide ion ($[ClC_6H_4O]^-$) with *m/z* 127, and for lornoxicam (Section 4.3.4), leading to the fragment ion with *m/z* 250 due to the loss of 2-isocyanatopyridine

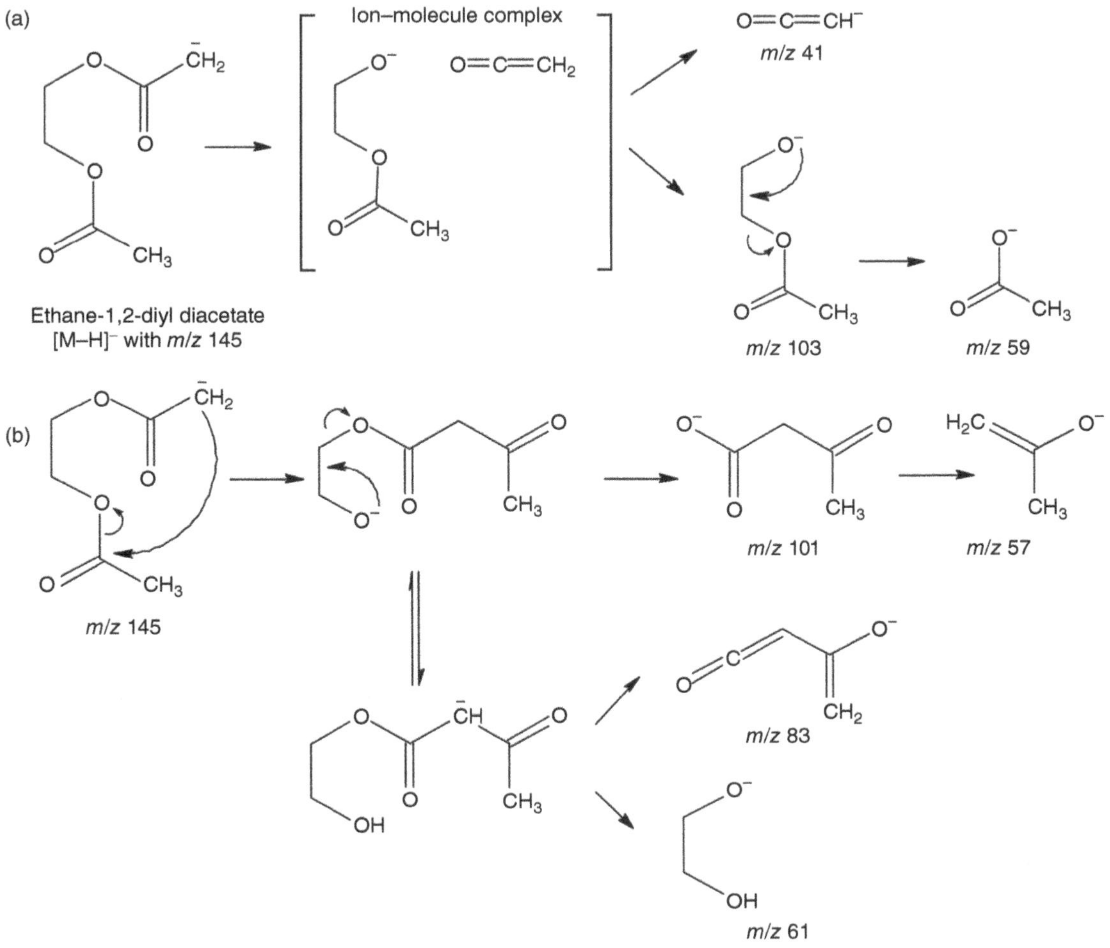

FIGURE 3.19 Two fragmentation pathways in high-energy CID of ethane-1,2-diyl diacetate, with (a) ion formation of a direct-cleavage pair of ions via an ion–molecule complex and (b) additional fragment ions after a Claisen-type rearrangement. (Source: Stroobant et al., 1995. Reproduced with permission of Springer.)

(C_5H_4N—N=C=O) (Figure 3.20, Eq. 3.37b). It is also seen for ACE inhibitors like ramipril (Section 4.1.3.2), and for leflunomide and teriflunomide (Section 4.5.7). Finally, fragmentation according to Eq. 3.37c can occur, as observed for lornoxicam (Section 4.3.4), leading to deprotonated 2-aminopyridine ([C_5H_4N—NH]$^-$) with *m/z* 93 (Figure 3.20). It is also seen for metoclopramide involving the loss of *N*,*N*-diethyl-2-(isocyanato)ethanamine (Section 4.4.5) and for leflunomide and teriflunomide involving the loss of 1-isocyanato-4-(trifluoromethyl)benzene (Section 4.5.7).

In compounds containing an *N*-carbamoyl sulfonamide moiety ($R^1NHCONHSO_2R^2$), as in antidiabetic drugs such as carbutamide, tolbutamide, and gliclazide (Section 4.4.1), the loss of a substituted isocyanate ($R^1N=C=O$) is observed, according to Eq. 3.38a. This step is often followed by the loss of sulfur dioxide (SO_2) via a rearrangement of the NH_2 group to R^2, according to Eq. 3.38b.

$$[R^1NHCONHSO_2R^2\text{–H}]^- \rightarrow$$

$$R^1N=C=O + [H_2NSO_2R^2\text{–H}]^- \qquad (3.38a)$$

$$[H_2NSO_2R^2\text{–H}]^- \rightarrow [H_2NR^2\text{–H}]^- + SO_2 \qquad (3.38b)$$

For carbutamide, this results in deprotonated 4-aminobenzene-1-sulfonamide with *m/z* 171 ([H_2N—$C_6H_4SO_2NH_2$–H]$^-$) and deprotonated benzene-1,4-diamine with *m/z* 107 ([$C_6H_4(NH_2)_2$–H]$^-$), while for tolbutamide and gliclazide deprotonated 4-methylbenzenesulfonamide with *m/z* 170 ([CH_3—C_6H_4—SO_2NH_2–H]$^-$) and deprotonated 4-methylaniline with *m/z* 106 ([$CH_3C_6H_4NH$]$^-$) are observed (Section 4.4.1).

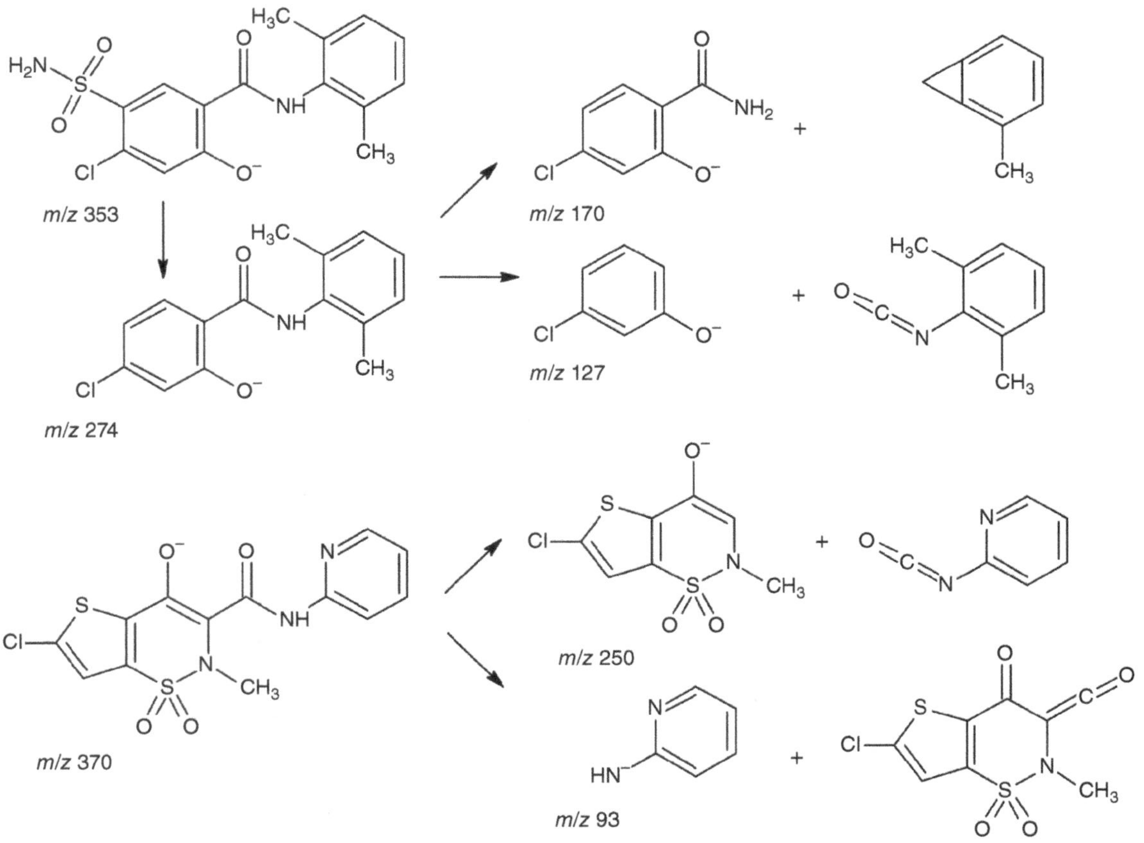

FIGURE 3.20 Fragmentation pathways for deprotonated amides, illustrated for xipamide ($[M-H]^-$ with *m/z* 353) and lornoxicam ($[M-H]^-$ with *m/z* 370).

In the fragmentation of benzoylphenylurea herbicides, cleavages are observed in the urea moiety of either of the N—C bonds of the carbonyl group with charge retention on either N-atom (Section 4.11.5.2). Barbiturates show common losses of hydrogen isocyanate (HN═C═O) and 87 Da (C_2HNO_3) as well as a common fragment ion with *m/z* 85 ($[C_2HN_2O_2]^-$, that is, the iminolate ion of 1,3-diazetidine-2,4-dione) (Section 4.2.6).

In the identification and amino-acid sequencing of peptides, fragmentation of $[M+H]^+$ is applied in the great majority of cases (Section 3.5.3). However, fragmentation of $[M-H]^-$ of a peptide may provide complementary information (Bowie et al., 2002; Bilusich & Bowie, 2009). The most prominent fragmentation reactions are the so-called α- and β-backbone cleavages (Figure 3.21). These involve cleavage of the peptide bond and results in two (complementary) fragment ions; the ions from the α-pathway are more abundant than those from the β-pathway, even in peptides where the C-terminal is amidated. A second series of fragment ions arise from the so-called β′-pathway, which involves a nucleophilic attack of the carboxylate ion (—COO^-, or —$CONH^-$ for an C-terminal amidated peptide) on any of the carbonyl groups in the peptide backbone, resulting in a cyclic intermediate, from which one or more amino-acid residues is eliminated, for example, a loss of 113 Da is observed due to the elimination of a Leu residue. The β′-cleavage is observed in peptides that lack Ser, Thr, Asp, Asn, Glu, or Gln and is in competition with the β-cleavage. Interestingly, the amino-acid pairs Asp/Asn and Glu/Gln are involved in other types of backbone cleavages via a nucleophilic attack from their side chain. Asp/Asn initiate backbone cleavages of the Asp/Asn N—C bond, whereas Glu/Gln direct similar cleavages of the Glu/Gln N—C bond via a lactone intermediate. Ser and Thr show characteristic substituent-chain fragmentation (Bowie et al., 2002).

3.7.7 Sulfonamides and Related Sulfones

Sulfonamides are prone to negative-ion formation. Thiazide diuretics have two sulfonamide moieties, one as part of a six-membered ring, and the other as a phenyl substituent (Section 4.1.4.1), both of which play an important role in the fragmentation: combined losses of hydrogen cyanide (HCN) and sulfur dioxide (SO_2), and substituted nitriles (R^3CN, from the C^3-position in the benzothiadiazine ring) and SO_2

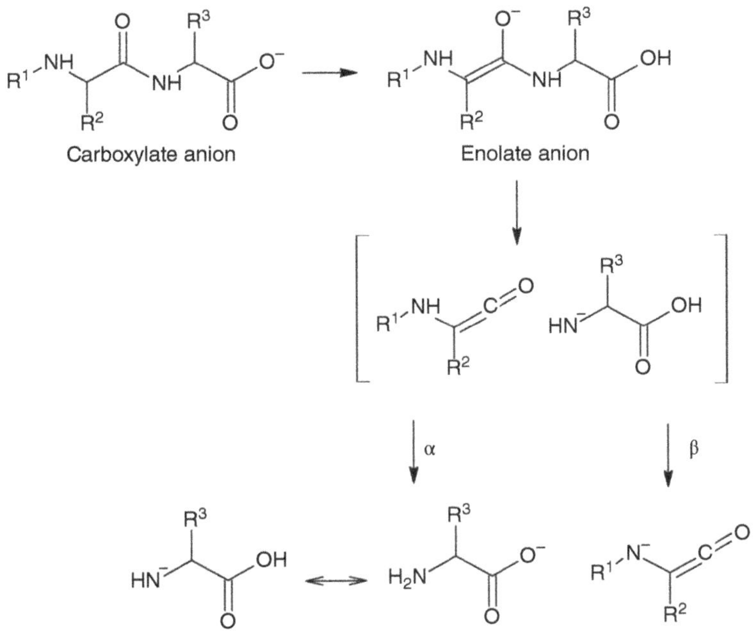

FIGURE 3.21 Fragmentation scheme of [M–H]$^-$ of a peptide.

are common. A fragment ion consistent with a deprotonated iminosulfene ($[NSO_2]^-$) with *m/z* 78 is also commonly observed. The loss of SO_2 from the phenyl sulfonamide moiety is also frequently observed, involving a rearrangement of the amino (—NH_2) group to the phenyl ring (Garcia et al., 2002). This is also observed for the COX-2 inhibitor valdecoxib (Section 4.3.4) and as secondary fragmentation in sulfonylurea antidiabetics (Section 4.4.1). H/D-exchange experiments on the fragmentation mechanisms of some thiazide diuretics have shown that there is molecular isomerization and the formation of ion–dipole complexes prior to fragmentation (Garcia et al., 2002).

$[NSO_2]^-$ with *m/z* 78 is observed for the thiazide diuretics (Section 4.1.4.1), whereas deprotonated sulfuramidous acid ($[H_2NSO_2]^-$) with *m/z* 80 is observed for the loop diuretics bumetanide and piretanide (Section 4.1.4.2). Sulfonamide antibiotics show the aminobenzenesulfonyl anion ($[H_2N—C_6H_4SO_2]^-$) with *m/z* 156 and secondary fragments thereof involving the loss of sulfur monoxide (SO), SO_2, or SO_2 and the cyano radical ($^\bullet$CN) (Section 4.8.1). A similar cleavage in the sulfonamide moiety of bosentan produces the 4-*tert*-butylbenzenesulfonyl anion ($[(CH_3)_3C—C_6H_4SO_2]^-$) with *m/z* 197 (Section 4.1.6.3). Losses of the iminosulfene ($HNSO_2$) and/or the sulfuramidous acid radical ($H_2NSO_2^\bullet$) are observed for chlorothiazide (Section 4.1.4.1) and for the COX-2 inhibitor valdecoxib (Section 4.3.4). Loss of a sulfonyl radical ($R^1SO_2^\bullet$) from the aromatic ring is observed for sildenafil (where R^1 is 1-methylpiperazine) (Section 4.1.6.4). *N*-Aryl methylsulfonamides ($ArNHSO_2CH_3$) like nimesulide (Section 4.3.4) may show losses of sulfene (CH_2SO_2) at low collision energy and a methylsulfonyl radical ($CH_3SO_2^\bullet$) at high collision energies. The methylsulfonyl anion ($[CH_3SO_2]^-$) with *m/z* 79 may also be observed.

Compounds with a methylenesulfonic acid group (—CH_2SO_3H), such as metamizole (Section 4.3.3.5), show the hydrogen sulfite anion ($[HSO_3]^-$) with *m/z* 81 (more abundant at low collision energy) and *m/z* 80 due to oxosulfane dioxide radical anion ($[SO_3]^{-\bullet}$) (more abundant at high collision energy). The loss of $H_2CSO_3^\bullet$ (94 Da) and a fragment ion with *m/z* 94 due to a methyliumsulfonate radical anion ($[H_2CSO_3]^{-\bullet}$) may also be observed.

Aromatic sulfate conjugates arising from Phase II drug metabolism show a characteristic loss of sulfur trioxide (SO_3) and a fragment ion with *m/z* 97 due to the hydrogen sulfate anion ($[HSO_4]^-$). In aliphatic sulfate conjugates, as observed in steroids for instance, the loss of sulfuric acid (H_2SO_4, 98 Da) and the fragment ion with *m/z* 97 are observed along with the less abundant $[SO_3]^{-\bullet}$ with *m/z* 80 and the sulfate radical anion ($[SO_4]^{-\bullet}$) with *m/z* 96.

3.7.8 Halogenated Compounds

Aliphatic halogenated compounds (R^1X, where X is F, Cl, Br, or I) show losses of hydrogen halides (HX) (Eq. 3.39a). Halogenated aromatic compounds (Ar^1X, where X is Cl, Br, or I) may show the loss of HX (Eq. 3.39b) and/or a halogen radical ($X^\bullet$) (Eq. 3.39c). Fluorinated aromatic compounds (Ar^1F) only show HF losses (Eq. 3.39b); no radical losses are seen. The alkyl or aryl group should contain a substituent that can be deprotonated and, in the case of HX loss, a substituent that can provide the required H-atom.

$$[R^1CH_2CH_2X\text{–}H]^- \rightarrow [R^1CH{=}CH_2\text{–}H]^- + HX \quad (X = F, Cl, Br, \text{or } I) \tag{3.39a}$$

$$[Ar^1X\text{–}H]^- \rightarrow [Ar^1\text{–}2H]^- + HX \quad (X = F, Cl, Br, \text{or } I) \tag{3.39b}$$

$$[Ar^1X\text{–}H]^- \rightarrow [Ar^1\text{–}H]^{-\bullet} + X^\bullet \quad (X = Cl, Br, \text{or } I) \tag{3.39c}$$

The loss of hydrogen fluoride (HF) is observed in fluorinated penicillin β-lactam antibiotics (Section 4.8.3.1) and fluorinated benzoylphenylurea herbicides (Section 4.11.5.2). In trifluoromethylaryl compounds, up to three consecutive HF losses can be observed, as is the case for polythiazide (Section 4.1.4.1), flufenamic acid and floctafenamic acid (Section 4.3.3), niflumic acid (Section 4.3.3.1), floctafenine (Section 4.3.3.2), and leflunomide and teriflunomide (Section 4.5.7). Proposing a structure for such fragment ions is not always straightforward, requiring molecular rearrangements in some cases. The loss of trifluoromethane (HCF_3) is observed for mefloquine (Section 4.10.2.4) and efavirenz (Section 4.10.3.2).

The loss of hydrogen chloride (HCl) is observed for chlorinated thiazide diuretics such as buthiazide and trichlormethiazide (Section 4.1.4.1), for diclofenac and related chlorinated compounds (Section 4.3.3), for chlorinated penicillin β-lactam antibiotics (Section 4.8.3.1), and for chlorinated phenols, for example, disinfectants and pesticides and their degradation products (Table 4.10.2 and Section 4.11.8). Competition between the loss of chlorine radical ($Cl^\bullet$) and of HCl is observed in the secondary fragmentation of tolfenamic acid (Section 4.3.3.1) and desloratadine (Section 4.5.2).

For brominated and iodated compounds, there is often a competition between the loss of bromine or iodine radicals ($Br^\bullet$ or $I^\bullet$) and of hydrogen bromide or hydrogen iodide (HBr or HI). Aliphatic brominated compounds show the loss of HBr, for example, for brominated barbiturates (Section 4.2.6) and for bromural and acebromural (Section 4.2.8). Aromatic brominated compounds show the loss of $Br^\bullet$ like for bromoxynil (Section 4.11.8) or the loss of HBr like for the fungicide dibromosalicylamide (Section 4.9.4). The loss of $I^\bullet$ is observed for iodine-containing X-ray contrast agents (Section 4.5.8), with even up to three consecutive losses of $I^\bullet$ being observed for acetrizoic acid (Section 4.5.8), and for the herbicide ioxynil (Section 4.11.8). In addition, bromide (Br^-) and iodide (I^-) ions with *m/z* 79 and 127, respectively, are frequently observed. Br^- is for instance observed for bromine-containing barbiturates (Section 4.2.6), bromural and acebromural (Section 4.2.8), the fungicides 5-bromo-4′-chlorosalicylamide and dibromosalicylamide (Section 4.9.4), and the antiviral drug brivudine (Section 4.10.3.4). I^- is observed for iodine-containing X-ray contrast agents (Section 4.5.8) and for the antiviral drug idoxuridine (Section 4.10.3.4).

3.7.9 Miscellaneous Compound Classes

Nitroaryl compounds are frequently analyzed in negative-ion mode. They show losses of a nitrosyl radical ($NO^\bullet$) and/or a nitryl radical ($NO_2^\bullet$). Examples are the fungicides 2,4-dinitrophenol and dichloran (Section 4.9.4) or several pesticides (Section 4.11.8).

A P—O $C^{5'}$ bond cleavage of the phosphoester moiety in nucleotides results in $[PO_3]^-$ and dihydrogen phosphate ($[H_2PO_4]^-$) ions with *m/z* 79 and 97, respectively. Cleavage of the phosphoester P—O $C^{5'}$ ribose bonds in fludarabine phosphate leads to fragment ions with charge retention on either one of the resulting products (Section 4.5.6.1); the *m/z* values of the two fragment ions add up to *m/z* ([M−H]−1), indicating a direct cleavage.

3.8 FRAGMENTATION OF METAL-ION CATIONIZED MOLECULES

Although in the data collection on fragmentation of drugs, pesticides, and related compounds in Chapter 4 attention is predominantly paid to the fragmentation of $[M+H]^+$ and $[M-H]^-$, it is useful to briefly discuss the fragmentation of (alkali) metal-ion cationized molecules ($[M+Metal]^+$). In positive-ion mode, in addition to $[M+H]^+$, other adduct ions may be generated like $[M+Na]^+$ or $[M+K]^+$, mostly due to residual Na^+ or K^+ in the sample. By the addition of particular cations to the sample, for example, Li^+ or Ag^+, adduct formation may be directed. In coordination-ESI-MS (CIS-MS), Ag^+ cationization is applied to achieve ionization of compounds that are not readily ionized under conventional ESI conditions (Bayer et al., 1997).

For most of the compounds discussed in Chapter 4, fragmentation of $[M+Na]^+$ or $[M+Li]^+$ would give little information. Frequently, no fragmentation occurs, except for the dissociation of the adduct ion itself, thus resulting in $[Na]^+$ with *m/z* 23 and $[Li]^+$ with *m/z* 7. However, for many biomolecules, fragmentation of $[M+Na]^+$, $[M+Li]^+$, or other metal-cationized molecules may provide different fragmentation characteristics, which can assist in structure elucidation. Some examples are briefly reviewed.

The fragmentation of protonated macrolide antibiotics like erythromycin A has been compared with their alkali-cationized molecules (Cerny et al., 1994). Protonation of erythromycin A takes place at the dimethylamino group of the C^6-aminosugar substituent of the 14-membered lactone. The fragment ions observed are consistent with the loss of the sugar substituents or with the loss of the lactone aglycone (Section 4.8.8). The $Alkali^+$ is localized on the highly oxygenated lactone aglycone. This results in a series of high-*m/z* value fragment ions, consistent with small-molecule neutral losses from the sugar substituents, cross-ring fragmentation of the sugar substituent, and finally loss of the complete sugar units with charge retention on the lactone aglycone (Cerny et al., 1994).

Due to easy chelation of the Alkali$^+$ with vicinal O-atoms, for example, hydroxy groups, typically in a tridentate interaction (Suzuki et al., 2009), [M+Alkali]$^+$ are readily observed for glycosides such as flavonoid glycosides (Cuyckens & Claeys, 2004; De Rijke et al., 2006; March & Brodbelt, 2008; Vukics & Guttman, 2010), saponins (Song et al., 2004) and iridoid glycosides (Madhusudanan et al., 2000; Es-Safi et al., 2007). Fragmentation of [M+Alkali]$^+$ often results in different information compared to fragmentation of [M+H]$^+$. This can be illustrated with the product-ion mass spectrum of globularin, an iridoid glycoside (Figure 3.22) (Es-Safi et al., 2007). Both [M+H]$^+$ and [M+Na]$^+$ with *m/z* 493 and 515, respectively, show the loss of the glucose moiety, either as glucose ($C_6H_{12}O_6$) or as dehydroglucose ($C_6H_{10}O_5$), although with lower abundance for [M+Na]$^+$. For [M+H]$^+$, this is followed by the loss of cinnamic acid ($C_6H_5CH{=}CHCOOH$). The protonated cinnamic acid with *m/z* 149 and the cynnamoyl ion ([$C_6H_5CH{=}CHC{=}O$]$^+$) with *m/z* 131 are observed as well. For [M+Na]$^+$, the ions with *m/z* 185 and 203 correspond to sodiated dehydroglucose ([$C_6H_{10}O_5$+Na]$^+$) and glucose ([$C_6H_{12}O_6$+Na]$^+$), respectively. Thus, H^+ is preferentially kept by the aglycone, whereas the Na^+ is preferentially kept by the glucose-containing fragment. Cross-ring fragmentation in the iridoid skeleton in [M+Na]$^+$ leads to fragment ions with *m/z* 283, containing the cinnamic acid moiety, and *m/z* 255, containing the glucose moiety (Es-Safi et al., 2007).

Metal-ion cationization with other than alkali ions has been extensively investigated for the structural characterization of flavonoid glycosides (March & Brodbelt, 2008; Satterfield & Brodbelt, 2000; Satterfield & Brodbelt, 2001; Pikulski & Brodbelt, 2003; Pikulski et al., 2007). Ion complexes like [Flavonoid−H+Metal+Ligand]$^+$ were generated with transition metal ions such as Cu^{2+} and Co^{2+}, and auxiliary neutral ligands such as 2,2′-bipyridine or 4,7-diphenyl-1,10-phenanthroline. In addition to up to a 100-fold response improvement compared to [M+H]$^+$, such complexes can be used to differentiate between 1,2- and

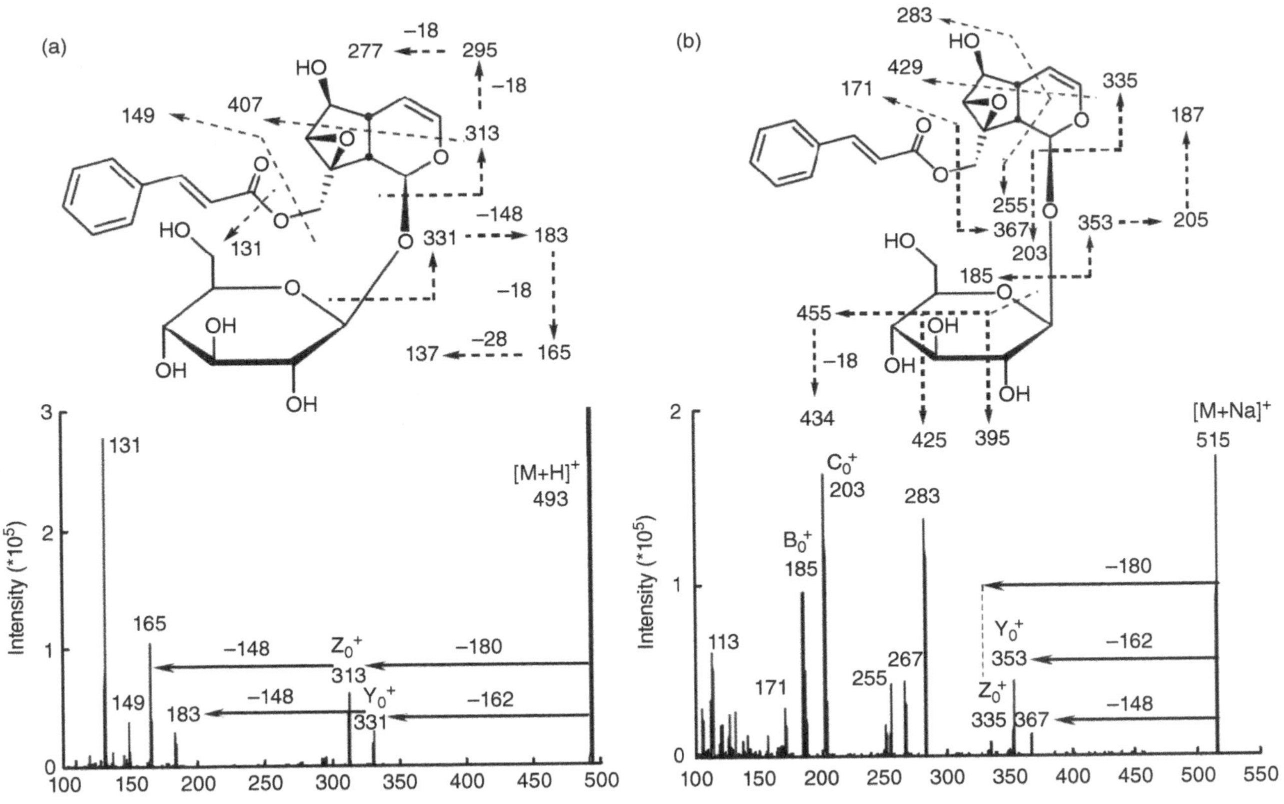

FIGURE 3.22 Comparison of the fragmentation of the iridoid glycoside globularin using (a) the ESI-generated [M+H]$^+$ as precursor ion and (b) the ESI-generated [M+Na]$^+$ as precursor ion. (Source: Es-Safi et al., 2007. Reproduced with permission of Wiley.)

1,6-linked disaccharide isomers of flavonoid glycosides (Satterfield & Brodbelt, 2001; Pikulski & Brodbelt, 2003). The research resulted in a toolbox with enhanced structure elucidation capabilities for flavonoids and their glycosides based on ESI-MS, ion-trap MS^n, and tunable transition metal–ligand complexation (Pikulski et al., 2007).

The fragmentation of oligosaccharides, glycans, and glycopeptides has been reviewed in detail (Zaia, 2004). Differences in the fragmentation characteristics between $[M+H]^+$, $[M-H]^-$, and $[M+Alkali]^+$ have been rationalized in terms of charge-induced cleavages by a localized charge in the case of (de)protonation and a CRF mechanism for $[M+Alkali]^+$. It was established that $[M+Alkali]^+$ are more prone to cross-ring cleavages than $[M+H]^+$ (Hofmeister et al., 1991; Lemoine et al., 1991; Asam & Glish, 1997). Thus, apart from the cleavage of the glycosidic bond, also observed with $[M+H]^+$, cross-ring cleavages are more readily observed for $[M+Alkali]^+$. This enables differentiation between the different glycosidic linkages (Zaia, 2004).

In various ways, $Alkali^+$-cationization also plays an important role in the characterization of lipids (Murphy & Axelsen, 2011; Cajka & Fiehn, 2014). Mono-, di-, and triacylglycerols are not readily ionized in ESI-MS, unless a lithium or an ammonium salt is added to the mobile phase resulting in $[M+Li]^+$ or $[M+NH_4]^+$ (Cajka & Fiehn, 2014). Fragmentation of $[M+H]^+$ of triacylglycerols is briefly discussed in Section 3.6.4. When $[M+Li]^+$ of a (16:0/18:1/18:0)-triacylglycerol (palmitoyl, oleoyl, and stearoyl; $[M+Li]^+$ with *m/z* 867) is fragmented, six instead of three peaks are observed with *m/z* values in the mid-region of the spectrum (Eq. 3.40; FA = RCOO).

$$[C_3H_5\text{–}FA^1(16:0)/FA^2(18:1)/FA^3(18:0)+Li]^+ \rightarrow [C_3H_4\text{–}FA^2(18:1)/FA^3(18:0)+H]^+ + FA^1(16:0)COOLi \quad (3.40a)$$

$$[C_3H_5\text{–}FA^1(16:0)/FA^2(18:1)/FA^3(18:0)+Li]^+ \rightarrow [C_3H_4\text{–}FA^2(18:1)/FA^3(18:0)+Li]^+ + FA^1(16:0)COOH \quad (3.40b)$$

$$[C_3H_5\text{–}FA^1(16:0)/FA^2(18:1)/FA^3(18:0)+Li]^+ \rightarrow [C_3H_4\text{–}FA^2(18:1)/FA^3(18:0)] + [FA^1(16:0)COOH+Li]^+ \quad (3.40c)$$

$$[C_3H_5\text{–}FA^1(16:0)/FA^2(18:1)/FA^3(18:0)+Li]^+ \rightarrow [C_3H_5OLi\text{–}FA^2(18:1)/FA^3(18:0)] + [FA^1(16:0)\text{—}C{=}O]^+ \quad (3.40d)$$

They occur in pairs with an *m/z* difference of 6. They can be attributed to the loss of either the lithium salt of the fatty acid (R^1COOLi) or the fatty acid (R^1COOH), that is, the ions with *m/z* 605 and 611 are due to the loss of the sn-1 palmitic acid (16:0), according to Eqs 3.40a and 3.40b, respectively. The less abundant ions with *m/z* 585 and 579 are due to the loss of the sn-2 oleic acid (18:1) and its lithium salt, respectively. In addition, lithiated fatty acids ($[R^1COOH+Li]^+$), according to Eq. 3.40c, and the fatty acid acylium ions ($[R^1C{=}O]^+$), according to Eq. 3.40d, are observed, for example, with *m/z* 291 and 267, respectively, for the sn-3 stearic acid (18:0) (*m/z* difference of 24). Ions due to the combined losses of either the sn-1 or the sn-3 fatty acid and the sn-2 fatty acid may be observed as well. This elimination involves the loss of a free fatty acid and an α,β-unsaturated fatty acid, thus leading to ions with *m/z* 331 involving the loss of the sn-1 palmitic acid and the sn-2 oleic acid, and with *m/z* 303 involving the loss of the sn-3 stearic acid and the sn-2 oleic acid (Hsu & Turk, 1999a).

An important contribution to the field of lipid analysis is the discovery of CRF in $[M-H]^-$ and $[(M-H+Li)+Li]^+$ (cf. Section 2.8) of fatty acids (Jensen et al., 1985; Adams & Gross, 1986). High-energy CID of FAB-MS-generated $[(M-H+Li)+Li]^+$ results in C—C bond cleavages along the alkyl chain yielding a neutral alkene, H_2, and an unsaturated fatty acid product ion, according to Eq. 3.41.

$$[RCH_2CH_2CH_2CH_2(CH_2)_nCOOLi+Li]^+ \rightarrow RCH{=}CH_2 + H_2 + [H_2C{=}CH(CH_2)_nCOOLi+Li]^+ \quad (3.41)$$

For a saturated fatty acid, a series of ions with an *m/z* difference of 14 (methylene (CH_2)) is observed. In an unsaturated fatty acid, the presence of a double bond interrupts this regular pattern and results in an *m/z* difference of 54. The cleavage of the allylic bond is favored over the cleavage of both the double and the vinylic bonds. CRF in high-energy CID has become an important tool in the determination of double-bond positions in fatty acids (Contado & Adams, 1991; Gross, 2000). Although CRF, as observed in high-energy CID, is not observed in ESI-MS with low-energy CID in a TQ instrument, a wealth of structural information can still be obtained, enabling discrimination between isomeric unsaturated fatty acids (Hsu & Turk, 1999b; Levery et al., 2000; Hsu et al., 2002). The fragmentation of various glycerophospholipid classes after ESI-MS and low-energy CID in both TQ MS–MS and ion-trap MS^n, using $[M+H]^+$, $[M+Li]^+$, and/or $[(M-H+Li)+Li]^+$ as precursor ion in the positive-ion mode and using $[M-H]^-$ as precursor ion in the negative-ion mode, has been reviewed in considerable detail (Hsu & Turk, 2009).

The fragmentation of $[M+Na]^+$ of peptides is significantly different from that of $[M+H]^+$. Complexation of Na^+ at the C-terminal carboxylic group polarizes the carbonyl

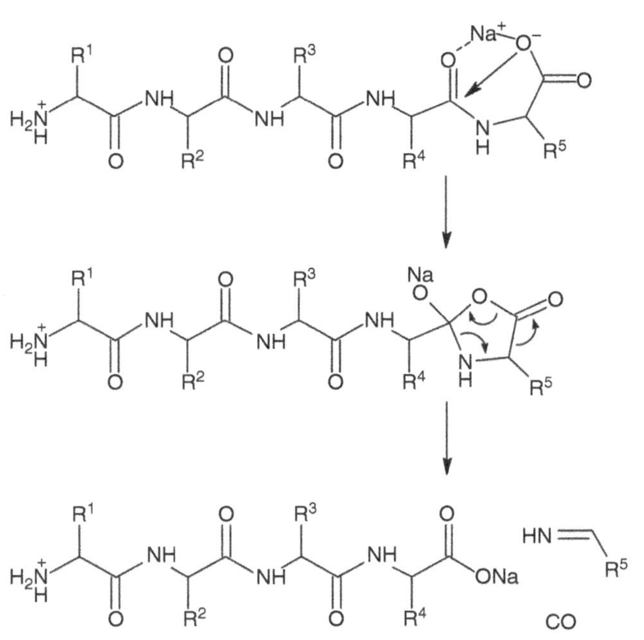

FIGURE 3.23 Proposed mechanism for the loss of the C-terminal amino-acid residue from $[M+Na]^+$ of a peptide. (Source: Grese et al., 1989. Reproduced with permission of American Chemical Society.)

bond of the adjacent amino-acid residue, allowing a nucleophilic attack by the negative O-atom. This catalyzes the hydrolysis of the amide bond, which leads to the loss of the C-terminal amino-acid residue, as CO and an imine (HN=CHR, where R is the side chain of the C-terminal amino-acid) (Figure 3.23). In effect, rearrangement of a hydroxy (OH) of the C-terminal amino acid takes place (Grese et al., 1989; Kulik et al., 1989). Thus, the $[b_{n-1}+Na+OH]^+$ ion is generated. In combination with MS^n in an ion-trap instrument, this enables C-terminal amino-acid sequencing.

Interesting fragmentation behavior is observed for CIS-MS-generated $[M+Ag]^+$ of amines, diamines, aminocarboxylic acids, and alkyl benzyl ethers (Grewal et al., 2000; Shi et al., 2004; Schäfer et al., 2009). Both Ag^+-containing and non-Ag^+-containing fragment ions are observed. The latter can be considered as iminium ions, ($[R^1R^2C{=}NH_2]^+$ or $[R^1CH{=}NH_2]^+$), generated by the loss of silver hydride (AgH) due to a 1,2-elimination. If methyl (CH_3—) and phenyl (C_6H_5—) substituents are present at the α-*C* to the amino-*N*, the loss of silver methide ($AgCH_3$) and silver phenide (AgC_6H_5) can be observed (Grewal et al., 2000; Shi et al., 2004). The hypothesis of 1,2-elimination has been tested by [D]-labeling studies (Schäfer et al., 2009). The losses of AgH or $AgCH_3$ have also been observed in the characterization of Ag^+-cationized ferrocenyl catalyst complexes (Martha et al., 2010).

3.9 GENERATION OF ODD-ELECTRON IONS IN ESI-MS, APCI-MS, AND APPI-MS

In both positive-ion and negative-ion ESI-MS, APCI-MS, and APPI-MS, mostly EE^+/EE^- ions, for example, $[M+H]^+$ or $[M-H]^-$, are generated. However, over the years, there has been considerable interest in the formation of $OE^{+\bullet}$ or $OE^{-\bullet}$ under ESI conditions (Vessecchi et al., 2007). Since an ESI source can be considered a controlled-current electrolytic flow cell, it should be possible to generate molecular ions $M^{+\bullet}$ or $M^{-\bullet}$ by oxidation or reduction of neutral molecules, respectively. Electrochemical processes in the ESI source have been extensively investigated (Van Berkel et al., 1992, 1998), with special interest in the formation of $M^{+\bullet}$. Topics related to the electrochemical processes in ESI have been discussed by others as well (Diehl & Karst, 2002). However, it seems that under electrochemical conditions favoring $M^{+\bullet}$ the proton concentration increases as well, thus stimulating $[M+H]^+$ formation (Van Berkel et al., 1997). A similar situation is observed in the development of atmospheric-pressure photoionization (APPI) for LC–MS (Robb et al., 2000). Photoionization could generate $M^{+\bullet}$, whereas the aqueous conditions in the reversed-phase mobile phases used in LC–MS analysis favor ion–molecule reactions that lead to the generation of $[M+H]^+$ rather than $M^{+\bullet}$ (Syage, 2004) (Section 1.2.5.3).

Nevertheless, $M^{+\bullet}$ or $M^{-\bullet}$ may be observed in ESI for a number of compound classes including polycyclic aromatic hydrocarbons, polyenes such as carotenoids, metallocenes, quinones, and fullerenes (Vessecchi et al., 2007).

Compounds that readily show $OE^{+\bullet}$ along with EE^+ under ESI-MS and APCI-MS conditions are carotenoids (van Breemen, 1995; van Breemen et al., 1996; Guaratini et al., 2007; van Breemen et al., 2012; Bijttebier et al., 2013). Carotenoids are naturally occurring compounds with antioxidant properties that are predominantly found in photosynthesizing organisms such as green plants and algae. Three subclasses of carotenoids can be discriminated: carotenes, xanthophylls, and xanthophyll esters. Carotenoids are tetraterpenoids, produced from 8 isoprene molecules; they contain 40 C-atoms. Whereas carotenes are non-oxygen-containing hydrocarbons, the xanthophylls do contain oxygen (as a hydroxy, epoxide, and/or keto group). Xanthophylls with a hydroxy group may be conjugated or esterified with fatty acids or glycosides. Carotenes show both $M^{+\bullet}$ and $[M+H]^+$, together with some characteristic fragments in positive-ion mode, and $[M-H]^-$ and $M^{-\bullet}$ in negative-ion mode. The ratio between $OE^{+\bullet}$ and EE^+ can be correlated to the water content of the mobile-phase (Bijttebier et al., 2013). Xanthophylls and their esters favor $[M+H]^+$ formation. Product-ion mass spectra of carotenoids frequently show fragmentation characteristics related to both the $OE^{+\bullet}$ and EE^+ precursor ion.

In positive-ion mode, α-carotene with $M^{+\bullet}$ (with *m/z* 536) as precursor ion shows the loss of *iso*-butene

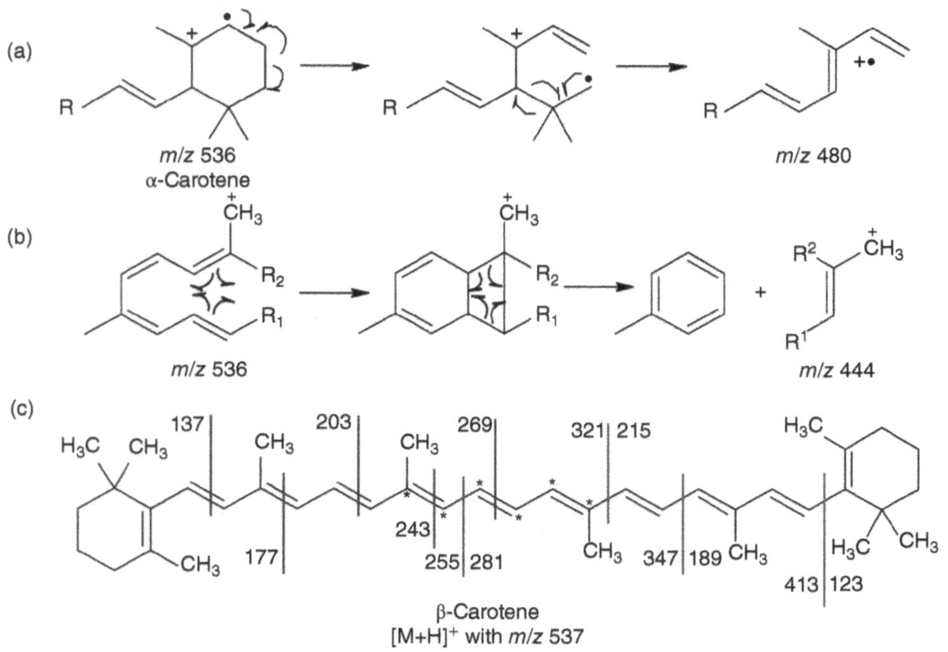

FIGURE 3.24 Fragmentation of carotenes: (a) retro-Diels–Alder fragmentation of the ionone ring of the ESI-generated $M^{+\bullet}$ of α-carotene, showing the loss of C_4H_8, (b) Edmunds–Johnstone fragmentation mechanism for $M^{+\bullet}$, leading to the loss of $C_6H_5CH_3$, and (c) various fragment ions observed from $[M+H]^+$ of β-carotene (* indicates positions of [^{13}C]-labels in [$^{13}C_6$]-β-carotene).

($(CH_3)_2C{=}CH_2$) due to an RDA fragmentation of the six-membered ionone ring, leading to an ion with *m/z* 480 (Figure 3.24a). The RDA fragmentation reaction on the β-carotene isomer would result in the loss of ethene (C_2H_4). In addition, an ion with *m/z* 444 is observed, which is consistent with a loss of toluene ($C_6H_5CH_3$) from the structure backbone. Combined loss of $(CH_3)_2C{=}CH_2$ and $C_6H_5CH_3$ results in a fragment ion with *m/z* 388. These losses are more abundant for $M^{+\bullet}$ than for $[M+H]^+$. The loss of $C_6H_5CH_3$ arises from the structure backbone C-atoms between C^{10} and $C^{10'}$ in α- or β-carotene (or between C^6 and $C^{6'}$ in linear carotenes). This loss can be explained by an Edmunds–Johnstone mechanism involving a [2+2]-cyclization via a four-membered ring transition state, with subsequent homolytic rupture of the newly formed cyclobutane σ-bonds (Figure 3.24b) (Edmunds & Johnstone, 1965). In some cases, the loss of xylene ($CH_3C_6H_4CH_3$) is also observed, that is, from C^8 to C^{14} or from $C^{8'}$ to $C^{14'}$.

The fragmentation of β-carotene ($[M+H]^+$ with *m/z* 537) as precursor ion is different. A series of fragments are observed consistent with cleavages at the various double bonds of the structure backbone (Figure 3.24c) (Andreoli et al., 2004; van Breemen et al., 2012). Apart from the losses indicated, fragment ions consistent with the loss of methylcyclopentadiene ($C_5H_5CH_3$, 80 Da) or $C_6H_5CH_3$ (92 Da) are observed. By studying [$^{13}C_6$]-labeled β-carotene, with the position of the labels indicated by the asterisks in Figure 3.24c, it was demonstrated that the $C_5H_5CH_3$ or $C_6H_5CH_3$ losses involve backbone carbons C^{11} to C^{15} and C^{10} to C^{15}, respectively. For the [$^{13}C_6$]-labeled analog, losses of 83 and 95 Da were observed (van Breemen et al., 2012). This is somewhat in contrast to proposals made regarding the $C_6H_5CH_3$ loss in $M^{+\bullet}$, discussed earlier. Differentiation between α- and β-carotene with $[M+H]^+$ as precursor ion can be based on the occurrence of the fragment ion with *m/z* 123 (the ionone ring, $[C_9H_{15}]^+$), which is not observed for β-carotene (Figure 3.24c).

In negative-ion mode, α- and β-carotene with $M^{-\bullet}$ as precursor ion show losses of the cyclopentadienyl methyl radical ($C_5H_5CH_2^{\bullet}$), of $C_6H_5CH_3$, and of $CH_3C_6H_4CH_3$. In α-carotene, a RDA fragmentation leads to the loss of $(CH_3)_2C{=}CH_2$, which is not observed for β-carotene. It is interesting to note that the MS–MS spectra with $M^{+\bullet}$ or $M^{-\bullet}$ as precursor ion show less fragmentation than the spectra with $[M+H]^+$ as precursor ion (van Breemen et al., 2012).

3.10 USEFUL TABLES

A practical way to summarize part of the information given in this and the next chapter is to provide tables with characteristic neutral losses (Tables 3.3 and 3.5) and

TABLE 3.3 Small neutral losses in positive-ion mode.

Neutral Loss (Da)		Formula	From Which Functional Group and/or Compound Class the Neutral Loss May Originate
15	15.0235	$^{\bullet}CH_3$	Aromatic methoxy-ethers, quaternary ammonium compounds, *N*-methyl anilines
16	15.9949	O	Atomic oxygen, from *N*-oxide
16	16.0313	CH_4	Quaternary ammonium compounds, aromatic dimethoxy-ethers
17	17.0027	$HO^{\bullet}$	Hydroxyl radical, from *N*-oxide, nitroaryl compounds
17	17.0266	NH_3	Primary amines and amides, cyclic amines (piperidine, piperazine), amino acids
18	18.0106	H_2O	Very common, from alcohols, aldehydes, ketones, carboxylic acids, *N*-oxides
20	20.0062	HF	Fluorine-containing compounds
26	26.0157	C_2H_2	Ethyne, from aromatic rings
27	27.0109	HCN	Aromatic nitriles, *N*-heteroaromatic compounds, (substituted) anilines
28	27.9949	CO	Very common, for example, from aliphatic acylium ions, cyclic ketones, phenols, aromatic aldehydes
28	28.0062	N_2	Tetrazole derivatives
28	28.0313	C_2H_4	Ethene, common, from compounds with *N*- or *O*-ethyl substituents, for example, ethers, amines, esters; diethyl phosphates and phosphoro(di)thioate pesticides
29	29.0027	$^{\bullet}CHO$	Formyl radical
29	29.0266	$HN{=}CH_2$	Methanimine, for example, from secondary amines, piperazine ring
29	29.0391	$^{\bullet}C_2H_5$	Ethyl radical
30	29.9742	$O_2 + H_2$	Thermally induced conversion of NO_2 to NH_2 in nitroaryl compounds
30	29.9980	$^{\bullet}NO$	Nitrosyl radical from nitroaryl compounds
30	30.0106	$H_2C{=}O$	Formaldehyde, for example, from aromatic methoxy compounds, benzyl alcohols
31	31.0058	HNO	Nitroxyl from nitroaryl compounds
31	31.0184	$CH_3O^{\bullet}$	Aromatic methoxy compounds
31	31.0422	CH_3NH_2	Aliphatic methylamino substituent, piperidine, *N*-methyl-piperazine
32	31.9721	S	Atomic sulfur, from phenothiazine
32	32.0262	CH_3OH	Common, from methyl ethers and esters, dimethyl phosphate and phosphoro(di)thioate pesticides
33	33.0340	$H_2O + {}^{\bullet}CH_3$	Cathinone designer drugs
34	33.9877	H_2S	Thiol-containing compounds
35	34.9683	$Cl^{\bullet}$	Chlorine-containing compounds
35	35.0371	$H_2O + NH_3$	Loss of water and ammonia
36	35.9767	HCl	Chlorine-containing compounds
36	36.0211	$2 \times H_2O$	Dihydroxy compounds
41	41.0266	$H_3C{-}C{\equiv}N$	Methyl *N*-heteroaromatic compounds
42	42.0106	$H_2C{=}C{=}O$	Ethenone, common, for example, from acetate, *N*-or *O*-acetyl derivatives
42	42.0218	$HN{=}C{=}NH$	Methanediimine, common, from *N*-heteroaromatic compounds, for example, 2-amine-1,3,5-triazines; guanidine derivatives
42	42.0470	$CH_3CH{=}CH_2$	Propene, from *N*- or *O*-(*n*- or *iso*-)propyl substituents, for example, ethers, amines, esters, amides; *iso*-propyl-phenyl derivatives
43	43.0058	$HN{=}C{=}O$	Hydrogen isocyanate, for example, from cyclic amides, carbamates, xanthines
43	43.0171	HN_3	Tetrazole derivatives
43	43.0422	$H_3C{-}N{=}CH_2$	Secondary *N*-methyl amines, piperazine
43	43.0548	$^{\bullet}C_3H_7$	Propyl or *iso*-propyl radical
44	43.9898	CO_2	Very common, from anhydrides, aryl carboxylic acids, carbamates
44	44.0136	$H_2NCO^{\bullet}$	Carbamoyl radical
44	44.0262	CH_3CHO	Ethanal, from *N*-CH_2CH_2OH derivatives
45	45.0215	$H_2O + HCN$; $HCONH_2$	Cathinones and some FUBINACA designer drugs; formamide, for example, from dibenzodiazepines
45	45.0579	$(CH_3)_2NH$; $CH_3CH_2NH_2$	Aliphatic *N,N*-dimethyl amino and *N*-ethyl amino substituents, piperidine

TABLE 3.3 *(Continued)*

Neutral Loss (Da)		Formula	From Which Functional Group and/or Compound Class the Neutral Loss May Originate
46	45.9924	$^{\bullet}NO_2$	Nitryl radical from nitroaryl compounds
46	46.0055	HCOOH	Common, from amino acids, aliphatic carboxylic acids
46	46.0419	C_2H_5OH	Ethyl ethers or esters
47	46.9950	$CH_3S^{\bullet}$	Aromatic methyl thioethers
47	47.0007	HNO_2	Rare: nitroaryl compounds
48	47.9670	SO	Aromatic sulfonyl ion
48	48.0034	CH_3SH	Thioethers
50	49.9923	CH_3Cl	
50	49.9968	CF_2	Trifluoromethyl derivatives
54	54.0470	C_4H_6	
56	55.9898	C_2O_2	Ethenedione, from benzodiazepine with an N^1-desmethyl-α-hydroxyacetamide skeleton
56	56.0262	$CH_3CH{=}C{=}O$	From amides
56	56.0626	C_4H_8; $2\times C_2H_4$	Butene, from *N*- or *O*-(*n*, *sec*, *iso*, *tert*-)butyl substituents, for example, ethers, esters, amines, amides; *tert*-butyl-phenyl derivatives; diethyl phosphates and phosphoro(di)thioate pesticides
57	57.0215	$H_3C{-}N{=}C{=}O$	*N*-Methyl carbamates, xanthines
57	57.0215	C_2H_3NO	Gly residue mass
57	57.0579	C_3H_7N	*N*-Methyl-piperazine, aliphatic *N*-methyl, *N*-ethyl amino substituents
58	58.0419	$(CH_3)_2C{=}O$	Acetone from D-ring of steroids
59	59.0484	$HN{=}C(NH_2)_2$	Guanidine derivatives
59	59.0735	$C_3H_7NH_2$	(*iso*-)Propylamine derivatives, for example, β-blockers
60	60.0211	CH_3COOH	Acetic acid ester, combined loss of $H_2C{=}C{=}O$ and water
60	60.0581	C_3H_7OH	(*iso*-)Propanol, from (*iso*-)propyl ethers or esters, or loss of water and C_3H_6
64	63.9441	S_2	Aromatic disulfides
64	63.9619	SO_2	Aromatic sulfoxides and sulfonamides, sulfonic acids, sulfonates
66	65.9776	$H_2 + SO_2$	Sulfonamide antibiotics
68	68.0262	C_4H_4O	From D-ring in hydroxy-ethenyl 3-keto-Δ^4 steroids
68	68.0374	$C_3H_4N_2$	Imidazole derivatives
69	69.0327	$C_2H_3N_3$	Triazole derivatives
70	70.0351	$C_3H_6N_2$	Pyrazoline derivatives
71	71.0007	$HN{=}C{=}O + CO$	Xanthines
71	71.0371	C_3H_5NO	Ala residue mass
71	71.0735	C_4H_9N	Pyrrolidinophenone designer drugs
73	73.0528	C_3H_7NO	*N'*,*N'*-Dimethyl phenylurea herbicides
73	73.0891	$C_4H_{11}N$	*N*-*tert*-Butyl derivatives
74	74.0368	$HCOOCH_2CH_3$	Ethyl ester ACE inhibitors
74	74.0732	$C_4H_8 + H_2O$; C_4H_9OH; $2\times C_2H_4 + H_2O$	*N*-*tert*-Butanol derivatives, for example, β-blockers and β-adrenergic receptor agonists, diethyl phosphates and phosphoro(di)thioate pesticides
75	75.0320	$C_2H_5NO_2$	Glycine or glutathione conjugates
77	77.0841	$C_3H_7NH_2 + H_2O$	(*n*- or *iso*-)Propylamine and water, for example, β-blockers
78	78.0470	C_6H_6	Phenyl derivatives, for example, benzodiazepines
79	78.9178	$Br^{\bullet}$	Bromine-containing compounds
79	79.0422	C_5H_5N	Pyridine derivatives
80	79.9256	HBr	Bromine-containing compounds
80	79.9806	$H_2NSO_2^{\bullet}$	Sulfuramidous acid radical
81	80.9641	$HSO_3^{\bullet}$	Sulfonic acids
82	82.0783	C_6H_{10}	Cyclohexane derivatives
84	84.0575	C_5H_8O	From 3-keto-Δ^1-steroid (A-ring C^1–C^4 and C^{19})
85	85.0164	$CH_3{-}N{=}C{=}O + CO$	Xanthines
85	85.0891	$C_5H_{11}N$	Piperidine derivatives
87	87.0320	$C_3H_5NO_2$	Ser residue mass
87	87.0684	C_4H_9NO	Morpholine derivatives
87	87.1048	$C_5H_{13}N$	Phenothiazines with 3-(*N*,*N*-dimethylamino)-2-methylpropyl substituent

TABLE 3.3 *(Continued)*

Neutral Loss (Da)		Formula	From Which Functional Group and/or Compound Class the Neutral Loss May Originate
91	91.0997	$C_4H_9NH_2 + H_2O$	*tert*-Butylamine and water, for example, from β-blockers and β-adrenergic receptor agonists with *N-tert*-butyl substituent
93	93.0579	C_6H_7N	*N*-Aniline derivatives, for example, sulfonamide antibiotics
97	97.0528	C_5H_7NO	Pro residue mass
98	98.0190	C_5H_6S	Methyl-thiophene
99	99.0684	C_5H_9NO	Val residue mass
100	100.1000	$C_5H_{12}N_2$	Phenothiazines with a 3-(4-methylpiperazin-1-yl)propyl substituent
101	101.0477	$C_4H_7NO_2$	Thr residue mass
103	103.0092	C_4H_5NOS	Cys residue mass
104	103.0422	$C_6H_5—C{\equiv}N$	Benzonitrile, for example, from benzodiazepines
104	104.0626	$C_6H_5—CH{=}CH_2$	Ethylbenzene derivatives
112	112.0080	C_6H_5Cl	Chlorobenzene derivatives, for example, benzodiazepines
113	113.0841	$C_6H_{11}NO$	Leu or Ile residue mass
114	114.0429	$C_4H_6N_2O_2$	Asn residue mass
115	115.0269	$C_4H_5NO_3$	Asp residue mass
115	115.0997	$C_6H_{13}NO$	β-Blockers with a 1-(*iso*-propylamino)-3-phenoxypropan-2-ol skeleton
119	119.0371	$C_6H_5—N{=}C{=}O$	Phenyl isocyanate
121	121.0198	$C_3H_7NO_2S$	Cysteine conjugates
121	121.0328	$FC_6H_4—C{\equiv}N$	Fluorobenzonitrile, for example, from fluorinated benzodiazepines
122	122.0732	$CH_3O—C_6H_4CH_3$	Methoxytoluene, for example, from 25X-NBOMe designer drugs
126	126.0082	$C_2H_7O_4P$	Dimethyl phosphate pesticides
127	126.9045	$I^{\bullet}$	Iodine-containing compounds
128	127.9123	HI	Iodine-containing compounds
128	128.0586	$C_5H_7NO_3$	Gln residue mass
128	128.0626	$C_{10}H_8$	Naphthoylindole class synthetic cannabinoids
128	128.0950	$C_6H_{12}N_2O$	Lys residue mass
128	128.1313	$C_7H_{16}N_2$	Phenothiazines with a 3-(4-methylpiperazin-1-yl)propyl substituent
129	129.0426	$C_5H_7NO_3$	Pyroglutamic acid from glutathione conjugates
129	129.0426	$C_5H_7NO_3$	Glu residue mass
131	131.0405	C_5H_9NOS	Met residue mass
136	136.0289	$C_4H_9O_3P$	Aryl diethylphosphate and phosphorothioate substituents (after thiono–thiolo rearrangement)
137	137.0032	$ClC_6H_4—C{\equiv}N$	Chlorobenzonitrile, for example, from chlorinated benzodiazepines
137	137.0589	$C_6H_7N_3O$	His residue mass
137	137.0841	$C_8H_{11}NO$	1-(2-Methoxyphenyl)methanamine, from 25X-NBOMe designer drugs
142	141.9854	$C_2H_7O_3PS$	Dimethyl phosphorothioate pesticides
146	146.0691	$C_5H_{10}N_2O_3$	Glutamine from glutathione or GluCys conjugates
147	147.0684	C_9H_9NO	Phe residue mass
152	152.0061	$C_4H_9O_2PS$	Diethyl phosphorothioate pesticides
154	154.0395	$C_4H_{11}O_4P$	Aliphatic diethyl phosphate substituents
155	155.0041	$HN{=}C_6H_4{=}SO_2$	Sulfonamide antibiotics
156	156.0535	$C_6H_8N_2O_3$	Cephalosporin β-lactam antibiotics
156	156.1011	$C_6H_{12}N_4O$	Arg residue mass
157	157.0197	$C_6H_7NO_2S$	Cephalosporin β-lactam antibiotics
158	157.9625	$C_2H_7O_2PS_2$	Dimethyl phosphorodithioate pesticides
158	158.0943	$C_8H_{14}O_3$	Macrolide antibiotics with cladinose substituent
159	159.0354	$C_6H_9NO_2S$	Penicillin β-lactam antibiotics
162	162.0528	$C_6H_{10}O_5$	From hexose glycosides
163	163.0303	$C_5H_9NO_3S$	*N*-Acetylcysteine conjugates
163	163.0633	$C_9H_9NO_2$	Tyr residue mass
164	164.0637	$C_{10}H_9FO$	From butyrophenones
176	176.0256	$C_5H_8N_2O_3S$	Glycylcysteinyl conjugates
176	176.0321	$C_6H_8O_6$	Glucuronic acid conjugates

TABLE 3.3 (*Continued*)

Neutral Loss (Da)		Formula	From Which Functional Group and/or Compound Class the Neutral Loss May Originate
183	183.0756	$C_6H_9N_5O_2$	Sulfonylurea herbicides with 4-methoxy-6-methyl-triazinyl substituent
186	185.9938	$C_4H_{11}O_2PS_2$	Aliphatic diethyl phosphorodithioate substituents
186	186.0793	$C_{11}H_{10}N_2O$	Trp residue mass
188	187.9796	$C_8H_6Cl_2O$	Some imidazole fungicides
194	194.0427	$C_6H_{10}O_7$	(Benzylic or aliphatic) glucuronic acid conjugates
198	198.0753	$C_7H_{10}N_4O_3$	Sulfonylurea herbicides with 4,6-dimethoxy-pyrimidinyl substituent
273	273.0961	$C_{10}H_{15}N_3O_6$	Glutathione conjugates
307	307.0838	$C_{10}H_{17}N_3O_6S$	Glutathione conjugates

characteristic products ions (Tables 3.4 and 3.6). Separate sets of tables have been made for the positive-ion (Tables 3.3 and 3.4) and the negative-ion modes (Tables 3.5 and 3.6). For each neutral loss and characteristic fragment ion, the tables provide nominal mass, accurate mass and elemental composition, as well as an indication of the typical compound classes for which such losses are observed, in some cases very specific.

TABLE 3.4 Characteristic fragment ions in positive-ion mode.

m/z		Formula	From Which Functional Group and/or Compound Class the Fragment Ion May Originate
15	15.0229	$[CH_3]^+$	Methyl cation
18	18.0338	$[NH_3+H]^+$	Protonated ammonia
30	30.0338	$[H_2C{=}NH+H]^+$	Protonated methanimine
32	32.0495	$[CH_3NH_2+H]^+$	Protonated methylamine
43	43.0178	$[CH_3C{=}O]^+$	Acetyl cation
43	43.0542	$[C_3H_7]^+$	*iso*-Propyl cation
44	44.0495	$[C_2H_5N+H]^+$	Protonated ethanimine
46	46.0651	$[C_2H_5NH+H]^+$	Protonated dimethylamine or ethylamine from alkyl *N,N*-dimethyl or *N*-ethyl amino substituent
56	56.0495	$[C_3H_5N+H]^+$	β-Blockers
57	57.0699	$[(CH_3)_3C]^+$	*tert*-Butyl carbocation, from compounds with *tert*-butyl substituent, for example, ethers, esters, or amines
58	58.0287	$[C_2H_4NO]^+$	*N*-Methylcarbamoyl cation or protonated methyl isocyanate
58	58.0651	$[C_3H_8N]^+$	*N,N*-Dimethyliminium ion from compounds with alkyl *N,N*-dimethylamino substituents, for example, some phenothiazines, tricyclic antidepressants, pheniramines; protonated *N*-ethyl-methaneimine
60	60.0556	$[CH_5N_3+H]^+$	Guanidine derivatives
60	60.0808	$[C_3H_9N+H]^+$	Protonated propylamine from *N*-(*n*- or *iso*-)-propyl derivatives
65	64.9787	$[HOP{=}O+H]^+$	Dimethyl phosphate and diethyl phosphorothioate pesticides
65	65.0386	$[C_5H_5]^+$	Cyclopentadienyl cation, from substituted phenyl derivatives, for example, amphetamines; fragment of tropylium ion (*m/z* 91)
68	68.0243	$[C_2H_2N_3]^+$	1,3,5-Triazines
69	69.0447	$[C_3H_5N_2]^+$	Imidazole derivatives
70	70.0400	$[C_2H_4N_3]^+$	Triazole derivatives
70	70.0651	$[C_4H_8N]^+$	Dihydropyrrolium ion, for example, from pyrrolidinophenone designer drugs; from phenothiazines with a 3-(4-methylpiperazin-1-yl)propyl substituent; Pro immonium ion
72	72.0444	$[(CH_3)_2N{=}C{=}O]^+$	*N,N*-Dimethylcarbamoyl cation, from *N,N*-dimethylureas, for example, phenylurea herbicides
72	72.0808	$[C_4H_{10}N]^+$	*N,N*-Dimethyl-ethyliminium ion; protonated *N*-*iso*-propylmethanimine from β-blockers; Val immonium ion; from pyrrolidinophenone designer drugs
74	74.0600	$[C_3H_8NO]^+$	β-Blockers with *N*-*iso*-propyl or with *N*-*tert*-butyl substituents; Thr immonium ion

TABLE 3.4 *(Continued)*

m/z		Formula	From Which Functional Group and/or Compound Class the Fragment Ion May Originate
75	75.0553	$[C_2H_7N_2O]^+$	Methoxy-1,3,5-triazines
77	77.0386	$[C_6H_5]^+$	Phenylium ion, from phenyl derivatives
79	78.9943	$[CH_3OP{=}O{+}H]^+$	Dimethyl phosphate and phosphoro(di)thioate pesticides
79	79.0058	$[CH_4ClN_2]^+$	Chloro-1,3,5-triazines
80	80.0495	$[C_5H_5N{+}H]^+$	Protonated pyridine
81	80.9736	$[(HO)_2P{=}O]^+$	Dimethyl and diethyl phosphate and diethyl phosphorothioate pesticides
81	81.0447	$[C_4H_5N_2]^+$	(1*H*-Pyrazol-4-yl)methylium ion from methyl-imidazole derivatives and stanozolol
82	82.0651	$[C_5H_8N]^+$	From cocaine and its metabolites
86	86.0349	$[C_2H_4N_3O]^+$	Methoxy-1,3,5-triazines
86	86.0600	$[C_4H_8NO]^+$	*N*-Ethyl-*N*-methylcarbamoyl cation
86	86.0964	$[C_5H_{12}N]^+$	Compounds with 3-(*N,N*-dimethylamino)propyl substituent, for example, phenothiazines; Leu or Ile immonium ion
91	91.0542	$[C_7H_7]^+$	Tropylium ion (benzyl cation) from benzyl derivatives
92	92.0495	$[H_2NC_6H_4]^+$	Sulfonamide antibiotics; acetaminophen; sulfonylurea antidiabetics
93	93.0335	$[HOC_6H_4]^+$	Hydroxyphenylium ion, from phenol derivatives
95	94.9893	$[CH_3O\ HO\ P{=}O]^+$	Dimethyl phosphate pesticides
95	95.0292	$[FC_6H_4]^+$	Fluorophenylium ion, from fluorophenyl derivatives, for example, butyrophenones
95	95.0491	$[C_6H_5OH{+}H]^+$	Phenol derivatives
97	96.9508	$[(HO)_2P{=}S]^+$	Diethyl phosphoro(di)thioate pesticides
97	97.0106	$[C_5H_5S]^+$	Thenyl cation
97	97.0648	$[C_6H_9O]^+$	From 3-keto-Δ^4 steroid, for example, anabolic steroids, progestogens, or corticosteroids
98	98.0839	$[C_5H_{10}N_5]^{+\bullet}$	Compounds with a 3-(4-methylpiperazin-1-yl)propyl substituent, for example, phenothiazines
98	98.0964	$[C_6H_{12}N]^+$	β-Blockers with *N-iso*-propyl substituent; also 1-methylidenepiperidinium ion or similar
99	98.9842	$[(HO)_3P{=}O{+}H]^+$	Diethyl phosphate pesticides
99	99.0917	$[C_5H_{11}N_2]^+$	4-Methyl-2,3,4,5-tetrahydropyrazin CH_3 ium ion, for example, from sildenafil
100	100.0505	$[C_3H_6N_3O]^+$	Methoxy-1,3,5-triazines
100	100.0995	$[C_5H_{12}N_2]^{+\bullet}$	4-Methyl-2,3,4,5-tetrahydropyrazinium ion, for example, from sildenafil
100	100.1121	$[C_6H_{14}N]^+$	Compounds with 3-(*N,N*-dimethylamino)-2-methylpropyl substituent, for example, phenothiazines
101	101.0709	$[C_4H_9N_2O]^+$	Gln immonium ion
102	102.0550	$[C_4H_8NO_2]^+$	ACE inhibitors (acid form)
104	104.0010	$[C_2H_3ClN_3]^+$	Chloro-1,3,5-triazines
104	104.0495	$[C_6H_5C{\equiv}N{+}H]^+$	Protonated benzonitrile, for example, from benzodiazepines
104	104.0528	$[C_4H_{10}NS]^+$	Met immonium ion
105	105.0335	$[C_6H_5{-}C{=}O]^+$	Benzoyl derivatives, for example, cocaine and its metabolites
105	105.0447	$[C_6H_5N_2]^+$	Carbendazim
105	105.0699	$[C_6H_5CHCH_3]^+$; $[C_8H_9]^+$	Dimethylphenyl or ethylbenzene derivatives, for example, from 3,4-methylenedioxyamphetamines
107	107.0491	$[HOC_6H_4CH_2]^+$	Hydroxybenzyl derivatives
108	108.0444	$[O{=}C_6H_4{=}NH_2]^+$	Sulfonamide and sulfonylurea derivatives
109	109.0049	$[(CH_3O)_2P{=}O]^+$	Dimethyl phosphate and phosphorothioate pesticides
109	109.0448	$[FC_6H_5CH_2]^+$	Fluorobenzyl derivatives
109	109.0648	$[C_7H_9O]^+$	From 3-keto-Δ^4 steroids, for example, anabolic steroids, progestogens, or corticosteroids
110	110.0461	$[C_3H_4N_5]^+$	1,3,5-Triazines
110	110.0713	$[C_5H_8N_3]^+$	His immonium ion
111	110.9664	$[(CH_3O)\ HO\ P{=}S]^+$	Dimethyl phosphoro(di)thioate pesticides
111	110.9996	$[ClC_6H_4]^+$	Chlorophenyl derivatives

TABLE 3.4 (*Continued*)

m/z		Formula	From Which Functional Group and/or Compound Class the Fragment Ion May Originate
112	112.1121	$[C_7H_{14}N]^+$	β-Blockers with *N-tert*-butyl substituent
113	113.1073	$[C_6H_{13}N_2]^+$	Compounds with an *N*-propyl-*N′*-methyl-piperazine substituent, for example, phenothiazines
114	114.0372	$[C_5H_8NS]^+$	β-Lactam antibiotics
116	116.0277	$[C_3H_6N_3S]^+$	Methylthio-1,3,5-triazines
116	116.0495	$[N{\equiv}C{-}C_6H_4CH_2]^+$	Cyanobenzyl derivatives
116	116.1070	$[C_6H_{14}NO]^+$	β-Blockers with *N-iso*-propyl substituent
119	119.0855	$[C_9H_{11}]^+$	α,α-(Dimethyl)benzyl cation or α-(ethyl)benzyl cation
120	120.0444	$[C_6H_5{-}N{=}C{=}O{+}H]^+$	Aminobenzoyl derivatives; protonated phenyl isocyanate
120	120.0808	$[C_8H_{10}N]^+$	Phe immonium ion
121	121.0284	$[HOC_6H_4{-}C{=}O]^+$	Hydroxybenzoyl ion
121	121.0648	$[C_8H_9O]^+$	From 3-keto-$\Delta^{1,4}$ steroids, for example, anabolic or corticosteroids; methoxybenzyl derivatives; from 25X-NBOMe designer drugs
121	121.1012	$[C_9H_{13}]^+$	(1*H*-Pyrazol-4-yl)methylium ion, from stanozolol
122	122.0401	$[FC_6H_4{-}C{\equiv}N{+}H]^+$	Protonated fluorobenzonitrile, for example, from fluorinated benzodiazepines
122	122.0964	$[C_8H_{12}N]^+$	Dimethylaniline derivatives
123	123.0241	$[FC_6H_4{-}C{=}O]^+$	Fluorobenzoyl derivatives
123	123.0804	$[C_8H_{11}O]^+$	From 3-keto-Δ^4 steroids, for example, anabolic steroids, progestogens, or corticosteroids
125	124.9821	$[(CH_3O)_2P{=}S]^+$ $[CH_3CH_2O\ HO\ P{=}S{+}H]^+$	Dimethyl and diethyl phosphoro(di)thioate pesticides
125	125.0153	$[ClC_6H_4CH_2]^+$	Chlorobenzyl derivatives
126	126.0120	$[C_4H_4N_3S]^+$	Cephalosporin β-lactam antibiotics with a 2-(2-amino-1,3-thiazol-4-yl)-2-(methoxyimino)acetyl substituent
126	126.1277	$[C_8H_{16}N]^+$	1-(Pyrrolidin-1-yl)but-1-ylium ion from pyrrolidinophenone designer drugs
127	127.0155	$[(CH_3O)_2\ HO\ P{=}O{+}H]^+$	Dimethyl and diethyl phosphate pesticides
127	127.0542	$[C_{10}H_7]^+$	Naphthalene-1-ylium ion, for example, from naphthoylindole class synthetic cannabinoids
128	128.0944	$[C_6H_{12}N_2O]^+$	Compounds with a 3-[4-(2-hydroxyethyl)piperazin-1-yl]propyl substituent, for example, phenothiazines
130	130.0863	$[C_6H_{12}NO_2]^+$	ACE inhibitors (ethyl ester form)
130	130.1226	$[C_7H_{16}NO]^+$	β-Blockers with *N-tert*-butyl substituent
130	130.1590	$[(C_4H_9)_2NH_2{+}H]^+$	Protonated dibutylamine
131	131.0491	$[C_9H_7O]^+$	(2-Aminopropyl)benzofuran designer drugs
132	132.0556	$[C_7H_6N_3]^+$	Carbendazim
133	133.0648	$[C_9H_9O]^+$	(2-Aminopropyl)dihydrobenzofuran designer drugs
134	134.0964	$[C_9H_{12}N]^+$	ACE inhibitors enalapril, quinapril, and ramipril
135	135.0441	$[C_8H_7O_2]^+$	3,4-Methylenedioxyamphetamines
135	135.0804	$[C_9H_{11}O]^+$	From 3-keto-$\Delta^{1,4}$ steroids
136	136.0757	$[C_8H_{10}NO]^+$	Tyr immonium ion
138	138.0105	$[C_5H_5ClN]^+$	Protonated chlorobenzonitrile, for example, from chlorinated benzodiazepines
139	138.9945	$[ClC_6H_4{-}C{=}O]^+$	Chlorobenzoyl derivatives
141	141.0771	$[C_5H_9N_4O]^+$	Sulfonylurea herbicides with 4-methoxy-6-methyltriazinyl substituent
141	141.1386	$[C_8H_{17}N_2]^+$	Compounds with a 3-(4-methylpiperazin-1-yl)propyl substituent, for example, phenothiazines
143	142.9385	$[(HO)_2\ P{=}S\ SCH_2]^+$	Diethyl phosphorodithioate pesticides
143	143.1179	$[C_7H_{15}N_2O]^+$	Compounds with a with a 3-[4-(2-hydroxyethyl)piperazin-1-yl]propyl substituent, for example, phenothiazines
146	146.0228	$[C_3H_5ClN_5]^+$	Chloro-1,3,5-triazines
147	147.0804	$[C_{10}H_{11}O]^+$	From 3-keto-$\Delta^{1,4}$ steroids, for example, corticosteroids
150	150.0913	$[C_9H_{12}NO]^+$	Cocaine and its metabolites
151	151.0754	$[(CH_3O)_2C_6H_3CH_2]^+$	Dimethoxybenzyl derivatives

TABLE 3.4 (*Continued*)

m/z		Formula	From Which Functional Group and/or Compound Class the Fragment Ion May Originate
153	153.0134	$[(CH_3CH_2O)_2P{=}S]^+$	Diethyl phosphoro(di)thioate pesticides
155	154.9926	$[(CH_3O)_2\ P{=}S\ OCH_2]^+$	Dimethyl phosphorothioate pesticides
155	155.0468	$[(CH_3CH_2O)_2\ HO\ P{=}O{+}H]^+$	Diethyl phosphate pesticides
155	155.0491	$[C_{11}H_7O]^+$	Naphthoylindole class synthetic cannabinoids
156	156.0114	$[H_2N{-}C_6H_4{-}SO_2]^+$	Sulfonamide and sulfonylurea derivatives
156	156.0226	$[C_5H_6N_3OS]^+$	Cephalosporin β-lactam antibiotics with a 2-(2-amino-1,3-thiazol-4-yl)-2-(methoxyimino)acetyl substituent
158	158.0270	$[C_6H_8NO_2S]^+$	Cephalosporin β-lactam antibiotics
158	158.1176	$[C_8H_{16}NO_2]^+$	Deoxydesosamine, from some macrolide antibiotics
159	158.9698	$[(CH_3CH_2O)\ HO\ HS\ P{=}S{+}H]^+$	Diethyl phosphorodithioate pesticides
159	158.9763	$[Cl_2C_6H_3CH_2]^+$	Dichlorobenzyl derivatives
159	159.0804	$[C_{11}H_{11}O]^+$	(2-Aminopropyl)-benzofuran designer drugs
159	159.0917	$[C_{10}H_{11}N_2]^+$	Trp immonium ion
160	160.0427	$[C_6H_{10}NO_2S]^+$	Penicillin β-lactam antibiotics
160	160.1121	$[C_{11}H_{14}N]^+$	ACE inhibitors
161	161.0961	$[C_{11}H_{13}O]^+$	(2-Aminopropyl)dihydrobenzofuran designer drugs
163	163.0754	$[C_{10}H_{11}O_2]^+$	3,4-Methylenedioxyamphetamines
164	164.0832	$[C_{10}H_{12}O_2]^+$	25X-NBOMe designer drugs
165	165.0214	$[C_8H_6ClN_2]^+$	Chlorinated benzodiazepines with an N^1-desmethyl acetamide skeleton
165	165.0710	$[C_{10}H_{10}FO]^+$	From butyrophenones
165	165.0910	$[C_{10}H_{13}O_2]^+$	Verapamil
167	167.0564	$[C_6H_7N_4O_2]^+$	Sulfonylurea herbicides with 4-methoxy-6-methyltriazinyl substituent
167	167.0855	$[C_{13}H_{11}]^+$	α-(Phenyl)benzyl cation, for example, from histamine antagonists
168	168.0808	$[C_{12}H_{10}N]^+$	α-(Pyridinyl)benzyl cation, for example, from histamine antagonists
169	168.9647	$[BrC_6H_4CH_2]^+$	Bromobenzyl derivatives
171	170.9698	$[(CH_3O)_2\ P{=}S\ SCH_2]^+$	Dimethyl and diethyl phosphorodithioate pesticides
171	171.0804	$[C_{12}H_{11}O]^+$	From 3-keto-$\Delta^{1,4}$ steroids, for example, corticosteroids
171	171.1492	$[C_9H_{19}N_2O]^+$	Compounds with a 3-(2-hydroxyethylpiperazin-1-yl)propyl substituent, for example, phenothiazines
172	171.9721	$[C_7H_4Cl_2N]^+$	Protonated dichlorobenzonitrile
173	173.0961	$[C_{12}H_{13}O]^+$	From 3-keto-$\Delta^{1,4}$ steroids
178	178.0321	$[C_9H_8NOS]^+$	Diltiazem
179	179.0941	$[C_{10}H_{13}NO_2]^+$	25X-NBOMe designer drugs
180	180.0808	$[C_{13}H_{10}N]^+$	Non-ring-substituted phenothiazines
181	181.0859	$[(CH_3O)_3C_6H_2CH_2]^+$	Trimethoxybenzyl derivatives
181	181.1012	$[C_{14}H_{13}]^+$	α-(2-Methylphenyl)benzyl cation, for example, from histamine antagonists
182	182.0560	$[C_7H_8N_3O_3]^+$	Sulfonylurea herbicides with 4,6-dimethoxypyrimidinyl substituent
182	182.0964	$[C_{13}H_{12}N]^+$	α-(Methyl)-*α*-(2-pyridyl)benzyl cation, for example, from histamine antagonists
192	192.0808	$[C_{14}H_{10}N]^+$	Angiotensin II receptor antagonist (sartans)
195	195.0652	$[C_{10}H_{11}O_4]^+$	Vasodilators dilazep and hexobendine
198	198.0372	$[C_{12}H_8NS]^+$	Non-ring-substituted phenothiazines
199	199.0011	$[(CH_3CH_2O)_2\ P{=}S\ SCH_2]^+$	Diethyl phosphorodithioate pesticides
200	200.0528	$[C_{12}H_{19}NS]^+$	From non-ring-substituted phenothiazines
201	201.0466	$[C_{13}H_{10}Cl]^+$	α-(4-Chlorophenyl)benzyl cation, for example, from histamine antagonists
202	202.4180	$[C_{12}H_9ClN]^+$	α-(2-Pyridyl-4-chloro)benzyl cation, for example, from histamine antagonists
203	203.1543	$[C_{13}H_{19}N_2]^+$	(1*H*-Pyrazol-4-yl)methylium ion from stanozolol
206	206.1176	$[C_{12}H_{16}NO_2]^+$	From ACE inhibitors (acid form)
207	207.0917	$[C_{14}H_{11}N_2]^+$	From angiotensin II receptor antagonists (sartans)
212	212.0528	$[C_{13}H_{10}NS]^+$	From non-ring-substituted phenothiazines
214	214.0418	$[C_{13}H_9ClNS]^+$	From chlorine-substituted phenothiazines

TABLE 3.4 (*Continued*)

m/z		Formula	From Which Functional Group and/or Compound Class the Fragment Ion May Originate
215	215.0622	$[C_{14}H_{12}Cl]^+$	α-(4-Chlorophenyl)-*α*-(methyl)benzyl cation, for example, from histamine antagonists
234	234.0139	$[C_{12}H_9ClNS]^+$	Chlorine-substituted phenothiazines
234	234.1489	$[C_{14}H_{20}NO_2]^+$	ACE inhibitors (ethyl ester form)
235	235.0978	$[C_{14}H_{11}N_4]^+$	Angiotensin II receptor antagonist (sartans)
240	240.0841	$[C_{15}H_{14}NS]^+$	Non-ring-substituted phenothiazines
241	241.0390	$[C_8H_9N_4O_3S]^+$	Cephalosporin β-lactam antibiotics with a 2-(2-amino-1,3-thiazol-4-yl)-2-(methoxyimino)acetyl substituent
245	244.9960	$[C_{13}H_{10}Br]^+$	α-(4-Bromophenyl)benzyl cation, for example, from histamine antagonists
246	245.9913	$[C_{12}H_9BrN]^+$	α-(2-Pyridyl-4-bromo)benzyl cation, for example, from histamine antagonists
246	246.0139	$[C_{13}H_9ClNS]^+$	Chlorine-substituted phenothiazines
253	253.1071	$[C_{13}H_{17}O_5]^+$	Vasodilators dilazep and hexobendine
274	274.0452	$[C_{15}H_{13}ClNS]^+$	Chlorine-substituted phenothiazines

TABLE 3.5 Small neutral losses in negative-ion mode.

Neutral Loss (Da)		Formula	From Which Functional Group and/or Compound Class the Neutral Loss May Originate
15	15.0235	$^{\bullet}CH_3$	Aromatic methoxy compounds
18	18.0106	H_2O	Common, from alcohols, aldehydes, ketones, carboxylic acids
20	20.0062	HF	Fluorine-containing compounds
27	27.0109	HCN	Hydrogen cyanide, for example, from aromatic nitriles, thiazide diuretics
28	27.9949	CO	Phenol derivatives
28	28.0313	C_2H_4	Ethene, from compounds with *N*- or *O*-ethyl substituents, for example, ethers, amines, esters
30	29.9980	$^{\bullet}NO$	Nitrosyl radical from nitroaryl compounds
30	30.0106	$H_2C{=}O$	Formaldehyde, for example, from corticosteroids
31	31.0058	HNO	Nitroxyl from nitroaryl compounds
32	32.0262	CH_3OH	Methyl ethers and esters
34	33.9877	H_2S	Hydrogen sulfide, from thiols
35	34.9683	$Cl^{\bullet}$	Chlorine-containing compounds
36	35.9767	HCl	Chlorine-containing compounds
36	36.0211	$2 \times H_2O$	Dihydroxy compounds
42	42.0106	$H_2C{=}C{=}O$	Ethenone, from acetates, *N*- or *O*-acetyl derivatives
42	42.0470	C_3H_6	Propene, from *N*- or *O*-(*n*- or *iso*-)propyl substituents, for example, ethers, amines or esters
43	43.0058	$HN{=}C{=}O$	Hydrogen isocyanate, for example, from cyclic amides, barbiturates
44	43.9898	CO_2	Aryl carboxylic acids, β-lactam antibiotics
44	44.0136	$H_2NCO^{\bullet}$	Carbamoyl radical, for example, from nitrazepam, oxazepam
44	44.0262	CH_3CHO	Ethanal, for example, from *N*-CH_2CH_2OH derivatives
46	45.9924	$^{\bullet}NO_2$	Nitryl radical from nitroaryl compounds
46	46.0055	HCOOH	Aliphatic carboxylic acids
46	46.0419	C_2H_5OH	Ethyl ethers or esters
48	47.9670	SO	Aromatic sulfonyl anions
56	55.9898	$O{=}C{=}C{=}O$	Ethenedione, for example, from various NSAIDs
58	58.0055	$C_2H_2O_2$	Oxiran-2-one, for example, from phenoxyacetic acids
58	58.0419	$(CH_3)_2C{=}O$	Acetone, for example, from 3-keto-$\Delta^{1,4}$-glucocorticosteroids
60	60.0211	CH_3COOH	Acetic acid from aliphatic-*O*-acetyl substituents; from glucuronic acid conjugates
60	60.0581	C_3H_7OH	(*iso*-)Propyl ethers or esters
64	63.9619	SO_2	Aromatic sulfoxides, sulfonic acids, sulfonates, and thiazide diuretics

TABLE 3.5 (*Continued*)

Neutral Loss (Da)		Formula	From Which Functional Group and/or Compound Class the Neutral Loss May Originate
69	68.9958	$^{\bullet}CF_3$	Aromatic trifluoromethyl derivatives
70	70.0030	HCF_3	Trifluoromethyl derivatives
76	76.0161	$H_2C{=}O + HCOOH$	Loss of $H_2C{=}O$ from $[M{+}HCOO]^-$ of corticosteroids
78	77.9776	$H_2S + CO_2$	Cephalosporin β-lactam antibiotics
78	78.0470	C_6H_6	Benzene, from phenyl derivatives
79	78.9178	$Br^{\bullet}$	Bromine-containing compounds
79	78.9728	$HNSO_2$	Sulfonamide diuretics
80	79.9568	SO_3	Steroid sulfate conjugates
80	79.9256	HBr	Bromine-containing compounds
80	79.9806	$H_2NSO_2^{\bullet}$	Sulfamoyl radical, for example, from thiazide diuretics
87	86.9956	C_2HNO_3	Barbiturates
90	90.0137	$H_2C{=}O + CH_3COOH$	Loss of $H_2C{=}O$ from $[M{+}CH_3COO]^-$ of corticosteroids
91	90.9728	$HCN + SO_2$	Thiazide diuretics
91	91.0548	$C_6H_5CH_2^{\bullet}$	Benzyl radical, from benzyl derivatives
98	97.9674	H_2SO_4	Steroid sulfate conjugates
104	104.0474	$C_4H_8O_3$	Loss of $(H_3C)_2C{=}O$ from $[M{+}HCOO]^-$ of acetonide corticosteroids
113	113.0477	$C_5H_7NO_2$	Cephalosporin β-lactam antibiotics
118	118.0630	$C_5H_{10}O_3$	Loss of $(H_3C)_2C{=}O$ from $[M{+}CH_3COO]^-$ of acetonide corticosteroids
119	119.0371	$C_6H_5{-}N{=}C{=}O$	Phenyl isocyanate
123	123.0326	$C_6H_5NO_2$	Nitrobenzene derivatives
124	123.9748	$C_2H_5O_2PS$	Dimethyl phosphoro(di)thioate pesticides
127	126.9045	$I^{\bullet}$	Iodine-containing compounds
128	127.9123	HI	Iodine-containing compounds
141	141.0426	$C_6H_7NO_3$	Penicillin β-lactam antibiotics
152	152.0061	$C_4H_9O_2PS$	Diethyl phosphoro(di)thioate pesticides
157	157.0198	$C_6H_7NO_2S$	Cephalosporin β-lactam antibiotics
176	176.0321	$C_6H_8O_6$	Glucuronic acid conjugates
183	183.0132	$C_8H_3F_2NO_2$	Benzoylphenylureas
194	194.0421	$C_6H_{10}O_7$	Aliphatic glucuronic acid conjugates

TABLE 3.6 Characteristic product ions in negative-ion mode.

m/z		Formula	From Which Functional Group and/or Compound Class the Fragment Ion May Originate
58	57.9757	$[N{=}C{=}S]^-$	Thioisocyanate, for example, from thiobarbiturates
59	59.0139	$[CH_3COO]^-$	Acetate
64	64.0319	$[C_5H_4]^{-\bullet}$	Cyclopentadiene radical anion, for example, from sulfonamide antibiotics
65	65.0397	$[C_5H_5]^-$	Cyclopentadiene anion, phenol derivatives
69	68.9958	$[CF_3]^-$	Trifluoromethyl compounds
77	77.0397	$[C_6H_5]^-$	Phenide ion, 2-methyl-4-chloro-phenoxy acids
78	77.9655	$[NSO_2]^-$	Deprotonated iminosulfene, from thiazide diuretics
79	78.9591	$[PO_3]^-$	Phosphate-containing compounds
79	78.9859	$[CH_3SO_2]^-$	Methylsulfonyl anion, for example, from cyclooxygenase-2 enzyme (COX-2) inhibitors
79	78.9189	Br^-	Bromine-containing compounds
79	79.0553	$[C_6H_7]^-$	Cannabinoids
80	79.9574	$[SO_3]^{-\bullet}$	Sulfonates; sulfate conjugates
80	79.9812	$[SO_2NH_2]^-$	From sulfonamide diuretics
81	80.9652	$[HSO_3]^-$	Hydrogen sulfite, from sulfonates
85	85.0044	$[C_2HN_2O_2]^-$	Barbiturates
85	85.0295	$[C_4H_5O_2]^-$	Glucuronic acid conjugates
92	92.0268	$[C_6H_4O]^{-\bullet}$	Loss of $^{\bullet}NO$ from *m/z* 122 ($[O_2N{-}C_6H_4]^-$)
92	92.0506	$[C_6H_5NH]^-$	Anilinide anion, from NSAIDs and sulfonamide antibiotics
93	93.0346	$[C_6H_5O]^-$	Phenoxide ion, from phenoxy derivatives, for example, NSAIDs

TABLE 3.6 (*Continued*)

m/z		Formula	From Which Functional Group and/or Compound Class the Fragment Ion May Originate
93	93.0458	$[C_5H_4N—NH]^-$	Cyclooxygenase-2 enzyme (COX-2) inhibitors
96	95.9523	$[SO_4]^{-\bullet}$	Sulfate radical anion, from steroid sulfate conjugates
97	96.9601	$[HSO_4]^-$	Steroid sulfate conjugates
97	96.9696	$[H_2PO_4]^-$	Phosphate-containing compounds
106	106.0662	$[C_7H_8N]^-$	Sulfonylurea antidiabetics
107	107.0502	$[C_7H_7O]^-$	Cannabinoids
108	108.0455	$[H_2N—C_6H_4O]^-$	Aminophenoxide ion, for example, from sulfonamide antibiotics
112	112.0404	$[C_5H_6NO_2]^-$	Cephalosporin β-lactam antibiotics
113	113.0208	$[C_6H_3F_2]^-$	Difluorobenzene derivatives
113	113.0244	$[C_5H_5O_3]^-$	Glucuronic acid conjugates
122	122.0248	$[O_2N—C_6H_4]^-$	Nitrobenzene derivatives
126	126.0116	$[ClC_6H_3—NH_2]^-$	Thiazide diuretics with Cl-substituent on C^6
127	126.9050	I^-	Iodine-containing compounds
127	126.9945	$[C_6H_4ClO]^-$	Chlorophenol derivatives
143	143.0502	$[C_{10}H_6O–H]^-$	Deprotonated 2-naphthol, from estrogens
145	144.9617	$[C_6H_3Cl_2]^-$	Dichlorobenzene derivatives
145	145.0659	$[C_{10}H_9O]^-$	Estrogens
141	149.9781	$[(CH_3O)_2P{=}SO]^-$	Dimethyl phosphorothioate pesticides
156	156.0125	$[H_2N—C_6H_4—SO_2]^-$	Aminobenzyl sulfonyl anion, for example, from sulfonamide antibiotics
156	156.0266	$[C_7H_4F_2NO]^-$	Difluorobenzylamide anion, from benzoylphenylureas
157	156.9552	$[(CH_3O)_2P{=}SS]^-$	Dimethyl phosphodithioate pesticides
160	160.0380	$[F_3C—C_6H_3—NH_2]^-$	Thiazide diuretics with CF_3-substituent on C^6
169	169.0094	$[(CH_3CH_2O)_2P{=}SO]^-$	Diethyl phosphothioate pesticides
169	169.0659	$[C_{12}H_{10}O–H]^-$	Deprotonated 6-vinyl-2-naphthol, from estrogens
170	170.0281	$[C_7H_8NO_2S]^-$	Sulfonylurea antidiabetics
175	175.0248	$[C_6H_7O_6]^-$	Glucuronic acid conjugates
183	183.0815	$[C_{13}H_{12}O–H]^-$	Deprotonated 6-allyl-2-naphthol, from estrogens
185	184.9865	$[(CH_3CH_2O)_2P{=}SS]^-$	Diethyl phosphodithioate pesticides
190	189.9735	$[ClC_6H_3—SO_2NH_2]^-$	Thiazide diuretics with Cl-substituent on C^6
205	204.9844	$[C_6H_6ClN_2O_2S]^-$	Thiazide diuretics with Cl-substituent on C^6 (loss of R^3—CN + SO_2)
224	223.9999	$[F_3C—C_6H_3—SO_2NH_2]^-$	Thiazide diuretics with CF_3-substituent on C^6
239	239.0108	$[C_7H_6F_3N_2O_2S]^-$	Thiazide diuretics with CF_3-substituent on C^6 (loss of R^3—CN+SO_2)
239	239.1441	$[C_{17}H_{20}O–H]^-$	Estrogens
269	268.9463	$[C_6H_6ClN_2O_4S_2]^-$	Thiazide diuretics with Cl-substituent on C^6 (loss of R^3—CN)
303	302.9727	$[C_7H_6F_3N_2O_4S_2]^-$	Thiazide diuretics with CF_3-substituent on C^6 (loss of R^3—CN)

REFERENCES

Abrankó L, Szilvássy B. 2015. Mass spectrometric profiling of flavonoid glycoconjugates possessing isomeric aglycones. J Mass Spectrom 50: 71–80.

Adams J, Gross ML. 1986. Energy requirement for remote charge site ion decompositions and structural information from collisional activation of alkali metal cationized fatty alcohols. J Am Chem Soc 108: 6915–6921.

Ambihapathy K, Yalcin T, Leung H-W, Harrison AG. 1997. Pathways to immonium ions in the fragmentation to protonated peptides. J Mass Spectrom 32: 209–215.

Anari MR, Bakhtiar R, Franklin RB, Pearson PG, Baillie TA. 2003. Study of the fragmentation mechanism of protonated 6-hydroxychlorzoxazone: Application in simultaneous analysis of CYP2E1 activity with major human cytochrome P450s. Anal Chem 75: 469–478.

Andreoli R, Manini P, Poli D, Bergamaschi E, Mutti A, Niessen WMA. 2004. Development of a simplified method for the simultaneous determination of retinol, α-tocopherol, and β-carotene in serum by liquid chromatography-tandem mass spectrometry with atmospheric pressure chemical ionization. Anal Bioanal Chem 378: 987–994.

Arpino PJ, Guiochon G. 1982. Optimization of the instrumental parameters of a combined LC–MS, coupled by an interface for DLI. III. Why the solvent should not be removed in LC–MS interfacing methods. J Chromatogr 251: 153–164.

Asam MR, Glish GL. 1997. Tandem mass spectrometry of alkali cationized polysaccharides in a quadrupole ion trap, J Am Soc Mass Spectrom 8: 987–995.

Audier HE, Milliet A, Perret C, Tabet JC, Varenne P. 1978. Mécanismes de fragmentation en spectrométrie de masse par ionisation chimique. III. Ions formés par cyclisation. Org Mass Spectrom 13: 315–318.

Bandu ML, Watkins KR, Bretthauer ML, Moore CA, Desaire H. 2004. Prediction of MS/MS data. 1. A focus on pharmaceuticals containing carboxylic acids. Anal Chem 76: 1746–1753.

Bayer E, Gfrörer P, Rentel C. 1997. Coordination-ionspray-MS (CIS-MS), a universal detection and characterization method for direct coupling with separation techniques. Angew Chem Int Ed 38: 992–995.

Benoit FN, Harrison AG. 1976. Hydrogen migrations in mass spectrometry. I. The loss of olefin from phenyl-*n*-propyl ether following electron impact ionization and chemical ionization. Org Mass Spectrom 11: 599–608.

Benoit FN, Harrison AG. 1978. Hydrogen migrations in mass spectrometry. V. Loss of olefin from *n*-propyl esters following chemical ionization. Org Mass Spectrom 13: 128–132.

Biemann K. 1988. Contributions of mass spectrometry to peptide and protein structure. Biomed Environ Mass Spectrom 16: 99–111.

Bigler L, Hesse M. 1995. Neighboring group participation in the electrospray ionization tandem mass spectra of polyamine toxins of spiders. Part 1: α, ω-diaminoalkane compounds. J Am Soc Mass Spectrom 6, 634–637.

Bijlsma L, Sancho JV, Hernández F, Niessen WMA. 2011. Fragmentation pathways of drugs of abuse and their metabolites based on QTOF MS/MS and MS^E accurate-mass spectra. J Mass Spectrom 46: 865–875.

Bijttebier SK, D'Hondt E, Hermans N, Apers S, Voorspoels S. 2013. Unravelling ionization and fragmentation pathways of carotenoids using orbitrap technology: A first step towards identification of unknowns. J Mass Spectrom 48: 740–754.

Bilusich D, Bowie JH. 2009. Fragmentations of $(M–H)^-$ anions of underivatised peptides. Part 2: Characteristic cleavages of Ser and Cys and of disulfides and other post-translational modifications, together with some unusual internal processes. Mass Spectrom Rev 28: 20–34.

Bleiholder C, Osburn S, Williams TD, Suhai S, Van Stipdonk M, Harrison AG, Paizs B. 2008. Sequence-scrambling fragmentation pathways of protonated peptides. J Am Chem Soc 130: 17774–17789.

Bowen RD, Stapleton BJ, Williams DH. 1978. Non-concerted unimolecular reactions of ions in the gas phase: Isomerisation of weakly coordinated carbonium ions. J Chem Soc Chem Commun 24–26.

Bowen RD, Harrison AG. 1981. Loss of methyl radical from some small immonium ions: Unusual violation of the even-electron rule. Org Mass Spectrom 16: 180–182.

Bowie JH. 1990. The fragmentation of even-electron organic negative ions. Mass Spectrom Rev 9: 349–379.

Bowie JH, Brinkworth CS, Dua S. 2002. Collision-induced fragmentations of the $(M–H)^-$ parent anions of underivatized peptides: An aid to structure determination and some unusual negative ion cleavages. Mass Spectrom Rev 21: 87–107.

Boyd R, Somogyi A. 2010. The mobile proton hypothesis in fragmentation of protonated peptides: A perspective. J Am Soc Mass Spectrom 21: 1275–1278.

Brophy JJ, Nelsen D, Shannon JS, Middleton S. 1979. Electron impact and chemical ionization mass spectra of acyl ureas. Org Mass Spectrom 14: 379–386.

Byrdwell WC. 2001. Atmospheric-pressure chemical ionization mass spectrometry for analysis of lipids. Lipids 36: 327–346.

Cajka T, Fiehn O. 2014. Comprehensive analysis of lipids in biological systems by liquid chromatography-mass spectrometry. Trends Anal Chem 61: 192–206.

Cerny RL, MacMillan DK, Gross ML, Mallams AK, Pramanik BN. 1994. Fast-atom bombardment and tandem mass spectrometry of macrolide antibiotics. J Am Soc Mass Spectrom 5: 151–158.

Chen H, Wang H, Chen Y, Zhang H. 2005. Liquid chromatography-tandem mass spectrometry analysis of anisodamine and its phase I and II metabolites in rat urine. J Chromatogr B 824: 21–29.

Contado MJ, Adams J. 1991. Collision-induced dissociations and B/E linked scans for structural determination of modified fatty acid esters. Anal Chim Acta 246: 187–197.

Cren-Olivé C, Déprez S, Lebrun S, Coddeville B, Rolando C. 2000. Characterization of methylation site of monomethylflavan-3-ols by liquid chromatography/electrospray ionization tandem mass spectrometry. Rapid Commun Mass Spectrom 14: 2312–2319.

Crombie RA, Harrison AG. 1988. Unimolecular and collision-induced fragmentation of protonated nitroarenes. Org Mass Spectrom 23: 327–333

Crotti S, Stella L, Munari I, Massaccesi F, Cotarca L, Forcato M, Traldi P. 2007. Claisen rearrangement induced by low-energy collision of ESI-generated, protonated benzyloxy indoles. J Mass Spectrom 42: 1562–1568.

Cuyckens F, Claeys M. 2004. Mass spectrometry in the structural analysis of flavonoids. J Mass Spectrom 39: 1–15.

Delatour T, Richoz J, Vouros P, Turesky RJ. 2002. Simultaneous determination of 3-nitrotyrosine and tyrosine in plasma proteins of rats and assessment of artifactual tyrosine nitration. J Chromatogr B 779: 189–199.

Deng P, Zhan Y, Chen X, Zhong D. 2012. Derivatization methods for quantitative bioanalysis by LC-MS/MS. Bioanalysis 4: 49–69.

De Rijke E, Out P, Niessen WMA, Ariese F, Gooijer C, Brinkman UATh. 2006. Analytical separation and detection methods for flavonoids: An overview. J Chromatogr A 1112: 31–63.

Diakiw V, Shannon JS, Lacey MJ, MacDonald CG. 1979. Skeletal rearrangements of protonated molecular ions of some ethers. Org Mass Spectrom 14: 58.

Diehl G, Karst U, 2002. On-line electrochemistry–MS and related techniques. Anal Bioanal Chem 373: 390–398.

Djerassi C, Fenselau C. 1965. Mass spectrometry in structural and stereochemical problems. LXXXV. The nature of the cyclic transition state in hydrogen rearrangements of aliphatic amines. J Am Chem Soc 87: 5752–5756.

Domon B, Costello CE. 1988. A systematic nomenclature for carbohydrate fragmentations in FAB-MS/MS spectra of glycoconjugates. Glycoconjugate J 5: 397–409.

Edmunds FS, Johnstone RAW. 1965. Constituents of cigarette smoke. Part IX. The pyrolysis of polyenes and the formation of aromatic hydrocarbons. J Chem Soc 2892–2897.

Es-Safi NE, Kerhoas L, Ducrot PH. 2007. Fragmentation study of iridoid glucosides through positive and negative electrospray ionization, collision-induced dissociation and tandem mass spectrometry. Rapid Commun Mass Spectrom 21: 1165–1275.

Falck D, Kool J, Honing M, Niessen WMA. 2013. Tandem mass spectrometry study of p38a kinase inhibitors and related substances. J Mass Spectrom 48: 718–731.

Field FH. 1968. Chemical ionization mass spectrometry. VIII. Alkenes and alkynes. J Am Chem Soc 90: 5649–5656.

Field FH. 1972. Chemical Ionization Mass Spectrometry. in: Maccoll A (Ed.): Mass Spectrometry. Butterworths & Co Publishers Ltd, London, pp. 133–181. ISBN 978-0-408702-66-9.

Garcia P, Popot MA, Fournier F, Bonnaire Y, Tabet JC. 2002. Gas-phase behaviour of negative ions produced from thiazidic diuretics under electrospray conditions. J Mass Spectrom 37: 940–953.

Gerlich M, Neumann S. 2013. MetFusion: Integration of compound identification strategies. J Mass Spectrom 48: 291–298.

Grese RP, Cerny RL, Gross ML. 1989. Metal ion-peptide interactions in the gas phase: A tandem mass spectrometry study of alkali metal cationized peptides. J Am Chem Soc 111: 2835–2842.

Grewal RN, Rodriquez CF, Shoeib T, Chu IK, Tu Y-P, Hopkinson AC, Siu KWM. 2000. Elimination of AgR (R = H, CH_3, C_6H_5) from collisionally-activated argentinated amines. Eur J Mass Spectrom 6: 187–192.

Griffiths WJ. 2003. Tandem mass spectrometry in the study of fatty acids, bile acids, and steroids. Mass Spectrom Rev 22: 81–152.

Gross ML. 2000. Charge-remote fragmentation: An account of research on mechanisms and applications. Int J Mass Spectrom 200: 611–624.

Guaratini T, Gates PJ, Pinto E, Colepicolo P, Lopes NP. 2007. Differential ionisation of natural antioxidant polyenes in electrospray and nanospray mass spectrometry. Rapid Commun Mass Spectrom 21: 3842–3848.

Guy PA, Gremaud E, Richoz J, Turesky RJ. 2000. Quantitative analysis of mutagenic heterocyclic aromatic amines in cooked meat using liquid chromatography-atmospheric pressure chemical ionisation tandem mass spectrometry. J Chromatogr A 883: 89–102.

Ham BM. 2008. Even electron mass spectrometry with biomolecule applications, John Wiley & Sons, Ltd., Hoboken, NJ. ISBN 978-0-470-11802-3.

Harrison AG. 1992. Chemical ionization mass spectrometry, 2nd ed. CRC Press, Boca Raton, FL. ISBN 978-0-849-34254-7.

Harrison AG, Young AB, Bleiholder C, Suhai S, Paizs B. 2006. Scrambling of sequence information in collision-induced dissociation of peptides. J Am Chem Soc 128: 10364–10365.

Harrison AG. 2009. Ciclization of peptide b_9 ions. J Am Soc Mass Spectrom 20: 2248–2253.

Hill AW, Mortishire-Smith RJ. 2005. Automated assignment of high-resolution collisionally activated dissociation mass spectra using a systematic bond disconnection approach. Rapid Commun Mass Spectrom 19: 3111–3118.

Hofmeister GE, Zhou Z, Leary JA. 1991. Linkage position determination in lithium-cationized disaccharides: Tandem mass spectrometry and semiempirical calculations. J Am Chem Soc 113: 5964–5970.

Holčapek M, Jirasko R, Lisa M. 2010. Basic rules for the interpretation of atmospheric pressure ionization mass spectra of small molecules, J Chromatogr A, 1217: 3908–3921.

Hsu F-F, Turk J. 1999a. Structural characterization of triacylglycerols as lithiated adducts by electrospray ionization mass spectrometry using low-energy collisionally activated dissociation on a triple stage quadrupole instrument. J Am Soc Mass Spectrom 10: 587–599.

Hsu F-F, Turk J. 1999b. Distinction among isomeric unsaturated fatty acids as lithium adducts by ESI-MS using low energy CID on a triple stage quadrupole instrument. J Am Soc Mass Spectrom 10: 600–612.

Hsu F-F, Turk J, Stewart ME, Downing DT. 2002. Structural studies on ceramides as lithiated adducts by low energy collisional-activated dissociation tandem mass spectrometry with electrospray ionization. J Am Soc Mass Spectrom 13: 680–695.

Hsu F-F, Turk J. 2009. Electrospray ionization with low-energy collisionally activated dissociation tandem mass spectrometry of glycerophospholipids: Mechanisms of fragmentation and structural characterization. J Chromatogr B 877: 2673–2695.

Hunt DF, Yates JR, III,, Shabanowitz J, Winston S, Hauer CR. 1986. Protein sequencing by tandem mass spectrometry. Proc Natl Acad Sci U S A 83: 6233–6237.

Hunter EPL, Lias SG. 1998. Evaluated gas phase basicities and proton affinities of molecules: An update, J Phys Chem Ref Data 27: 413–656.

Ichikawa H, Harrison AG. 1978. Hydrogen migrations in mass spectrometry. VI. the chemical ionization mass spectra of substituted benzoic acids and benzyl alcohols. Org Mass Spectrom 13: 389–396.

Jensen NJ, Tomer KB, Gross ML. 1985. Gas-phase decompositions occurring remote to a charge site. J Am Chem Soc 107: 1863–1868.

Jensen ON, Podtelejnikov AV, Mann M. 1997. Identification of the components of simple protein mixtures by high-accuracy peptide mass mapping and database searching. Anal Chem 69: 4741–4750.

Jia C, Qi W, He Z. 2007. Cyclization reaction of peptide fragment ions during multistage collisionally activated decomposition: An inducement to lose internal amino-acid residues. J Am Soc Mass Spectrom 18: 663–678.

Johnson RS, Martin SA, Biemann K. 1988. Collision-induced fragmentation of $(M+H)^+$ ions of peptides. Side chain specific sequence ions. Int J Mass Spectrom Ion Processes 86: 137–154.

Julian RR, Ly T, Finaldi A-M, Morton TH. 2007. Dissociation of a protonated secondary amine in the gas phase via an ion–neutral complex. Int J Mass Spectrom 265: 302–307.

Justino GC, Borges CM, Florêncio MH. 2009. Electrospray ionization tandem mass spectrometry fragmentation of protonated flavone and flavonol aglycones: A re-examination. Rapid Commun Mass Spectrom 23: 237–248.

Karancsi T, Slégel P. 1999. Reliable molecular mass determination of aromatic nitro compounds: Elimination of gas-phase reduction occurring during atmospheric pressure chemical ionization. J Mass Spectrom 34: 975–977.

Karni M, Mandelbaum A. 1980. The 'even-electron' rule. Org Mass Spectrom 15 (1980) 53–64.

Kasper PT, Rojas-Chertó M, Mistrik R, Reijmers T, Hankemeier T, Vreeken RJ. 2012. Fragmentation trees for the structural characterisation of metabolites. Rapid Commun Mass Spectrom, 26: 2275–2286.

Kerwin JL, Wiens AM, Ericsson LH. 1996. Identification of fatty acids by electrospray mass spectrometry and tandem mass spectrometry. J Mass Spectrom 31: 184–192.

Kind T, Fiehn O. 2007. Seven Golden Rules for heuristic filtering of molecular formulas obtained by accurate mass spectrometry. BMC Bioinf 8: 105.

Kind T, Fiehn O. 2010. Advances in structure elucidation of small molecules using mass spectrometry. Bioanal Rev 2: 23–60.

Kingston DGI, Bursey JT, Bursey MM. 1974. Intramolecular hydrogen transfer in mass spectra. II. McLafferty rearrangement and related reactions. Chem Rev 74: 215–242.

Kingston EE, Shannon JS, Diakiw V, Lacey MJ. 1981. Skeletal rearrangements on chemical ionization of dibenzyl ether and derivatives. Org Mass Spectrom 16: 428–440.

Kingston EE, Shannon JS, Lacey MJ. 1983. Rearrangements in chemical ionization mass spectra. Org Mass Spectrom 18: 183–192.

Kinter M, Sherman NE. 2000. Protein sequencing and identification using tandem mass spectrometry. Wiley Interscience, NY. ISBN 978-0-47132-249-8.

Klagkou K, Pullen F, Harrison M, Organ A, Firth A, Langley GJ. 2003. Fragmentation pathways of sulphonamides under electrospray tandem mass spectrometric conditions. Rapid Commun Mass Spectrom 17: 2373–2379.

Kuck D. 1990a. Mass spectrometry of alkylbenzenes and related compounds. Part I. Gas-phase ion chemistry of alkylbenzene radical cations. Mass Spectrom Rev 9: 187–233.

Kuck D. 1990b. Mass spectrometry of alkylbenzenes and related compounds. Part 11. Gas phase ion chemistry of protonated alkylbenzenes (alkylbenzenium ions). Mass Spectrom Rev 9: 583–630.

Kulik W, Heerma W, Terlouw JK. 1989. A novel fragmentation process in the fast-atom bombardment/tandem mass spectra of peptides cationized with Na^+, determining the identity of the C-terminal amino acid. Rapid Commun Mass Spectrom 3: 276–279.

Kuuranne T, Kotiaho T, Pedersen-Bjergaard S, Einar Rasmussen K, Leinonen A, Westwood S, Kostiainen R. 2003. Feasibility of a liquid-phase microextraction sample clean-up and liquid chromatographic/mass spectrometric screening method for selected anabolic steroid glucuronides in biological samples. J Mass Spectrom 38: 16–26.

Lee VW-M, Li H, Lau T-C, Siu KWM. 1998. Structures of b and a product ions from the fragmentation of argentinated peptides. J Am Chem Soc 120: 7302–7309.

Lemoine J, Strecker G, Leroy Y, Fournet B, Ricart G. 1991. Collisional-activation tandem mass spectrometry of sodium adduct ions of methylated oligosaccharides: Sequence analysis and discrimination between alpha-NeuAc-(2–3) and alpha-NeuAc-(2–6) linkages. Carbohydr Res 221: 209–217.

Levery SB, Toledo MS, Doong RL, Straus AH. 2000. Takahashi, H.K.: Comparative analysis of ceramide structural modification found in fungal cerebrosides by electrospray tandem mass spectrometry with low energy collision-induced dissociation of Li^+ adduct ions. Rapid Commun Mass Spectrom 14: 551–563.

Levsen K, Schiebel HM, Terlouw JK, Jobst KJ, Elend M, Preiss A, Thiele A, Ingedoh A. 2007. Even-electron ions: A systematic study of the neutral species lost in the dissociation of quasi-molecular ions. J Mass Spectrom 42: 1024–1044.

Lias SG, Liebman JF, Levin RD. 1984. Evaluated gas phase basicities and proton affinities of molecules; heats of formation of protonated molecules. J Phys Chem Ref Data 13: 695–808.

Lias SG, Bartmess JE, Liebman JF, Holmes JL, Levin RD, Mallard WG. 1988. Gas phase ion and neutral thermochemistry. J Phys Chem Ref Data 17 (Suppl 1): 1–268.

Liebler DC. 2002. Introduction to proteomics: Tools for the new biology. Humana Press Inc., Totowa, NJ. ISBN 978-0-89603-991-9.

Liu DQ, Hop CECA. 2005. Strategies for characterization of drug metabolites using liquid chromatography–tandem mass spectrometry in conjunction with chemical derivatization and on-line H/D exchange approaches. J Pharm Biomed Anal 37: 1–18.

Ma YL, Li QM, Van den Heuvel H, Claeys M. 1997. Characterization of flavone and flavonol aglycones by collision-induced dissociation tandem mass spectrometry. Rapid Commun Mass Spectrom 11: 1357–1364.

Ma YL, Verdernikova I, Van den Heuvel H, Claeys M. 2000. Internal glucose residue loss in protonated O-diglycosyl flavonoids upon low-energy CID. J Am Soc Mass Spectrom 11: 136–144.

Ma YL, Cuyckens F, Van den Heuvel H, Claeys M. 2001. MS methods for the characterization and differentiation of isomeric O-diglycosyl flavonoids. Phytochem Anal 12: 159–165.

Madhusudanan KP, Mathad VT, Raj SK, Bhaduri AP. 2000. Characterization of iridoids by fast atom bombardment mass spectrometry followed by collision-induced dissociation of $[M+Li]^+$ ions. J Mass Spectrom 35: 321–329.

March R, Brodbelt J. 2008. Analysis of flavonoids: Tandem mass spectrometry, computational methods, and NMR. J Mass Spectrom 43: 1581–1617.

Martha CT, van Zeist W-J, Bickelhaupt FM, Irth H, Niessen WMA. 2010. Tandem mass spectrometry of silver-adducted ferrocenyl catalyst complexes in continuous-flow reaction detection systems. J Mass Spectrom 45: 1332–1343.

McLafferty FW. 1959. Mass spectrometric analysis. Molecular rearrangements. Anal Chem 31: 82–87.

McLafferty FW, Bente, III, PF, Kornfeld R, Tsai S-C, Howe I. 1973. Metastable ion characteristics. XXII. Collisional activation spectra of organic ions. J Am Chem Soc 95: 2120–2129.

McLafferty FW. 1980. Unimolecular decompositions of even-electron ions. Org Mass Spectrom 15: 114–121.

McLafferty FW, Tureček F. 1993. Interpretation of mass spectra, 4th ed. University Science Books, Mill Valley, CA. ISBN 978-0-935702-25-3.

Morton TH. 1980. Ion–molecule complexes in unimolecular fragmentations of gaseous cations. Alkyl phenyl ether molecular ions. J Am Chem Soc 102: 1596–1602.

Mouls L, Aubagnac JL, Martinez J, Enjalbal C. 2007. Low energy peptide fragmentations in an ESI-Q–TOF type mass spectrometer. J Proteome Res 6: 1378–1391.

Murphy RC, Axelsen PH. 2011. Mass spectrometric analysis of long-chain lipids. Mass Spectrom Rev 30: 579–599.

Niessen WMA. 1998. Analysis of antibiotics by liquid chromatography–mass spectrometry. J Chromatogr A 812: 53–76.

Niessen WMA. 2000 Structure Elucidation by LC-MS, Foreword. Analusis 28: 885– 887.

Niessen WMA. 2005. Mass spectrometry of antibiotics. in: Nibbering NMM (ed.): Encyclopedia of mass spectrometry. Volume 4: Fundamentals of and applications of organic (and organometallic) compounds. Elsevier Ltd, Oxford. pp. 822–837. ISBN 978-0-08-043846-7.

Niessen WMA. 2010. Group-specific fragmentation of pesticides and related compounds in liquid chromatography–tandem mass spectrometry. J Chromatogr A 1217: 4061–4070.

Niessen WMA. 2011. Fragmentation of toxicologically relevant drugs in positive-ion liquid chromatography–tandem mass spectrometry. Mass Spectrom Rev, 30: 626–663.

Niessen WMA. 2012. Fragmentation of toxicologically relevant drugs in negative-ion liquid chromatography–tandem mass spectrometry. Mass Spectrom Rev, 31: 626–665.

Niessen WMA, Honing M. 2015. Mass spectrometry strategies in the assignment of molecular structure; breaking chemical bonds before bringing the pieces of the puzzle together. Ch. 4 in Cid MM and Bravo J (Eds.), Tools in structure determination of organic molecules and complexes. VCH-Wiley, pp. 105–144. ISBN: 978-3-527-33336-3.

Paizs B, Suhai S. 2005. Fragmentation pathways of protonated peptides. Mass Spectrom Rev 24: 508–548.

Papayannopoulos IA. 1995. The interpretation of collision-induced dissociation tandem mass spectra of peptides. Mass Spectrom Rev 14: 49–73.

Pelander A, Tyrkkö E, Ojanperä I.2009. In silico methods for predicting metabolism and mass fragmentation applied to quetiapine in liquid chromatography/time-of-flight mass spectrometry urine drug screening. Rapid Commun Mass Spectrom 23: 506–514.

Pikulski M, Brodbelt JS. 2003. Differentiation of flavonoid glycoside isomers by using metal complexation and electrospray ionization mass spectrometry. J Am Soc Mass Spectrom 14: 1437–1453.

Pikulski M, Aguilar A, Brodbelt JS. 2007. Tunable transition metal-ligand complexation for enhanced elucidation of flavonoid diglycosides by electrospray ionization mass spectrometry. J Am Soc Mass Spectrom 18: 422–431.

Pluskal T, Uehara T, Yanagida M. 2012. Highly accurate chemical formula prediction tool utilizing high-resolution mass spectra, MS/MS fragmentation, heuristic rules, and isotope pattern matching. Anal Chem 84: 4396–4403.

Qin X-Z. 2001. Tandem mass spectrum of a farnesyl transferase inhibitor. Gas-phase rearrangements involving imidazole. J Mass Spectrom 36: 911–917.

Reddy PN, Srikanth R, Venkateswarlu N, Rao RN, Srinivas R. 2005. Electrospray ionization tandem mass spectrometric study of three isomeric substituted aromatic sulfonic acids; differentiation via ortho effects. Rapid Commun Mass Spectrom 19: 72–76.

Reiner EJ, Harrison AG, Bowen RD. 1989. Collision-induced dissociation mass spectra of protonated alkyl amines. Can J Chem 67: 2081–2088.

Robb DB, Covey TR, Bruins AP. 2000. Atmospheric pressure photoionization: An ionization method for liquid chromatography-mass spectrometry. Anal Chem 72: 3653–3659.

Roepstorff P, Fohlmann J. 1984. Proposal for a common nomenclature for sequence ions in mass spectra of peptides. Biomed Mass Spectrom 11: 601.

Satterfield M, Brodbelt JS. 2000. Enhanced detection of flavonoids by metal complexation and electrospray ionization mass spectrometry. Anal Chem 72: 5898–5906.

Satterfield M, Brodbelt JS. 2001. Structural characterization of flavonoid glycosides by collisionally activated dissociation of metal complexes. J Am Soc Mass Spectrom 12: 537–549.

Schäfer M, Dreiocker F, Budzikiewicz H. 2009. Collision-induced loss of AgH from Ag^+ adducts of alkylamines, aminocarboxylic acids and alkyl benzyl ethers leads exclusively to thermodynamically favored product ions. J Mass Spectrom 44: 278–284.

Schwarzenberg A, Ichou F, Cole RB, Machuron-Mandard X, Junot C, Lesage D, Tabet JC. 2013. Identification tree based on fragmentation rules for structure elucidation of organophosphorus esters by electrospray mass spectrometry. J Mass Spectrom 48: 576–586.

Shah RP, Garg A, Putlur SP, Wagh S, Kumar V, Rao V, Singh S, Mandlekar S, Desikan S. 2013. Practical and economical implementation of online H/D exchange in LC-MS. Anal Chem 85: 10904–10912.

Shi T, Zhao J, Shoeib T, Siu KWM, Hopkinson AC. 2004. Fragmentation of singly charged silver/α,ω-diaminoalkane complexes: Competition between the loss of H_2 and AgH molecules. Eur J Mass Spectrom 2004: 931–940.

Sigsby ML, Day RJ, Cooks RG. 1979. Fragmentation of even electron ions: Protonated amines and esters. Org Mass Spectrom 14: 556–561.

Smith RD, Loo JA, Edmonds CG, Barinaga CJ, Udseth HR. 1990. New developments in biochemical mass spectrometry: Electrospray ionization. Anal Chem 62: 882–899.

Smith RD, Loo JA, Ogorzalek Loo RR, Busman M, Udseth HR. 1991. Principles and practice of electrospray ionization-mass spectrometry for large polypeptides and proteins. Mass Spectrom Rev 10: 359–452.

Smith RM. 2004. Understanding mass spectra. A basic approach, 2nd ed. Wiley Interscience, Hoboken, NJ. ISBN 978-0-471-42949-4.

Song F, Cui M, Liu Z, Yu B, Liu S. 2004. Multiple-stage tandem mass spectrometry for differentiation of isomeric saponins. Rapid Commun Mass Spectrom 18: 2241–2248.

Stroobant V, Rozenberg R, El Bouabsa M, Deffense E, de Hoffmann E. 1995. Fragmentation of conjugate bases of esters derived from multifunctional alcohols including triacylglycerols. J Am Soc Mass Spectrom 6: 498–506.

Suzuki H, Kameyama A, Tachibana K, Narimatsu H, Fukui K. 2009. Computationally and experimentally derived general rules for fragmentation of various glycosyl bonds in sodium adduct oligosaccharides. Anal Chem 81: 1108–1120.

Syage JA. 2004. Mechanism of $[M+H]^+$ formation in photoionization. J Am Soc Mass Spectrom 15: 1521–1533.

Thurman EM, Ferrer I, Pozo ÓJ, Sancho JV, Hernández F. 2007. The even-electron rule in electrospray mass spectra of pesticides. Rapid Commun Mass Spectrom 21: 3855–3868.

Tureček F, Hanuš V. 1984. Retro-Diels–Alder reaction in mass spectrometry. Mass Spectrom Rev 3 (1984) 85–152.

Upthagrove AL, Hackett M, Nelson WL. 1999. Fragmentation pathways of selectively labeled propranolol using electrospray ionization on an ion trap mass spectrometer and comparison with ions formed by electron impact. Rapid Commun Mass Spectrom 13: 534–541.

Van Berkel GJ, McLuckey SA, Glish GL. 1992. Electrochemical origin of radical cations observed in electrospray ionization mass spectra. Anal Chem 64: 1586–1593.

Van Berkel GJ, Zhou F, Aronson JT. 1997. Changes in bulk solution pH caused by the inherent controlled-current electrolytic process of an electrospray ion source. Int J Mass Spectrom Ion Processes 162: 55–67.

Van Berkel GJ, Quirke JME, Tigani RA, Dilley AS, Covey TR. 1998. Derivatization for electrospray ionization mass spectrometry. 3. Electrochemically ionizable derivatives. Anal Chem 70: 1544–1554.

van Breemen RB. 1995. Electrospray liquid chromatography–mass spectrometry of carotenoids. Anal Chem 67: 2004–2009.

van Breemen RB, Huang C-R, Tan Y, Sander LC, Schilling AB. 1996. Liquid chromatography/mass spectrometry of carotenoids using atmospheric pressure chemical ionization. J Mass Spectrom 31: 975–981.

van Breemen RB, Dong L, Pajkovic ND. 2012. Atmospheric pressure chemical ionization tandem mass spectrometry of carotenoids. Int J Mass Spectrom 312: 163–172.

van der Hooft JJ, Vervoort J, Bino RJ, Beekwilder J, de Vos RC. 2011. Polyphenol identification based on systematic and robust high-resolution accurate mass spectrometry fragmentation. Anal Chem 83: 409–416.

Veith HJ. 1983. Mass spectrometry of ammonium and iminium salts. Mass Spectrom Rev 2: 419–446.

Vessecchi R, Crotti AEM, Guaratini T, Colepicolo P, Galembeck SE, Lopes NP. 2007. Radical ion generation processes of organic compounds in electrospray ionization mass spectrometry. Mini-Rev Org Chem 4: 75–87.

Vestal ML. 1983. Ionization techniques for nonvolatile molecules. Mass Spectrom Rev 2: 447–480.

Vukics V, Guttman A. 2010. Structural characterization of flavonoid glycosides by multi-stage mass spectrometry. Mass Spectrom Rev 29: 1–16.

Wagner W, Heimbach H, Levsen K. 1980. Gaseous odd- and even-electron ions. Int J Mass Spectrom Ion Processes 36: 125–142.

Wang H-Y, Zhang X, Guo Y-L, Dong X-C, Tang Q-H, Lu L. 2005. Sulfonamide bond cleavage in benzenesulfonamides and rearrangement of the resulting *p*-aminophenylsulfonyl cations: Application to a 2-pyrimidinyloxybenzylaminobenzenesulfonamide herbicide, Rapid Commun Mass Spectrom 19: 1696–1702.

Weeks DP, Field FH. 1970. Chemical ionization mass spectrometry. XI. Reactions of methoxymethyl formate and methoxymethyl acetate with methane and isobutane. J Am Chem Soc 92: 1600–1605.

Weissberg A, Dagan S. 2011. Interpretation of ESI(+)-MS–MS spectra—Towards the identification of "unknowns", Int J Mass Spectrom 299: 158–168.

Weissberg A, Madmon M, Dagan S. 2016. Structural identification of compounds containing tertiary amine side chains using ESI-MS^3 combined with fragmentation pattern matching to chemical analogues – Benzimidazole derivatives as a case study. Int J Mass Spectrom 394 (2016) 9–21.

Williams TM, Kind AJ, Houghton E, Hill DW. 1999. Electrospray collision-induced dissociation of testosterone and testosterone hydroxy analogs. J Mass Spectrom 34: 206–216.

Wojakowska A, Perkowski J, Góral T, Stobiecki M. 2013. Structural characterization of flavonoid glycosides from leaves of wheat (*Triticum aestivum L.*) using LC/MS/MS profiling of the target compounds. J Mass Spectrom 48: 329–339.

Wolfender J-L, Rodriguez S, Hostettmann K. 1998. Liquid chromatography coupled to mass spectrometry and nuclear magnetic resonance spectroscopy for the screening of plant constituents. J Chromatogr A 794: 299–316.

Wright P, Alex A, Pullen F. 2016. Predicting collision-induced dissociation mass spectra: Understanding the role of the mobile proton in small molecule fragmentation. Rapid Commun Mass Spectrom 30: 1163–1175.

Xia YQ, Miller JD, Bakhtiar R, Franklin RB, Liu DQ. 2003. Use of a quadrupole linear ion trap mass spectrometer in metabolite identification and bioanalysis. Rapid Commun Mass Spectrom 17: 1137–1145.

Yalcin T, Khouw C, Csizmadia IG, Peterson MR, Harrison AG. 1995. Why are b ions stable species in peptide spectra? J Am Soc Mass Spectrom 6: 1165–1174.

Yalcin T, Csizmadia IG, Peterson MB, Harrison AG. 1996. The structure and fragmentation of B-n ($b \geq 3$) ions in peptide spectra. J Am Soc Mass Spectrom 7: 233–242.

Yates, III, JR, Eng JK, McCormack AL, Schieltz D. 1995. Methods to correlate tandem mass spectra of modified peptides to amino acid sequences in the protein database. Anal Chem 67: 1426–1438.

Yates, III, JR. 1998. Mass spectrometry and the age of proteome. J Mass Spectrom 33: 1–19.

Yinon J, Laschever M. 1981. Reduction of Trinitroaromatic Compounds in Water by Chemical Ionization Mass Spectrometry. Org Mass Spectrom 16: 264–266.

Zaia J. 2004. Mass spectrometry of oligosaccharides. Mass Spectrom Rev 23: 161–227.

Zhang Y, Fonslow BR, Shan B, Baek MC, Yates, III, JR. 2013. Protein analysis by shotgun/bottom-up proteomics. Chem Rev 113: 2343–2394.

Zhurov KO, Fornelli L, Woodrich MD, Laskay ÜA, Tsybin YO. 2013. Principles of electron capture and transfer dissociation mass spectrometry applied to peptide and protein structure analysis. Chem Soc Rev 42: 5014–5030.

Zollinger M, Seibl J. 1985. McLafferty reactions in even-electron ions. Org Mass Spectrom 20: 649–661.

4

FRAGMENTATION OF DRUGS AND PESTICIDES

Interpretation of MS–MS Mass Spectra of Drugs and Pesticides, First Edition. Wilfried M. A. Niessen and Ricardo A. Correa C.

4.1

FRAGMENTATION OF DRUGS FOR CARDIOVASCULAR DISEASES AND HYPERTENSION

Antihypertonic or antihypertensive compounds are used to treat high blood pressure and are also prescribed in relation to various cardiac diseases. There are many classes of antihypertensive compounds, which lower blood pressure by different mechanisms. In this section, the interpretation of product-ion mass spectra of the various compound classes of antihypertensive drugs as well as other drugs related to cardiovascular diseases is discussed. Most compounds can be readily analyzed in positive-ion mode, whereas some compound classes are also prone to negative-ion formation.

4.1.1 β-BLOCKERS OR β-ADRENERGIC ANTAGONISTS

β-Blockers, also called β-adrenergic blocking agents, β-adrenergic antagonists, or β-antagonists, are a class of drugs particularly used for the management of cardiac arrhythmias, cardioprotection after myocardial infarction, and hypertension. As they reduce blood pressure and trembling, their use in sports is prohibited by the World Anti-Doping Agency (WADA). Multiresidue methods for β-blocker analysis have been reported in relation to therapeutic drug monitoring (Li et al., 2007), toxicology (Johnson & Lewis, 2006), sports doping control (Gergov et al., 2000; Thevis & Schänzer, 2005; Mazzarino et al., 2008; Pujos et al., 2009), food safety analysis (Zhang et al., 2009), and environmental analysis (Hernando et al., 2007; Lee et al., 2007).

β-Blockers have a general structure (albeit not exclusively) of 1-alkylamino-3-phenoxypropan-2-ol, with alkyl being *iso*-propyl such as in atenolol and propranolol, *tert*-butyl such as in timolol or talinolol, or other substituents such as in labetalol, which has the general structure of 2-aminoethanol (Figure 4.1.1). Individual compounds show further substitution or changes in the substituent of the aminopropanol skeleton, for example, a naphthoxy rather than a phenoxy in propranolol or a 4-(morpholin-4-yl)-1,2,5-thiadiazol-3-yl group in timolol.

General features in the fragmentation of all β-blockers in positive-ion mode comprise loss of water, cleavage of either of the C—N bonds of the amine moiety with charge retention on either resulting fragment, and combinations thereof (Figure 4.1.2). These C—N bond cleavages lead to a series of complementary ions, characterized by the sum of their *m/z* values being equal to *m/z* ([M+H]+1) (Section 3.5.2). Additional cleavage of the 2-propanol C^3—O ether bond (if present) may be observed, with charge retention on the 1-alkylamino-propan-2-ol fragment and, if sufficient proton affinity is present, charge retention on the corresponding ether oxygen containing fragment such as in timolol, as illustrated in Figure 4.1.2. Please note that additional compound-specific fragments may be observed due to losses from the substituents on the aromatic rings; these types of losses are not discussed here.

With *N-iso*-propyl-substituted compounds like propranolol ($[M+H]^+$ with *m/z* 260), one expects as the primary fragment ions the *iso*-propyl cation with *m/z* 43 ($[(CH_3)_2CH]^+$), protonated *iso*-propylamine ($[(CH_3)_2CHNH_2+H]^+$) with *m/z* 60, and the primary (1°) 2-hydroxy-3-[(propan-2-yl)amino]propyl carbocation ($[(CH_3)_2CHNHCH_2CHOHCH_2]^+$) with *m/z* 116, and their complementary ions with *m/z* ([M+H]−42), *m/z* ([M+H]−59), and *m/z* ([M+H]−115), respectively (Figure 4.1.2). For propranolol, these would be the ions with *m/z* 218, *m/z* 201, and *m/z* 145. Due to the low proton affinity of 1-naphthol, the ion with *m/z* 145 is not observed (Section 3.5.2). Secondary fragmentation of the ion with *m/z* 116 leads to an ion with *m/z* 98 due to a water loss, the 1-amino-2-hydroxypropan-3-yl carbocation with *m/z* 74 ($[H_2NCH_2CHOHCH_2]^+$) due to the loss of propene ($CH_3CH{=}CH_2$), and an ion with *m/z* 56 due to the loss of both water and $CH_3CH{=}CH_2$. Similar secondary fragmentation from $[M+H]^+$ leads to the ions with *m/z* 200 due to

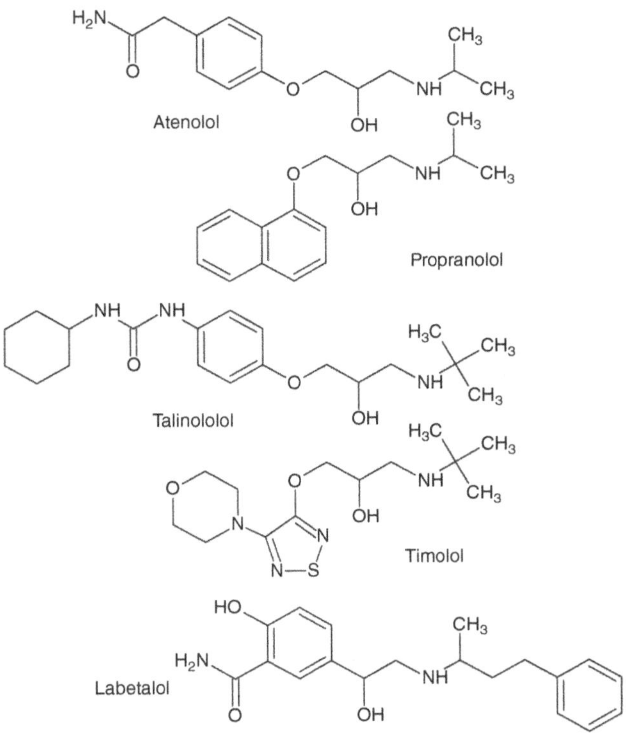

FIGURE 4.1.1 Typical structures of β-blockers.

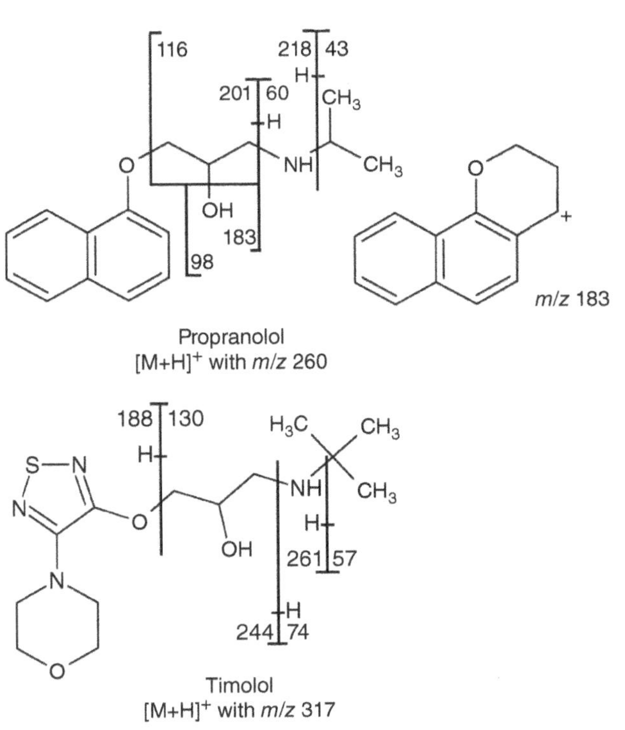

FIGURE 4.1.2 Fragmentation schemes for protonated propranolol and timolol.

losses of water and $CH_3CH{=}CH_2$ and with *m/z* 183 due to losses of water and *iso*-propylamine ($(CH_3)_2CHNH_2$). The spectra of 17 compounds of this class, available in our library collection, were searched for the presence of these ions (Table 4.1.1; a somewhat similar table was reported by others (Lee et al., 2007)). From the data in Table 4.1.1, one may conclude that the fragment ions with low *m/z* values are observed for most of the compounds, whereas losses of $(CH_3)_2CHNH_2$ or water and $CH_3CH{=}CH_2$ are relatively rare. For most compounds, either the ion with *m/z* 116 or the ion due to the loss of 77 Da is most abundant at low collision energy. In Table 4.1.1, the observation of an ion with *m/z* 72 is also indicated for most compounds. This ion is due to a β,γ-C—C cleavage relative to the amine group, resulting in protonated *N-iso*-propylmethanimine ($[(CH_3)_2CHN{=}CH_2{+}H]^+$). These observations are consistent with interpretations based on deuterium ([D]) labeling studies and accurate-mass determination (Upthagrove et al., 1999; Kumar et al., 2008).

In the same way, the fragmentation of *tert*-butyl-substituted compounds such as timolol ($[M{+}H]^+$ with *m/z* 317) may be considered (Figure 4.1.2). The characteristic fragment ions observed for 10 compounds in this class, available in our library collection, are summarized in Table 4.1.2. One expects as the primary fragment ions the *tert*-butyl cation with *m/z* 57 ($[(CH_3)_3C]^+$), protonated *tert*-butylamine ($[(CH_3)_3CNH_2{+}H]^+$) with *m/z* 74, and the primary (1°) 2-hydroxy-3-[(*tert*-butyl)amino]propyl carbocation ($[(CH_3)_3CNHCH_2CHOHCH_2]^+$) with *m/z* 130, and their complementary ions with *m/z* ([M+H]−56), *m/z* ([M+H]−73), and *m/z* ([M+H]−129), respectively (Figure 4.1.2). For these compounds, ions with *m/z* 130, which are structurally similar to the above-described ion with *m/z* 116, are not observed, indicating that the *tert*-butyl group is more readily lost than the *iso*-propyl group. The fragment ion with *m/z* 74, due to the loss of 2-methylpropene ($(CH_3)_2C{=}CH_2$) from the ion with *m/z* 130, is observed for all compounds.

Recently, the primary loss of either $CH_3CH{=}CH_2$ leading to an ion with *m/z* ([M+H]−42) or $(CH_3)_2C{=}CH_2$ leading to an ion with *m/z* ([M+H]−56) has been investigated in more detail using an ion-trap MS^n system (Holman et al., 2011). It was found that similar to electron ionization (EI) MS, the larger alkyl group is lost more readily, thus resulting in a more abundant fragment ion. Using *ab initio* calculations, this behavior could be correlated to the longer C—N bond with the *tert*-butyl instead of the *iso*-propyl substituent (Holman et al., 2011). Similar behavior is observed in our library collection based on collision-induced dissociation (CID) in a collision cell rather than in an ion trap. In the spectra of the same set of model compounds, the relative abundance of the ion with *m/z* ([M+H]−56) ranged from 10% to 75%, whereas either the ion with *m/z* ([M+H]−42) was absent or its relative abundance did not exceed 3%.

TABLE 4.1.1 Fragmentation of β-blockers with a 1-(*iso*-propylamino)-3-phenoxypropan-2-ol skeleton.

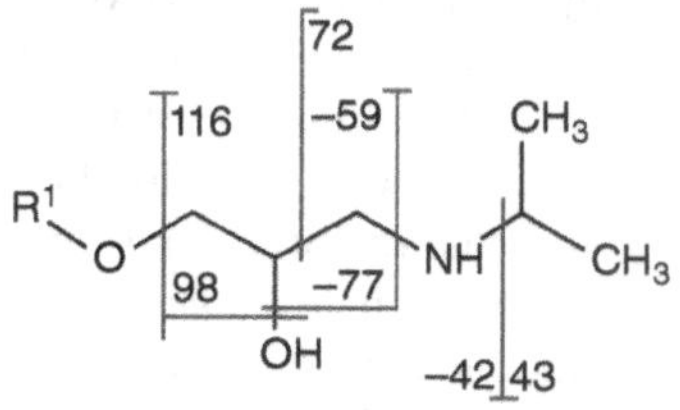

m/z values	[M+H]+	*m/z* 43	*m/z* 56	*m/z* 72	*m/z* 74	*m/z* 98	*m/z* 116	−18 −w	−42 $-C_3H_6$	−59 $-i$PrNH$_2$	−60 −water $-C_3H_6$	−77 −water $-i$PrNH$_2$
Acebutolol	337	x	x	x	x	x	x	319			277	260
Alprenolol	250	x	x	x	x	x	x	232	208		190	173
Atenolol	267	x	x	x	x	x	x	249	225	208	207	190
Befunolol	292	x				x	x	274	250		232	215
Betaxolol	308	x	x	x	x	x	x	290	266			231
Bisoprolol	326	x	x	x	x	x	x	308				
Carazolol	299	x	x	x	x	x	x		257			222
Diacetolol	309		x	x	x	x	x	291			249	232
Esmolol	296	x	x	x	x	x	x	278	254			219
Mepindolol	263	x	x	x	x	x	x		221			186
Metipranolol	310	x	x	x	x	x	x		268			233
Metoprolol	268	x	x	x	x	x	x		226			191
Oxprenolol	266	x	x	x	x	x	x	248			206	189
Pindolol	249	x	x	x	x	x	x		207	190		172
Practolol	267		x	x	x	x	x		225		207	190
Propranolol	260	x	x	x	x	x	x	242	218	201	200	183
Toliprolol	224	x	x	x	x	x	x	206	182		164	147

TABLE 4.1.2 Fragmentation of β-blockers with a 1-(*tert*-butylamino)-3-phenoxypropan-2-ol skeleton.

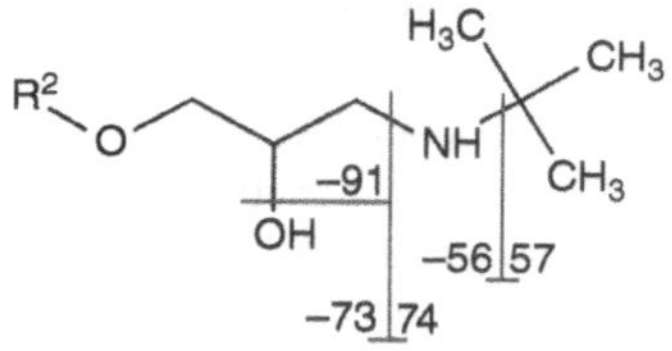

	[M+H]+	*m/z* 56	*m/z* 57	*m/z* 74	−18 −water	−56 $-C_4H_8$	−73 $-t$BuNH$_2$	−74 −water $-C_4H_8$	−91 −water $-t$BuNH$_2$
Bunitrolol	249	x	x	x		193		175	158
Bupranolol	272	x	x	x		216		198	181
Carteolol	293	x	x	x		237			202
Celiprolol	380	x	x	x	362	324	307	306	289
Levobunolol	292	x	x	x		236	219	218	201
Nadolol	310	x	x	x		254		236	219
Penbutolol	292	x	x	x		236			201
Talinolol	364	x	x	x		308			273
Tertatolol	296	x	x	x		240		222	205
Timolol	317	x	x	x		261	244	243	

The fragment ion due to the loss of alkylamine, that is, *iso*-propylamine ($(CH_3)_2CHNH_2$) (Table 4.1.1) or *tert*-butylamine ($(CH_3)_3CNH_2$) (Table 4.1.2), is relatively insignificant and often not present. It has been suggested that the loss of water in β-blockers is not due to a 1,2-elimination, but rather a 1,3-elimination, resulting in the formation of an aziridine ring via neighboring group participation, which would block the cleavage of the α-C—N bond relative to the hydroxy-substituted C-atom (Upthagrove et al., 1999). Thus, the loss of 77 Da (water and *iso*-propylamine) and 91 Da (water and *tert*-butylamine) should be the result of initial amine loss followed by the loss of water rather than vice versa.

Although β-blockers are generally analyzed in positive-ion mode, negative-ion MS–MS spectra are available in our library collection for acebutolol, carvedilol, celiprolol, isoproterenol, and labetalol. Hardly any common fragmentation routes could be observed for these five compounds. Acebutolol ($[M-H]^-$ with *m/z* 335), carvedilol ($[M-H]^-$ with *m/z* 405), and celiprolol ($[M-H]^-$ with *m/z* 378) have an aromatic-aliphatic ether moiety as a common feature (Figure 4.1.3). For all three compounds, cleavage of the aliphatic C—O ether bond is observed with charge retention on the aromatic substituent group, and the loss of the aliphatic group (2-propanol skeleton) as a radical. The resulting fragment ions with *m/z* 219 ($[C_{12}H_{13}NO_3]^{-\bullet}$) for acebutolol, with *m/z* 181 ($[C_{12}H_7NO]^{-\bullet}$) for carvedilol, and with *m/z* 248 ($[C_{13}H_{16}N_2O_3]^{-\bullet}$) for celiprolol show further fragmentation, for example, the loss of an *iso*-propyl radical ($^\bullet C_3H_7$) to an ion with *m/z* 176 for acebutolol, and the loss of a diethylamino radical $(C_2H_5)_2N^\bullet$ to an ion with *m/z* 176 for celiprolol. Further studies with accurate-mass determination are required for this compound class.

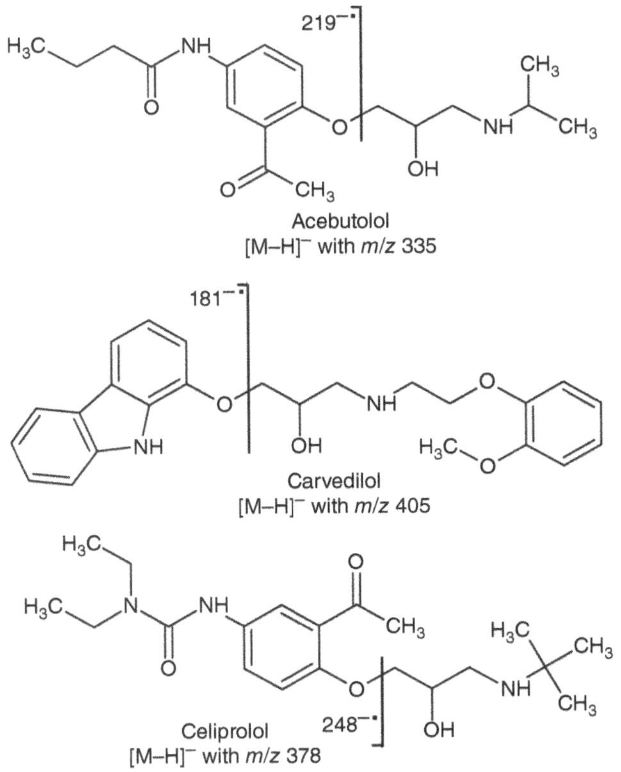

FIGURE 4.1.3 Fragmentation schemes for the deprotonated β-blockers acebutolol, carvedilol, and celiprolol.

4.1.2 DIHYDROPYRIDINE CALCIUM ANTAGONISTS

Dihydropyridine calcium channel blockers (CCBs) are a class of drugs that disrupt Ca^{2+}-flux through calcium channels. This class of compounds is easily identified by the suffix -dipine. Their main clinical effect is to reduce systemic vascular resistance and arterial pressure in individuals with hypertension. In GC–MS, dipines containing a nitro group can be analyzed in electron-capture negative ionization, resulting in molecular anions $M^{-\bullet}$ (Ehrhardt & Ziegler, 1988). Multiresidue LC–MS methods for this class of compounds have been reported in the positive-ion mode (Mueller et al., 2004; Baranda et al., 2005), although LC–MS analysis in negative-ion mode has been reported for some individual dipines (Tian et al., 2006; Lee et al., 2008; Kong et al., 2010). A comparison between electrospray ionization (ESI) and atmospheric-pressure photoionization (APPI) revealed that the positive-ion mode is preferred in ESI, whereas the negative-ion mode is more favorable in APPI (Pereira et al., 2008). The general structure of the dihydropyridine CCBs or dipines and the substituent groups of 10 representative compounds are shown in Table 4.1.3a (Mueller et al., 2004).

In positive-ion MS–MS, most dipines show losses of an alcohol or an alkene derived from the ester alkyl chains at the 3- and 5-positions (Table 4.1.3b) of the dihydropyridine ring. In most cases, the loss of the largest aliphatic chain yields the more abundant fragment ion (Mueller et al., 2004). The resulting fragment ions are also used as product ions in selected-reaction monitoring (SRM) in multiresidue LC–MS quantification methods (Mueller et al., 2004). After the loss of the alcohol, the resulting fragment is an acylium ion. The fragmentation behavior can be illustrated with the fragmentation of nimodipine, which contains a 2-methoxyethyl moiety at the 3-position (R_1) and an *iso*-propyl moiety at the 5-position (R_2), as shown in Table 4.1.3 (Qiu et al., 2004; Migliorança et al., 2005). Nimodipine ($[M+H]^+$ with *m/z* 419) shows the losses of 2-methoxyethanol ($HOCH_2CH_2OCH_3$) to an ion with *m/z* 343 and *iso*-propylalcohol ($(CH_3)_2CHOH$) to an ion with 359 corresponding to the loss of the ester groups R_1 and R_2, respectively (Table 4.1.3). The more abundant fragment ion with *m/z* 343 shows a subsequent loss of propene

($CH_3CH{=}CH_2$) from the R_2 group to a fragment ion with *m/z* 301, from which water is lost to an ion with *m/z* 283. This sequence of events, from $[M+H]^+$ with *m/z* 419 via the acylium ion with *m/z* 343 and the acylium ion with *m/z* 301 to the ion with *m/z* 283, raises some interesting mechanistic questions. The loss of $CH_3CH{=}CH_2$ from the ion with *m/z* 343 can in principle be explained as a charge-remote fragmentation, resulting in an ion with a carboxylic (—COOH) group at the 5-position and the acylium ion (—C=O$^+$) at the 3-position, whereas the subsequent loss of water to an ion

TABLE 4.1.3 Dihydropyridine calcium channel blockers.

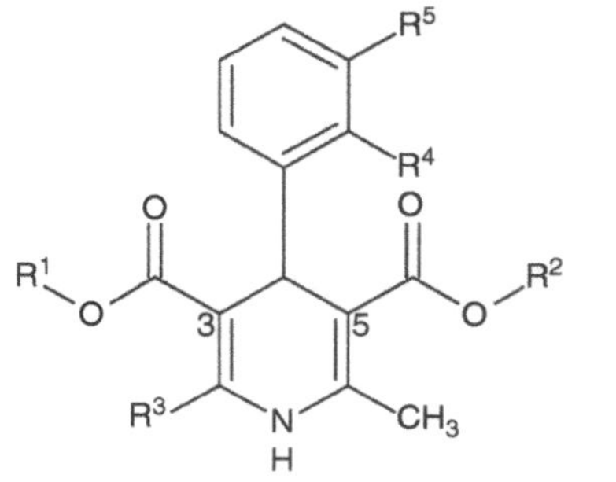

(a) Structures of the Compounds Studied

Substance	R_1	R_2	R_3	R_4	R_5
Amlodipine	$—CH_2CH_3$	$—CH_3$	$—CH_2O(CH_2)_2NH_2$	—Cl	—H
Aranidipine	$—CH_2COCH_3$	$—CH_3$	$—CH_3$	$—NO_2$	—H
Benidipine	$—C_5H_9N—CH_2C_6H_5$	$—CH_3$	$—CH_3$	—H	$—NO_2$
Cilnidipine	$—CH_2CHCHC_6H_5$	$—CH_2CH_2OCH_3$	$—CH_3$	—H	$—NO_2$
Felodipine	$—CH_2CH_3$	$—CH_3$	$—CH_3$	—Cl	—Cl
Isradipine	$—CH_3$	$—CH(CH_3)_2$	$—CH_3$	=N—O—N=	
Lacidipine	$—CH_2CH_3$	$—CH_2CH_3$	$—CH_3$	$—CHCHCOOC(CH_3)_3$	—H
Nicardipine	$—(CH_2)_2N(CH_3)CH_2C_6H_5$	$—CH_3$	$—CH_3$	—H	$—NO_2$
Nifedipine	$—CH_3$	$—CH_3$	$—CH_3$	$—NO_2$	—H
Nilvadipine	$—CH_3$	$—CH(CH_3)_2$	—C≡N	—H	$—NO_2$
Nimodipine	$—CH_2CH_2OCH_3$	$—CH(CH_3)_2$	$—CH_3$	—H	$—NO_2$
Nisoldipine	$—CH_2CH(CH_3)_2$	$—CH_3$	$—CH_3$	$—NO_2$	—H
Nitrendipine	$—CH_2CH_3$	$—CH_3$	$—CH_3$	—H	$—NO_2$

(b) Characteristic Fragmentation Observed in Positive-Ion Mode

Substance	$[M+H]^+$	Loss of R_1		Loss of R_2		Loss of Both R_1 and R_2
	m/z	*m/z*	Neutral Lost	*m/z*	Neutral Lost	*m/z*
Amlodipine	409			377	CH_3OH	
Felodipine	384	356	C_2H_4	352	CH_3OH	324
		338	CH_3CH_2OH			
Isradipine	372	330	$CH_2{=}CHCH_3$	340	CH_3OH	298
		312	$(CH_3)_2CHOH$			
Lacidipine	456	410	CH_3CH_2OH	410	CH_3CH_2OH	364
Nicardipine	480	315	$HO(CH_2)_2N(CH_3)CH_2C_6H_5$	448	CH_3OH	
Nifedipine	347	315	CH_3OH	315	CH_3OH	
Nivaldipine	386	354	CH_3OH	344	$CH_2{=}CHCH_3$	
				326	$(CH_3)_2CHOH$	
Nimodipine	419	343	$HOCH_2CH_2OCH_3$	359	$(CH_3)_2CHOH$	283
Nisoldipine	389	315	$HOCH_2CH(CH_3)_2$	357	CH_3OH	
Nitrendipine	361	315	CH_3CH_2OH	329	CH_3OH	

(continued)

TABLE 4.1.3 (*Continued*)

(c) Characteristic Fragmentation Observed in Negative-Ion Mode

Substance	$[M-H]^-$	C—C Cleavage between Rings		R^1 or R^2 Ester Aliphatic Chain Losses	Other Fragments
	m/z	*m/z*	*m/z*	*m/z*	*m/z*
Aranidipine	387	122	264	355: $-CH_3OH$ 313: $-CH_3COCH_2OH$	
Benidipine	504	122	381		
Cilnidipine	491	122			
Felodipine	382	145	236	350: $-CH_3OH$ 336: $-CH_3CH_2OH$	314: $-CH_3OH-HCl$ 300: $-CH_3CH_2OH-HCl$
Isradipine	370	119	250	310: $-CH_3CH_2OH$	
Lacidipine	454		250	408: $-CH_3CH_2OH$	380: $-(CH_3)_3COH$ from R^4 147: $[C_6H_5CH{=}CHCOO]^-$ 101: $[C_8H_5]^-$
Nicardipine	478	122	355	446: $-CH_3OH$	
Nifedipine	345	122	222		
Nivaldipine	384	122	261	324: $-(CH_3)_2CHOH$	
Nimodipine	417	122	294	357: $-(CH_3)_2CHOH$ 341: $-CH_3O(CH_2)_2OH$	
Nitrendipine	359	122	236	327: $-CH_3OH$ 313: $-CH_3CH_2OH$	208: 236 $-C_2H_4$

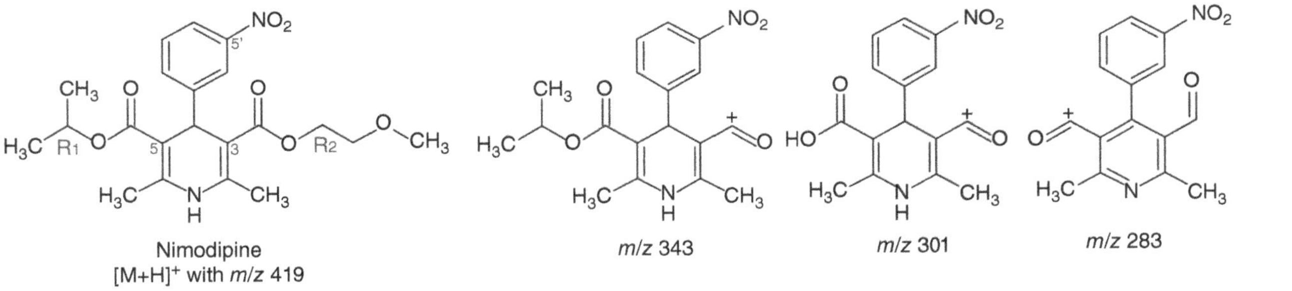

FIGURE 4.1.4 Proposed structures for the fragment ions of protonated nimodipine.

with *m/z* 283 demands charge migration and hydrogen rearrangement, resulting in the aromatization from a dihydropyridine ring to a pyridine ring (Figure 4.1.4). Accurate-mass data show that the fragment ion with *m/z* 255 is due to the loss of the nitryl radical ($^\bullet NO_2$) rather than the loss of formic acid (HCOOH) from the ion with *m/z* 301. The ester aliphatic group losses from the 3- and 5-positions of the dihydropyridine ring (R_1 and R_2 groups), that is, the fragment ions with *m/z* 359, *m/z* 343, and *m/z* 301 for nimodipine, are specified for all dipines in Table 4.1.3b. The fragment ion corresponding to the ion with *m/z* 301 in nimodipine is not observed if there is a methyl ester in the 3- and/or 5-positions.

Amlodipine ($[M+H]^+$ with *m/z* 409) also shows fragments due to loss of the R_1 and R_2 ester aliphatic substituents from the 3- and 5-position of the dihydropyridine ring, but the product-ion mass spectrum is dominated by the loss of ammonia (NH_3) from the (aminoethoxy)methyl group ($—CH_2OCH_2CH_2NH_2$) at the 2-position, and the subsequent fragmentation initiated by this loss, leading to the fragment ion with *m/z* 294 ($[C_{15}H_{17}ClNO_2]^+$), containing a morpholine-3-ylidene substituent, as shown in Figure 4.1.5 (Yasuda et al., 1996). This fragmentation route is not observed in analogs of amlodipine where the primary amino group is substituted by a dimethyl or an acetyl group, which would sterically hinder the formation of the morpholine ring (Yasuda et al., 1996). After intramolecular proton transfer to the dihydropyridine-N-atom in $[M+H]^+$, another ring cleavage involving the loss of 171 Da ($C_8H_{13}NO_3$) leads to the fragment ion with *m/z* 238 ($[C_{12}H_{13}ClNO_2]^+$). The structures and the proposed mechanisms for the formation of ions with *m/z* 294 and *m/z* 238 are outlined in Figure 4.1.5. In ion-trap MS^2, the fragment ions with *m/z* 238 and *m/z* 294 are not observed (Suchanova et al., 2006).

For negative-ion MS–MS, the class-specific fragmentation of the dihydropyridines is illustrated for nitrendipine ($[M-H]^-$ with *m/z* 359) in Figure 4.1.6. The cleavage of

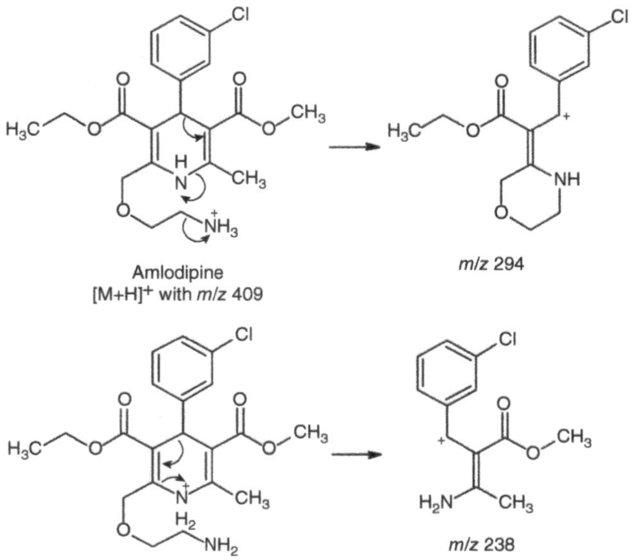

FIGURE 4.1.5 Proposed structures and fragmentation mechanisms for the formation of the ions with *m/z* 294 and 238 of protonated amlodipine.

the C—C bond between the substituted phenyl and the dihydropyridine rings results in two characteristic fragment ions, that is, the nitrophenide ion ($[O_2N—C_6H_4]^-$) with *m/z* 122 for nitrosubstituted compounds (*m/z* values for other substituents in Table 4.1.3c) and *m/z* 236 ($[C_{12}H_{14}NO_4]^-$) for nitrendipine (*m/z* values for other dipines in Table 4.1.3c). The *m/z* of the two fragments sum up to *m/z* ([M−H]−1) (Section 3.7.2). The $[O_2N—C_6H_4]^-$ with *m/z* 122 gives a secondary fragment ion, that is, the cyclohex-2,5-dien-4-ide-3-yl-one radical anion with *m/z* 92 ($[C_6H_4O]^{-\bullet}$) due to the loss of the nitrosyl radical ($^{\bullet}NO$). An alternative interpretation of the ion with *m/z* 92 as a methylidenide ion ($[CH_2C_5H_4N]^-$) (Kong et al., 2010) is probably not correct, as its formation would require the loss of a methyl substituent from the dihydropyridine ring (and its aromatization). In addition, the dipines show losses of R^1 or R^2 ester aliphatic substituents as alcohols, for nitrendipine that means losses of methanol (CH_3OH) or ethanol (CH_3CH_2OH) (Figure 4.1.6). Secondary losses of the R^1 or R^2 ester aliphatic groups from the ion with *m/z* 236 (due to loss of the substituted phenyl ring) may be observed as well, for example, the ion with *m/z* 208 for nitrendipine, involving the loss of ethene (C_2H_4). Similar losses for other dipines are given in Table 4.1.3c. For both felodipine and lacidipine, fragment ions are observed due to losses of (part of) the R^4 (or R^5) substituent, for example, loss of hydrogen chloride (HCl) for felodipine and loss of *tert*-butanol ($(CH_3)_3COH$) for lacidipine (Table 4.1.3c).

4.1.3 ANGIOTENSIN-CONVERTING ENZYME INHIBITORS

Angiotensin-converting enzyme inhibitors (ACE inhibitors) are pharmaceuticals that are primarily used in the treatment of hypertension and congestive heart failure. ACE inhibitors can be subdivided into three groups based on their molecular structure: (1) thiol-containing agents such as captopril, (2) dicarboxylate-containing agents such as enalapril, and (3) phosphinate-containing agents such as fosinopril. Captopril was the first ACE inhibitor developed (1975) and was one of the earliest successes of structure–activity relationship (SAR)-based drug design. Due to its thiol group (—SH), captopril shows adverse drug reactions. Therefore, other

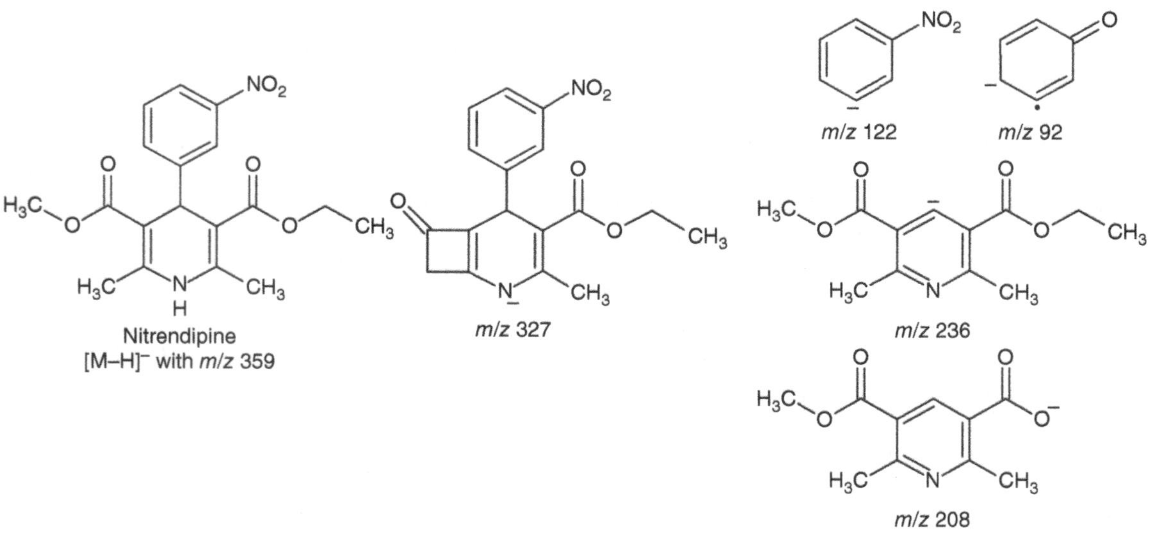

FIGURE 4.1.6 Proposed structures for the fragment ions of deprotonated nitrendipine.

classes of ACE inhibitors were designed without a thiol group. Most ACE inhibitors are administered as prodrugs (ethyl esters) to improve oral bioavailability. Due to the carboxylic acid group, ACE inhibitors are amenable to negative-ion formation, although they can also be analyzed in positive-ion mode.

4.1.3.1 Thiol-Containing ACE Inhibitors: Captopril

The positive-ion fragmentation of captopril ($[M+H]^+$ with *m/z* 218) is outlined in Figure 4.1.7. The most abundant fragment ion with *m/z* 116 ($[C_5H_{10}NO_2+H]^+$) results from the loss of carbon monoxide (CO) and 2-propene-1-thiol ($H_2C{=}CHCH_2SH$). Charge retention on the complementary acylium ion, that is, the 2-methyl-3-sulfanylpropanoyl cation ($[HSCH_2CHCH_3C{=}O]^+$) with *m/z* 103, is also observed as well as an ion with *m/z* 75 due to the loss of CO from the ion with *m/z* 103. Alternatively, the loss of formic acid (HCOOH) from $[M+H]^+$ is observed. A combination of these losses leads to the protonated 3,4-dihydro-2*H*-pyrrole ($[C_4H_7N+H]^+$) with *m/z* 70.

In negative-ion mode, captopril ($[M-H]^-$ with *m/z* 216) shows the loss of hydrogen sulfide (H_2S) to an ion with *m/z* 182, the loss of methanethial ($H_2C{=}S$) to an ion with *m/z* 170, and the loss of the complete *N*-sulfur-containing substituent of the pyrrolidine ring to deprotonated proline ($[C_5H_9NO_2-H]^-$) with *m/z* 114 (Figure 4.1.7).

4.1.3.2 Dicarboxylate ACE Inhibitors

The dicarboxylate ACE inhibitors are administered as monoethyl-ester prodrugs, for example, ramipril, which are metabolized to the active compound, for example, ramiprilat. Using ramipril ($[M+H]^+$ with *m/z* 417; $[M-H]^-$ with *m/z* 415) as example, characteristic class-specific fragments for dicarboxylate ACE inhibitors in both positive-ion and negative-ion fragmentations are outlined in Table 4.1.4 (Bhardwaj & Singh, 2008; Parekh et al., 2008; Gupta et al., 2011).

In positive-ion mode (Table 4.1.4a), the ramipril fragment ion with *m/z* 343 (F_1) is due to the loss of $HCOOCH_2CH_3$. The fragment ion with *m/z* 234 (F_2) is due to a cleavage of the C—C bond of the α-aminocarbonyl functionality, with charge retention on the amino group bearing fragment ion, that is, ($[C_{14}H_{19}NO_2+H]^+$). The fragment ion with *m/z* 160 (F_3; $[C_{11}H_{14}N]^+$) results from a combination of these two losses. In addition, a fragment ion with *m/z* 130 (F_4) is observed, which can be considered as secondary fragmentation of the ion with *m/z* 234 involving the loss of styrene ($C_6H_5CH{=}CH_2$). The same fragmentation routes are observed for related compounds (Table 4.1.4a).

In negative-ion mode (Table 4.1.4b), the ramipril fragment ion with *m/z* 341 (F_1) is due to the loss of $HCOOCH_2CH_3$, in a similar manner to the positive-ion mode. Other fragmentation reactions in negative-ion mode are different. The five cleavage sites, indicated for ramipril in Table 4.1.4b, are not always observed for all dicarboxylate ACE inhibitors studied, which could be related to the collision energy applied. Additional fragment ions for specific compounds are indicated and interpreted as well.

4.1.3.3 Phosphinate-Containing ACE Inhibitors: Fosinopril

The fragmentation of fosinopril ($[M+H]^+$ with *m/z* 564) involves the loss of the phosphinic acid ester group, that is, 2-methylprop-1-en-1-yl propanoate ($CH_3CH_2COOCH{=}C(CH_3)_2$) resulting in fragment ions with *m/z* 436 and with *m/z* 418 after further loss of water (Figure 4.1.8). Secondary loss of formic acid (HCOOH) from the ion with *m/z* 436 leads to a fragment ion with *m/z* 390. At higher collision energies, protonated 5-cyclohexyl-3,4-dihydro-2*H*-pyrrole with *m/z* 152 ($[C_{10}H_{17}N+H]^+$) is most abundant (Figure 4.1.8).

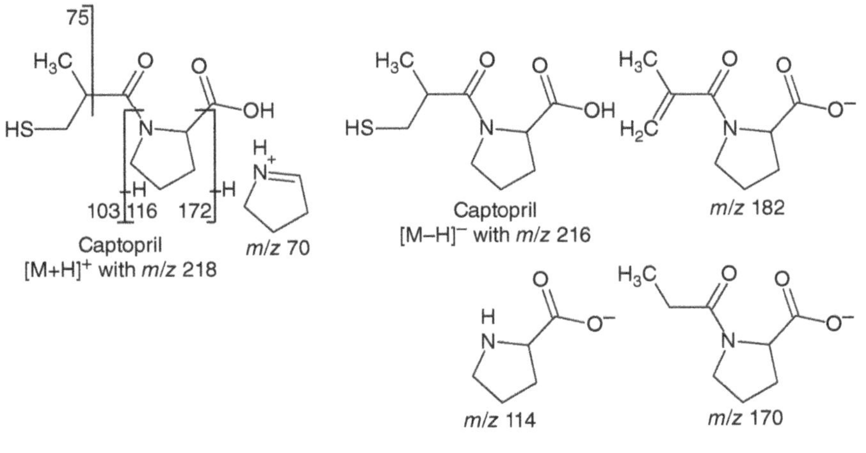

FIGURE 4.1.7 Fragmentation schemes for protonated and deprotonated captopril.

TABLE 4.1.4 Fragmentation of angiotensin II converting enzyme (ACE) inhibitors.

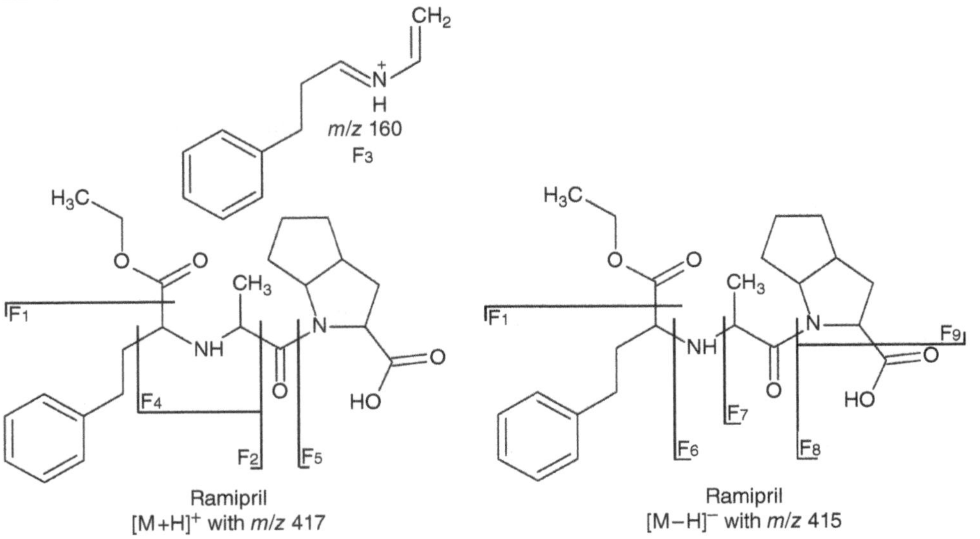

(a) Class-Specific Fragmentation Observed in Positive-Ion Mode

Compound	$[M+H]^+$ (*m/z*)	F_1 (*m/z*)	F_2 (*m/z*)	F_3 (*m/z*)	F_4 (*m/z*)	F_5 (*m/z*)
Enalapril	377	303	234	160	130	116
Enalaprilat	349	303	206	160	102	116
Imidapril	406	332	234	160	130	145
Moexipril	499	425	234	160	130	238
Perindopril	369	295	172	98		170
Ramipril	417	343	234	160	130	156
Ramiprilat	389	343	206	160	102	156
Quinapril	439	365	234	160	130	178
Quinaprilat	411	365	206	160	102	178
Spirapril	467	393	234	160	130	206
Trandopril	431	357	234	160	130	170

(b) Class-Specific Fragmentation Observed in Negative-Ion Mode

Compound	$[M-H]^-$ (*m/z*)	F_1 (*m/z*)	F_6 (*m/z*)	F_7 (*m/z*)	F_8 (*m/z*)	F_9 (*m/z*)	Other Fragments
Benazepril	423	349	231				*m/z* 174: possibly F_6 − HN=CHCHO
Imidapril	404		212		143	99	*m/z* 360: loss of CO_2
Lisinopril	404				114	68	*m/z* 386: loss of H_2O *m/z* 289: complementary to *m/z* 114
Quinapril	437	363		232	176		*m/z* 319: $F_1 - CO_2$ *m/z* 188: $F_7 - CO_2$ *m/z* 347: loss of C_2H_5OH and CO_2
Ramipril	415	341		210	154		*m/z* 166: $F_7 - CO_2$
Ramiprilat	387	341	225		154		

4.1.4 DIURETIC DRUGS

A diuretic is any drug that elevates the miction rate and thus provides a means of induced diuresis. As such, diuretics are frequently used in the treatment of hypertension. They are misused in sports to achieve rapid weight loss or to mask the use of other banned substances. There are several categories of diuretics: (1) thiazide-type diuretics derived from

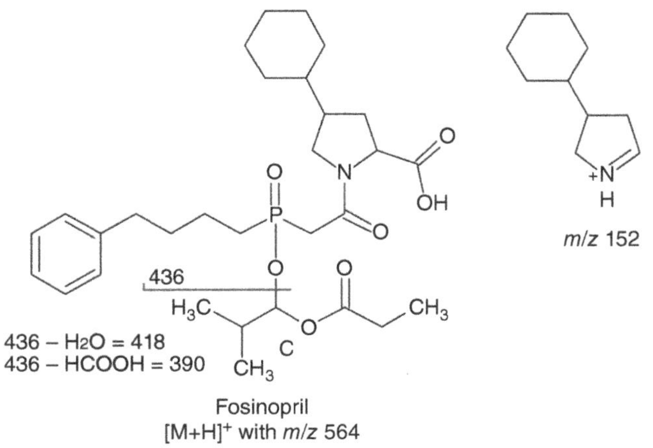

FIGURE 4.1.8 Fragmentation scheme for protonated fosinopril.

benzothiadiazine such as hydrochlorothiazide, benzthiazide, buthiazide, and indapamide, (2) high-ceiling loop diuretics such as torsemide and bumetanide, (3) epithelial sodium channel blockers such as amiloride and triamterene, and (4) carbonic anhydrase inhibitors such as acetazolamide. Most of these compounds have (cyclic) sulfonamide groups, and thus are readily amenable to negative-ion formation, mostly using ESI but sometimes using APCI (Qin et al., 2003). Multiresidue analysis of diuretic drugs using mostly negative-ion LC–MS is primarily directed at sports doping analysis (Deventer et al., 2002, 2009; Goebel et al., 2004; Thevis & Schänzer, 2005; Mazzarino et al., 2008; Tsai & Lee, 2008; Deventer et al., 2009; Wong et al., 2011). The fragmentation of diuretic drugs in negative-ion mode has been studied extensively (Garcia et al., 2002; Thevis & Schänzer, 2007; Giancotti et al., 2008).

4.1.4.1 Thiazide Diuretic Drugs

Thiazide diuretics are derived from 2,4-benzothiadiazine. Three general structures can be further discerned showing differences in fragmentation behavior: hydrochlorothiazide (Group 1), chlorothiazide (Group 2), and methyclothiazide (Group 3) are typical examples (Table 4.1.5).

The Group 1 compounds have a C^3—N^4 single bond and no methyl substituent at N^2. In negative-ion mode, hydrochlorothiazide and related compounds (althiazide, buthiazide, cyclopenthiazide, ethiazide) and the trifluoromethyl analogs (bendroflumethiazide and hydroflumethiazide) show a characteristic loss of hydrogen cyanide (HCN), followed by a loss of sulfur dioxide (SO_2). For hydrochlorothiazide ($[M-H]^-$ with *m/z* 296) and hydroflumethiazide ($[M-H]^-$ with *m/z* 330), the loss of HCN can be readily understood. However, for structure analogs with a substituents at the C^3-position (buthiazide, cyclopenthiazide, ethiazide, bendroflumethiazide), the loss of HCN requires a rearrangement of the substituent, most likely to the N-atom at the 4-position of the 2,4-benzothiadiazine ring (Figure 4.1.9). Two other class-specific fragment ions are observed for these compounds. The ions with *m/z* 269 (compounds with Cl at the 6-position) or *m/z* 303 (compounds with CF_3 at the 6-position) involve the loss of R^3—C≡N, and in turn undergo the loss of SO_2 to fragment ions with *m/z* 205 ($[C_6H_6ClN_2O_2S]^-$) or *m/z* 239 ($(C_7H_6F_3N_2O_2S]^-$), respectively. In addition, three common fragment ions are observed: an ion with *m/z* 190 ($[ClC_6H_3SO_2NH_2]^-$) or *m/z* 224 ($[CF_3C_6H_3$—$SO_2NH_2]^-$), an ion with *m/z* 126 ($[ClC_6H_3NH_2]^-$) or *m/z* 160 ($[CF_3C_6H_3NH_2]^-$), and the deprotonated iminosulfene ($[NSO_2]^-$) with *m/z* 78 (Figure 4.1.9) (Garcia et al., 2002; Thevis & Schänzer, 2007). A mechanism for the formation of most of these fragment ions has been proposed (Garcia et al., 2002).

Next to these class-specific fragment ions, individual compounds provide some compound-specific fragment ions. Althiazide ($[M-H]^-$ with *m/z* 382) shows the loss of 41 Da, attributed to the unlikely loss of acetonitrile ($CH_3C≡N$) (Garcia et al., 2002), but most likely due to the loss of an allyl radical ($^{\bullet}C_3H_5$) from the thioether group R^3. The fragmentation of bendroflumethiazide ($[M-H]^-$ with *m/z* 420) has been investigated in more detail (Giancotti et al., 2008). It shows the loss of a benzyl radical ($C_6H_5CH_2^{\bullet}$) or toluene ($C_6H_5CH_3$) leading to fragment ions with *m/z* 329 and *m/z* 328, respectively. Two consecutive losses of hydrogen fluoride (HF) from the ion with *m/z* 329 result in a fragment ion with *m/z* 289. Buthiazide ($[M-H]^-$ with *m/z* 352) shows the loss of hydrogen chloride (HCl). Cyclopenthiazide ($[M-H]^-$ with *m/z* 378) shows the loss of cyclopentadiene (C_5H_6) from the R^3-side chain. Hydroflumethiazide ($[M-H]^-$ with *m/z* 330) shows a fragment ion with *m/z* 266 due to the loss of SO_2. Methyclothiazide ($[M-H]^-$ with *m/z* 358) shows the loss of HCl leading to an ion with *m/z* 322. Trichlormethiazide ($[M-H]^-$ with *m/z* 378) show the losses of one or two HCl molecules to produce ions with *m/z* 342 or *m/z* 306, respectively (Garcia et al., 2002).

In positive-ion MS–MS, the Group 1 compounds show complex and extensive fragmentation, which is difficult to interpret without accurate-mass data. Some general fragmentation characteristics have been illustrated with buthiazide ($[M+H]^+$ with *m/z* 354) in Figure 4.1.10. The major fragment ions are due to cleavages of the benzothiadiazine ring, resulting in the fragment ions with *m/z* 269 (Figure 4.1.10), *m/z* 253, and *m/z* 205 (due to the loss of SO_2 from the fragment ion with *m/z* 269).

The Group 2 compounds (benzthiazide and chlorothiazide) have a C^3—N^4 double bond. This double bond effectively blocks all the characteristic negative-ion fragmentation observed for the Group 1 compounds (without this double bond). Benzthiazide ($[M-H]^-$ with *m/z* 430) first shows two fragment ions related to the loss of the (benzylsulfanyl)methyl group R^3 (—$CH_2SCH_2C_6H_5$):

TABLE 4.1.5 Structures and negative-ion fragmentation of the benzothiadiazine diuretics studied.

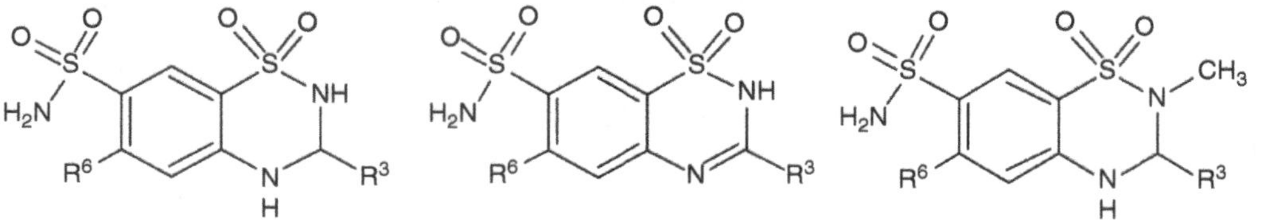

Group 1 Group 2 Group 3

Group	Name	R^3	R^6	M
1	Althiazide	$—CH_2SCH_2CH{=}CH_2$	—Cl	382.9835
1	Bendroflumethiazide	$—CH_2C_6H_5$	$—CF_3$	421.0377
1	Buthiazide	$—CH_2CH(CH_3)_2$	—Cl	353.0270
1	Cyclopenthiazide	$—CH_2C_5H_9$	—Cl	379.0427
1	Cyclothiazide	—Norborn-2-en-5-yl (C_7H_9)	—Cl	389.0270
1	Ethiazide	$—CH_2CH_3$	—Cl	324.9958
1	Hydrochlorothiazide	—H	—Cl	296.9644
1	Hydroflumethiazide	—H	$—CF_3$	330.9908
1	Trichlormethiazide	$—CHCl_2$	—Cl	378.9022
2	Benzthiazide	$—CH_2SCH_2C_6H_5$	—Cl	430.9834
2	Chlorothiazide	—H	—Cl	294.9488
3	Methyclothiazide	$—CH_2Cl$	—Cl	358.9568
3	Polythiazide	$—CH_2SCH_2CF_3$	—Cl	438.9708

Group	Name	$[M-H]^-$	$-HCN$ $-SO_2$	$-RC{\equiv}N$	$-RC{\equiv}N$ $-SO_2$	$C_6H_5R^6N^-$	SO_2N^-
1	Althiazide	382		269	205		78
1	Bendroflumethiazide	420		303	239	160	78
1	Buthiazide	352	261	269	205	126	78
1	Cyclopenthiazide	378	287	269	205	126	78
1	Cyclothiazide	388		269	205		78
1	Ethiazide	324	233	269	205	126	78
1	Hydrochlorothiazide	296	205	269	205	126	78
1	Hydroflumethiazide	330	239	303	239	160	78
1	Trichlormethiazide	378					
2	Benzthiazide	430	339				78
2	Chlorothiazide	294					78
3	Methyclothiazide	358					78
3	Polythiazide	438					

Three groups are indicated: (1) no double bond between C^3 and N^4; (2) double bond between C^3 and N^4; (3) no double bond between C^3 and N^4 and a methyl substituent at N^2.

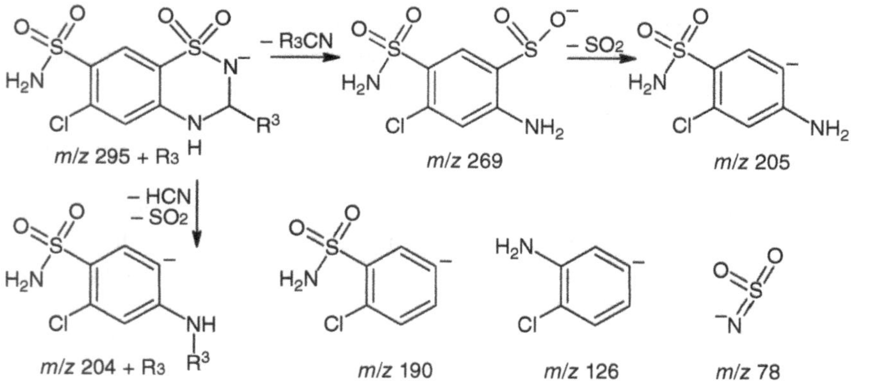

FIGURE 4.1.9 General fragmentation pathways for deprotonated thiazide diuretic drugs.

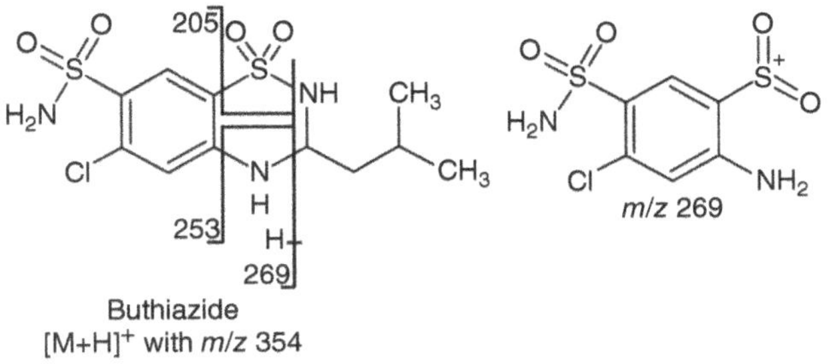

FIGURE 4.1.10 Fragmentation scheme for protonated buthiazide.

an ion with m/z 339 due to the loss of $C_6H_5CH_2^{\bullet}$, and an ion with m/z 308 due to the loss of phenylmethanethial (C_6H_5CHS), followed by the loss of a sulfamoyl radical ($NH_2SO_2^{\bullet}$) to an ion with m/z 228, most likely involving the sulfonamide group at the 7-position (Garcia et al., 2002). Chlorothiazide ($[M-H]^-$ with m/z 294) shows a fragment ion with m/z 214 due to the loss of $NH_2SO_2^{\bullet}$ as well, followed by the loss of a chlorine radical ($Cl^{\bullet}$) from the 6-position to an ion with m/z 179 ($[C_7H_3N_2O_2S]^-$).

Positive-ion MS–MS data are also available for the Group 2 compounds benzthiazide and chlorothiazide. The major fragment of benzthiazide ($[M+H]^+$ with m/z 432) is the tropylium ion ($[C_7H_7]^+$) with m/z 91 derived from the R^3 group (—$CH_2SCH_2C_6H_5$), whereas chlorothiazide ($[M+H]^+$ with m/z 296) yields fragment ions related to the benzothiadiazine structure. There appear to be three fragmentation pathways: (a) subsequent losses of ammonia (NH_3) and SO_2 leading to fragment ions with m/z 279 and m/z 215 ($[C_7H_4ClN_2O_2S]^+$), respectively; (b) rearrangement of the C^7 sulfonamide group (—SO_2NH_2) resulting in a fragment ion with m/z 231 (effective loss of H_2NOS), which subsequently loses hydrogen cyanide (HCN) to an ion with m/z 204 ($[C_6H_3ClNO_3S]^+$); (c) loss of both SO_2 and HCN from the sulfonamide group of the thiadiazine ring to an ion with m/z 205 ($[C_6H_6ClN_2O_2S]^+$). Proposed fragmentation pathways are given in Figure 4.1.11.

The Group 3 compounds (methyclothiazide and polythiazide) differ from the Group 1 compounds by a methyl substituent at N^2. These two compounds have little in common in their negative-ion fragmentation patterns, however. Methyclothiazide ($[M-H]^-$ with m/z 358) shows the loss of HCl, followed by two consecutive SO_2 losses (Garcia et al., 2002). Polythiazide ($[M-H]^-$ with m/z 438) shows three consecutive HF losses, as well as losses from the C^3 (2,2,2-trifluoroethylsulfanyl)methyl group ($CF_3CH_2SCH_2$—). Polythiazide shares with benzthiazide the same —CH_2SCH_2— structural feature at C^3, resulting in similar fragmentation, that is, the loss of a 2,2,2-trifluoroethyl radical ($^{\bullet}CH_2CF_3$) and the loss of trifluoroethanethial (CF_3CHS) to ions with m/z 355 and m/z 324, respectively.

4.1.4.2 Other Sulfonamide Thiazide-Like Diuretic Drugs

Clofenamide ($[M-H]^-$ with m/z 269) is a low-ceiling sulfonamide diuretic. Its fragmentation in negative-ion MS–MS can be described by sequential small-molecule losses, partly involving rearrangements in the sulfonamide groups. Thus, a fragment ion with m/z 233 is observed due to the loss of hydrogen chloride (HCl); m/z 205 due to the loss of sulfur dioxide (SO_2); m/z 169 due to the loss of HCl and SO_2 in either order; m/z 141 due to the loss of two times SO_2; m/z 121 due to the loss of SO_2, HCl, and SO; m/z 105 due to HCl loss

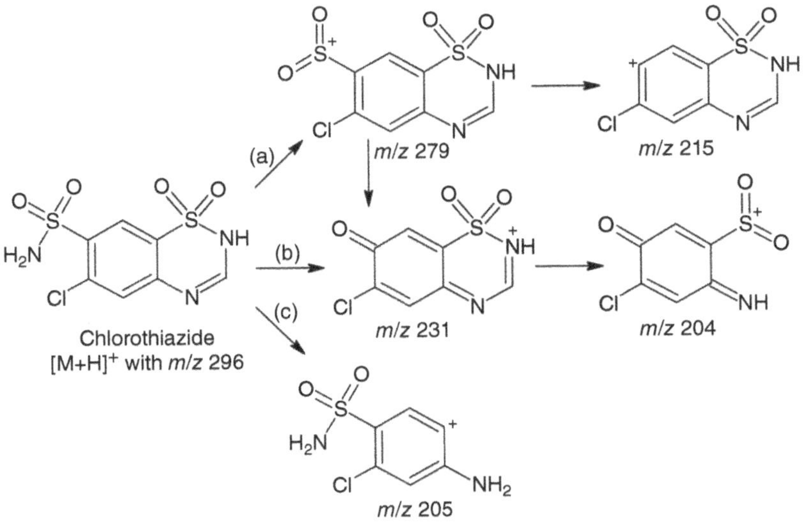

FIGURE 4.1.11 Fragmentation pathways for protonated chlorothiazide.

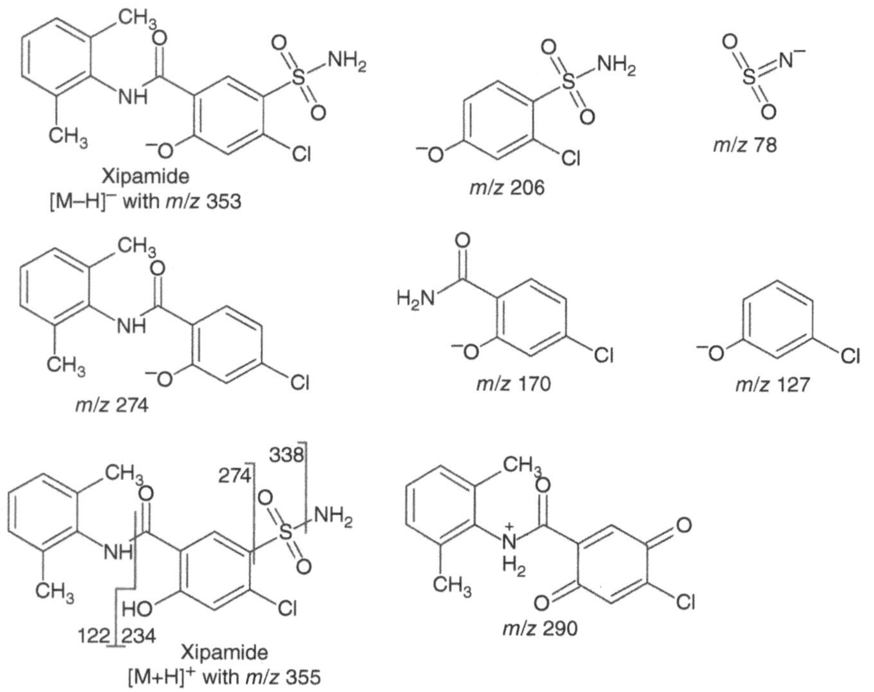

FIGURE 4.1.12 Fragmentation schemes for deprotonated and protonated xipamide.

from *m/z* 141; and the deprotonated iminosulfene ($[NSO_2]^-$) with *m/z* 78.

Although three sulfamoyl benzamide diuretics (clopamide, indapamide, xipamide) share common structural features, their fragmentation behavior is quite different. Clopamide ($[M-H^-$ with *m/z* 344) shows fragment ions with *m/z* 308 due the loss of HCl and with *m/z* 280 due to the loss of SO_2 resulting from a rearrangement of the sulfonamide group. Xipamide ($[M-H]^-$ with *m/z* 353), on the other hand, shows the loss of iminosulfene ($HNSO_2$) to an ion with *m/z* 274 (Figure 4.1.12). With clopamide and indapamide ($[M-H]^-$ with *m/z* 364), the radical anion with *m/z* 189 ($[ClC_6H_3—SO_2NH]^{-\bullet}$) is observed, whereas with xipamide the presence of an additional C^2 hydroxy group on the benzenesulfonamide ring results in an even-electron ion (EE^-) with *m/z* 206 ($[ClC_6H_4O—SO_2NH]^-$). Interestingly, an EE^- fragment with *m/z* 190 ($[ClC_6H_4—SO_2NH]^-$) is observed in the fragmentation of chlorthalidone ($[M-H]^-$ with *m/z* 337), whereas the odd-electron ($OE^{-\bullet}$) fragment with *m/z* 189 is observed again for the benzene-1,3-disulfonamide mefruside ($[M-H]^-$ with *m/z* 381), which, like clopamide, also shows the loss of HCl or SO_2 to ions with *m/z* 345 and *m/z* 317, respectively.

In positive-ion mode, xipamide ($[M+H]^+$ with *m/z* 355) shows losses of ammonia (NH_3), SO_2 and NH_3, and sulfur monoxide (SO) and NH_3 (involving a rearrangement in the SO_2 group) to ions with *m/z* 338, *m/z* 274, and *m/z* 290, respectively. In addition, two complementary ions are observed due to cleavage of the amide C—N bond, that is, the protonated 2,6-dimethylaniline ($[C_8H_{11}N+H]^+$) with *m/z* 122 and an acylium ion with *m/z* 234 ($[C_6H_5ClNO_3S—C{=}O]^+$) (Figure 4.1.12).

4.1.4.3 Loop Diuretic Drugs

Loop diuretics act on the Na^+–K^+–$2Cl^-$ transporter in the nephron's ascending limb of the loop of Henle, and by competing for the Cl^- binding site they inhibit sodium and chloride reabsorption. They are especially effective in patients with impaired kidney function.

As an example of a high-ceiling loop diuretic, the positive-ion fragmentation of torasemide ($[M+H]^+$ with *m/z* 349) is outlined in Figure 4.1.13. Characteristic cleavages on either side of the carbonyl group in the urea moiety result in an acylium ion with *m/z* 290 ($[C_{12}H_{12}N_3O_2S—C{=}O]^+$) due to the loss of *iso*-propylamine ($(CH_3)_2CHNH_2$) and a protonated [4-(3-methylanilino)pyridin-3-yl]sulfonamide ($[C_{12}H_{13}N_3O_2S+H]^+$) with *m/z* 264. Cleavage of the C—S bond between the pyridine ring and the SO_2 group results in the 4-(3-methylanilino)pyridin-3-ium ion with *m/z* 183 ($[C_{12}H_{11}N_2]^+$).

Other loop diuretics are preferentially analyzed in negative-ion mode. The negative-ion fragmentation of furosemide ($[M-H]^-$ with *m/z* 329) has been investigated using a linear-ion-trap–orbitrap (LIT–orbitrap) hybrid instrument (Giancotti et al., 2008). The initial loss of carbon

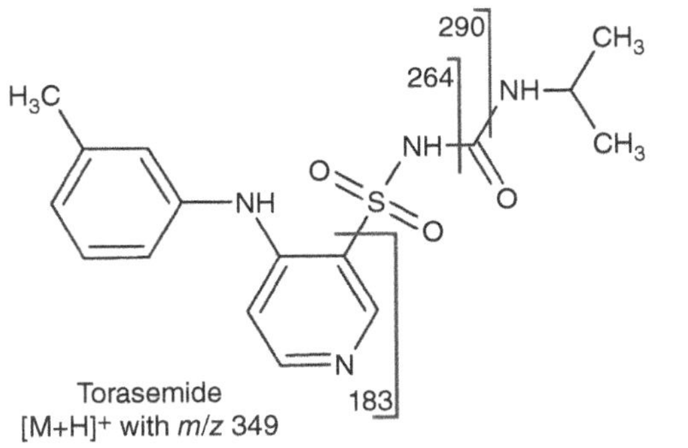

FIGURE 4.1.13 Fragmentation scheme for protonated torasemide.

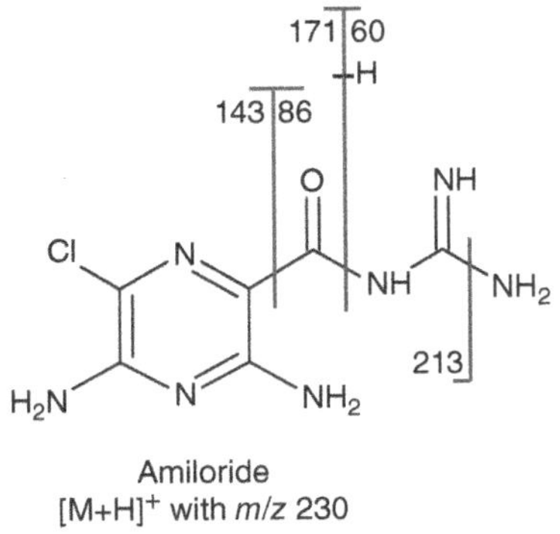

FIGURE 4.1.14 Fragmentation scheme for protonated amiloride.

dioxide (CO_2) leads to an ion with *m/z* 285, which is also observed for the other sulfamoylbenzoic acid diuretics bumetanide ($[M-H]^-$ with *m/z* 363) and piretanide ($[M-H]^-$ with *m/z* 361). Subsequently, furosemide shows a minor (≈1%) loss of a sulfamoyl radical ($NH_2SO_2^•$) to an ion with *m/z* 205.030 ($[C_{11}H_8ClNO]^{-•}$) and a major (≈99%) loss of C_5H_4O from the 2-methyl-furanamine group to an ion with *m/z* 204.984 ($[C_6H_6ClN_2O_2S]^-$). The major fragment with *m/z* 205 shows secondary fragmentation involving the loss of iminosulfene ($NHSO_2$) to an ion with *m/z* 126 ($[ClC_6H_4NH]^-$) and the complementary deprotonated iminosulfene ($[NSO_2]^-$) with *m/z* 78 (Giancotti et al., 2008). In collision-cell CID, mostly the same fragment ions are observed for furosemide. Piretanide shows the loss of 92 Da (C_6H_4O) from both $[M-H]^-$ and $[M-H-CO_2]^-$, resulting in fragment ions with *m/z* 269 and *m/z* 225, respectively. Loss of SO_2 from the ion with *m/z* 269 via a rearrangement in the sulfonamide group results in the fragment ion with *m/z* 205. Bumetanide and piretanide produce a fragment ion with *m/z* 80 (deprotonated sulfuramidous acid, $[SO_2NH_2]^-$) rather than the ion with *m/z* 78 ($[NSO_2]^-$).

The fragmentation of ethacrynic acid ($[M-H]^-$ with *m/z* 301) has been investigated using a LIT–orbitrap hybrid instrument (Giancotti et al., 2008). Three consecutive losses are observed resulting in fragment ions with *m/z* 243 due to the loss of oxiran-2-one ($C_2H_2O_2$), with *m/z* 207 due to the loss of hydrogen chloride (HCl), and with *m/z* 192 due to the loss of a methyl radical ($^•CH_3$).

4.1.4.4 Other (Non-sulfonamide) Diuretic Drugs

The major fragment ions of cicletanine ($[M-H]^-$ with *m/z* 260) with *m/z* 232 and *m/z* 196 are due to consecutive losses of CO and HCl. An ion with *m/z* 121, most likely the deprotonated radical anion of 2,4-dimethylpyridin-3-ol ($[C_7H_7NO]^{-•}$), is observed as well. In positive-ion mode, losses of CO and a formyl radical ($^•CHO$) as well as a 4-chlorobenzyl cation with *m/z* 125 ($[ClC_6H_4CH_2]^+$) are observed.

The fragmentation of amiloride ($[M+H]^+$ with *m/z* 230), an epithelial sodium channel blocker diuretic drug, is outlined in Figure 4.1.14. It shows cleavages of the C—C and C—N bonds of the carbonyl functional group and possible charge retention on either fragment. Note that the cleavage of the C—C bond between the carbonyl and the pyrazine ring is a direct cleavage, that is, the *m/z* values of two fragments equal *m/z* ($[M+H]-1$) (Section 3.5.4), whereas the cleavage of the C—N bond between the carbonyl and the guanidine moiety results in complementary ions, that is, with a sum of the fragments *m/z* values equal to *m/z* ($[M+H]+1$) (Section 3.5.2). The fragment ion with charge retention on the pyrazine ring (*m/z* 143) shows secondary fragmentation involving the loss of a chlorine radical ($Cl^•$) to an ion with *m/z* 108, or the loss of hydrogen cyanide (HCN) to an ion with *m/z* 116. The identity of most of these fragments was confirmed by accurate-mass determination in a LIT–orbitrap hybrid instrument (Giancotti et al., 2008).

4.1.4.5 Carbonic Anhydrase Inhibitors

Carbonic anhydrase inhibitors are a class of drugs that suppress the activity of carbonic anhydrase. They have a variable clinical use as antiglaucoma agents, as diuretics, as antiepileptics and in the treatment of ulcers. Acetozolamide and methazolamide are used as diuretics. The positive-ion fragmentation of acetazolamide ($[M+H]^+$ with *m/z* 223) involves the loss of ammonia (NH_3) from the sulfonamide group to an ion with *m/z* 206, the loss of ethenone ($H_2C{=}C{=}O$) from the acetamide group to an ion with *m/z* 181, or the loss of both groups to an ion with *m/z* 164. The subsequent loss of sulfur dioxide (SO_2) from the ion with *m/z* 164 results in an ion with *m/z* 100. Methazolamide ($[M+H]^+$ with *m/z* 237) yields only two fragments, that is, ions with *m/z* 195 due to the loss of $H_2C{=}C{=}O$ and with *m/z* 57 due to the protonated methylcyanamide $[CH_3NH{-}C{\equiv}N{+}H]^+$ derived from the 1,3,4-thiadiazole ring and the C^2 amide N-atom.

4.1.5 ANGIOTENSIN II RECEPTOR ANTAGONISTS

Angiotensin II receptor antagonists, also known as angiotensin receptor blockers or sartans, are drugs that modulate the renin–angiotensin–aldosterone system. Their main use is found in hypertension, kidney damage due to diabetes mellitus, and congestive heart failure. Multiresidue analysis of angiotensin II receptor antagonists using LC–MS in human plasma has been reported (Ferreirós et al., 2007).

Losartan, candesartan, irbesartan, olmesartan, and valsartan have a [2′-(1*H*-tetrazol-5-yl)[1,1′-biphenyl]-4-yl]methyl group as a common structure element. In positive-ion mode, this group leads to three common substituted benzyl cations with *m/z* 235 ($[C_7H_5N_4—C_6H_4CH_2]^+$), *m/z* 207 ($[C_7H_5N_2—C_6H_4CH_2]^+$), and *m/z* 192 ($[C_7H_4N—C_6H_4CH_2]^+$). The structures for some sartans and the proposed structures of their class-specific fragment ions are shown in Figure 4.1.15 (Zhao et al., 1999). The identity of these fragments was confirmed using accurate-mass determination on a time-of-flight (TOF) MS instrument (Shah et al., 2010). Additional fragmentation of losartan ($[M+H]^+$ with *m/z* 423) involves the loss of water and of water and N_2 to ions with *m/z* 405 and 377, respectively (Zhao et al., 1999). For irbesartan ($[M+H]^+$ with *m/z* 429), the loss of both N_2 and hydrogen azide (HN_3) from the tetrazole ring to ions with *m/z* 401 and 386 is observed (Shah et al., 2010). In addition, various compound-specific fragments are observed involving the various substituents. With irbesartan, the complementary fragment to the ion with *m/z* 235 is observed with *m/z* 195 ($[C_{11}H_{18}N_2O+H]^+$). With valsartan ($[M+H]^+$, *m/z* 436), compound-specific fragment ions with *m/z* 306 ($C_{18}H_{20}N_5]^+$) and *m/z* 291 ($[C_{17}H_{17}N_5]^{+\bullet}$) are due to the loss of pentanal ($C_5H_{10}O$) and carbon dioxide (CO_2) and of a subsequent loss of a methyl radical ($^\bullet CH_3$), respectively. Alternative interpretations, that is, the loss of the amide *N*-substituent isopentanoic acid ($(CH_3)_2CHCH_2COOH$) along with either N_2 or HN_3 from the tetrazole ring (Lu et al., 2009a) or the loss of the tetrazole ring and one of the phenyl rings (Koseki et al., 2007), do not fit to the elemental compositions for the fragments derived from accurate-mass data.

Due to the presence of a carboxylic acid group and/or the tetrazole ring (with $pK_a \approx 4.9$), some sartans can also be analyzed in negative-ion mode. The main fragment ion in the negative-ion MS–MS spectra of losartan ($[M-H]^-$ with *m/z* 421) and valsartan ($[M-H]^-$ with *m/z* 434) is the 9*H*-fluorene-2-yl-methyl carbanion ($[C_{14}H_{11}]^-$) with *m/z*

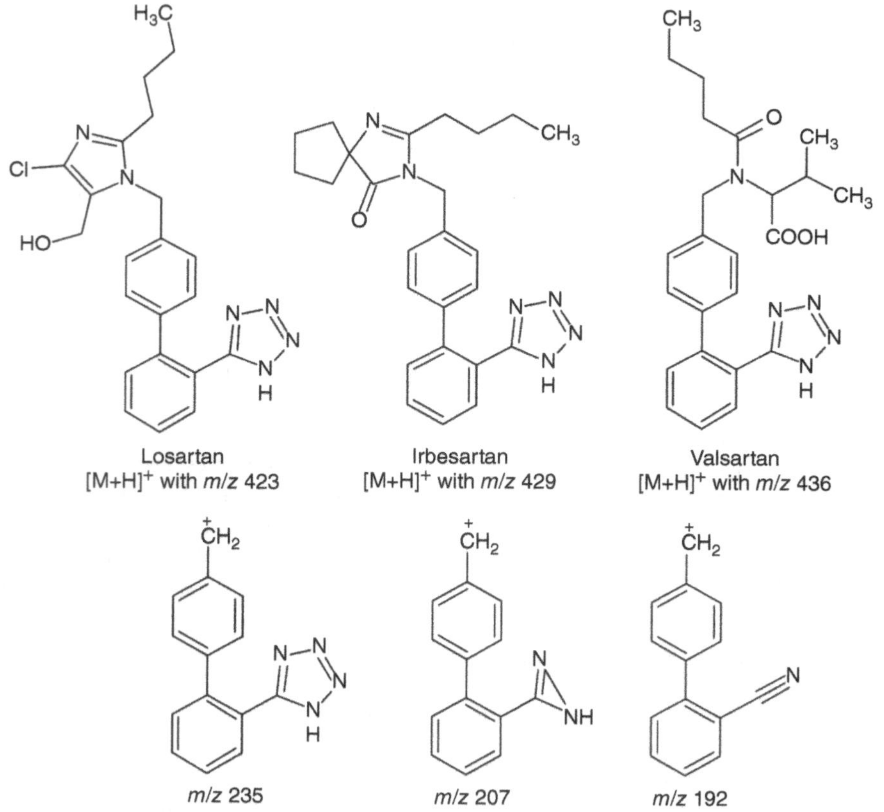

FIGURE 4.1.15 Structures and characteristic class-specific fragment ions for protonated angiotensin II receptor antagonists (sartans).

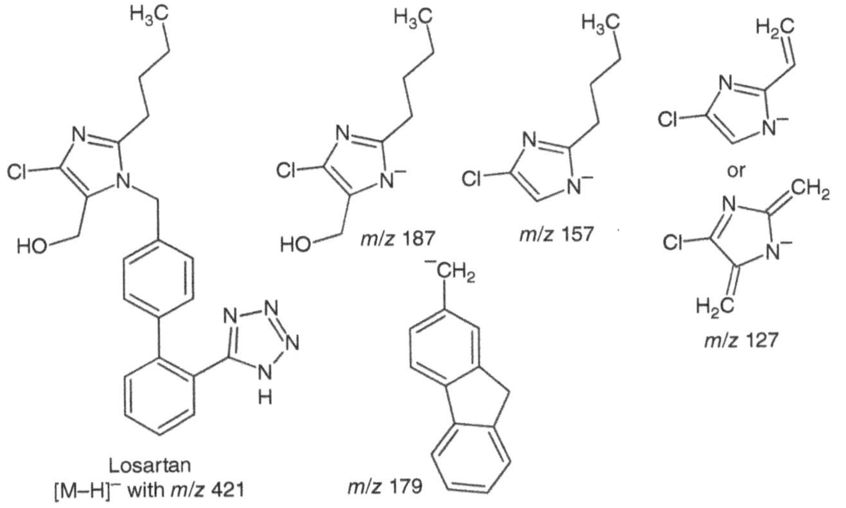

FIGURE 4.1.16 Proposed structures for the fragment ions of the deprotonated angiotensin II receptor antagonist losartan.

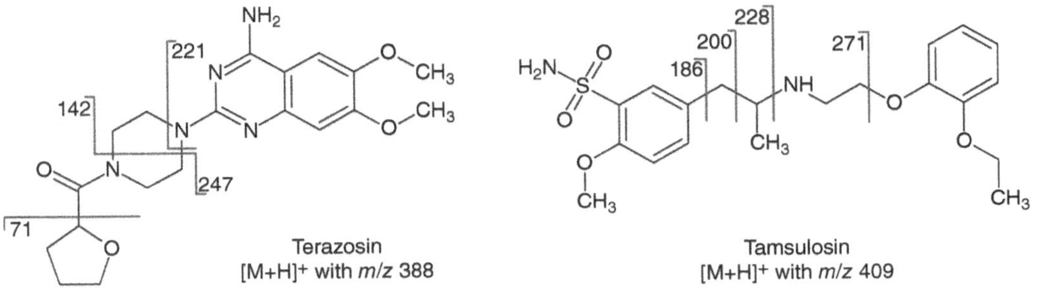

FIGURE 4.1.17 Fragmentation schemes for protonated terazosin and tamsulosin.

179, which originates from the 1,1′-biphenyl group. Other fragment ions of losartan involve the loss of the 1,1′-biphenyl group with its substituents and charge retention on the substituted imidazole group, resulting in a fragment ion with *m/z* 187 ($[C_8H_{13}ClN_2O-H]^-$), as well as subsequent secondary fragmentation thereof (Figure 4.1.16).

In telmisartan ($[M-H]^-$ with *m/z* 513), the tetrazole substituent is replaced by a carboxylic acid group. The fragmentation involves the loss of CO_2 to an ion with *m/z* 469 followed by the loss of a ([1,1′-biphenyl]-4-yl)methyl radical ($^{\bullet}C_{13}H_{11}$) to a radical anion with *m/z* 302 ($[C_{19}H_{18}N_4]^{-\bullet}$), which in turn loses $^{\bullet}CH_3$ to an ion with *m/z* 287 (Gupta et al., 2011).

4.1.6 OTHER ANTIHYPERTENSIVE COMPOUNDS

Along with the compound classes discussed in the previous sections, there are a number of other classes of antihypertensive compounds, including α_1-adrenoceptor antagonists, α_2-adrenoceptor agonists, centrally acting adrenergic drugs, and vasodilators. Multiresidue analysis of various classes of antiarrhythmic compounds in human plasma has been reported using positive-ion LC–MS (Li et al., 2007). The MS–MS fragmentation of some of the chemical classes involved is discussed here.

4.1.6.1 α_1-Adrenoceptor Antagonists

The α_1-adrenoceptor antagonists comprise compounds such as alfuzosin, doxazosin, prazosin, silodosin, tamsulosin, terazosin, trimazosin, and urapidil. They are used to treat not only hypertension but also benign prostatic hyperplasia. Except for alfuzosin, silodosin, and tamsulosin, these compounds are characterized by an *N,N′*-substituted piperazine ring, which plays an important role in the fragmentation. The typical fragmentation of these compounds in positive-ion mode is illustrated for terazosin ($[M+H]^+$ with *m/z* 388) in Figure 4.1.17. Fragmentation of the piperazine ring results in two complementary fragments with *m/z* 247 ($[C_{12}H_{14}N_4O_2+H]^+$) and *m/z* 142 ($[C_7H_{11}NO_2+H]^+$). The fragment ion with *m/z* 221 results from a two-step cleavage involving both piperazine C–N^4 bonds. Similar cleavages of the piperazine ring are observed for prazosin, trimazosin, and doxazosin. Multistage MS^n fragmentation of tamsulosin ($[M+H]^+$ with *m/z* 409) leads to a series of fragments, as outlined in Figure 4.1.17 (Nageswara Rao et al., 2008). Some fragments in the MS^4 spectrum are not well understood yet.

4.1.6.2 α_2-Adrenoceptor Agonists and Central Adrenergic Drugs

This compound class comprises a number of molecules containing a guanidine group ($HN{=}C(NH_2)_2$) such as guanabenz, guanoxan, and guanfacine. Characteristic fragmentation in positive-ion MS–MS involves the loss of ammonia (NH_3), the loss of methanediimine ($HN{=}C{=}NH$), and/or the loss of $HN{=}C(NH_2)_2$. Fragment ions with *m/z* 60, corresponding to protonated guanidine, and with *m/z* 58, corresponding to protonated diaziridin-3-imine ($[CH_3N_3+H]^+$), are also observed. For guanabenz ($[M+H]^+$ with *m/z* 231), this results in fragment ions with *m/z* 214, *m/z* 189, and *m/z* 172 (corresponding to protonated 2,6-dichlorobenzonitrile, $[Cl_2C_6H_3{-}C{\equiv}N+H]^+$), *m/z* 60, and *m/z* 58. The ions with *m/z* 60 and *m/z* 172 are complementary ions (Section 3.5.2). For guanoxan ($[M+H]^+$ with *m/z* 208), fragment ions are observed with *m/z* 191, *m/z* 166, *m/z* 149, *m/z* 58, and the protonated N^2-allylguanidine ($[C_4H_9N_3+H]^+$) with *m/z* 100. For guanfacine ($[M+H]^+$ with *m/z* 246), fragment ions are observed with *m/z* 204 due to the loss of $HN{=}C{=}NH$, *m/z* 60 due to protonated guanidine, and *m/z* 159 due to the 2,6-dichlorobenzyl cation ($[Cl_2C_6H_3CH_2]^+$).

The fragmentation of clonidine ($[M+1]^+$ with *m/z* 231) and moxonidine ($[M+1]^+$ with *m/z* 242) does not result in structure-informative fragmentation. The most abundant fragment ion is the 2-aminoethyl carbocation ($[H_2NCH_2CH_2]^+$) or protonated ethanimine ($[CH_3CH{=}NH+H]^+$) with *m/z* 44, derived from the imidazole ring. Phentolamine ($[M+H]^+$ with *m/z* 282) yields an intense *N*-(3-hydroxyphenyl)-*N*-(4-methylphenyl) methaniminium ion with *m/z* 212 ($[C_{14}H_{14}NO]^+$) due to the loss of the 4,5-dihydro-1*H*-imidazole ring.

Extensive fragmentation is observed for the adrenergic receptor antagonist urapidil ($[M+H]^+$ with *m/z* 388), mainly involving cleavages at various positions of the 1,3-diaminopropane skeleton of the molecule (Figure 4.1.18). Cleavage at the piperazine ring results in two fragments: the $OE^{+\bullet}$ 4-(2-methoxyphenyl)piperazin-2-ylium-1-yl radical cation ($[C_{11}H_{14}N_2O]^{+\bullet}$) with *m/z* 190 and the EE^+ 6-(1-amino-3-ylium-propanyl)-1,3-dimethylpyrimidine-2,4-dione ion ($[C_9H_{14}N_3O_2]^+$) with *m/z* 196. Other major EE^+ fragments are observed with *m/z* 141 ($[C_6H_8N_2O_2+H]^+$), *m/z* 205 ($[C_{12}H_{17}N_2O]^+$), *m/z* 233 ($[C_{14}H_{21}N_2O]^+$), and *m/z* 248 ($[C_{14}H_{21}N_3O+H]^+$) (Figure 4.1.18). These fragments were confirmed by accurate-mass data.

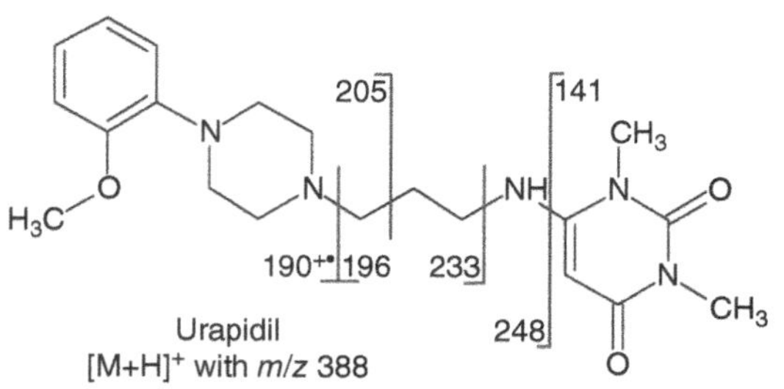

FIGURE 4.1.18 Fragmentation scheme for protonated urapidil.

4.1.6.3 Vasodilators

Vasodilation refers to the widening of blood vessels, which leads to a decrease in blood pressure; compounds with this effect are termed vasodilators. Chemically, vasodilators show great differences in structure from a small molecule such as (pyridin-2-yl)methanol ($[M+H]^+$ with *m/z* 110) to a large molecule such as inositol niacinate ($[M+H]^+$ with *m/z* 811). Most of the compounds in our mass spectral library collection show readily interpretable fragmentation. The fragmentation of butalamine ($[M+H]^+$ with *m/z* 317) may serve as an example (Figure 4.1.19). Initially, two major fragments are formed, that is, a carbocation with *m/z* 188 ($[C_{10}H_{10}N_3O]^+$) due to the loss of dibutylamine ($(C_4H_9)_2NH$) and an acylium ion with *m/z* 199 ($[(C_4H_9)_2NCH_2CH_2NH{-}C{=}O]^+$), resulting from cleavages in the oxadiazole ring. At higher collision energies, additional fragments are observed (Figure 4.1.19). A fragment ion with *m/z* 143 is due to secondary fragmentation of the ion with *m/z* 199 (loss of butene, C_4H_8). Another secondary fragment is the ion with *m/z* 87 ($[C_3H_7N_2O]^+$) (Figure 4.1.19). These fragments were confirmed by accurate-mass data.

The vasodilator ajmalicine ($[M+H]^+$ with *m/z* 353), an alkaloid from *Catharanthus roseus* roots, shows two major fragments, both resulting from cleavages in the quinolizidine ring, that is, ions with *m/z* 144 ($[C_{10}H_{10}N]^+$) and *m/z* 210 ($[C_{11}H_{15}NO_3+H]^+$) (Figure 4.1.19) (Lu et al., 2009b; Ferreres et al., 2010). Additionally, both $[M+H]^+$ and the fragment ion with *m/z* 210 show a loss of methanol (CH_3OH) from the methyl ester group.

The vasodilators dilazep ($[M+H]^+$ with *m/z* 605) and hexobendine ($[M+H]^+$ with *m/z* 593) have symmetric structures with propyl-3,4,5-trimethoxybenzoates attached to a diamine, which is a diazepine ring in dilazep and an ethylene diamine in hexobendine. Both compounds show the (3,4,5-trimethoxybenzoyl cation ($[(CH_3O)_3C_6H_2{-}C{=}O]^+$) with *m/z* 195 and the 3-[(3,4,5-trimethoxybenzoyl)oxy]prop-1-yl carbocation ($[(CH_3O)_3C_6H_2{-}COOCH_2CH_2CH_2]^+$) with *m/z* 253.

Some vasodilators contain a sulfonamide group like bosentan or a sulfonylimine group like diazoxide and can thus be analyzed in both positive-ion and negative-ion

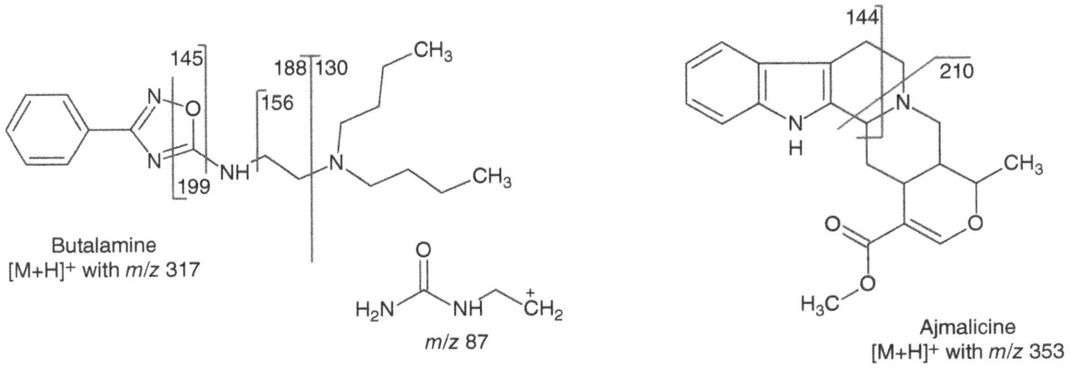

FIGURE 4.1.19 Fragmentation schemes for protonated butalamine and ajmalicine.

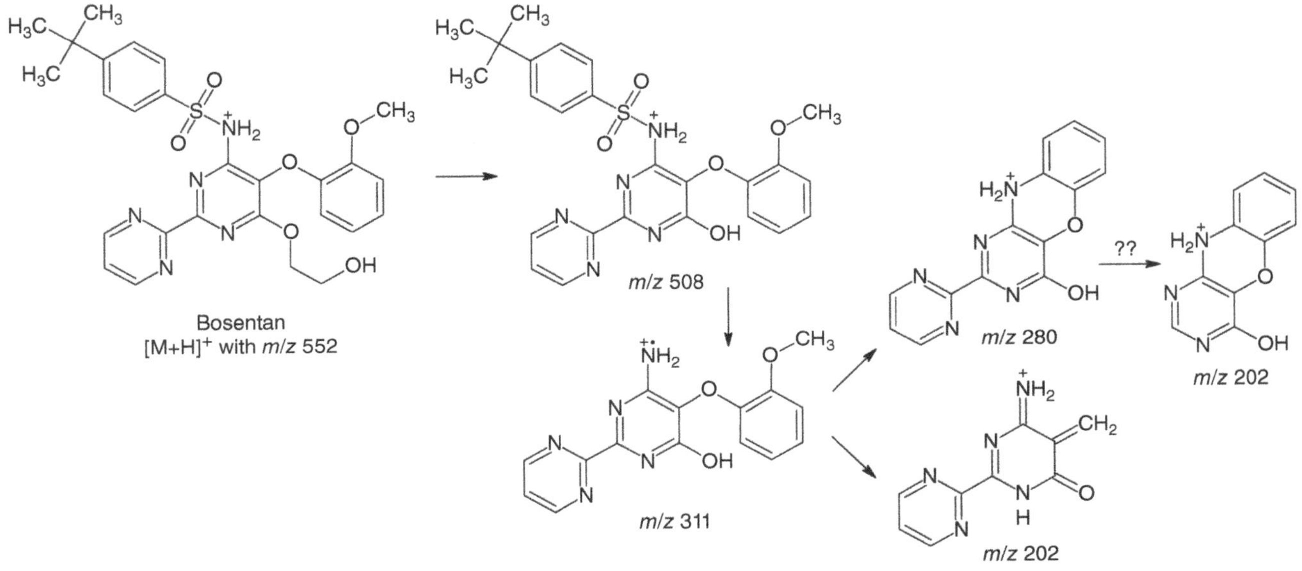

FIGURE 4.1.20 Fragmentation pathways for protonated bosentan.

modes. In negative-ion mode, bosentan ($[M-H]^-$ with *m/z* 550), an endothelin receptor antagonist used against pulmonary artery hypertension, shows only one major fragment ion, that is, the 4-*tert*-butylbenzene-1-sulfonyl anion ($[(CH_3)_3C—C_6H_4—SO_2]^-$) with *m/z* 197. In positive-ion mode, the fragmentation of the bosentan ($[M+H]^+$ with *m/z* 552) is more complicated. It was initially elucidated using H/D exchange, pseudo-MS^3, and accurate-mass measurement on a sector instrument (Hopfgartner et al., 1996). Subsequently, similar data was obtained using an early Q–TOF hybrid instrument (Hopfgartner et al., 1999). Proposed structures for the fragments are given in Figure 4.1.20 (Hopfgartner et al., 1999). The loss of ethanal (CH_3CHO) results in a fragment ion with *m/z* 508. This ion shows the loss of a 4-(*tert*-butyl)benzenesulfonyl radical ($(CH_3)_3C—C_6H_4—SO_2^•$) to an ion with *m/z* 311, which in turn shows the loss of a methoxy radical ($CH_3O^•$) to an ion with *m/z* 280. The ion with *m/z* 202 is due to a rearrangement of the ion with *m/z* 311 with a loss of a 2-hydroxyphenoxyl radical ($HOC_6H_4—O^•$), and not due to the (unlikely) loss of the pyrimidine ring as $C_4H_2N_2$ from the ion with *m/z* 280. Accurate-mass determination shows that the ion with *m/z* 202 has an elemental composition of $[C_9H_8N_5O]^+$, consistent with the radical loss just mentioned and not $[C_{10}H_8N_3O_2]^+$, which would be consistent with the unlikely fragmentation route shown in Figure 4.1.20.

Diazoxide, a potassium channel activator that is used as a vasodilator in the treatment of acute hypertension, shows some structure similarities with thiazide diuretic drugs. In negative-ion mode, next to a weak loss of hydrogen chloride (HCl) from $[M-H]^-$, diazoxide ($[M-H]^-$ with *m/z* 229) shows the loss of sulfur dioxide (SO_2) to an ion with *m/z* 165, followed by a loss of HCl to an ion with *m/z* 129. In addition, an ion with *m/z* 141 is observed, which is probably due to

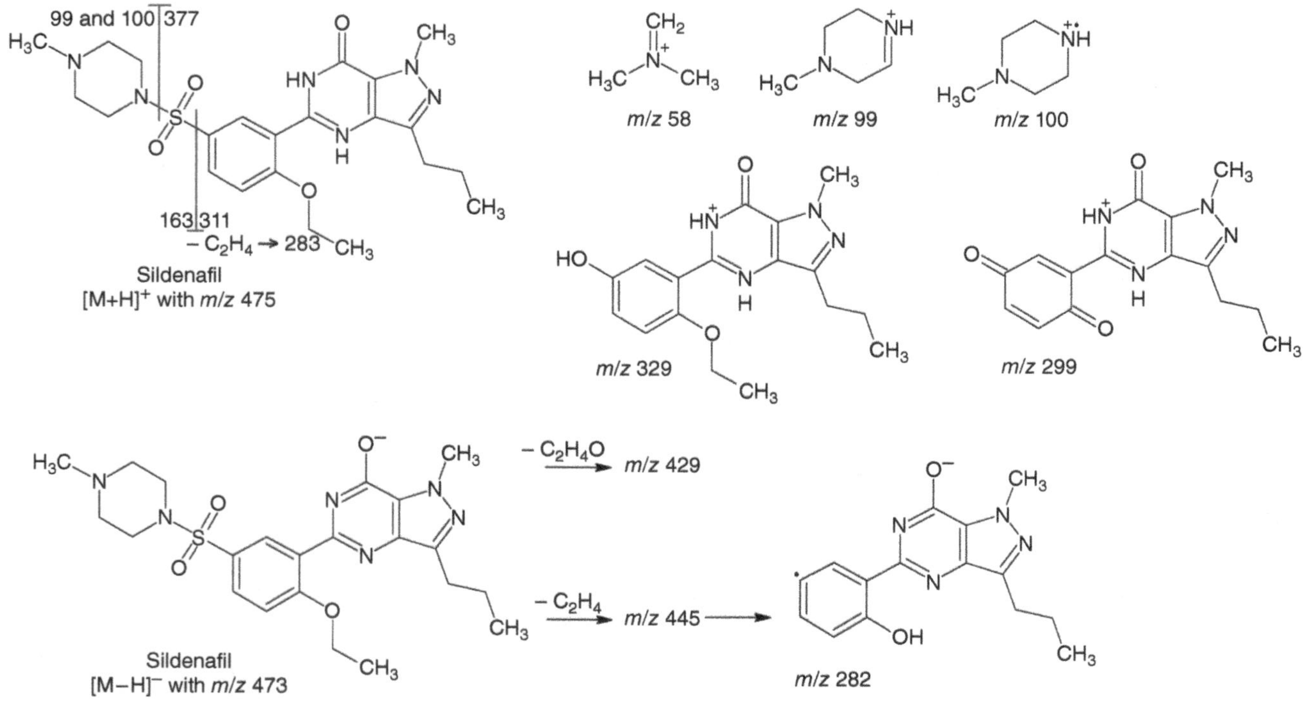

FIGURE 4.1.21 Fragmentation schemes for protonated and deprotonated sildenafil.

the loss of sulfur monoxide (SO), requiring a rearrangement in the SO_2 moiety as seen more often (Section 3.7.7), and a vinylideneamino radical ($H_2C{=}C{=}N^•$). In positive-ion mode, the fragmentation of diazoxide ($[M+H]^+$ with *m/z* 231) is different. The loss of acetonitrile ($CH_3C{\equiv}N$) results in a fragment ion with *m/z* 190, which subsequently shows the loss of SO to an ion with *m/z* 142 or of SO_2 to an ion with *m/z* 126. The latter shows a further loss of HCl to an ion with *m/z* 90.

4.1.6.4 Phosphodiesterase-5 Inhibitors

The vasodilator sildenafil (Viagra) is a phosphodiesterase-5 inhibitor, which is primarily used for the treatment of erectile dysfunction. Sildenafil and related compounds such as vardenafil (Levitra) and tadalafil (Cialis) are frequently found as adulterants in "natural" herbal medicines (Bogusz et al., 2006; Gratz et al., 2006). A multiresidue LC–MS method for the detection of 80 common synthetic adulterants from various pharmacological classes in herbal remedies has been described (Bogusz et al., 2006). A negative-ion LC–MS–MS multiresidue screening method for sildenafil, vardenafil, acetildenafil, and some of their analogs was reported (Ng et al., 2010). The method included the acquisition of TOF-MS data to confirm the identity of the fragment ions. The positive-ion fragmentation of sildenafil and six related compounds was studied with the help of accurate-mass measurement using a Fourier-transform ion-cyclotron resonance mass spectrometer (FT-ICR-MS) (Gratz et al., 2006).

In positive-ion mode, abundant fragments of sildenafil ($[M+H]^+$ with *m/z* 475) result from cleavages of the C—S and N—S bonds of the sulfonyl (SO_2) group (Figure 4.1.21). The C—S bond cleavage results in a fragment ion with *m/z* 311 ($[C_{17}H_{19}N_4O_2]^+$) due to the loss of 4-methylpiperazine-1-sulfinic acid, and a fragment ion with *m/z* 163 ($[H_3C{-}C_4H_8N_2{-}SO_2]^+$). The C—S bond cleavage is a direct cleavage, where the sum of the *m/z* values of the fragment ions add up to *m/z* ($[M+H]-1$) (Section 3.5.4). The N—S bond cleavage leads to two complementary ions (Section 3.5.2), that is, the ion with *m/z* 99 ($[C_5H_{10}N_2+H]^+$) due to charge retention on the 1-methylpiperazine ring and the ion with *m/z* 377 ($[C_{17}H_{20}N_4O_4S+H]^+$) involving a H-rearrangement (Figure 4.1.21). Next to an EE^+ ion with *m/z* 99, an $OE^{+•}$ ion with *m/z* 100 ($[C_5H_{12}N_2]^{+•}$) is observed (Section 3.5.6). Accurate-mass data confirm that the fragment ion with *m/z* 283 ($[C_{15}H_{15}N_4O_2]^+$) results from the loss of ethene (C_2H_4) from the fragment ion with *m/z* 311, and not from the (unlikely) cleavage of the C—C bond between the benzene and the pyrimidinone rings with charge retention on the benzene ring. The fragment ion with *m/z* 377 shows secondary fragmentation involving a rearrangement in the SO_2 group, resulting in the loss of sulfur monoxide (SO) to a fragment ion with *m/z* 329 and subsequent loss of ethane (C_2H_6) to a fragment ion with *m/z* 299 ($[C_{15}H_{14}N_4O_3+H]^+$) (Figure 4.1.21). The SO_2

rearrangement is frequently observed in compounds with a sulfonamide group (Section 3.7.8), for example, in sulfonamide antibiotics (Section 4.8.1) (Niessen, 1998; Klagkou et al., 2003; Sun et al., 2008). Secondary fragmentation of the 1-methylpiperazine ring (*m/z* 99 or *m/z* 100) leads to the *N,N*-dimethylmethaniminium ion ($[(CH_3)_2N{=}CH_2]^+$) with *m/z* 58. The positional isomers vardenafil (Ku et al., 2009; Gratz et al., 2006) and aildenafil (Wang et al., 2007) ($[M+H]^+$ with *m/z* 489) show similar fragmentation to sildenafil because they also contain the sulfonamide group (R_2NSO_2—), whereas tadalafil due to its different chemical structure shows different fragmentation routes (Gratz et al., 2006).

In negative-ion MS–MS, sildenafil ($[M-H]^-$ with *m/z* 473) and some of its analogs such as vardenafil and acetildenafil show a common neutral loss of ethene (C_2H_4) from the ethyl ether group and a common fragment ion with *m/z* 282 ($[C_{15}H_{14}N_4O_2]^{-\bullet}$) due to the subsequent loss of 4-methylpiperazine sulfonyl radical ($C_5H_{11}N_2$—$SO_2^\bullet$), or the corresponding substituent for analog compounds (Ng et al., 2010). The fragmentation scheme is outlined for sildenafil in Figure 4.1.21. Minor fragments are the ions with *m/z* 429 due to the loss of ethanal (CH_3CHO) and with *m/z* 310 due to the loss of $C_5H_{11}N_2$—$SO_2^\bullet$ directly from $[M-H]^-$. Interestingly, acetildenafil and piperazinonafil follow the same fragmentation route although they do not contain the SO_2 group, whereas isopiperazinonafil gives different fragment ions, that is, a cleavage comprising the piperazine ring rather than the benzene ring (Wollein et al., 2011).

Pentoxifylline, diprophylline, and other methylxanthines act as nonselective phosphodiesterase-5 inhibitors. Pentoxifylline ($[M+H]^+$ with *m/z* 279) shows the loss of the 5-oxohexyl N^1 (purine base) substituent with charge retention on the purine to a fragment ion with *m/z* 181 ($[C_7H_8N_4O_2+H]^+$), that is, protonated theobromine. The complementary fragment ion with *m/z* 99 ($[C_6H_{11}O]^+$) with charge retention on the oxohexyl substituent is observed as well. The ion with *m/z* 181 shows the characteristic fragmentation of the xanthine structure, that is, the formation of fragment ions with *m/z* 138 and 110 due to the losses of hydrogen isocyanate (HN=C=O) and a subsequent loss of carbon monoxide (CO), respectively (Section 4.2.8).

4.1.7 ANTIARRHYTHMIC AGENTS

Antiarrhythmic agents are drugs used to treat an irregular heartbeat. A wide variety of compounds are applied for this purpose, for example, (1) Na^+-channel blocking agents such as quinidine, procainamide (and its *N*-acetyl metabolite acecainide), disopyramide, lidocaine, phenytoin, mexiletine, tocainide, flecainide, propafenone, moricizine; (2) β-blockers such as propranolol, esmolol, timolol, metoprolol, atenolol, bisoprolol (Tables 4.1.1 and 4.1.2); (3) K^+-channel blockers such as amiodarone, sotalol, ibutilide, dofetilide, dronedarone; (4) Ca^{2+}-channel blockers such as verapamil and diltiazem; and (5) other compounds such as adenosine and digoxin. The fragmentation of some members of these classes is discussed in the following. A multiresidue method for the analysis of 10 antiarrhythmic drugs has been reported (Li et al., 2007).

4.1.7.1 Na^+-Channel Blockers

Mexiletine ($[M+H]^+$ with *m/z* 180) shows several fragment ions, for example, protonated *iso*-propenylamine ($[C_3H_7N+H]^+$) with *m/z* 58, the 1,3-dimethylphenyl cation ($[C_8H_9]^+$) with *m/z* 105, and the 1,6-dimethylphenoxy cation ($[C_8H_9O]^+$) with *m/z* 121. Tocainide ($[M+H]^+$ with *m/z* 192), an analog of the local anesthetic lidocaine discussed in Section 4.2.5, shows a different fragmentation: the major fragment is the protonated 1,6-dimethylaniline ($[C_8H_9NH_2+H]^+$) with *m/z* 122.

Acecainide, flecainide, and procainamide are benzamide analogs. The major fragment ions of acecainide ($[M+H]^+$ with *m/z* 278) and of procainamide ($[M+H]^+$ with *m/z* 236) are due to the loss of diethylamine ($(C_2H_5)_2NH$), resulting in ions with *m/z* 205 and 163, respectively. The cleavage of the amide bond with charge retention on the acylium ion is also observed, leading to the 4-acetamidobenzoyl cation ($[CH_3CONHC_6H_4$—$C{=}O]^+$) with *m/z* 162 or the 4-aminobenzoyl cation ($[H_2N$—C_6H_4—$C{=}O]^+$) with *m/z* 120, respectively. Further fragmentation results in an ion with *m/z* 120 due to the loss of ethenone ($H_2C{=}C{=}O$) for acecainide, and an ion with *m/z* 92 due to the loss of carbon monoxide (CO) for procainamide. Flecainide ($[M+H]^+$ with *m/z* 415) shows the loss of ammonia (NH_3) from the piperidine ring to an ion with *m/z* 398. The cleavage at the amide bond leads to 2,5-bis(2,2,2-trifluoroethoxy)benzoyl cation with *m/z* 301 ($[(CF_3CH_2O)_2C_6H_3$—$C{=}O]^+$). The complementary ion with *m/z* 115 is not observed; the (piperidin-2-yl)methylium

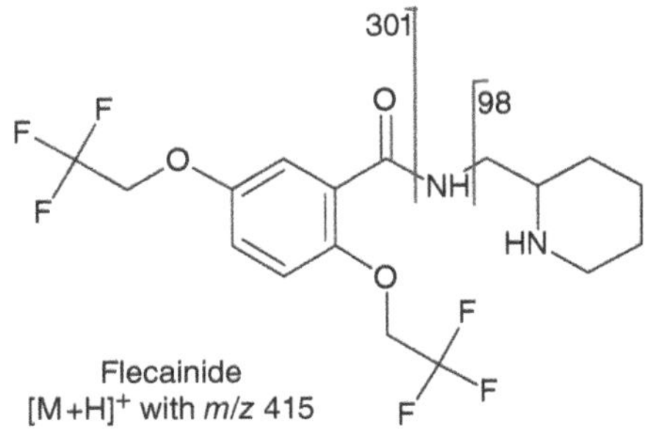

FIGURE 4.1.22 Fragmentation scheme for protonated flecainide.

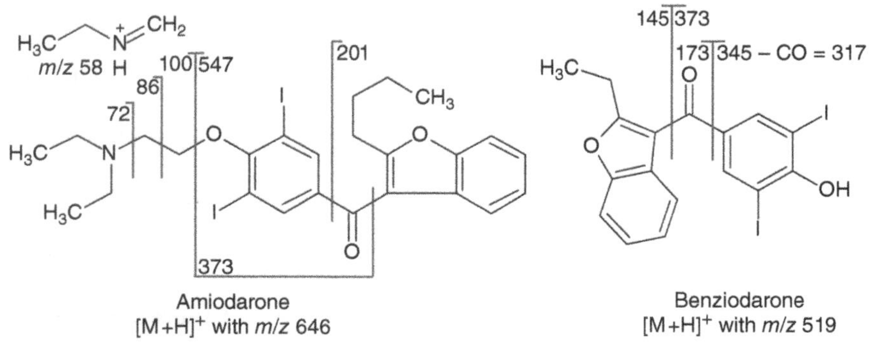

FIGURE 4.1.23 Fragmentation schemes for protonated amiodarone and benziodarone.

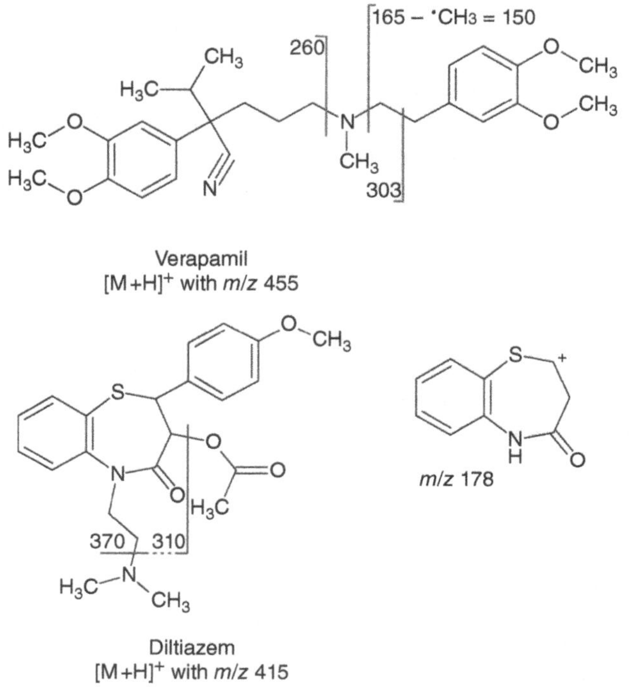

FIGURE 4.1.24 Fragmentation schemes for protonated verapamil and diltiazim.

ion ($[C_5H_{10}N{-}CH_2]^+$) with *m/z* 98 is observed instead (Figure 4.1.22).

Disopyramide ($[M+H]^+$ with *m/z* 340) shows fragment ions with *m/z* 239 ($[C_{15}H_{15}N_2O]^+$) due to the loss of di-*iso*-propylamine ($((CH_3)_2CH)_2NH$) and with *m/z* 195 ($[C_{14}H_{13}N]^{+\bullet}$) due to the subsequent loss of a carbamoyl radical ($H_2NCO^\bullet$). Similar fragments are observed for the des-*iso*-propyl metabolite.

4.1.7.2 K$^+$-Channel Blockers

The major fragment ions of amiodarone ($[M+H]^+$ with *m/z* 646) result from cleavages in the β-amino ether group. This leads to a sequence of ions, that is, the 2-(diethylamino)eth-1-yl carbocation ($[(CH_3CH_2)_2NCH_2CH_2]^+$) with *m/z* 100, the *N,N*-diethylmethaniminium ion ($[(CH_3CH_2)_2N{=}CH_2]^+$) with 86, the protonated diethylamine ($[(CH_3CH_2)_2NH+H]^+$) with *m/z* 72, and the *N*-ethylmethaniminium ion ($[CH_3CH_2NH{=}CH_2]^+$) with *m/z* 58. Less abundant ions are the acylium ions derived from the benzofuran ring with *m/z* 201 ($[C_{12}H_{13}O{-}C{=}O]^+$) and from the diiodobenzene ring, that is, the 4-hydroxy-3,5-diiodobenzoyl cation with *m/z* 373 ($[C_6H_3I_2O{-}C{=}O]^+$) (Figure 4.1.23). Similarly, the desethyl metabolite of amiodarone ($[M+H]^+$ with *m/z* 618) shows abundant fragment ions with *m/z* 72, that is, the 2-(ethylamino)eth-1-yl carbocation ($[CH_3CH_2NHCH_2CH_2]^+$), and *m/z* 58, that is, $[CH_3CH_2NH{=}CH_2]^+$, as well as an ion with *m/z* 547 ($[C_{19}H_{16}I_2O_3+H]^+$) due to the loss of ethyl(vinyl)amine ($CH_3CH_2NHCH{=}CH_2$) from the β-amino ether group. The fragment ions with *m/z* 201 and 373 are also observed (Kuhn et al., 2010). In the product-ion mass spectrum of benziodarone ($[M+H]^+$ with *m/z* 519), two sets of fragment ions are observed resulting from C—C bond cleavages at the carbonyl group. Both sets of fragment ions involve a direct cleavage reaction, i.e., the sum of the *m/z* values of the fragment ions add up to *m/z* ($[M+H]-1$) (Section 3.5.4). These sets of ions are, the 4-hydroxy-3,5-diiodobenzoyl cation ($[I_2C_6H_3O{-}C{=}O]^+$) with *m/z* 373 and the 2-ethyl-1-benzofuran-3-yl cation with *m/z* 145 ($[C_{10}H_9O]^+$), and the 4-hydroxy-3,5-diiodophenyl cation ($[I_2C_6H_3O]^+$) with *m/z* 345 and the (2-ethyl-1-benzofuran-3-yl)acylium ion with *m/z* 173 ($[C_{10}H_9O{-}C{=}O]^+$) (Figure 4.1.23).

4.1.7.3 Ca^{2+}-Channel Blockers

Next to the dihydropyridine calcium antagonists (Section 4.1.2), there are a number of other compounds with similar function, especially phenylalkylamines like verapamil and benzothiazepines like diltiazem.

Verapamil ($[M+H]^+$ with *m/z* 455) shows four fragment ions (Figure 4.1.24). The carbocations with *m/z* 165 ($[(CH_3O)_3C_6H_3CH_2CH_2]^+$) and *m/z* 260 ($[C_{16}H_{22}NO_2]^+$)

are due to C—N bond cleavages. The ion with *m/z* 150 is due to the loss of a methyl radical ($^{\bullet}CH_3$) from the ion with *m/z* 165. The less abundant iminium ion with *m/z* 303 ($[C_{18}H_{27}N_2O_2]^+$) is due to α,β-C—C bond cleavage (relative to the amine N-atom) (Rousu et al., 2010). The metabolism of verapamil has been extensively investigated (Walles et al., 2003).

Diltiazem ($[M+H]^+$ with *m/z* 415) shows subsequent losses of dimethylamine ($(CH_3)_2NH$) and acetic acid (CH_3COOH) to ions with *m/z* 370 and 310, respectively. The base peak is an ion with *m/z* 178 ($[C_9H_8NOS]^+$) derived from the benzothiazocinone ring (Figure 4.1.24). Multiresidue LC–MS analysis of diltiazem and 11 of its Phase I metabolites has been described, where the fragment ion with *m/z* 178 is used as a common fragment for all Phase I metabolites (Molden et al., 2003).

REFERENCES

Baranda AB, Mueller CA, Alonso RM, Jiménez RM, Weinmann W. 2005. Quantitative Determination of the calcium channel antagonist amlodipine, lercanidipine, nitrendipine, felodipine and lacidipine in human plasma using liquid chromatography–tandem mass spectrometry. Ther Drug Monit 27: 44–52.

Bhardwaj SP, Singh S. 2008. Study of forced degradation behavior of enalapril maleate by LC and LC–MS and development of a validated stability-indicating assay method. J Pharm Biomed Anal 46: 113–120.

Bogusz MJ, Hassan H, Al-Enazi E, Ibrahim Z, Al-Tufail M. 2006. Application of LC–ESI-MS–MS for detection of synthetic adulterants in herbal remedies. J Pharm Biomed Anal 41: 554–564.

Deventer K, Delbeke FT, Roels K, Van Eenoo P. 2002. Screening for 18 diuretics and probenecid in doping analysis by liquid chromatography–tandem mass spectrometry. Biomed Chromatogr 16: 529–535.

Deventer K, Pozo ÓJ, Van Eenoo P, Delbeke FT. 2009. Qualitative detection of diuretics and acidic metabolites of other doping agents in human urine by high-performance liquid chromatography–tandem mass spectrometry: Comparison between liquid–liquid extraction and direct injection, J Chromatogr A 1216: 5819–5827.

Ehrhardt JD, Ziegler JM. 1988. Negative ion mass spectra of dihydropyridine calcium-channel blockers. Biomed Environ Mass Spectrom 15: 525–528.

Ferreirós N, Dresen S, Alonso RM, Weinmann W. 2007. Validated quantitation of angiotensin II receptor antagonists (ARA-II) in human plasma by liquid-chromatography–tandem mass spectrometry using minimum sample clean-up and investigation of ion suppression. Ther Drug Monit 29: 824–834.

Ferreres F, Pereira DM, Valentão P, Oliveira JMA, Faria J, Gaspar L, Sottomayor M, Andrade PB. 2010. Simple and reproducible HPLC–DAD–ESI-MS/MS analysis of alkaloids in *Catharanthus roseus* roots, J Pharm Biomed Anal 51: 65–69.

Garcia P, Popot MA, Fournier F, Bonnaire Y, Tabet JC. 2002. Gas-phase behaviour of negative ions produced from thiazidic diuretics under electrospray conditions. J Mass Spectrom 37: 940–953.

Gergov M, Robson JN, Duchoslav E, Ojanperä I. 2000. Automated liquid chromatographic/tandem mass spectrometric method for screening β-blocking drugs in urine. J Mass Spectrom 35: 912–918.

Giancotti V, Medana C, Aigotti R, Pazzi M, Baiocchi C. 2008. LC-high-resolution multiple stage spectrometric analysis of diuretic compounds Unusual mass fragmentation pathways. J Pharm Biomed Anal 48: 462–466.

Goebel C, Trout GJ, Kazlauskas R. 2004. Rapid screening methods for diuretics in doping control using automated solid-phase extraction and liquid chromatography–electrospray tandem mass spectrometry. Anal Chim Acta 502: 65–74.

Gratz SR, Gamble BM, Flurer RA. 2006. Accurate mass measurement using Fourier transform ion cyclotron resonance mass spectrometry for structure elucidation of designer drug analogs of tadalafil, vardenafil and sildenafil in herbal and pharmaceutical matrices. Rapid Commun Mass Spectrom 20: 2317–2327.

Gupta VK, Jain R, Lukram O, Agarwal S, Dwivedi A. 2011. Simultaneous determination of ramipril, ramiprilat and telmisartan in human plasma using liquid chromatography tandem mass spectrometry. Talanta 83: 709–716.

Hernando MD, Gómez MJ, Agüera A, Fernández-Alba AR. 2007. LC–MS analysis of basic pharmaceuticals (beta-blockers and anti-ulcer agents) in wastewater and surface water, Trends Anal Chem 26: 581–594.

Holman SW, Wright P, Langley GJ. 2011. The low-energy collision-induced dissociation product ion spectra of protonated beta-blockers reveal an analogy to fragmentation behaviour under electron ionisation conditions. J Mass Spectrom 46: 1182–1185.

Hopfgartner G, Vetter W, Meister W, Ramuz H. 1996. Fragmentation of Bosentan ® (Ro 47-0203) in ionspray mass spectrometry after collision-induced dissociation at low energy: A case of radical fragmentation of an even-electron ion. J Mass Spectrom 31: 69–76.

Hopfgartner G, Chernushevich IV, Covey T, Plomley JB, Bonner R. 1999. Exact mass measurement of product ions for the structural elucidation of drug metabolites with a tandem quadrupole orthogonal-acceleration time-of-flight mass spectrometer. J Am Soc Mass Spectrom 10: 1305–1314.

Johnson RD, Lewis RJ. 2006. Quantitation of atenolol, metoprolol, and propranolol in postmortem human fluid and tissue specimens via LC/APCI-MS. Forensic Sci Int 156: 106–117.

Klagkou K, Pullen F, Harrison M, Organ A, Firth A, Langley GJ. 2003. Fragmentation pathways of sulphonamides under electrospray tandem mass spectrometric conditions. Rapid Commun Mass Spectrom 17: 2373–2379.

Kong D, Li S, Zhang X, Gu J, Liu M, Meng Y, Fu Y, La X, Xue G, Zhang L, Wang Q. 2010. Simultaneous determination of m-nisoldipine and its three metabolites in rat plasma by liquid chromatography–mass spectrometry. J Chromatogr B 878: 2989–2996.

Koseki N, Kawashita H, Hara H, Niina M, Tanaka M, Kawai R, Nagae Y, Masuda N. 2007. Development and validation of a method for quantitative determination of valsartan in human plasma by liquid chromatography–tandem mass spectrometry. J Pharm Biomed Anal 43: 1769–1774.

Ku HY, Shon JH, Liu KH, Shin JG, Bae SK. 2009. Liquid chromatography/tandem mass spectrometry method for the simultaneous determination of vardenafil and its major metabolite, N-desethylvardenafil, in human plasma: Application to a pharmacokinetic study. J Chromatogr B 877: 95–100.

Kuhn, J Götting C, Kleesiek K. 2010. Simultaneous measurement of amiodarone and desethylamiodarone in human plasma and serum by stable isotope dilution liquid chromatography–tandem mass spectrometry assay. J Pharm Biomed Anal 51: 210–216.

Kumar V, Malik S, Singh S. 2008. Polypill for the treatment of cardiovascular diseases part 2. LC–MS/TOF characterization of interaction/degradation products of atenolol/lisinopril and aspirin, and mechanisms of formation thereof. J Pharm Biomed Anal 48: 619–628.

Lee HB, Sarafin K, Peart TE. 2007. Determination of beta-blockers and beta2-agonists in sewage by solid-phase extraction and liquid chromatography–tandem mass spectrometry. J Chromatogr A 1148: 158–167.

Lee HW, Seo JH, Lee HS, Jeong SY, Cho YW, Lee KT. 2008. Development of a liquid chromatography/negative-ion electrospray tandem mass spectrometry assay for the determination of cilnidipine in human plasma and its application to a bioequivalence study. J Chromatogr B 862: 246–251.

Li S, Liu G, Jia J, Liu Y, Pan C, Yu C, Cai Y, Ren J. 2007. Simultaneous determination of ten antiarrhythmic drugs and a metabolite in human plasma by liquid chromatography–tandem mass spectrometry. J Chromatogr B 847: 174–181.

Lu CY, Chang YM, Tseng WL, Feng CH, Lu CY. 2009a. Analysis of angiotensin II receptor antagonist and protein markers at microliter level plasma by LC–MS/MS. J Pharm Biomed Anal 49: 123–128.

Lu S, Tran BN, Nelsen JL, Aldous KM. 2009b. Quantitative analysis of mitragynine in human urine by high performance liquid chromatography–tandem mass spectrometry. J Chromatogr B 877: 2499–2505.

Mazzarino M, de la Torre X, Botrè F. 2008. A screening method for the simultaneous detection of glucocorticoids, diuretics, stimulants, anti-oestrogens, beta-adrenergic drugs and anabolic steroids in human urine by LC–ESI-MS/MS. Anal Bioanal Chem 392: 681–698.

Migliorança LH, Barrientos-Astigarraga RE, Schug BS, Blume HH, Pereira AS, De Nucci G. 2005. Felodipine quantification in human plasma by high-performance liquid chromatography coupled to tandem mass spectrometry. J Chromatogr 814: 217–223.

Molden E, Helen Bøe G, Christensen H, Reubsaet L. 2003. High-performance liquid chromatography–mass spectrometry analysis of diltiazem and 11 of its phase I metabolites in human plasma. J Pharm Biomed Anal 33: 275–285.

Mueller CA, González AB, Weinmann W. 2004. Screening for dihydropyridine calcium channel blockers in plasma by automated solid-phase extraction and liquid chromatography/tandem mass spectrometry. J Mass Spectrom 39: 639–646.

Nageswara Rao R, Kumar Talluri MV, Narasa Raju A, Shinde DD, Ramanjaneyulu GS. 2008. Development of a validated RP-LC/ESI-MS–MS method for separation, identification and determination of related substances of tamsulosin in bulk drugs and formulations. J Pharm Biomed Anal 46: 94–103.

Ng CS, Law TY, Cheung YK, Ng PC, Choi KK. 2010. Development of a screening method for the detection of analogues of sildenafil and vardenafil by the use of liquid chromatograph coupled with triple quadrupole linear ion trap mass spectrometer. Anal Methods 2: 890–896.

Niessen WMA. 1998. Analysis of antibiotics by liquid chromatography–mass spectrometry. J Chromatogr A 812: 53–76.

Parekh SA, Pudage A, Joshi SS, Vaidya VV, Gomes NA, Kamat SS. 2008. Simultaneous determination of hydrochlorothiazide, quinapril and quinaprilat in human plasma by liquid chromatography–tandem mass spectrometry. J Chromatogr B 873: 59–69.

Pereira AS, Bicalho B, Ilha JO, De Nucci G. 2008. Analysis of dihydropyridine calcium channel blockers using negative ion photoionization mass spectrometry. J Chromatogr Sci 46: 35–41.

Pujos E, Cren-Olivé C, Paisse O, Flament-Waton MM, Grenier-Loustalot MF. 2009. Comparison of the analysis of β-blockers by different techniques. J Chromatogr B 877: 4007–4014.

Qin Y, Wang XB, Wang C, Zhao M, Wu MT, Xu YX, Peng SQ. 2003. Application of high-performance liquid chromatography–mass spectrometry to detection of diuretics in human urine. J Chromatogr B 794: 193–203.

Qiu F, Chen X, Li X, Zhong D. 2004. Determination of nimodipine in human plasma by a sensitive and selective liquid chromatography–tandem mass spectrometry method. J Chromatogr B 802: 291–297.

Rousu T, Herttuainen J, Tolonen A. 2010.Comparison of triple quadrupole, hybrid linear ion trap triple quadrupole, time-of-flight and LTQ-Orbitrap mass spectrometers in drug discovery phase metabolite screening and identification in vitro – amitriptyline and verapamil as model compounds. Rapid Commun Mass Spectrom 24: 939–957.

Shah RP, Sahu A, Singh S. 2010. Identification and characterization of degradation products of irbesartan using LC–MS/TOF, MS(*n*), on-line H/D exchange and LC-NMR. J Pharm Biomed Anal 51: 1037–1046.

Suchanova B, Sispera L, Wsol V. 2006. Liquid chromatography–tandem mass spectrometry in chiral study of amlodipine biotransformation in rat hepatocytes. Anal Chim Acta 573–574: 273–283.

Sun M, Dai W, Liu DQ. 2008. Fragmentation of aromatic sulfonamides in electrospray ionization mass spectrometry: Elimination of SO_2 via rearrangement. J Mass Spectrom 43: 383–393.

Thevis M, Schänzer W. 2005. Examples of doping control analysis by liquid chromatography–tandem mass spectrometry: Ephedrines, beta-receptor blocking agents, diuretics, sympathomimetics, and cross-linked hemoglobins. J Chromatogr Sci 43: 22–31.

Thevis M, Schänzer W. 2007. Mass spectrometry in sports drug testing: Structure characterization and analytical assays. Mass Spectrom Rev 26: 79–107.

Tian L, Jiang J, Huang Y, Xu L, Liu H, Li Y. 2006.Determination of aranidipine and its active metabolite in human plasma by liquid chromatography/negative electrospray ionization tandem mass spectrometry. Rapid Commun Mass Spectrom 20: 2871–2877.

Tsai TF, Lee MR. 2008. Liquid-phase microextration combined with liquid chromatography–electrospray tandem mass spectrometry for detecting diuretics in urine. Talanta 75: 658–665.

Upthagrove AL, Hackett M, Nelson WL. 1999. Fragmentation pathways of selectively labeled propranolol using electrospray ionization on an ion trap mass spectrometer and comparison with ions formed by electron impact. Rapid Commun Mass Spectrom 13: 534–541.

Walles M, Thum T, Levsen K, Borlak J. 2003. Metabolism of verapamil: 24 new phase I and phase II metabolites identified in cell cultures of rat hepatocytes by liquid chromatography–tandem mass spectrometry. J Chromatogr B 798: 265–274.

Wang J, Jiang Y, Wang Y, Zhao X, Cui Y, Gu J. 2007. Liquid chromatography tandem mass spectrometry assay to determine the pharmacokinetics of aildenafil in human plasma. J Pharm Biomed Anal 44: 231–235.

Wollein U, Eisenreich W, Schramek N. 2011. Identification of novel sildenafil-analogues in an adulterated herbal food supplement. J Pharm Biomed Anal 56: 705–712.

Yasuda T, Tanaka M, Iba K. 1996. Quantitative determination of amlodipine in serum by liquid chromatography with atmospheric-pressure chemical ionization tandem mass spectrometry. J Mass Spectrom 31: 879–884.

Zhang J, Shao B, Yin J, Wu Y, Duan H. 2009. Simultaneous detection of residues of β-adrenergic receptor blockers and sedatives in animal tissue by high-performance liquid chromatography–tandem mass spectrometry. J Chromatogr B 877: 1915–1922.

Zhao Z, Wang Q, Tsai EW, Qin XZ, Ip D. 1999. Identification of losartan degradates in stressed tablets by LC–MS and LC–MS/MS. J Pharm Biomed Anal 20: 129–136.

4.2

FRAGMENTATION OF PSYCHOTROPIC OR PSYCHOACTIVE DRUGS

A psychotropic compound, psychoactive drug or psychopharmaceutical, is a chemical substance that acts primarily upon the central nervous system (CNS) where it alters brain function, resulting in changes in perception, mood, consciousness, and/or behavior. These drugs may be used recreationally to purposefully alter one's consciousness or therapeutically as medications. Most classes of psychotropic drugs are analyzed in positive-ion mode, except for the barbiturates that do not give significant response in positive-ion mode.

4.2.1 PHENOTHIAZINES

Phenothiazines are frequently used as neuroleptic (antipsychotic) drugs, for example, chlorpromazine and fluphenazine, which are also suitable for the treatment of mental disorders by acting mainly as inhibitors of dopamine D_1 and D_2 receptors. In addition, some phenothiazines are used as antihistaminic drugs, for example, diethazine, dimetotiazine, dioxoprothazine, and triethylperazine (Section 4.5.2), or as sedatives, for example, acepromazine and aceprometazine. Multiresidue analysis of phenothiazines in human bodily fluids has been reported (Kumazawa et al., 2000; Kratzsch et al., 2003). The identity of three product ions of some phenothiazines, frequently used in SRM during veterinary residue analysis, has been elucidated (Nuñez et al., 2015).

10*H*-Phenothiazine is a heterotricyclic compound with molecular formula $C_{12}H_9NS$. One of the two benzene rings may be substituted at the 2-position, for instance with a Cl-atom as in chlorpromazine, prochlorperazine, and perphenazine, or a trifluoromethyl group (—CF_3) as in fluphenazine and trifluoperazine. Most phenothiazine antipsychotic drugs can be mainly classified into three groups that differ with respect to the *N*-substituent of the 10*H*-phenothiazine heterotricyclic system: (a) compounds with a 3-(4-methylpiperazin-1-yl)propyl substituent such as perazine, prochlorperazine, and trifluoperazine, or with a 3-[4-(2-hydroxyethyl)piperazin-1-yl]propyl substituent such as perphenazine and fluperphenazine; (b) compounds with an substituent such as 3-(*N,N*-dimethylamino)propyl such as promazine and chlorpromazine, or 3-(*N,N*-dimethylamino)-2-methylpropyl such as alimemazine; and (c) compounds with a 2-(1-methyl-2-piperidinyl)ethyl substituent such as thioridazine. In addition, some other structures are available such as a 3-(4-hydroxypiperidin-1-yl)propyl N^{10}-substituent in periciazine or a 2-(*N,N*-dimethylamino)propyl substituent in promethazine.

Characteristic fragmentation patterns of promazine ($[M+H]^+$ with *m/z* 285) and perazine ($[M+H]^+$ with *m/z* 340) by collision-induced dissociation (CID) in tandem mass spectrometry (MS–MS) are shown in Figure 4.2.1. Cleavages occur at various sites of the propane chain linking the dimethylamino or *N*-methyl-piperazinyl groups, respectively, and the heterotricyclic phenothiazine (McClean et al., 2000).

The cleavage of the phenothiazine C—N bond occurs mostly with charge retention on the resulting aliphatic amine, leading to the 3-(dimethylamino)prop-1-yl cation ($[C_5H_{12}N]^+$) with *m/z* 86 in promazine, the 3-(dimethylamino)-2-methylprop-1-yl cation ($[C_6H_{14}N]^+$) with *m/z* 100 in alimemazine ($[M+H]^+$ with *m/z* 299), the 3-(4-methylpiperazin-1-yl)prop-1-yl carbocation ($[C_8H_{17}N_2]^+$) with *m/z* 141 in perazine, and the 3-[4-(2-hydroxyethyl)piperazin-1-yl]prop-1-yl carbocation ($[C_9H_{19}N_2O]^+$) with *m/z* 171 in perphenazine ($[M+H]^+$ with *m/z* 404). Less abundant ions result from charge retention on the phenothiazine ring system either as a protonated phenothiazine with *m/z* (200+{2-substituent−1}), as a phenothiazine radical cation with *m/z* (199+{2-substituent−1}), or as a phenothiazin-5-ium ion with *m/z* (198+{2-substituent−1}) (Table 4.2.1a,b).

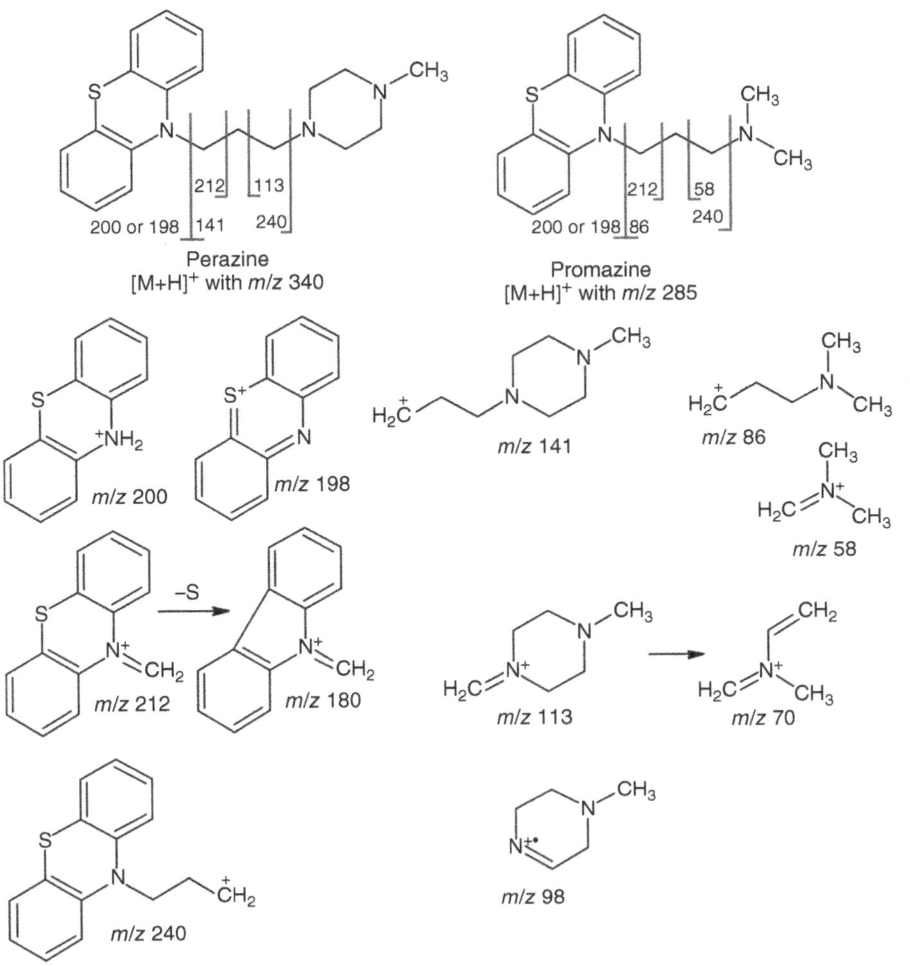

FIGURE 4.2.1 Fragmentation schemes for protonated perazine and promazine.

The cleavage of the propane-chain C^1–C^2 bond results in an iminium ion (10-methylidene phenothiazinium ion with *m/z* (212+{2-substituent−1}). The loss of elemental S with ring contraction to 9-methylidene carbazolium ion from the iminium ion is frequently observed to an ion with *m/z* (180+{2-substituent−1}) (Figure 4.2.1).

The cleavage of the propane-chain C^2—C^3 bond of perazine and promazine results in an abundant iminium ion, that is, the 1-methylidene-4-methylpiperazinium ion ($[C_6H_{13}N_2]^+$) with *m/z* 113 or the *N,N*-dimethylmethaniminium ion ($[(CH_3)_2N{=}CH_2]^+$) with *m/z* 58, respectively (Figure 4.2.1). Secondary fragmentation of the iminium ion with *m/z* 113 may lead to the *N*-methyl-*N*-vinylmethaniminium ion ($[C_4H_8N]^+$) with *m/z* 70 due to the loss of *N*-methylmethanimine ($CH_3N{=}CH_2$).

The cleavage of the propane-chain C^3—N bond to the N^1-piperazinyl or *N*-dimethylamino groups results in ions with *m/z* (240+{2-substituent−1}) due to charge retention on the phenothiazine containing fragment. In some cases, the cleavage of the propane-chain C^3—N bond to the N^1-piperazinyl group leads to a radical cation with *m/z* 98 ($[C_5H_{10}N_2]^{+\bullet}$) (Figure 4.2.1). For compounds with a 3-[4-(2-hydroxyethyl)piperazin-1-yl]propyl substituent, a similar radical cation with *m/z* 128 ($[C_6H_{12}N_2O]^{+\bullet}$) is observed. Characteristic fragment ions observed for typical examples from the three groups indicated earlier are given in Table 4.2.1.

The cleavages leading to the characteristic fragments with *m/z* 141, 113, and 212 were also described by others (Kumazawa et al., 2000; McClean et al., 2000). A recently reported secondary fragmentation of the ion with *m/z* 240 to a fragment ion with *m/z* 166 ($[C_9H_{12}NS]^+$) was not observed in our set of library spectra (Wen & Zhou, 2009). With phenothiazines containing a 2-Cl-substituent, the loss of chlorine radical ($Cl^\bullet$) may occur (Wen & Zhou, 2009).

In ion-trap MS^n of chlorpromazine ($[M+H]^+$ with *m/z* 319), subsequent cleavages occur in subsequent stages of MS^n, that is, the loss of dimethylamine ($(CH_3)_2NH$) in MS^2,

TABLE 4.2.1 Characteristic fragments of various classes of phenothiazines.

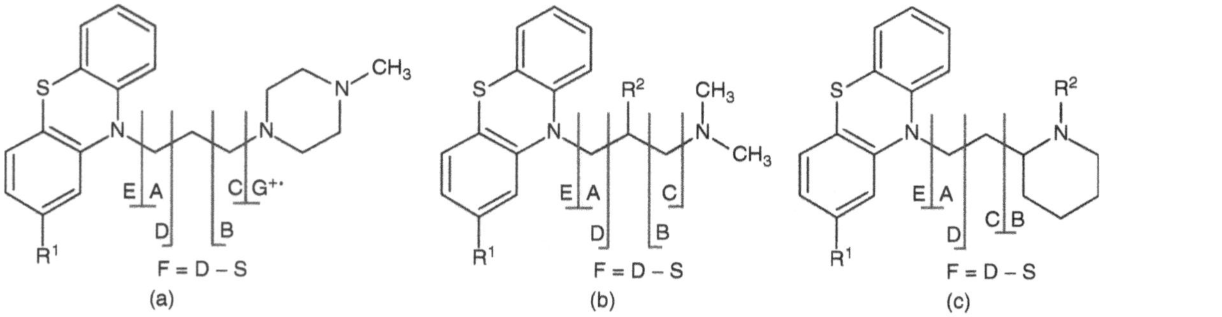

(a) Compounds with a 3-(4-methylpiperazin-1-yl)propyl or a 3-[4-(2-hydroxyethyl)piperazin-1-yl]propyl Substituent

Compound	R^1	$[M+H]^+$	A	B	C	D	E	F	$G^{+\bullet}$
Compounds with a 3-(4-methylpiperazin-1-yl)propyl substituent									
Perazine	—H	340	141	113	240	212	199	180	98
Prochlorperazine	—Cl	374	141	113	274	246	233	214	98
Thiethylperazine*	$—SCH_2CH_3$	400	141	113	300	272	259	240	98
Trifluoperazine	$—CF_3$	408	141	113	308	280		248	98
Butaperazine	$—C_4H_7O$	410	141	113	310	282		250	98
Thioproperazine	$—SO_2N(CH_3)_2$	447	141	113		319		287	98
Compounds with a 3-[4-(2-hydroxyethyl)piperazin-1-yl]propyl substituent									
Perphenazine	—Cl	404	171	143		246		214	128
Fluphenazine	$—CF_3$	438	171	143	308	280		248	128

(b) Compounds with a 3-(*N*,*N*-dimethylamino)propyl or a 3-(*N*,*N*-dimethylamino)-2-methylpropyl Substituent

Compound	R^1	R^2	$[M+H]^+$	A	B	C	D	E	F
Compounds with a 3-(*N*,*N*-dimethylamino)propyl substituent									
Promazine	—H	—H	285	86	58	240	212	199	180
Chlorpromazine	—Cl	—H	319	86	58	274	246		
Triflupromazine	$—CF_3$	—H	353	86		308	280		248
Compounds with a 3-(*N*,*N*-dimethylamino)-2-methylpropyl substituent									
Alimemazine	—H	$—CH_3$	299	100	58	254	212	199	180
Cyamemazine	—C≡N	$—CH_3$	324	100	58	279	237		205
Levomepromazine	$—OCH_3$	$—CH_3$	329	100	58	284	242	229	210
Oxomemazine*,†	$—SO_2$	$—CH_3$	331	100	58		244		180
Methiomeprazine	$—SCH_3$	$—CH_3$	345	100	58	300	258		226

(c) Compounds with a 2-(1-methyl-2-piperidinyl)ethyl Substituent

Compound	R^1	R^2	$[M+H]^+$	A	B	C	D	E	F
Northioridazine	$—SCH_3$	—H	357	112	84		258	245	226
Thioridazine	$—SCH_3$	$—CH_3$	371	126	98		258		226
Mesoridazine	$—SOCH_3$	$—CH_3$	387	126	98		274		
Sulforidazine	$—SO_2CH_3$	$—CH_3$	403	126	98		290		258

*Phenothiazine used as histamine antagonist (cf. Section 4.5.2).
†Oxememazine contains SO_2 rather than an S-atom in the phenothiazine ring (cf. Section 4.5.2).

of ethene (C_2H_4) in MS^3, and of elemental S in MS^4 (Joyce et al., 2004). Similar fragmentation routes are observed for phenothiazines with different *N*-substitution.

Phenothiazines are extensively metabolized involving sulfoxidation, aromatic ring hydroxylation, *N*-demethylation, and *N*-oxidation. In sulfoxidated metabolites, the loss of sulfur dioxide (SO_2) rather than elemental S is observed.

4.2.2 OTHER CLASSES OF NEUROLEPTIC DRUGS

Along with the butyrophenones and the thioxanthenes, the phenothiazines constitute the three compound classes of the so-called first-generation antipsychotics. Subsequently, the second-generation or so-called atypical antipsychotic drugs were developed, including risperidone, and clozapine and other dibenzazepine analogs. Even though they often encompass a wide range of receptor targets, both first- and second-generation antipsychotics tend to block receptors in the dopamine pathways in the brain. Multiresidue LC–MS analysis of various classes of neuroleptic drugs, including phenothiazines, has been described using single-quadrupole (SQ) MS, with in-source CID and library searching (Kratzsch et al., 2003), and SRM in a tandem-quadrupole (TQ) instrument (Kirchherr & Kühn-Velten, 2006; Saar et al., 2009). In order to improve confidence of identity during analysis, three SRM transitions per compound rather than one have been applied (Saar et al., 2009).

Risperidone ($[M+H]^+$ with *m/z* 411) is a second-generation antipsychotic drug used in the treatment of schizophrenia and other states associated with bipolar disorder and irritability in autistic children. In MS–MS, risperidone shows only one major fragment ion with *m/z* 191 ($[C_{11}H_{15}N_2O]^+$) (Figure 4.2.2) and a minor secondary fragment ion with *m/z* 163 due to subsequent loss of carbon monoxide (CO) (McClean et al., 2000). Further fragmentation may be observed using MS^n in an ion-trap instrument (McClean et al., 2000). The major metabolites, 7- and 9-hydroxyrisperidone, show the same cleavage, thus leading to a major fragment ion with *m/z* 207.

Pyritinol ($[M+H]^+$ with *m/z* 369), formed by binding two pyridoxine (vitamin B_6) molecules via a disulfide bridge (—S—S—), shows unusual fragmentation: two $OE^{+\bullet}$ fragments are generated with *m/z* 153 ($[C_8H_{11}NO_2]^{+\bullet}$) and *m/z* 217 ($[C_8H_{11}NO_2S_2]^{+\bullet}$) differing by S_2 (Figure 4.2.2). Both fragments show the loss of water to ions with *m/z* 135 and 199, respectively.

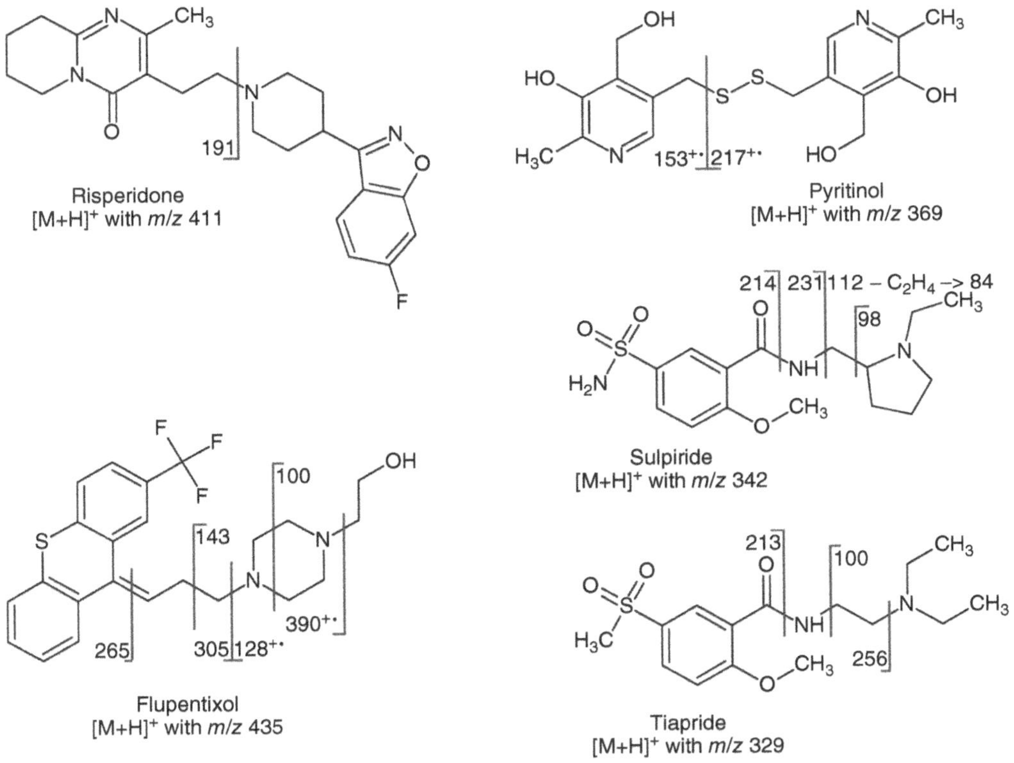

FIGURE 4.2.2 Fragmentation schemes for protonated risperiodone, pyritinol, flupentixol, sulpiride, and tiapride.

Amisulpride, sulpiride, and tiapride are atypical antipsychotics with a sulfonyl group (—SO_2—). The fragmentation of sulpiride ($[M+H]^+$ with *m/z* 342) and tiapride ($[M+H]^+$ with *m/z* 329), outlined in Figure 4.2.2, mainly involves fragmentation reactions common to amide and amine functional groups (Sections 3.6.5 and 3.6.6), that is, cleavages of amine and amide C—N bonds. With tiapride, however, secondary fragmentation of the ions with *m/z* 256 ($[C_{11}H_{14}NO_4S]^+$) and *m/z* 213 ($[C_9H_9O_4S]^+$) involving the loss of a methanesulfonyl radical ($CH_3SO_2^{\bullet}$) results in the formation of two radical cations with *m/z* 177 and 134, respectively.

The thioxanthene antipsychotics flupentixol ($[M+H]^+$ with *m/z* 435) and clopenthixol ($[M+H]^+$ with *m/z* 401) have very similar chemical structures, differing only by the C^2-thioxanthene substituent, that is, a trifluoromethyl and a Cl-atom, respectively. Their fragmentation involves the unexpected initial loss of 2-hydroxyethyl radical ($^{\bullet}CH_2CH_2OH$) from the N^4-atom of the piperazine ring, generating fragment ions with *m/z* 390 ($[C_{21}H_{21}F_3N_2S]^{+\bullet}$) and *m/z* 356 ($[C_{20}H_{21}ClN_2S]^{+\bullet}$), respectively (Figure 4.2.2). Additionally, two major fragment ions containing the thioxanthene ring system are observed with *m/z* 305 ($[C_{17}H_{12}F_3S]^+$) and *m/z* 265 ($[C_{14}H_9F_3S]^+$) for flupentixol and with *m/z* 271 ($[C_{16}H_{12}ClS]^+$) and *m/z* 231 ($[C_{13}H_8ClS]^+$) for clopenthixol. Moreover, other fragments contain the piperazine ring, such as the EE^+ ion [4-(2-hydroxyethyl)piperazin-1-yl]methylium ion ($[C_7H_{15}N_2O]^+$) with *m/z* 143 and the 4-(2-hydroxyethyl)piperazin-1-ium-1-yl radical cation ($[C_6H_{12}N_2O]^{+\bullet}$) with *m/z* 128. The ion-trap MS^n fragmentation of flupentixol has also been repored (McClean et al., 2000).

4.2.2.1 Butyrophenones

Butyrophenones are used to treat psychiatric disorders such as schizophrenia; they also act as antiemetics, however. Typical examples include azaperone, benperidol, bromperidol, droperidol, fluanisone, haloperidol, melperone, moperone, and trifluperidol. The characteristic fragmentation of the butyrophenones is illustrated for haloperidol ($[M+H]^+$ with *m/z* 376) in Figure 4.2.3. Characteristic fragment ions, observed for all butyrophenones available in our library collection, are the 4-fluorophenylium ion ($[FC_6H_4]^+$) with *m/z* 95, the 4-fluorobenzoyl ion ($[FC_6H_4—C{=}O]^+$) with *m/z* 123, and the 1-(4-fluorophenyl)-butan-1-one-4-yl carbocation ($[FC_6H_4—COCH_2CH_2CH_2]^+$) with *m/z* 165. A fragment ion complementary to the ion with *m/z* 165 may also be observed, that is, an ion with *m/z* 212 (and with *m/z* 194 ($[C_{11}H_{13}ClN]^+$) after loss of water) for haloperidol, an ion with *m/z* 100 for melperone ($[M+H]^+$ with *m/z* 264), an ion with *m/z* 246 (or with *m/z* 228 after loss of water) for trifluperidol ($[M+H]^+$ with *m/z* 410), an ion with *m/z* 193 for fluanisone ($[M+H]^+$ with *m/z* 357), and an ion with *m/z* 218 for benperidol ($[M+H]^+$ with *m/z* 382). For bromperidol ($[M+H]^+$ with *m/z* 420) and moperone ($[M+H]^+$ with *m/z* 356), only a fragment ion analogous to the ion with *m/z* 194 for haloperidol is observed, that is, the ions with *m/z* 238 and with *m/z* 174, respectively. Azaperone ($[M+H]^+$ with *m/z* 328) does not show the anticipated fragment ion complementary to the ion with *m/z* 165; it shows two ions with *m/z* 147 ($[C_9H_{11}N_2]^+$) and *m/z* 121 ($[C_7H_9N]^+$) instead, which may be considered secondary fragments of the expected ion with *m/z* 164 after the loss of ammonia (NH_3) and ethanimine ($CH_3CH{=}NH_2$).

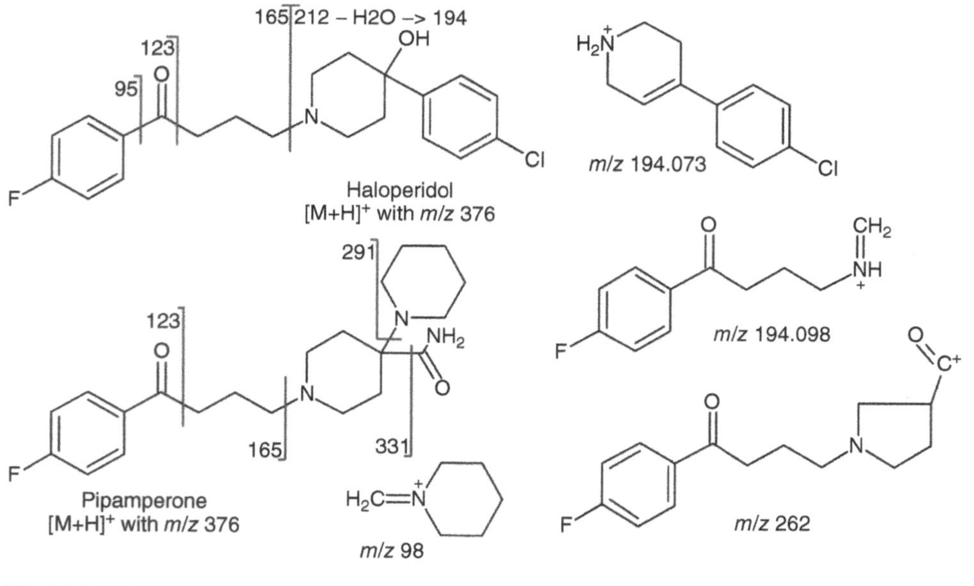

FIGURE 4.2.3 Fragmentation schemes and proposed structures for some of the fragment ions of protonated haloperidol and pipamperone.

The fragmentation of the [D_4]-labeled 4-fluorophenyl analog of haloperidol reveals that the haloperidol fragment ion with *m/z* 194 actually consists of two isobaric ions: an ion with *m/z* 194.073 ([$C_{11}H_{13}ClN$]$^+$), as described earlier, and an ion with *m/z* 194.098 ([$C_{11}H_{13}FNO$]$^+$), which in the [D_4]-labeled analog is observed with *m/z* 198.121 ([$C_{11}H_9D_4FNO$]$^+$). The ion with *m/z* 194.098 is also observed for droperidol, bromperidol, moperone, and trifluperidol. Proposed structures for both isobaric ions are given in Figure 4.2.3.

Pipamperone ([M+H]$^+$ with *m/z* 376) is an isobar of haloperidol. However, next to the common fragment ions with *m/z* 165 and 123, pipamperone shows several discriminative fragments ions, such as the ion with *m/z* 291 due to the loss of piperidine ($C_5H_{11}N$) and the 1-methylidenepiperidinium ion ([$C_6H_{12}N$]$^+$) with *m/z* 98 (Figure 4.2.3). The fragment ion with *m/z* 194.098, discussed for haloperidol, is observed as well.

4.2.2.2 Dibenzazepines and Related Structures

The 10,11-dihydrodibenzazepine structure is a common feature of some tricyclic antidepressants including imipramine, lofepramine, and trimipramine (Section 4.2.3.1). Related structures of second-generation antipsychotics have a disubstituted 5*H*-dibenzo[1,4]diazepine ring system, for example, clozapine ([M+H]$^+$ with *m/z* 327), or an oxazepine or a thiazepine ring instead of the azepine ring system, as in loxapine ([M+H]$^+$ with *m/z* 328) and quetiapine ([M+H]$^+$ with *m/z* 384), respectively. Furthermore, they have a 4-*N*-substituted piperazin-1-yl group attached to C^6 of the azepine ring forming a formamidine moiety (—N—C=N—).

Characteristic fragmentation involving the piperazine ring is illustrated for clozapine in Figure 4.2.4. Losses of methylamine (CH_3NH_2), methylvinylamine ($CH_3NHCH{=}CH_2$), and 1-methylpiperazine ($C_5H_{12}N_2$) lead to fragment ions with *m/z* 296, *m/z* 270, and *m/z* 227 ([$C_{13}H_8ClN_2$]$^+$), respectively. For both clozapine and loxapine, the subsequent loss of chlorine radical ($Cl^\bullet$) is observed, leading to fragment ions with *m/z* 192 and 193, respectively. The *N*-desmethyl metabolite of clozapine ([M+H]$^+$ with *m/z* 313) shows mostly the same fragment ions with *m/z* 296, *m/z* 270, *m/z* 227, and *m/z* 192. A fragment ion with *m/z* 70 ([C_4H_8N]$^+$) is observed rather than the ion with *m/z* 84 ([$C_5H_{10}N$]$^+$). The clozapine *N*-oxide metabolite ([M+H]$^+$ with *m/z* 343) shows quite a different fragmentation. Next to the losses of a hydroxyl radical ($OH^\bullet$) and water, characteristic for *N*-oxides, ions are observed with *m/z* 299 due to a loss of C_2H_4O (possibly consistent with the loss of ethanal, CH_3CHO), with *m/z* 256 due to a loss of C_4H_9NO (possibly consistent with the loss of ethylvinylhydroxylamine), and with *m/z* 243 due to a loss of $C_5H_{10}NO^\bullet$ (possibly consistent with the loss of the 1-methyl-1-oxo-pyrrolidinyl radical).

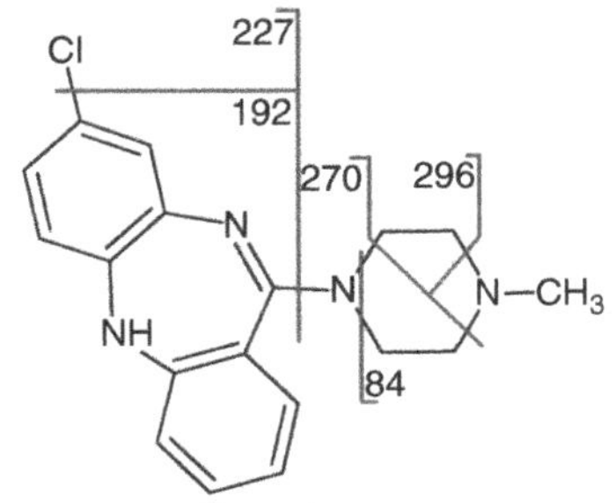

Clozapine
[M+H]$^+$ with *m/z* 327

FIGURE 4.2.4 Fragmentation scheme for protonated clozapine.

4.2.3 ANTIDEPRESSANTS

Tricyclic antidepressants (TCAs) and tetracyclic antidepressants (TeCAs) are heterocyclic compounds that were introduced in the late 1950s and 1970s, respectively. In clinical use, the TCAs have nowadays been largely replaced by newer types of antidepressants such as the selective serotonin reuptake inhibitors (SSRIs) and serotonin-norepinephrine reuptake inhibitors (SNRIs). Multiresidue LC–MS methods have been described for the analysis of antidepressants in plasma and other bodily fluids for therapeutic drug monitoring (Kollroser & Schober, 2002; Kirchherr & Kühn-Velten, 2006; Sauvage et al., 2006; de Castro et al., 2007; de Castro et al., 2008; Breaud et al., 2010) as well as in wastewater and water from sewage treatment plants for environmental analysis (Lajeunesse et al., 2008).

4.2.3.1 Tricyclic Antidepressants

There are various structurally different classes of TCAs. Trimipramine ([M+H]$^+$ with *m/z* 295) and lofepramine ([M+H]$^+$ with *m/z* 419) are derived from the 10,11-dihydrodibenzazepine tricyclic system, amitriptyline ([M+H]$^+$ with *m/z* 278) from the 10,11-dihydrodibenzocycloheptane tricyclic system, and protriptyline ([M+H]$^+$ with *m/z* 264) from the dibenzocyclohep-10-ene tricyclic system. Alternative structures are also available, such as those related to oxepine, for example, doxepin ([M+H]$^+$ with *m/z* 280), or thiepine, for example, dosulepin ([M+H]$^+$ with *m/z* 296).

To some extent, the fragmentation of TCAs is similar to that of phenothiazines (Section 4.2.1). Characteristic to all TCAs with a 3-dimethylamino substituent in the propane chain (trimipramine, amitriptyline) is the loss of dimethylamine ($(CH_3)_2NH$) (Kollroser & Schober, 2002) and the formation of an *N,N*-dimethylmethaniminium ion ([$(CH_3)_2N{=}CH_2$]$^+$) with *m/z* 58 due to a cleavage of the C^2—C^3 bond of the propane chain (Figure 4.2.5 for amitriptyline). In compounds with a 3-methylamino substituent (nortriptyline, protriptyline), the loss of methylamine

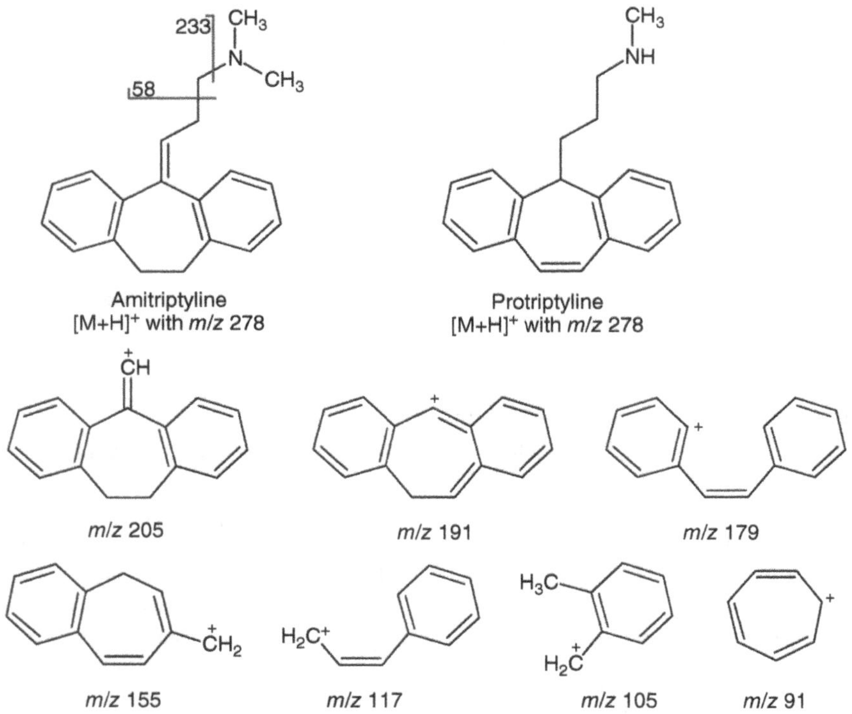

FIGURE 4.2.5 Fragmentation scheme and proposed structures for some fragment ions of protonated amitriptyline and protriptyline.

(CH_3NH_2) is observed. The fragment ion with *m/z* 58 and those resulting from the loss of (di)methylamine are generally used in SRM for quantification of these drugs in biological matrices (Kollroser & Schober, 2002; de Castro et al., 2007; de Castro et al., 2008; Lajeunesse et al., 2008). One may question the selectivity of such a transition (Allen, 2006). For 10,11-dihydrodibenzazepines, for example, trimpramine, lofepramine, imipramine ($[M+H]^+$ with *m/z* 280), and clomipramine ($[M+H]^+$ with *m/z* 315), the cleavage of the propane-chain C—N azepine bond leads to a fragment ion with charge retention on the substituent chain, that is, ions with *m/z* 100 ($[C_6H_{14}N]^+$), *m/z* 224 ($[C_{12}H_{15}ClNO]^+$), *m/z* 86 ($[C_5H_{12}N]^+$), and *m/z* 252 ($[C_{14}H_{26}N_3O]^+$), respectively. The complementary fragment is not observed although low-abundant ions with *m/z* 193 ($[C_{14}H_{11}N]^{+\bullet}$) and/or *m/z* 192 ($[C_{14}H_{10}N]^+$) are observed at high collision energies.

For amitriptyline, fragmentation of the 10,11-dihydrodibenzocycloheptane skeleton is observed at higher collision energies. Proposed structures for the more abundant fragment ions are given in Figure 4.2.5. The molecular formulae of these fragment ions have been confirmed by accurate-mass data. Similar fragment ions are observed for the dibenzocyclohept-10-ene tricyclic system, as in protriptyline (Figure 4.2.5).

The most abundant fragment ions of opipramol ($[M+H]^+$ with *m/z* 364), with a 3-[4-(2-hydroxyethyl)piperazin-1-yl]propyl substituent, thus similar to the phenothiazine perphenazine (Section 4.2.1), are the same as the ones observed for perphenazine. Cleavage of the propane chain C—N dibenzazepin bond results in an ion with *m/z* 171 ($[C_9H_{19}N_2O]^+$) and cleavage of the propane chain C^2—C^3 bond in the 1-methylidene-4-(2-hydroxyethyl)piperazinium ion with *m/z* 143 ($[C_7H_{15}N_2O]^+$) (Section 4.2.1).

4.2.3.2 Tetracyclic Antidepressants

The fragmentation of the TeCAs mianserin ($[M+H]^+$ with *m/z* 265) and mirtazepine ($[M+H]^+$ with *m/z* 266) is characterized by typical losses from the 4-methyl-substituted piperazine ring, that is, losses of methylamine (CH_3NH_2), of *N*-methylmethanimime, ($CH_3N{=}CH_2$), and of *N*-methylethanimine ($CH_3N{=}CHCH_3$) (Figure 4.2.6). Mirtazepine generates an additional fragment ion with *m/z* 195, which is due to the loss of dimethylvinylamine ($(CH_3)_2NCH{=}CH_2$). The latter fragment is the most abundant fragment ion of mirtazepine in ion-trap MS^2. The molecular formulae of these fragment ions were confirmed using high-resolution MS on a quadrupole–time-of-flight (Q–TOF) hybrid instrument (Smyth et al., 2006).

The TeCA maprotiline ($[M+H]^+$ with *m/z* 278) is isomeric to the TCA amitriptyline. Although these compounds have a number of fragment ions in common, their discrimination appears to be possible based on the primary

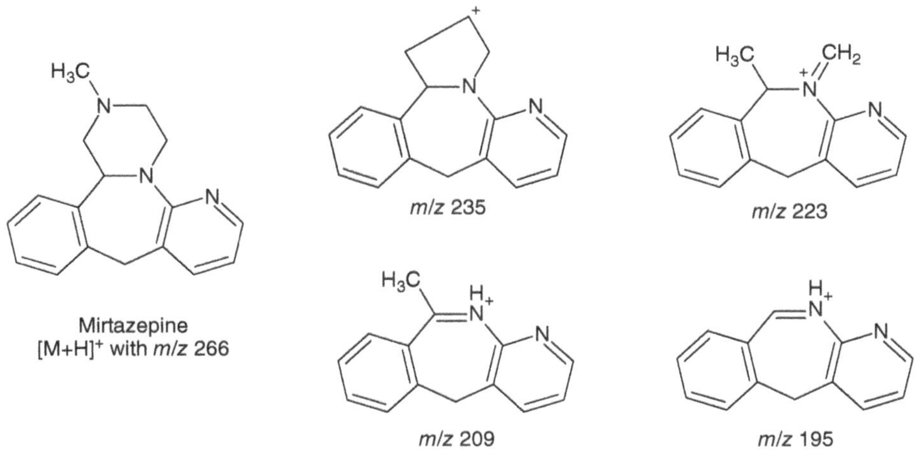

FIGURE 4.2.6 Proposed structures for some of the fragment ions of protonated mirtazepine.

loss of dimethylamine ($(CH_3)_2NH$) to a fragment ion with *m/z* 233 for amitriptyline and the primary loss of ethane (C_2H_4) due to a retro-Diels–Alder fragmentation in maprotiline.

4.2.3.3 Selective Serotonin Reuptake Inhibitors

Apart from TCAs and TeCAs, there are three other classes of antidepressant drugs: SSRIs, monoamine oxidase inhibitors (MAOIs), and SNRIs. SSRIs are modern antidepressants with less adverse effects than TCAs. They increase the levels of the neurotransmitter serotonin in the brain, which is known to play an important role in mood states. From a chemistry point of view, these drugs do not share clear common structural features. Therefore, class-specific fragmentation cannot be defined. Fragmentation of the SSRIs citalopram, fluoxetine, paroxetine, and sertraline was studied using ion-trap MS^n and accurate-mass determination on a Q–TOF instrument (Smyth et al., 2006).

In ion-trap MS^2, citalopram ($[M+H]^+$ with *m/z* 325) generates an abundant fragment ion with *m/z* 262 due to the loss of water and dimethylamine ($(CH_3)_2NH$), and two less abundant ions with *m/z* 307 due to water loss and *m/z* 280 due to $(CH_3)_2NH$ loss (Smyth et al., 2006; Raman et al., 2009). The loss of water most likely results in ring opening of the furan ring (Raman et al., 2009), although an alternative structure was proposed involving ring closure (Smyth et al., 2006) (A and B in Figure 4.2.7). Further fragmentation of the ion with *m/z* 262 in MS^3 results in an ion with *m/z* 234 due to the loss of ethene (C_2H_4) and with *m/z* 215 ($[C_{16}H_9N]^+$) in MS^4 due to the subsequent loss of a fluorine radical ($F^\bullet$) (Smyth et al., 2006). This interpretation is in agreement with the Q–TOF data (Smyth et al., 2006). More extensive fragmentation of citalopram is observed in a TQ instrument (Jiang et al., 2010). Next to the fragments already reported and a number of additional minor fragments, the most abundant fragment is the 4-fluorobenzyl cation ($[FC_6H_4CH_2]^+$) with *m/z* 109, while the 4-cyanobenzyl cation ($[N{\equiv}C{-}C_6H_4CH_2]^+$) with *m/z* 116 is also readily observed (Figure 4.2.7). The SRM transition *m/z* 325 > 109 is applied for the quantitative analysis of citalopram (de Castro et al., 2007; de Castro et al., 2008; Jiang et al., 2010). The fragment ions with low *m/z* observed in the TQ data are not observed in the ion-trap MS^n spectra, possibly due to the inability to trap fragment ions with *m/z* values less than ≈25% of the *m/z* value of the precursor ion (Hakala et al., 2006).

Paroxetine ($[M+H]^+$ with *m/z* 330) is reported to form product ions with *m/z* 220 and 70 (Massaroti et al., 2005). The fragment ion with *m/z* 220 ($[C_{12}H_{14}NO_3]^+$) was erroneously attributed to the loss of 4-fluorostyrene ($FC_6H_4{-}CH{=}CH_2$, 122 Da), whereas in fact it should involve the loss of 4-fluorotoluene ($FC_6H_4{-}CH_3$, 110 Da). This fragmentation pattern is not in agreement with our library data, which shows carbocations with *m/z* 192 ($[C_{12}H_{15}FN]^+$), *m/z* 178 ($[C_{11}H_{13}FN]^+$), *m/z* 163 ($[C_{11}H_{12}F]^+$), *m/z* 123 ($[C_8H_8F]^+$; the α-(methyl)-4-fluorobenzyl cation), and *m/z* 70 ($[C_4H_8N]^+$), and an oxonium ion with *m/z* 151 ($C_8H_7O_3]^+$) as the major fragments. Proposed structures for these fragments are given in Figure 4.2.7; molecular formulae are in agreement with accurate-mass data. The [4-(4-fluorophenyl)piperidin-3-yl]methylium ion with *m/z* 192 is also the most abundant fragment in ion-trap MS^2, along with less abundant ions with *m/z* 313 (loss of ammonia, NH_3), *m/z* 151, and *m/z* 123 (Smyth et al., 2006).

Characteristic fragmentation of fluoxetine ($[M+H]^+$ with *m/z* 310), also known as Prozac, involves the loss of 4-trifluoromethylphenol ($CF_3C_6H_4OH$) leading to the α-[2-(methylamino)ethyl]benzyl cation with *m/z* 148 ($[C_{10}H_{14}N]^+$). At higher collision energies, secondary fragmentation of this ion results in the α-(vinyl)benzyl cation with *m/z* 117 ($[C_6H_5CHCH{=}CH_2]^+$) due to the

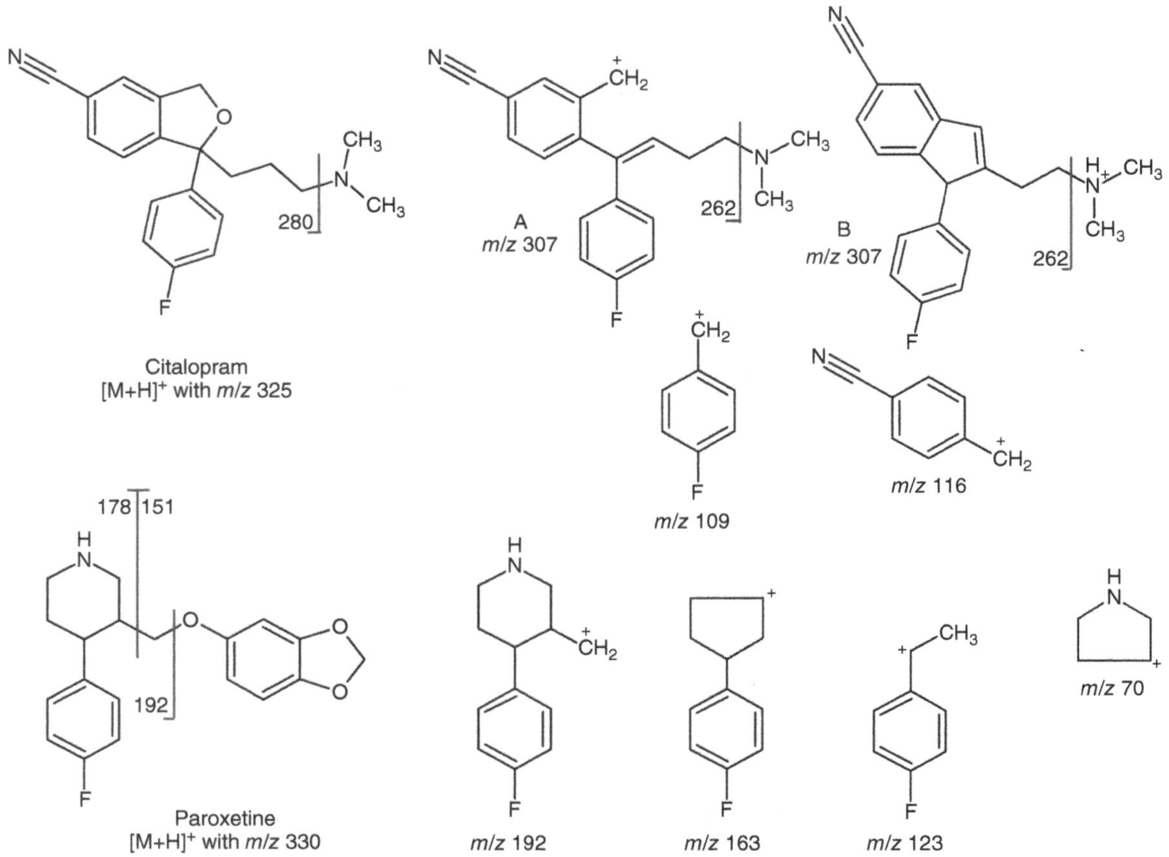

FIGURE 4.2.7 Fragmentation schemes and proposed structures for some of the fragment ions of protonated citalopram and paroxetine.

loss of CH_3NH_2 and the tropylium cation with *m/z* 91 ($[C_7H_7]^+$). At low *m/z*, the *N*-methylmethaniminium ion ($[CH_3NH{=}CH_2]^+$) with *m/z* 44 is observed (Sutherland et al., 2001; Smyth et al., 2006). The molecular formulae were confirmed by accurate-mass data.

4.2.3.4 Monoamine Oxidase Inhibitors and Serotonin-Norepinephrine Reuptake Inhibitors

Monoamine oxidase inhibitors (MAOIs) are prescribed only in special cases due to possible dietary and drug–drug interactions. The fragmentation of the MAOI moclobemide ($[M+H]^+$ with *m/z* 269) is outlined in Figure 4.2.8; it involves straightforward cleavages of the C—N bonds of the amine and amide functional groups to ions with *m/z* 182 due to the loss of morpholine (C_4H_9NO), the 4-chlorobenzoyl cation ($[ClC_6H_4{-}C{=}O]^+$) with *m/z* 139, the primary (1°) 2-(morpholin-4-yl)ethyl carbocation ($[C_6H_{12}NO]^+$) with *m/z* 114, and the 4-chlorophenylium ion ($[ClC_6H_4]^+$ with *m/z* 111.

Similar to SSRIs in their mode of action, the SNRIs are modern antidepressants with less adverse effects than TCAs and MAOIs. Fragmentation of the SNRI venlafaxine ($[M+H]^+$ with *m/z* 278) was studied using ion-trap MS^n and accurate-mass determination on a Q–TOF instrument (Smyth et al., 2006), and in a TQ instrument (Bhatt et al., 2005, Patel et al., 2008) (Figure 4.2.8). In ion-trap MS^n, a water loss to an ion with *m/z* 260 was observed in MS^2, a subsequent loss of dimethylamine ($(CH_3)_2NH$) to *m/z* 215 in MS^3, and the loss of butene (C_4H_8) in MS^4, resulting in an ion with *m/z* 159 ($[C_{11}H_{11}O]^+$) (Smyth et al., 2006). The latter ion is hardly observed in TQ MS–MS spectra, indicating that multistage MS^n and ion-trap CID can yield structure information different from collision-cell CID in a TQ instrument. Apart from fragments due to water and subsequent $(CH_3)_2NH$ losses, four additional fragment ions are observed in the TQ product-ion mass spectrum (Bhatt et al., 2005, Patel et al., 2008), that is, the *N,N*-dimethylmethaniminium ion ($[(CH_3)_2N{=}CH_2]^+$) with *m/z* 58, the 4-methoxybenzyl cation ($[CH_3O{-}C_6H_4CH_2]^+$) with *m/z* 121, *m/z* 147 ($[C_{10}H_{11}O]^+$), and the α-(cyclohexyl) benzyl cation with *m/z* 173 ($[C_{13}H_{17}]^+$), for which structure proposals are provided in Figure 4.2.8.

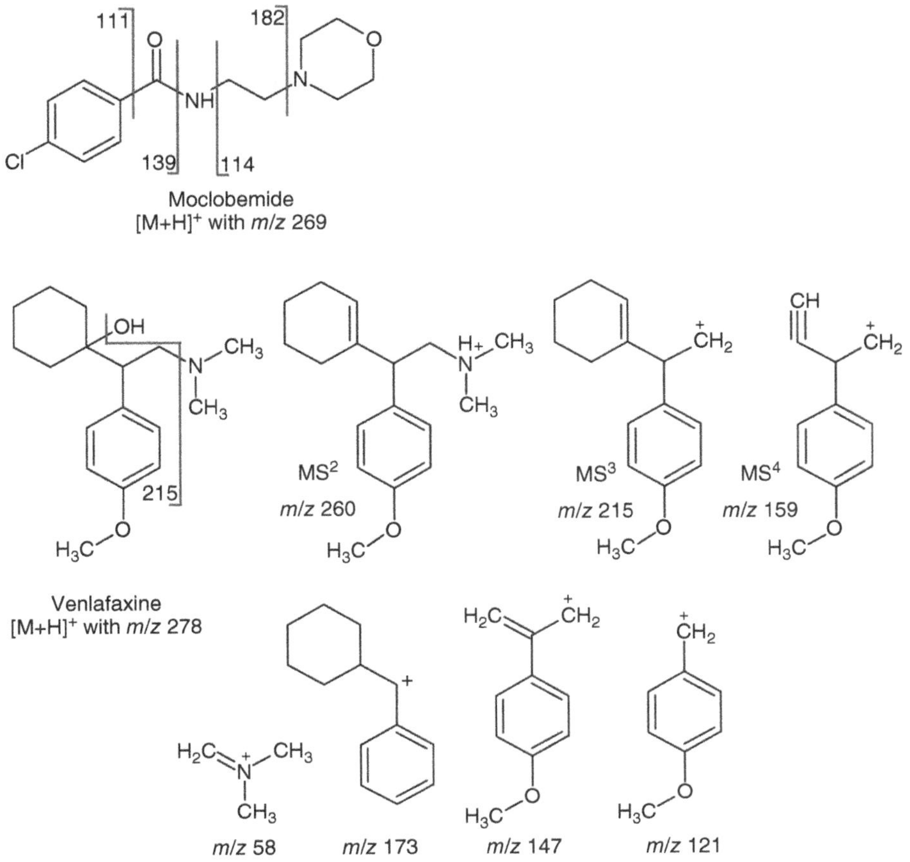

FIGURE 4.2.8 Fragmentation schemes and proposed structures for some of the fragment ions of protonated moclobemide and venlafaxine. For venlafaxine, both ion-trap MS^n fragment ions and TQ MS–MS fragment ions are shown.

4.2.4 BENZODIAZEPINES

Benzodiazepines are psychoactive drugs that enhance the effect of the neurotransmitter γ-aminobutyric acid (GABA), which results in sedative, hypnotic, anxiolytic, anticonvulsant, muscle relaxant, and/or amnesic action. As such, they are useful in a variety of conditions such as alcohol dependence, seizures, anxiety, panic, agitation, and insomnia. Benzodiazepines are commonly misused and taken in combination with other drugs of abuse. When combined with other CNS depressants such as alcohol and opiates, the potential for toxicity increases. Multiresidue LC–MS analysis of benzodiazepines has been frequently reported, using either selected-ion monitoring (SIM) on SQ instruments (Kratzsch et al., 2004; Ishida et al., 2009), SRM on TQ instruments (Villain et al., 2005; Marin et al., 2008; Badawi et al., 2009; Nakamura et al., 2009; Marin & McMillin, 2010; Glover & Allen, 2010), SRM on ion-trap instruments (Smink et al., 2004), or accurate-mass determination on TOF instruments (Hayashida et al., 2009; Nielsen et al., 2010). The benzodiazepines are analyzed in matrices such as whole blood, plasma, serum, urine, hair, saliva, and meconium.

The core chemical structure of benzodiazepines, as the name suggests, is a benzene ring fused to a diazepine. The drugs are 1,4-benzodiazepines with different substituents attached to the main structure, which influence the binding of the molecule to the GABA receptor, thereby modulating their pharmacological properties. Many of the benzodiazepine drugs contain a 5-phenyl-1,3-dihydro-[1,4]-benzodiazepin-2-one core structure, for example, diazepam with an N^1-methyl and a 7-Cl substituent, oxazepam with a 3-hydroxy and a 7-Cl substituent, or alprazolam with a methyltriazole ring fused to the diazepine.

The fragmentation of benzodiazepines has been systematically studied using ion-trap and Q–TOF instruments (Smyth et al., 2000; Smyth et al., 2004) and using a quadrupole–linear-ion-trap (Q–LIT) hybrid instrument (Risoli et al., 2007). The fragmentation of the following benzodiazepine subclasses is discussed here: compounds containing an N^1-methylacetamide skeleton in the diazepine ring (diazepam), compounds containing an N^1-desmethylacetamide skeleton in the diazepine ring (nordiazepam), compounds containing an N^1-desmethyl-α-hydroxyacetamide skeleton in the diazepine ring (oxazepam), compounds containing an N^1-methyl-α-hydroxyacetamide skeleton (temazepam), and compounds with other N^1-substituents than methyl (prazepam and halazepam).

Ion-trap MS^n of diazepam ($[M+H]^+$ with *m/z* 285, containing an N^1-methylacetamide skeleton in the diazepine ring) shows the loss of carbon monoxide (CO) with diazepine ring contraction to an ion with *m/z* 257 in MS^2, followed by the loss of either a chlorine radical ($Cl^•$) to an ion with *m/z* 222 or methanimine ($H_2C{=}NH$) to an ion with *m/z* 228 in MS^3. In MS^4, the ion with *m/z* 228 shows the loss of $Cl^•$ to an ion with *m/z* 193 (Smyth et al., 2000). The fragment ions with *m/z* 257, 228, 222, and 193 are also observed in TQ MS–MS spectra. The ion with *m/z* 241 ($[C_{14}H_{10}ClN_2]^+$) is due to the loss of ethanal (CH_3CHO). The ion with *m/z* 154 ($[C_8H_9ClN]^+$) is due to the loss of benzonitrile ($C_6H_5{-}C{\equiv}N$) from the ion with *m/z* 257 (Badawi et al., 2009; Nakamura et al., 2009). The tropylium ion ($[C_7H_7]^+$), that is, the benzyl cation ($[C_6H_5CH_2]^+$) with *m/z* 91 is also observed. Proposed structures for these fragment ions are given in Figure 4.2.9 (Risoli et al., 2007). Related compounds with the N^1-methylacetamide skeleton in the diazepine ring show the same fragmentation behavior, that is, loss of CO or CH_3CHO from $[M+H]^+$, as well as low-abundant ions due to the loss of $H_2C{=}NH$ or the loss of $C_6H_5{-}C{\equiv}N$ (or its analog with additional phenyl substituents) from the $[M+H-CO]^+$-ion. Thus, the loss of fluorobenzonitrile ($FC_6H_4{-}C{\equiv}N$) is observed for fludiazepam ($[M+H]^+$ with *m/z* 303), flunitrazepam ($[M+H]^+$ with *m/z* 314), zolazepam ($[M+H]^+$ with *m/z* 287), and 7-aminoflunitrazepam ($[M+H]^+$ with *m/z* 284), whereas the loss of chlorobenzonitrile ($ClC_6H_4{-}C{\equiv}N$) is observed for clotiazepam ($[M+H]^+$ with *m/z* 319). Tetrazepam ($[M+H]^+$ with *m/z* 289) does not show a similar loss, that is, the loss of 1-cyanocyclohexene ($C_6H_9{-}C{\equiv}N$). In flunitrazepam, additional fragment ions are observed due to the loss of a nitryl radical ($^•NO_2$), that is, from $[M+H]^+$ to an ion with *m/z* 268, from $[M+H-CO]^+$ to *m/z* 240, from $[M+H-CO-H_2C{=}NH]^+$ to *m/z* 211, and from $[M+H\text{-}CO\text{-}FC_6H_4{-}C{\equiv}N]^+$ to *m/z* 119. The loss of the $^•NO_2$ is also the primary fragmentation of flunitrazepam in ion-trap MS^2, with a subsequent loss of a formyl radical ($^•CHO$) in MS^3, resulting in an ion with *m/z* 239 (Smyth et al., 2004). The same ion with *m/z* 239 is observed at high collision energy in collision-cell CID, along with the ion with *m/z* 211, which can be considered as a secondary fragment of the ion with *m/z* 268, involving the loss of methylisocyanate ($CH_3{-}N{=}C{=}O$). The fragmentation of the 7-amino-, 3-hydroxy-, and *N*-desmethyl metabolites of flunitrazepam has also been studied (Smyth et al., 2000; Smyth et al., 2004) (see later).

The fragmentation scheme of nordiazepam ($[M+H]^+$ with *m/z* 271, containing an N^1-desmethyl acetamide skeleton in the diazepine ring) is shown in Figure 4.2.10. The primary losses lead to the ions with *m/z* 243 due to CO loss, *m/z* 226 due to the loss of formamide ($HCONH_2$), and *m/z* 193 due

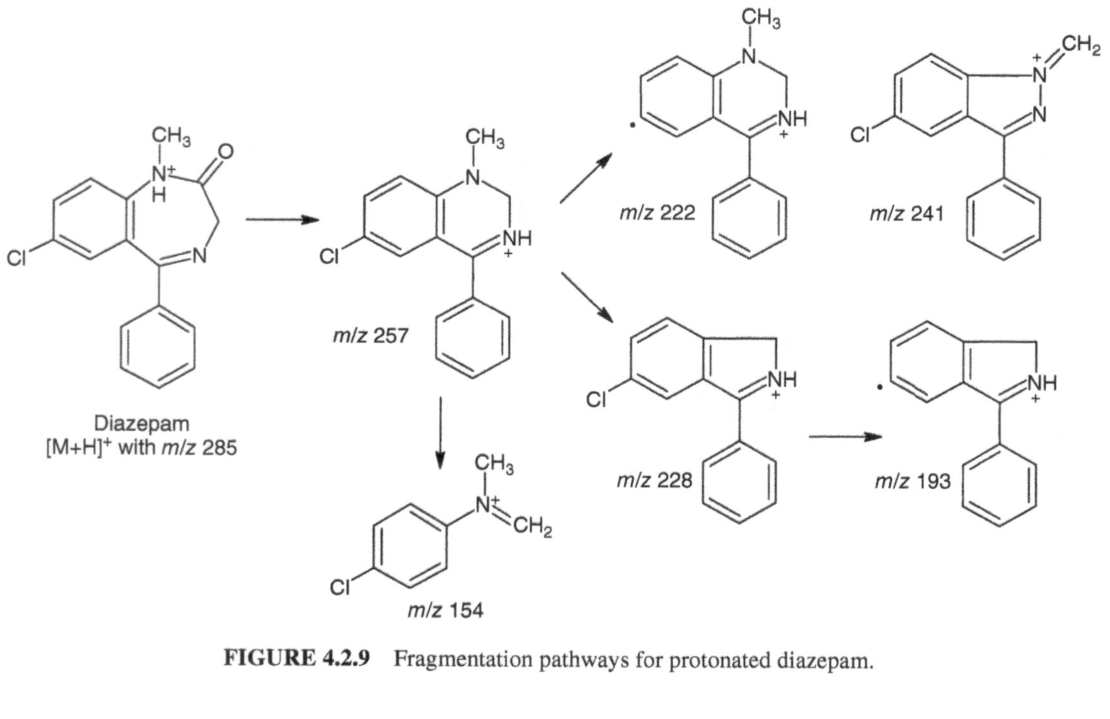

FIGURE 4.2.9 Fragmentation pathways for protonated diazepam.

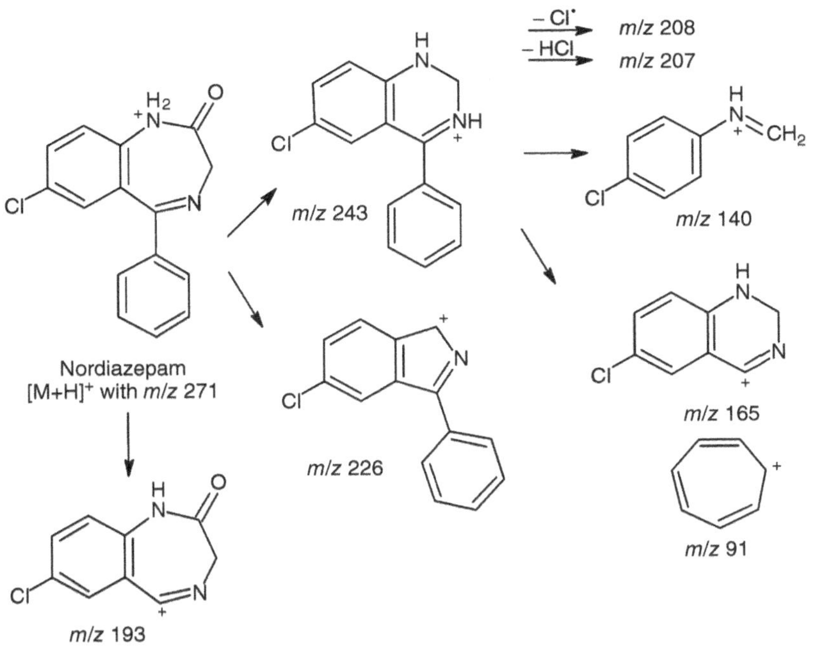

FIGURE 4.2.10 Fragmentation pathways for protonated nordiazepam.

to the loss of benzene (C_6H_6). From the ion with *m/z* 243, several secondary losses are observed, leading to fragment ions with *m/z* 208 due to the loss of Cl•, *m/z* 207 due to the loss of hydrogen chloride (HCl), *m/z* 165 due to the loss of C_6H_6, and *m/z* 140 due to the loss of C_6H_5—C≡N (Badawi et al., 2009; Nakamura et al., 2009). Protonated benzonitrile ([C_6H_5—C≡N+H]$^+$) with *m/z* 104 and $[C_7H_7]^+$ with *m/z* 91 are also observed.

Product-ion mass spectral data for 12 benzodiazepines with an N^1-desmethyl acetamide skeleton in the diazepine ring are available in our library collection. None of the others show all the fragment ions observed for nordiazepam. Often some of the ions are missing, while in addition other mostly compound-specific fragment ions are observed. An overview of the *m/z* values of the observed fragment ions for all 12 compounds is given in Table 4.2.2. The loss of CO is observed for all 12 compounds. The loss of $HCONH_2$ is only observed for 7-aminonitrazepam ([M+H]$^+$ with *m/z* 252) and its 7-acetamide analog (7-acetamidenitrazepam; [M+H]$^+$ with *m/z* 294) after initial loss of ethenone (H_2C=C=O) from the acetamide group. The primary loss of C_6H_6 is only observed for 7-aminonitrazepam and 4′-hydroxynordiazepam ([M+H]$^+$ with *m/z* 287), where it involves the loss of phenol (C_6H_5OH). Primary loss of C_6H_5—C≡N, FC_6H_4—C≡N, and ClC_6H_4—C≡N, not seen for nordiazepam, is observed for 7-aminonitrazepam, *nor*-aminoflunitrazepam, and 7-aminoclonazepam, respectively.

The secondary loss of Cl• and/or HCl is observed for all compounds with a Cl-substituent, that is, for 7-aminoclonazepam ([M+H]$^+$ with *m/z* 286), 4-hydroxynordiazepam, desalkylflurazepam ([M+H]$^+$ with *m/z* 289), delorazepam ([M+H]$^+$ with *m/z* 305), and clonazepam ([M+H]$^+$ with *m/z* 316) after initial loss of •NO_2. The secondary loss of C_6H_6 is observed for 7-aminonitrazepam, for 7-acetamidenitrazepam after initial loss of H_2C=C=O thereof, and for nitrazepam ([M+H]$^+$ with *m/z* 282), whereas a secondary loss of C_6H_5OH, fluorobenzene (C_6H_5F), and chlorobenzene (C_6H_5Cl) is observed for 4′-hydroxynordiazepam, desalkylflurazepam, and delorazepam, respectively. The secondary loss of C_6H_5—C≡N is observed for 7-aminonitrazepam and 7-acetamidenitrazepam, both prior to and after initial loss of H_2C=C=O. Nor-aminoflunitrazepam ([M+H]$^+$ with *m/z* 270) and desalkylflurazepam show secondary loss of FC_6H_4—C≡N; 4′-hydroxynordiazepam of hydroxybenzonitrile (HOC_6H_4—C≡N); and 7-aminoclonazepam, delorazepam, and fenazepam ([M+H]$^+$ with *m/z* 349) of ClC_6H_4—C≡N.

The tropylium ion ($[C_7H_7]^+$) with *m/z* 91 is observed for nitrazepam and 7-acetamidenitrazepam, whereas hydroxy- and fluorobenzyl cations with *m/z* 107 ($[HOC_6H_4CH_2]^+$) and with *m/z* 109 ($[FC_6H_4CH_2]^+$) are observed for 4′-hydroxynordiazepam and desalkylflurazepam, respectively. As indicated in Table 4.2.2, [C_6H_5—C≡N+H]$^+$ with *m/z* 104 is also observed for several of these compounds. Additional fragment ions are observed, for

TABLE 4.2.2 Fragmentation of benzodiazepines containing the N^1-desmethyl acetamide skeleton in the diazepine ring: *m/z* values of the ions observed.

	$[M+H]^+$	Primary Losses			
		$-CO$	$-HCONH_2$	$-C_6H_5X$	$-XC_6H_4CN$
Nordiazepam	271	243	226	193 (X = H)	
7-Aminonitrazepam	252	224	207	174 (X = H)	149 (X = H)
nor-Aminoflunitrazepam	270	242			149 (X = F)
Nitrazepam	282	254			
7-Aminoclonazepam	286	222 (after −HCl)			149 (X = Cl)
4′-Hydroxynordiazepam	287	259		193 (X = OH)	
Desalkylflurazepam	289	261			
7-Acetamidenitrazepam after $H_2C{=}C{=}O$ loss	294	266	207		
Delorazepam	305	277			
Clonazepam	316	288			
Fenazepam	349	321			
Bromazepam	316	288			

	$[M+H-CO]^+$	Secondary losses from $[M+H-CO]^+$			Other ions
		$-X^\bullet$	$-C_6H_5X$	$-XC_6H_4CN$	
Nordiazepam	243	208 (X = Cl)	165 (X = H)	140 (X = H)	91, 104
7-Aminonitrazepam	224		146 (X = H)	121 (X = H)	94, 104, 180
nor-Aminoflunitrazepam	242			121 (X = F)	195 250 (270−HF) 222 (250−CO)
Nitrazepam	254	208 (X = NO_2)	176 (X = H)		91, 104, 180
7-Aminoclonazepam				121 (X = Cl)	94, 104, 195 250 (286−HCl) 222 (250−CO)
4′-Hydroxynordiazepam	259	224 (X = Cl)	165 (X = OH)	140 (X = OH)	107
Desalkylflurazepam	261	226 (X = Cl)	165 (X = F)		104, 109
7-Acetamidenitrazepam after $H_2C{=}C{=}O$ loss	266		146 (X = H)	163 (X = H) 121 (X = H)	91, 94, 180
Delorazepam	277	242 (X = Cl)	165 (X = Cl)		
Clonazepam	288	207 (X = Cl) (after $-^\bullet NO_2$)			104
Fenazepam	321	242 (X = Br)		184 (X = Cl)	158, 206 285 (321−HCl)
Bromazepam	288	209 (X = Br)			209 ($-Br^\bullet$) 261, 182

example, protonated aniline with *m/z* 94 ($[C_6H_5NH_2{+}H]^+$) for 7-aminonitrazepam, 7-acetamidenitrazepam, and 7-aminoclonazepam; an ion with *m/z* 180 ($[C_{13}H_{10}N]^+$) for 7-aminonitrazepam, 7-acetamidenitrazepam (after initial $H_2C{=}C{=}O$ loss), as well as for nitrazepam; and an ion with *m/z* 195 ($[C_{13}H_{11}N_2]^+$) for *nor*-aminoflunitrazepam and 7-aminoclonazepam. The fragment ions with *m/z* 180 and 195 can be formed along various fragmentation pathways.

The fragmentation of two other compounds with an N^1-desmethyl acetamide skeleton in the diazepine ring requires some special attention. Apart from the loss of CO to an ion with *m/z* 288 (Table 4.2.2), bromazepam ($[M+H]^+$ with *m/z* 316) shows a fragment ion with *m/z* 209 due to the consecutive losses of CO and a bromine radical ($Br^\bullet$); *m/z* 261 due to the combined loss of hydrogen cyanide (HCN) and CO; and *m/z* 182 due to the combined loss of HCN, CO, and $Br^\bullet$. Apart from the class-specific fragment ions shown in Table 4.2.2, fenazepam ($[M+H]^+$ with *m/z* 349) shows a fragment ion with *m/z* 158 ($[C_9H_6N_2O]^{+\bullet}$), which is due to the combined loss of C_6H_5Cl and $Br^\bullet$, and an ion with *m/z* 206 ($[C_{14}H_{10}N_2]^{+\bullet}$), which is due to the combined loss of CO, HCl, and $Br^\bullet$.

Oxazepam ($[M+H]^+$ with *m/z* 287) is a metabolite of diazepam with an N^1-desmethyl-α-hydroxyacetamide skeleton. In ion-trap MS^n, the loss of water to an ion with *m/z* 269 in MS^2 and of CO to an ion with *m/z* 241 in MS^3 are observed. In MS^4, the ion with *m/z* 241 shows losses of C_6H_6 to an ion with *m/z* 163, of C_6H_5—C≡N to an ion with *m/z* 138, of $Cl^•$ to an ion with *m/z* 206, and of HCN to an ion with *m/z* 214 (Smyth et al., 2000). In a TQ product-ion mass spectrum, apart from these ions, $[C_6H_5—C≡N+H]^+$ with *m/z* 104 and an ion with *m/z* 231, consistent with the loss of ethenedione (O═C═C═O), are observed (Risoli et al., 2007). In addition, an ion is observed with *m/z* 166 ($[C_8H_5ClNO]^+$) due to the combined losses of water and C_6H_5—C≡N. Structure proposals for the fragment ions of oxazepam are shown in Figure 4.2.11. Two other compounds featuring an N^1-desmethyl-α-hydroxyacetamide skeleton are available in our library collection, that is, 3-hydroxynorflurazepam ($[M+H]^+$ with *m/z* 305) and lorazepam ($[M+H]^+$ with *m/z* 321). They show similar fragmentation characteristics to oxazepam; the *m/z* values of their fragment ions are summarized in Table 4.2.3.

Temazepam and lormetazepam contain an N^1-methyl-α-hydroxyacetamide skeleton in the diazepine ring. Temazepam ($[M+H]^+$ with *m/z* 301) shows two primary fragment ions, that is, ions with *m/z* 283 due to the loss of water and with *m/z* 239 due to the loss of ethane-1,2-diol ($HOCH_2CH_2OH$). Secondary fragmentation of the ion with *m/z* 283 results in ions with *m/z* 255 due to the loss of CO and with *m/z* 180 due to loss of C_6H_5—C≡N. Further fragmentation of the ion with *m/z* 255 results in an ion with *m/z* 228 due to the loss of HCN, which in turn shows the loss of $Cl^•$ to an ion *m/z* 193 ($[C_{14}H_{11}N]^{+•}$), and an ion with *m/z* 177 due to the loss of C_6H_6. The chlorobenzyl cation ($[ClC_6H_4CH_2]^+$) with *m/z* 125 is also observed. Lormetazepam ($[M+H]^+$ with *m/z* 335) shows fragment ions consistent with most of the neutral losses observed for temazepam, except for the ion with *m/z* 180 due to losses of water and ClC_6H_4—C≡N. The ion with *m/z* 177 is due the loss of C_6H_5Cl rather than the loss of C_6H_6 (observed with temazepam). In addition, ions are observed with *m/z* 253 due to losses of water, $HCO^•$, and $Cl^•$; with *m/z* 250 consistent with losses of O═C═C═O and HN═CH_2; and with *m/z* 213 consistent with loss of O═C═C═O, methanamine (H_2NCH_3), and $Cl^•$.

Sulazepam ($[M+H]^+$ with *m/z* 301) and quazepam ($[M+H]^+$ with *m/z* 387), both containing an N^1-substituted ethanethioamide skeleton in the diazepine ring, show characteristic radical losses, that is, the loss of the sulfanyl radical ($HS^•$) to ions with *m/z* 268 and 354, respectively. Sulazepam shows the loss of the *N*-methylthioformamidyl radical ($CH_3NCHS^•$) to ions with *m/z* 227 and a subsequent loss of $Cl^•$ to an ion with *m/z* 192. At higher collision

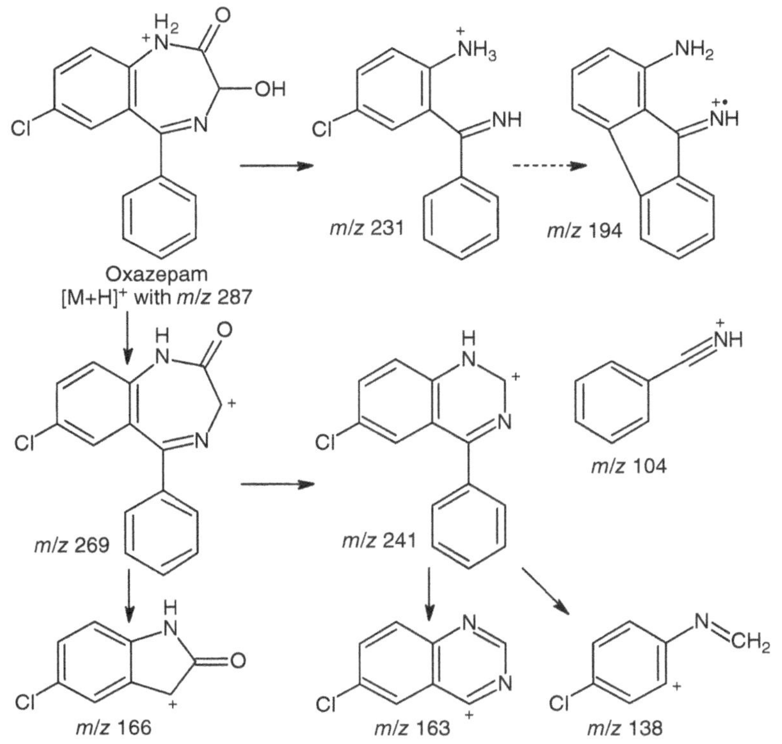

FIGURE 4.2.11 Fragmentation pathways for protonated oxazepam. The ion with *m/z* 194 is not observed for oxazepam, but it is observed for its analogue compounds (Section 4.2.2).

TABLE 4.2.3 Fragmentation of benzodiazepines containing the N^1-desmethyl-α-hydroxy acetamide skeleton in the diazepine ring: *m/z* values of the ions observed.

	Oxazepam	3-Hydroxynorflurazepam	Lorazepam
$[M+H]^+$	287	305	321
$[M+H-H_2O]^+$	269	287	303
Loss of $XC_6H_4C{\equiv}N$	166 (X = H)	166 (X = F)	
$[M+H-H_2O-CO]^+$	241	259	275
Loss of HCN	214		
Loss of Cl•	206		
Loss of C_6H_5X	163 (X = H)	163 (X = F)	163 (X = Cl)
Loss of $XC_6H_4C{\equiv}N$	138 (X = H)	138 (X = F)	138 (X = Cl)
$[M+H-O{=}C{=}C{=}O]^+$	231	249	265
Loss of HX		229 (X = F)	229 (X = Cl)
Loss of HX and Cl•		194 (X = F)	194 (X = Cl)
$[XC_6H_4C{\equiv}N+H]^+$	104 (X = H)	122 (X = F)	138 (X = Cl)

energy, complex fragmentation patterns are observed resulting from secondary fragmentation of $OE^{+\bullet}$ fragment ions (and in quazepam also due to the N^1-1,1,1-trifluoroethyl substituent).

Even with accurate-mass data available, the fragmentation of a methyltriazolobenzodiazepine like alprazolam is difficult to understand. Based on the derived molecular formulae for the fragments of alprazolam ($[M+H]^+$ with *m/z* 309), structures are proposed for most of the fragments observed (Figure 4.2.12). However, in some cases, similar alternative structures could be drawn.

In negative-ion MS–MS, nitrazepam ($[M-H]^-$ with *m/z* 280) shows subsequent losses of CO to *m/z* 252, a nitrosyl radical (•NO) to *m/z* 222 or $^\bullet NO_2$ to *m/z* 206, and CO to *m/z* 194; the ion with *m/z* 166 needs further study (no accurate-mass data available). The nitrazepam 3-hydroxy metabolite ($[M-H]^-$ with *m/z* 296) shows somewhat different fragmentation, that is, either the loss of CO to *m/z* 268 or the loss of hydrogen isocyanate (HN=C=O) or the carbamoyl radical ($H_2NCO^\bullet$) to *m/z* 253 or 252, respectively. The latter two ions show the loss of nitroxyl (HNO) or •NO, respectively, leading to an ion with *m/z* 222, which in turn shows the loss of CO to *m/z* 196. The ion with *m/z* 166, described earlier, is observed as well.

The glucuronic acid conjugate of oxazepam ($[M-H]^-$ with *m/z* 461) shows the characteristic loss of

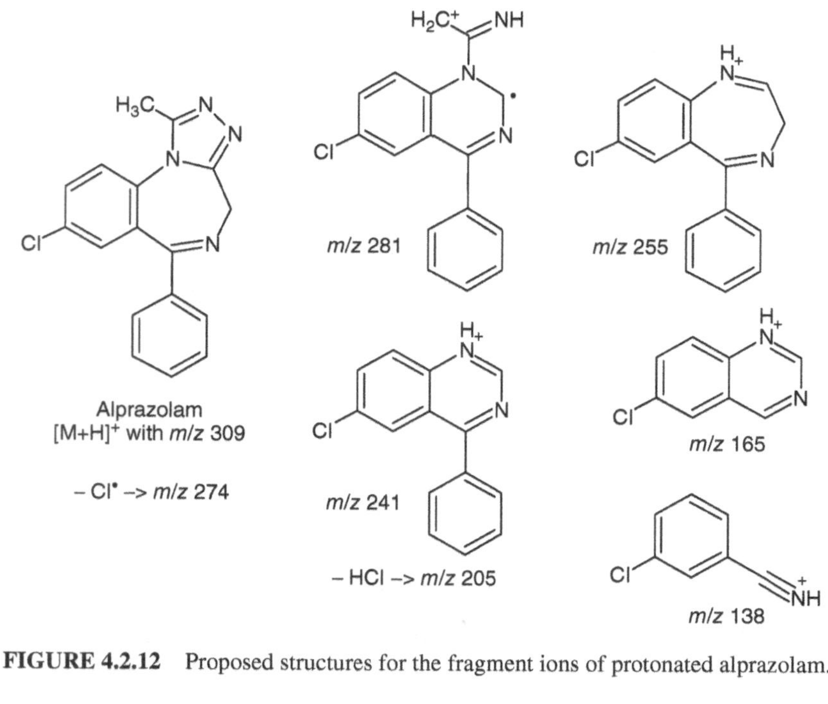

FIGURE 4.2.12 Proposed structures for the fragment ions of protonated alprazolam.

dehydroglucuronide ($C_6H_8O_6$) to *m/z* 285, as well as some characteristic secondary fragments of the dehydroglucuronate anion, for example, the ions with *m/z* 113 and 85 (Section 3.7.3) (Kuuranne et al., 2000). The ion with *m/z* 285 (deprotonated oxazepam) shows similar fragmentation to 3-hydroxynitrazepam, differing only by the C^7 substituent with a nitro group instead of a Cl-atom as in oxazepam, that is, loss of CO to an ion with *m/z* 257 or loss of HN=C=O or $H_2NCO^•$ to ions with *m/z* 242 or *m/z* 241, respectively.

4.2.5 LOCAL ANESTHETICS

A local anesthetic drug causes reversible anesthesia and a loss of nociception. There are two classes of local anesthetics: the more frequently used aminoamides like lidocaine and the aminoesters like procaine. Multiresidue LC–MS analysis of the aminoamide anesthetics bupivacaine, mepivacaine, ropivacaine, and prilocaine in human serum has been described (Koehler et al., 2005).

In MS–MS, the aminoamide anesthetics show the loss of the corresponding *N*-phenylformamide molecule with charge retention on the resulting *N,N*-diethylmethaniminium ion ($[(CH_3CH_2)_2N{=}CH_2]^+$) with *m/z* 86, as in lidocaine ($[M+H]^+$ with *m/z* 235) (Figure 4.2.13). Secondary fragments from the iminium ion may be observed, as for instance to the *N*-ethylmethaniminium ion ($[CH_3CH_2N{=}CH_2]^+$) with *m/z* 58 for lidocaine.

In contrast to aminoamide anesthetics, the aminoester anesthetics show more fragmentation covering all functionalities of the molecule. Procaine ($[M+H]^+$ with *m/z* 237) shows the 4-aminophenylium ion ($[H_2N{-}C_6H_4]^+$) with *m/z* 92, the 4-aminobenzoyl cation ($[H_2N{-}C_6H_4{-}C{=}O]^+$) with *m/z* 120, the 2-(4-aminobenzoate)ethyl carbocation ($[H_2N{-}C_6H_5{-}COOCH_2CH_2]^+$) with *m/z* 164, the 2-(*N,N*-diethylamino)ethyl carbocation ($[(CH_3CH_2)_2NCH_2CH_2]^+$) with *m/z* 100, and the 2-(*N*-ethylamino)ethyl carbocation with *m/z* 72 ($[CH_3CH_2NHCH_2CH_2]^+$) (Figure 4.2.13).

4.2.6 BARBITURATES

Barbiturates are derivatives of barbituric acid. They act as CNS depressants and produce a variety of effects ranging from mild sedation to total anesthesia. They are also effective as anxiolytics, hypnotics, and anticonvulsants.

Barbiturates can readily be analyzed by negative-ion ESI, with the negative charge most likely on the O^4-atom, due to an amide-iminol tautomerism. A toxicological screening method for some barbiturates and nonsteroidal anti-inflammatory drugs (NSAIDs) in serum based on detection of their $[M-H]^-$ has been reported (Hori et al., 2006). A number of barbiturates were among the 140 compounds analyzed in a multiresidue screening method for equine urine using positive-ion/negative-ion switching in ultra-performance LC with TQ MS–MS (Wong et al., 2011). Analysis of barbiturates in serum using supercritical fluid chromatography (SFC) in combination with ESI-MS–MS has also been described (Spell et al., 1998).

For 18 out of the 38 barbiturates studied, losses of hydrogen isocyanate (HN=C=O, 43 Da) and 87 Da (C_2HNO_3) are observed as well as a common fragment ion with *m/z* 85. The loss of 87 Da involves either the loss of CO_2 from $[M-H-43]^-$ or the loss of 1-hydroxyaziridine-2,3-dione (C_2HNO_3) from $[M-H]^-$, thus in both cases requiring the shift of an O-atom. The proposed structure of the fragment ion with *m/z* 85 ($[C_2HN_2O_2]^-$) is the 1,3-diazetidine-2,4-dione iminolate tautomer. Proposed structures for the fragment ions of amobarbital ($[M-H]^-$ with *m/z* 225) are shown in Figure 4.2.14. The loss of 43 Da (HN=C=O) is very common: it is observed for 28 out of the 38 barbiturates studied. In the case of benzobarbital ($[M-H]^-$ with *m/z* 335) with an N^1-benzoyl substituent, the loss of benzoyl isocyanate ($C_6H_5CON{=}C{=}O$) is observed instead of HN=C=O. For 23 out of the 38 barbiturates, the fragment ion with *m/z* 85 is missing. Although not observed in our library collection, the isocyanate ion ($[NCO]^-$) with *m/z* 42 has been reported as well (Spell et al., 1998). Surprisingly, hexobarbital, metharbital, methylphenobarbital, and thiamylal do not yield any fragment ions. Thiobutabarbital ($[M-H]^-$ with *m/z* 227) and thiopental ($[M-H]^-$ with *m/z* 241) with a thioketone on the C^2-atom only show one fragment ion, that is, the isothiocyanate ion ($[NCS]^-$) with *m/z* 58.

Bromine-containing barbiturates, for example, brallobarbital ($[M-H]^-$ with *m/z* 285), butallylonal ($[M-H]^-$ with *m/z* 301), and propallylonal ($[M-H]^-$ with *m/z* 287), show

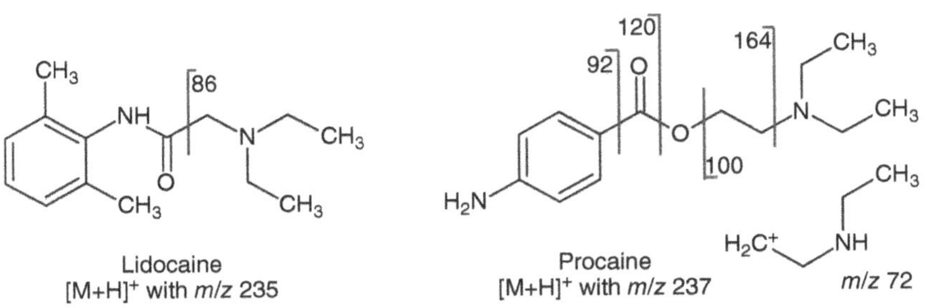

FIGURE 4.2.13 Fragmentation schemes for protonated lidocaine and procaine.

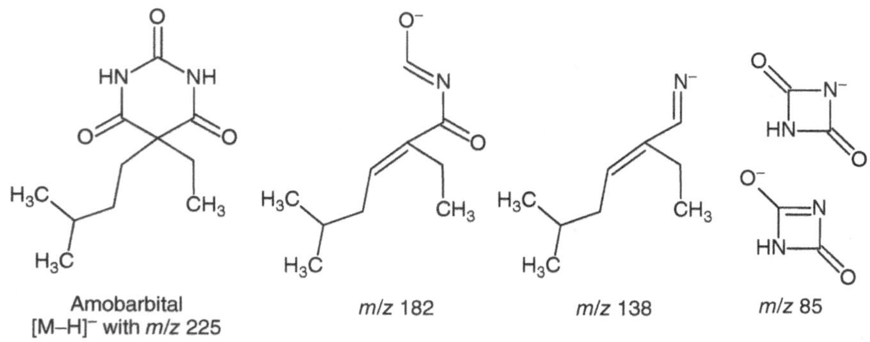

FIGURE 4.2.14 Proposed structures for the fragment ions of deprotonated amobarbital.

the loss of hydrogen bromide (HBr), the loss of HN═C═O from both $[M-H]^-$ and $[M-H-HBr]^-$, and the formation of bromide ion (Br^-) with *m/z* 79. Narcobarbital ($[M-H]^-$ with *m/z* 301) shows Br^- with *m/z* 79 as its only fragment ion.

4.2.7 ANTICONVULSANT DRUGS

Anticonvulsant drugs comprise a wide variety of compound classes used to prevent the stimulation of the brain that causes seizure activity, as is the case in epilepsy. These compounds are also being increasingly used in the treatment of bipolar disorders. Some examples of anticonvulsant drugs are barbiturates like pyrimidine-2,4,6-triones, for example, phenobarbital; and pyrimidine-4,6-diones, for example, primidone; benzodiazepines like clobazam and clonazepam (Section 4.2.4); carboxamides like carbamazepine; valproates and valproylamides; hydantoins like phenytoin; propionamides like beclamide; succinimides like mesuximide; and triazines like lamotrigine.

Popular antiepileptic drugs are carbamazepine and its derivative oxcarbazepine. With a dibenzoazepine skeleton, these compounds resemble tricyclic antidepressants (Section 4.2.3.1). The fragmentation of carbamazepine ($[M+H]^+$ with *m/z* 237) involves losses from the carboxamide (—$CONH_2$) group, that is, the loss of ammonia (NH_3) to an ion with *m/z* 220 or the loss of hydrogen isocyanate (HN═C═O) to *m/z* 194, as well as the loss of NH_3 and carbon monoxide (CO) to *m/z* 192. In oxcarbazepine (carbamazepine-10,11-epoxide, $[M+H]^+$ with *m/z* 253), the same losses are observed as for carbamazepine, thus resulting in fragment ions with *m/z* 236, *m/z* 210, and *m/z* 208. The latter two ions show the loss of CO with the ion with *m/z* 180 being more abundant than that of *m/z* 182.

Valproic acid is a widely used broad-spectrum antiepileptic drug. In negative-ion MS–MS, valproic acid ($[M-H]^-$ with *m/z* 143) does not produce any fragments. In LC–MS–MS, the SRM transition *m/z* 143 > 143 is widely used. Some of its metabolites show fragmentation, for example, 3-hydroxy-valproic acid ($[M-H]^-$ with *m/z* 159) shows the loss of propanal (CH_3CH_2CHO) to a fragment ion with *m/z* 101, the isomeric 5-hydroxy-valproic acid shows the loss of formic acid (HCOOH) to *m/z* 113, whereas 4-ene-valproic acid ($[M-H]^-$ with *m/z* 141) does not form fragment ions (Gao et al., 2011).

Phenytoin ($[M-H]^-$ with *m/z* 251) shows the cyanophenide anion with *m/z* 102 ($[N{\equiv}C{-}C_6H_4]^-$) and the loss of HN═C═O to an ion with *m/z* 208. In positive-ion mode, phenytoin ($[M+H]^+$ with *m/z* 253) shows consecutive losses of CO to an ion with *m/z* 225, of HN═C═O to *m/z* 182, and of benzene (C_6H_6) to the protonated benzonitrile ($[C_6H_5{-}C{\equiv}N{+}H]^+$) with *m/z* 104 (Figure 4.2.15).

MPPH ($[M-H]^-$ with *m/z* 265, 5-(4-methylphenyl)-5-phenylhydantoin), which is used as internal standard in the analysis of phenytoin, shows fragments due to the subsequent losses of HN═C═O and CO to ions with *m/z* 222 and 194, respectively. In addition, $[N{\equiv}C{-}C_6H_4]^-$ with *m/z* 102 and the 4-cyanotoluene aryl anion ($[CH_3{-}C_6H_3{-}C{\equiv}N]^-$) with *m/z* 116 are observed.

Topiramate ($[M-H]^-$ with *m/z* 338) yields two fragment ions, that is, the sulfamate anion ($[H_2N{-}SO_3]^-$) with *m/z* 96 and the deprotonated iminosulfene ($[NSO_2]^-$) with *m/z* 78. Ethotoin ($[M-H]^-$ with *m/z* 203) shows two major fragment ions, one with *m/z* 174 ($[C_9H_6N_2O_2]^{-\bullet}$) due to the loss of an ethyl radical ($^\bullet C_2H_5$) and another with *m/z* 54 ($[C_3H_4N]^-$).

The major fragment ion of beclamide ($[M+H]^+$ with *m/z* 198) is the tropylium ion ($[C_7H_7]^+$) with *m/z* 91 due to the loss of 3-chloropropanamide ($ClCH_2CH_2CONH_2$). Sultiame ($[M+H]^+$ with *m/z* 291) shows loss of NH_3 to an ion with *m/z* 274, loss of sulfur dioxide (SO_2) to *m/z* 227, as well as a combined loss of SO_2 and propene ($CH_3CH{=}CH_2$) to a fragment ion with *m/z* 185 ($[H_2NSO_2{-}C_6H_4{-}N{=}CH_2{+}H]^+$). At higher collision energy, the *N*-phenylmethaniminiumyl radical cation with *m/z* 105 ($[C_6H_5{-}N{=}CH_2]^{+\bullet}$) is observed (Figure 4.2.15).

Primidone ($[M+H]^+$ with *m/z* 219) shows subsequent losses of methylisocyanate ($CH_3{-}N{=}C{=}O$) to an ion with *m/z* 162 and of HN═C═O to the α-(ethyl)benzyl cation ($[C_6H_5CHCH_2CH_3]^+$) with *m/z* 119, as well as the

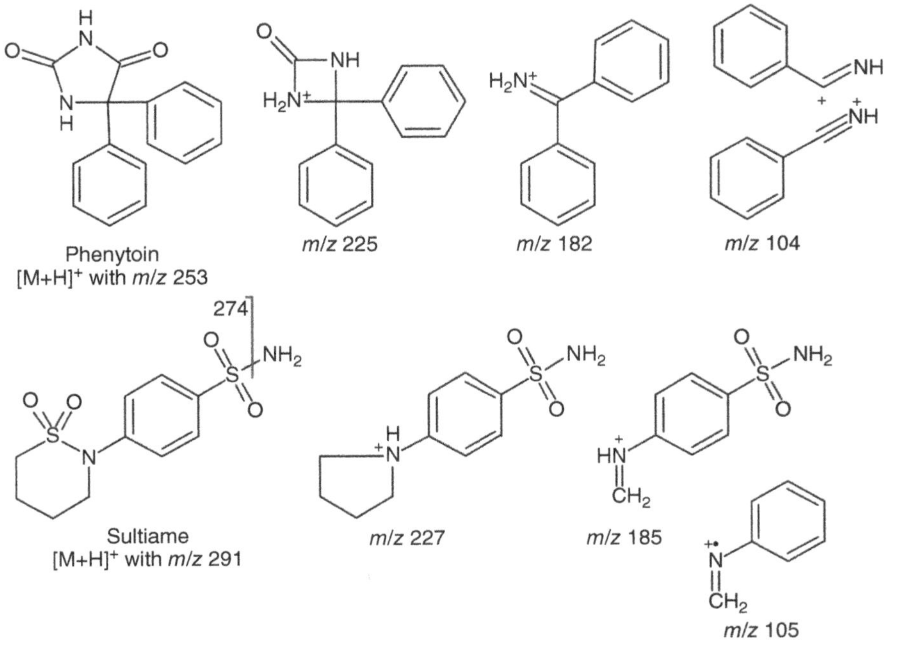

FIGURE 4.2.15 Proposed structures for the fragment ions of protonated phenytoin and sultiame.

formation of $[C_7H_7]^+$ (*m/z* 91) (Figure 4.2.16). Mesuximide ($[M+H]^+$ with *m/z* 204) shows the loss of CO to an ion with *m/z* 176, followed by the loss of methylamine (CH_3NH_2) to *m/z* 145. Furthermore, a fragment ion with *m/z* 126 ($[C_6H_9NO_2]^+$) due to the loss of C_6H_6 is observed, as well as the α,α-(dimethyl)benzyl cation ($[C_6H_5C(CH_3)_2]^+$) with *m/z* 119 and $[C_7H_7]^+$ with *m/z* 91 (Figure 4.2.16).

4.2.8 OTHER PSYCHOTROPIC DRUGS

There are numerous other psychotropic drugs. Among them are various classes of drugs of abuse, many of which are illicit, such as amphetamine and related compounds, cocaine and related compounds, and opiates, all of which are discussed separately in Section 4.7.

Methylphenidate ($[M+H]^+$ with *m/z* 234) or ritalin is a CNS stimulant that is used in the treatment of attention-deficit disorders, with or without hyperactivity, as well as narcolepsy. The major fragment is the tetrahydropyridinium ion with *m/z* 84 ($[C_5H_{10}N]^+$), which at higher collision energies shows the loss of ethene (C_2H_4) via a retro-Diels–Alder fragmentation to an ion with *m/z* 56.

Zolpidem (an imidazopyridine acetamide) and zopiclone (a pyrrolopyrazinone acetamide) are non-benzodiazepine hypnotics prescribed for the short-term treatment of insomnia, as well as some other CNS disorders. Zolpidem ($[M+H]^+$ with *m/z* 308) primarily shows the loss of dimethylamine ($(CH_3)_2NH$) to an ion with *m/z* 263 followed by the loss of carbon monoxide (CO) to an ion with *m/z* 235. Zopiclone ($[M+H]^+$ with *m/z* 389) is easily fragmented, generating fragment ions with *m/z* 345, 263, 245, 217, 112, and 99. Accurate-mass data show that the ion with *m/z* 345 results from a rearrangement involving the loss of carbon dioxide (CO_2). The ion with *m/z* 263 ($[C_{11}H_7ClN_4O_2]^+$) results from the loss of CO and the 4-methylpiperazine group as $C_5H_{10}N_2$, while the ions with *m/z* 245 and 217 are due to secondary fragmentation of the ion with *m/z* 263, involving the loss of water and water and CO, respectively (Figure 4.2.17). The ion with *m/z* 112 is due to 3-chloropyridinium ion ($[ClC_5H_3N]^+$) and the ion with *m/z* 99 is due to the protonated *N*-methyl tetrahydropyrazine ($C_5H_{11}N_2^+$).

Baclofen ($[M+H]^+$ with *m/z* 214) is a derivative of γ-aminobutyric acid (GABA), primarily used to treat spasticity and is currently under investigation for the treatment of alcoholism. MS–MS analysis of baclofen results in fragment ions with *m/z* 197 due to the loss of ammonia (NH_3), *m/z* 196 due to the loss of water, *m/z* 179 due to a combined loss of NH_3 and water, *m/z* 151 due to the loss of CO from the ion with *m/z* 179, which in turn shows the loss of a chlorine radical ($Cl^\bullet$) or of hydrogen chloride (HCl) to the fragment ions with *m/z* 116 and 115, respectively.

The methylxanthine alkaloid caffeine ($[M+H]^+$ with *m/z* 195) is probably the most widely used compound worldwide with CNS stimulant action. The fragmentation of caffeine results in two major fragment ions: the ion with *m/z* 138 is due to the loss of methylisocyanate (CH_3—N=C=O) and the ion with *m/z* 110 results from an additional loss of

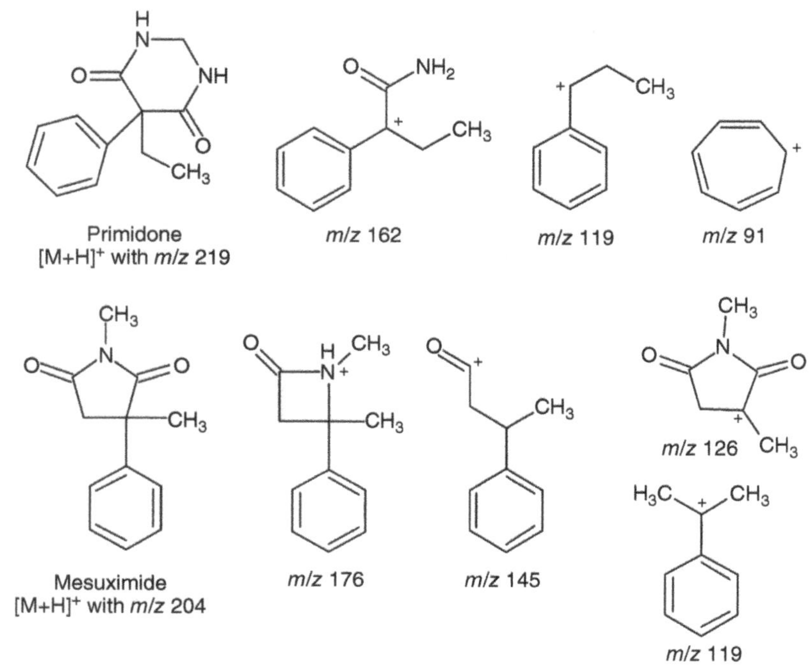

FIGURE 4.2.16 Proposed structures for the fragment ions of protonated primidone and mesuximide.

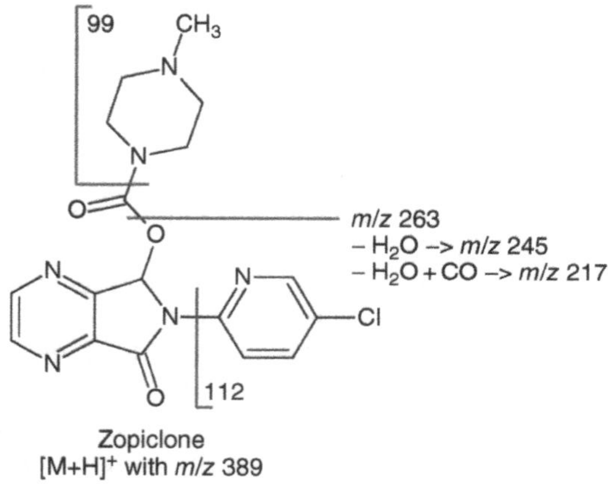

FIGURE 4.2.17 Fragmentation scheme for protonated zopiclone.

CO. The formation of the ion with *m/z* 138 can be envisaged as a retro-Diels–Alder fragmentation. At higher collision energies, additional fragment ions are observed, including an ion with *m/z* 123 due to loss of a methyl radical ($^{\bullet}CH_3$) from the ion with *m/z* 138.

The sedative and hypnotic compounds bromural (bromovalerylurea, $[M-H]^-$ with *m/z* 221) and acebromural ($[M-H]^-$ with *m/z* 277) show similar fragmentation, involving the loss of hydrogen bromide (HBr) and the formation of the bromide anion (Br^-) with *m/z* 79.

REFERENCES

Allen KR. 2006. Interference by venlafaxine ingestion in the detection of tramadol by liquid chromatography linked to tandem mass spectrometry for the screening of illicit drugs in human urine. Clin Toxicol (Phila) 44: 147–153.

Badawi N, Simonsen KW, Steentoft A, Bernhoft IM, Linnet K. 2009. Simultaneous screening and quantification of 29 drugs of abuse in oral fluid by solid-phase extraction and ultra-performance LC–MS/MS, Clin Chem 55: 2004–2018.

Bhatt J, Jangid A, Venkatesh G, Subbaiah G, Singh S. 2005. Liquid chromatography-tandem mass spectrometry (LC–MS–MS) method for simultaneous determination of venlafaxine and its active metabolite O-desmethyl venlafaxine in human plasma. J Chromatogr B 829: 75–81.

Breaud AR, Harlan R, Di Bussolo JM, McMillin GA, Clarke W. 2010. A rapid and fully-automated method for the quantitation of tricyclic antidepressants in serum using turbulent-flow liquid chromatography-tandem mass spectrometry. Clin Chim Acta 411: 825–832.

de Castro A, Ramírez Fernandez Mdel M, Laloup M, Samyn N, De Boeck G, Wood M, Maes V, López-Rivadulla M. 2007. High-throughput on-line solid-phase extraction-liquid chromatography–tandem mass spectrometry method for the simultaneous analysis of 14 antidepressants and their metabolites in plasma. J Chromatogr A 1160: 3–12.

de Castro A, Concheiro M, Quintela O, Cruz A, López-Rivadulla M. 2008. LC–MS/MS method for the determination of nine antidepressants and some of their main metabolites in oral fluid and plasma. Study of correlation between venlafaxine concentrations in both matrices. J Pharm Biomed Anal 48: 183–193.

Gao S, Miao H, Tao X, Jiang B, Xiao Y, Cai F, Yun Y, Li J, Chen W. 2011. LC–MS/MS method for simultaneous determination of valproic acid and major metabolites in human plasma. J Chromatogr B 879: 1939–1944.

Glover SJ, Allen KR. 2010. Measurement of benzodiazepines in urine by liquid chromatography-tandem mass spectrometry: Confirmation of samples screened by immunoassay. Ann Clin Biochem 47: 111–117.

Hakala KS, Kostiainen R, Ketola RA. 2006. Feasibility of different mass spectrometric techniques and programs for automated metabolite profiling of tramadol in human urine. Rapid Commun Mass Spectrom 20: 2081–2090.

Hayashida M, Takino M, Terada M, Kurisaki E, Kudo K, Ohno Y. 2009. Time-of-flight mass spectrometry (TOF-MS) exact mass database for benzodiazepine screening. Leg Med (Tokyo) Suppl 1: S423–S425.

Hori Y, Fujisawa M, Shimada K, Hirose Y, Yoshioka T. 2006. Method for screening and quantitative determination of serum levels of salicylic Acid, acetaminophen, theophylline, phenobarbital, bromvalerylurea, pentobarbital, and amobarbital using liquid chromatography/electrospray mass spectrometry. Biol Pharm Bull 29: 7–13.

Ishida T, Kudo K, Hayashida M, Ikeda N. 2009. Rapid and quantitative screening method for 43 benzodiazepines and their metabolites, zolpidem and zopiclone in human plasma by liquid chromatography/mass spectrometry with a small particle column. J Chromatogr B 877: 2652–2657.

Jiang T, Rong Z, Peng L, Chen B, Xie Y, Chen C, Sun J, Xu Y, Lu Y, Chen H. 2010. Simultaneous determination of citalopram and its metabolite in human plasma by LC–MS/MS applied to pharmacokinetic study. J Chromatogr B 878: 615–619.

Joyce C, Smyth WF, Ramachandran VN, O'Kane E, Coulter DJ. 2004.The characterisation of selected drugs with amine-containing side chains using electrospray ionisation and ion trap mass spectrometry and their determination by HPLC-ESI-MS. J Pharm Biomed Anal 36: 465–476.

Kirchherr H, Kühn-Velten WN. 2006. Quantitative determination of forty-eight antidepressants and antipsychotics in human serum by HPLC tandem mass spectrometry: A multi-level, single-sample approach. J Chromatogr B 843: 100–113.

Koehler A, Oertel R, Kirch W. 2005. Simultaneous determination of bupivacaine, mepivacain, prilocaine and ropivacain in human serum by liquid chromatography-tandem mass spectrometry. J Chromatogr A 1088: 126–130.

Kollroser M, Schober C. 2002. Simultaneous determination of seven tricyclic antidepressant drugs in human plasma by direct-injection HPLC-APCI-MS-MS with an ion trap detector. Ther Drug Monit 24: 537–544.

Kratzsch C, Peters FT, Kraemer T, Weber AA, Maurer HH. 2003. Screening, library-assisted identification and validated quantification of fifteen neuroleptics and three of their metabolites in plasma by liquid chromatography/mass spectrometry with atmospheric pressure chemical ionization. J Mass Spectrom 38: 283–295.

Kratzsch C, Tenberken O, Peters FT, Weber AA, Kraemer T, Maurer HH. 2004. Screening, library-assisted identification and validated quantification of 23 benzodiazepines, flumazenil, zaleplone, zolpidem and zopiclone in plasma by liquid chromatography/mass spectrometry with atmospheric pressure chemical ionization. J Mass Spectrom 39: 856–872.

Kumazawa T, Seno H, Watanabe-Suzuki K, Hattori H, Ishii A, Sato K, Suzuki O. 2000. Determination of phenothiazines in human body fluids by solid-phase microextraction and liquid chromatography/tandem mass spectrometry. J Mass Spectrom 35: 1091–1099.

Kuuranne T, Vahermo M, Leinonen A, Kostianen R. 2000. Electrospray and atmospheric pressure chemical ionization tandem mass spectrometric behavior of eight anabolic steroid glucuronides. J Am Soc Mass Spectrom 11: 722–730.

Lajeunesse A, Gagnon C, Sauvé S. 2008. Determination of basic antidepressants and their N-desmethyl metabolites in raw sewage and wastewater using solid-phase extraction and liquid chromatography-tandem mass spectrometry. Anal Chem 80: 5325–5333.

Marin SJ, Coles R, Merrell M, McMillin GA. 2008. Quantitation of benzodiazepines in urine, serum, plasma, and meconium by LC–MS–MS. J Anal Toxicol 32: 491–498.

Marin SJ, McMillin GA. 2010. LC–MS/MS analysis of 13 benzodiazepines and metabolites in urine, serum, plasma, and meconium. Methods Mol Biol 603: 89–105.

Massaroti P, Cassiano NM, Duarte LF, Campos DR, Marchioretto MA, Bernasconi G, Calafatti S, Barros FA, Meurer EC, Pedrazzoli J. 2005. Validation of a selective method for determination of paroxetine in human plasma by LC–MS/MS. J Pharm Pharm Sci 8: 340–347.

McClean S, O'Kane EJ, Smyth WF. 2000. Electrospray ionisation-mass spectrometric characterisation of selected anti-psychotic drugs and their detection and determination in human hair samples by liquid chromatography-tandem mass spectrometry. J Chromatogr B 740: 141–157.

Nakamura M, Ohmori T, Itoh Y, Terashita M, Hirano K. 2009. Simultaneous determination of benzodiazepines and their metabolites in human serum by liquid chromatography–tandem mass spectrometry using a high-resolution octadecyl silica column compatible with aqueous compounds LC–MS/MS assay for benzodiazepines. Biomed Chromatogr 23: 357–364.

Nielsen MK, Johansen SS, Dalsgaard PW, Linnet K. 2010. Simultaneous screening and quantification of 52 common pharmaceuticals and drugs of abuse in hair using UPLC-TOF-MS. Forensic Sci Int 196: 85–92.

Nuñez A, Lehotay SJ, Geis-Asteggiante L. 2015. Structural characterization of product ions by electrospray ionization and quadrupole time-of-flight mass spectrometry to support regulatory analysis of veterinary drug residues in foods. Part 2: Benzimidazoles, nitromidazoles, phenothiazines, and mectins. Rapid Commun Mass Spectrom 29: 719–729.

Patel BN, Sharma N, Sanyal M, Shrivastav PS. 2008. Liquid chromatography tandem mass spectrometry assay for the simultaneous determination of venlafaxine and O-desmethylvenlafaxine in human plasma and its application to a bioequivalence study. J Pharm Biomed Anal 47: 603–611.

Raman B, Sharma BA, Ghugare PD, Karmuse PP, Kumar A. 2009. Semi-preparative isolation and structural elucidation of an impurity in citalopram by LC/MS/MS. J Pharm Biomed Anal 50: 377–383.

Risoli A, Cheng JB, Verkerk UH, Zhao J, Ragno G, Hopkinson AC, Siu KW. 2007. Gas-phase fragmentation of protonated benzodiazepines. Rapid Commun Mass Spectrom 21: 2273–2281.

Saar E, Gerostamoulos D, Drummer OH, Beyer J. 2009. Comparison of extraction efficiencies and LC–MS–MS matrix effects using LLE and SPE methods for 19 antipsychotics in human blood. Anal Bioanal Chem 393: 727–734.

Sauvage FL, Gaulier JM, Lachâtre G, Marquet P. 2006. A fully automated turbulent-flow liquid chromatography-tandem mass spectrometry technique for monitoring antidepressants in human serum. Ther Drug Monit 28: 123–130.

Smink BE, Brandsma JE, Dijkhuizen A, Lusthof KJ, de Gier JJ, Egberts AC, Uges DR. 2004. Quantitative analysis of 33 benzodiazepines, metabolites and benzodiazepine-like substances in whole blood by liquid chromatography-(tandem) mass spectrometry. J Chromatogr B 811: 13–20.

Smyth WF, McClean S, Ramachandran VN. 2000. A study of the electrospray ionisation of pharmacologically significant 1,4-benzodiazepines and their subsequent fragmentation using an ion-trap mass spectrometer. Rapid Commun Mass Spectrom 14: 2061–2069.

Smyth WF, Joyce C, Ramachandran VN, O'Kane E, Coulter D. 2004. Characterisation of selected hypnotic drugs and their metabolites using electrospray ionisation with ion trap mass spectrometry and with quadrupole time-of-flight mass spectrometry and their determination by liquid chromatography-electrospray ionisation–ion trap mass spectrometry. Anal Chim Acta 506: 203–214.

Smyth WF, Leslie JC, McClean S, Hannigan B, McKenna HP, Doherty B, Joyce C, O'Kane E. 2006. The characterisation of selected antidepressant drugs using electrospray ionisation with ion trap mass spectrometry and with quadrupole time-of-flight mass spectrometry and their determination by high-performance liquid chromatography/electrospray ionisation tandem mass spectrometry. Rapid Commun Mass Spectrom 20: 1637–1642.

Spell JC, Srinivasan K, Stewart JT, Bartlett MG. 1998. Supercritical fluid extraction and negative ion electrospray liquid chromatography tandem mass spectrometry analysis of phenobarbital, butalbital, pentobarbital and thiopental in human serum. Rapid Commun Mass Spectrom 12: 890–894.

Sutherland FC, Badenhorst D, de Jager AD, Scanes T, Hundt HK, Swart KJ, Hundt AF. 2001. Sensitive liquid chromatographic-tandem mass spectrometric method for the determination of fluoxetine and its primary active metabolite norfluoxetine in human plasma. J Chromatogr A 914: 45–51.

Villain M, Concheiro M, Cirimele V, Kintz P. 2005. Screening method for benzodiazepines and hypnotics in hair at pg/mg level by liquid chromatography–mass spectrometry/mass spectrometry. J Chromatogr B 825: 72–78.

Wen B, Zhou M. 2009. Metabolic activation of the phenothiazine antipsychotics chlorpromazine and thioridazine to electrophilic iminoquinone species in human liver microsomes and recombinant P450s. Chem Biol Interact 181: 220–226.

Wong CH, Tang FP, Wan TS. 2011. A broad-spectrum equine urine screening method for free and enzyme-hydrolysed conjugated drugs with ultra-performance liquid chromatography/tandem mass spectrometry. Anal Chim Acta 697: 48–60.

4.3

FRAGMENTATION OF ANALGESIC, ANTIPYRETIC, AND ANTI-INFLAMMATORY DRUGS

Analgesics, antipyretic, and antiphlogistic drugs are used to relieve pain, lower body temperature, and for their anti-inflammatory properties, respectively. These compounds can have one or more of those properties, reason for which they are used as antirheumatic drugs. Analgesics, antipyretic, and anti-inflammatory drugs act in various ways on the peripheral and CNS. Important drug classes are acetaminophen (also called paracetamol), salicylic acid derivatives, nonsteroidal anti-inflammatory drugs (NSAIDs), selective cyclooxygenase-2 enzyme (COX-2) inhibitors, and opiates. Since the opiates are compounds primarily related to the structure of morphine, they are discussed in Section 4.7.5.

Multiresidue analysis of analgesic drugs has been reported for plasma (Suenami et al., 2006a,b; Vinci et al., 2006), bovine muscle and milk (Van Hoof et al., 2004; Gallo et al., 2008; Dowling et al., 2009; Malone et al., 2009; Dowling et al., 2010; Dubreil-Chéneau et al., 2011), and surface and wastewater (Farré et al., 2007; Petrović et al., 2005).

4.3.1 ACETAMINOPHEN

Acetaminophen (*N*-(4-hydroxyphenyl)acetamide, paracetamol, $[M+H]^+$ with *m/z* 152) is a widely used over-the-counter analgesic and antipyretic drug. Important fragment ions are the protonated 4-aminophenol ($[H_2N—C_6H_4OH+H]^+$) with *m/z* 110, which is due to the loss of ethenone ($H_2C{=}C{=}O$) from the acetyl group, and its complementary acylium ion ($[CH_3—C{=}O]^+$) with *m/z* 43. Secondary fragmentation of the ion with *m/z* 110 involves the loss of ammonia (NH_3) or water leading to the 4-hydroxyphenylium ion with *m/z* 93 and the 4-aminophenylium ion with *m/z* 92, respectively, and further to the cyclopentadienyl cation ($[C_5H_5]^+$) with *m/z* 65 after the loss of carbon monoxide (CO) and hydrogen cyanide (HCN), respectively (Figure 4.3.1). Acetaminophen is also widely studied in relation to its ability to form reactive

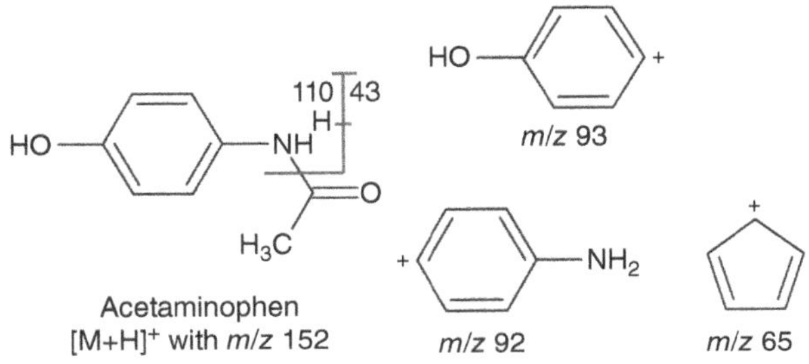

FIGURE 4.3.1 Fragmentation scheme for protonated acetaminophen (paracetamol).

quinoneimine ($O{=}C_6H_4{=}NH$) metabolites (Evans et al., 2004).

4.3.2 SALICYLIC ACID DERIVATIVES

Aspirin (acetylsalicylic acid or 2-(acetoxy)benzoic acid) was the first discovered NSAID. Because of its undesirable side effects in gastrointestinal ulcers, it is nowadays primarily used for its antiplatelet activity, that is, the inhibition of thromboxane production, which helps to prevent myocardial infarction (heart attacks), strokes, and blood clots. In positive-ion MS–MS, acetylsalicylic acid ($[M+H]^+$ with *m/z* 181) shows losses of water to an acylium ion with *m/z* 163, of ethenone ($H_2C{=}C{=}O$) to form the protonated 2-hydroxybenzoic acid with *m/z* 139, and a combination of these two losses to the 2-hydroxybenzoyl cation ($[HOC_6H_4C{=}O]^+$) with *m/z* 121 (Figure 4.3.2) (Williams et al., 2006). In negative-ion MS–MS, acetylsalicylic acid ($[M-H]^-$ with *m/z* 179) shows three main fragment ions, that is, 2-hydroxybenzoate with *m/z* 137 due to the loss of $H_2C{=}C{=}O$, the phenoxide ion ($[C_6H_5O]^-$) with *m/z* 93 due to the subsequent loss of carbon dioxide (CO_2), and the acetate anion ($[CH_3COO]^-$) with *m/z* 59 (Figure 4.3.2).

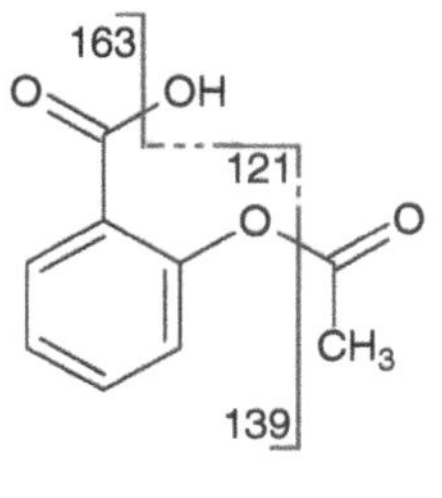

Acetylsalicylic acid
$[M+H]^+$ with *m/z* 181

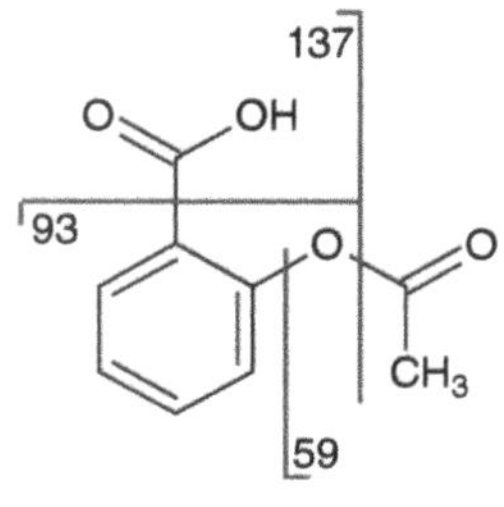

Acetylsalicylic acid
$[M-H]^-$ with *m/z* 179

FIGURE 4.3.2 Fragmentation schemes for protonated and deprotonated acetylsalicylic acid (aspirin).

Similar fragmentation is observed for related compounds. Salicylic acid ($[M-H]^-$ with *m/z* 137) shows subsequent losses of CO_2 and carbon monoxide (CO) to fragment ions with *m/z* 93 and 65, respectively.

The positive-ion product-ion mass spectra of a number of related compounds present in the library collection were also studied. 5-Aminosalicylic acid (mesalamine, $[M+H]^+$ with *m/z* 154) shows losses of water or formic acid (HCOOH) to form ions with *m/z* 136 and 108, respectively. At higher collision energies, the 3-aminocyclopentadienyl cation ($[H_2NC_5H_4]^+$) with *m/z* 80 is formed as a result of the loss of HCOOH and CO. Salicylamide ($[M+H]^+$ with *m/z* 138) shows fragment ions with *m/z* 121, *m/z* 93, and *m/z* 65 ($[C_5H_5]^+$) due to subsequent losses of ammonia (NH_3), CO, and CO. In positive-ion MS–MS, *N-iso*-propyl salicylamide ($[M+H]^+$ with *m/z* 180) first shows the loss of propene ($CH_3CH{=}CH_2$) to an ion with *m/z* 138, and further losses similar to what is seen for salicylamide. In negative-ion MS–MS, *N-iso*-propyl salicylamide ($[M-H]^-$ with *m/z* 178) shows subsequent losses of $CH_3CH{=}CH_2$ and hydrogen isocyanate (HN=C=O) to ions with *m/z* 136 and 93, respectively. Ethenzamide (2-ethyoxybenzamide, $[M+H]^+$ with *m/z* 166) shows subsequent losses of NH_3 and ethene (C_2H_4) to an ion with *m/z* 121, which shows further losses of CO and CO to the cyclopentadienyl cation ($[C_5H_5]^+$) with *m/z* 65. Salsalate ($[M+H]^+$ with *m/z* 259) shows the loss of salicylic acid to the 2-hydroxybenzoyl cation with *m/z* 121 and further fragmentation as described for salicylamide. In negative-ion MS–MS, salsalate ($[M-H]^-$ with *m/z* 257) shows the 2-hydroxybenzoate anion with *m/z* 137 and the ion with *m/z* 93 due to subsequent loss of CO_2.

Benorilate ($[M+H]^+$ with *m/z* 314) is an ester-linked co-drug of acetaminophen and aspirin. In MS–MS, it shows the loss of $H_2C{=}C{=}O$ either from the *N*-acetyl or from the *O*-acetyl substituent. In addition, $[HOC_6H_4{-}C{=}O]^+$ with *m/z* 121 and protonated acetaminophen with *m/z* 152 are formed, as well as protonated 4-aminophenol ($[H_2N{-}C_6H_4OH+H]^+$) with *m/z* 110, due the loss of $H_2C{=}C{=}O$ from the ion with *m/z* 152.

Mesalamine or 5-aminosalicylic acid ($[M-H]^-$ with *m/z* 152) shows the loss of CO_2 or of the formyloxyl radical ($HCO_2^\bullet$) to form fragment ions with *m/z* 108 and 107, respectively. At lower collision energy, the ion with *m/z* 108 is more abundant, and at higher collision energy the ion with *m/z* 107. Diflusinal ($[M-H]^-$ with *m/z* 249) shows subsequent losses of CO_2 and hydrogen fluoride (HF) and also the characteristic salicylate fragment ions with *m/z* 137 and 93.

Olsalazine is an *N*-quinoneimine derivative of 4-aminosalicylic acid, which is used in the treatment of inflammatory bowel diseases such as ulcerative colitis. In negative-ion MS–MS, olsalazine ($[M-H]^-$ with *m/z* 301) shows the loss of water to an ion with *m/z* 283, the loss of HCOOH to *m/z* 255, subsequent loss of CO_2 to *m/z* 211 and of CO to *m/z* 183 (Figure 4.3.3). In addition, the ion cyclohexa-2,5-dienenone-4-iminide ion ($O{=}C_6H_4{=}N^-$) with *m/z* 106 is observed, as well as the phenoxide-4-yl radical anion ($[C_6H_4O]^{-\bullet}$) with *m/z* 92, and the cyclopenta-2,4-dieneiminide ion ($[C_5H_4N]^-$) with *m/z* 78. Finally, a rearrangement reaction leads to the fragment ions with *m/z* 154 ($[C_{11}H_6O]^{-\bullet}$) and *m/z* 155 ($[C_{11}H_7O]^-$), respectively (Figure 4.3.3).

4.3.3 NONSTEROIDAL ANTI-INFLAMMATORY DRUGS

4.3.3.1 *N,N*-Diphenylamine or *N*-Phenyl-*N*-Pyridinylamine Derivatives

NSAIDs such as diclofenac and aceclofenac derive from 2-(anilinophenyl)acetic acid; tolfenamic acid, flufenamic acid, mefenamic acid, meclofenamic acid from *N*-phenylanthranilic acid; while flunixin and niflumic acid are N^2-anilino substituted derivatives of nicotinic acid. Most compounds can be analyzed in both positive-ion and negative-ion modes, although in multiresidue LC–MS analysis, the negative-ion mode is mostly preferred (Van Hoof et al., 2004; Farré et al., 2007). Like acetaminophen, diclofenac is extensively studied in relation to its ability to form reactive quinoneimine metabolites (Evans et al., 2004, Dieckhaus et al., 2005).

In positive-ion mode, diclofenac ($[M+H]^+$ with *m/z* 296) shows subsequent losses of water, carbon monoxide (CO), and either a chlorine radical ($Cl^\bullet$) or hydrogen chloride (HCl) to fragment ions with *m/z* 278, *m/z* 250, *m/z* 215, and *m/z* 214, respectively. In negative-ion MS–MS, diclofenac ($[M-H]^-$ with *m/z* 294) shows subsequent losses of carbon dioxide (CO_2), HCl and HCl to fragment ions with *m/z* 250, *m/z* 214, and *m/z* 178, respectively (Marchese et al., 2003; Kosjek et al., 2008). Flufenamic acid ($[M-H]^-$ with *m/z* 280) and floctafenamic acid ($[M-H]^-$ with *m/z* 331, with a 4-quinolinyl substituent instead of a phenyl) show after CO_2 loss three subsequent hydrogen fluoride (HF)

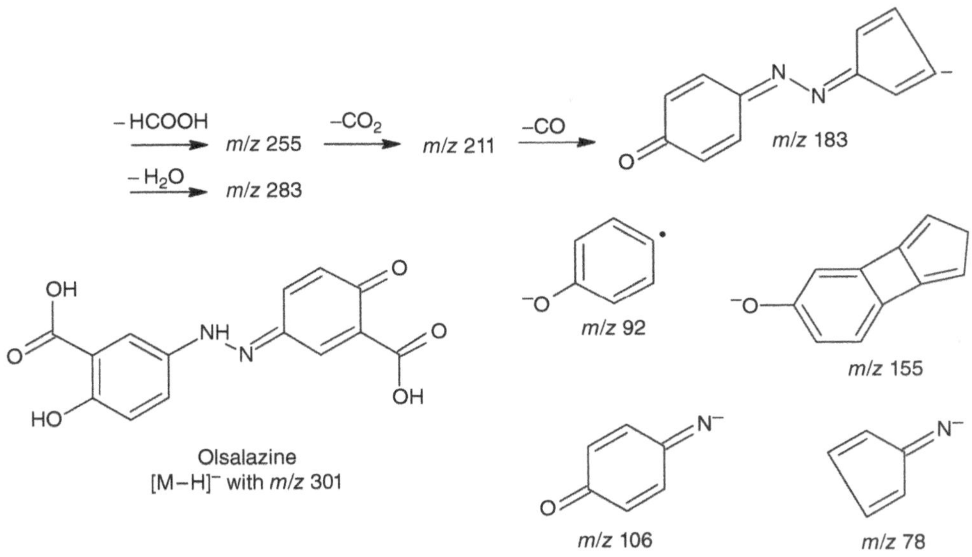

FIGURE 4.3.3 Proposed structures for some of the fragment ions of deprotonated olsalazine.

losses to ions with *m/z* 176 ([$C_{13}H_6N]^-$) and *m/z* 227 ([$C_{16}H_7N_2]^-$), respectively. Alclofenac ([M−H]$^-$ with *m/z* 225) shows subsequent losses of CO_2 and an allyl radical ($^\bullet C_3H_5$) to fragment ions with *m/z* 181 and 140, respectively. In meclofenamic acid ([M−H]$^-$ with *m/z* 294), competing primary losses of CO_2 and HCl are followed by other subsequent losses. Thus, fragment ions with *m/z* 258 (loss of HCl), *m/z* 250 (loss of CO_2), *m/z* 214 (loss of HCl and CO_2), and *m/z* 178 (loss of two times HCl and of CO_2) are observed.

Tolfenamic acid ([M−H]$^-$ with *m/z* 260) shows more complex fragmentation. Whereas at low collision energy only the loss of CO_2 to a fragment ion with *m/z* 216 is observed, at higher collision energy the loss of formic acid (HCOOH) to a fragment ion with *m/z* 214 is observed as well. Both these fragment ions show secondary fragmentation, for example, losses of methane (CH_4), HCl, and combinations of CH_4 and HCl from the ion with *m/z* 216 to fragment ions with *m/z* 200, *m/z* 180, and *m/z* 164, respectively, and losses of a chlorine radical ($Cl^\bullet$) and HCl from the ion with *m/z* 214 to fragment ions with *m/z* 179 and *m/z* 178, respectively. In addition, a phenide ion ([$C_6H_5]^-$) with *m/z* 77 and a fragment ion with *m/z* 116 ([$C_8H_6N]^-$) are found. Mefenamic acid ([M−H]$^-$ with *m/z* 240) shows losses of both CO_2 and HCOOH to fragment ions with *m/z* 196 and 194, respectively, and the loss of CH_4 from the ion with *m/z* 196. In addition, deprotonated aniline ([$C_6H_5NH]^-$) with *m/z* 92, deprotonated 2,3-dimethylaniline with *m/z* 120, and deprotonated 2- or 3-methylaniline with *m/z* 105 are observed (Figure 4.3.4). In negative-ion MS–MS, niflumic acid ([M−H]$^-$ with *m/z* 281) shows the loss of CO_2 and three subsequent losses of HF, resulting in fragment ions with *m/z* 237, *m/z* 217, *m/z* 197, and *m/z* 177, respectively. The pyidin-2-aminide ion ([C_5H_4N—$NH]^-$) with *m/z* 93 is also observed. Similar features are observed in ion-trap MS^2 spectra of these compounds (Vinci et al., 2006).

In positive-ion MS–MS, meclofenamic acid ([M+H]$^+$ with *m/z* 296) shows subsequent losses of water and $Cl^\bullet$ to form an ion with *m/z* 243. Niflumic acid ([M+H]$^+$ with *m/z* 283) shows losses of water and HF to an ion with *m/z* 245. Mefenamic acid ([M+H]$^+$ with *m/z* 242) shows losses of water and a methyl radical ($^\bullet CH_3$) to *m/z* 209. The cleavage of the phenyl C—N bond does not occur because the required H-rearrangement to the N-atom is not likely to occur from a phenyl ring (Section 3.6.5).

In positive-ion MS–MS, aceclofenac ([M+H]$^+$ with *m/z* 354) shows the loss of the ester substituent (hydroxyacetic acid ($HOCH_2COOH$) leading to an acylium ion with *m/z* 278, which shows subsequent losses of CO and $Cl^\bullet$ to fragment ions with *m/z* 250 and 215, respectively. In negative-ion MS–MS, aceclofenac ([M−H]$^-$ with *m/z* 352) also shows the loss of the ester substituent (oxiran-2-one, $C_2H_2O_2$) to the deprotonated dichlofenac with *m/z* 294, which shows subsequent losses of CO_2 and two times HCl to an ion with *m/z* 178. However, the base peak in the negative-ion product-ion mass spectrum of aceclofenac is the deprotonated hydroxyacetic acid ($HOCH_2COO^-$) with *m/z* 75.

4.3.3.2 Anthranilic Acid Derivatives

Glafenine and floctafenine are *N*-(4-quinolinyl) derivatives of anthranilic acid. In positive-ion MS–MS, glafenine ([M+H]$^+$ with *m/z* 373) and floctafenine ([M+H]$^+$ with *m/z* 407) first show the loss of the 2,3-dihydroxypropyl ester alkyl chain. This substituent is predominantly lost as 1,2,3-trihydroxypropane, because that loss results in the formation of an acylium ion ([RC=O]$^+$), that is, to form

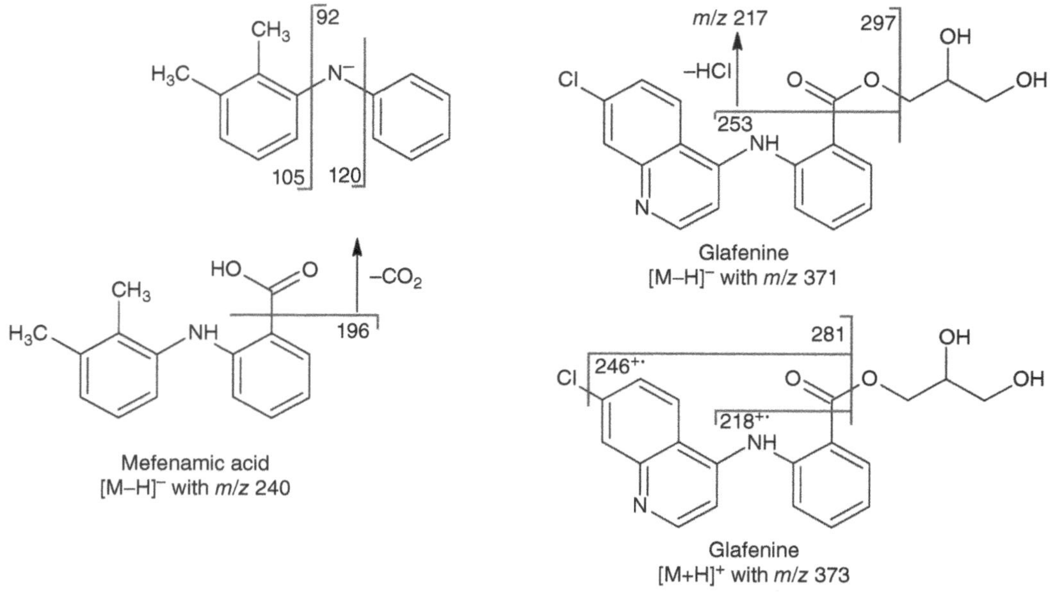

FIGURE 4.3.4 Fragmentation schemes for deprotonated mefenamic acid and for protonated and deprotonated glafenine.

an ion with *m/z* 281 for glafenine and with *m/z* 315 for floctafenine. In glafenine, this primary loss is followed by the loss of carbon monoxide (CO), a chlorine radical (Cl•), or both to produce ions with *m/z* 253, *m/z* 246, and *m/z* 218, respectively. Floctafenine shows the subsequent loss of CO and hydrogen fluoride (HF). In negative-ion mode, glafenine ($[M-H]^-$ with *m/z* 371) and floctafenine ($[M-H]^-$ with *m/z* 405) show the loss of the ester alkyl chain (prop-2-ene-1,2-diol or 1-hydroxypropan-2-one, $C_3H_6O_2$) to produce an ion with *m/z* 297 and 331, respectively. Subsequent loss of carbon dioxide (CO_2) and hydrogen chloride (HCl) leads to an ion with *m/z* 217 for glafenine. After losing CO_2, floctafenine can show HF loss up to three times. The fragmentation of glafenine is outlined in Figure 4.3.4.

4.3.3.3 Phenylpropanoic Acid Derivatives

Another group of NSAID can be considered as derivatives of 2-aryl propanoic acid, for example, carprofen, fenoprofen, flurbiprofen, ibuprofen, ketoprofen, and naproxen. In both positive-ion and negative-ion MS–MS, most of these compounds show competing losses of carbon dioxide (CO_2) and formic acid (HCOOH).

In positive-ion MS–MS, ketoprofen ($[M+H]^+$ with *m/z* 255) shows an abundant loss of HCOOH to an ion with *m/z* 209, followed by the loss of either benzene (C_6H_6) to the 3-vinylbenzoyl cation with *m/z* 131 ($[H_2C{=}CH{-}C_6H_4{-}C{=}O]^+$) or a methyl radical (•$CH_3$) to an ion with *m/z* 194. In addition, the phenyl acylium ion ($[C_6H_5{-}C{=}O]^+$) with *m/z* 105 and the phenylium ion ($[C_6H_5]^+$) with *m/z* 77 are found. In negative-ion MS–MS, ketoprofen ($[M-H]^-$ with *m/z* 253) shows the loss of CO_2 to form an ion with *m/z* 209 (Kosjek et al., 2011). An additional fragment ion with *m/z* 197, consistent with the loss of ethene-1,2-dione (O=C=C=O), has been reported as well (Marchese et al., 2003).

In positive-ion MS–MS, naproxen ($[M+H]^+$ with *m/z* 231) shows a loss of HCOOH followed by a loss of a •CH_3, whereas in negative-ion MS–MS, naproxen ($[M-H]^-$ with *m/z* 229) shows fragment ions with *m/z* 185, *m/z* 169, and *m/z* 170, which are consistent with losses of CO_2 and subsequent loss of either methane (CH_4) or •CH_3. Ketorolac ($[M+H]^+$ with *m/z* 256, a formic acid derivative with a benzoylpyrrolizine substituent group) shows a predominant loss of HCOOH to an ion with *m/z* 210, a minor loss of C_6H_6, and an abundant benzoyl cation ($[C_6H_5{-}C{=}O]^+$) with *m/z* 105.

In negative-ion MS–MS, carprofen ($[M-H]^-$ with *m/z* 272, with a 6-chloro-9*H*-carbazole-2-yl substituent) shows the loss of CO_2 to *m/z* 228 and of HCOOH to *m/z* 226. Both ions show subsequent loss of hydrogen chloride (HCl). Fenoprofen ($[M-H]^-$ with *m/z* 241) shows primarily a loss of CO_2, as well as a phenoxide ion ($[C_6H_5O]^-$) with *m/z* 93. An additional fragment with *m/z* 185 ($[C_{13}H_{13}O]^-$) has been observed in a Q–TOF product-ion mass spectrum, and even though it is not readily understood, it is consistent with the loss of C_2O_2. Flurbiprofen ($[M-H]^-$ with *m/z* 243) shows subsequent losses of CO_2, hydrogen fluoride (HF), and ethene (C_2H_4) or C_6H_6 to fragment ions with *m/z* 199, *m/z* 179, *m/z* 151, and *m/z* 101. In addition, the loss of HCOOH followed by the loss of HF leads to fragment ions with 197 and 177, respectively. Ibuprofen ($[M-H]^-$ with *m/z* 205) is reported to primarily show the loss of CO_2 (Ferrer

& Thurman, 2005), whereas in ion-trap MS^2 the loss of HCOOH is observed (Vinci et al., 2006). The loss of both CO_2 and HCOOH from $[M-H]^-$ of ibuprofen has also been observed by others (Marchese et al., 2003). Interestingly, such fragmentation differences were not observed for the other compounds studied; mostly CO_2 losses were observed, as described earlier (Vinci et al., 2006).

Pirprofen ($[M+H]^+$ with *m/z* 252) shows fragment ions with *m/z* 210 due to the loss of propene ($CH_3CH{=}CH_2$) from the 2,5-dihydropyrrole ring, with *m/z* 206 due to the loss of HCOOH, as well as two radical ions, that is, *m/z* 179 due to the loss of propan-2-yl-1-carboxylic acid radical ($CH_3{}^\bullet CHCOOH$) from the *iso*-propanoic acid skeleton, and *m/z* 151 due to a subsequent loss of C_2H_4 from the 2,5-dihydropyrrole ring. Tiaprofenic acid ($[M+H]^+$ with *m/z* 261) shows three fragment ions, that is, an ion with *m/z* 215 due to the loss of HCOOH, the [5-(1-carboxyethyl)thiophen-2-yl]acylium ion ($[C_7H_7O_2S{-}C{=}O]^+$) with *m/z* 183 due to the loss of C_6H_6, and the benzoyl cation ($[C_6H_5{-}C{=}O]^+$) with *m/z* 105.

4.3.3.4 Pyrazolone Compounds

NSAIDs of the pyrazolone class, for example, propyphenazone, morazone, and aminophenazone, are analyzed in positive-ion mode. Propyphenazone ($[M+H]^+$ with *m/z* 231) yields three major fragments and a variety of less abundant ones. Based on accurate-mass data, the fragment ion with *m/z* 201 is consistent with the subsequent loss of two methyl radicals ($^\bullet CH_3$), whereas the ion with *m/z* 189 is due to the loss of the *iso*-propyl group as propene ($CH_3CH{=}CH_2$); the protonated *N*-methylethanimine ($[CH_3N{=}CHCH_3{+}H]^+$) with *m/z* 58 is due to a pyrazolone ring cleavage. Morazone ($[M+H]^+$ with *m/z* 378) yields just one major fragment ion with *m/z* 201 ($[C_{12}H_{13}N_2O]^+$) due to the loss of 3-methyl-2-phenylmorpholine. The major fragments of aminophenazone ($[M+H]^+$ with *m/z* 232) are the protonated *N,N*-dimethyl-2-(methaniminyl)prop-1-en-1-amine ($[C_6H_{12}N_2{+}H]^+$) with *m/z* 113 due to the loss of phenyl isocyanate ($C_6H_5{-}N{=}C{=}O$) and the *N,N*-dimethyl-2-(methaniminyl)prop-2-en-1-iminium ion ($[C_6H_{11}N_2]^+$) with *m/z* 111 due to the loss of *N*-phenylformamide ($C_6H_5{-}NHCHO$). Both ions are due to cleavages in the pyrazolone ring. At higher collision energies, secondary fragments of these two ions are observed. It may be concluded that no common trends were found in the fragmentation of these pyrazolone compounds.

The most abundant fragment ions of metamizole ($[M-H]^-$ with *m/z* 310, also called dipyrone) are the ions with *m/z* 191 due to the loss of phenyl isocyanate ($C_6H_5{-}N{=}C{=}O$), with *m/z* 175 due to a subsequent loss of methane (CH_4), with *m/z* 81 (the hydrogen sulfonate anion $[HSO_3]^-$) at low collision energy, and with *m/z* 80 (sulfonate radical anion ($[SO_3]^{-\bullet}$) at higher collision energy. Minor fragments are the ion with *m/z* 295 due to the loss of $^\bullet CH_3$ and the ions with *m/z* 216 due to the loss of $H_2CSO_3{}^\bullet$ and *m/z* 94 due to $[H_2CSO_3]^{-\bullet}$, both involving the *N*-methylene sulfonic acid moiety.

4.3.3.5 Butazone NSAID, Pyrazolidine-3,5-Dione Derivatives

A third group of NSAIDs have a pyrazolidine-3,5-dione group, for example, phenylbutazone and its metabolite oxyphenylbutazone, mofebutazone, and suxibuzone. The butazone NSAIDs can be analyzed in both positive-ion and negative-ion modes.

Proposed structures of the major positive-ion fragments of phenylbutazone ($[M+H]^+$ with *m/z* 309) are given in Figure 4.3.5. The (1,2-diphenylhydrazinyl)acylium ion ($[C_6H_5NH{-}C_6H_5N{-}C{=}O]^+$) with *m/z* 211 is due to the loss of 2-hexenal. The ion with *m/z* 190 ($[C_{12}H_{16}NO]^+$) is due to the loss of phenyl isocyanate ($C_6H_5{-}N{=}C{=}O$). The ion with *m/z* 188 ($[C_{12}H_{14}NO]^+$) is due to the loss of *N*-phenylformamide ($C_6H_5{-}NHCHO$); ring closure to a 2-piperidinone ring is proposed. Carbon monoxide (CO) extrusion from the ion with *m/z* 188 results in the ion with *m/z* 160. The protonated phenyl isocyanate (or anilinoacylium ion) with *m/z* 120 is also observed.

Suxibuzone ($[M+H]^+$ with *m/z* 439) shows the loss of water to an ion with *m/z* 421, the loss of (carboxymethyl)ketene ($HOOHCH_2CH{=}C{=}O$) to an ion with *m/z* 339, and the complementary acylium ion with *m/z* 101 ($[HOOCCH_2CH_2C{=}O]^+$). Two secondary ions are produced from *m/z* 339, that is, an ion with *m/z* 321 due to the loss of water and an ion with *m/z* 309 (protonated phenylbutazone) due to the loss of formaldehyde ($H_2C{=}O$). In addition, the molecular ion of the hexyl succinate ($[C_{10}H_{18}O_4]^{+\bullet}$) with *m/z* 202 is observed and also the ion with *m/z* 160, which is also found for phenylbutazone (Figure 4.3.5).

Interestingly, both mofebutazone ($[M+H]^+$ with *m/z* 233; with one instead of two *N*-phenyl substituents) and oxyphenylbutazone ($[M+H]^+$ with *m/z* 325; with a 4-hydroxy group on the N^1-phenyl substituent) show quite different fragmentation than phenylbutazone. Mofebutazone shows the loss of propane (C_3H_8) from the butyl substituent of the 1-pyrazolidine-3,5-dione group to form a fragment ion with *m/z* 189. The molecular ion of phenylhydrazine ($[C_6H_5{-}NH{-}NH_2]^{+\bullet}$) with *m/z* 108 seems to result from the loss of CO and the hex-2-enoyl radical ($C_6H_9O^\bullet$). Oxyphenylbutazone shows an ion with *m/z* 297 due to the loss of CO, an ion with *m/z* 232 due to the loss of aniline ($C_6H_5{-}NH_2$), and an ion with *m/z* 204 due to the loss of both CO and $C_6H_5{-}NH_2$. In addition, oxyphenylbutazone has three fragments in common with phenylbutazone. These are the ions with *m/z* 120, *m/z* 160, and *m/z* 227 (a +16 (O-atom) shift relative to *m/z* 211 due to the N^1-4-hydroxy O-atom of the phenyl substituent) (Figure 4.3.5).

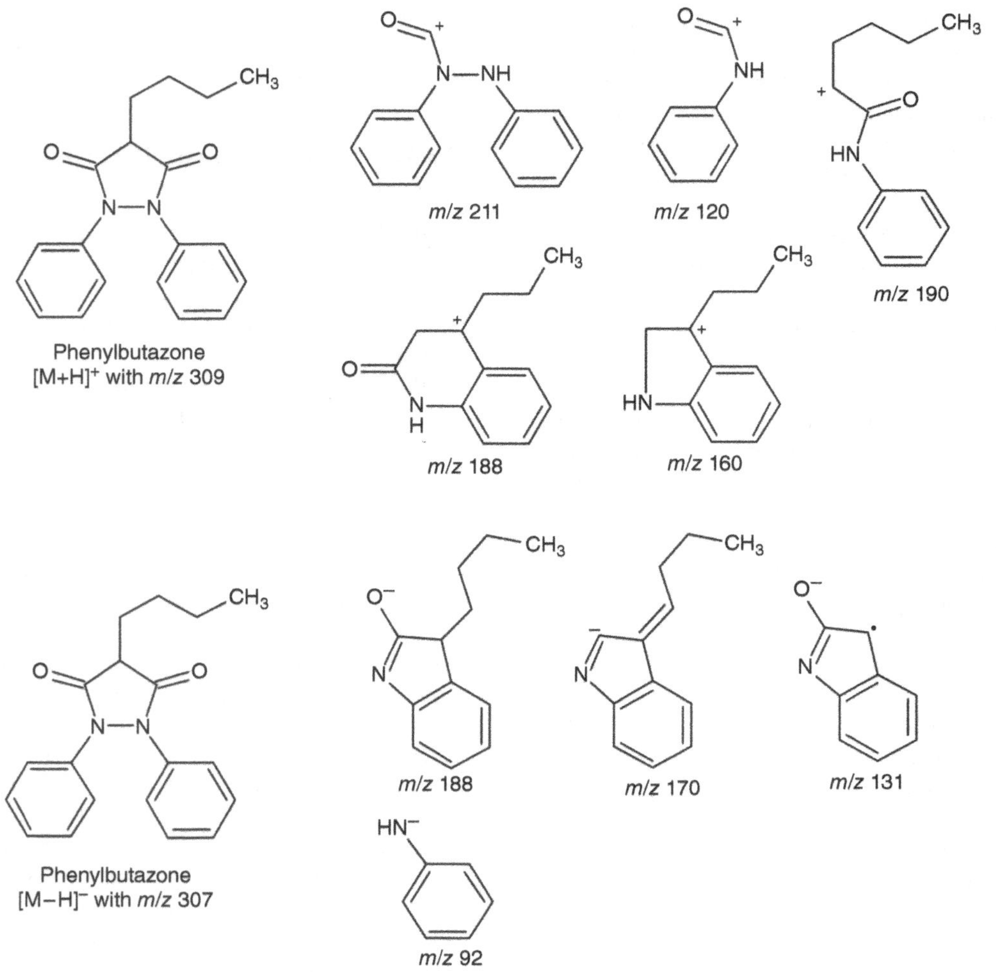

FIGURE 4.3.5 Proposed structures for the fragment ions of protonated and deprotonated phenylbutazone.

In negative-ion mode, phenylbutazone ($[M-H]^-$ with *m/z* 307) shows the loss of C_6H_5—N=C=O to produce a substituted 3*H*-indole fragment ion with *m/z* 188 ($[C_{12}H_{14}NO]^-$). Secondary fragmentation of this ion involves the loss of water to an ion with *m/z* 170 ($[C_{12}H_{12}N]^-$) or the homolytic loss of a butyl radical ($^{\bullet}C_4H_9$) to an ion with *m/z* 131 ($[C_8H_5NO]^{-\bullet}$), both of which conserve the 3*H*-indole ring structure. The anilinide ion ($[C_6H_5NH]^-$) with *m/z* 92 can be produced along different routes. Structure proposals for these fragment ions are given in Figure 4.3.5. With oxyphenylbutazone ($[M-H]^-$ with *m/z* 323), the fragment ions with *m/z* 147, *m/z* 186, and *m/z* 204 show a +16 (O) shift relative to the phenylbutazone fragment ions with *m/z* 131, *m/z* 170, and *m/z* 188, respectively. This is due to the N^1-4-hydroxy O-atom of the phenyl substituent. Mofebutazone ($[M-H]^-$ with *m/z* 231) shows $[C_6H_5NH]^-$ with *m/z* 92, the phenyldiazenide anion ($[C_6H_5$—N=N$]^-$) with *m/z* 105, and an ion with *m/z* 133 ($[C_7H_5N_2O]^-$), which can be described as deprotonated 3*H*-indazol-3-ol.

4.3.4 COX-2 INHIBITORS

The cyclooxygenase-2 enzyme (COX-2) inhibitors piroxicam ($[M+H]^+$ with *m/z* 332), tenoxicam ($[M+H]^+$ with *m/z* 338), isoxicam ($[M+H]^+$ with *m/z* 336), and meloxicam ($[M+H]^+$ with *m/z* 352) are NSAIDs belonging to the oxicam group. Most oxicams are benzothiazine carboxamide derivatives, but tenoxicam is a thienothiazine carboxamide derivative. In product-ion mass spectra, these compounds show three characteristic fragment ions with charge retention either on the 2-pyridinyl containing fragment (piroxicam and tenoxicam), that is, protonated 2-aminopyridine ($[C_5H_4N$—$NH_2+H]^+$) with *m/z* 95, protonated 2-isocyanatopyridine

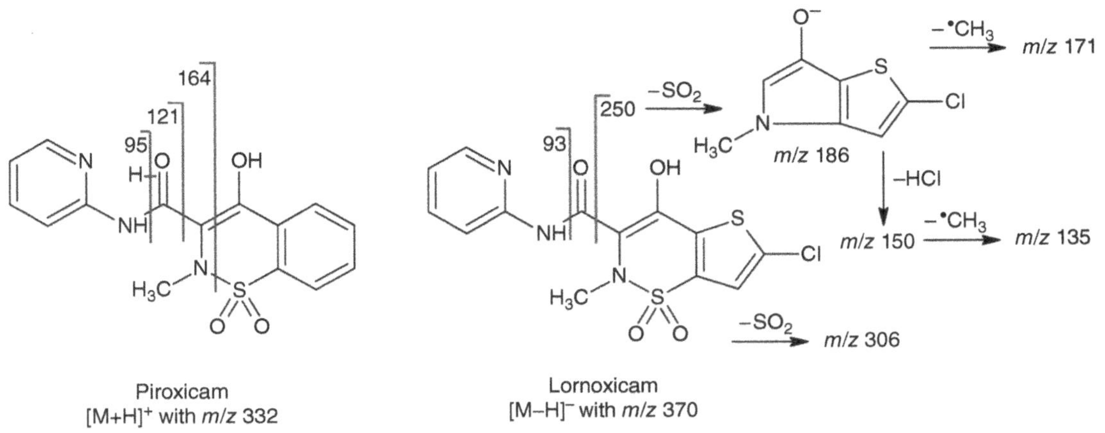

FIGURE 4.3.6 Fragmentation schemes for protonated piroxicam and deprotonated lornoxicam.

($[C_5H_4N{-}N{=}C{=}O+H]^+$) with *m/z* 121, and protonated 2-(methylimino)-*N*-(pyridin-2-yl)acetamide ($[C_8H_9N_3O+H]^+$) with *m/z* 164, or on the 5-methyl-1,2-oxazol-3-yl group (isoxicam) and the 5-methyl-1,3-thiazol-2-yl group (meloxicam) of the molecule, that is, the 5-methyl-3*H*-1,3-thiazolium cation ($[C_4H_5NS]^+$) with *m/z* 99, the protonated 3-isocyanato-5-methyl-1,2-oxazole with *m/z* 125 ($[C_5H_4N_2O_2+H]^+$), and the protonated 2-(methylimino)-*N*-(5-methyl-1,2-oxazol-3-yl)acetamide with *m/z* 168 ($[C_7H_9N_3O_2+H]^+$). The fragmentation for piroxicam is outlined in Figure 4.3.6 (Ji et al., 2005).

In negative-ion mode, the COX-2 inhibitors isoxicam ($[M-H]^-$ with *m/z* 334), lornoxicam ($[M-H]^-$ with *m/z* 370), piroxicam ($[M-H]^-$ with *m/z* 330), and tenoxicam ($[M-H]^-$ with *m/z* 336) show similar fragmentation as well. The loss of sulfur dioxide (SO_2) from the sulfonamide group leads to ions with *m/z* 270, *m/z* 306, *m/z* 266, and *m/z* 272, respectively. Cleavage of the C—C bond of the acetamide, that is, the loss of 3-isocyanato-5-methyl-1,2-oxazole for isoxicam, the loss of 2-isocyanatopyridine for lornoxicam, piroxicam, and tenoxicam, with charge retention on the benzothiazine group, results in fragment ions with *m/z* 210, *m/z* 250, *m/z* 210, and *m/z* 216, respectively. These fragments show secondary fragmentation by subsequent loss of SO_2 and a methyl radical ($^\bullet CH_3$), which for isoxicam and piroxicam results in common fragment ions with *m/z* 146 and 131. The general fragmentation routes for lornoxicam are outlined in Figure 4.3.6. In addition to the fragment ions with *m/z* 186 and 171 due to secondary losses of SO_2 and $CH_3^\bullet$ from the ion with *m/z* 250, lornoxicam shows an additional secondary route due to the loss of hydrogen chloride (HCl) to fragment ions with *m/z* 150 and 135. Lornoxicam, piroxicam, and tenoxicam show a common fragment ion, the deprotonated 2-pyridinylamine ($[C_5H_4N{-}NH]^-$) with *m/z* 93. The same fragmentation reaction for isoxicam results in deprotonated (5-methyl-1,2-oxazol-3-yl)amine with *m/z* 97 ($[C_4H_6N_2O-H]^-$).

Rofecoxib ($[M-H]^-$ with *m/z* 313, a 2-furanone derivative) shows two subsequent losses of carbon monoxide (CO) to fragment ions with *m/z* 285 and 257, respectively (Chavez-Eng et al., 2002). Both fragment ions show subsequent loss of SO_2 to ions with *m/z* 221 and 193. In addition, the methylsulfonyl anion ($[CH_3SO_2]^-$) with *m/z* 79 is observed. Extensive fragmentation is observed in the negative-ion product-ion mass spectrum of celecoxib ($[M-H]^-$ with *m/z* 380) (Zhang et al., 2000). The loss of SO_2 results in a fragment ion with *m/z* 316, which shows several secondary fragmentation routes, for example, three consecutive losses of hydrogen fluoride (HF) to fragment ions with *m/z* 296, *m/z* 276, and *m/z* 256, the loss of the trifluoromethyl radical ($^\bullet CF_3$) to a fragment ion with *m/z* 247, and the loss of 1*H*-benzazirine (C_6H_5N, 91 Da) to a fragment ion with *m/z* 225. The deprotonated sulfuramidous acid ($[H_2NSO_2]^-$) with *m/z* 80 and an abundant trifluoromethanide anion with *m/z* 69 ($[CF_3]^-$) are also observed (Zhang et al., 2000).

In positive-ion mode, valdecoxib ($[M+H]^+$ with *m/z* 315) shows a variety of fragments. Often, both $OE^{+\bullet}$ and EE^+ fragments are found with similar relative abundance. Primary losses involve the loss of ethenone ($H_2C{=}C{=}O$) from the 1,2-oxazole ring to an ion with *m/z* 273, the loss of the sulfuramidous acid radical ($H_2NSO_2^\bullet$) or sulfuramidous acid (H_2NSO_2H) to ions with *m/z* 235 and 234, respectively, and the loss of (most likely) $H_2NSO_2^\bullet$ and a methyl radical ($^\bullet CH_3$) to an ion with *m/z* 220. Structure proposals for some of the other fragment ions are given in Figure 4.3.7. Next to the EE^+ fragment ions with *m/z* 192 and 131, $OE^{+\bullet}$ fragment ions with *m/z* 193 and 191 and with *m/z* 132 are observed.

Our interpretation of the negative-ion product-ion mass spectrum of valdecoxib ($[M-H]^-$ with *m/z* 313) differs from the one given in the literature (Zhang et al., 2003). A comparison is given in Figure 4.3.7. Valdecoxib shows

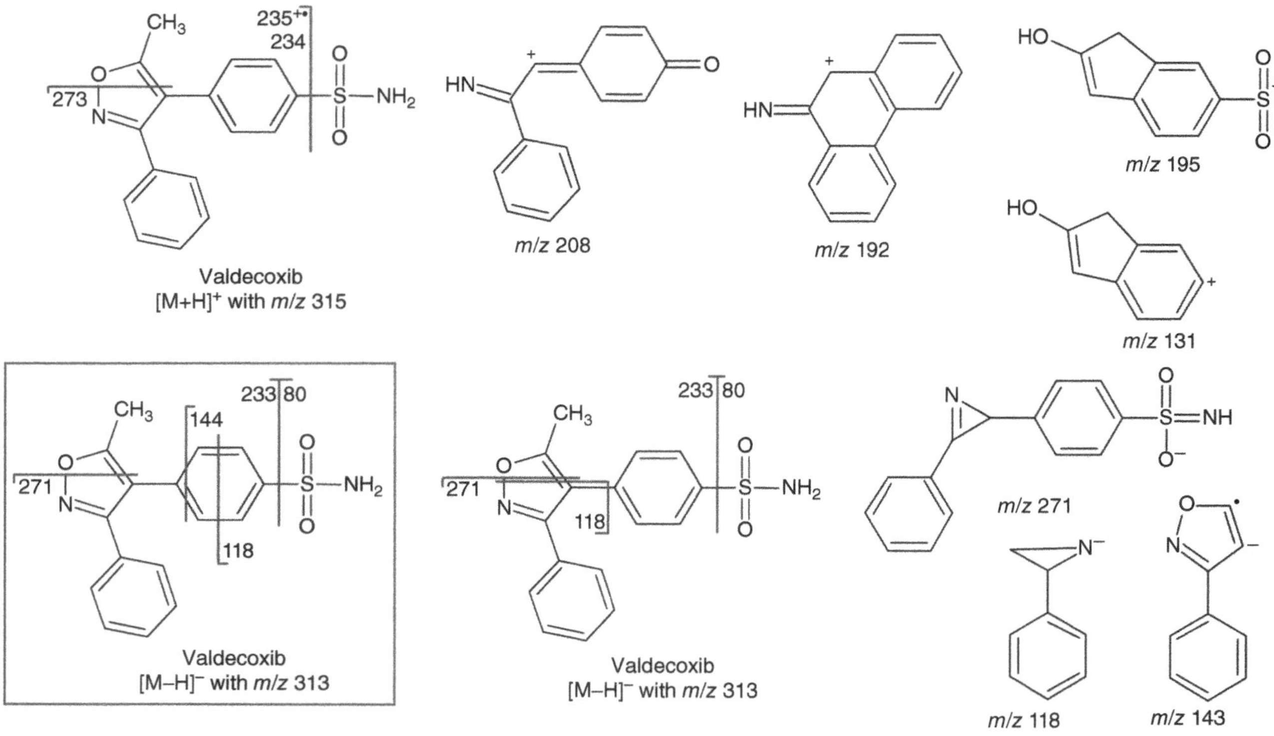

FIGURE 4.3.7 Fragmentation schemes and proposed structures for some of the fragment ions of protonated and deprotonated valdecoxib. The framed structure shows the interpretation of the negative-ion product-ion mass spectrum proposed elsewhere (Zhang et al., 2003). (Source: Adapted from Niessen, 2012.)

minor fragments due to the losses of $H_2C{=}C{=}O$, SO_2, iminosulfene ($HNSO_2$), and $H_2NSO_2^{\bullet}$ to fragment ions with *m/z* 271, *m/z* 249, *m/z* 234, and *m/z* 233, respectively. The base peak, however, is due to a fragment ion with *m/z* 118, most likely deprotonated 2-phenylaziridine ($[C_8H_9N{-}H]^-$), although $[C_3H_3SO_2NH]^-$ has been suggested as an alternative due to cleavages of the benzene ring (Zhang et al., 2003). No accurate-mass data were available to discriminate between these two possibilities. For valdecoxib metabolites, a novel intramolecular S_N2 rearrangement in the 1,2-oxazole ring was described (Zhang et al., 2004), which leads to quite different fragmentation routes for the metabolites.

The COX-2 inhibitor nimesulide ($[M{-}H]^-$ with *m/z* 307) shows the loss of sulfene (H_2CSO_2) to a fragment ion with *m/z* 229 and the methylsulfonyl anion with *m/z* 79 ($[CH_3SO_2]^-$). At higher collision energies, extensive fragmentation occurs, for example, the formation of the nitrophenyl anion ($[C_6H_4{-}NO_2]^-$) with *m/z* 122, an ion with *m/z* 198 due to losses of the methanesulfonyl radical ($CH_3SO_2^{\bullet}$) and the nitrosyl radical ($^{\bullet}NO$) from $[M{-}H]^-$, and the ions with *m/z* 170 and 142 due to two subsequent losses of carbon monoxide (CO) from the ion with *m/z* 198. The formation of the ion with *m/z* 142 requires a skeletal rearrangement of the molecular structure.

REFERENCES

Chavez-Eng CM, Constanzer ML, Matuszewski BK. 2002. High-performance liquid chromatographic–tandem mass spectrometric evaluation and determination of stable isotope labeled analogs of rofecoxib in human plasma samples from oral bioavailability studies. J Chromatogr B 767: 117–129.

Dieckhaus CM, Fernández-Metzler CL, King R, Krolikowski PH, Baillie TA. 2005. Negative ion tandem mass spectrometry for the detection of glutathione conjugates. Chem Res Toxicol 18: 630–638.

Dowling G, Gallo P, Malone E, Regan L. 2009. Rapid confirmatory analysis of non-steroidal anti-inflammatory drugs in bovine milk by rapid resolution liquid chromatography tandem mass spectrometry. J Chromatogr A 1216: 8117–8131.

Dowling G, Malone E, Harbison T, Martin S. 2010. Analytical strategy for the determination of non-steroidal anti-inflammatory drugs in plasma and improved analytical strategy for the determination of authorized and non-authorized non-steroidal anti-inflammatory drugs in milk by LC–MS/MS. Food Addit Contam A 27: 962–982.

Dubreil-Chéneau E, Pirotais Y, Bessiral M, Roudaut B, Verdon E. 2011. Development and validation of a confirmatory method for the determination of 12 non-steroidal anti-inflammatory drugs in milk using liquid chromatography–tandem mass spectrometry. J Chromatogr A 1218: 6292–6301.

Evans DC, Watt AP, Nicoll-Griffith DA, Baillie TA. 2004. Drug-protein adducts: An industry perspective on minimizing the potential for drug bioactivation in drug discovery and development. Chem Res Toxicol 17: 3–16.

Farré M, Petrovic M, Barceló D. 2007. Recently developed GC/MS and LC/MS methods for determining NSAIDs in water samples. Anal Bioanal Chem 387: 1203–1214.

Ferrer I, Thurman EM. 2005. Measuring the mass of an electron by LC/TOF-MS: A study of "twin ions". Anal Chem 77: 3394–3400.

Gallo P, Fabbrocino S, Vinci F, Fiori M, Danese V, Serpe L. 2008. Confirmatory identification of sixteen non-steroidal anti-inflammatory drug residues in raw milk by liquid chromatography coupled with ion trap mass spectrometry. Rapid Commun Mass Spectrom 22: 841–854.

Ji HY, Lee HW, Kim YH, Jeong DW, Lee HS. 2005. Simultaneous determination of piroxicam, meloxicam and tenoxicam in human plasma by liquid chromatography with tandem mass spectrometry. J Chromatogr B 826: 214–219.

Kosjek T, Zigon D, Kralj B, Heath E. 2008. The use of quadrupole-time-of-flight mass spectrometer for the elucidation of diclofenac biotransformation products in wastewater. J Chromatogr A 1215: 57–63.

Kosjek T, Perko S, Heath E, Kralj B, Žigon D. 2011. Application of complementary mass spectrometric techniques to the identification of ketoprofen phototransformation products. J Mass Spectrom 46: 391–401.

Malone EM, Dowling G, Elliott CT, Kennedy DG, Regan L. 2009. Development of a rapid, multi-class method for the confirmatory analysis of anti-inflammatory drugs in bovine milk using liquid chromatography tandem mass spectrometry. J Chromatogr A 1216: 8132–8140.

Marchese S, Gentili A, Perret D, Ascenzo GD, Pastori F. 2003. Quadrupole time-of-flight versus triple-quadrupole mass spectrometry for the determination of non-steroidal anti-inflammatory drugs in surface water by liquid chromatography/tandem mass spectrometry. Rapid Commun Mass Spectrom 17: 879–886.

Niessen WMA. 2012. Fragmentation of toxicologically relevant drugs in negative-ion liquid chromatography – tandem mass spectrometry. Mass Spectrom Rev, 31: 626–665.

Petrović M, Hernando MD, Díaz-Cruz MS, Barceló D. 2005. Liquid chromatography–tandem mass spectrometry for the analysis of pharmaceutical residues in environmental samples: A review. J Chromatogr A 1067: 1–14.

Suenami K, Wah Lim L, Takeuchi T, Sasajima Y, Sato K, Takekoshi Y, Kanno S. 2006a. Direct determination of non-steroidal anti-inflammatory drugs by column-switching LC-MS. J Sep Sci 29: 2725–2732.

Suenami K, Lim LW, Takeuchi T, Sasajima Y, Sato K, Takekoshi Y, Kanno S. 2006b. Rapid and simultaneous determination of nonsteroidal anti-inflammatory drugs in human plasma by LC–MS with solid-phase extraction. Anal Bioanal Chem 384: 1501–1505.

Van Hoof N, De Wasch K, Poelmans S, Noppe H, De Brabander HF. 2004. Multi-residue liquid chromatography/tandem mass spectrometry method for the detection of non-steroidal anti-inflammatory drugs in bovine muscle: Optimisation of ion trap parameters. Rapid Commun Mass Spectrom 18: 2823–2829.

Vinci F, Fabbrocino S, Fiori M, Serpe L, Gallo P. 2006. Determination of fourteen non-steroidal anti-inflammatory drugs in animal serum and plasma by liquid chromatography/mass spectrometry. Rapid Commun Mass Spectrom 20: 3412–3420.

Williams JP, Nibbering NMM, Green BN, Patel VJ, Scrivens JH. 2006. Collision-induced fragmentation pathways including odd-electron ion formation from desorption electrospray ionisation generated protonated and deprotonated drugs derived from tandem accurate mass spectrometry. J Mass Spectrom 41: 1277–1286.

Zhang JY, Wang Y, Dudkowski C, Yang D, Chang M, Yuan J, Paulson SK, Breau AP. 2000. Characterization of metabolites of Celecoxib in rabbits by liquid chromatography/tandem mass spectrometry. J Mass Spectrom 35: 1259–1270.

Zhang JY, Fast DM, Breau AP. 2003. Determination of valdecoxib and its metabolites in human urine by automated solid-phase extraction-liquid chromatography–tandem mass spectrometry. J Chromatogr B 785: 123–134.

Zhang JY, Xu F, Breau AP. 2004. Collision-induced dissociation of valdecoxib metabolites: A novel rearrangement involving an isoxazole ring. J Mass Spectrom 39: 295–302.

4.4

FRAGMENTATION OF DRUGS RELATED TO DIGESTION AND THE GASTROINTESTINAL TRACT

4.4.1 ANTIDIABETIC DRUGS

Antidiabetic drugs or oral hypoglycemic or antihyperglycemic agents are compounds effective against diabetes mellitus by way of lowering blood sugar levels. There are different classes of antidiabetic drugs. They are prescribed depending on the nature of the diabetes, the age and social condition of the patient, among other factors. Product-ion mass spectra for two classes of antidiabetic drugs are discussed here: (1) secretagogues, including sulfonylureas and meglitinides, and (2) sensitizers, including biguanides and thiazolidinediones. Multiresidue analysis of antidiabetic agents has been described for analysis in plasma (Maurer et al., 2002; Wang & Miksa, 2007) and in equine urine (Ho et al., 2004).

4.4.1.1 First-Generation Sulfonylurea Antidiabetics

First-generation sulfonylurea antidiabetic agents include compounds such as carbutamide, tolazamide, and tolbutamide. Characteristic fragmentation of these compounds in positive-ion mode involves cleavages of the sulfonyl (SO_2) group S—N and S—C bonds in the sulfonamide moiety. Characteristic sulfonamide-type fragmentation is observed (Sections 3.6.6 and 3.7.7). For carbutamide ($[M+H]^+$ with *m/z* 272, Figure 4.4.1), cleavage of the S—N bond results in the (4-aminobenzene)sulfonyl cation ($[H_2NC_6H_4SO_2]^+$) with *m/z* 156 and cleavage of the S—C bond results in the 4-aminophenylium ion with *m/z* 92. A fragment ion with *m/z* 108 is also observed, which is due to the loss of sulfur monoxide (SO) from the fragment with *m/z* 156, after rearrangement in the SO_2 group (Section 3.6.7). In addition, protonated butylamine ($[C_4H_9NH_2+H]^+$) with *m/z* 74 and the butyl carbocation ($[C_4H_9]^+$) with *m/z* 57 are observed. Tolazamide and tolbutamide show similar fragmentation. In negative-ion mode, for carbutamide ($[M-H]^-$ with *m/z* 270), deprotonated (4-aminobenzene)sulfonamide ($[H_2N—C_6H_4—SO_2NH]^-$) with *m/z* 171 is formed, which gives secondary fragmentation to a deprotonated benzenediamine ($[H_2N—C_6H_4NH]^-$) with *m/z* 107 involving the loss of sulfur dioxide (SO_2) and rearrangement of the NH-group (Figure 4.4.1). Similar fragmentation in chlorpropamide ($[M-H]^-$ with *m/z* 275) and tolbutamide ($[M-H]^-$ with *m/z* 269) results in fragment ions with *m/z* 190 and 126, and *m/z* 170 and 106, respectively.

4.4.1.2 Second-Generation Sulfonylurea Antidiabetics

The second-generation sulfonylurea agents include glipizide, glyburide (glibenclamide), glimepiride, gliclazide, gliquidone, and glibornuride. As an example, the class-specific fragmentation of these compounds is outlined for glyburide ($[M+H]^+$ with *m/z* 494, glibenclamide) in Figure 4.4.2. The identity of the fragments has been confirmed by accurate-mass determination on a Q–TOF instrument (Radjenović et al., 2008). Again, the cleavages of the S—C and S—N bonds on the SO_2 group of the sulfonamide moiety are important fragmentation routes, yielding fragment ions with *m/z* 288 and 352, respectively. The loss of sulfur monoxide (SO) from the ion with *m/z* 352 results in an ion with *m/z* 304. Other compounds of this class show similar class-specific fragmentation, as summarized in Table 4.4.1. In negative-ion mode, gliclazide ($[M-H]^-$ with *m/z* 322) and glibornuride ($[M-H]^-$ with *m/z* 365) show fragmentation similar to carbutamide (Figure 4.4.1). Thus, deprotonated 4-methylbenzenesulfonamide ($[CH_3—C_6H_4—SO_2NH]^-$) with *m/z* 170 and deprotonated 4-methylaniline ($[CH_3—C_6H_4NH]^-$) with *m/z* 106 are formed. The latter ion is due to the loss of sulfur dioxide (SO_2) and rearrangement of the NH-group. The negative-ion fragmentation of other second-generation sulfonylurea agents glyburide ($[M-H]^-$ with *m/z* 492, glibenclamide), glimepiride ($[M-H]^-$ with *m/z* 489), gliquidone ($[M-H]^-$ with *m/z* 526), and glysoxepide ($[M-H]^-$ with *m/z* 448)

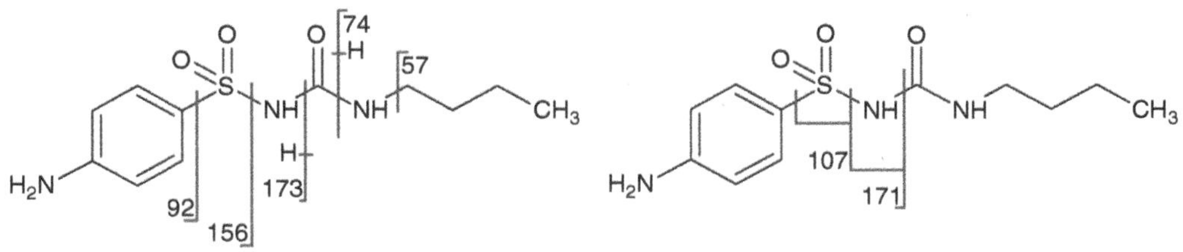

Carbutamide
[M+H]$^+$ with *m/z* 272

Carbutamide
[M–H]$^-$ with *m/z* 270

FIGURE 4.4.1 Fragmentation schemes for protonated and deprotonated carbutamide.

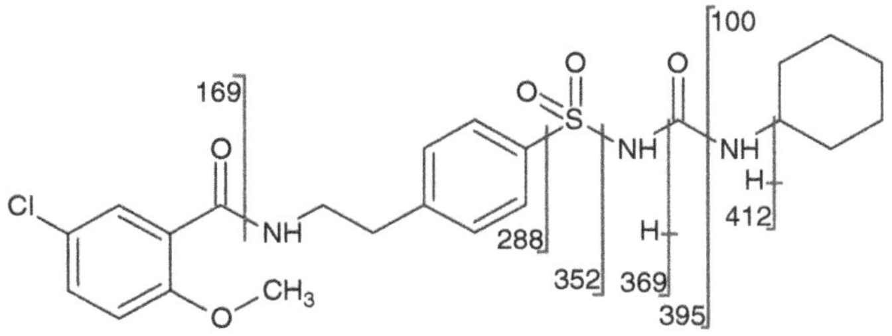

Glyburide
[M+H]$^+$ with *m/z* 494

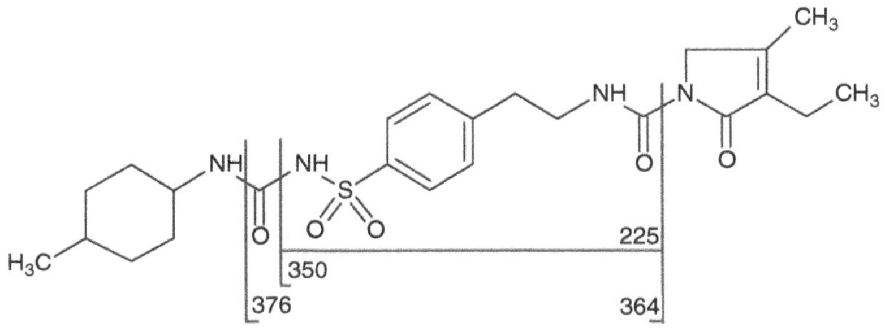

Glimepiride
[M–H]$^-$ with *m/z* 489

FIGURE 4.4.2 Fragmentation schemes for protonated glyburide and deprotonated glimepiride.

TABLE 4.4.1 Class-specific fragmentation for second-generation sulfonylurea antidiabetics.

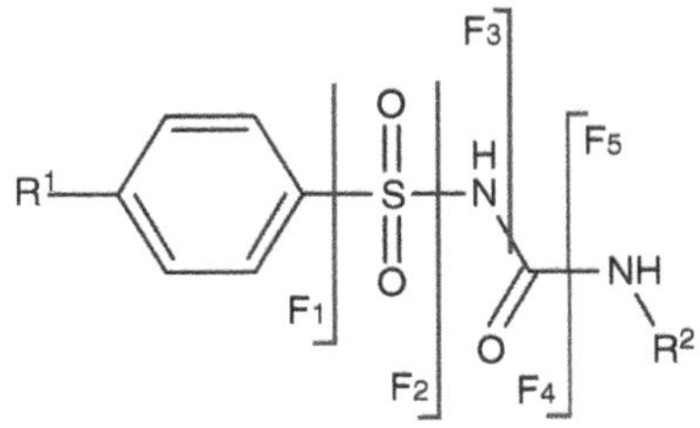

Compound	[M+H]$^+$	F_1	F_2	F_3	F_4	F_5
Glyburide, glibencamide	494	288	352	369	395	100
Gliclazide	324	91	155			127
Glibornuride	367	91				170
Glipizide	446		304	321	347	100
Glisoxepide	450		293	310		115
Glimepiride	491		335	352	378	114
Gliquidone	528	322	386	403		100

is only slightly different. The general behavior is demonstrated for glimepiride in Figure 4.4.2 (Bansal et al., 2008). Thus, cleavage of the carbonyl C—N bond to the sulfonamide group yields a fragment ion with a general structure $[R^1SO_2NH]^-$, that is, the fragment ion with *m/z* 350 for glimepiride, *m/z* 367 for glyburide, *m/z* 401 for gliquidone, and *m/z* 308 for glisoxepide. The latter compound also forms the deprotonated 4-(2-isocyanatoethyl)benzenesulfonamide with *m/z* 225 (identical to glimepiride). In addition, a 5-methyl-1,2-oxazol-3-ide ion with *m/z* 82 ($[C_4H_4NO]^-$) is formed for glisoxepide.

4.4.1.3 Other Antidiabetic Drugs

Extensive fragmentation is observed for the biguanide antidiabetic drug metformin ($[M+H]^+$ with *m/z* 130), with the formation of the complementary ions, that is, protonated dimethylaminonitrile with *m/z* 71 ($[(CH_3)_2N{-}C{\equiv}N+H]^+$) and protonated guanidine with *m/z* 60 ($[H_2NCNHNH_2+H]^+$) being most abundant (Figure 4.4.3). In addition, the fragmentation of metformin yields two other pairs of complementary ions, that is, protonated dimethylamine with *m/z* 46 ($[(CH_3)_2NH+H]^+$) and protonated *N*-cyanoguanidine ($[H_2NCNHNH{-}C{\equiv}N+H]^+$) with *m/z* 85, and protonated *N,N*-dimethylguanidine ($[(CH_3)_2NCNHNH_2+H]^+$) with *m/z* 88 and protonated methanediimine ($HN{=}C{=}NH+H]^+$) with *m/z* 43. Similar fragmentation is observed for phenformin ($[M+H]^+$ with *m/z* 206) (Wang et al., 2004).

The meglitinides are compounds used to treat type 2 diabetes. They are called insulin secretagogues because they stimulate (pro)insulin secretion by the pancreatic β-cell. Although nateglinide and repaglinide have clear structure differences, both of them are α-amino acid derivatives. They show extensive fragmentation. Nateglinide ($[M+H]^+$ with *m/z* 318) shows ions with *m/z* 300 due to the loss of water and *m/z* 272 due to the loss of formic acid (HCOOH). Cleavage of the amide N—C bond results in two complementary ions, that is, the protonated 2-amino-3-phenylpropanoic acid ($[C_9H_{11}NO_2+H]^+$) with *m/z* 166 and the 4-*iso*-propylcyclohexyl acylium ion ($[C_{10}H_{17}O]^+$) with *m/z* 153 (Figure 4.4.3). The ion with *m/z* 166 shows secondary fragmentation to *m/z* 120 due to the loss of HCOOH. The ion with *m/z* 153 shows secondary fragmentation to an ion with *m/z* 125 ($[C_9H_{17}]^+$) due to the loss of carbon monoxide (CO), which in turn shows fragment ions with *m/z* 83 ($[C_6H_{11}]^+$), *m/z* 69 ($[C_5H_9]^+$), and *m/z* 57 ($[C_4H_9]^+$). Repaglinide ($[M+H]^+$ with *m/z* 453) does not show primary fragment ions due to the cleavage of the amide N—C bond, but it does show the secondary fragmentation product of one of the resulting fragments: the ion with *m/z* 162 ($[C_{11}H_{16}N]^+$) is due to the loss of piperidine ($C_5H_{11}N$) from the (nonobserved) ion with *m/z* 247 ($[C_{16}H_{27}N_2]^+$). The protonated piperidine is observed with *m/z* 86. Cleavage of the amide N—C bond to the (*iso*-butyl)-2-(1-pyperidinyl)benzyl group results in the α-(*iso*-butyl)-2-(1-pyperidinyl)benzyl cation with *m/z* 230 ($[C_{16}H_{24}N]^+$). Finally, the loss of 1-phenylpiperidine ($C_{11}H_{15}N$) from $[M+H]^+$ results in an ion with *m/z* 292.

The 1,3-thiazolidine-2,4-dione or glitazone anti-type 2 diabetics include compounds such as rosiglitazone, pioglitazone, and troglitazone. In their product-ion mass spectra, these compounds show one major fragment due to the cleavage of the ether-O—C-aliphatic substituent bond. For rosiglitazone ($[M+H]^+$ with *m/z* 358), this fragmentation results in the 2-[*N*-methyl-*N*-(pyridin-2-yl)]ethyl carbocation ($C_5H_4N{-}NCH_3CH_2CH_2]^+$) with *m/z* 135 (Figure 4.4.3), for pioglitazone ($[M+H]^+$ with *m/z* 357) in the 2-(5-ethylpyridin-2-yl)ethyl carbocation

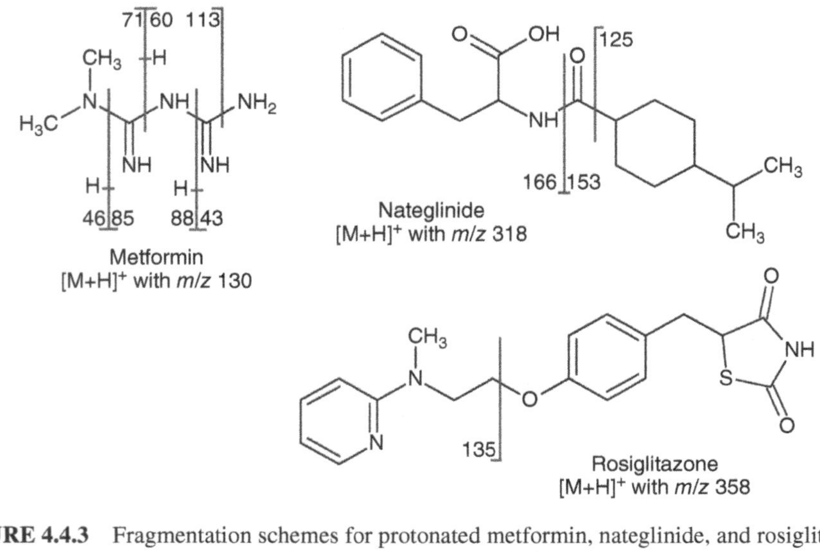

FIGURE 4.4.3 Fragmentation schemes for protonated metformin, nateglinide, and rosiglitazone.

($[C_2H_5—C_5H_4N—CH_2CH_2]^+$) with *m/z* 134. Troglitazone is studied because of its ability for form reactive metabolites (Kassahun et al., 2001; Dieckhaus et al., 2005).

4.4.2 ANTIULCER DRUGS

There are two major classes of antiulcer agents used in the treatment of various gastric diseases: H_2-histamine antagonist and proton-pump inhibitors. The use of proton-pump inhibitors has in fact largely superseded the use of H_2-receptor antagonists. Multiresidue analysis of antiulcer drugs, both H_2-histamine antagonist and proton-pump inhibitors, has been described (Chung et al., 2004).

4.4.2.1 Proton-Pump Inhibitors

Characteristic fragmentation of the proton-pump inhibitors involves the cleavage of the C—S bond between the benzimidazole heterocycle and the sulfoxide (SO) group with charge retention on the SO. The resulting fragment ions are found with *m/z* 198, 242, 200, and 252 for omeprazole ($[M+H]^+$ with *m/z* 346), rabeprazole ($[M+H]^+$ with *m/z* 360), pantoprazole ($[M+H]^+$ with *m/z* 384), and lansoprazole ($[M+H]^+$ with *m/z* 370), respectively. They are generally used in SRM for quantitative analysis. The fragmentation of omeprazole is outlined in Figure 4.4.4 (Boix et al., 2013). In addition to the characteristic fragment ion with *m/z* 198 for omeprazole, the complementary ion is also observed, that is, the protonated 5-methoxy-1*H*-benzimidazole ($[C_8H_8N_2O+H]^+$) with *m/z* 149. Cleavage of the S—C bond to the pyridine-containing group results in a radical cation, the molecular ion of 4-methoxy-2,3,5-trimethylpyridine ($[C_9H_{13}NO]^{+\bullet}$) with *m/z* 151 for omeprazole, and similar radical cations for the related compounds, that is, the molecular ion of 4-(3-methoxypropoxy)-2,3-dimethylpyridine ($[C_{11}H_{17}NO_2]^{+\bullet}$) with *m/z* 195 for rabeprazole, of 3,4-dimethoxy-2-methylpyridine ($[C_8H_{11}NO_2]^{+\bullet}$) with *m/z* 153 for pantoprazole, and of 2,3-dimethyl-4-(2,2,2-trifluoroethoxy)pyridine ($[C_9H_{10}F_3NO]^{+\bullet}$) with *m/z* 205 for lansoprazole. In the negative-ion product-ion mass spectrum of omeprazole ($[M-H]^-$ with *m/z* 344), two major fragment ions are observed involving cleavages of either of the C—S bonds to the SO group with charge retention on the benzimidazole heterocycle, that is, the 5-methoxybenzimidazol-1-ide ion ($[C_8H_7N_2O]^-$) with *m/z* 147 and the radical anion with *m/z* 194 ($[C_8H_6N_2O_2S]^{-\bullet}$). The ion with *m/z* 179 is due to the loss of a methyl radical ($^\bullet CH_3$) from the fragment ion with *m/z* 194.

4.4.2.2 H_2-Histamine Antagonists

H_2-histamine antagonists are used to reduce the secretion of gastric acid and are prescribed in the treatment of various gastric diseases. The multiresidue LC–MS of several H_2 antagonists has been reported for horse urine (Chung et al., 2004), in waste and surface water (Hernando et al., 2007), and in plasma (Sun et al., 2009). In these studies, target compounds are cimetidine ($[M+H]^+$ with *m/z* 253), famotidine ($[M+H]^+$ with *m/z* 338), lafutidine ($[M+H]^+$ with *m/z*

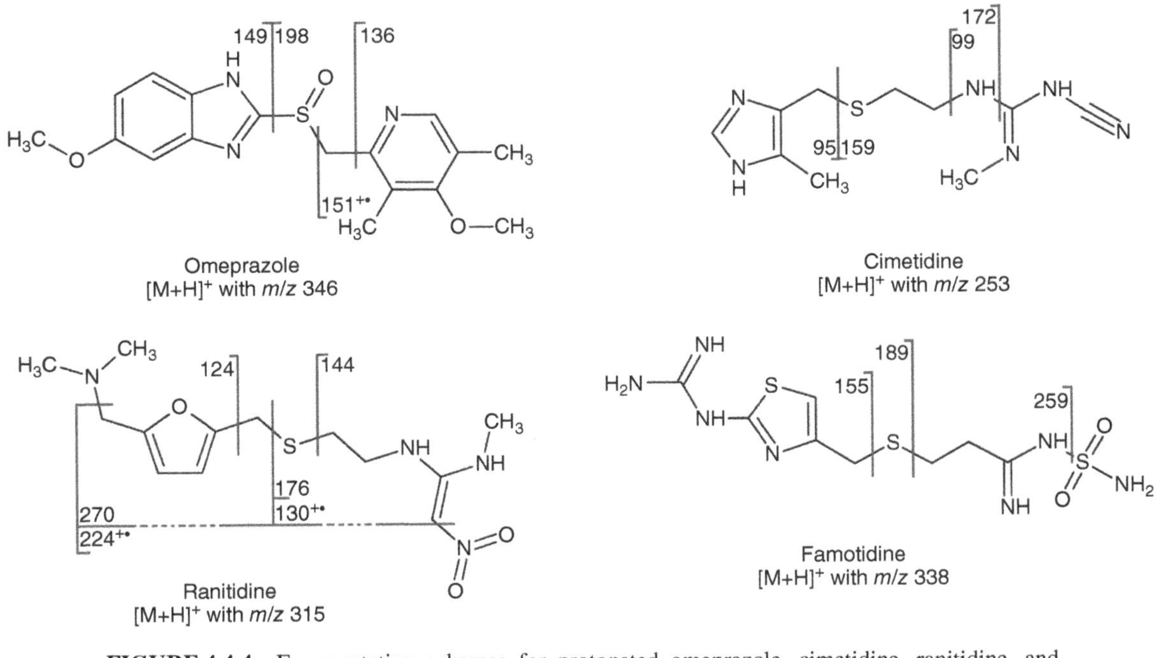

FIGURE 4.4.4 Fragmentation schemes for protonated omeprazole, cimetidine, ranitidine, and famotidine.

432), nizatidine ($[M+H]^+$ with *m/z* 332), and/or ranitidine ($[M+H]^+$ with *m/z* 315). The analysis of ranitidine played an important role in the history of LC–MS, because in the 1980s ranitidine served as a benchmark for LC–MS interface performance (Tomer & Parker, 1989).

Except for lafutidine (with a β-amidosulfoxide group), these compounds contain a thioether linkage, that is, $R^1CH_2SCH_2CH_2R^2$, which is prone to fragmentation. Cleavage of the R^1CH_2—$SCH_2CH_2R^2$ bond accompanied by a H-rearrangement results in the most abundant fragment ion with *m/z* 189 ($[C_5H_8N_4S_2+H]^+$) for famotidine, *m/z* 176 ($[C_5H_{11}N_3O_2S+H]^+$) for ranitidine, and *m/z* 159 ($[C_5H_{10}N_4S+H]^+$) for cimetidine (Figure 4.4.4). For famotidine, additional fragment ions are observed with *m/z* 259 due to the loss of iminosulfene ($HNSO_2$) and with *m/z* 155 due to the cleavage of the R^1CH_2S—$CH_2CH_2R^2$ bond (or the loss of H_2S from the ion with *m/z* 189) (Figure 4.4.4) (Qin et al., 1994). Additional fragment ions for ranitidine are due to losses of dimethylamine ($(CH_3)_2NH$) to a fragment ion with *m/z* 270 followed by the loss of a nitryl radical ($^{\bullet}NO_2$) to the radical cation with *m/z* 224. The loss of $^{\bullet}NO_2$ from the fragment ion with *m/z* 176 leads to the radical cation with *m/z* 130 (Hernando et al., 2007; Chung et al., 2004). Two other fragments include the 5-[(dimethylamino)methyl]furan-2-ylium ion ($[C_7H_{10}NO]^+$) with *m/z* 124 due to cleavage of the C—C bond to the furan ring-C^2 (R^1—$CH_2SCH_2CH_2R^2$ bond), and the ion with *m/z* 144 ($[C_5H_{10}N_3O_2]^+$) due to cleavage of the R^1CH_2S—$CH_2CH_2R^2$ bond (Figure 4.4.4). The fragmentation of cimetidine ($[M+H]^+$ with *m/z* 253) is also outlined in Figure 4.4.4. Next to the fragments indicated, additional fragment ions are observed, for example, an ion with *m/z* 211 due to the loss of methanediimine (HN=C=NH) from $[M+H]^+$, an ion with *m/z* 117 due to loss of HN=C=NH from the fragment with *m/z* 159, and an ion with *m/z* 82 due to the loss of NH_3 from the ion with *m/z* 99. MS^n fragmentation of lafutidine and its metabolites after microsomal incubation has been studied using a LIT–orbitrap hybrid instrument (Wang et al., 2008).

4.4.3 LIPID-LOWERING AGENTS

Raised or abnormally high levels of lipids and/or lipoproteins in the blood (hyperlipidemia), currently quite common due to obesity prevalence in many segments of the population, is considered a high risk factor for cardiovascular diseases. General treatment involves, along with dietary modification, the use of statins and/or fibrates. Often used in combination with simvastatin, ezetimibe ($[M-H]^-$ with *m/z* 409) is another antihyperlipidemic agent, which in its negative-ion product-ion mass spectrum shows only one major fragment ion with *m/z* 271 ($[C_{17}H_{16}FO_2]^-$) due to a cleavage of the carbonyl C—C^3 and the N^1—C^2 bonds of the β-lactam ring, thereby losing 4-fluorophenylisocyanate (FC_6H_4—N=C=O).

4.4.3.1 Fibrates

The fibrates are agonists of the peroxisome proliferator-activated receptor α (PPAR-α) and are used in the treatment of hypercholesterolemia (high blood cholesterol level). This compound class is frequently analyzed in (municipal) wastewater for environmental screening of pharmaceutical compounds (Garcia-Ac et al., 2009; Sousa et al., 2011). In positive-ion MS–MS, bezafibrate ($[M+H]^+$ with *m/z* 362) and fenofibrate ($[M+H]^+$ with *m/z* 361) show fragments due to cleavage of the aryl-O—C-alkyl bond of the ether linkage, which results in the protonated 4-[2-(4-chlorobenzamido)ethyl]phenol ($[C_{15}H_{14}ClNO_2+H]^+$) with *m/z* 276 due to the loss of 2-methylprop-2-enoic acid ($C_4H_6O_2$) for bezafibrate and in protonated 4-chloro-4′-hydroxybenzophenone ($[C_{13}H_9ClO_2+H]^+$) with *m/z* 233 due to the loss of *iso*-propyl 2-methylprop-2-enoate ($C_7H_{12}O_2$) for fenofibrate (Figure 4.4.5). In addition, common fragment ions are observed, that is, the 4-chlorophenylium ion ($[ClC_6H_4]^+$) with *m/z* 111 and the 4-chlorobenzoyl cation (ClC_6H_4—C=O]$^+$) with *m/z* 139. In addition, the α-(methyl)-4-hydroxybenzyl cation ([HO—$C_6H_4CHCH_3]^+$) with *m/z* 121 for bezafibrate is isobaric with the 4-hydroxybenzoyl cation ([HO—C_6H_4—C=O]$^+$) from fenofibrate (Figure 4.4.5).

The negative-ion product-ion mass spectrum of bezafribrate ($[M-H]^-$ with *m/z* 360) shows the deprotonated 4-[2-(4-chlorobenzamido)ethyl]phenol ion ($[C_{15}H_{14}ClNO_2-H]^-$) with *m/z* 274 and the deprotonated 2-methylprop-2-enoic acid ($[C_4H_6O_2-H]^-$) with *m/z* 85 due to cleavage of the aryl-O—C-alkyl bond of the ether moiety. The *m/z* values of these two ions add up to *m/z* ($[M-H]-1$) (Section 3.7.2). The loss of carbon dioxide (CO_2) from $[M-H]^-$ leads to a fragment ion with *m/z* 316. In addition, deprotonated 4-chlorobenzamide ([ClC_6H_4—$CONH_2-H]^-$) with *m/z* 154 and the radical anion with *m/z* 119 due to the loss of Cl$^{\bullet}$ from the ion with *m/z* 154 are observed (Figure 4.4.5). The cleavage of the aryl-O—C-alkyl ether linkage is also responsible for the major fragment ions in the negative-ion product-ion mass spectra of ciprofibrate ($[M-H]^-$ with *m/z* 288) and gemfibrozil ($[M-H]^-$ with *m/z* 249), that is, deprotonated 2-methylprop-2-enoic acid ($[C_4H_6O_2-H]^-$) with *m/z* 85 and the deprotonated 2,5-dimethylphenol ion with *m/z* 121 ($[(CH_3)_2C_6H_4O-H]^-$), respectively.

Etiroxate ($[M-H]^-$ with *m/z* 817.7) shows four major fragment ions (Figure 4.4.6). The cleavage of the hydroxy-diiodoaryl-O—C-diiodoaryl ether linkage leads to the 2,5-diiodo-4-oxyl-phenoxide radical anion ($[^{\bullet}OC_6H_2I_2O]^-$) with *m/z* 359.8. The radical anion

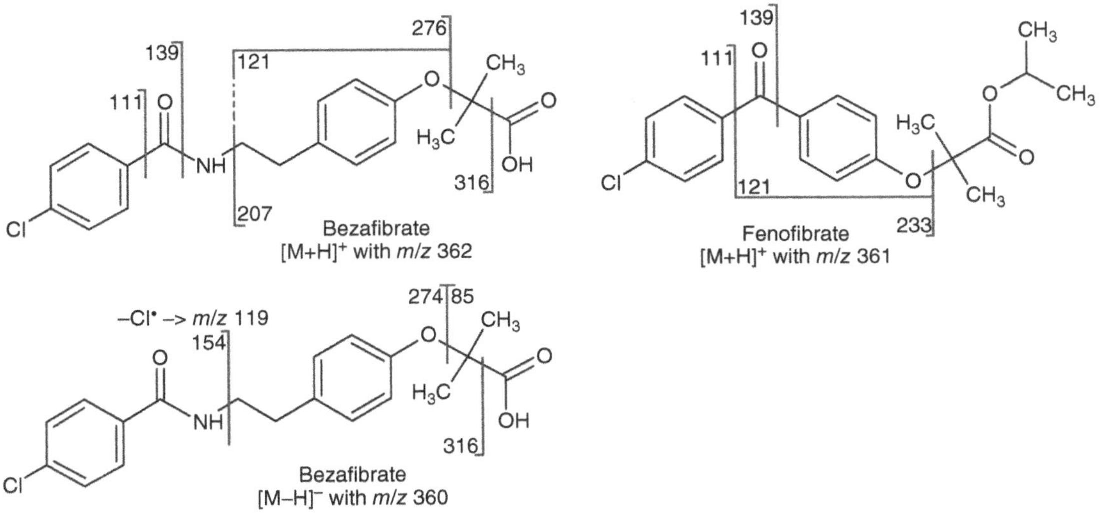

FIGURE 4.4.5 Fragmentation schemes for protonated and deprotonated bezafibrate and for protonated fenofibrate.

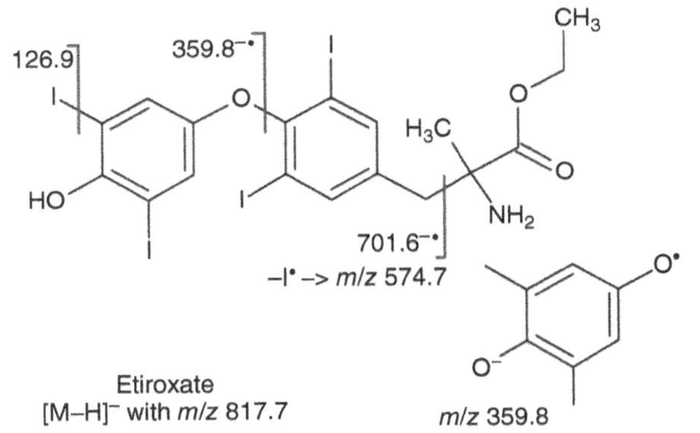

FIGURE 4.4.6 Fragmentation scheme for deprotonated etiroxate.

with *m/z* 701.7 is due to the loss of the ester ethyl 2-aminopropan-2-yl-oate radical ($CH_3{}^\bullet CNH_2COOC_2H_5$). The ion with *m/z* 701.7 shows a loss of iodine radical ($I^\bullet$) to an ion with *m/z* 574.8; the iodide anion (I^-) with *m/z* 126.9 is also observed.

4.4.3.2 Statins

Statins (or HMG-CoA reductase inhibitors) are a class of drugs used to lower plasma cholesterol levels by inhibition of the enzyme 3-hydroxy-3-methylglutaryl-coenzyme-A reductase, which is the rate-limiting-step enzyme of the mevalonate pathway of cholesterol synthesis. Inhibiting this enzyme in the liver results in a decrease in the synthesis of cholesterol and an increase in the synthesis of low-density lipoprotein (LDL) receptors, which in turn leads to an increase in the clearance of LDL from the bloodstream. Members of this class comprise synthetic compounds, such as atorvastatin, cerivastatin, fluvastatin, pitavastatin, and rosovastatin, and fermentation-derived compounds, such as lovastatin, mevastatin, and pravastatin. The most widely used statin, simvastatin, is a synthetic derivate of a fermentation product. These compounds can exist as either a 3-hydroxy lactone, analyzed in positive-ion mode, or a ring-opened dihydroxy acid, analyzed in negative-ion mode.

The positive-ion product-ion mass spectrum of simvastatin ($[M+H]^+$ with *m/z* 419) has been studied in detail using TQ, ion-trap, and Q–TOF instruments (Wang et al., 2001; Vuletić et al., 2005). The initial fragmentation involves losses of the 2,2-dimethylbutanoate group to the fragment ions with *m/z* 321 and 303, that is, involving the loss of 2,2-dimethylbut-3-enal ($C_6H_{10}O$) or the loss of 2,2-dimethylbutanoic acid ($C_6H_{12}O_2$), respectively (Figure 4.4.7). Further fragment ions are best considered as secondary fragmentation products of the ion with *m/z* 303 ($[C_{19}H_{27}O_3]^+$). A loss of water results in a fragment ion with *m/z* 285. This could either involve the opening of the δ-lactone ring or the loss of its hydroxy substituent, resulting in an α,β-double bond formation. The loss of acetic acid (CH_3COOH) results in an ion with *m/z* 243. The subsequent loss of ethanal (CH_3CHO) from the ion with *m/z* 243 results in a fragment ion with *m/z* 199 ($[C_{15}H_{19}]^+$), which can be written as a tricyclic structure (Figure 4.4.7). The loss of *iso*-butene (C_4H_8) from *m/z* 199 results in a fragment ion with *m/z* 143 ($[C_{11}H_{11}]^+$, *m/z* 143.086) (Wang et al., 2001). An alternative structure for the ion with *m/z* 143 has been suggested, involving the cleavage of the hexahydronaphthalene-C—C-ethyl-δ-lactone bond with charge retention on the lactone ring ($[C_7H_{11}O_3]^+$, *m/z* 143.070) (Figure 4.4.7). Accurate-mass data indicate that

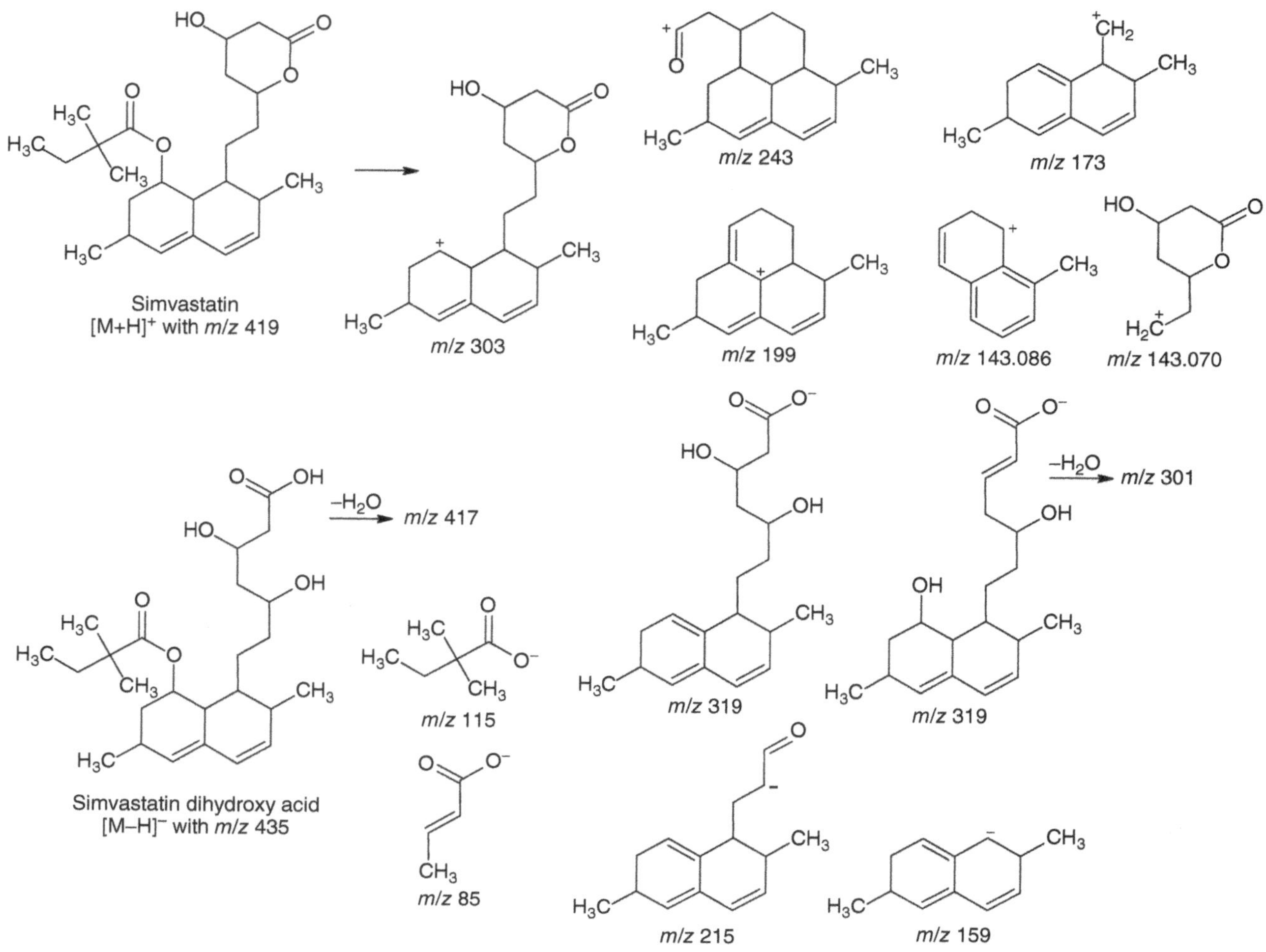

FIGURE 4.4.7 Proposed structures for the fragment ions of protonated simvastatin and deprotonated simvastatin dihydroxy acid.

both fragment ions are observed: the ion with $[C_7H_{11}O_3]^+$ at low collision energy and the ion with $[C_{11}H_{11}]^+$ at higher collision energy. This is according to expectations based on the fragmentation pathways of both ions, that is, a single 3°–2° C—C bond cleavage to form the ion with *m/z* 143.070, but a series of consecutive fragmentation reactions to form the ion with *m/z* 143.086. Furthermore, fragment ions are observed with *m/z* 267 (loss of water from the ion with *m/z* 285), *m/z* 225 (loss of water from the ion with *m/z* 243), and *m/z* 173 ($[C_{13}H_{17}]^+$) (Figure 4.4.7). Similar fragments are observed for related structures, such as lovastatin ($[M+H]^+$ with *m/z* 405), mevastatin ($[M+H]^+$ with *m/z* 391), and pravastatin ($[M+H]^+$ with *m/z* 425).

The negative-ion MS–MS spectra of the dihydroxy acids of simvastatin, lovastatin, pravastatin, and some oxidized related compounds were studied in detail using H/D-exchange and ^{18}O-labeling experiments (Qin, 2003). The two most abundant fragments for simvastatin dihydroxy acid ($[M-H]^-$ with *m/z* 435) are the ions with *m/z* 319 ($[C_{19}H_{27}O_4]^-$, two possible positional isomers) due to the loss of the ester substituent group as 2,2-dimethylbutanoic acid and 2,2-dimethylbutanoate with *m/z* 115 (Figure 4.4.7). The *m/z* values of these two ions add up to *m/z* ([M−H]−1) (Section 3.7.2). Next to ions due to the loss of water from the ions with *m/z* 435 and 319, three other fragment ions are observed, which can be considered secondary fragments of the ion with *m/z* 319, that is, ions with *m/z* 215, *m/z* 159, and *m/z* 85, all involving cleavages in the dihydroxy carboxylic acid group (Figure 4.4.7). Similar fragments are observed for lovastatin and pravastatin dihydroxy acids (Qin, 2003).

The other statins have different structures. As an example, the fragment ions in the positive-ion product-ion mass spectrum of atorvastatin ($[M+H]^+$ with *m/z* 559) are interpreted in Figure 4.4.8. Initial fragmentation results in losses of aniline ($C_6H_5NH_2$) or phenylisocyanate (C_6H_5—N=C=O) to form fragment ions with *m/z* 466 and 440, respectively. Both

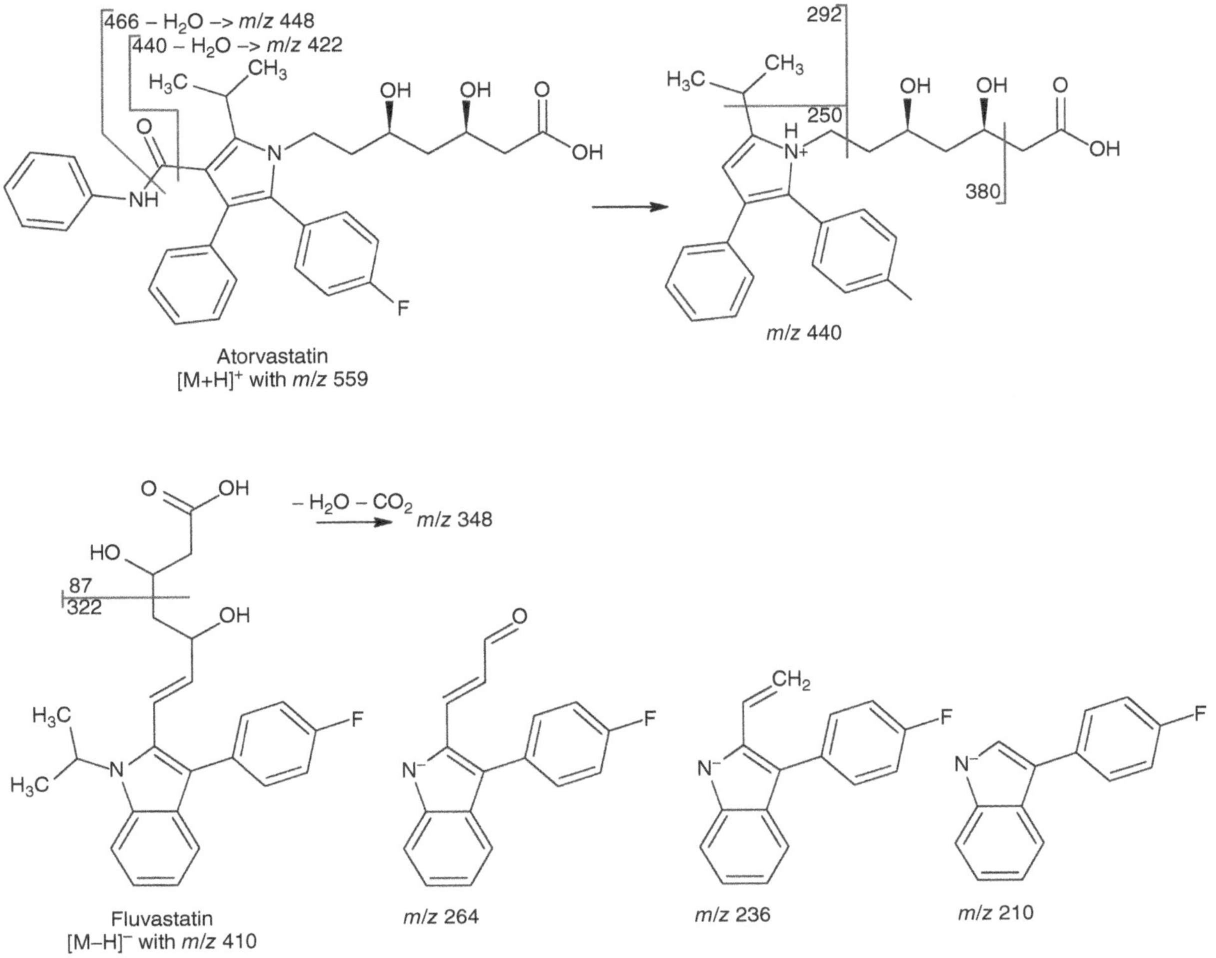

FIGURE 4.4.8 Fragmentation schemes for protonated atorvastatin and deprotonated fluvastatin.

fragments show the loss of water to ions with *m/z* 448 and 422, respectively. Further fragmentation may be considered as secondary fragmentation of the ion with *m/z* 440. The loss of acetic acid (CH_3COOH) results in a fragment ion with *m/z* 380, and the loss of 3,5-dihydroxyhexanoic acid ($C_6H_{12}O_4$) in an iminium ion with *m/z* 292 (Shah et al., 2008). The 2-(4-fluorophenyl)-1-methylidene-3-phenyl-1*H*-pyrrolium ion ($[C_{17}H_{13}FN]^+$) with *m/z* 250 results from the subsequent loss of propene ($CH_3CH{=}CH_2$) from the C^2-atom of the pyrrole ring.

Quite extensive fragmentation is observed in negative-ion MS–MS of fluvastatin ($[M-H]^-$ with *m/z* 410) (Figure 4.4.8). The combined loss of water and carbon dioxide (CO_2) results in an abundant fragment ion with *m/z* 348. The loss of 3-oxopropanoic acid ($C_3H_4O_3$) results in a fragment ion with *m/z* 322, while the deprotonated 3-oxopropanoic acid with *m/z* 87 is also observed. The *m/z* values of these two ions add up to *m/z* ($[M-H]-1$) (Section 3.7.2). The loss of propene ($CH_3CH{=}CH_2$) and other substituent losses lead to fragment ions with *m/z* 264, *m/z* 236, and finally the deprotonated 3-(4-fluorophenyl)-1*H*-indole ($[C_{14}H_{10}FN-H]^-$) with *m/z* 210 (Figure 4.4.8).

4.4.4 ANOREXIC DRUGS

Anorexic drugs are appetite-suppressing compounds, the use of which should result in weight loss. A wide variety of compounds may be listed as anorexic drugs, including a number of amphetamine derivatives, such as mefenorex, fenproporex, and dexfenfluramine. These compounds show fragmentation similar to amphetamine (Section 4.7.2.1), that is, formation of the α-(ethyl)benzyl cation ($[C_9H_{11}]^+$) with *m/z* 119 and of the tropylium ion ($[C_7H_7]^+$) with *m/z* 91. For dexfenfluramine ($[M+H]^+$ with *m/z* 232), which has a trifluoromethyl (—CF_3) substituent on the phenyl ring, these fragment ions are found with *m/z* 187 and 159 (*m/z* shift of 68). Norephedrine ($[M+H]^+$ with *m/z* 152; 2-amino-1-phenylpropan-1-ol) shows the loss of water to an ion with *m/z* 134, and subsequently the loss of ammonia

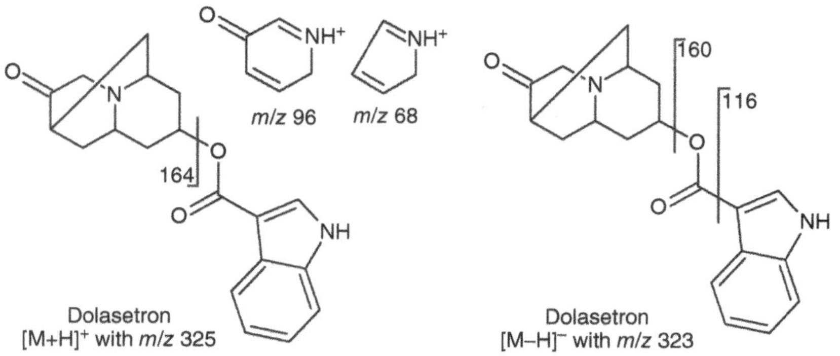

FIGURE 4.4.9 Fragmentation schemes for protonated and deprotonated dolasetron.

(NH_3) to the α-(vinyl)benzyl cation with *m/z* 117 ($[C_9H_9]^+$) and the formation of $[C_7H_7]^+$ with *m/z* 91. The ion with *m/z* 115 is probably better described as a ring closure to an 1*H*-inden-1-ylium ion rather than as an α-(ethynyl)benzyl cation ($[C_9H_7]^+$). Metamfepramone ($[M+H]^+$ with *m/z* 178, *N,N*-dimethyl cathinone, Section 4.7.2.4) shows three major fragment ions, that is, a 1-benzoylethyl carbocation ($[C_9H_9O]^+$) with *m/z* 133 due to the loss of dimethylamine ($(CH_3)_2NH$), the benzoyl cation with *m/z* 105 ($[C_6H_5{-}C{=}O]^+$), and the *N,N*-dimethylethaniminium ion ($[(CH_3)_2N{=}CHCH_3]^+$) with *m/z* 72.

The cyclobutane derivative sibutramine ($[M+H]^+$ with *m/z* 280) provides extensive fragmentation. Based on accurate-mass data, the most abundant ions with *m/z* 125, *m/z* 139, and *m/z* 153 can be interpreted as the 4-chlorobenzyl cation ($[ClC_6H_4CH_2]^+$), the α-(methyl)-4-chlorobenzyl cation ($[ClC_6H_4CHCH_3]^+$), and the α-(ethyl)-4-chlorobenzyl cation ($[ClC_6H_4CHCH_2CH_3]^+$), respectively. The [1-(4-chlorophenyl)cyclobutyl]methylium ion ($[C_{11}H_{12}Cl]^+$) with *m/z* 179 is due to losses of dimethylamine ($(CH_3)_2NH$) and *iso*-butene ($H_2C{=}C(CH_3)_2$) from $[M+H]^+$. Two minor fragments are the ions with *m/z* 97 ($[C_7H_{13}]^+$) and *m/z* 109 ($[C_8H_{13}]^+$).

4.4.5 ANTIEMETIC DRUGS

An antiemetic drug is used to treat vomiting and nausea. They are typically used to treat motion sickness and the side effects from chemotherapy directed against cancer. Our library collection contains negative-ion product-ion mass spectra for two classes of antiemetic drugs.

Dolasetron is a 5-hydroxytryptamine (5-HT_3) receptor antagonist antiemetic which acts by blocking serotonin receptors in the CNS and the gastrointestinal tract. Dolasetron ($[M+H]^+$) with *m/z* 325) and its related compounds are generally analyzed in positive-ion mode. The aminoketone azatricycloundecan-5-ylium ion with *m/z* 164 ($[C_{10}H_{14}NO]^+$) is due to the loss of 1*H*-indole-3-carboxylic acid. Secondary fragmentation of the ion with *m/z* 164 results in protonated 6*H*-pyridin-3-one with *m/z* 96 ($[C_5H_5NO+H]^+$) and protonated 2*H*-pyrrole with *m/z* 68 ($[C_4H_5N+H]^+$) (Figure 4.4.9). For tropisetron ($[M+H]^+$ with *m/z* 285), the loss of 1*H*-indole-3-carboxylic acid leads to a fragment ion with *m/z* 124 ($[C_8H_{14}N]^+$), which in turn forms a cycloheptadienylium ion ($[C_7H_9]^+$) with *m/z* 93 after the loss of methylamine (CH_3NH_2). For granisetron ($[M+H]^+$ with *m/z* 313), the loss of 1-methyl-indazole-3-carboxamide leads to a fragment ion with *m/z* 138 ($[C_9H_{16}N]^+$). Ondansetrone ($[M+H]^+$ with *m/z* 294) is a N^1-(methylcarbazol-4-one)-2-methylimidazole derivate. Its major fragment ions are the ion with *m/z* 212 ($[C_{14}H_{14}NO]^+$) due to the loss of 2-methyl-1*H*-imidazole ($C_4H_6N_2$), which is also observed as complementary ion with *m/z* 83 ($[C_4H_6N_2+H]^+$), the ion with *m/z* 184 due to the subsequent loss of carbon monoxide (CO), and the ion with *m/z* 170 ($[C_{13}H_{12}NO]^+$) due to the loss of CO and 1,2-dimethylimidazole from $[M+H]^+$. In its negative-ion product-ion mass spectrum, dolasetron ($[M-H]^-$ with *m/z* 323) shows the loss of the azatricycloundecanone group and the formation of the deprotonated 1*H*-indole-3-carboxylic acid ($[C_9H_7NO_2-H]^-$) with *m/z* 160, which in turn shows the loss of carbon dioxide (CO_2) to an ion with *m/z* 116 (Figure 4.4.9). The ions with *m/z* 160 and 114 also are the

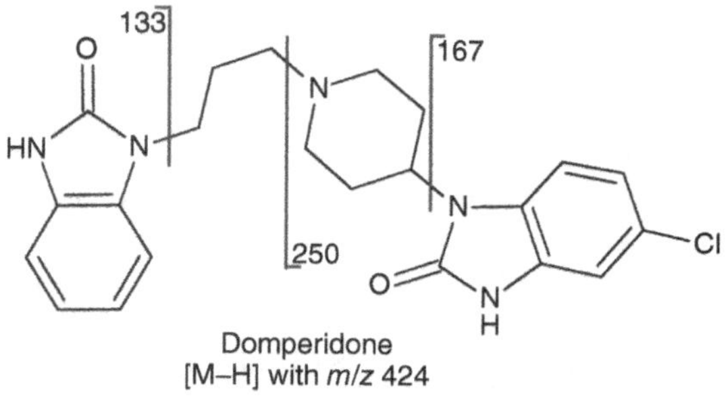

FIGURE 4.4.10 Fragmentation scheme for deprotonated domperidone.

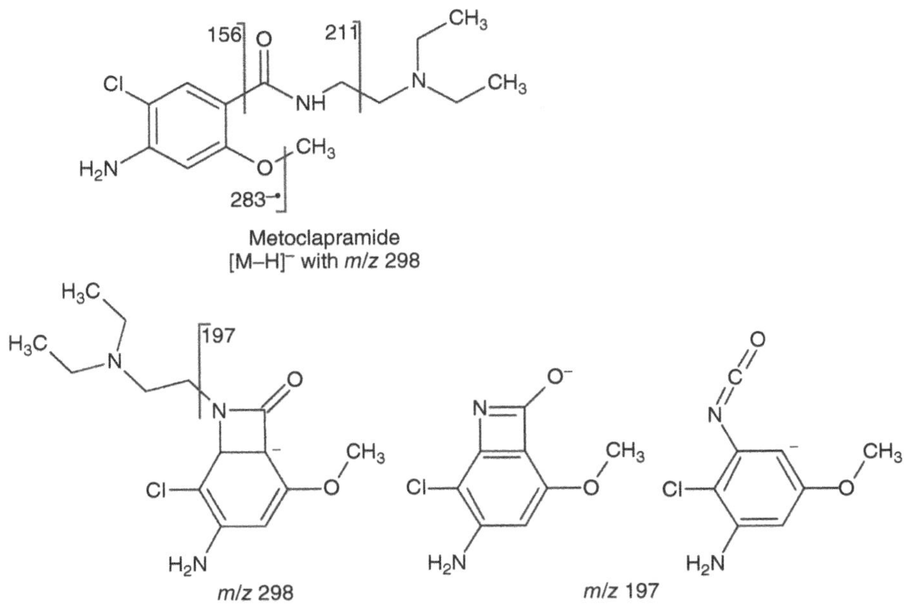

FIGURE 4.4.11 Fragmentation scheme for deprotonated metoclapramide.

most abundant fragment ions for tropisetron ($[M-H]^-$ with *m/z* 283) in negative-ion mode.

Another class of antiemetic drugs is dopamine antagonists such as domperidone and metoclopramide, which act on the CNS, and are used to treat nausea and vomiting associated with neoplastic disease, radiation sickness, and cytotoxic drugs. The major fragments in the negative-ion product-ion mass spectrum of domperidone ($[M-H]^-$ with *m/z* 424) are the deprotonated 1,3-dihydrobenzimidazol-2-one ($[C_7H_6N_2O-H]^-$) with *m/z* 133, the deprotonated 5-chloro-1,3-dihydrobenzimidazol-2-one ($[C_7H_5ClN_2O-H]^-$) with *m/z* 167, and the deprotonated 5-chloro-1-(pyperidin-4-yl)-benzimidazol-2-one ($[C_{12}H_{14}ClN_3O-H]^-$) with *m/z* 250 (Figure 4.4.10).

The negative-ion fragmentation of metoclopramide ($[M-H]^-$ with *m/z* 298) has been studied in detail using desorption-ESI on a Q–TOF hybrid instrument (Williams et al., 2006). Losses of a methyl radical ($^{\bullet}CH_3$) and of *N*-ethyl-*N*-methylethanamine ($CH_3N(CH_2CH_3)_2$) yield fragment ions with *m/z* 283 and 211, respectively. The loss of *N,N*-diethyl-2-(isocyanato)ethanamine ($(CH_2CH_3)_2NCH_2CH_2N{=}C{=}O$) leads to an ion with *m/z* 156 (Figure 4.4.11). The fragment ion with *m/z* 298 is considered to form a β-lactam-type intermediate, which after elimination of triethylamine ($(CH_2CH_3)_3N$) results in an ion with *m/z* 197. Subsequent ring opening of this ion leads to its structural isomer, the 3-isocyanato-substituted (—N=C=O) 2-chloro-5-methoxyaniline (*m/z* 197) (Figure 4.4.11) (Williams et al., 2006).

REFERENCES

Bansal G, Singh M, Jindal KC, Singh S. 2008. LC-UV-PDA and LC–MS studies to characterize degradation products of glimepiride. J Pharm Biomed Anal 48: 788–795.

Boix C, Ibáñez M, Sancho JV, Niessen WMA, Hernández F. 2013. Investigating the presence of omeprazole in waters by liquid chromatography coupled to low and high resolution mass spectrometry: Degradation experiments. J Mass Spectrom 48: 1091–1100.

Chung EW, HO ENM, Leung DKK, Tang FPW, Yiu KCH, Wan TSM. 2004. Detection of anti-ulcer drugs and their metabolites in horse urine by liquid chromatography–mass spectrometry. Chromatographia 59: S29–S38.

Dieckhaus CM, Fernández-Metzler CL, King R, Krolikowski PH, Baillie TA. 2005. Negative ion tandem mass spectrometry for the detection of glutathione conjugates. Chem Res Toxicol 18: 630–638.

Garcia-Ac A, Segura PA, Gagnon C, Sauvé S. 2009. Determination of bezafibrate, methotrexate, cyclophosphamide, orlistat and enalapril in waste and surface waters using on-line solid-phase extraction liquid chromatography coupled to polarity-switching electrospray tandem mass spectrometry. J Environ Monit 11: 830–838.

Hernando MD, Gómez MJ, Agüera A, Fernández-Alba AR. 2007. LC–MS analysis of basic pharmaceuticals (beta-blockers and anti-ulcer agents) in wastewater and surface water, Trends Anal Chem 26: 581–594.

Ho ENM, Yiu KCH, Wan TSM, Stewart BD, Watkins KL. 2004. Detection of anti-diabetics in equine plasma and urine by liquid chromatography–tandem mass spectrometry, J Chromatogr B 811: 65–73.

Kassahun K, Pearson PG, Tang W, McIntosh I, Leung K, Elmore C, Dean D, Wang R, Doss G, Baillie TA. 2001. Studies on the metabolism of troglitazone to reactive intermediates in vitro and in vivo. Evidence for novel biotransformation pathways involving quinone methide formation and thiazolidinedione ring scission. Chem Res Toxicol 14: 62–70.

Maurer HH, Kratzsch C, Kraemer T, Peters FT, Weber AA 2002 Screening, library-assisted identification and validated quantification of oral antidiabetics of the sulfonylurea-type in plasma by atmospheric pressure chemical ionization liquid chromatography–mass spectrometry, J Chromatogr B 773: 63–73.

Qin XZ, Ip DP, Chang KH, Dradransky PM, Brooks MA, Sakuma T. 1994. Pharmaceutical application of LC–MS. 1. Characterization of a famotidine degradate in a package screening study by LC–APCI MS. J Pharm Biomed Anal 12: 221–233.

Qin XZ. 2003. Collision-induced dissociation of the negative ions of simvastatin hydroxy acid and related species. J Mass Spectrom 38: 677–686.

Radjenović J, Pérez S, Petrović M, Barceló D. 2008. Identification and structural characterization of biodegradation products of atenolol and glibenclamide by liquid chromatography coupled to hybrid quadrupole time-of-flight and quadrupole ion trap mass spectrometry. J Chromatogr A 1210: 142–153.

Shah RP, Kumar V, Singh S. 2008. Liquid chromatography/mass spectrometric studies on atorvastatin and its stress degradation products. Rapid Commun Mass Spectrom 22: 613–622.

Sousa MA, Gonçalves C, Cunha E, Hajšlová J, Alpendurada MF. 2011. Cleanup strategies and advantages in the determination of several therapeutic classes of pharmaceuticals in wastewater samples by SPE-LC–MS/MS. Anal Bioanal Chem 399: 807–822.

Sun X, Tian Y, Zhang Z, Chen Y. 2009. A single LC–tandem mass spectrometry method for the simultaneous determination of four H2 antagonists in human plasma. J Chromatogr B 877: 3953–3959.

Tomer KB, Parker CE. 1989. Biochemical applications of liquid chromatography–mass spectrometry. J Chromatogr 492: 189–221.

Vuletić M, Cindrić M, Koruznjak JD. 2005. Identification of unknown impurities in simvastatin substance and tablets by liquid chromatography/tandem mass spectrometry. J Pharm Biomed Anal 37: 715–721.

Wang H, Wu Y, Zhao Z. 2001. Fragmentation study of simvastatin and lovastatin using electrospray ionization tandem mass spectrometry. J Mass Spectrom 36: 58–70.

Wang Y, Tang Y, Gu J, Fawcett JP, Bai X. 2004. Rapid and sensitive liquid chromatography–tandem mass spectrometric method for the quantitation of metformin in human plasma. J Chromatogr B 808: 215–219.

Wang M, Miksa IR. 2007. Multi-component plasma quantitation of anti-hyperglycemic pharmaceutical compounds using liquid chromatography–tandem mass spectrometry. J Chromatogr B 856: 318–327.

Wang Y, Chen X, Li Q, Zhong D. 2008. Characterization of metabolites of a novel histamine H2-receptor antagonist, lafutidine, in human liver microsomes by liquid chromatography coupled with ion trap mass spectrometry. Rapid Commun Mass Spectrom 22: 1843–1852.

Williams JP, Nibbering NMM, Green BN, Patel VJ, Scrivens JH. 2006. Collision-induced fragmentation pathways including odd-electron ion formation from desorption electrospray ionisation generated protonated and deprotonated drugs derived from tandem accurate mass spectrometry. J Mass Spectrom 41: 1277–1286.

4.5

FRAGMENTATION OF OTHER CLASSES OF DRUGS

4.5.1 β-ADRENERGIC RECEPTOR AGONISTS

The β-adrenergic receptor agonists such as bambuterol, clenbuterol, and salbutamol are structurally very similar to the β-blockers (Section 4.1.1). These compounds can impair the relaxation of bronchial muscle and are therefore used against asthma and chronic obstructive pulmonary disease. The general structure of the β-adrenergic agonists is 2-alkylamino-1-phenylethanol, with *tert*-butyl being the predominant alkyl group. The phenyl group may be substituted as well, for example, 4-amino-3,5-dichlorophenyl in clenbuterol.

The fragmentation of the β-adrenergic receptor agonists (Table 4.5.1) closely resembles that of β-blockers with a 1-(*tert*-butylamino)-3-phenoxypropan-2-ol skeleton (Section 4.1.1 and Table 4.1.2). Fragmentation of clenbuterol ($[M+H]^+$ with *m/z* 277) involves the loss of water to an ion with *m/z* 259, subsequent loss of 2-methylprop-1-ene ($(CH_3)_2C{=}CH_2$) to an ion with *m/z* 203, and subsequent losses of a chlorine radical ($Cl^\bullet$) and hydrogen chloride (HCl) to ions with *m/z* 168 and *m/z* 132, respectively (Doerge et al., 1993; Debrauwer & Bories, 1993). Carbuterol ($[M+H]^+$ with *m/z* 268) shows the loss of water to an ion with *m/z* 250, the loss of $(CH_3)_2C{=}CH_2$ to an ion with *m/z* 212, and the loss of both water and $(CH_3)_2C{=}CH_2$ to an ion with *m/z* 194. Secondary fragmentation of the ion with *m/z* 194 involves the loss of ammonia (NH_3) to an ion with *m/z* 177, the loss of hydrogen isocyanate (HN=C=O) to an ion with *m/z* 151, or the loss of both NH_3 and HN=C=O to an ion with *m/z* 134 ($[C_8H_8NO]^+$). Terbutaline ($[M+H]^+$ with *m/z* 226) also shows the loss of water to an ion with *m/z* 208, the loss of $(CH_3)_2C{=}CH_2$ to an ion with *m/z* 170, and the loss of both water and $(CH_3)_2C{=}CH_2$ to an ion with *m/z* 152. The same fragmentation is observed for tulobuterol, whereas salbutamol and pirbuterol show a second water loss as well (Table 4.5.1).

Fenoterol is a β-adrenergic agonists with a 2-alkylamino-1-phenylethanol skeleton, but with a (4-hydroxyphenyl) propan-2-yl rather than an *iso*-propyl or a *tert*-butyl substituent. The predicted positive-ion fragmentation of fenoterol ($[M+H]^+$ with *m/z* 304) is outlined in Figure 4.5.1. Cleavage of the α-C—N bond relative to the hydroxy-substituted C-atom would result in the two complementary fragment ions with *m/z* 153 and 152, whereas cleavage of the other C—N bond would result in the ions with *m/z* 170 and 135. Interestingly, the fragment ions with *m/z* 152 and 135 could also be obtained by secondary losses of water from the ions with *m/z* 170 and 153, respectively (Figure 4.5.1). However, accurate-mass data show that none of the ions resulting from the cleavage of the α-C—N bond relative to the hydroxy-substituted C-atom are actually present. This can be explained when assuming an abundant primary water loss via 1,3-elimination rather than via a 1,2-elimination, as described for the β-blocker propranolol (Section 4.1.1) (Upthagrove et al., 1999). Formation of an aziridine ring would block the cleavage of the α-C—N bond relative to the hydroxy-substituted C-atom. Further research is needed to prove this point.

Other structurally related drugs acting as bronchodilators are compounds such as ephedrine, etafedrine, isoprenaline, and dioxethedrin. For etafedrine ($[M+H]^+$ with *m/z* 194), apart from the loss of water, two complementary fragment ions (Section 3.5.2) are observed, that is, 1-phenylpropan-1-ol ($[C_9H_{11}O]^+$) with *m/z* 135 and protonated *N*-methylethanamine ($[C_3H_9N+H]^+$) with *m/z* 60.

4.5.2 HISTAMINE ANTAGONISTS

Histamine antagonists are agents that inhibit the release or action of histamine. The term antihistamine is usually reserved for compounds that act upon the H_1 histamine receptor. They are mostly used in the treatment against allergies. H_2 antagonists are used to reduce the secretion of gastric acid. They are prescribed in the treatment of various gastric diseases (Section 4.4.2.2). H_3 and H_4 receptor antagonists

TABLE 4.5.1 Fragmentation of β-adrenergic receptor agonists.

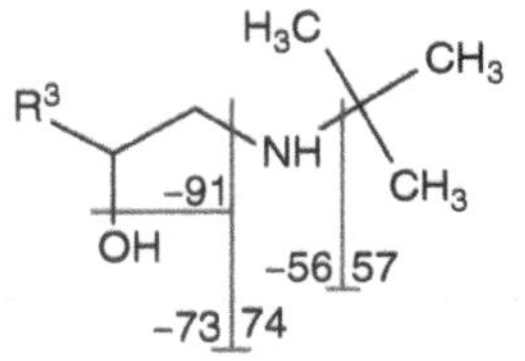

	$[M+H]^+$	*m/z* 57	−18 −water	−56 $-C_4H_8$	−74 −water $-C_4H_8$	−91 −water $-tBuNH_2$
Bambuterol	368	x		312	294	
Carbuterol	268	x	250	212	194	177
Clenbuterol	278				204	187
Pirbuterol	241	x	223	185	167	
Salbutamol	240	x	222		166	
Terbutaline	226	x	208	170	152	135
Tulobuterol	228	x	210	172	154	

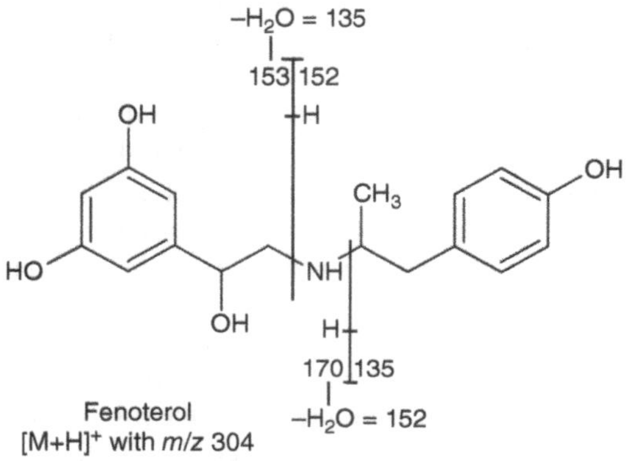

Fenoterol
$[M+H]^+$ with *m/z* 304

FIGURE 4.5.1 Predicted fragmentation scheme for protonated fenoterol. The fragment ions related to the indicated cleavage, that is, α-C—N bond relative to the hydroxy-substituted C-atom, are not observed.

are still under investigation, H_3 antagonists in relation to the treatment of conditions such as ADHD, Alzheimer's disease, and schizophrenia, and H_4 antagonists as anti-inflammatory and analgesic drugs. No examples of H_3 and H_4 antagonists are available in our spectral library collection.

From a structure point of view, the H_1 histamine antagonists include a variety of different compounds such as phenothiazines, pheniramines, diphenhydramines, 1-(diphenyl methyl)piperazines, and *N*′-phenyl-*N*′-methylthiopheneamines. A multiresidue analysis of 18 H_1 antihistamine drugs from different subclasses in blood using LC–MS in SRM mode has been reported (Gergov et al., 2001).

The phenothiazines with antihistamine properties show similar fragmentation behavior to the antipsychotic agents (Section 4.2.1). The compounds with a 10-*N*,*N*-dimethyl-2-propanamine substituent such as dimetotiazine, dioxoprotazine, and isothipendyl show cleavage of the phenothiazine C—N bond, mostly with charge retention on the resulting aliphatic amine, thus resulting in a 2-(dimethylamino)prop-1-yl carbocation ($[C_5H_{12}N]^+$) with *m/z* 86 for all three compounds. The cleavage of the C^1—C^2 bond in the propane chain to yield a 10-methylidenephenothiazinium ion or the *N*,*N*-(dimethyl)methylmethaniminium ion seems to be blocked by the presence of the 2-methyl substituent. Oxomemazine, with a 3-(*N*,*N*-dimethylamino)-2-methylpropyl substituent, and thiethylperazine, with a 3-(4-methylpiperazin-1-yl)propyl substituent, show the fragment ions common to their structural class (Table 4.2.1).

Pheniramine ($[M+H]^+$ with *m/z* 241) is an antihistamine with anticholinergic properties used to treat allergic conditions such as rhinitis (hay fever) or urticaria. Derivatives of pheniramine include chlorpheniramine ($[M+H]^+$ with *m/z* 275), brompheniramine ($[M+H]^+$ with *m/z* 319), and tolpropamine ($[M+H]^+$ with *m/z* 254). The halogenation of the phenyl substituent in pheniramine increases its potency, for example, chlorpheniramine and brompheniramine. The characteristic fragmentation of this class of compounds involves the loss of dimethyl amine ($(CH_3)_2NH$) and the loss of ethyldimethylamine ($(CH_3)_2NCH_2CH_3$), resulting in the formation of the secondary (2°) 2-pyridyl-4-substituted benzyl cation, where the 4-substituent is H in pheniramine (*m/z* 168), Cl in chlorpheniramine (*m/z* 202), and Br in brompheniramine (*m/z* 246). In the latter two cases, a subsequent loss of a chlorine ($Cl^\bullet$) or a bromine ($Br^\bullet$) radical is observed, respectively, to an ion with *m/z* 167 ($[C_{12}H_9N]^{+\bullet}$) (Figure 4.5.2). Tolpropamine having a 4-methylphenyl substituent instead of a 2-pyridyl leads to the 2° α-(4-methylphenyl)benzyl cation ($[C_{14}H_{13}]^+$) with *m/z* 181.

Related structures are the diphenhydramines, such as the substituted dialkyl ethers bromazine ($[M+H]^+$ with *m/z* 344), carbinoxamine ($[M+H]^+$ with *m/z* 291), chlorphenoxamine ($[M+H]^+$ with *m/z* 304), clemastine ($[M+H]^+$ with *m/z* 344, Figure 4.5.2), cloperastine ($[M+H]^+$ with *m/z* 330), diphenhydramine ($[M+H]^+$ with *m/z* 256), doxylamine ($[M+H]^+$ with *m/z* 271), orphenadrine ($[M+H]^+$ with *m/z* 240), and the substituted alkyl aryl ether etoloxamine. Except for etoloxamine, all these compounds show the loss of the ethanolamine group to produce the corresponding 2° or tertiary (3°) benzylic carbocation: Formation of the 2° α-(phenyl)benzyl cation ($[C_{13}H_{11}]^+$) with *m/z* 167 for diphenhydramine, the 2° α-(4-bromophenyl)benzyl cation ($[C_{13}H_{10}Br]^+$) with *m/z* 245 for bromazine, the 2° α-(2-pyridyl)-4-chlorobenzyl cation ($[C_{12}H_9ClN]^+$) with *m/z* 202 for carbinoxamine, the 3° α-(4-chlorophenyl)-α-(methyl)benzyl cation ($[C_{14}H_{12}Cl]^+$) with *m/z* 215 for both chlorphenoxamine and clemastine, the 2° α-(4-chlorophenyl)benzyl cation ($[C_{13}H_{10}Cl]^+$) with *m/z* 201 for cloperastine, the 3° α-(methyl)-α-(2-pyridyl)benzyl cation ($[C_{13}H_{12}N]^+$) with *m/z* 182 for doxylamine, and the 2° α-(2-methylphenyl)benzyl cation ($[C_{14}H_{13}]^+$) with *m/z* 181 for orphenadrine. In the case of bromazine and doxylamine, a subsequent loss of $Br^\bullet$ or a methyl radical ($^\bullet CH_3$) is observed, respectively. In the case of carbinoxamine, chlorphenoxamine, and clemastine, a subsequent loss of $Cl^\bullet$ is observed. With clemastine, the complementary fragment ion is observed, that is, protonated 2-(1-methylpyrrolidin-2-yl)ethanol ($[C_7H_{15}NO+H]^+$) with *m/z* 130. Etoloxamine ($[M+H]^+$ with *m/z* 284) behaves somewhat differently, with the loss of diethylamine ($(CH_2CH_3)_2NH$) to an ion with *m/z* 211, and the formation of the 2-ethyl carbocation of triethylamine ($[(CH_3CH_2)_2NCH_2CH_2]^+$) with *m/z* 100, and an ion with *m/z* 72 ($[C_4H_{10}N]^+$), which can be either of the two structural isomers: the *N*-ethylethaniminium ion ($[CH_3HC{=}NHC_2H_5]^+$) or the 2-ethyl carbocation of diethylamine ($[C_2H_5NHCH_2CH_2]^+$). The complementary ion to the ion with *m/z* 100, which would be expected as an ion with *m/z* 185, is not observed due to the (relatively) low proton affinity of 2-benzylphenol (Section 3.5.2).

Another class of related structures are derivatives of 1-(diphenylmethyl)piperazine, as for instance cetirizine, chlorcyclizine, cinnarizine, cyclizine, hydroxyzine, or meclozine. In MS–MS, the C—N bond between the diphenylmethyl group and the piperazine ring N^1 is cleaved, with charge retention on either of the resulting fragments. Charge retention at the 4-substituted piperazine ring of cinnarizine ($[M+H]^+$ with *m/z* 369, Figure 4.5.2) leads to an ion with *m/z* 201 ($[C_{13}H_{16}N_2+H]^+$), whereas charge retention on the diphenylmethyl group, as in cinnarizine and cyclizine ($[M+H]^+$ with *m/z* 267), results in the 2° α-(phenyl)benzyl cation ($[C_{13}H_{11}]^+$) with *m/z* 167. Compounds with 4-chloro substitution in one of the two phenyl rings such as in cetirizine, chlorcyclizine, hydroxyzine, and meclozine produce the 2° α-(4-chlorophenyl)benzyl cation ($[C_{13}H_{10}Cl]^+$) with *m/z* 201, which shows a subsequent loss of a $Cl^\bullet$ to a radical cation with *m/z* 166.

Yet another class of compounds is based on the *N'*-benzyl-*N'*-phenylethane-1,2-diamine structure, as in histapyrrodine, eventually with pyridinyl or pyrimidinyl instead of phenyl and/or substituents on the benzyl ring, as for instance bromtripelennamine, chlorpyramine, pyrilamine, thonzylamine, and tripelenamine ($[M+H]^+$ with *m/z* 256, Figure 4.5.3). Fragment ions are observed due to either the loss of $(CH_3)_2NH$ with *m/z* ($[M+H]-45$), for example, an ion with *m/z* 211 for tripelenamine, the 2-(dimethylamino)ethyl carbocation ($[(CH_3)_2NCH_2CH_2]^+$) with *m/z* 72, and the benzyl cation ($[C_6H_5CH_2]^+$) with *m/z* 91 for tripelenamine or the corresponding cation with a different substituent, for example, *m/z* 169 ($[BrC_6H_4CH_2]^+$) for bromtripelennamine, *m/z* 125 ($[ClC_6H_4CH_2]^+$) for chlorpyramine, and *m/z* 121 ($[CH_3O{-}C_6H_4CH_2]^+$) for pyrilamine and thonzylamine.

Structurally similar, but with a 2- or 3-thenyl group (methylthiophene) instead of an *N'*-benzyl substituent are methapyrilene, methaphenilene, and thenyldiamine. They show characteristic fragmentation involving the cleavage of the diamine N′—C bond to the thenyl group with charge retention on either resulting fragment. For methaphenilene ($[M+H]^+$ with *m/z* 261, Figure 4.5.3), this results in two complementary ions, that is, the thenylium ion ($[C_4H_3S{-}CH_2]^+$) with *m/z* 97 and the protonated *N'*-phenyl-substituted 2-imino-*N,N*-dimethylethan-1-amine with *m/z* ($[M+H]-98$). Loss of $HN(CH_3)_2$ from $[M+H]^+$ is also observed.

The quantitative bioanalysis of the tricyclic antihistamine loratadine and its active metabolite desloratadine ($[M+H]^+$ with *m/z* 311, Figure 4.5.4) has been extensively studied (e.g., Yang et al., 2003). In MS–MS, desloratadine shows the loss of ammonia (NH_3) to an ion with *m/z* 294 followed by the loss of $Cl^\bullet$ to an ion with *m/z* 259 ($[C_{19}H_{17}N]^{+\bullet}$) or the loss of HCl to an ion with *m/z* 258, or the loss of methanimine ($H_2C{=}NH$) from the 4-methylenepiperidine ring to an ion with *m/z* 282 (cf. Section 3.6.5). Similar losses from the ring are observed for azatadine ($[M+H]^+$ with *m/z* 291), which is the dechlorinated-*N*-methylpiperidine analog of desloratadine. Rupatadine ($[M+H]^+$ with *m/z* 416), which is a desloratadine analog with a (5-methyl-3-pyridinyl)methyl group at the piperidine nitrogen, shows somewhat similar fragmentation, that is, the loss of 3,5-dimethylpyridine (C_7H_9N) to an ion with *m/z* 309, the loss of *N*-methyl-(5-methylpyridin-3-yl)methanimine to an ion with *m/z* 282 ($[C_{18}H_{17}ClN]^+$), and the loss of (5-methylpyridin-3-yl)methanamine and $Cl^\bullet$ to an ion with *m/z* 259 ($[C_{19}H_{17}N]^{+\bullet}$); the protonated 3,5-dimethylpyridine

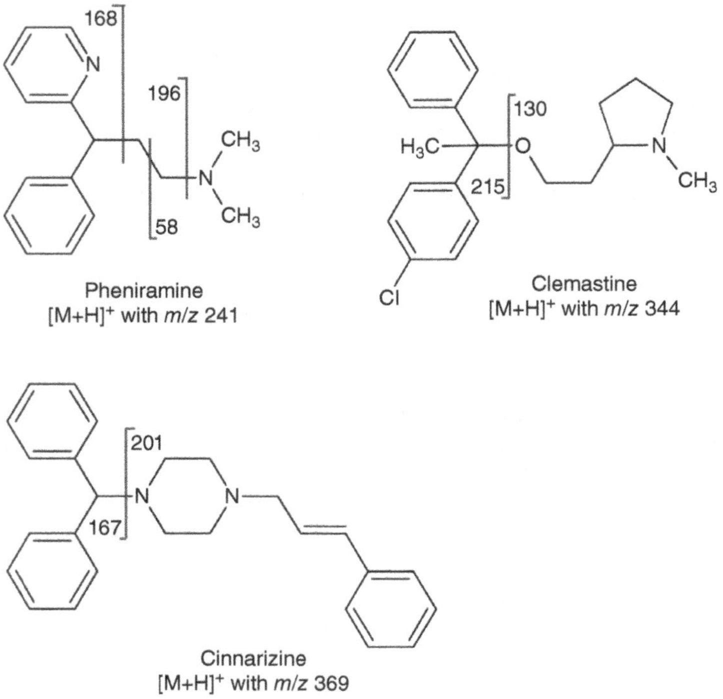

FIGURE 4.5.2 Fragmentation schemes for protonated pheniramine, clemastine, and cinnarizine.

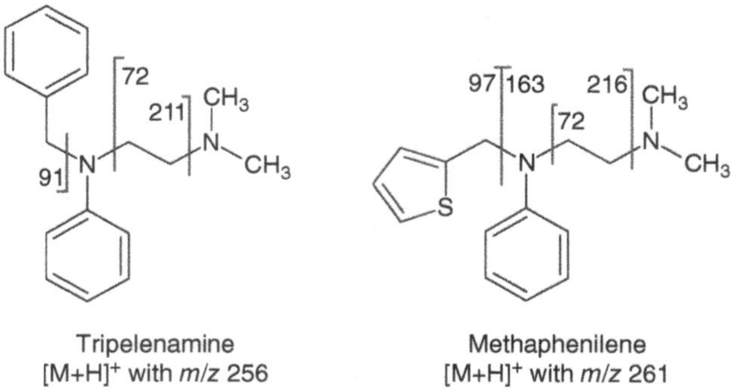

FIGURE 4.5.3 Fragmentation schemes for protonated tripelenamine and methaphenilene.

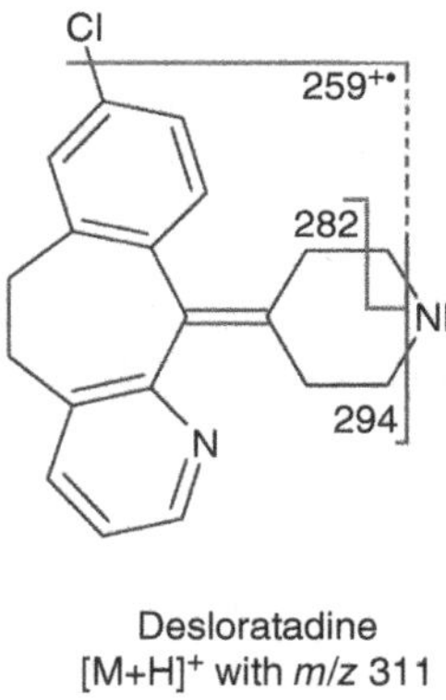

FIGURE 4.5.4 Fragmentation scheme for protonated desloratadine.

($[C_7H_9N+H]^+$) with m/z 108 is observed as well (Wen et al., 2009). The fragmentation of loratadine, desloratadine, and their metabolites has recently been studied in more detail using MS^2 and MS^3 on Q–LIT and LIT–orbitrap hybrid instruments (Picard et al., 2009; Chen et al., 2009).

4.5.3 ANTICHOLINERGIC AGENTS

An anticholinergic agent is a compound that blocks the neurotransmitter acetylcholine in the central and the peripheral nervous systems; a classic example is atropine. Anticholinergics inhibit parasympathetic nerve impulses by selectively blocking the binding of the neurotransmitter acetylcholine to its receptor in the nerve cells. They can be subdivided into three categories in accordance with their specific targets in the central and/or peripheral nervous system: antimuscarinic agents, ganglionic blockers, and neuromuscular blockers. Anticholinergic drugs are used in the treatment of a variety of conditions, such as gastrointestinal disorders such as gastritis, pylorospasm, diverticulitis, and ulcerative colitis, as well as genitourinary disorders, respiratory disorders, Parkinson's disease, and others. This class of compounds contains a high structural variety, including for instance 10*H*-phenothiazine-like structures as in profenamine and diethazine (Section 4.2.1). Some other examples are discussed in the following.

The major fragment ion of atropine ($[M+H]^+$ with *m/z* 290) is the 8-methyl-8-azabicyclo[3.2.1]oct-3-yl ion ($[C_8H_{14}N]^+$) with *m/z* 124 due to C—O bond cleavage between the heterocyclic substituent and the alkoxy O-atom of the ester moiety. Subsequent loss of methylamine (CH_3NH_2) results in a fragment ion with *m/z* 93 ($[C_7H_9]^+$); the tropylium ion ($[C_7H_7]^+$) with *m/z* 91 is also observed. In addition, losses of water and formaldehyde ($H_2C{=}O$) are found (Chen et al., 2006). In the case of scopolamine ($[M+H]^+$ with *m/z* 304), apart from the major fragment ion with *m/z* 138 resulting (as in atropine) from the heterocyclic group of the ester's moiety alkoxy group, a fragment due to the cleavage of the C—O bond between the acyl group and the alkoxy group is also observed, resulting in a protonated oxazatricyclic alcohol ($[C_8H_{13}NO_2+H]^+$) with *m/z* 156. The related dialkyl ether benzatropine ($[M+H]^+$ with *m/z* 308) shows just one major fragment ion, the 2° α-(phenyl)benzyl cation ($[C_{13}H_{11}]^+$) with *m/z* 167.

A number of anticholinergic compounds are derivatives of phenylacetic acid with additional substituents attached to the acid-C^2. With cyclohexyl substitution as in drofenine ($[M+H]^+$ with *m/z* 318), the two major fragments are an ion with *m/z* 245 due to the loss of diethylamine ($(C_2H_5)_2NH$) and the α-(ethoxycarbonyl)benzyl cation ($[C_{10}H_{11}O_2]^+$) with *m/z* 163 due to the subsequent loss of cyclohexene (C_6H_{12}). In addition, the 2-(diethylamino)ethyl carbocation ($[(C_2H_5)_2NCH_2CH_2]^+$) with *m/z* 100, $[C_7H_7]^+$ with *m/z* 91, and the 2° α-(cyclohexyl)benzyl cation ($[C_{13}H_{17}]^+$) with *m/z* 173 are observed (Figure 4.5.5). With cyclohexyl and hydroxy substituents, oxybutynin ($[M+H]^+$ with *m/z* 358) shows the loss of water to an ion with *m/z* 340, and the cleavage of the acid's C^2—C bond to the carbonyl group to form a 3° α-(cyclohexyl)-α-(hydroxy)benzyl cation ($[C_{13}H_{17}O]^+$) with *m/z* 189, followed by a water loss to *m/z* 171. Cleavage of the C—O ester bond leads to fragment ions with *m/z* 142 and 124 due to the protonated 4-diethylaminobut-2-ynol ($[C_8H_{15}NO+H]^+$) and water loss thereof, respectively (Figure 4.5.5). Propiverine ($[M+H]^+$ with *m/z* 368) with phenyl and propoxide substituents also shows similar fragmentation (Figure 4.5.5) in addition to losses of propene ($CH_3CH{=}CH_2$) and propanol ($CH_3CH_2CH_2OH$) (Oertel et al., 2007).

The two related structures triperiden ($[M+H]^+$ with *m/z* 312) and trihexyphenidyl ($[M+H]^+$ with *m/z* 302), which are based on a common 1-phenyl-3-(piperidin-1-yl)propan-1-ol backbone, show a major fragment ion with *m/z* 98, the 1-methylidenepiperidinium ion ($[C_6H_{12}N]^+$).

The quaternary ammonium compound propantheline (M^+ with *m/z* 368), containing a 9*H*-xanthene group, shows the loss of propene ($CH_3CH{=}CH_2$) to an ion with *m/z* 326, the

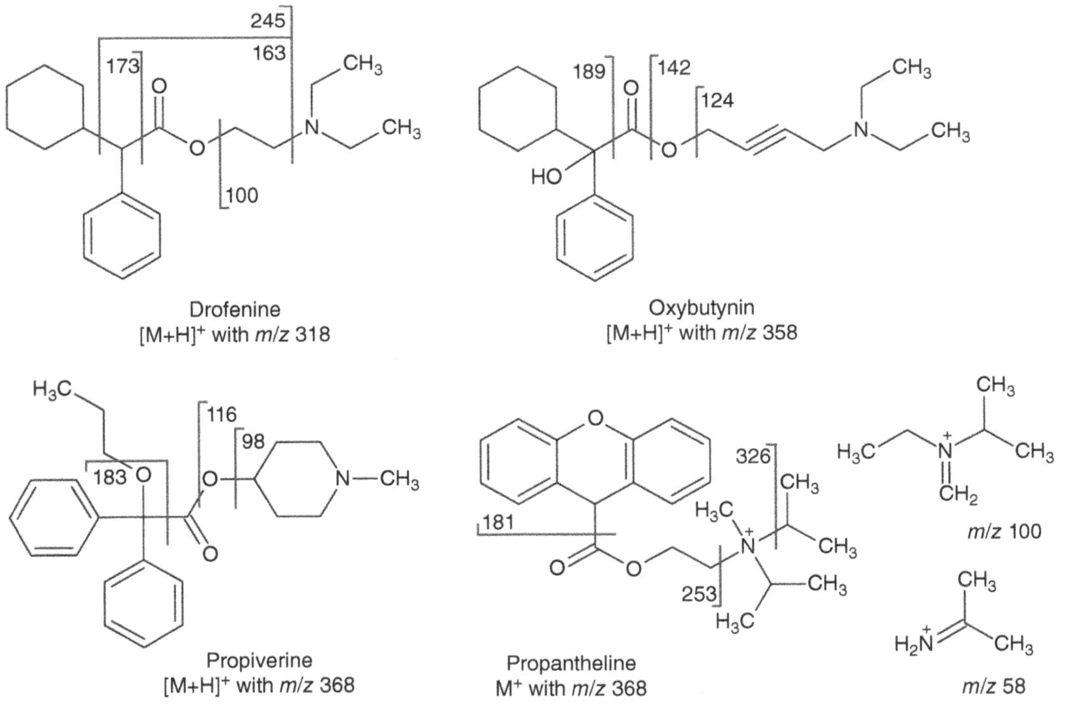

FIGURE 4.5.5 Fragmentation schemes for protonated drofenine, oxybutynin, and propiverine and for the propantheline cation.

loss of di-*iso*-propylmethylamine ($C_7H_{17}N$) to an ion with *m/z* 253. Further loss of ethene (C_2H_4) and carbon dioxide (CO_2) leads to the xanthenium cation ($[C_{13}H_9O]^+$) with *m/z* 181. In addition, the *N*-ethyl-*N*-(*iso*-propyl)methaniminium ion ($[C_6H_{14}N]^+$) with *m/z* 100 and the protonated propan-2-imine ($[C_3H_7N+H]^+$) with *m/z* 58 are observed, which are related to the di-*iso*-propylmethylamine group (Figure 4.5.5).

4.5.4 DRUGS AGAINST ALZHEIMER'S DISEASE: ACETYLCHOLINESTERASE INHIBITORS

Alzheimer's disease is a neurodegenerative disease attributed to a deficiency of cholinergic neurotransmission. Various drugs are under development for the treatment of Alzheimer's disease. Even though these drugs differ widely in structure, their mode of action is based on the inhibition of the enzyme acetylcholinesterase. LC–MS methods for the bioanalysis of these types of drugs have been reviewed (Azevedo Marques et al., 2011a; Ponnayyan Sulochana et al., 2014).

Tacrine was the first drug tested for the treatment of Alzheimer's disease. The fragment ions of tacrine ($[M+H]^+$ with *m/z* 199) are outlined in Figure 4.5.6. The major fragment ion with *m/z* 171 involves the loss of ethene (C_2H_4) via a retro-Diels–Alder fragmentation. The ion with *m/z* 183 is due to an unexpected loss of methane (CH_4); a structure based on an intramolecular cyclization is proposed. The fragment ion with *m/z* 144 ($[C_{10}H_9N+H]^+$) is most likely due to the loss of propionitrile ($CH_3CH_2C{\equiv}N$), resulting in a protonated 1-naphthylamine rather than the protonated 3-methylquinoline proposed previously (Azevedo Marques et al., 2010) (Figure 4.5.6). The latter would involve the unlikely combined losses of ammonia (NH_3) and C_3H_2.

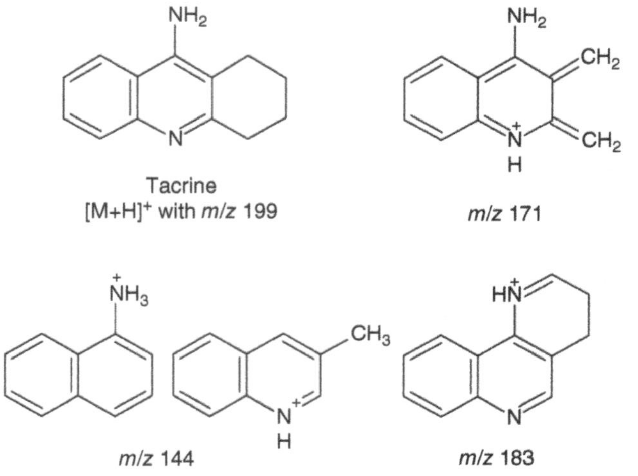

FIGURE 4.5.6 Proposed structures for the fragment ions of protonated tacrine.

Physostigmine ($[M+H]^+$ with *m/z* 276) is a naturally occurring alkaloid extracted from Calabar beans. Its complex fragmentation scheme is proposed in Figure 4.5.7. One would expect the *N*-methylcarbamate group to be the weakest spot in the molecule, and thus the predominant fragmentation site, yet extensive fragmentation of the *N*-methyl pyrrolidine group, is also observed as indicated by accurate-mass data. Loss of methylamine (CH_3NH_2) from the carbamate leads to a fragment ion with *m/z* 245. Fragmentation of the pyrrolidine ring leads to two fragment ions: the loss of *N*-methylmethanimine ($CH_3N{=}CH_2$) to form an ion with *m/z* 233 and the loss of *N*-methylethenamine ($CH_3NHCH{=}CH_2$) to an ion with *m/z* 219. Additionally, the combined fragmentation of the carbamate and the pyrrolidine ring leads to several fragment ions, including the radical cation with *m/z* 147 ($[C_9H_9NO]^{+\bullet}$), as shown in Figure 4.5.7.

Rivastigmine ($[M+H]^+$ with *m/z* 251) shows major fragment ions with *m/z* 206 due to the loss of dimethylamine ($(CH_3)_2NH$) from the phenyl substituent, and the *N*-ethyl-*N*-methylcarbamoyl cation ($[C_4H_8NO]^+$) with m/z 86, caused by fragmentation of the carbamate moiety. Structure proposals for some of the minor fragment ions are also shown in Figure 4.5.8 (Thomas et al., 2012; Thevis et al., 2006). Loss of ethene (C_2H_4) from the ion with *m/z* 206 leads to the ion with *m/z* 178 (Figure 4.5.8). Our proposed structure for this ion differs from a structure proposed elsewhere (Thevis et al., 2006); despite the fact that their accurate-mass data indicates an elemental composition $[C_{10}H_{12}NO_2]^+$ for the ion with *m/z* 178, thus involving the loss of C_2H_4 from the ion with *m/z* 206, a structure involving the loss of carbon monoxide (CO) is proposed, thus leading to an isobaric ion $[C_{11}H_{16}NO]^+$ (Thevis et al., 2006).

Two related acetylcholinesterase inhibitors, neostigmine and pyridostigmine, are quaternary ammonium compounds, which cannot cross the blood–brain barrier and thus cannot be used in the treatment of Alzheimer's disease. All major fragment ions of neostigmine (M^+ with *m/z* 223) and pyridostigmine (M^+ with *m/z* 181) derive from the fragmentation of the same carbamate moiety present in both compounds. The occurrence of the *N,N*-dimethylcarbamoyl cation ($[(CH_3)_2N{-}C{=}O]^+$) with *m/z* 72 is common to both molecules. Two other fragment ions are the ion with *m/z* 208 due to the loss of a methyl radical ($^\bullet CH_3$) in the case of neostigmine, and the ion with *m/z* 124 due to the loss of methyl isocyanate ($CH_3{-}N{=}C{=}O$) in the case of pyridostigmine. This latter fragmentation involves the rearrangement of a methyl (CH_3) group, leading either to the 1-methyl-3-methoxypyridinium ion, the 3-hydroxy-1,4-dimethylpyridinium ion, or the 3-hydroxy-1,2-dimethylpyridinium ion ($[C_7H_{10}NO]^+$) with *m/z* 124. The subsequent loss of a $^\bullet CH_3$ from the ion with *m/z* 124 to an ion with *m/z* 109 can result from either the methoxy group or the quaternary ammonium moiety (Section 3.5.7).

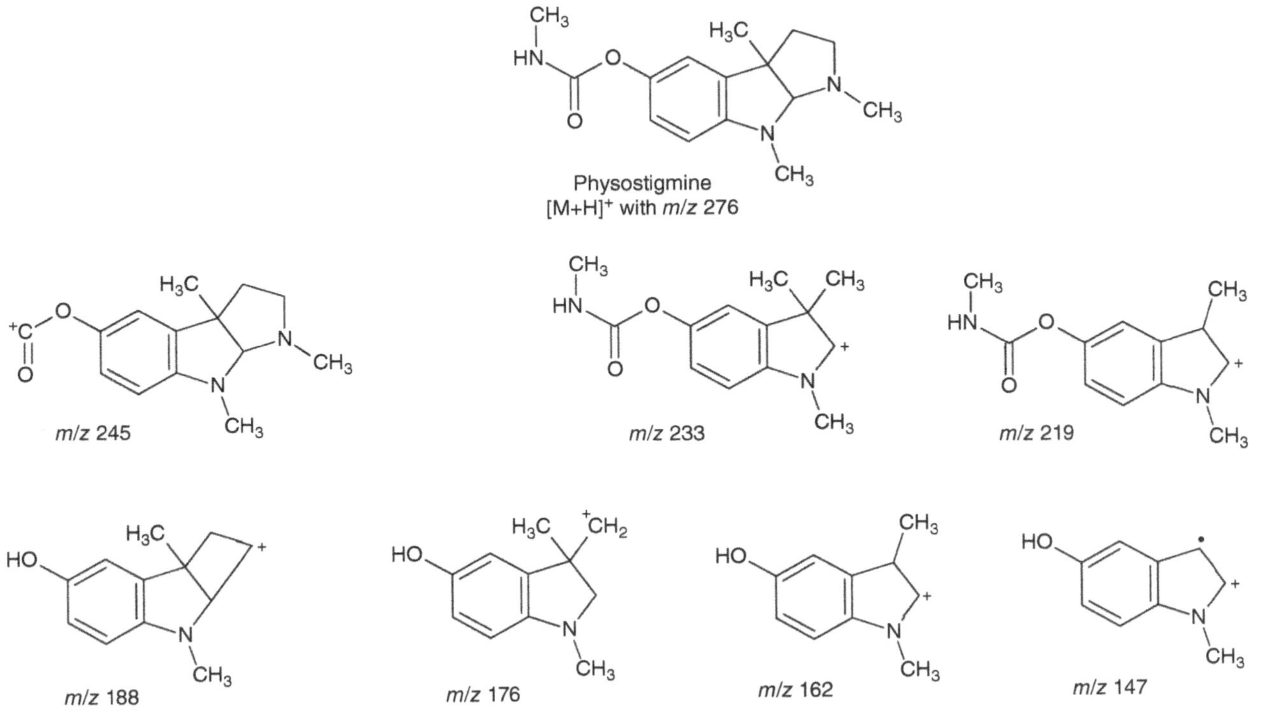

FIGURE 4.5.7 Proposed structures for the fragment ions of protonated physostigmine.

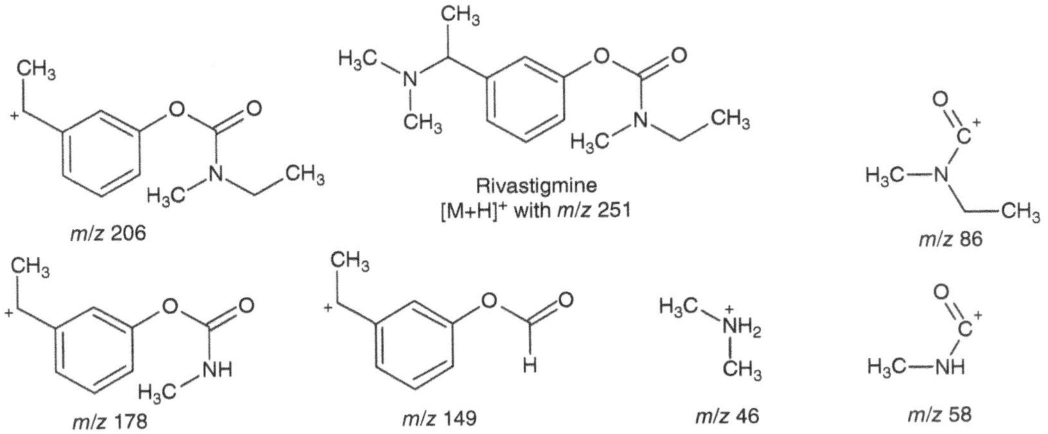

FIGURE 4.5.8 Proposed structures for the fragment ions of protonated rivastigmine.

The most abundant fragment ion for donepezil ($[M+H]^+$ with *m/z* 380) is the tropylium ion ($[C_7H_7]^+$) with *m/z* 91, resulting from the cleavage of the N—C bond between the N-atom of the pyrimidine ring and the benzylic C-atom. This ion is frequently used in the quantitative bioanalysis of donepezil when using SRM (Azevedo Marques et al., 2011a; Ponnayyan Sulochana et al., 2014). Based on accurate-mass data, structures have been proposed for some of the minor ions (Thevis et al., 2006) (Figure 4.5.9). The proposed structures are in agreement with data from a [D_7]-labeled analog used as an internal standard (Matsui et al., 1999).

Galantamine is an alkaloid isolated from the bulbs and flowers of *Galanthus caucasicus* and related Amaryllidaceae such as the *Narcissus poeticus* (daffodil). Considerable attention has been paid to the fragmentation of galantamine ($[M+H]^+$ with *m/z* 288) in MS–MS (Verhaeghe et al., 2003; Jegorov et al., 2006; Thevis et al., 2006; Maláková et al., 2007; Azevedo Marques et al., 2011b). However, there seems

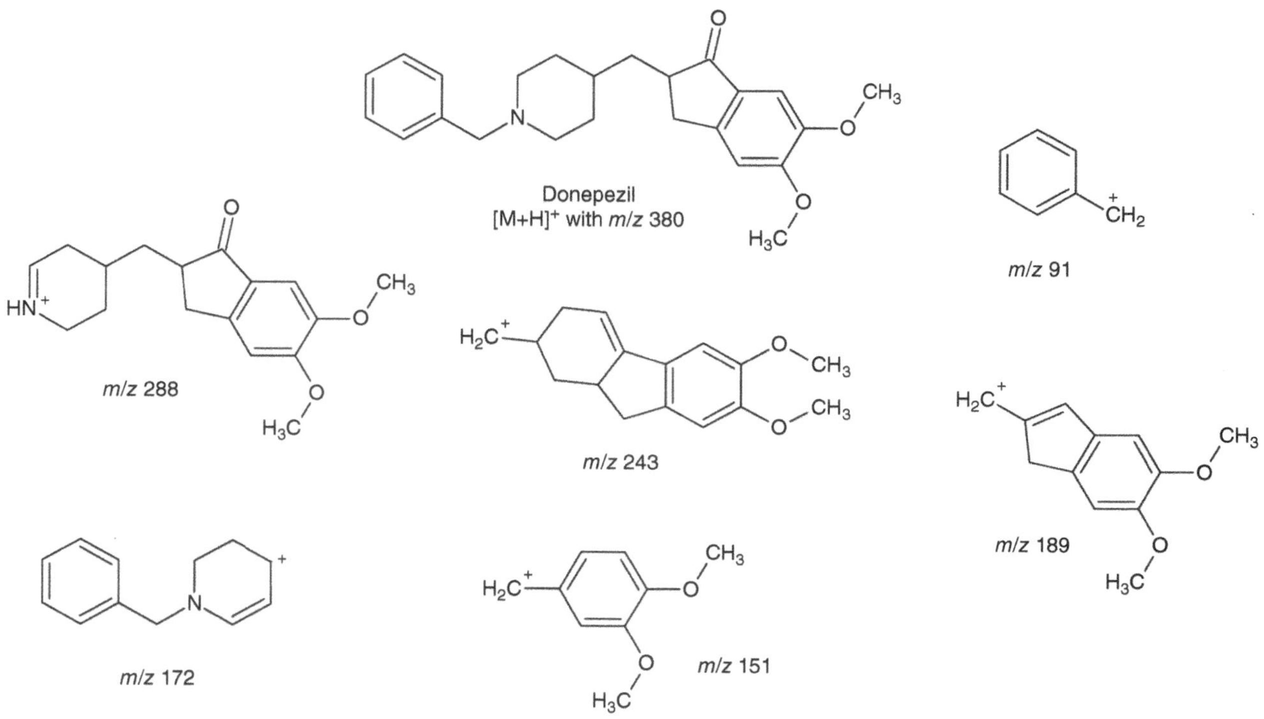

FIGURE 4.5.9 Proposed structures for the fragment ions of protonated donepezil. (Source: Partly based and redrawn from (Thevis et al., 2006) with permission of Wiley.)

to be no agreement on the identity of several fragment ions. A good starting point is the fragmentation scheme proposed by the group of Jegorov (Jegorov et al., 2006) (Figure 4.5.10), which is based on accurate-mass data acquired using an FT-ICR-MS instrument. Although there are some minor differences in the way the structures are drawn, the literature agrees on the proposed structures for the fragment ions with *m/z* 270 due to the loss of water, with *m/z* 257 due to the loss of methylamine (CH_3NH_2), and with *m/z* 239 due to losses of both water and CH_3NH_2. Possible structures for the ions with *m/z* 231, 213, and 225 were not reported by the authors (Figure 4.5.10). Based on the molecular formulae derived from accurate-mass data, the ion with *m/z* 231 is due to the loss of C_3H_7N, for example, *N*-methylethenamine, the ion with *m/z* 213 to a subsequent water loss, and the ion with *m/z* 225 due to the losses of both water and $(CH_3)_2NH$ (structure *m/z* 225-I in Figure 4.5.11). Except for the ion with *m/z* 225, this interpretation is in agreement with that of other authors (Verhaeghe et al., 2003; Thevis et al., 2006; Maláková et al., 2007). The ion with *m/z* 225 has also been described as a secondary fragmentation, involving the loss of methanol (CH_3OH) from the ion with *m/z* 257 (Thevis et al., 2006) (structure *m/z* 225-II in Figure 4.5.11). Interestingly, MS^3 data acquired using an IT-TOF instrument (Azevedo Marques, unpublished data) shows losses of a methyl radical ($^{\bullet}CH_3$), water, carbon monoxide (CO), and the combined losses of either water and CO or CO and formaldehyde ($H_2C{=}O$), which cannot be explained by either of the two proposed structures for the ion with *m/z* 225. The structure proposed for the ion with *m/z* 209 (Figure 4.5.10) seems rather unlikely, as it would involve loss of water from the hydrogenated dibenzofuran alcohol, the loss of the furan ring O-atom, as well as the loss of $(CH_3)_2NH$ from the hydrogenated azepine ring, including a ring skeletal rearrangement. A more likely route to the fragment ion with *m/z* 209 is the loss of $H_2C{=}O$ from the ion with *m/z* 239 (Figure 4.5.11). An alternative route to the ion with *m/z* 209 would involve (in either order) the loss of water and of methanol from the hydrogenated dibenzofuran ring, followed by a 1,5-*H* shift to the more stable benzylic cation, and a ring contraction of the hydrogenated azepin with concomitant extrusion of methanimine ($H_2C{=}NH$), although the intermediate fragment ions involved are not all observed. An alternative structure is also proposed for the ion with *m/z* 198: rather than the losses of water and the 2-ethyl radical of ethyldimethylamine ($^{\bullet}CH_2CH_2N(CH_3)_2$) (Figure 4.5.10), it could be envisaged as a result of the combined losses of ethylmethylamine (C_3H_9N) and a methoxy radical ($CH_3O^{\bullet}$) (Figure 4.5.11).

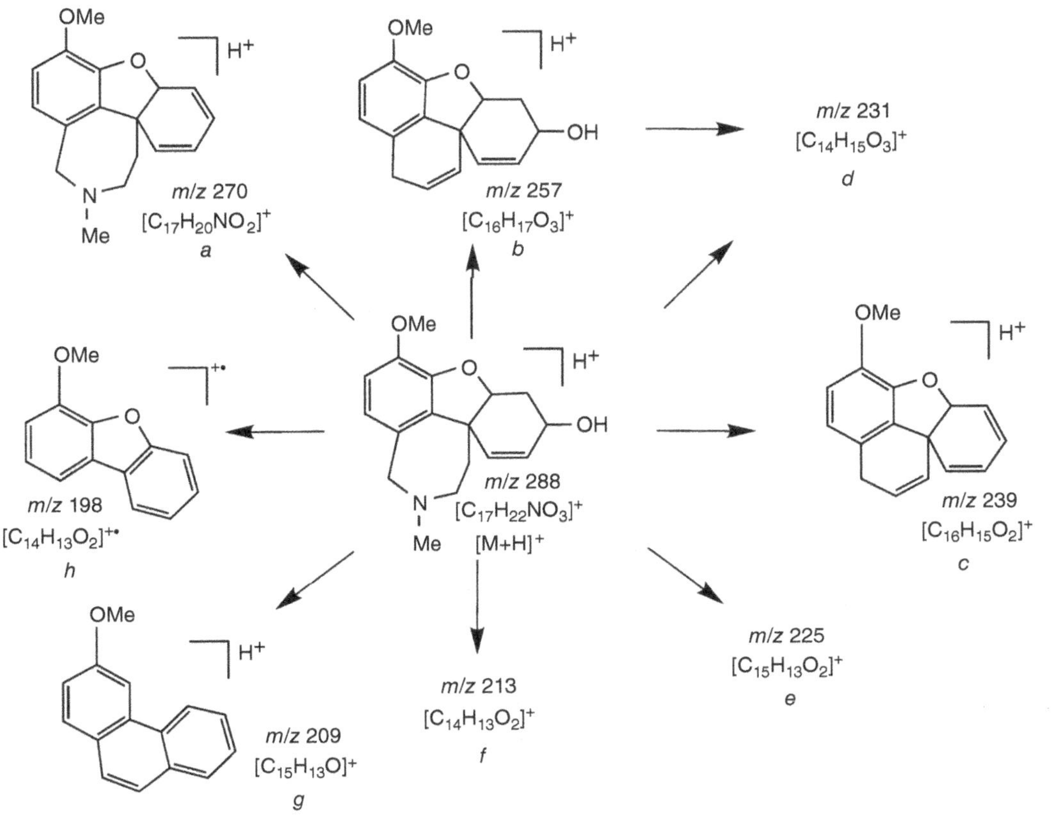

FIGURE 4.5.10 Fragmentation pathways for protonated galantamine as proposed by Jegorov et al. (Source: Jegorov et al., 2006. Reproduced with permission of Wiley.)

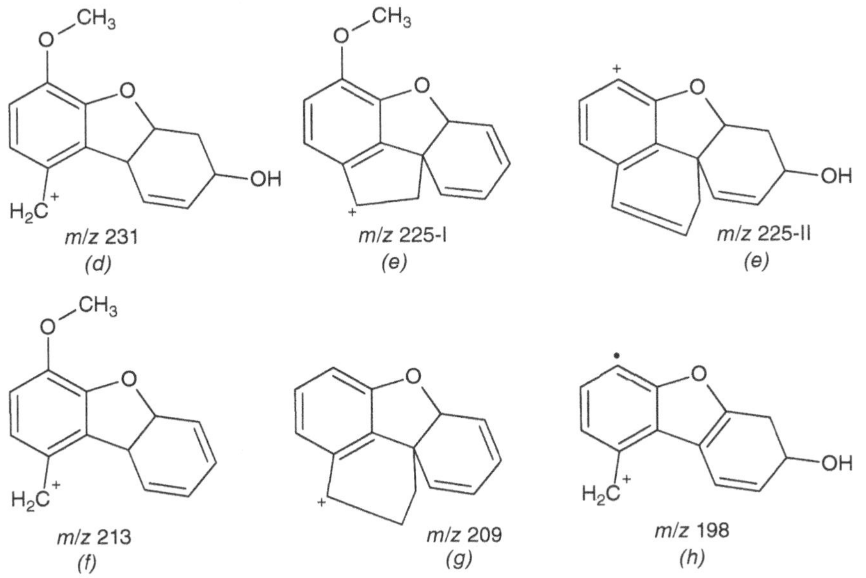

FIGURE 4.5.11 Alternative structure proposals for some of the fragments observed for protonated galantamine.

4.5.5 ANTIPARKINSONIAN DRUGS

The compounds that are indicated in the treatment of parkinsonisms do not belong to a particular structural class. Some of them show great resemblance to (or can even be considered to be) anticholinergic drugs (Section 4.5.3), for example, benzatropine, biperiden, budipine, procyclidine, and trihexyphenidyl. Benzatropine and trihexyphenidyl are discussed in Section 4.5.3. Biperiden ($[M+H]^+$ with *m/z* 312) shows the loss of water and a major fragment ion with *m/z* 98 due to the (piperidin-1-yl)methylium ion ($[C_6H_{12}N]^+$). Budipine ($[M+H]^+$ with *m/z* 294) shows fragment ions with *m/z* 238 due to the loss of *iso*-butene (C_4H_8) and the complementary ion with *m/z* 57 ($[C_4H_9]^+$), with *m/z* 160 due to the subsequent losses of *iso*-butene and benzene (C_6H_6), and the 2-(methaniminyl)ethyl carbocation ($[H_2C{=}NCH_2CH_2]^+$) with *m/z* 56. Procyclidine ($[M+H]^+$ with *m/z* 288) shows the loss of water and the (pyrrolidin-1-yl)methylium ion ($[C_5H_{10}N]^+$) with *m/z* 84.

Another type of antiparkinsonian drugs comprises L-DOPA (or levodopa) and the related structure carbidopa, as well as methamphetamine (Section 4.7.2.1) with its related structures like selegiline. The fragment ions of levodopa ($[M+H]^+$ with *m/z* 198) are shown in Figure 4.5.12. The fragmentation of carbidopa ($[M+H]^+$ with *m/z* 227) shows a similar pattern. For levodopa, it involves the loss of ammonia (NH_3) or formic acid (HCOOH) to ions with *m/z* 181 or *m/z* 152, respectively. In the case of carbidopa, the loss of hydrazine (H_2NNH_2) or HCOOH results in fragment ions with *m/z* 195 or *m/z* 181, respectively. The formation of the ion with *m/z* 139, protonated 3,4-dihydroxybenzaldehyde ($[C_7H_6O_3+H]^+$), is initiated by a 1,2-H-shift in the ion with *m/z* 195 resulting in the more stable benzyl cation. It involves the rearrangement of the hydroxy group of the acid moiety to the benzylic-C and the loss of 1-propen-1-one ($CH_3CH{=}C{=}O$). Similar fragmentation is observed in the amino acid tyrosine and its 3-nitro analog. The mechanism has been elucidated using [D]-labeling experiments (Delatour et al., 2002) (Section 3.5.8). In negative-ion mode, levodopa ($[M-H]^-$ with *m/z* 196) shows fragment ions with *m/z* 179 due to the loss of NH_3, *m/z* 152 due to the loss of carbon dioxide (CO_2), and the major fragment ion with *m/z* 135 due to the loss of both CO_2 and NH_3. A fragment ion with *m/z* 122 ($[C_7H_6O_2]^{-\bullet}$) is also observed.

The fragmentation of four other antiparkinsonian drugs, selegiline, ropinirole, cabergoline, and metixene, is outlined in Figure 4.5.13. The fragmentation of selegiline ($[M+H]^+$ with *m/z* 188) is very similar to that of amphetamines (Section 4.7.2.1). Relative to the phenyl group, it involves the cleavage of the α,β-C—C bond to form a tropylium ion ($[C_7H_7]^+$) with *m/z* 91 or the cleavage of the β-C—N bond, resulting in two complementary fragments, the α-(ethyl)benzyl cation ($[C_6H_5CHCH_2CH_3]^+$) with *m/z* 119 and the protonated 3-(methylamino)propyne ($[HC{\equiv}CCH_2NHCH_3+H]^+$) with *m/z* 70. Related fragmentation is observed for pholedrine ($[M+H]^+$ with *m/z* 166) and rasagiline ($[M+H]^+$ with *m/z* 172).

Ropinirole ($[M+H]^+$ with *m/z* 261) shows the loss of ethyldipropylamine ($(C_3H_7)_2NCH_2CH_3$) from the 1,3-dihydroindol-2-one group to an ion with *m/z* 132 ($[C_8H_6NO]^+$). The loss of dipropylamine ($(C_3H_7)_2NH$) leads to a fragment ion with *m/z* 160 ($[C_{10}H_{10}NO]^+$). The *N,N*-dipropylmethaniminium ion ($[H_2C{=}N(C_3H_7)_2]^+$) with *m/z* 114 is observed as well, which in turn shows the loss of ethene (C_2H_4) via a six-membered ring transition state (Section 3.4.2 and Figure 3.6) to the *N*-methyl-*N*-propylmethaniminium ion ($[(CH_3)(C_3H_7)N{=}CH_2]^+$) with *m/z* 86 (Figure 4.5.13).

The fragmentation of cabergoline ($[M+H]^+$ with *m/z* 452) can be described as a series of consecutive small-molecule losses involving the initial loss of *N*-vinylformamide ($H_2C{=}CHNHCHO$) to an ion with *m/z* 381, the loss of

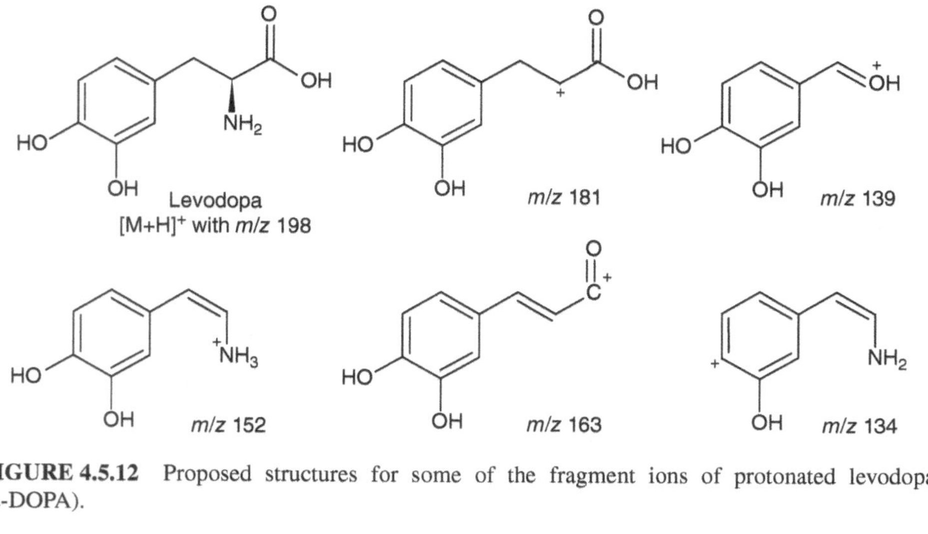

FIGURE 4.5.12 Proposed structures for some of the fragment ions of protonated levodopa (L-DOPA).

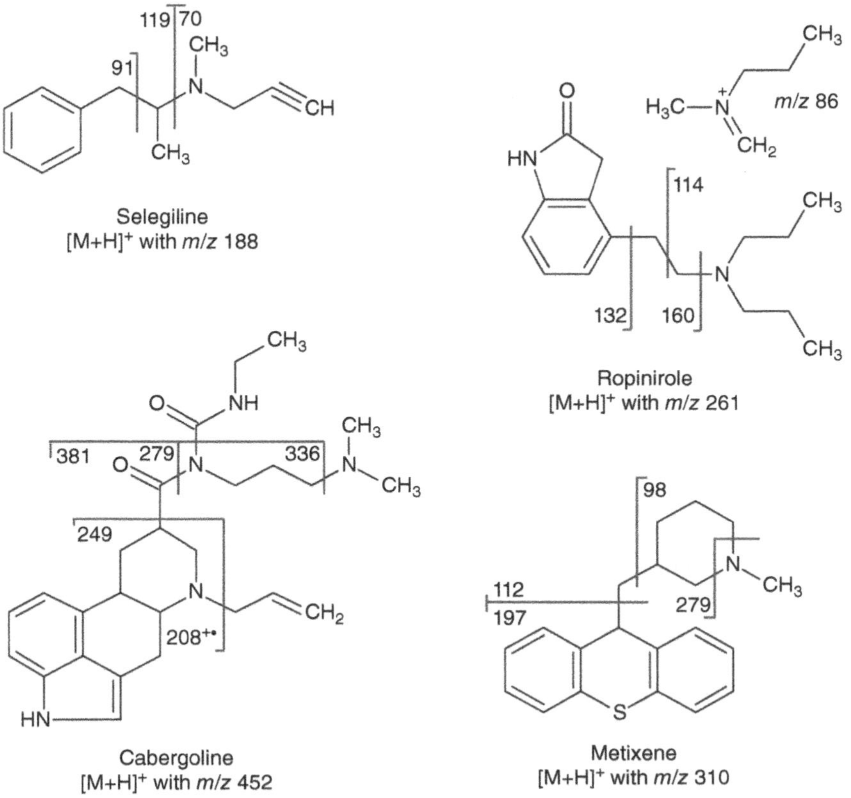

FIGURE 4.5.13 Fragmentation schemes for protonated selegiline, ropinirole, cabergoline, and metixene.

dimethylamine ($(CH_3)_2NH$) to an ion with *m/z* 336, further loss of either allylamine ($H_2C{=}CHCH_2NH_2$) to an ion with *m/z* 279 or *N*-propylformamide ($CH_3CH_2CH_2NHCHO$) to an ion with *m/z* 249, which in turn loses the allyl radical ($^{\bullet}CH_2CH{=}CH_2$) to an ion with *m/z* 208 (Figure 4.5.13) (Allievi & Dostert, 1998).

Metixene ($[M+H]^+$ with *m/z* 310) shows a minor fragment ion with *m/z* 279 due to the loss of methylamine (CH_3NH_2) and two major fragment ions due to a direct cleavage (Section 3.5.4) of the C^9—C bond of the thioxanthene group, thus resulting in the thioxanthenium ion ($[C_{13}H_9S]^+$) with *m/z* 197 and the (1-methylpiperidin-3-yl)methylium ion ($[C_7H_{14}N]^+$) with *m/z* 112 (Figure 4.5.13). The loss of propene ($CH_3CH{=}CH_2$) from the ion with *m/z* 112 results in the fragment ion with *m/z* 70, for example, the *N*-vinyl-*N*-methylmethaniminium ion or the protonated *N*-(allyl)methanimine ($[C_4H_8N]^+$).

Benserazide can be analyzed in both positive-ion and negative-ion modes. In positive-ion mode, benserazide ($[M+H]^+$ with *m/z* 258) shows two complementary fragment ions due to the cleavage of the C—N bond between the benzylic-C-atom and the hydrazine-N-atom, which leads to protonated 2-amino-3-hydroxypropanehydrazide ($[C_3H_9N_3O_2+H]^+$) with *m/z* 120 and the 1,2,3-trihydroxybenzyl cation ($(HO)_3C_6H_2CH_2]^+$) with *m/z* 139. The ion with *m/z* 120 undergoes secondary fragmentation to an ion with *m/z* 103 due to the loss of NH_3, and by subsequent losses of carbon monoxide (CO) or hydrogen isocyanate ($HN{=}C{=}O$) to the protonated 2-aminoacetamide ($[H_2NCH_2CONH_2+H]^+$) with *m/z* 75 or the protonated aminoacetaldehyde ($[H_2NCH_2CHO+H]^+$) with *m/z* 60, respectively. In negative-ion mode, benserazide ($[M-H]^-$ with *m/z* 256) shows two complementary fragment ions resulting from the same C—N bond cleavage as in positive-ion mode analysis. These ions with *m/z* 118 ($[C_3H_8N_3O_2]^-$) and *m/z* 137 ($[C_7H_5O_3]^-$) yield secondary fragment ions by the subsequent loss of diazene ($HN{=}NH$) to form an ion with *m/z* 88 or of CO to an ion with *m/z* 109, respectively.

4.5.6 ANTINEOPLASTIC AND CYTOSTATIC DRUGS

Antineoplastic drugs are used in the chemotherapy treatment of cancer. This involves various compound classes such as antimetabolites, cytotoxic antibiotics, alkylating agents, antimicrotubule agents, and topoisomerase inhibitors.

The analysis of antineoplastic drugs is relevant for a number of reasons. First of all, quantitative bioanalysis is relevant in drug development and therapeutic drug monitoring. Since these are highly toxic compounds, there are concerns with respect to laboratory safety for people preparing formulations and administering these types of drugs (Sottani et al., 2004; Nussbaumer et al., 2012). In addition, their presence in the (aquatic) environment calls for the development of multiresidue methods for the analysis of cytostatic and antineoplastic drugs (Kosjek & Heath, 2011; Negreira et al., 2013; Negreira et al., 2014).

4.5.6.1 Antimetabolites

Antimetabolites impede DNA and/or RNA synthesis. They often have a similar structure to the DNA or RNA building blocks, for example, nucleobases, nucleosides, or nucleotides. Similar compounds are also applied as antiviral agents (Section 4.10.3). These compounds are mostly analyzed in positive-ion mode, although some negative-ion MS–MS data are present in our spectral library collection as well.

In positive-ion mode, the modified nucleobases show small-molecule losses. 6-Mercaptopurine ($[M+H]^+$ with *m/z* 153), for instance, shows the loss of hydrogen cyanide (HCN) to an ion with *m/z* 126 or the loss of hydrogen isothiocyanate (HN=C=S) to an ion with *m/z* 94, which in turn shows the loss of HCN to an ion with *m/z* 67. Tioguanin ($[M+H]^+$ with *m/z* 168) shows the loss of HN=C=S to an ion with *m/z* 109, followed by the loss of ammonia (NH_3) to an ion with *m/z* 92.

In negative-ion mode, 6-mercaptopurine ($[M-H]^-$ with *m/z* 151) shows three major fragment ions: loss of HCN to an ion with *m/z* 124, the loss of hydrogen sulfide (H_2S) to an ion with *m/z* 117, and the loss of HN=C=S to an ion with *m/z* 92. Methylthiouracil ($[M-H]^-$ with *m/z* 141) produces an isothiocyanate fragment ion ($[N{=}C{=}S]^-$) with *m/z* 58. 5-Fluorouracil ($[M-H]^-$ with *m/z* 129) yields a fragment ion with *m/z* 86 due to the loss of hydrogen isocyanate (HN=C=O), possibly involving a retro-Diels–Alder mechanism (Section 3.5.6), which by the subsequent loss of HCN leads to an ion with *m/z* 59.

Nucleosides show primarily the cleavage of the glycosidic bond between the nucleobase and the anomeric $C^{1'}$ of (deoxy)ribose. In the positive-ion mode, charge retention is predominantly on the nucleobase. This is the case for tegafur ($[M+H]^+$ with *m/z* 201) producing protonated 5-fluorouracil ($[C_4H_3FN_2O_2+H]^+$) with *m/z* 131; for cytarabine ($[M+H]^+$ with *m/z* 244) and gemcitabine ($[M+H]^+$ with *m/z* 264), which share the same nucleobase, producing protonated cytosine ($[C_4H_5N_3O+H]^+$) with *m/z* 112; and for cladribine ($[M+H]^+$ with *m/z* 286) producing protonated 2-chloroadenine ($[C_5H_4ClN_5+H]^+$) with *m/z* 170. Capecitabine ($[M+H]^+$ with *m/z* 360) is a prodrug of 5-fluorouracil, which in positive-ion mode shows the loss of the 5′-deoxyribose group to an ion with *m/z* 244 (Montange et al., 2010). The negative-ion fragmentation of capecitabine ($[M-H]^-$ with *m/z* 358) results in an abundant fragment ion with *m/z* 154 due to the losses of the 5′-deoxyribose group (as the dehydrodeoxyribose ($C_5H_8O_3$)) and pentanol ($C_5H_{11}OH$), the latter originating from the alkyl chain of the ester group. In addition, several low-abundance fragment ions present are not well understood. The related substances 5′-deoxy-5-fluorocytidine ($[M-H]^-$ with *m/z* 244) and 5′-deoxy-5-fluorouridine ($[M-H]^-$ with *m/z* 245) show the loss of 5′-deoxyribose group to the main fragment ions, that is, deprotonated flucytosine ($[C_4H_4FN_3O-H]^-$) with *m/z* 128 and deprotonated fluorouracil ($[C_4H_3FN_2O_2-H]^-$) with *m/z* 129, respectively (Montange et al., 2010).

Nucleotides are analyzed in negative-ion mode as well. Fludarabine phosphate ($[M-H]^-$ with *m/z* 364) shows the $[PO_3]^-$ ion with *m/z* 79 and dihydrogen phosphate ($[H_2PO_4]^-$) with *m/z* 97. In addition, by the direct cleavage of the glycosidic bond, two fragment ions are produced: the deprotonated 2-fluoroadenine ($[C_5H_4FN_5-H]^-$) with *m/z* 152 and deprotonated dehydroribose 5′-phosphate ($[C_5H_9O_7P-H]^-$) with *m/z* 211. The *m/z* values of these two ions are equal to *m/z* ($[M-H]-1$) (Section 3.7.2).

Methotrexate and pemetrexed are antifolate antimetabolites, which interfere with nucleoside synthesis. The major fragment ion of methotrexate ($[M+H]^+$ with *m/z* 455) results from the amide C—N bond cleavage, leading to an acylium ion with *m/z* 308 ($[C_{15}H_{14}N_7O]^+$) due to the loss of 2-aminopentane-1,5-dioic acid ($C_5H_9NO_4$). The complementary fragment ions with *m/z* 175 and 134 may be considered as secondary fragments of the ion with *m/z* 308. They result from the C—N bond cleavage of the 4-methylamino group of the benzamide and the 6-methylpteridine group with charge retention on either fragment. Thus, the ion with *m/z* 175 is the (2,4-diaminopteridin-6-yl)methylium ion ($[C_7H_7N_6]^+$) and the ion with *m/z* 134 is the 4-(*N*-methylamino)phenyl acylium ion ($[(CH_3NH{-}C_6H_4{-}C{=}O]^+$). In negative-ion mode, pemetrexed ($[M-H]^-$ with *m/z* 426) shows fragment ions with *m/z* 408 due to the loss of water, *m/z* 382 due to the loss of carbon dioxide (CO_2), *m/z* 339 due to a subsequent loss of HN=C=O, and *m/z* 297 due to a subsequent loss of methanediimine (HN=C=NH).

4.5.6.2 Cytotoxic Antibiotics: Anthracyclines

Anthracyclines are derived from the bacterium *Streptomyces peucetius* and are widely prescribed in cancer chemotherapy because of their effectiveness against many types of carcinoma. Daunorubicin is produced naturally and other related compounds are derived from it. Its structure consists of the deoxyhexosamine daunosamine (4-amino-6-methyl-tetrahydropyran-2,5-diol) with an ether

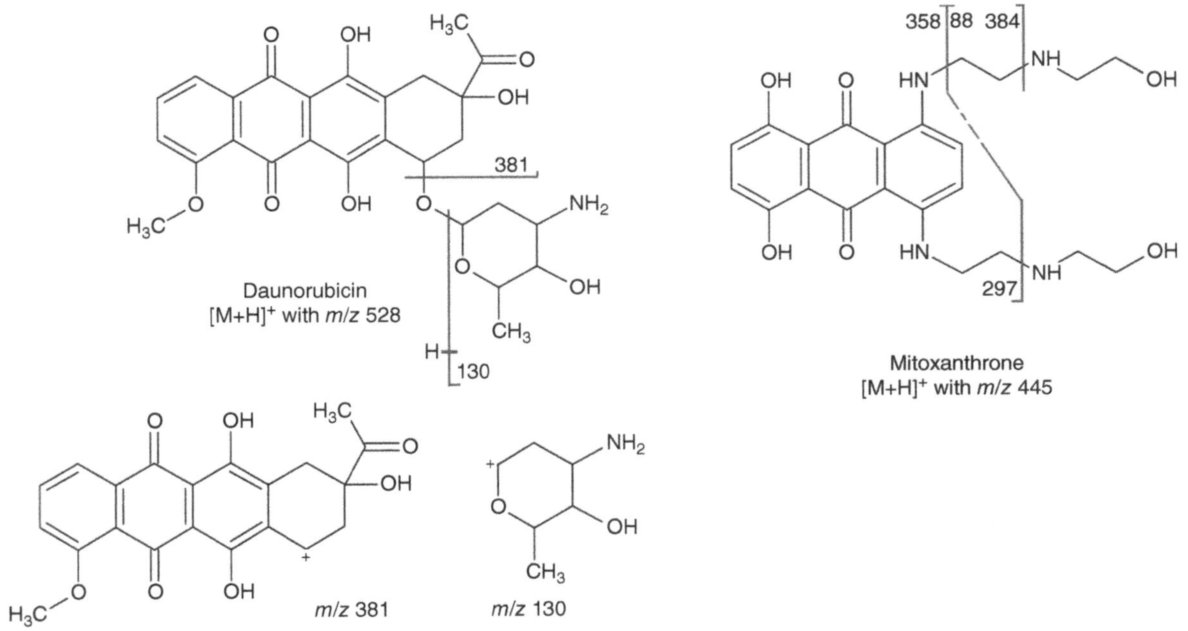

FIGURE 4.5.14 Proposed structures for the fragment ions of protonated anthracycline daunorubicin and fragmentation scheme for mitoxanthrone.

linkage to a substituted anthraquinone derivative. It is the exact nature of the substitution on this anthraquinone group that characterizes all anthracyclines sharing the same hexosamine daunosamine. The structure of daunorubicin ($[M+H]^+$ with *m/z* 528) and its two major fragment ions with *m/z* 381 and 130 are shown in Figure 4.5.14. These carbocations are the result of cleavage of one of the two C—O bonds of the ether linkage between the substituted anthraquinone group and daunosamine. Whereas the ion with *m/z* 381 ($[C_{21}H_{17}O_7]^+$) shows subsequent losses of water to an ion with *m/z* 363 and of ethenone ($H_2C{=}C{=}O$) to an ion with *m/z* 321, the ion with *m/z* 130 ($[C_6H_{12}NO_2]^+$) shows secondary fragmentation by the loss of either ethanal (CH_3CHO) or propen-1-ol ($CH_3CH{=}CHOH$) to ions with *m/z* 86 and 72, respectively (Sottani et al., 2004). Related compounds show analogous fragmentation. The above-mentioned cleavage of the ether linkages with the generation of two major fragments is quite general, for example, idarubicin ($[M+H]^+$ with *m/z* 498) shows fragment ions with *m/z* 351 ($[C_{20}H_{15}O_6]^+$) and *m/z* 130, epirubicin and its enantiomer doxorubicin ($[M+H]^+$ with *m/z* 544) ions with *m/z* 397 ($[C_{21}H_{17}O_8]^+$) and *m/z* 130, and zorubicin ($[M+H]^+$ with *m/z* 646) ions with *m/z* 499 ($[C_{28}H_{23}N_2O_7]^+$) and *m/z* 130. Aclarubicin ($[M+H]^+$ with *m/z* 812) first shows the loss of the 2-methylpyran dimer linked to $C^5$of daunosamine to an ion with *m/z* 570 [$(C_{30}H_{35}NO_{10}+H]^+$). Secondary fragmentation involving the cleavage of the two C—O bonds of the ether linkage between the substituted anthraquinone group and *N*,*N*-dimethyl daunosamine results in the two carbocations with *m/z* 395 ($[C_{22}H_{19}O_7]^+$) and *m/z* 158 ($[C_8H_{16}NO_2]^+$) analogous to daunorubicin. Mitoxanthrone ($[M+H]^+$ with *m/z* 445) is an anthracycline with a 5,8-bis[2-(2-hydroxyethylamino)ethylamino] substitution. Its fragmentation involves small-molecule losses due to cleavages of C—N bonds of the (2-hydroxyethylamino)ethylamino substituents. The loss of 2-aminoethanol (C_2H_7NO) results in an ion with *m/z* 384, the loss of 2-(ethylamino)ethanol ($C_4H_{11}NO$) in an ion with *m/z* 358, and the subsequent loss of another 2-aminoethanol to an ion with *m/z* 297 ($[C_{16}H_{13}N_2O_4]^+$). The 2-[(2-hydroxyethyl)amino]ethyl carbocation ($[C_4H_{10}NO]^+$) with *m/z* 88, complementary to the ion with *m/z* 358, is observed as well (Figure 4.5.14).

In negative-ion mode, daunorubicin ($[M-H]^-$ with *m/z* 526) shows the loss of daunosamine ($C_6H_{13}NO_3$) to an ion with *m/z* 379, followed by the loss of an acetyl radical ($CH_3C{\equiv}O^\bullet$) to an ion with *m/z* 336, followed by the loss of methyl radical ($^\bullet CH_3$) to an ion with *m/z* 321.

4.5.6.3 Alkylating Agents

Alkylating agents are the oldest class of compounds used in cancer chemotherapeutics. They interfere with DNA replication by alkylating the guanine base at the N^7-position. In cancer therapy, mostly nitrogen mustards, that is, compounds derived from bis(2-chloroethyl)amine ($HN(CH_2CH_2Cl)_2$), are used. This class comprises various subclasses.

The fragment ions observed in positive-ion MS–MS for the oxazaphosphinan-2-amine alkylating agents ifosfamide

($[M+H]^+$ with *m/z* 261), cyclophosphamide ($[M+H]^+$ with *m/z* 261), and trofosfamide (trosfamide, $[M+H]^+$ with *m/z* 323) are shown in Figure 4.5.15. Initial fragmentation involves either the oxazaphosphinan ring opening with the loss of ethene (C_2H_4) to ions with *m/z* 233, 233, and 295, respectively, or the cleavage of phophamide P—N bond to the 2-chloroethylamine group(s). The latter fragmentation leads to ions with charge retention on either the oxazephosphinan ring to ions with *m/z* 182 ($[C_5H_{10}ClNO_2]^+$), *m/z* 120 ($[C_3H_7NO_2P]^+$), and *m/z* 182 ($[C_5H_{10}ClNO_2]^+$), respectively, or the chloroethylamine group leading to ions with *m/z* 78 ($[ClCH_2CH{=}NH{+}H]^+$), *m/z* 140 ($[C_4H_7Cl_2N{+}H]^+$), and *m/z* 140 ($[C_4H_7Cl_2N{+}H]^+$), respectively (Figure 4.5.15). The loss of hydrogen chloride (HCl) from $[M+H]^+$ or from Cl-containing fragment ions is observed as well.

The urea derivative carmustin ($[M+H]^+$ with *m/z* 214, 1,3-bis(2-chloroethyl)-1-nitrosourea) shows fragment ions with *m/z* 151 due to the loss of 2-chloroethyl radical ($^\bullet CH_2CH_2Cl$), *m/z* 108 due to the subsequent loss of hydrogen isocyanate (HN=C=O), *m/z* 106 due to the losses of azanone (HN=O) and 2-chloroethanimine ($HN{=}CHCH_2Cl$), and with *m/z* 63 due to the 2-chloroethyl cation ($[ClCH_2CH_2]^+$). Another urea derivative is nimustine ($[M+H]^+$ with *m/z* 273), which shows an ion with *m/z* 165 due to the cleavage of the urea C—N bond between the carbonyl group and the disubstituted amino moiety with charge retention on the carbonyl and an ion with *m/z* 122 due to a subsequent loss of HN=C=O. The fragment ion with *m/z* 194 results from an intramolecular nucleophilic attack by the nitroso N-atom on the chloromethyl C-atom, leading to the loss of chloronitrosomethane (CH_2ClNO) and the production of a protonated substituted methanimine ($[C_8H_{11}N_5O{+}H]^+$) (Figure 4.5.16).

Chlorambucil ($[M+H]^+$ with *m/z* 304) shows the loss of $^\bullet CH_2CH_2Cl$ to a radical cation with *m/z* 241 ($[C_{12}H_{16}ClNO_2]^{+\bullet}$), which after radical rearrangement to the phenyl ring undergoes a McLafferty rearrangement (Section 3.3 and Figure 3.2) leading to the cleavage of the C^3—C^4 bond of the butanoic acid chain, with the loss of the propanoic-3-yl acid radical ($^\bullet CH_2CH_2COOH$), and the formation of an benzylic carbocation, that is, the 4-[(2-chloroethyl)amino]benzyl cation ($[C_9H_{11}ClN]^+$) with *m/z* 168, shown in Figure 4.5.17. Melphalan ($[M+H]^+$ with *m/z* 305) shows fragment ions due to the loss of ammonia (NH_3) to an ion with *m/z* 288, a benzylic carbocation with *m/z* 230 ($[C_{11}H_{14}Cl_2N]^+$) due to the C^2—C^3 cleavage of the propanoic acid chain, and the loss of aminoacetic acid (H_2NCH_2COOH), followed by the loss of either HCl to an ion with *m/z* 194 or chloroethene ($H_2C{=}CHCl$) to an ion with *m/z* 168, which in turn loses HCl to an ion with *m/z* 132 ($[C_9H_{10}N]^+$) (Figure 4.5.17). The ion with *m/z* 246, that is, protonated 4-[bis(2-chloroethyl)amino]benzaldehyde ($[C_{11}H_{13}Cl_2NO{+}H]^+$), results from the loss of NH_3 followed by H-rearrangement and subsequent loss of ethenone ($H_2C{=}C{=}O$), as is also observed for levodopa and (nitro)tyrosine (Sections 4.5.5 and 3.5.8, respectively).

Another subclass of alkylating agents are the aziridine-type compounds such as mitomycin ($[M+H]^+$ with *m/z* 335), which shows fragment ions with *m/z* 303 due to the loss of methanol (CH_3OH); with *m/z* 274 due to the loss of NH_3 and CO_2; with *m/z* 242 due to the loss of CH_3OH, NH_3, and CO_2; and with *m/z* 215 ($[C_{12}H_{11}N_2O_2]^+$) due to a loss of hydrogen cyanide (HCN) from the ion with *m/z* 242.

4.5.6.4 Antimicrotubule Agents

Antimicrotubule agents are plant-derived compounds that block cell division (mitosis) by interfering in the synthesis of microtubule proteins, thereby promoting apoptosis in cancer cell. Demecolcine ($[M+H]^+$ with *m/z* 372, colcemid) is derived from the alkaloid colchicine. It shows extensive fragmentation. Various substituent losses have been reported for colchicine (Li et al., 2002). Small-molecule losses result in fragment ions with *m/z* 341 due to the loss of methylamine (CH_3NH_2), *m/z* 340 due to the loss of methanol (CH_3OH), *m/z* 315 due to the loss of *N*-allylmethylamine (C_3H_7N), *m/z* 312 due to a loss of carbon monoxide (CO) from the ion with *m/z* 340, and *m/z* 310 due to the loss of methoxy radical ($CH_3O^\bullet$) from the ion with *m/z* 341. Combination of such losses results in the fragment ions with *m/z* 256 ($[C_{16}H_{16}O_3]^{+\bullet}$), *m/z* 251 ($[C_{14}H_{21}NO_3]^+$), *m/z* 235 ($[C_{13}H_{17}NO_3]^{+\bullet}$), and *m/z* 208 ($[C_{12}H_{16}O_{35}]^{+\bullet}$). The ion with *m/z* 181 corresponds to the 3,4,5-trimethoxybenzyl cation ($[(CH_3O)_3C_6H_2CH_2]^+$). Structures for these fragments have been proposed in Figure 4.5.18.

Paclitaxel ($[M+H]^+$ with *m/z* 854, initially called taxol) and the related docetaxel were isolated from the bark of *Taxus brevifolia* in the late 1960s. The fragmentation of paclitaxel has been studied in detail (Kerns et al., 1994; Royer et al., 1995). Consecutive losses of acetic acid (CH_3COOH), water, and benzamide ($C_6H_5CH_2CONH_2$) lead to fragment ions with *m/z* 794, 776, and 655, respectively. Complementary fragment ions with *m/z* 569 and 286 result from the cleavage of the C^{13}—O ester bond between the A-ring of taxane and the benzamide-containing α-hydroxy acetate group (Figure 4.5.19). Further fragments may be described by small-molecule losses from the ion with *m/z* 569. These include ions with *m/z* 551 due to the loss of water, with *m/z* 509 due to the loss of CH_3COOH, with *m/z* 447 due to the loss of benzoic acid (C_6H_5COOH), with *m/z* 387 due to losses of both CH_3COOH and C_6H_5COOH; a further loss of CH_3COOH leads to the ion *m/z* 327. At low *m/z*, fragment ions with *m/z* 105 and 122 are observed, which are due to the benzoyl cation ($[C_6H_5{-}C{=}O]^+$) and to protonated benzamide ($[C_6H_5{-}CONH_3]^+$), respectively.

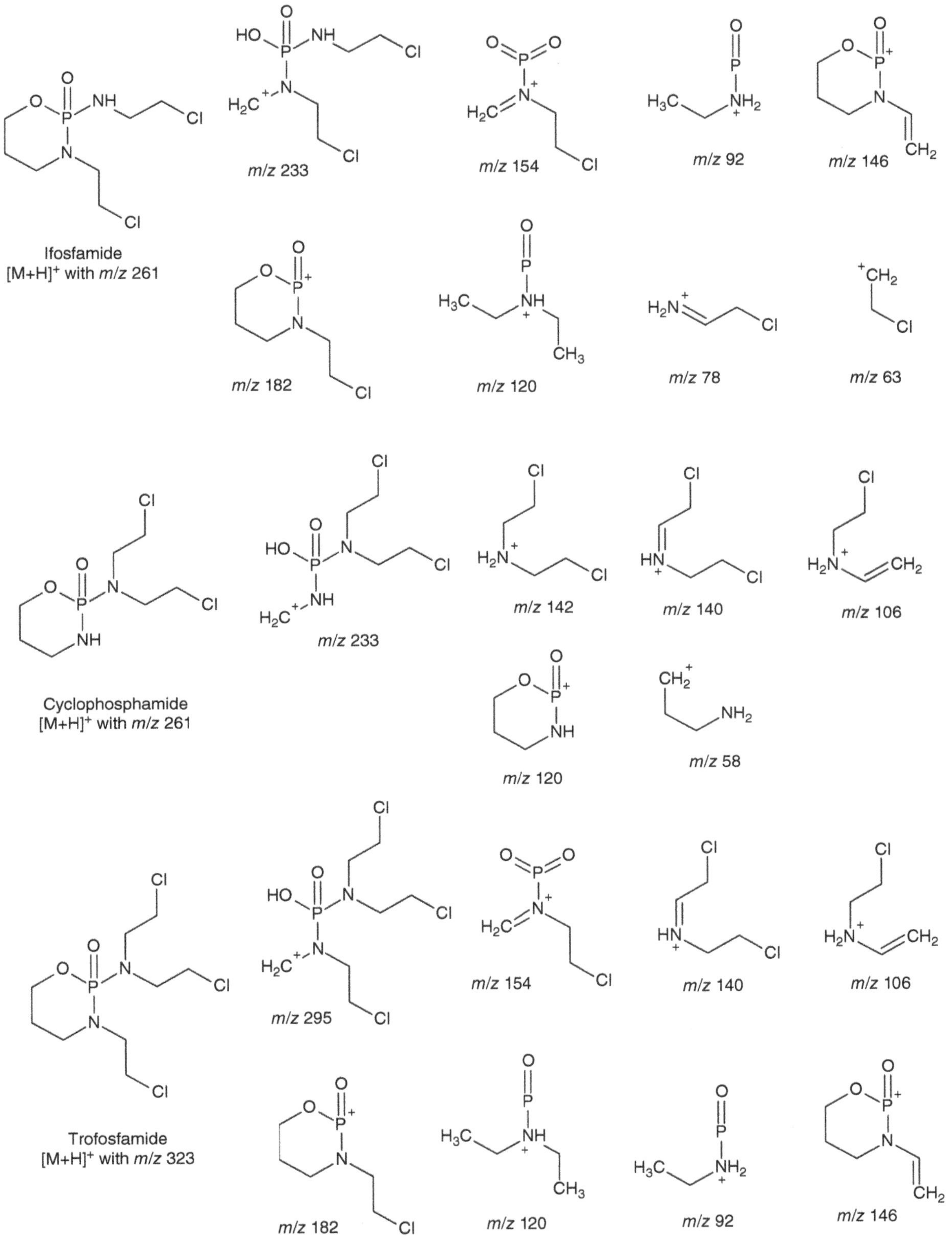

FIGURE 4.5.15 Proposed structures for the fragment ions of protonated ifosfamide, cyclophosphamide, and trofosfamide.

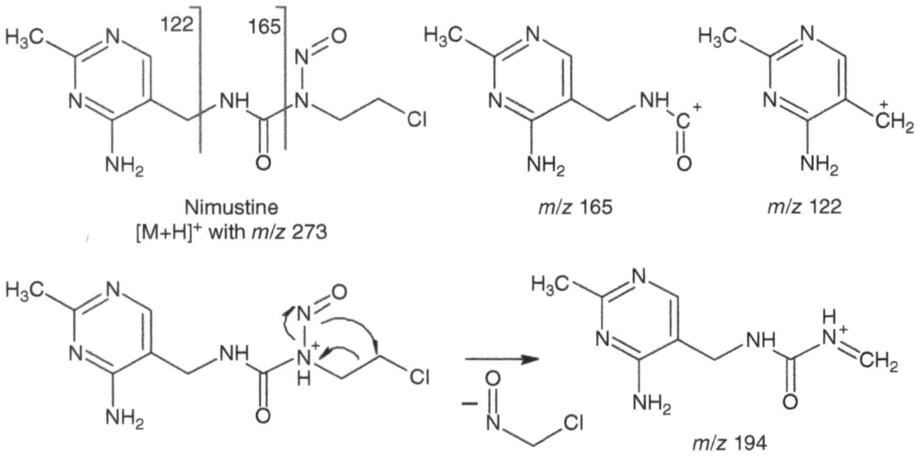

FIGURE 4.5.16 Proposed structures for the fragment ions of protonated nimustine.

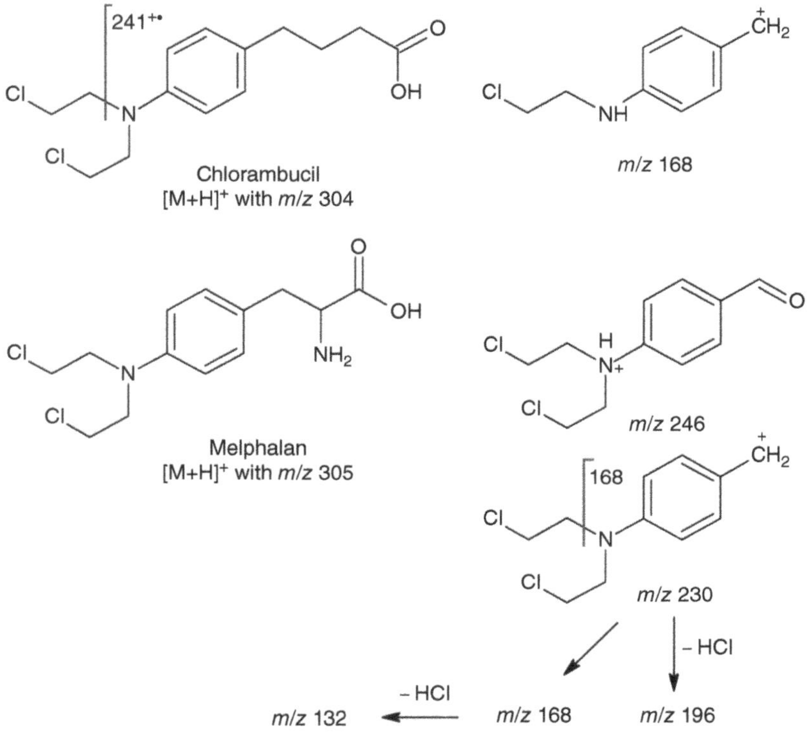

FIGURE 4.5.17 Fragmentation schemes for protonated chlorambucil and melphalan.

4.5.6.5 Topoisomerase Inhibitors

Teniposide ($[M+H]^+$ with *m/z* 657) is a derivative of the naturally occurring podophyllotoxin (a spindle toxin), whose mode of action is as a topoisomerase inhibitor. The major fragment ions of teniposide result from losses of either of the ring structures attached to C^1 and C^4 of the functionalized tetralin (Figure 4.5.20). This accounts for the fragment ions with *m/z* 503 and 383, whereas in the ion derived from tetraline with *m/z* 229 ($[C_{13}H_9O_4]^+$) both substituting ring structures have been lost. Further loss of carbon dioxide (CO_2) leads to the naphtho[2,3-d][1,3]dioxole-7-methylium ion with *m/z* 185 ($[C_{12}H_9O_2]^+$).

4.5.7 IMMUNOSUPPRESSIVE DRUGS

Immunosuppressive agents are drugs that inhibit or prevent the normal activity of the immune system, thereby averting

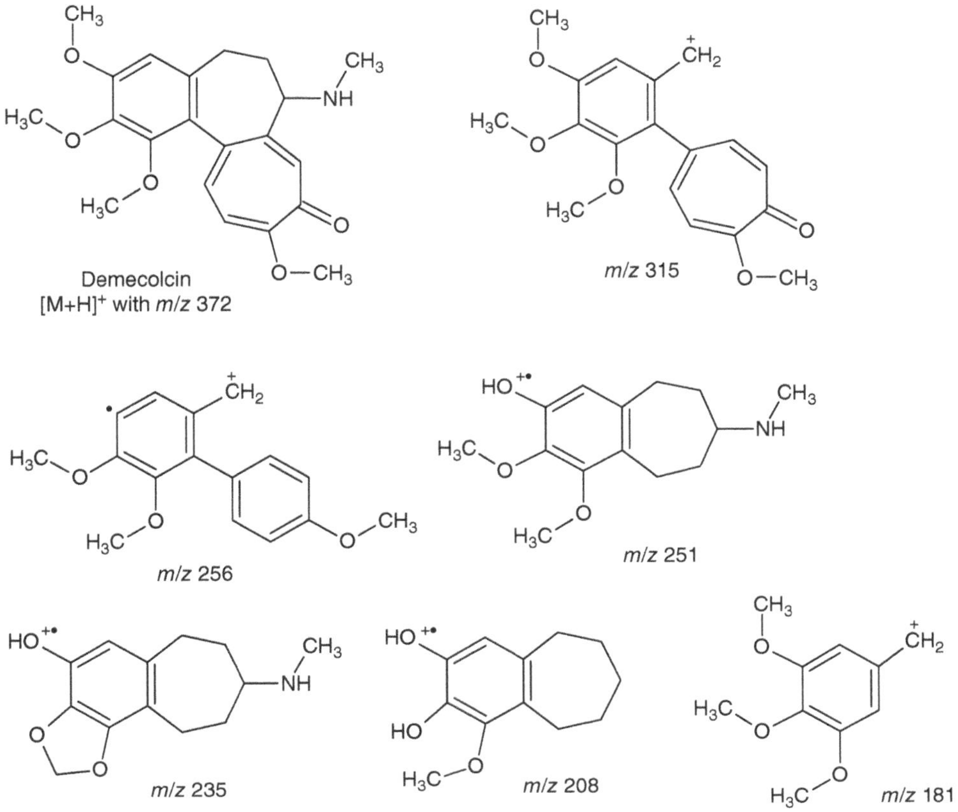

FIGURE 4.5.18 Proposed structures for the fragment ions of protonated demecolcin.

the rejection of transplanted organs or aiding in the treatment of various autoimmune diseases. There are various classes of immunosuppressive drugs: glucocorticosteroids (discussed in Section 4.6.6); cytostatic compounds like alkylating agents and antimetabolites (discussed in Section 4.5.6); and antibodies, interferons, and other therapeutic proteins that are beyond the scope of the present discussion on small-molecule drugs.

Ciclosporin A is a cyclic peptide that interferes in the activity and growth of T cells. In positive-ion mode, ciclosporin A is mostly analyzed as the ammoniated molecule ($[M+NH_4]^+$ with *m/z* 1219.9) and does not provide structure-specific fragmentation; often, the loss of ammonia (NH_3) is used in SRM (Keevil et al., 2002; Hinchliffe et al., 2011). In negative-ion MS–MS, ciclosporin A ($[M-H]^-$ with *m/z* 1200.8) shows an abundant fragment ion with *m/z* 1088.7 due to the loss of 2-methylhex-4-enal ($C_7H_{12}O$) from the C^{33} 1-hydroxy-2-methylhex-4-enyl peptidic bond substituent. This behavior was also observed in the thermospray ionization (TSI) LC–MS spectrum of ciclosporin A (Abián et al., 1992).

The macrolide immunosuppressants everolimus (SDZ-RAD), tacrolimus (FK-506), and sirolimus (rapamycin) show extensive fragmentation in positive-ion MS–MS. The fragmentation patterns of everolimus and sirolimus have been discussed in detail (Vidal et al., 1998; Gregory et al., 2006; Boernsen et al., 2007).

The immunosuppressive drug mycophenolic acid is a reversible inhibitor of inosine-5′-monophosphate dehydrogenase, an enzyme implicated in the *de novo* cellular synthesis of guanine nucleotides. It is frequently used in combination with ciclosporin or tacrolimus for treating transplant patients. Despite the presence of a carboxylic acid function, mycophenolic acid and its major glucuronic acid metabolite are analyzed as sodium or ammonium adducts in positive-ion mode (e.g., Kuhn et al., 2010). Protonated mycophenolic acid ($[M+H]^+$ with *m/z* 321) shows the loss of one or two water molecules to the ions with *m/z* 303 and 285, respectively, and the loss of formic acid (HCOOH) to an ion with *m/z* 275. Cleavage of C—C bonds in the 4-methylhex-4-enoic-6-yl acid substituent and charge retention on the 1(3*H*)-isobenzofuranone ring results in the fragment ions with *m/z* 207 ($[C_{11}H_{11}O_4]^+$) and *m/z* 195 ($[C_{10}H_{10}O_4+H]^+$) (Figure 4.5.21). In negative-ion mode, the glucuromycophenolic acid ($[M-H]^-$ with *m/z* 495) shows a fragment ion with *m/z* 319 due to the loss of dehydroglucuronic acid ($C_6H_8O_6$). The fragmentation of mycophenolic acid ($[M-H]^-$ with *m/z* 319) occurs on its 4-methylhex-4-enoic-6-yl acid

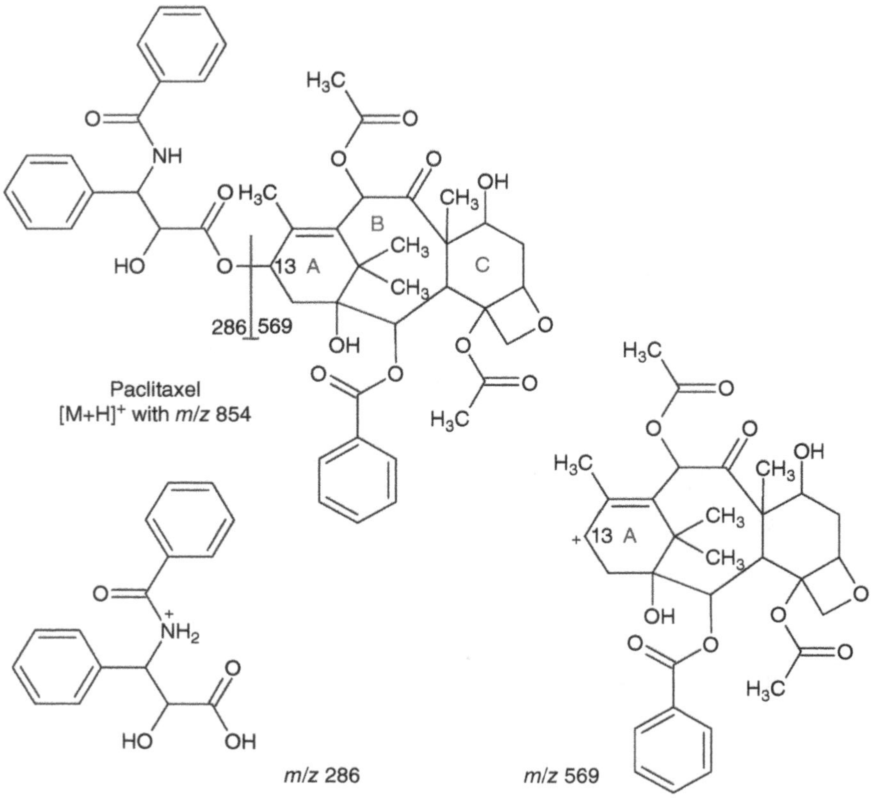

FIGURE 4.5.19 Fragmentation scheme for protonated paclitaxel.

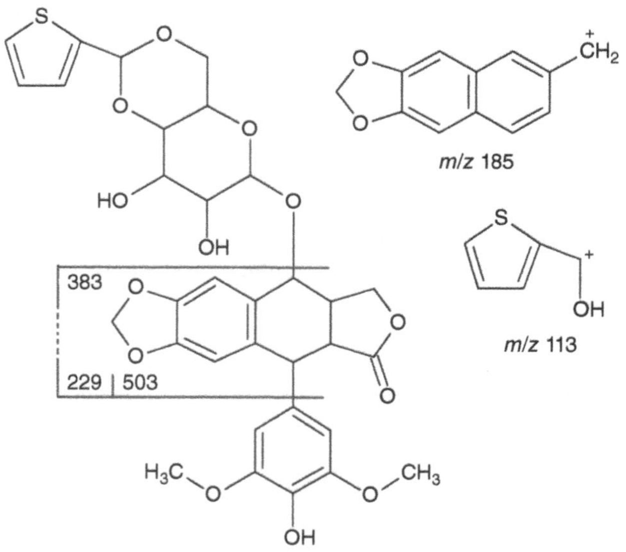

FIGURE 4.5.20 Fragmentation scheme for protonated teniposide.

substituent with the formation of ions with *m/z* 275 due to the loss of CO_2, *m/z* 245 due to the loss of propionic acid (C_2H_5COOH), and *m/z* 191 ($[C_{10}H_7O_4]^-$) due to the loss of 4-methylhex-4-enoic acid ($C_7H_{12}O_2$) (Figure 4.5.21).

The antirheumatic compound leflunomide ($[M-H]^-$ with *m/z* 269) and its isomeric immunosuppressive metabolite teriflunomide show the same fragment ions. Cleavage of the two C—C bonds to the carbonyl group leads to deprotonated 4-(trifluoromethyl)aniline ($[C_7H_6F_3N-H]^-$) with *m/z* 160 and the 1,2-oxazol-5-ide anion ($[C_4H_4NO]^-$) with *m/z* 82 (Figure 4.5.21). Secondary fragmentation of the ion with *m/z* 160 involves three consecutive losses of hydrogen fluoride (HF) to the ions with *m/z* 140, *m/z* 120, and *m/z* 100, whereas the ion with *m/z* 82 shows the subsequent loss of methane (CH_4) to a fragment ion with *m/z* 66 ($[C_3NO]^-$), for example, cyanoethynolate ($[N{\equiv}C—C{\equiv}C—O]^-$).

4.5.8 X-RAY CONTRAST AGENTS

Radiocontrast or X-ray contrast agents are compounds used to improve the visibility of internal bodily structures when using X-ray-based imaging techniques. The compounds present in our library collection all contain iodine. In

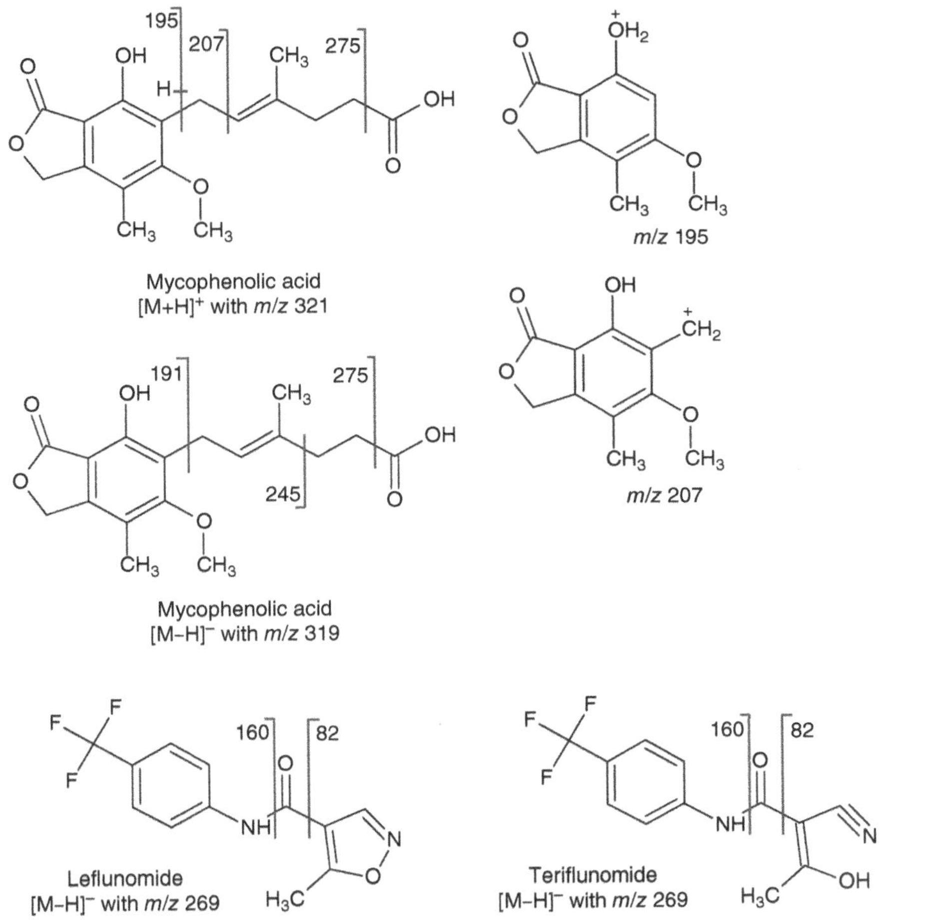

FIGURE 4.5.21 Fragmentation schemes for protonated mycophenolic acid and deprotonated mycophenolic acid, leflunomide, and teriflunomide.

negative-ion MS–MS, iodide (I^-) with *m/z* 126.9 is the most abundant fragment ion. This is about the only fragment ion observed for iobitridol ($[M-H]^-$ with *m/z* 833.9), iopromide ($[M-H]^-$ with *m/z* 789.9), and ioxaglic acid ($[M-H]^-$ with *m/z* 1267.6).

In positive-ion mode, iopromide ($[M+H]^+$ with *m/z* 791.9) shows fragment ions with *m/z* 773.9 due to the loss of water, with *m/z* 700.8 due to the loss of 3-aminopropane-1,2-diol ($C_3H_9NO_2$), and with *m/z* 686.8 due to the loss of 3-(methylamino)propane-1,2-diol ($C_4H_{11}NO_2$). The last two ions show two consecutive losses of hydrogen iodide (HI) to the ions with *m/z* 572.9 and 445.0 and with *m/z* 558.9 and 431.0, respectively, (Figure 4.5.22). Iocetamic acid ($[M+H]^+$ with *m/z* 614.8) shows fragment ions with *m/z* 596.8 due to the loss of water, *m/z* 554.8 due to the subsequent loss of the ethenone ($H_2C{=}C{=}O$), and *m/z* 498.8 due to a further loss of the methylethenone ($CH_3{-}CH{=}C{=}O$). The ion with *m/z* 596.8 shows two consecutive losses of an iodine radical ($I^\bullet$) to the ions with *m/z* 469.9 and 343.0, followed by a loss of $H_2C{=}C{=}O$ to the ions with *m/z* 427.9 and 300.9. The fragmentation of ioglicic acid ($[M+H]^+$ with *m/z* 671.8) can be partially described as a series of consecutive small-molecule losses. The ion with *m/z* 640.8 is due to the loss of methylamine (CH_3NH_2), the ion with *m/z* 612.8 due to a subsequent carbon monoxide (CO) loss, and the ion with *m/z* 583.7 due to either the subsequent loss of methanimine ($H_2C{=}NH$) or to the loss of 2-amino-*N*-methylacetamide ($C_3H_8N_2O$) from the intact protonated molecule (Figure 4.5.22). The ion with *m/z* 640.8 may show consecutive losses of HI and the iminoacetaldehyde (HN═CHCHO), thus accounting for the fragment ions with *m/z* 512.8 and 455.8, respectively. The ion with *m/z* 428.8 ($[C_9H_5NI_2O_3]^{+\bullet}$) is due to losses of $I^\bullet$ and CO from the ion with *m/z* 583.7. A further loss of $I^\bullet$ from the ion with *m/z* 428.8 results in an ion with *m/z* 301.9. Structure proposals for some of these ions are given in Figure 4.5.22.

Acetrizoic acid ($[M+H]^+$ with *m/z* 557.8) shows the loss of water to an ion with *m/z* 539.7, the loss of $I^\bullet$ to an ion with *m/z* 430.9, the losses of two $I^\bullet$ to an ion with *m/z* 303.9, and the losses of three $I^\bullet$ and CO_2 to an ion with *m/z* 133.

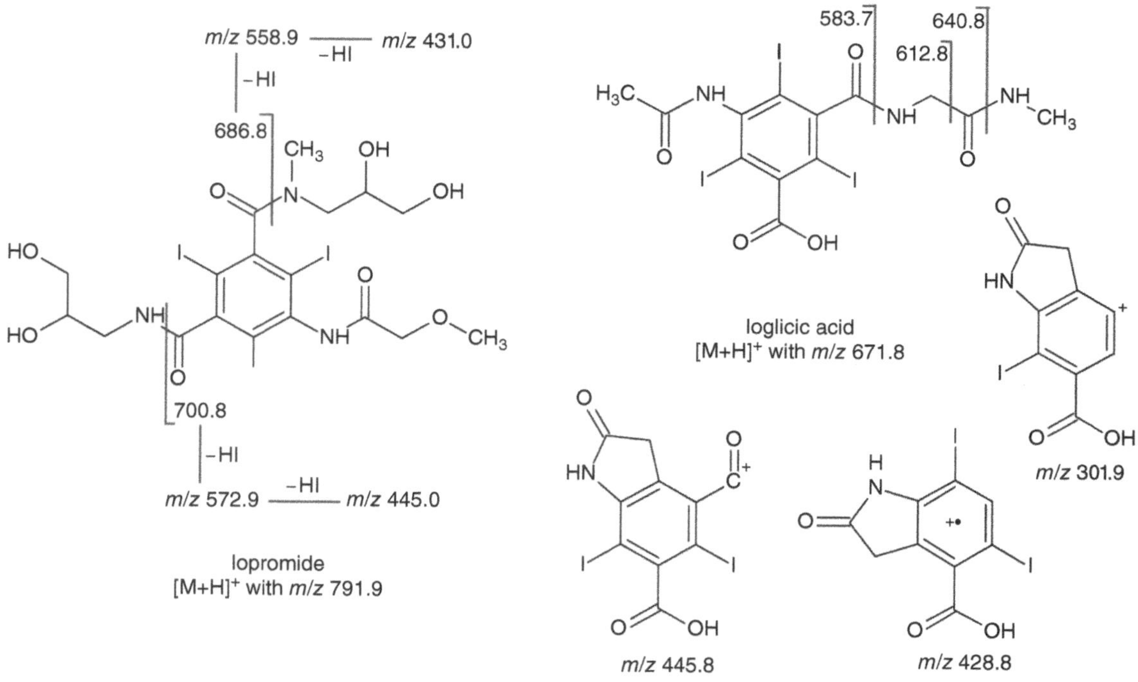

FIGURE 4.5.22 Fragmentation schemes for the protonated X-ray contrast agents iopromide and ioglicic acid.

The ion with *m/z* 430.9 shows the loss of $H_2C{=}C{=}O$ to an ion with *m/z* 388.8. The loss of formic acid (HCOOH) from the ion with *m/z* 303.9 results in an ion with *m/z* 257.9. In negative-ion mode, acetrizoic acid ($[M-H]^-$ with *m/z* 555.7) shows the loss of carbon dioxide (CO_2) and $H_2C{=}C{=}O$ to an ion with *m/z* 469.7, the loss of CO_2 and HI to an ion with *m/z* 383.8, and I^- with *m/z* 126.9. The structurally related metrizoic acid ($[M+H]^+$ with *m/z* 628.8) shows fragment ions with *m/z* 500.9 due to the loss of HI, with *m/z* 455.9 due to the loss of HCOOH and $I^\bullet$, with *m/z* 375.0 due to two $I^\bullet$ losses, which in turn either shows the loss of an acetyl radical ($CH_3C{=}O^\bullet$) to an ion with *m/z* 332.0 or the loss is HI to an ion with *m/z* 247.1. Adipiodone ($[M-H]^-$ with *m/z* 1138.5) shows fragments due to two consecutive losses of CO_2 and the loss of HI leading to ions with *m/z* 1094.5, *m/z* 1050.5, and *m/z* 922.6, respectively; I^- with *m/z* 127 is also observed. Propyliodone ($[M+H]^+$ with *m/z* 447.9) is based on a 3,5-diiodo-1*H*-pyridin-4-one ring instead of a 2,4,6-triiodo-benzene ring. It shows the loss of propene ($CH_3CH{=}CH_2$) to an ion with *m/z* 405.8, followed by consecutive losses of $I^\bullet$ to an ion with *m/z* 278.9 and of another $I^\bullet$ to an ion with *m/z* 152.0.

4.5.9 ANTICOAGULANTS AND RODENTICIDES

Anticoagulants are administered to prevent blood from clotting. They act as vitamin K antagonists, which is an essential element of the coagulation cascade. For this reason, they are used as medication for treating thrombotic disorders. Unexpectedly perhaps some anticoagulants are also widely used as rodenticides. Two compound classes can be distinguished: coumarins and indandiones. Multiresidue LC–MS analysis of anticoagulant rodenticides in various matrices has been described, using either SIM or SRM (Jin & Chen, 2006; Marek & Koskinen, 2007; Jin et al., 2008).

4.5.9.1 Coumarin Anticoagulants

Coumarin anticoagulants can be analyzed in both positive-ion and negative-ion modes. For the fragmentation in the positive-ion mode, coumafuryl ($[M+H]^+$ with *m/z* 299, fumarin) serves as an example. It shows fragment ions derived from either of the two substituent groups attached to the 3-methyl group of the 4-hydroxy-3-methylcoumarin ring. The loss of acetone (CH_3COCH_3) results in an ion with *m/z* 241 and the loss of the furan ring (C_4H_4O) in an ion with *m/z* 231. Cleavage of the C—C bond between 4-hydroxycoumarin and its 3-methyl substituent results in two complementary fragment ions: the protonated 4-hydroxycoumarin ($[C_9H_6O_3+H]^+$) with *m/z* 163, which is common to the compound class, and the 4-(2-furyl)butan-2-one-4-yl carbocation ($[C_8H_9O_2]^+$) with *m/z* 137 (Figure 4.5.23). The ion with *m/z* 163 may show a loss of ethenone ($H_2C{=}C{=}O$) to an ion with *m/z* 121 ($[C_7H_5O_2]^+$). Similar behavior is observed for other

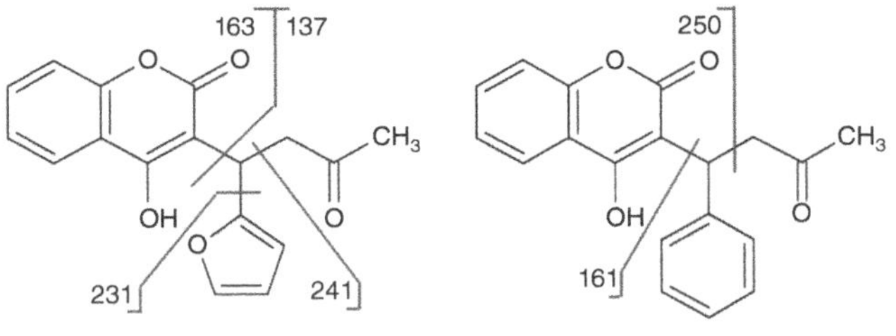

FIGURE 4.5.23 Fragmentation schemes for protonated coumafuryl and deprotonated warfarin.

molecules that share the same 4-hydroxycoumarin ring structure, as in coumatetralyl ($[M+H]^+$ with *m/z* 293), coumachlor ($[M+H]^+$ with *m/z* 343), warfarin ($[M+H]^+$ with *m/z* 309), and acenocoumarol ($[M+H]^+$ with *m/z* 354). In negative-ion mode, coumafuryl ($[M-H]^-$ with *m/z* 297), coumachlor ($[M-H]^-$ with *m/z* 341), warfarin ($[M-H]^-$ with *m/z* 307), and acenocoumarol ($[M-H]^-$ with *m/z* 352) show the loss of a propan-2-one-1-yl radical ($^{\bullet}CH_2COCH_3$) due to a homolytic C—C cleavage between the propan-2-one-1-yl substituent, which is common to all four compounds, and the 3-methyl substituent of coumarin. This results in deprotonated 4-hydroxycoumarin ($[C_9H_6O_3-H]^-$) with *m/z* 161, which in turn shows the loss of CO_2 to an ion with *m/z* 117 (for warfarin, Figure 4.5.23). Apart from the ion with *m/z* 295 due to the loss of $^{\bullet}CH_2COCH_3$, acenocoumarol also shows the loss of the nitrosyl radical ($^{\bullet}NO$) from its nitrobenzene ring to an ion with *m/z* 265.

4.5.9.2 Indanedione Anticoagulants

The indanedione anticoagulants such as pindone, chlorophacinone, clorindone, and phenindione are derivatives of 1,3-indandione (α,β-diketone). They can be analyzed in both positive-ion and negative-ion modes. In negative-ion mode, a class-specific fragment ion is observed due to the deprotonated 1,3-indandione ($[C_9H_5O_2]^-$) with *m/z* 145. Pindone ($[M+H]^+$ with *m/z* 231) shows fragment ions due to the loss of one or two water molecules to ions with *m/z* 213 or 195, respectively. The ions with *m/z* 185 and 171 are consistent with the loss of carbon monoxide (CO) and water and of CO and methane (CH_4). The ion with *m/z* 147 is protonated 1,3-indanedione ($[C_9H_6O_2+H]^+$), and the ion with *m/z* 57 is the *tert*-butyl cation ($[C_4H_9]^+$). In positive-ion mode, chlorophacinone ($[M+H]^+$ with *m/z* 375) shows fragment ions with *m/z* 357 due to the loss of water, with *m/z* 297 due to the loss of benzene (C_6H_6), and with *m/z* 263 due to the loss of chlorobenzene (C_6H_5Cl). The latter two fragment ions show a subsequent loss of CO to the ions with *m/z* 269 and 235, respectively. Next to the class-specific ion with *m/z* 145, chlorophacinone ($[M-H]^-$ with *m/z* 373) and diphenadione ($[M-H]^-$ with *m/z* 339) in negative-ion mode show diphenylmethanide anions with *m/z* 201 ($[ClC_6H_4—CH—C_6H_5]^-$) and *m/z* 167 ($[C_6H_5—CH—C_6H_5]^-$), respectively. The fragmentation of clorindione ($[M+H]^+$ with *m/z* 257) in positive-ion mode can be described as the result of consecutive small-molecule losses. One and two water losses result in the ions with *m/z* 239 and 221, the loss of water and CO in *m/z* 201, the loss of CO and hydrogen chloride (HCl) in *m/z* 193 ($[C_{14}H_9O]^+$), the loss of water, CO, and a chlorine radical ($Cl^{\bullet}$) in a radical cation with *m/z* 176 ($[C_{14}H_8]^{+\bullet}$), whereas the loss of HCl and two times CO leads to the ion with *m/z* 165 ($[C_{13}H_9]^+$). In negative-ion mode, clorindione ($[M-H]^-$ with *m/z* 255) shows a fragment ion with *m/z* 219 due to the loss of HCl and an intense fragment ion with *m/z* 169 (loss of 86 Da) that is not understood; accurate-mass data is missing. The ion with *m/z* 169 is also observed for the related fluindione ($[M-H]^-$ with *m/z* 239). Phenindione ($[M+H]^+$ with *m/z* 223) shows the loss of water to an ion with *m/z* 205, the loss of CO to an ion with *m/z* 195, and the loss of both water and CO to an ion with *m/z* 177. In negative-ion mode, phenindione ($[M-H]^-$ with *m/z* 221) shows an ion with *m/z* 193 due to the loss of CO, *m/z* 165 ($[C_{13}H_9]^-$) due to loss of two times CO, an ion with *m/z* 145 ($[C_9H_5O_2]^-$) due to the deprotonated 1,3-indandione, and an ion with *m/z* 117 ($[C_6H_4—CH{=}C{=}O]^-$).

4.5.9.3 Chloralose

Chloralose ($[M-H]^-$ with *m/z* 307) is a trichlorinated 1,2-acetal derivative of D-glucofuranose, which acts as a narcotic agent but is also used as a rodenticide. Its major fragments in negative-ion MS–MS are the ions with *m/z* 189 due to the loss of chloroform ($CHCl_3$) and an ion with *m/z* 161 due to a subsequent loss of CO. Other fragment ions observed are the trichloromethide ion (Cl_3C^-) with *m/z* 117, an ion with *m/z* 71 ($[C_3H_3O_2]^-$), and an ion with *m/z* 59 ($[C_2H_3O_2]^-$). The ion with *m/z* 101 ($[C_4H_5O_3]^-$) is due to secondary fragmentation of the ion with *m/z* 161, involving the loss of 1,2-ethenediol ($C_2H_4O_2$) from the C^5 substituent of the furanose.

4.5.10 CONCLUSIONS

In this section 4.5, a number of compound classes of several drugs has been taken together. Obviously, this list can be made considerably longer as many other drug classes are available. In some cases, known classes were left out because excellent reviews are available covering such a particular class. This is for instance the case for the nonsteroidal selective androgen receptor modulators (SARMs) (Thevis & Schänzer, 2008). In other cases, the therapeutic class of the drugs is clear, but compounds with widely differing structural features are used.

REFERENCES

Abián J, Stone A, Morrow MG, Creer MH, Fink LM, Lay JO Jr., 1992. Thermospray high-performance liquid chromatography/mass spectrometric determination of cyclosporins. Rapid Commun Mass Spectrom 6: 684–689.

Allievi C, Dostert P. 1998. Quantitative determination of cabergoline in human plasma using liquid chromatography combined with tandem mass spectrometry. Rapid Commun Mass Spectrom 12: 33–39.

Azevedo Marques L, Kool J, Lingeman H, Niessen WMA, Irth H. 2010. Production and on-line acetylcholinesterase bioactivity profiling of chemical and biological degradation products of tacrine. J Pharm Biomed Anal 53: 609–616.

Azevedo Marques L, Giera M, Lingeman H, Niessen WMA. 2011a. Analysis of acetylchlonesterase inhibitors: Bioanalysis, degradation and metabolism. Biomed Chromatogr 25: 278–299.

Azevedo Marques L, Maade I, de Kanter FJJ, Lingeman H, Irth H, Niessen WMA, Giera M. 2011b. Stability-indicating study of the anti-Alzheimer's drug galantamine hydrobromide. J Pharm Biomed Anal 55: 85–92.

Boernsen KO, Egge-Jacobsen W, Inverardi B, Strom T, Streit F, Schiebel HM, Benet LZ, Christians U. 2007. Assessment and validation of the MS/MS fragmentation patterns of the macrolide immunosuppressant everolimus. J Mass Spectrom 42: 793–802.

Chen G, Daaro I, Pramanik BN, Piwinski JJ. 2009. Structural characterization of *in vivo* rat liver microsomal metabolites of antihistamine desloratadine using LTQ–Orbitrap hybrid mass spectrometer in combination with online hydrogen/deuterium exchange HR-LC–MS. J Mass Spectrom 44: 203–213.

Chen H, Chen Y, Du P, Han F, Wang H, Zhang H. 2006. Sensitive and specific liquid chromatographic-tandem mass spectrometric assay for atropine and its eleven metabolites in rat urine. J Pharm Biomed Anal 40: 142–150.

Debrauwer L, Bories G. 1993. Determination of clenbuterol residues by liquid chromatography–electrospray mass spectrometry. Anal Chim Acta 275: 231–239.

Delatour T, Richoz J, Vouros P, Turesky RJ. 2002. Simultaneous determination of 3-nitrotyrosine and tyrosine in plasma proteins of rats and assessment of artifactual tyrosine nitration. J Chromatogr B 779: 189–199.

Doerge DR, Bajic S, Lowes S. 1993. Analysis of clenbuterol in human plasma using liquid chromatography/atmospheric-pressure chemical-ionization mass spectrometry. Rapid Commun Mass Spectrom 7: 462–464.

Gregory MA, Hong H, Lill RE, Gaisser S, Petkovic H, Low L, Sheehan LS, Carletti I, Ready SJ, Ward MJ, Kaja AL, Weston AJ, Challis IR, Leadlay PF, Martin CJ, Wilkinson B, Sheridan RM. 2006. Rapamycin biosynthesis: Elucidation of gene product function. Org Biomol Chem 4: 3565–3568.

Gergov M, Robson JN, Ojanperä I, Heinonen OP, Vuori E. 2001. Simultaneous screening and quantitation of 18 antihistamine drugs in blood by liquid chromatography ionspray tandem mass spectrometry. Forensic Sci Int 121: 108–115.

Hinchliffe E, Adaway JE, Keevil BG. 2011. Simultaneous measurement of cyclosporin A and tacrolimus from dried blood spots by ultra-high performance liquid chromatography tandem mass spectrometry. J Chromatogr B 883–884: 102–107.

Jegorov A, Buchta M, Sedmera P, Kuzma M, Havlicek V. 2006. Accurate product ion mass spectra of galanthamine derivatives. J Mass Spectrom 41: 544–548.

Jin MC, Chen XH. 2006. Rapid determination of three anticoagulant rodenticides in whole blood by liquid chromatography coupled with electrospray ionization mass spectrometry. Rapid Commun Mass Spectrom 20: 2741–2746.

Jin MC, Chen XH, Ye ML, Zhu Y. 2008. Analysis of indandione anticoagulant rodenticides in animal liver by eluent generator reagent free ion chromatography coupled with electrospray mass spectrometry. J Chromatogr A 1213: 77–82.

Keevil BG, Tierney DP, Cooper DP, Morris MR. 2002. Rapid liquid chromatography–tandem mass spectrometry for routine analysis of cyclosporin A over an extended concentration range. Clin Chem 48: 67–76.

Kerns EH, Volk KJ, Hill SE, Lee MS. 1994. Profiling taxanes in Taxus extracts using LC/MS and LC/MS/MS techniques. J Nat Prod 57: 1391–1403.

Kosjek T, Heath E. 2011. Occurrence, fate and determination of cytostatic pharmaceuticals in the environment, Trends Anal Chem 30: 1065–1087.

Kuhn, J Götting C, Kleesiek K. 2010. Simultaneous measurement of amiodarone and desethylamiodarone in human plasma and serum by stable isotope dilution liquid chromatography–tandem mass spectrometry assay. J Pharm Biomed Anal 51: 210–216.

Li W, Sun Y, Fitzloff JF, van Breemen RB. 2002. Evaluation of commercial ginkgo and echinacea dietary supplements for colchicine using liquid chromatography–tandem mass spectrometry. Chem Res Toxicol 15: 1174–1178.

Maláková J, Nobilis M, Svoboda Z, Lísa M, Holčapek M, Kvtina J, Klimeš J, Palička V. 2007. High-performance liquid chromatographic method with UV photodiode-array, fluorescence and mass spectrometric detection for simultaneous determination of galantamine and its phase I metabolites in biological samples. J Chromatogr B 853: 265–274.

Marek LJ, Koskinen WC. 2007. Multiresidue analysis of seven anticoagulant rodenticides by high-performance liquid chromatography/electrospray/mass spectrometry. J Agric Food Chem 55: 571–576.

Matsui K, Oda Y, Nakata H, Yoshimura T. 1999. Simultaneous determination of donepezil (aricept) enantiomers in human plasma by liquid chromatography–electrospray tandem mass spectrometry. J Chromatogr B 729: 147–155.

Montange D, Bérard M, Demarchi M, Muret P, Piédoux S, Kantelip JP, Royer B. 2010. An APCI LC–MS/MS method for routine determination of capecitabine and its metabolites in human plasma. J Mass Spectrom 45: 670–677.

Negreira N, Mastroianni N, López de Alda M, Barceló D. 2013. Multianalyte determination of 24 cytostatics and metabolites by liquid chromatography–electrospray-tandem mass spectrometry and study of their stability and optimum storage conditions in aqueous solution. Talanta 116: 290–299.

Negreira N, de Alda ML, Barceló D. 2014. Cytostatic drugs and metabolites in municipal and hospital wastewaters in Spain:

Filtration, occurrence, and environmental risk. Sci Total Environ 497–498: 68–77.

Nussbaumer S, Geiser L, Sadeghipour F, Hochstrasser D, Bonnabry P, Veuthey JL, Fleury-Souverain S. 2012. Wipe sampling procedure coupled to LC–MS/MS analysis for the simultaneous determination of 10 cytotoxic drugs on different surfaces. Anal Bioanal Chem 402: 2499–2509.

Oertel R, Kilian B, Siegmund W, Kirch W. 2007. Determination of propiverine and its metabolites in rat samples by liquid chromatography–tandem mass spectrometry. J Chromatogr A 1149: 121–126.

Picard N, Dridi D, Sauvage FL, Boughattas NA, Marquet P. 2009. General unknown screening procedure for the characterization of human drug metabolites: Application to loratadine phase I metabolism. J Sep Sci 32: 2209–2217.

Ponnayyan Sulochana S, Sharma K, Mullangi R, Sukumaran SK. 2014. Review of the validated HPLC and LC–MS/MS methods for determination of drugs used in clinical practice for Alzheimer's disease. Biomed Chromatogr 28: 1431–1490.

Royer I, Alvinerie P, Armand JP, Ho LK, Wright M, Monsarrat B. 1995. Paclitaxel metabolites in human plasma and urine: Identification of 6 alpha-hydroxytaxol, 7-epitaxol and taxol hydrolysis products using liquid chromatography/atmospheric-pressure chemical ionization mass spectrometry. Rapid Commun Mass Spectrom 9: 495–502.

Sottani C, Tranfo G, Bettinelli M, Faranda P, Spagnoli M, Minoia C. 2004. Trace determination of anthracyclines in urine: A new high-performance liquid chromatography/tandem mass spectrometry method for assessing exposure of hospital personnel. Rapid Commun Mass Spectrom 18: 2426–2436.

Thevis M, Wilkens F, Geyer H, Schänzer W. 2006. Determination of therapeutics with growth-hormone secretagogue activity in human urine for doping control purposes. Rapid Commun Mass Spectrom 20: 3393–3402.

Thevis M, Schänzer W. 2008. Mass spectrometry of selective androgen receptor modulators. J Mass Spectrom 43: 865–876.

Thomas S, Shandilya S, Bharati A, Paul SK, Agarwal A, Mathela CS. 2012. Identification, characterization and quantification of new impurities by LC-ESI/MS/MS and LC-UV methods in rivastigmine tartrate active pharmaceutical ingredient. J Pharm Biomed Anal 57: 39–51.

Upthagrove AL, Hackett M, Nelson WL. 1999. Fragmentation pathways of selectively labeled propranolol using electrospray ionization on an ion trap mass spectrometer and comparison with ions formed by electron impact. Rapid Commun Mass Spectrom 13: 534–541.

Verhaeghe T, Diels L, de Vries R, De Meulder M, de Jong J. 2003. Development and validation of a liquid chromatographic–tandem mass spectrometric method for the determination of galantamine in human heparinised plasma. J Chromatogr B 789: 337–346.

Vidal C, Kirchner GI, Sewing K-Fr. 1998. Structural elucidation by electrospray mass spectrometry: An approach to the in vitro metabolism of the macrolide immunosuppressant SDZ RAD. J Am Soc Mass Spectrom 9: 1267–1274.

Wen J, Hong Z, Wu Y, Wei H, Fan G, Wu Y. 2009. Simultaneous determination of rupatadine and its metabolite desloratadine in human plasma by a sensitive LC–MS/MS method: Application to the pharmacokinetic study in healthy Chinese volunteers. J Pharm Biomed Anal 49: 347–353.

Yang L, Clement RP, Kantesaria B, Reyderman L , Beaudry F, Grandmaison C, Di Donato L, Masse R, Rudewicz PJ. 2003. Validation of a sensitive and automated 96-well solid-phase extraction liquid chromatography–tandem mass spectrometry method for the determination of desloratadine and 3-hydroxydesloratadine in human plasma. J Chromatogr B 792: 229–240.

4.6

FRAGMENTATION OF STEROIDS

4.6.1 INTRODUCTION

Steroids are characterized by the sterane core that is composed of 17 carbon atoms bound together to form four fused rings: three cyclohexane rings (designated as the A-, B-, and C-rings) and one cyclopentane ring (the D-ring). Carbon numbering has been standardized (Figure 4.6.1). There are hundreds of steroids and related compounds found in animals, plants, and fungi that vary by the functional groups attached to the rings, as well as by the oxidation state of the ring *C*-atoms. In mammals, steroids acts as hormones classified as (1) sex steroids such as androgens, estrogens, and progestagens that produce sex differences and support reproduction, (2) corticosteroids including glucocorticoids, which regulate many aspects of metabolism and immune functions, and mineralocorticoids, which help maintain blood volume and control renal excretion of electrolytes, (3) anabolic steroids that interact with androgen receptors to increase muscle and bone synthesis, and (4) neurosteroids that are synthesized in the brain or in an endocrine gland and then reach the brain via the bloodstream, where they alter neuronal excitability. In addition, synthetic steroid drugs have been made resembling naturally occurring steroids, but showing favorable properties as drugs.

Cholesterol also contains a sterane core. Cholesterol modulates the fluidity of cell membranes and is the principal constituent of the plaques implicated in atherosclerosis. Bile acids and vitamin D and their metabolites are also considered to be steroids. In addition, there are brassinolides, phyto(ecdy)sterols, steroid alkaloids, and saponins in plants; ergosterols in fungi and protozoa; and ecdysteroids in insects. There are many steroid drugs derived from any one of these groups.

Steroids generally show poor ionization characteristics in ESI-MS, as they lack basic or acidic moieties for direct protonation or deprotonation (Ma & Kim, 1997; Pozo et al., 2007). Different strategies have been explored to deal with this drawback. Both ESI-MS and APCI-MS can be applied for the ionization of steroids. Depending on the compound class, the LC–MS analysis can be achieved in positive-ion mode, in negative-ion mode, or in both modes. The so-called 3-keto-Δ^4-steroids like testosterone (with a C^3-carbonyl (C=O) group and a C^4—C^5-double bond) are preferably analyzed in positive-ion mode, whereas estrogens can only be analyzed in negative-ion mode. Corticosteroids are frequently analyzed in the negative-ion mode as formate or acetate adducts, that is, $[M+HCOO]^-$ or $[M+CH_3COO]^-$ with *m/z* (M+45) or *m/z* (M+59), respectively (Cui et al., 2006; Pozo et al., 2009a; Dusi et al., 2011; Chen et al., 2011).

In order to assess optimum ionization conditions in ESI-MS, adduct formation of anabolic steroids in positive-ion ESI-MS was investigated (Pozo et al., 2007). Different mobile-phase additives were applied, that is, formic acid, Na^+, and NH_4^+. Different behavior was also observed for either methanol or acetonitrile as organic solvent, as also reported by others (Ma & Kim, 1997). Seven groups of anabolic steroids were discriminated. (1) Steroids containing N-atoms like stanozolol are readily ionized to form $[M+H]^+$. (2) Steroids with a conjugated keto group like testosterone form $[M+H]^+$, but $[M+Na]^+$ may also be observed. Addition of 0.1 mM Na^+ resulted in $[M+Na+CH_3OH]^+$ for most steroids in this group, when methanol was present, and $[M+H]^+$ when acetonitrile was present. Adduct formation can be reduced by the addition of ammonium formate. (3) Steroids with an unconjugated keto group show $[M+Na+CH_3OH]^+$, with methanol present, and only a poor response of $[M+H+CH_3CN]^+$, when acetonitrile is present. (4) Hydroxyandrostenes mostly show either protonated, ammoniated, or sodiated molecules after a water loss. (5) Hydroxyandrostanes only provide response with acetonitrile as organic solvent: often both $[M+H-2H_2O]^+$ and $[M+H+CH_3CN-H_2O]^+$ are observed. (6) 6,17-Dihydroxy anabolic steroids show the formation of $[M+HCOO]^-$ in negative-ion mode similar to many corticosteroids. (7) 1,3-Diketo steroids show keto–enol tautomerism and can thus be analyzed as $[M-H]^-$ in negative-ion mode, whereas

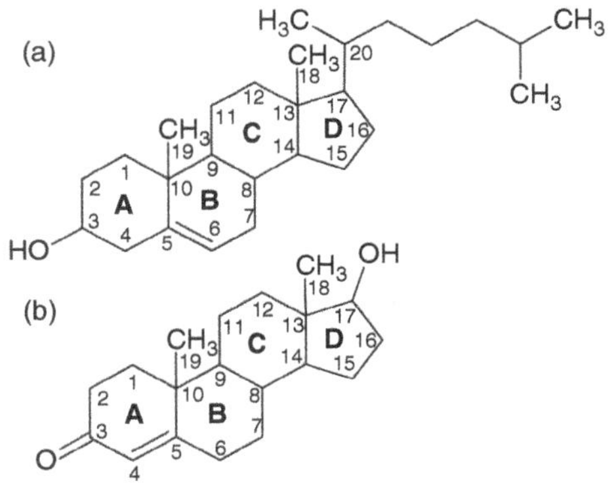

FIGURE 4.6.1 General structure and carbon numbering in steroids with (a) cholesterol and (b) testosterone.

1,2-diketo steroids are not readily ionized, although in positive-ion mode ions with *m/z* (M+149) have been observed, which are due to adduct formation with a protonated phthalic anhydride ($[C_8H_4O_3+H]^+$) (Pozo et al., 2007).

ESI-MS, APCI-MS, and APPI-MS were optimized for the analysis of anabolic steroids with respect to mobile-phase composition and ion-source parameters. The three ionization methods provided similar specificity and detection limits. In the end, ESI-MS was preferred using a methanol/water gradient with 5 mM ammonium acetate and 0.01% acetic acid (Leinonen et al., 2002). In other studies, APPI was preferred for the LC–MS analysis of steroids, for example, in the analysis of esterified phytosterols in human serum (Lembcke et al., 2005), 17α-ethinylestradiol in hepatocytes (Li et al., 2008), and of estradiol in human serum and endometrial tissue (Keski-Rahkonen et al., 2013). More recently, coordination electrospray ionization (CIS-MS) involving Ag^+-cationization has been proposed for the residue analysis of anabolic androgenic steroids (Kim et al., 2014).

Another strategy to improve the ionization of steroids is derivatization, where steroids are converted, for instance, to acetate esters, oximes or methyloximes, ferrocene boronates, Girard P hydrazones, 4-toluenesulfonhydrazones, phenylboronates, which can all be analyzed in positive-ion mode, or to sulfate, pentafluorobenzyl ether, or 2-nitro-4-trifluoromethyl phenylhydrozone derivatives, which can be analyzed in negative-ion mode (Higashi & Shimada, 2004). New steroid derivatization agents are reported regularly. The fragmentation of derivatized steroids is not discussed here.

LC–MS–MS in SRM mode is widely applied in quantitative steroid analysis, both in various biological and pharmacological applications and in relation to food safety. In most cases, intense product ions are selected without paying attention to the identity of the ions. Three SRM transitions per compound were reported for the positive-ion LC–MS analysis of 36 anabolic steroids (Joos & Van Ryckeghem, 1999). Numerous examples of multiresidue analysis of steroids by LC–MS in SRM mode are available in the literature (e.g., Deventer & Delbeke, 2003; Pozo et al., 2008b; Leporati et al., 2014). Next to sports doping and food safety applications, steroid analysis is also important in clinical endocrinology applications (e.g., Cho et al., 2009; Rauh, 2010; Kushnir et al., 2011). Given the developments in the bioanalytical field, comparison between TQ or Q–LIT instruments and high-resolution accurate-mass instruments, that is, TOF, Q–TOF hybrid, and ion-trap–orbitrap hybrid instruments, has also been investigated (Thevis et al., 2005b; Pozo et al., 2011)

4.6.2 FRAGMENTATION OF STEROIDS

The interpretation of low-resolution product-ion mass spectra of steroids is rather difficult. Many C—C bonds in a steroid are of similar strength, which upon fragmentation can give rise to numerous isomeric and isobaric fragments, for which in turn different structures may be proposed, involving different parts of the molecule. To some extent, one may be helped in the interpretation of steroid product-ion mass spectra by the extensive tables for structure elucidation of steroids generated for EI-MS (Von Unruh & Spiteller, 1970a,b,c; Von Unruh et al., 1970).

A nomenclature system was proposed to annotate steroid product-ion mass spectra (Griffiths et al., 1996). The basics of the nomenclature rules are demonstrated for testosterone in Figure 4.6.2. Fragmentation in the B-ring results in b- or B-fragments. Lower-case fonts are used for fragments with charge retention on the 3-keto group and upper-case fonts for fragment ions with charge retention on the part containing C^{17}. Subscripts are used to indicate which bond is broken; superscripts are used to annotate hydrogens. This nomenclature system is not very widely used.

In an early LC–MS report, the fragmentation of 60 steroids from different classes was studied under APCI-MS conditions (Kobayashi et al., 1993). Relatively simple MS spectra were obtained, featuring small-molecule losses, for example, losses of water, two water molecules, methanol (CH_3OH), ethanal (CH_3CHO), and 2-hydroxyethanal

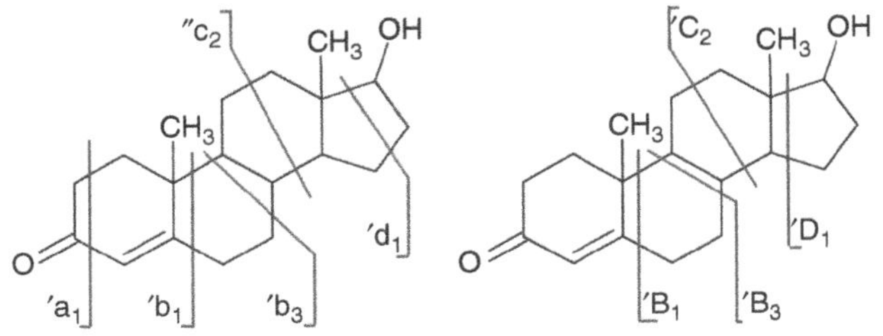

FIGURE 4.6.2 Proposed nomenclature for steroid fragment ions. (Source: Griffiths et al., 1996. Reproduced with permission of Wiley.)

($HOCH_2CHO$). No MS–MS was performed. For 3-keto-Δ^4-steroids, $[M+H]^+$ was observed for most compounds, accompanied by fragment ions due to losses of water if a hydroxy group was present. In cortisol and related compounds, featuring a C^{17}-OH or C^{17}-($COCH_2OH$) group, the base peak often is an ion with *m/z* ([M+H]−60), involving the loss of $HOCH_2CHO$. For 3-OH-Δ^4- and 3-OH-Δ^5-steroids, the ion with *m/z* ([M+H]−18] was the base peak.

The recent advent of readily accessible instruments for high-resolution accurate-mass mass spectrometry (HRAM-MS) somewhat simplifies the interpretation because at least the elemental composition of the fragments can unambiguously be determined. An excellent illustration of this is a study on fragmentation characteristics of anabolic steroids in relation to sports doping control, providing useful tables with characteristic fragment ions (Pozo et al., 2008a).

In an elaborate paper, the fragmentation of fatty acids, bile acids, steroids, and steroid conjugates has been reviewed (Griffiths, 2003). Both high-energy CID and low-energy CID were considered. The current discussion concentrates on low-energy CID, that is, collision-cell CID obtained in TQ and Q–TOF MS–MS instruments and ion-trap CID in ion-trap MS^n systems.

4.6.3 FRAGMENTATION IN 3-KETO-Δ^4-STEROIDS

An important structure element in many steroids from different classes, that is, both anabolics, corticosteroids, and progestogens, is the 3-keto-Δ^4-group. The fragmentation of 3-keto-Δ^4-steroids has been studied in detail, for example, with respect to the fragmentation of testosterone (Williams et al., 1999; Liu et al., 2000).

In MS–MS, testosterone ($[M+H]^+$ with *m/z* 289) shows two consecutive water losses in addition to three major fragment ions with *m/z* 97, *m/z* 109, and *m/z* 123. In a study on the ESI-MS analysis of testosterone esters, structure proposals were made for the latter two ions, with the ion with *m/z* 109 ($[C_7H_9O]^+$) being the A-ring with C^{19} (Figure 4.6.3a) and the one with *m/z* 123 ($[C_8H_{11}O]^+$) being the A-ring with C^{19} and C^6 (Shackleton et al., 1997) (Figure 4.6.3b). Series of hydroxylated and [D]-labeled analogs were used to elucidate the identity of the three fragment ions (Williams et al., 1999). The rationale behind the use of [D]-labeled standards for fragmentation studies is that if the label is kept in the ion, the C-atom is kept as well. Given the frequently observed hydrogen rearrangement reactions in the fragmentation of $[M+H]^+$ (Section 3.5.2), one should take great care in such observations. Often, at least two experiments have to be performed with different labeling to confirm the presence of a particular C-atom in a fragment ion.

When analyzing the [1,2-D_2]-analog (Figure 4.6.1 for carbon numbering), both labels are kept. However, upon fragmenting the [2,2,4,6,6-D_5]-analog, one label is lost. This

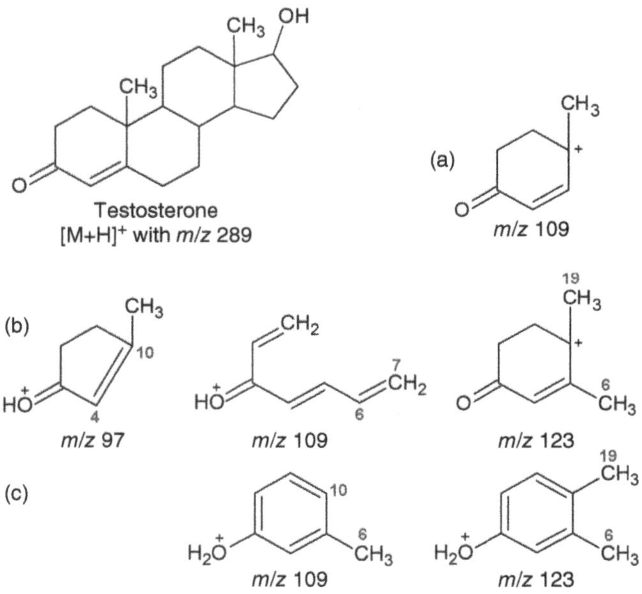

FIGURE 4.6.3 Proposed structures for the fragment ions of protonated testosterone.

can be interpreted in two ways: (1) C^2 and C^4 are in the ion with *m/z* 109 and one of the labels from C^6 is rearranged to the A-ring, (2) C^2, C^4, and C^6 are in the ion with *m/z* 109 and one of the C^6-labels is shifted to the neutral lost. Subsequent analysis of the [7-D]- and [6,7-D_2]-analogs confirms that both C^6 and C^7 are in the ion with *m/z* 109. This would mean that most likely C^{10} and C^{19} are not in the ion, which is confirmed by fragmentation of the [19,19,19-D_3]-analog: all three labels are lost. In treating these data, the original discussion was somewhat simplified because reviewing the complete data set demonstrates that the fragmentation may actually be more complicated. But the general conclusion is that the identity of the ion with *m/z* 109 involves C^1 through C^7 ($[C_7H_9O]^+$), with the charge stabilized by resonance (Williams et al., 1999) (Figure 4.6.3b).

In another study, this result was questioned because of the easy scrambling of [D]-labels during fragmentation (Liu et al., 2000). An alternative 3-methylphenol structure was proposed, involving C^6 and C^{10} and not C^7 (Figure 4.6.3). The controversy can only be solved by experiments with [^{13}C]-labeling. On the other hand, the loss of the C^{19}-substituent, as indicated by the [19,19,19-D_3]-analog, as well as the occurrence of the fragment ion with *m/z* 109 for 19-nortestosterone, strongly suggest that the 3-methylphenol structure is not the correct one.

Along the same lines, the identity of the ions with *m/z* 97 ($[C_6H_9O]^+$) and 123 ($[C_8H_{11}O]^+$) was elucidated (Figure 4.6.3b). Fragmentation mechanisms were proposed for the formation of the ions with *m/z* 109 and 123 (Williams et al., 1999), whereas a mechanism for *m/z* 97 was proposed elsewhere (Pozo et al., 2008a).

The influence of hydroxy groups at various positions of testosterone on the formation of the fragment ions with *m/z* 97, 109, and 123 was also investigated. These fragments show lower relative abundance with hydroxy groups at 2α-, 2β-, 6α-, 6β-, 14α-, and 19-positions. Similar relative abundances (compared to testosterone) are observed for analogs with a hydroxy group at 15β-, 16α-, and 16β-positions. Hydroxylation at 11α- or 11β-position suppresses the formation of the ions with *m/z* 97 and 109, but enhances the formation of the ions with *m/z* 123 and *m/z* 121 ($[C_8H_9O]^+$). The ion with *m/z* 121 is also abundant with 2α- and 2β-hydroxy analogs. Hydroxylation at the 7α-, 15α-, and 18-positions reduced the relative abundance of the ions with *m/z* 97 and 109. With androst-4-ene-3,17-dione, the fragment ions with *m/z* 97, 109, and 123 are also observed with similar abundance as for testosterone (Williams et al., 1999).

4.6.4 ANABOLIC STEROIDS

The fragmentation of the 3-keto-Δ^4-steroid testosterone, discussed earlier (Figure 4.6.3), provides quite some information on the fragmentation of related anabolic steroids. The identified fragment ions with *m/z* 97 ($[C_6H_9O]^+$), *m/z* 109 ($[C_7H_9O]^+$), and *m/z* 123 ($[C_8H_{11}O]^+$) (Section 4.6.3) are characteristic fragments for 3-keto-Δ^4-steroids (Griffiths, 2003).

In an early LC–MS–MS paper, the fragmentation of 3-keto-$\Delta^{1,4}$-steroid methandrostenolone ($[M+H]^+$ with *m/z* 301) and its metabolism was described (Edlund et al., 1989). Apart from a fragment with *m/z* 283 due to water loss, fragment ions (carbocations) are observed with *m/z* 121 ($[C_8H_9O]^+$) and *m/z* 135 ($[C_9H_{11}O]^+$), corresponding to the A-ring with C^6 and with C^6 and C^7, respectively; with *m/z* 173 ($[C_{12}H_{13}O]^+$), corresponding to the AB-ring and a C-atom from the C-ring; and with *m/z* 149 ($[C_{11}H_{17}]^+$), corresponding to the dehydrated CD-ring. The same fragments are observed for the 16-hydroxy metabolite, except for the ion with *m/z* 149, which is now observed with *m/z* 147 ($[C_{11}H_{15}]^+$), due to an additional water loss (Edlund et al., 1989). In a more recent study, the fragmentation of the C^6-, C^{11}-, and C^{15}-hydroxylated analogs of methandrostenolone has been investigated in order to correlate structure changes with differences in fragmentation behavior (Musharraf et al., 2013). In the C^6-, C^{11}-, and C^{15}-hydroxylated analogs, the carbocation with *m/z* 121 ($[C_8H_9O]^+$) is still observed, while carbocations with *m/z* 147 ($[C_{11}H_{15}]^+$), *m/z* 171 ($[C_{12}H_{11}O]^+$; due to AB-rings and C^{11}), and *m/z* 173 are now observed with higher abundance. In the C^{15}-analog, the ion with *m/z* 121 is the base peak, indicating that structural features in the D-ring may influence the cleavage of the B-ring. C^{11}-hydroxy analogs may show the loss of water from the C^{11}-position, resulting in a C^{10}—C^{11} double bond, which via a retro-Diels–Alter fragmentation will result in a carbocation with *m/z* 95 in the 17-hydroxy compound ($[C_7H_{11}]^+$; after water loss from the C^{17}-position) or the ion *m/z* 97 in the 17-keto compound ($[C_6H_8O+H]^+$) (Musharraf et al., 2013).

Using data from TQ, ion-trap, and FT-ICR-MS instruments, the fragmentation pathways of some anabolic steroids were elucidated (Guan et al., 2006). Fragmentation pathways were derived using Mass Frontier software. The 3-keto-Δ^4-steroids testosterone, nandrolone, normethandrolone, and mibolerone; the 3-keto, $\Delta^{1,4}$-steroids boldenone and methandrostenolone; the 3-keto-$\Delta^{4,9,11}$-steroids tetrahydrogestrinone (THG) and trenbolone; and the 1-methyl-3-keto-Δ^4-steroid methenolone served as model compounds. The product-ion spectra of some of these compounds, that is, for boldenone, methandrostenolone, testosterone, and methenolone, are dominated by only a few fragment ions, indicating energetically favored fragmentation pathways. In the product-ion spectra of the other model compounds, many product ions are observed, indicating many competing fragmentation pathways (Guan et al., 2006).

Comparison of the fragment ions of 3-keto-Δ^4-steroids testosterone and mibolerone ($[M+H]^+$ with *m/z* 303, with a methyl substituent at the C^7-position) are of some interest. Both produce a fragment ion with *m/z* 109, that is, an ion with formula $[C_7H_9O]^+$ for testosterone, but a carbocation with $[C_8H_{13}]^+$ for mibolerone, which is not related to the A-ring, but contains C^7+methyl substituent, C^8, C^{12} up to and including C^{15} and C^{18} (Guan et al., 2006).

In another study, the fragmentation of 41 anabolic steroids was investigated using TQ and Q–TOF instruments (Pozo et al., 2008a). Some compounds show a limited number of abundant fragment ions, whereas others show extensive fragmentation, for example, with more than 25 ions with higher than 15% relative abundance. As it is nearly impossible to interpret all these ions, the attention was especially focused at the fragment ions with high *m/z* and at the fragment ions with higher than 60% relative abundance at higher collision energy (30 eV). Based on this study, characteristic neutral losses and product ions were tabulated to assist in the structure elucidation of anabolic steroids (Pozo et al., 2008a). Some of the results are discussed in the following.

The 3-keto-Δ^4-steroids showed a number of water losses corresponding to the number of O-atoms present. Neutral loss of acetone (CH_3COCH_3) from the D-ring was observed for steroids with 17-methyl substituents. Most model compounds showed fragment ions with *m/z* 97 ($[C_6H_9O]^+$) and 109 ($[C_7H_9O]^+$), or analogs thereof, for example, ions with *m/z* 113 ($[C_6H_9O_2]^+$) and 125 ($[C_7H_9O_2]^+$) for oxymesterone and 4-hydroxytestosterone (with a C^4-hydroxy substituent) or *m/z* 143 ($[C_7H_8ClO]^+$), but no equivalent to the ion with *m/z* 97 for norclostebol (with a C^4-Cl substituent). The 19-*nor*- and 4-hydroxy steroids showed a neutral loss of 2-butanone (C_4H_8O) or butanol (C_4H_9OH), involving the loss of the D-ring including C^{18}. For norclostebol

([M+H]$^+$ with *m/z* 309), two isobaric ions with *m/z* 237 are observed, that is, one due to the loss of either C_4H_8O, that is, $[C_{14}H_{18}ClO]^+$ with *m/z* 237.104, and one due to the loss two times water and hydrogen chloride (HCl), that is, $[C_{18}H_{21}]^+$ with *m/z* 237.164 (Pozo et al., 2008a).

In the context of a metabolite identification study, the fragmentation of the 3-keto-Δ^4-steroid fluoxymesterone ([M+H]$^+$ with *m/z* 337) was studied (Pozo et al., 2008a; Pozo et al., 2008c). The presence of the 11-hydroxy group, the associated water loss, and the easy loss of hydrogen fluoride (HF) results in additional double bonds (between C^8 and C^9, and between C^{11} and C^{12}) in the steroid skeleton, which significantly influence the fragmentation. It suppresses the formation of the characteristic fragment ions with *m/z* 97 and 109. Mainly fragment ions involving small-molecule losses are observed, that is, *m/z* 317 due to HF loss, *m/z* 299 due to a subsequent water loss from the C^{11} position, *m/z* 281 due to a subsequent water loss from the C^3 position, *m/z* 263 due to the third water loss, that is, from the C^{17}-position, *m/z* 241 due to D-ring-related loss of acetone (CH_3COCH_3) from the ion with *m/z* 299, and *m/z* 233 due to CH_3COCH_3 loss from *m/z* 281 (Pozo et al., 2008c). Further fragments of fluoxymesterone were reported elsewhere (Pozo et al., 2008a). These are the carbocations with *m/z* 181 ($[C_{14}H_{13}]^+$) involving the ABC-rings; *m/z* 157 ($[C_{12}H_{13}]^+$) and *m/z* 145 ($[C_{11}H_{13}]^+$) both involving the AB-rings; *m/z* 131 ($[C_{10}H_{11}]^+$) involving the BC-rings; and *m/z* 123 ($[C_8H_{11}O]^+$) involving the A-ring, as also described for testosterone (Figure 4.6.3b).

To facilitate the identification of steroid metabolites, the fragmentation of four anabolic steroids (17-methyltestosterone, methandrostenolone, *cis*-androsterone, and adrenosterone) and some of their metabolites was studied (Musharraf et al., 2013). In 17-methyltestosterone, hydroxylation at C^{15}- and C^{16}-positions has little effect on the fragmentation, whereas hydroxylation at C^6-, C^{11}-, and C^9-positions results in changes in the fragmentation behavior. To some extent, this was also described for hydroxylated testosterone (Williams et al., 1999) and progesterone (Kang et al., 2004) (Sections 4.6.3 and 4.6.5).

In MS–MS, 17-methyltestosterone ([M+H]$^+$ with *m/z* 303) yields the ions with *m/z* 285 and 267 due to two subsequent water losses, the fragment ions with *m/z* 97 ($[C_6H_9O]^+$) and *m/z* 109 ($[C_7H_9O]^+$) (Section 4.6.3), and an ion with *m/z* 245 due to the loss of CH_3COCH_3 from the D-ring. In the C^6- and C^{11}-hydroxy analogs, the ions with *m/z* 97 and 109 are less abundant. Three subsequent water losses as well as a subsequent CH_3COCH_3 loss are observed. Many fragment ions with low *m/z* are observed. The C^6- and C^{11}-hydroxy analogs can be differentiated from the more abundant AB-ring fragments with *m/z* 147, 159, and 185 for the C^6-analog and with *m/z* 149, 161, and 187 for the C^{11}-analog. The *m/z* difference of 2 is due to the fact that upon water loss of the C^6-analog, a C^6—C^7 double bond is formed in the B-ring, whereas with the C^{11}-analog, a C^{11}—C^{12} double bond is formed in the C-ring (Musharraf et al., 2013). Similarly, *cis*-androsterone and adrenosterone and some of their metabolites were studied, but with these compounds less pronounced differences were found (Musharraf et al., 2013).

The fragmentation of 3-keto-Δ^1-steroids is significantly different. A number of water losses corresponding to the number of O-atoms present is observed. The loss of 84 Da (C_5H_8O) is also observed, eventually in combination with a water loss (102 Da). This is attributed to the loss of the A-ring (C^1—C^4) and C^{19} (Thevis et al., 2005b; Pozo et al., 2008a). For Δ^1-testosterone ([M+H]$^+$ with *m/z* 289), this results in fragment ions with *m/z* 205 ($[C_{14}H_{21}O]^+$) and *m/z* 187 ($[C_{14}H_{19}]^+$). In addition, fragment ions with *m/z* 145 and 131 are observed. From the occurrence of the ion with *m/z* 145 with other 3-keto-Δ^1-steroids, it may be concluded that this carbocation relates to the BC-ring ($[C_{11}H_{13}]^+$) (Pozo et al., 2008a).

With unconjugated 3-keto-steroids such as dihydrotestosterone (DHT) and mesterolone, the loss of 76 Da is observed, which is attributed to the loss of water from the C^{17}-position and of CH_3COCH_3 from the A-ring (C^2—C^4). No common fragment ions with low *m/z* are observed (Pozo et al., 2008a).

The 3-keto-$\Delta^{1,4}$-steroid boldenone ([M+H]$^+$ with *m/z* 287) also shows the fragment ions (carbocations) with *m/z* 121 ($[C_8H_9O]^+$), 135 ($[C_9H_{11}O]^+$), 173 ($[C_{12}H_{13}O]^+$), and both *m/z* 147 ($[C_{10}H_{11}O]^+$) due to a B-ring cleavage and charge retention at the keto group and *m/z* 149 ($[C_{11}H_{17}]^+$) due to dehydration in the CD-rings (Weidolf et al., 1988). The determination of accurate *m/z* for the product ions demands that some of the earlier interpretations must be adapted. For boldenone, for instance, the fragment ions with *m/z* 121 and 135 were attributed to A-ring-related fragments, similar to what is observed for methandrostenolone (Weidolf et al., 1988; Edlund et al., 1989). However, accurate-mass data show that this is true for the ion with *m/z* 121 ($[C_8H_9O]^+$), but not for the ion with *m/z* 135 ($[C_{10}H_{15}]^+$), which in fact results from the dehydrated CD-rings, similar to the ion with *m/z* 149 for methandrostenolone (Guan et al., 2006).

Anabolic steroids play an important role in sports doping. Their use is prohibited by the International Olympic Committee (IOC) and WADA. In order to trick the drug screening procedures of the regulatory laboratories, designer steroid drugs have been developed and used, for example, tetrahydrogestrinone (THG) (Catlin et al., 2004). This calls for the adaptation of the screening procedures and strategies to identify such designer steroids when found. To this end, the fragmentation pathways of 21 steroids with 3-keto-$\Delta^{4,9,11}$-, 3-keto-Δ^4-, and 3-keto-Δ^1-skeletons, some of which carry atypical substituents (—CH_2—CH_3, —CH=CH_2, —CH_2—CH=CH_2, etc.), were investigated

(Thevis et al., 2005b). The gestrinone analogs with a 3-keto-$\Delta^{4,9,11}$-skeleton and a C^{13} ethyl group show fragment ions with *m/z* 241 ($[C_{17}H_{21}O]^+$), due to loss of C^{16} and C^{17} and their substituents, and *m/z* 199 ($[C_{14}H_{15}O]^+$), due to the loss of the D-ring and the C^{13}-substituent. The 3-keto-$\Delta^{4,9,11}$-steroid trenbolone ($[M+H]^+$ with *m/z* 271) shows fragment ions with *m/z* 253 due to the loss of H_2O, *m/z* 227 ($[C_{16}H_{19}O]^+$) due to the loss of ethanal (CH_3CHO), probably involving the D-ring, and *m/z* 199 ($[C_{14}H_{15}O]^+$) due to C_4H_8O loss.

The 3-keto-$\Delta^{4,9,11}$-steroids such as gestrinone and THG show just one water loss (Pozo et al., 2008a). They also yield fragment ions due to the loss of both water and the C^{13}-substituent, either as ethene (C_2H_4) or as an ethyl radical ($^{\bullet}CH_2CH_3$) with gestrinone and THG and as a methyl radical ($^{\bullet}CH_3$) with 17α-trenbolone. Radical losses seem to be favored by conjugated double bonds. The $^{\bullet}CH_3$ losses from the C^{10} or C^{13} position (or both) are also observed for the two 3-keto-$\Delta^{4,9(11)}$-diene steroids studied (Pozo et al., 2008a). In addition, D-ring fragmentation similar to the 3-keto-Δ^4-steroids is observed for the 3-keto-$\Delta^{4,9,11}$-steroids, but irrespective of the C^{17}-substitution. Thus, gestrinone shows the loss of but-3-yn-2-one (C_4H_4O), THG of C_4H_8O, and 17α-trenbolone of CH_3CHO. Gestrinone and 17α-trenbolone also show an abundant fragment ion with *m/z* 199 ($[C_{14}H_{15}O]^+$) due to the ABC-rings (Pozo et al., 2008a). This is in agreement with an interpretation reported elsewhere (Thevis et al., 2005b).

A steroid that gained a lot of attention due to sports doping affairs is stanozolol and its metabolites. These compounds show extensive fragmentation in collision-cell CID. Next to the loss of water (and two water losses for the 4β-hydroxy-, the 3′-hydroxy-, and the 16β-hydroxy metabolites), not resulting in abundant fragment ions (Pozo et al., 2008a), stanozolol ($[M+H]^+$ with *m/z* 329) shows fragments with *m/z* 121 and 229, whereas the 16β-hydroxy metabolite yields fragment ions with *m/z* 159 and 227 (Mück & Henion, 1990; De Brabander et al., 1998). Many other fragments of the 3′-hydroxy-, the 4β-hydroxy-, and the 16β-hydroxy metabolites of stanozolol have been tentatively identified (Mück & Henion, 1990). In another study, based on APCI rather than ESI, only small-*m/z* fragment ions are observed, that is, ions with *m/z* 81, 95, 107, and 121 for both stanozolol and its 16β-hydroxy metabolite (Draisci et al., 2001a). As under MS^2 conditions in an ion-trap, stanozolol and its 16β-hydroxy metabolite give extensive fragmentation (Van de Wiele et al., 2000), a phenylboronic acid derivative is produced, which shows far less fragmentation. Based on the accurate-mass data in our library collection, some fragments of stanozolol ($[M+H]^+$ with *m/z* 329) can be tentatively identified: the ion with *m/z* 81 is the (1*H*-pyrazol-4-yl)methylium ion ($[C_4H_5N_2]^+$), the carbocation with *m/z* 121 ($[C_9H_{13}]^+$) could be the dehydrated D-ring with two C-atoms from the C-ring or the A-ring with C^{19} and two C-atoms from the B-ring, the carbocation with *m/z* 203 ($[C_{13}H_{19}N_2]^+$) is the pyrazole-AB-rings with one C-atom from the C-ring, and the carbocation with *m/z* 229 ($[C_{17}H_{25}]^+$) is the dehydrated D-ring with C- and B-ring and with one C-atom from the A-ring (Figure 4.6.4). This is in agreement with the fragment structures proposed by others (Mück & Henion, 1990; McKinney et al., 2004). A detailed study on the fragmentation of stanozolol, using Q–LIT and LIT–orbitrap hybrid instruments, stable isotope labeling, H/D-exchange experiments, and MS^3, was also reported (Thevis et al., 2005a). Based on these results, fragmentation mechanisms were proposed, partly charge driven, partly due to charge-remote fragmentation. Special attention was paid to the ions with *m/z* 81 ($[C_4H_5N_2]^+$), *m/z* 95, *m/z* 135 ($[C_8H_{10}N_2+H]^+$), *m/z* 119 ($[C_7H_6N_2+H]^+$), and *m/z* 91 ($C_7H_7]^+$) (Figure 4.6.4) (Thevis et al., 2005a). Two isobaric carbocations with *m/z* 95, one with $[C_5H_7N_2]^+$, thus containing the pyrazole ring, and one with $[C_7H_{11}]^+$ due to the A-ring and C19 (Figure 4.6.4), were reported for stanozolol (Pozo et al., 2008a). Detection and structure elucidation of a total of 19 stanozolol metabolites using TQ and Q–TOF instruments have also been reported (Pozo et al., 2009b).

Structure proposals for the fragments of the 3-keto-Δ^4-steroid norethisterone, based on molecular formulae derived from accurate-mass data using an ion-trap–TOF instrument, are given in Figure 4.6.5. These data were acquired in the course of a metabolic study with on-line activity assessment of the metabolites against the estrogen receptors ERα and ERβ (de Vlieger et al., 2010). For most fragments, unambiguous structure proposals can be made. Note that due to restrictions of the ion-trap in trapping fragment ions with low *m/z*, some expected fragment ions, especially the ion with *m/z* 109, are missing.

So far, fragmentation of anabolic steroids under low and medium collision energy conditions has been discussed. It has been demonstrated that at high collision energies in TQ instruments less specific fragment ions are observed, that is, the phenyl ion ($[C_6H_5]^+$) with *m/z* 77, the tropylium ion ($[C_7H_7]^+$) with *m/z* 91, and the 2-methylbenzyl cation ($[C_8H_9]^+$) with *m/z* 105, respectively. These ions may serve for screening purposes by precursor-ion analysis (PIA) in TQ instruments (Pozo et al., 2008b).

4.6.5 PROGESTOGENS

The progestin progesterone ($[M+H]^+$ with *m/z* 315) shows fragment ions with *m/z* 297 due to the loss of water, *m/z* 279 due to two losses of water, *m/z* 255 due to the subsequent loss of ethenone ($H_2C{=}C{=}O$) from the ion with *m/z* 297, and with *m/z* 97 ($[C_6H_9O]^+$), *m/z* 109 ($[C_7H_9O]^+$), and *m/z* 123 ($[C_8H_{11}O]^+$), characteristic for a 3-keto-Δ^4-steroid (Kang et al., 2004). The effects of hydroxylation of progesterone

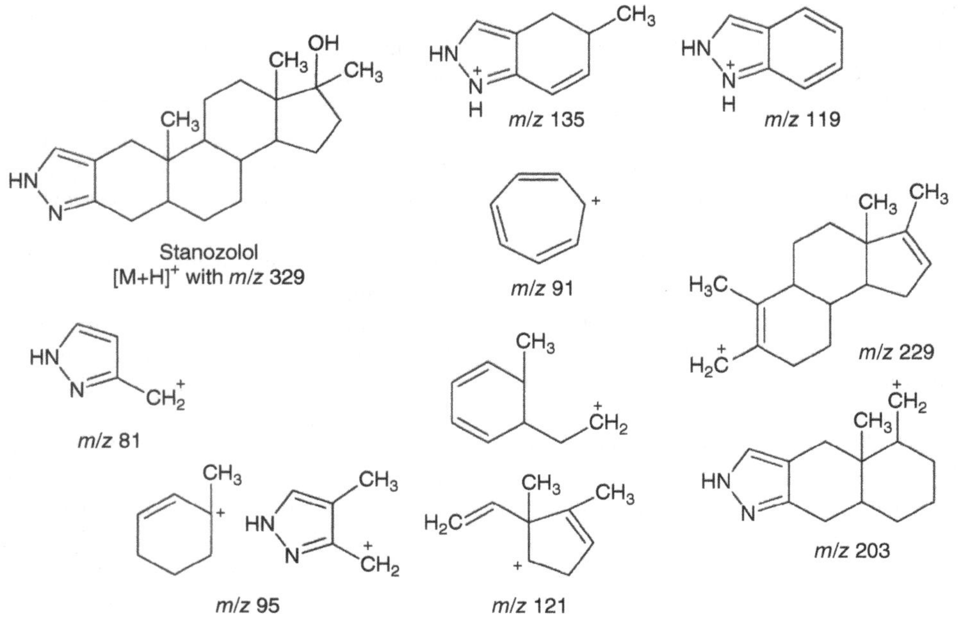

FIGURE 4.6.4 Proposed structures for the fragment ions of protonated stanozolol.

on its fragmentation behavior have been studied (Kang et al., 2004). At the high-*m/z* end, similar losses are observed for hydroxyprogesterones. With respect to the low-*m/z* end, almost similar spectra to that of progesterone were obtained for 15β-, 16α-, 17α-, and 21-hydroxyprogesterone, whereas with 2α-, 6β, 11α-, 11β-, 19-hydroxyprogestrone, extensive fragmentation is observed, with *m/z* 121 ($[C_8H_9O]^+$) being the most abundant ion for some compounds. The 11α-, 11β-hydroxy-, or 11-keto-substitution suppresses the formation of the fragment ions with *m/z* 97, 109, and 123. For 7β-hydroxyprogesterone, the ion with *m/z* 97 is far more abundant than any other fragment ion. With 9α-hydroxyprogesterone, the ion with *m/z* 137 ($[C_9H_{13}O]^+$), due to the A-ring with C^{19}, C^6, and C^7, is especially abundant. Seven hydroxyprogesterones (2α-, 6β-, 7β-, 9α-, 11α-, 11β-, and 19-) could be readily discriminated by comparison of the relative abundance of some eight fragment ions. The hydroxylation position at the D-ring (in 15β-, 16α-, 17α-, and 21-hydroxyprogesterones) could not be determined from the product-ion mass spectra (Kang et al., 2004).

Medroxyprogesterone acetate ($[M+H]^+$ with *m/z* 387) shows fragment ions with *m/z* 327 due to the loss of acetic acid (CH_3COOH), *m/z* 309 and 285 due to subsequent loss of water or $H_2C{=}C{=}O$, and *m/z* 267 due to the loss of both (Kim & Kim, 2001). The fragment ion with *m/z* 205 ($[C_{14}H_{21}O]^+$) is due to the AB-ring with two C's from the C-ring and the ion with *m/z* 97 ($[C_6H_9O]^+$) is probably similar to that of testosterone (Figure 4.6.3b). The fragment ion with *m/z* 123 ($[C_8H_{11}O]^+$), also observed for testosterone, must arise from a different fragmentation route, as an additional methyl substituent is present at the C^6-position. In the low-*m/z* range, the spectra of medroxyprogesterone acetate and medroxyprogesterone are identical, as expected.

4.6.6 CORTICOSTEROIDS

Corticosteroids are a class of steroid hormones that are produced in the adrenal cortex. They are involved in a wide range of physiologic systems such as stress response, immune response and regulation of inflammation, carbohydrate metabolism, protein catabolism, blood electrolyte levels, and behavior. Some common natural corticosteroid hormones are corticosterone ($C_{21}H_{30}O_4$), cortisone ($C_{21}H_{28}O_5$), and its isomer aldosterone.

Corticosteroids can be grouped into four compound classes based on chemical structure: (1) hydrocortisones such as cortisone and prednisone, (2) acetonides such as triamcinolone acetonide and budesonide, (3) betamethasones such as betamethasone and dexamethasone, and (4) esters such as hydrocortisone-17-valerate and dexamethasone acetate. Structures of the corticosteroids studied in this context are summarized in Table 4.6.1.

LC–MS analysis of corticosteroids is important because of their use as anti-inflammatory and antiallergic drugs and also in relation to their illegal use in sports doping, veterinary medicine, and as growth promoters. Multiresidue LC–MS analysis of corticosteroids has been described, using either positive-ion or negative-ion mode, (Deventer & Delbeke, 2003; Cui et al., 2006; Touber et al., 2007; Cho et al., 2009; Dusi et al., 2011; Chen et al., 2011). The fragmentation

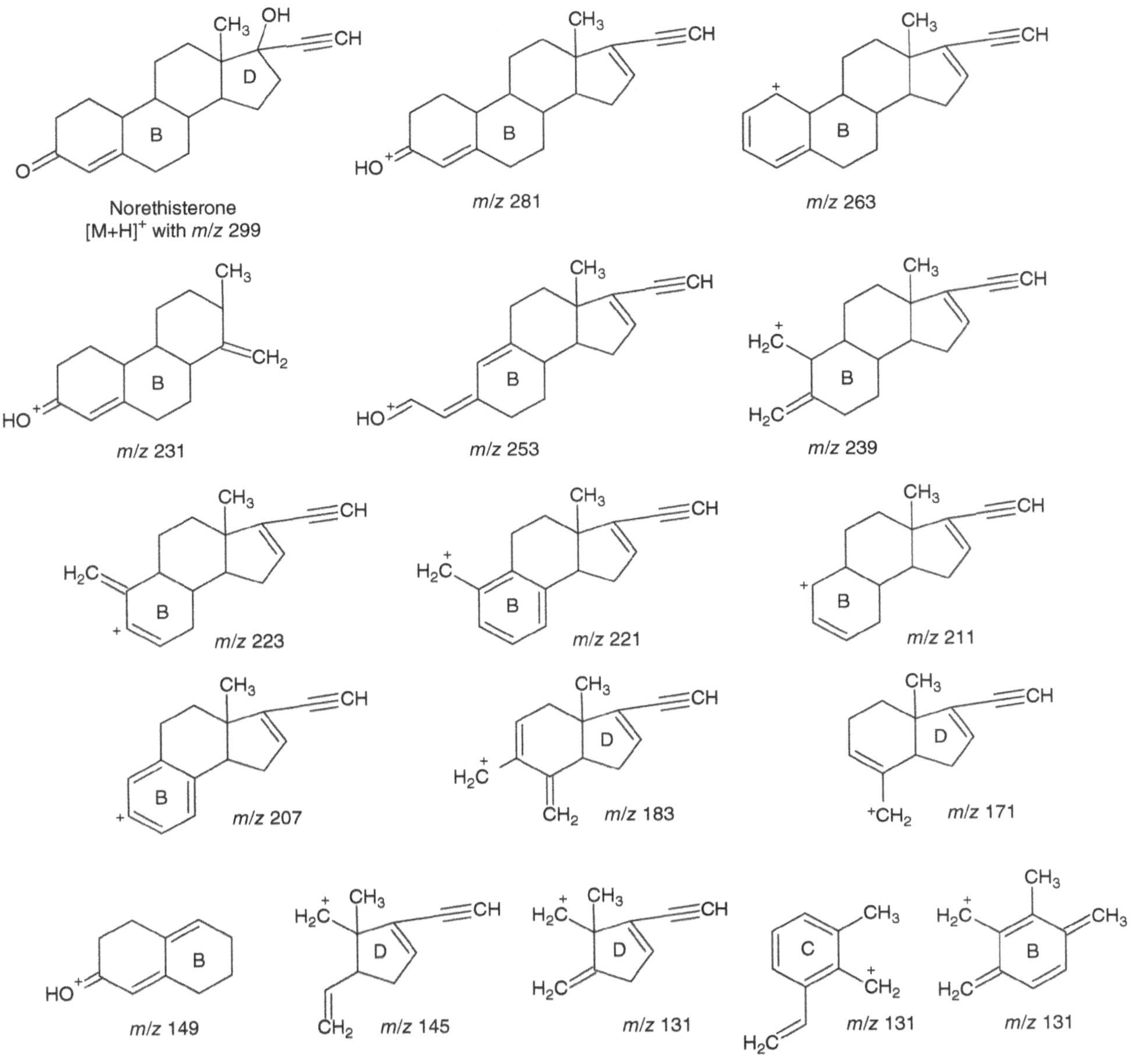

FIGURE 4.6.5 Proposed structures for the fragment ions of norethisterone in positive-ion MS^n in an ion-trap–time-of-flight hybrid instrument.

of corticosteroids in positive-ion and negative-ion modes is treated separately.

Similar to many anabolic steroids, many corticosteroids have a 3-keto-Δ^4-skeleton. Therefore, one would expect to observe the fragment ions with *m/z* 97 ($[C_6H_9O]^+$), *m/z* 109 ($[C_7H_9O]^+$), and *m/z* 123 ($[C_8H_{11}O]^+$), characteristic for a 3-keto-Δ^4-steroid. However, as already demonstrated for 11-hydroxytestosterone (Williams et al., 1999; Liu et al., 2000), an additional 11-hydroxy- or 11-keto group (Table 4.6.1) reduces the abundance of these fragment ions. The 3,11-diketo-$\Delta^{1,4}$-steroid prednisone ($[M+H]^+$ with *m/z* 359) shows fragment ions with *m/z* 341 due to water loss, *m/z* 323 due to another water loss, *m/z* 313 due to the loss of water and carbon monoxide (CO), *m/z* 295 due to two water losses and a CO loss, *m/z* 267 ($[C_{18}H_{19}O_2]^+$), which is probably due to the loss of C^{17} and its side chains, and *m/z* 147 ($[C_{10}H_{11}O]^+$) (see later).

The fragmentation of betamethasone ($[M+H]^+$ with *m/z* 393) under TSI-MS and MS–MS conditions has been studied (Polettini et al., 1998). Under TSI-MS conditions, fragment ions are observed with *m/z* 375 due to the loss of water, *m/z* 373 due to the loss of hydrogen fluoride (HF), *m/z* 363 due to the loss of formaldehyde ($H_2C{=}O$), and *m/z* 333 due to the loss of ethanal (CH_3CHO). The ion with

TABLE 4.6.1 Structures of glucocorticosteroids.

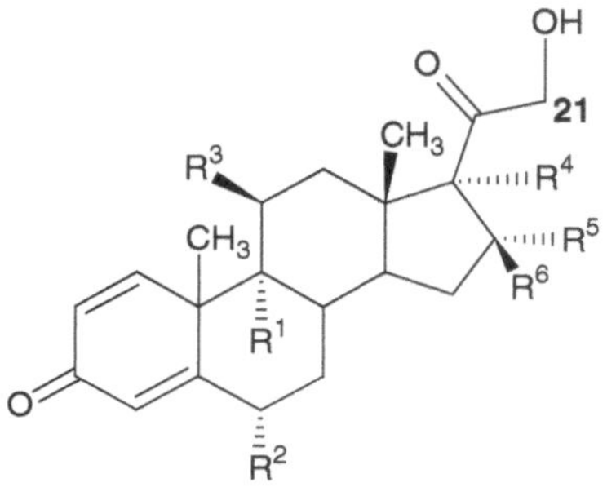

Compound	[M−H]$^-$	Δ^1	R^1	R^2	R^3	R^4	R^5	R^6
Prednisone	357	Yes	—H	—H	=O	—OH	—H	—H
Prednisolone	359	Yes	—H	—H	—OH	—OH	—H	—H
Cortisone	359	No	—H	—H	=O	—OH	—H	—H
Methylprednisolone	373	Yes	—H	—CH_3	—OH	—OH	—H	—H
Fluprednisolone	377	Yes	—H	—F	—OH	—OH	—H	—H
Fludrocortisone	379	No	—F	—H	—OH	—OH	—H	—H
Dexamethasone	391	Yes	—F	—H	—OH	—OH	—CH_3	—H
Betamethasone	391	Yes	—F	—H	—OH	—OH	—H	—CH_3
Triamcinolone	393	Yes	—F	—H	—OH	—OH	—OH	—H
Beclomethasone	407	Yes	—Cl	—H	—OH	—OH	—H	—CH_3
Flumethasone	409	Yes	—F	—F	—OH	—OH	—CH_3	—H
Desonide	415	Yes	—H	—H	—OH	—$OCH(CH_3)_2O$—		—H
Budesonide	429	Yes	—H	—H	—OH	—$OC(C_3H_7)O$—		—H
Triamcinolone acetonide	433	Yes	—F	—H	—OH	—$OCH(CH_3)_2O$—		—H
Flunisolide	433	Yes	—H	—F	—OH	—$OCH(CH_3)_2O$—		—H
Halometasone*	443	Yes	—F	—F	—OH	—OH	—CH_3	—H
Fluocinolone acetonide	451	Yes	—F	—F	—OH	—$OCH(CH_3)_2O$—		—H

*With —Cl at C2.

m/z 333 is used as precursor ion in TSI-MS–MS, resulting in fragment ions with, among others, *m/z* 313 due to the loss of HF, *m/z* 295 due to combined loss of HF and water, and *m/z* 171 and 147 (see later) (Polettini et al., 1998). In another study (Antignac et al., 2000), using ESI-MS–MS, a similar series of small-molecule losses is observed, that is, ions with *m/z* 373 due to the loss of HF, *m/z* 355, 337, and 319 due to three subsequent water losses. Additional fragment ions are observed with *m/z* 279 ($[C_{20}H_{22}O{+}H]^+$), which involves losses of HF, two times water from the C^{11}- and C^{17}-position, and the loss of 1,2-ethanedial (OHCCHO) from the C^{17}-position, *m/z* 237 ($[C_{17}H_{16}O{+}H]^+$) due to a subsequent loss of propene ($CH_3CH{=}CH_2$) from the D-ring, and the carbocations with *m/z* 171 ($[C_{12}H_{11}O]^+$), *m/z* 147 ($[C_{10}H_{11}O]^+$), and *m/z* 121 ($[C_8H_9O]^+$). Structure proposals for these fragments are given in Figure 4.6.6. These are characteristic fragments for many 3-keto-$\Delta^{1,4}$-corticosteroids.

Considerable attention has been paid to the differentiation between the epimers betamethasone and dexamethasone. Under some conditions, chromatographic separation can be achieved (Draisci et al., 2001b; Van Den Hauwe et al., 2001). Multistage MS^n in ion-trap instrument has also been proposed (De Wasch et al., 2001). Alternatively, the use of MS–MS on a TQ instrument in combination with multivariate statistics has been proposed (Antignac et al., 2002). Higher relative abundance of the fragment ions with *m/z* 121 and 279 for betamethasone and with *m/z* 161 and 237 for dexamethasone has also been indicated as a way to discriminate between these two epimers and their phosphate, acetate, valerate, and propionate esters (Arthur et al., 2004).

The fragmentation of beclomethasone dipropionate ($[M{+}H]^+$ with *m/z* 521), its 17- or 21-propionate metabolite ($[M{+}H]^+$ with *m/z* 465), and beclomethasone ($[M{+}H]^+$ with *m/z* 409) has been studied (Guan et al., 2003). In order to discriminate between two subsequent water losses and the loss of hydrogen chloride (HCl), the ^{37}Cl-isotope peak was selected as precursor ion, leading to a loss of either 36 Da (two times water) or 38 Da ($H^{37}Cl$). Beclomethasone dipropionate (precursor ion with *m/z* 523) shows fragment ions along two initial fragmentation pathways. Along one line, ions are observed with *m/z* 505 due to the loss of

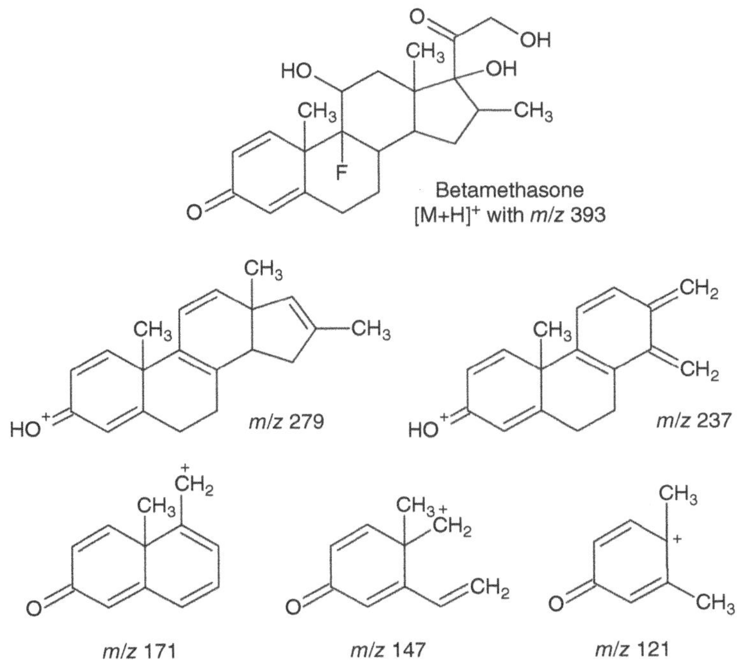

FIGURE 4.6.6 Proposed structures for the fragment ions of protonated betamethasone, showing characteristic fragment ions of 3-keto-$\Delta^{1,4}$-corticosteroids.

water from the 11-OH group, *m/z* 431 due to subsequent loss of propionic acid (CH_3CH_2COOH), and either *m/z* 357 due to the subsequent loss of another CH_3CH_2COOH or *m/z* 393 due to subsequent loss of $H^{37}Cl$. Along the other line, ions are observed with *m/z* 411 due to the subsequent loss of $H^{37}Cl$ and CH_3CH_2COOH, and either *m/z* 337 due to the loss of another CH_3CH_2COOH or *m/z* 393 due to a subsequent loss of water. The same initial fragmentation pathways are observed for the beclomethasone 17- or 21-propionate metabolites (precursor ions with *m/z* 467 (^{37}Cl-isotope)), thus either the loss of water followed by the loss of CH_3CH_2COOH to an ion with *m/z* 375 or the loss of $H^{37}Cl$ and propionic acid to an ion with 355 followed by the loss of water to an ion with *m/z* 337.

Beclomethasone itself (precursor ion with *m/z* 411 (^{37}Cl-isotope); Figure 4.6.7) shows a series of water losses and a $H^{37}Cl$ loss, thus fragment ions with *m/z* 373 due to $H^{37}Cl$ loss with subsequent losses of water to ions with *m/z* 335, *m/z* 337, *m/z* 319, and *m/z* 301, and fragment ions with *m/z* 393 and 375 due to water losses, which by subsequent $H^{37}Cl$ loss result in the ions with *m/z* 355 and 337 mentioned earlier (Guan et al., 2003). Additional common fragments are the ions with *m/z* 279 ($[C_{20}H_{22}O+H]^+$), which involves losses of $H^{37}Cl$, and two times water from the C^{11}- and C^{17}-positions, and the loss of OHCCHO from the C^{17}-position, and *m/z* 237 ($[C_{17}H_{16}O+H]^+$) due to a subsequent loss of $CH_3CH{=}CH_2$ from the D-ring (Figure 4.6.7).

The major fragment ions of the glucocorticosteroid fluticasone-17-propionate ($[M+H]^+$ with *m/z* 501) are the ions with *m/z* 483 due to water loss, *m/z* 481 due to hydrogen fluoride (HF) loss, *m/z* 333 due to the loss of the two C^{17}-substituents, *m/z* 313 and 293 due to two subsequent HF losses, and the ion with *m/z* 275 and 265, which could be considered as secondary losses of water and CO, respectively, from the ion with *m/z* 293. An ion with *m/z* 295 involves the loss of HF, the two C^{17}-substituents and two C's from the D-ring. Finally, the ion with *m/z* 121 ($[C_8H_9O]^+$) is characteristic for a 3-keto-$\Delta^{1,4}$-steroid (Section 4.6.4), while the ion with *m/z* 205 ($[C_{12}H_{10}FO_2]^+$) corresponds to an acylium ion involving the AB-ring and the C^{11}-keto group (after HF loss).

The fragmentation of budesonide ($[M+H]^+$ with *m/z* 431) results in fragment ions with *m/z* 413, 395, and 377 due to three consecutive water losses, *m/z* 341 due to the loss of butane-1,1-diol ($C_4H_{10}O_2$) from the 2-propyl-1,3-dioxolane side group at C^{16} and C^{17}, *m/z* 323, *m/z* 305, *m/z* 295, and *m/z* 277 due to subsequent losses of water, two times water, water and CO, and two times water and CO, respectively. In addition, fragment ions with *m/z* 121 ($[C_8H_9O]^+$), *m/z* 147 ($[C_{10}H_{11}O]^+$), and *m/z* 173 ($[C_{12}H_{13}O]^+$) are observed (Hou et al., 2005). In-source CID of 23 other corticosteroids with 3-keto-Δ^4- and 3-keto-$\Delta^{1,4}$-skeletons, with or without fluorine atoms, has been studied (Hou et al., 2005). Attention has been paid to water losses and, if applicable, HF loss. HF loss seems to occur prior to water loss. The fragment ions

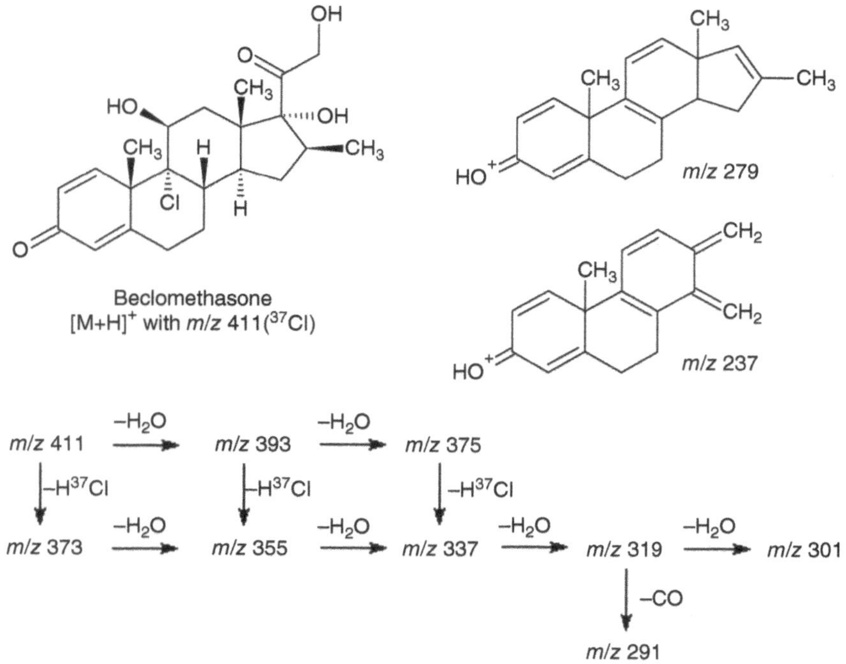

FIGURE 4.6.7 Fragmentation pathways for protonated beclomethasone.

described for budesonide, that is, *m/z* 121 ([C_8H_9O]$^+$), *m/z* 147 ([$C_{10}H_{11}O$]$^+$), and *m/z* 173 ([$C_{12}H_{13}O$]$^+$), are frequently observed for other 3-keto-$\Delta^{1,4}$-steroids (Section 4.6.4) (Hou et al., 2005).

In a sports doping related screening study, three characteristic fragment ions for 15 glucocorticosteroids have been applied, both in SRM and for PIA (Section 1.5.3). These are the fragment ions with *m/z* 121 ([C_8H_9O]$^+$), *m/z* 237 ([$C_{17}H_{16}O$+H]$^+$; ABC-rings plus C^{18} and C^{15}), and *m/z* 279 ([$C_{20}H_{22}O$+H]$^+$, ABCD-rings and C^{16} methyl substituent) (Figure 4.6.6), which are thus formed after loss of eventual substituents from C^6, C^9, C^{11}, and C^{17} positions (Mazzarino et al., 2008). PIA was also applied in the screening for 6-methylprednisolone acetate and its metabolites in rat urine, using a carbocation with *m/z* 161 ([$C_{11}H_{13}O$]$^+$, part of AB-rings and C^{19}) as the common product ion (Panusa et al., 2010). In this way, six metabolites were found in rat urine.

In negative-ion ESI-MS, the corticosteroids are usually observed as formate or acetate adducts, that is, [M+HCOO]$^-$ or [M+CH_3COO]$^-$ with *m/z* (M+45) or *m/z* (M+59), respectively (Cui et al., 2006; Pozo et al., 2009a; Dusi et al., 2011; Chen et al., 2011).

The fragmentation of corticosteroids in negative-ion MS–MS involves the loss of $H_2C{=}O$ from the C^{21}-position (Table 4.6.1) and its hydroxy group (Antignac et al., 2000; Deventer & Delbeke, 2003; Pozo et al., 2009a), except for corticosteroids with an acetonide function at the D-ring (Pozo et al., 2009a). The loss of $H_2C{=}O$ is observed as a neutral loss of 76 or 90 Da from [M+HCOO]$^-$ or [M+CH_3COO]$^-$, respectively. For acetonide corticosteroids, the loss of 104 or 118 Da is observed from [M+HCOO]$^-$ or [M+CH_3COO]$^-$, respectively, which is consistent with the loss of acetone (CH_3COCH_3) and formic acid (HCOOH) or acetic acid (CH_3COOH), respectively. In budesonide, which has a different side chain in the acetonide group, the loss of 118 Da is observed from the [M+HCOO]$^-$ (Pozo et al., 2009a). The identity of the fragment ions with high *m/z* in the negative-ion product-ion mass spectra of 17 representative corticosteroids is summarized in Table 4.6.2, where for clarity the *m/z* values of [M−H]$^-$ have been specified rather than that of [M+HCOO]$^-$ or [M+CH_3COO]$^-$. In addition to the common neutral losses indicated earlier, other small-molecule losses are observed, involving losses of a methyl radical ($^\bullet CH_3$), water, HF, CO, and/or HCl. The loss of CO from [M+HCOO−HCOOH−$H_2C{=}O$]$^-$ is specifically observed for corticosteroids with a keto rather than a hydroxy group at C^{11}. Nonspecific loss of 46 Da (water and CO) is also frequently observed from [M+HCOO−HCOOH−$H_2C{=}O$]$^-$ (Pozo et al., 2009a). Beclomethasone ([M+HCOO]$^-$ with *m/z* 453) shows fragment ions with *m/z* 407 due to loss of HCOOH, thus [M−H]$^-$; with *m/z* 377 due to the characteristic loss of 76 Da (HCOOH and $H_2C{=}O$); with *m/z* 341 due to subsequent loss of HCl; and with *m/z* 297 due to subsequent loss of carbon dioxide (CO_2) (Guan et al., 2003). For prednisolone ([M+HCOO]$^-$ with *m/z* 405) and methylprednisolone ([M+HCOO]$^-$ with *m/z*

TABLE 4.6.2 Fragmentation of 17 representative corticosteroids in negative-ion mode (structures in Table 4.6.1).

Compound	$[M-H]^-$	$-H_2C{=}O$	*m/z*	Interpretation
Prednisone	357	327	299	327 – CO
			285	327 – $H_2C{=}C{=}O$
Prednisolone	359	329	313	359 – H_2O – CO
			295	313 – H_2O
			280	295 – $^{\bullet}CH_3$
			187	Proposed: $A{+}B{+}C^{11}{-}O^-$
Cortisone	359	329	301	329 – CO
Methylprednisolone	373	343	327	373 – H_2O – CO
			309	327 – H_2O
			294	309 – $^{\bullet}CH_3$
			201	Proposed: $A{+}B{+}C^{11}{-}O^-$
Fluprednisolone	377	347	331	377 – H_2O – CO
			327	347 – HF
			313	331 – H_2O
			298	313 – $^{\bullet}CH_3$
Fludrocortisone	379	349	333	379 – H_2O – CO
			313	333 – HF
			295	313 – H_2O
			280	295 – $^{\bullet}CH_3$
Dexamethasone	391	361	345	391 – H_2O – CO
			325	345 – HF
			307	325 – H_2O
			292	307 – $^{\bullet}CH_3$
			277	292 – $^{\bullet}CH_3$ (possibly)
Betamethasone	391	361	345	391 – H_2O – CO
			325	345 – HF
			307	325 – H_2O
			292	307 – $^{\bullet}CH_3$
			277	292 – $^{\bullet}CH_3$ (possibly)
Triamcinolone	393	363	345	363 – H_2O
			325	345 – HF
			310	325 – $^{\bullet}CH_3$
Beclomethasone	407	377	341	377 – HCl
			297	341 – CO_2
Flumethasone	409	379	363	409 – H_2O – CO
			343	363 – HF
			328	343 – $^{\bullet}CH_3$
			325	343 – H_2O
			305	325 – HF
Desonide	415	No	357	425 – $(CH_3)_2C{=}O$
			339	357 – H_2O
Budesonide	429	No	357	429 – C_3H_7CHO
			339	357 – H_2O
Triamcinolone acetonide	433	No	413	433 – HF
			375	433 – $(CH_3)_2C{=}O$
			357	375 – H_2O
			337	357 – HF
Flunisolide	433	No	375	433 – $(CH_3)_2C{=}O$
			357	375 – H_2O
Halometasone	443	413	377	413 – HCl
			362	377 – $^{\bullet}CH_3$
			357	377 – HF
			342	357 – $^{\bullet}CH_3$
Fluocinolone acetonide	451	No	431	451 – HF
			393	451 – $(CH_3)_2C{=}O$
			373	393 – HF
			355	373 – H_2O

419), fragment ions with *m/z* 187 ([$C_{12}H_{12}O_2$−H]$^-$) and *m/z* 201 ([$C_{13}H_{14}O_2$−H]$^-$) are observed, which are attributed to the loss of the D-ring and a large part of the C-ring, leaving the A- and B-ring with C^{11} and the O-atom attached to it, bearing the negative charge. In aldosterone ([M−H]$^-$ with *m/z* 359), lacking the double bond at C^1, a similar fragment ion is observed with *m/z* 189 ([$C_{12}H_{14}O_2$−H]$^-$) (Turpeinen et al., 2008).

4.6.7 ESTROGENS

Estrogens form a group of compounds that are important as the primary female sex hormones in the estrous cycle of humans and other animals. The three major naturally occurring estrogens in women are estrone, estradiol, and estriol. Estrogen drugs are used as oral contraceptives and in estrogen replacement therapy for postmenopausal women. In environmental science, estrogens are among a wide range of endocrine-disrupting compounds (EDCs), which may cause male reproductive dysfunction to wildlife (Sosa-Ferrera et al., 2013).

Multiresidue LC–MS analysis of estrogens is mainly done using SRM in TQ instruments for clinical, pharmaceutical, and environmental applications (e.g., Díaz-Cruz et al., 2003; Pedrouzo et al., 2009; Malone et al., 2010; Wang et al., 2010). Estrogens are generally analyzed as [M−H]$^-$ in negative-ion mode. Characteristic fragment ions for some model estrogen compounds are summarized in Table 4.6.3. Molecular formulae of the fragment ions are confirmed by accurate-mass data. Structure proposals for some of the fragment ions, illustrated for estradiol, are given in Figure 4.6.8 (Croley et al., 2000; Griffiths, 2003; Sun et al., 2005; Lampinen-Salomonsson et al., 2006; Pedrouzo et al., 2009; Wooding et al., 2013). Note that these fragments most likely result from charge-remote fragmentation of the phenolic steroid anion.

The negative-ion fragmentation of estradiol ([M−H]$^-$ with *m/z* 271) has been studied in detail (Wooding et al., 2013), also taking advantage of two labeled analogs, that is, a [D_4]-labeled analog (with D-labels at C^2, C^4, and C^{16}) and a [$^{13}C_6$]-labeled analog (with [^{13}C]-labels at the D-ring and C^{18}). A wide variety of fragment ions is observed, including the deprotonated 2-naphthol ([$C_{10}H_8O$−H]$^-$) with

TABLE 4.6.3 Fragmentation of estrogen in negative-ion MS–MS.

Compound	[M−H]$^-$	Fragment ions (*m/z*)	Interpretation
Dienestrol	265	249	265 − CH_4
		235	265 − C_2H_6
		221	265 − CH_4 − C_2H_4
		147	[$C_{10}H_{11}O$]$^-$
		117	[C_8H_5O]$^-$, thus: HC≡C—C_6H_4—O^-
		93	[C_6H_5O]$^-$
Diethylstilbestrol	267	251	267 − CH_4
		237	267 − C_2H_6
		221	267 − CH_4 − C_2H_4
		209	267 − C_2H_6 − C_2H_4
		93	[C_6H_5O]$^-$
Estrone	269	159	[$C_{11}H_{11}O$]$^-$ (A+B+1×C from C)
		145	[$C_{10}H_9O$]$^-$ (A+B)
		143	[$C_{10}H_7O$]$^-$ (A+B)
Estradiol (17α- and 17β-)	271	253	271 − H_2O
		239	271 − CH_3OH
		183	[$C_{13}H_{11}O$]$^-$ (A+B+3×C from C)
		145	[$C_{10}H_9O$]$^-$ (A+B)
		143	[$C_{10}H_7O$]$^-$ (A+B)
Estriol	287	171	[$C_{12}H_{11}O$]$^-$ (A+B+2×C from C)
		145	[$C_{10}H_9O$]$^-$ (A+B)
17α-Ethinylestradiol	295	277	295 − H_2O
		251	277 − HC≡CH
		159	[$C_{11}H_{11}O$]$^-$ (A+B+1×C from C)
		145	[$C_{10}H_9O$]$^-$ (A+B)
		143	[$C_{10}H_7O$]$^-$ (A+B)
Norethindrone	299	171	[$C_{12}H_{11}O$]$^-$ (A+B+2×C from C)
		145	[$C_{10}H_9O$]$^-$ (A+B)

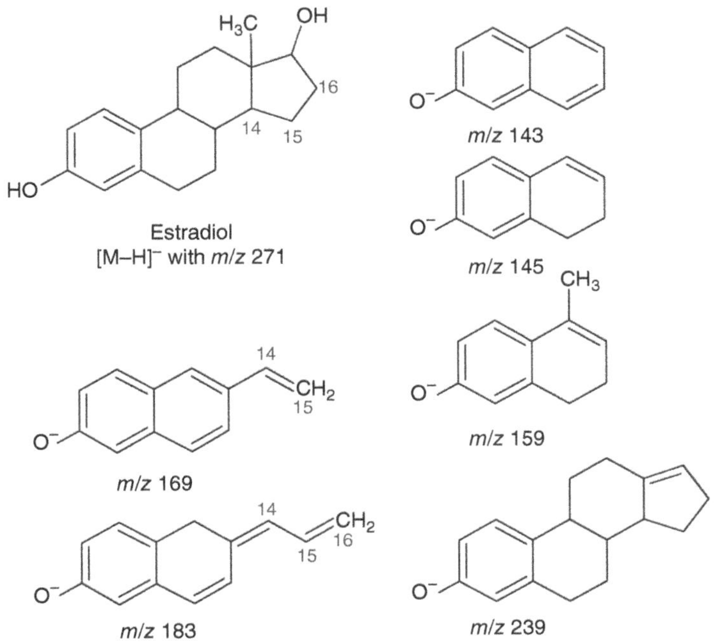

FIGURE 4.6.8 Proposed structures for the fragment ions of deprotonated estrogens, illustrated for estradiol.

m/z 143, the ion with *m/z* 145 ($[C_{10}H_{10}O{-}H]^-$), the deprotonated 6-vinyl-2-naphthol ($[C_{12}H_{10}O{-}H]^-$) with *m/z* 169, the deprotonated 6-allyl-2-naphthol ($[C_{13}H_{12}O{-}H]^-$) with *m/z* 183, and *m/z* 239 ($[C_{17}H_{20}O{-}H]^-$) (Figure 4.6.8). The major fragment ions with *m/z* 169 ($[C_{12}H_{10}O{-}H]^-$) and *m/z* 183 ($[C_{13}H_{12}O{-}H]^-$) are formed from $[M{-}H]^-$ directly. These ions are also observed for estrone and estriol. The use of labeled analogs revealed, perhaps somewhat surprising, that the ion with *m/z* 183 is due to the AB-rings and C^{14}, C^{15}, and C^{16}, as three [^{13}C]-labels and all four [D]-labels are kept. The ion with *m/z* 169 is due to the AB-rings and C^{14} and C^{15}, as two [^{13}C]- and two [D]-labels are kept (Wooding et al., 2013). The ion with *m/z* 239 apparently involves two radical losses, that is, a methyl radical ($^{\bullet}CH_3$) and a hydroxy radical ($^{\bullet}OH$), as all four [D]-labels and five out of six [^{13}C]-labels are kept.

The fragmentation of estrone ($[M{+}H]^+$ with *m/z* 271; $[M{-}H]^-$ with *m/z* 269) was studied using estrone methyl ester and a [D_4]-estrone analog (with [D]-labels at C^2, C^4, and C^{16}) to assist in elucidating the fragmentation pathways (Bourcier et al., 2010). In negative-ion mode, the major fragment is the ion with *m/z* 145 ($[C_{10}H_{10}O{-}H]^-$, AB-rings, Figure 4.6.8). In positive-ion mode, a variety of fragment ions are observed (Figure 4.6.9 for the fragmentation pathways). Common fragment ions for estrone, 17α-estradiol, estriol, and 17α-ethynylestradiol are the carbocations with *m/z* 133 ($[C_9H_9O]^+$), *m/z* 159 ($[C_{11}H_{11}O]^+$), *m/z* 183 ($[C_{13}H_{11}O]^+$), and the ion due to the loss of water. The loss of water was found not to involve 1,2-elimination using a H (or D) from the C^2, C^4, and C^{16} positions. Therefore, rearrangement of the C^{18}-methyl group from C^{13} to C^{17} is proposed. The water loss can then either arise as a 1,3-elimination (via a four-center mechanism) resulting in a methylidene ($=CH_2$) at C^{17} or involve the C^{13}-H-atom, after H-migration from C^{14}. Both fragment ions are stable tertiary carbocations conjugated with a C=C double bond (Figure 4.6.9). The former route shows two subsequent losses of ethene (C_2H_4) to fragment ions with *m/z* 225 ($[C_{16}H_{17}O]^+$) and *m/z* 197 ($[C_{14}H_{13}O]^+$), whereas the latter route results in three fragment ions, that is, with *m/z* 159 ($[C_{11}H_{11}O]^+$; loss of C_7H_{10}, e.g., 2,3-dimethylcyclopenta-1,3-diene), *m/z* 211 ($[C_{15}H_{15}O]^+$; loss of propene ($CH_3CH{=}CH_2$)), and *m/z* 183 ($[C_{13}H_{11}O]^+$; subsequent loss of C_2H_4) (Figure 4.6.9). Two different mechanisms were proposed for the formation of the ion with *m/z* 213 ($[C_{15}H_{17}O]^+$; due to loss of C_3H_6O from D-ring) and the two subsequent C_2H_4 losses to the ions with *m/z* 185 ($[C_{13}H_{13}O]^+$) and *m/z* 157 ($[C_{11}H_9O]^+$) (Bourcier et al., 2010).

Estrogen esters such as estradiol-17-acetate or estradiol-17-valerate show fragment ions due to the loss of the ester chain as acid, as well as a fragment ion due to charge retention on the carboxylic acid group. For example, estradiol-17-valerate ($[M{-}H]^-$ with *m/z* 355) shows fragment ions with *m/z* 253 due to the loss of valeric acid (C_4H_9COOH) and with *m/z* 101 due to the valerate anion ($[C_4H_9COO]^-$).

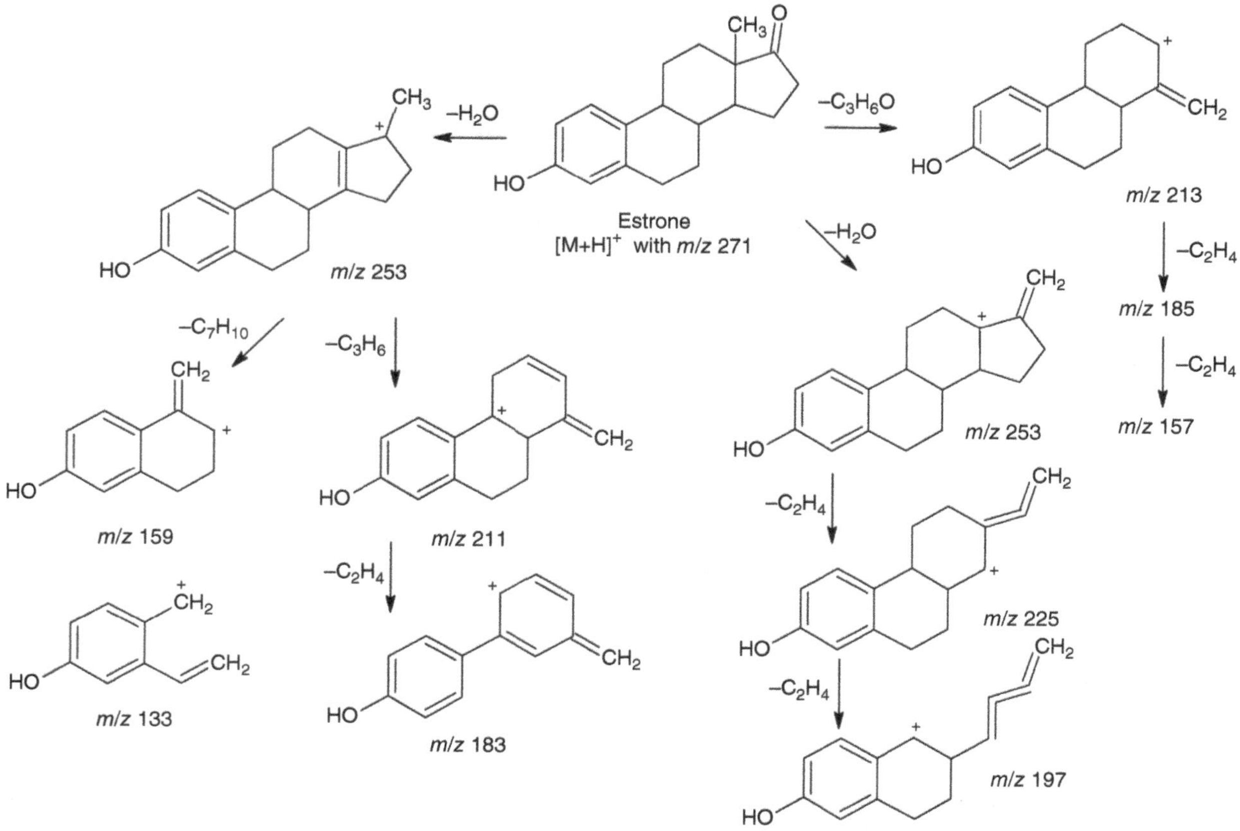

FIGURE 4.6.9 Proposed structures for the fragment ions of protonated estrone.

4.6.8 STEROID CONJUGATES

Important sulfate and glucuronic acid conjugates in the human body are the result of steroid metabolism. Both types of conjugates can be analyzed in either positive-ion or negative-ion mode.

In negative-ion mode, estrogen 3-sulfates or steroid disulfates show the characteristic loss of sulfur trioxide (SO_3). In addition, hydrogen sulfate ($[HSO_4]^-$) with *m/z* 97 may be observed. In positive-ion mode, the loss of dihydrogen sulfate (H_2SO_4) may be observed next to the loss of SO_3. Thus, estrone-3-sulfate ($[M-H]^-$ with *m/z* 349) shows fragment ions with *m/z* 269 due to the unconjugated estrone anion and with *m/z* 80 ($[SO_3]^{-\bullet}$) (Weidolf et al., 1988; Murray et al., 1996; Griffiths, 2003). Aliphatic 17β-sulfates of nortestosterone, testosterone, and boldenone show the hydrogen sulfate ($[HSO_4]^-$) with *m/z* 97 as the most abundant ion, next to radical anions with *m/z* 80 ($[SO_3]^{-\bullet}$) and 96 ($[SO_4]^{-\bullet}$) (Weidolf et al., 1988).

In both positive-ion and negative-ion modes, steroid glucuronides show characteristic losses of dehydroglucuronic acid ($C_6H_8O_6$, 176 Da) or glucuronic acid ($C_6H_{10}O_7$, 194 Da), the latter only from aliphatic glucuronides. In negative-ion mode, additional losses may be the loss of acetic acid (CH_3COOH) and a gluconolactone ($C_6H_{10}O_6$), whereas in positive-ion mode, steroids with an additional hydroxy or keto group may show the loss of 212 Da ($C_6H_{12}O_8$) (Kuuranne et al., 2000; Griffiths, 2003). In negative-ion mode, the deprotonated dehydroglucuronic acid ($[C_6H_7O_6]^-$) with *m/z* 175 may be observed, which may show secondary fragmentation, involving the loss of water to *m/z* 157 ($[C_6H_5O_5]^-$), the subsequent loss of carbon dioxide (CO_2) to deprotonated oxane-3,5-dione ($[C_5H_5O_3]^-$) with *m/z* 113, and the subsequent loss of carbon monoxide (CO) to deprotonated 2,5-dihydrofuran-3-ol ($[C_4H_5O_2]^-$) with *m/z* 85. Estrone-3-glucuronide ($[M-H]^-$ with *m/z* 445) shows fragment ions with *m/z* 269 due to the unconjugated estrone anion and with *m/z* 175 ($[C_6H_7O_6]^-$).

Equilin sulfate and the isomeric steroids estrone sulfate, 17α-dihydroequilin sulfate, and 17β-dihydroequilin sulfate are the primary constituents in hormone replacement therapies for traumatic brain injuries. Multistage MS^n in an ion-trap–time-of-flight hybrid instrument was used to discriminate between these components. The epimers 17α- and 17β-dihydroequilin sulfate could be distinguished at MS^4 level (Tedmon et al., 2013).

The negative-ion fragmentation of 18 tetrahydrocorticosteroid sulfates has been studied using LIT and TQ instruments (Mitamura et al., 2014). In MS–MS, 5β-tetrahydrocortisol-3-sulfate ($[M-H]^-$ with *m/z* 445) shows an abundant hydrogen sulfate ion (HSO_4^-) with *m/z* 97 and minor fragments with *m/z* 427 due the loss of water, *m/z* 415 due to the loss of formaldehyde ($H_2C{=}O$), and *m/z* 385 due to the loss of water and ethenone ($H_2C{=}C{=}O$). In the 5α-analog, the major fragment ion is the ion with *m/z* 415, whereas the relative abundance of the ions with *m/z* 97 and 427 is far less. Such difference between 5α- and 5β-analogs is not observed for the tetrahydrocortisol-21-sulfates, for which the most abundant fragment ion is due to the loss of $H_2C{=}O$ and SO_3. The tetrahydrocortisol-3,21-disulfates show abundant $[M-2H]^{2-}$ ions with *m/z* 262, along with $[M-H]^-$ ions with *m/z* 525. The singly-charged ions show loss of SO_3 to an ion with *m/z* 445 and SO_3 and water (or H_2SO_4) to *m/z* 427, while the doubly-charged ions predominantly show the loss of dihydrogen sulfate (H_2SO_4) to *m/z* 427 and the ion with *m/z* 97 (HSO_4^-) (Mitamura et al., 2014).

REFERENCES

Antignac JP, Le Bizec B, Monteau F, Poulain F, André F. 2000. Collision-induced dissociation of corticosteroids in electrospray tandem mass spectrometry and development of a screening method by high performance liquid chromatography/tandem mass spectrometry. Rapid Commun Mass Spectrom 14: 33–39.

Antignac JP, Le Bizec B, Monteau F, André F. 2002. Differentiation of betamethasone and dexamethasone using liquid chromatography/positive electrospray tandem mass spectrometry and multivariate statistical analysis. J Mass Spectrom 37: 69–75.

Arthur KE, Wolff JC, Carrier DJ. 2004. Analysis of betamethasone, dexamethasone and related compounds by liquid chromatography/electrospray mass spectrometry. Rapid Commun Mass Spectrom 18: 678–684.

Bourcier S, Poisson C, Souissi Y, Kinani S, Bouchonnet S, Sablier M. 2010. Elucidation of the decomposition pathways of protonated and deprotonated estrone ions: Application to the identification of photolysis products. Rapid Commun Mass Spectrom 24: 2999–3010.

Catlin DH, Sekera MH, Ahrens BD, Starcevic B, Chang YC, Hatton CK. 2004. Tetrahydrogestrinone: Discovery, synthesis, and detection in urine. Rapid Commun Mass Spectrom 18: 1245–1249.

Chen D, Tao Y, Liu Z, Zhang H, Liu Z, Wang Y, Huang L, Pan Y, Peng D, Dai M, Wang X, Yuan Z. 2011. Development of a liquid chromatography–tandem mass spectrometry with pressurized liquid extraction for determination of glucocorticoid residues in edible tissues. J Chromatogr B 879: 174–180.

Cho HJ, Kim JD, Lee WY, Chung BC, Choi MH. 2009. Quantitative metabolic profiling of 21 endogenous corticosteroids in urine by liquid chromatography–triple quadrupole-mass spectrometry. Anal Chim Acta 632: 101–108.

Croley TR, Hughes RJ, Koenig BG, Metcalfe CD, March RE. 2000. Mass spectrometry applied to the analysis of estrogens in the environment. Rapid Commun Mass Spectrom 14: 1087–1093.

Cui X, Shao B, Zhao R, Yang Y, Hu J, Tu X. 2006. Simultaneous determination of seventeen glucocorticoids residues in milk and eggs by ultra-performance liquid chromatography/electrospray tandem mass spectrometry. Rapid Commun Mass Spectrom 20: 2355–2364.

De Brabander HF, De Wasch K, van Ginkel LA, Sterk SS, Blokland MH, Delahaut P, Taillieu X, Dubois M, Arts CJ, van Baak MJ, Gramberg LG, Schilt R, van Bennekom EO, Courtheyn D, Vercammen J, Witkamp RF. 1998. Multi-laboratory study of the analysis and kinetics of stanozolol and its metabolites in treated calves. Analyst 123: 2599–2604.

Deventer K, Delbeke FT. 2003. Validation of a screening method for corticosteroids in doping analysis by liquid chromatography/tandem mass spectrometry. Rapid Commun Mass Spectrom 17: 2107–2114.

de Vlieger JS, Kolkman AJ, Ampt KAM, Commandeur JNM, Vermeulen NPE, Kool J, Wijmenga SS, Niessen WMA, Irth H, Honing M. 2010. Determination and identification of estrogenic compounds generated with biosynthetic enzymes using hyphenated screening assays, high resolution mass spectrometry and off-line NMR. J Chromatogr B 878: 667–674.

De Wasch K, De Brabander HF, Van de Wiele M, Vercammen J, Courtheyn D, Impens S. 2001. Differentiation between dexamethasone and betamethasone in a mixture using multiple mass spectrometry. J Chromatogr A 926: 79–86.

Díaz-Cruz MS, López de Alda MJ, López R, Barceló D. 2003. Determination of estrogens and progestogens by mass spectrometric techniques (GC/MS, LC/MS and LC/MS/MS). J Mass Spectrom 38: 917–923.

Draisci R, Palleschi L, Marchiafava C, Ferretti E, Delli Quadri F. 2001a. Confirmatory analysis of residues of stanozolol and its major metabolite in bovine urine by liquid chromatography–tandem mass spectrometry. J Chromatogr A 926: 69–77.

Draisci R, Marchiafava C, Palleschi L, Cammarata P, Cavalli S. 2001b. Accelerated solvent extraction and liquid chromatography–tandem mass spectrometry quantitation of corticosteroid residues in bovine liver. J Chromatogr B 753: 217–223.

Dusi G, Gasparini M, Curatolo M, Assini W, Bozzoni E, Tognoli N, Ferretti E. 2011. Development and validation of a liquid chromatography–tandem mass spectrometry method for the simultaneous determination of nine corticosteroid residues in bovine liver samples. Anal Chim Acta 700: 49–57.

Edlund PO, Bowers L, Henion JD. 1989. Determination of methandrostenolone and its metabolites in equine plasma and urine by coupled-column liquid chromatography with ultraviolet detection and confirmation by tandem mass spectrometry. J Chromatogr 487: 341–356.

Griffiths WJ, Brown A, Reimendal R, Yang Y, Zhang J, Sjövall J. 1996. A comparison of fast-atom bombardment and electrospray as methods of ionization in the study of sulphated- and sulphonated-lipids by tandem mass spectrometry. Rapid Commun Mass Spectrom 10: 1169–1174.

Griffiths WJ. 2003. Tandem mass spectrometry in the study of fatty acids, bile acids, and steroids. Mass Spectrom Rev 22: 81–152.

Guan F, Uboh C, Soma L, Hess A, Luo Y, Tsang DS. 2003. Sensitive liquid chromatographic/tandem mass spectrometric method for the determination of beclomethasone dipropionate and its metabolites in equine plasma and urine. J Mass Spectrom 38: 823–838.

Guan F, Soma LR, Luo Y, Uboh CE, Peterman S. 2006. Collision-induced dissociation pathways of anabolic steroids by electrospray ionization tandem mass spectrometry. J Am Soc Mass Spectrom 17: 477–489.

Higashi T, Shimada K. 2004. Derivatization of neutral steroids to enhance their detection characteristics in liquid chromatography–mass spectrometry. Anal Bioanal Chem 378: 875–882.

Hou S, Hindle M, Byron PR. 2005. Chromatographic and mass spectral characterization of budesonide and a series of structurally related corticosteroids using LC–MS. J Pharm Biomed Anal 39: 196–205.

Joos PE, Van Ryckeghem M. 1999. Liquid chromatography–tandem mass spectrometry of some anabolic steroids. Anal Chem 71: 4701–4710.

Kang MJ, Lisurek M, Bernhardt R, Hartmann RW. 2004. Use of high-performance liquid chromatography/electrospray ionization collision-induced dissociation mass spectrometry for structural identification of monohydroxylated progesterones. Rapid Commun Mass Spectrom 18: 2795–2800.

Keski-Rahkonen P, Huhtinen K, Desai R, Harwood DT, Handelsman DJ, Poutanen M, Auriola S. 2013. LC–MS analysis of estradiol in human serum and endometrial tissue: Comparison of electrospray ionization, atmospheric pressure chemical ionization and atmospheric pressure photoionization. J Mass Spectrom 48: 1050–1058.

Kim SM, Kim DH. 2001. Quantitative determination of medroxyprogesterone acetate in plasma by liquid chromatography/electrospray ion trap mass spectrometry. Rapid Commun Mass Spectrom 15: 2041–2045.

Kim SH, Cha EJ, Lee KM, Kim HJ, Kwon OS, Lee J. 2014. Simultaneous ionization and analysis of 84 anabolic androgenic steroids in human urine using liquid chromatography-silver ion coordination ionspray/triple-quadrupole mass spectrometry. Drug Test Anal 6: 1174–1185.

Kobayashi Y, Saiki K, Watanabe F. 1993. Characteristics of mass fragmentation of steroids by atmospheric pressure chemical ionization-mass spectrometry. Biol Pharm Bull 16: 1175–1178.

Kushnir MM, Rockwood AL, Roberts WL, Yue B, Bergquist J, Meikle AW. 2011. Liquid chromatography tandem mass spectrometry for analysis of steroids in clinical laboratories. Clin Biochem 44: 77–88.

Kuuranne T, Vahermo M, Leinonen A, Kostianen R. 2000. Electrospray and atmospheric pressure chemical ionization tandem mass spectrometric behavior of eight anabolic steroid glucuronides. J Am Soc Mass Spectrom 11: 722–730.

Lampinen-Salomonsson M, Bondesson U, Petersson C, Hedeland M. 2006. Differentiation of estriol glucuronide isomers by chemical derivatization and electrospray tandem mass spectrometry. Rapid Commun Mass Spectrom 20: 1429–1440.

Leinonen A, Kuuranne T, Kostiainen R. 2002. Liquid chromatography/mass spectrometry in anabolic steroid analysis – optimization and comparison of three ionization techniques: Electrospray ionization, atmospheric pressure chemical ionization and atmospheric pressure photoionization. J Mass Spectrom 37: 693–698.

Lembcke J, Ceglarek U, Fiedler GM, Baumann S, Leichtle A, Thiery J. 2005. Rapid quantification of free and esterified phytosterols in human serum using APPI-LC–MS/MS. J Lipid Res 46: 21–26.

Leporati M, Bergoglio M, Capra P, Bozzetta E, Abete MC, Vincenti M. 2014. Development, validation and application to real samples of a multiresidue LC–MS/MS method for determination of β2-agonists and anabolic steroids in bovine hair. J Mass Spectrom 49: 936–946.

Li F, Hsieh Y, Korfmacher WA. 2008. High-performance liquid chromatography-atmospheric pressure photoionization/tandem mass spectrometry for the detection of 17alpha-ethinylestradiol in hepatocytes. J Chromatogr B 870: 186–191.

Liu S, Sjövall J, Griffiths WJ. 2000. Analysis of oxosteroids by nano-electrospray mass spectrometry of their oximes. Rapid Commun Mass Spectrom 14: 390–400.

Ma Y-C, Kim H-Y. 1997. Determination of steroids by liquid chromatography–mass spectrometry. J Am Soc Mass Spectrom 8: 1010–1020.

Malone EM, Elliott CT, Kennedy DG, Regan L. 2010. Rapid confirmatory method for the determination of sixteen synthetic growth promoters and bisphenol A in bovine milk using dispersive solid-phase extraction and liquid chromatography–tandem mass spectrometry. J Chromatogr B 878: 1077–1084.

Mazzarino M, Turi S, Botrè F. 2008. A screening method for the detection of synthetic glucocorticosteroids in human urine by liquid chromatography–mass spectrometry based on class-characteristic fragmentation pathways. Anal Bioanal Chem 390: 1389–1402.

McKinney AR, Suann CJ, Dunstan AJ, Mulley SL, Ridley DD, Stenhouse AM. 2004. Detection of stanozolol and its metabolites in equine urine by liquid chromatography–electrospray ionization ion trap mass spectrometry. J Chromatogr B 811: 75–83.

Mitamura K, Satoh (née Okihara) R, Kamibayashi M, Sato K, Iida T, Ikegawa S. 2014. Simultaneous determination of 18 tetrahydrocorticosteroid sulfates in human urine by liquid chromatography/electrospray ionization-tandem mass spectrometry. Steroids 85: 18–29.

Mück WM, Henion JD. 1990. High-performance liquid chromatography/tandem mass spectrometry: Its use for the identification of stanozolol and its major metabolites in human and equine urine. Biomed Environ Mass Spectrom 19: 37–51.

Murray S, Rendell NB, Taylor GW. 1996. Microbore high-performance liquid chromatography–electrospray ionisation mass spectrometry of steroid sulphates. J Chromatogr A 738: 191–199.

Musharraf SG, Ali A, Khan NT, Yousuf M, Choudhary MI, Atta-ur-Rahman. 2013. Tandem mass spectrometry approach for the investigation of the steroidal metabolism: Structure-fragmentation relationship (SFR) in anabolic steroids and their metabolites by ESI-MS/MS analysis. Steroids 78: 171–181.

Panusa A, Aldini G, Orioli M, Vistoli G, Rossoni G, Carini M. 2010. A sensitive and specific precursor ion scanning approach in liquid chromatography/electrospray ionization tandem mass spectrometry to detect methylprednisolone acetate and its metabolites in rat urine. Rapid Commun Mass Spectrom 24: 1583–1594.

Pedrouzo M, Borrull F, Pocurull E, Marcé RM. 2009. Estrogens and their conjugates: Determination in water samples by solid-phase extraction and liquid chromatography–tandem mass spectrometry. Talanta 78: 1327–1331.

Polettini A, Marrubini Bouland G, Montagna M. 1998. Development of a coupled-column liquid chromatographic–tandem mass spectrometric method for the direct determination of betamethasone in urine. J Chromatogr B 713: 339–352.

Pozo ÓJ, Van Eenoo P, Deventer K, Delbeke FT. 2007. Ionization of anabolic steroids by adduct formation in liquid chromatography electrospray mass spectrometry. J Mass Spectrom 42: 497–516.

Pozo ÓJ, Van Eenoo P, Deventer K, Grimalt S, Sancho JV, Hernández F, Delbeke FT. 2008a. Collision-induced dissociation of 3-keto anabolic steroids and related compounds after electrospray ionization. Considerations for structural elucidation. Rapid Commun Mass Spectrom 22: 4009–4024.

Pozo ÓJ, Deventer K, Eenoo PV, Delbeke FT. 2008b. Efficient approach for the comprehensive detection of unknown anabolic steroids and metabolites in human urine by liquid chromatography–electrospray-tandem mass spectrometry. Anal Chem 80: 1709–1720.

Pozo ÓJ, Van Thuyne W, Deventer K, Van Eenoo P, Delbeke FT. 2008c. Elucidation of urinary metabolites of fluoxymesterone by liquid chromatography–tandem mass spectrometry and gas chromatography–mass spectrometry. J Mass Spectrom 43: 394–408.

Pozo ÓJ, Ventura R, Monfort N, Segura J, Delbeke FT. 2009a. Evaluation of different scan methods for the urinary detection of corticosteroid metabolites by liquid chromatography tandem mass spectrometry. J Mass Spectrom 44: 929–944.

Pozo ÓJ, Van Eenoo P, Deventer K, Lootens L, Grimalt S, Sancho JV, Hernández F, Meuleman P, Leroux-Roels G, Delbeke FT. 2009b. Detection and structural investigation of metabolites of stanozolol in human urine by liquid chromatography tandem mass spectrometry. Steroids 74: 837–852.

Pozo ÓJ, Van Eenoo P, Deventer K, Elbardissy H, Grimalt S, Sancho JV, Hernández F, Ventura R, Delbeke FT. 2011. Comparison between triple quadrupole, time of flight and hybrid quadrupole time of flight analysers coupled to liquid chromatography for the detection of anabolic steroids in doping control analysis. Anal Chim Acta 684: 98–111.

Rauh M. 2010. Steroid measurement with LC–MS/MS. Application examples in pediatrics. J Steroid Biochem Mol Biol 121: 520–527

Shackleton CH, Chuang H, Kim J, de la Torre X, Segura J. 1997. Electrospray mass spectrometry of testosterone esters: Potential for use in doping control. Steroids 62: 523–529.

Sosa-Ferrera Z, Mahugo-Santana C, Santana-Rodríguez JJ. 2013. Analytical methodologies for the determination of endocrine disrupting compounds in biological and environmental samples. BioMed Res Int 2013: 674838.

Sun Y, Gu C, Liu X, Liang W, Yao P, Bolton JL, van Breemen RB. 2005. Ultrafiltration tandem mass spectrometry of estrogens for characterization of structure and affinity for human estrogen receptors. J Am Soc Mass Spectrom 16: 271–279.

Tedmon L, Barnes JS, Nguyen HP, Schug KA. 2013. Differentiating isobaric steroid hormone metabolites using multi-stage tandem mass spectrometry. J Am Soc Mass Spectrom 24: 399–409.

Thevis M, Makarov AA, Horning S, Schänzer W. 2005a. Mass spectrometry of stanozolol and its analogues using electrospray ionization and collision-induced dissociation with quadrupole-linear ion trap and linear ion trap-orbitrap hybrid mass analyzers. Rapid Commun Mass Spectrom 19: 3369–3378.

Thevis M, Bommerich U, Opfermann G, Schänzer W. 2005b. Characterization of chemically modified steroids for doping control purposes by electrospray ionization tandem mass spectrometry. J Mass Spectrom 40: 494–502.

Touber ME, van Engelen MC, Georgakopoulus C, van Rhijn JA, Nielen MW. 2007. Multi-detection of corticosteroids in sports doping and veterinary control using high-resolution liquid chromatography/time-of-flight mass spectrometry. Anal Chim Acta 586: 137–146

Turpeinen U, Hämäläinen E, Stenman UH. 2008. Determination of aldosterone in serum by liquid chromatography–tandem mass spectrometry. J Chromatogr B 862: 113–118.

Van Den Hauwe O, Perez JC, Claereboudt J, Van Peteghem C. 2001. Simultaneous determination of betamethasone and dexamethasone residues in bovine liver by liquid chromatography/tandem mass spectrometry. Rapid Commun Mass Spectrom 15: 857–861.

Van de Wiele M, De Wasch K, Vercammen J, Courtheyn D, De Brabander HF, Impens S. 2000. Determination of 16beta-hydroxystanozolol in urine and faeces by liquid chromatography–multiple mass spectrometry. J Chromatogr A 904: 203–209.

Von Unruh G, Spriteller-Friedmann M, Spiteller G. 1970. Schlüsselbruchstücke und schlüsseldifferenzen als kriterium bei der datenerfassung von massenspektren. Tetrahedron 26: 3039–3044.

Von Unruh G, Spiteller G. 1970a. Tabellen zur massenspektrometrischen strukturaufklärung von steroiden II: Schlüsselbruchstücke von freien steroiden. Tetrahedron 26: 3329–3346.

Von Unruh G, Spiteller G. 1970b. Tabellen zur massenspektrometrischen strukturaufklärung von steroiden—III: Schlüsseldifferenzen von freien steroiden. Tetrahedron 26: 3289–3301.

Von Unruh G, Spiteller G. 1970c. Tabellen zur massenspektrometrischen strukturaufklärung von steroiden—IV: Schlüsselbruchstücke und schlüsseldifferenzen von steroidderivaten. Tetrahedron 26: 3303–3311.

Wang QL, Zhang AZ, Pan X, Chen LR. 2010. Simultaneous determination of sex hormones in egg products by ZnCl2 depositing lipid, solid-phase extraction and ultra-performance liquid chromatography/electrospray ionization tandem mass spectrometry. Anal Chim Acta 678: 108–116.

Weidolf LO, Lee ED, Henion JD. 1988. Determination of boldenone sulfoconjugate and related steroid sulfates in equine urine by high-performance liquid chromatography/tandem mass spectrometry. Biomed Environ Mass Spectrom 15: 283–289.

Williams TM, Kind AJ, Houghton E, Hill DW. 1999. Electrospray collision-induced dissociation of testosterone and testosterone hydroxy analogs. J Mass Spectrom 34: 206–216.

Wooding KM, Barkley RM, Hankin JA, Johnson CA, Bradford AP, Santoro N, Murphy RC. 2013. Mechanism of formation of the major estradiol product ions following collisional activation of the molecular anion in a tandem quadrupole mass spectrometer. J Am Soc Mass Spectrom 24: 1451–1455.

4.7

FRAGMENTATION OF DRUGS OF ABUSE

4.7.1 INTRODUCTION

The term "drugs of abuse" or "illicit drugs" is not very clear, as abuse or misuse of drugs can occur with many classes of drugs. When anabolic steroids and/or some classes of antibiotics are used for growth promotion in veterinary applications, this use can be considered as abuse. The use of compounds on the WADA doping list by (professional) athletes can be considered as abuse. However, the term "drugs of abuse" generally refers to the use of illegal or illicit substances of CNS action, such as amphetamines, barbiturates, cannabis-related substances, cocaine, benzodiazepines, methaqualone, opioids, and ethanol. In this chapter, the MS–MS fragmentation of several of these drug classes is described. Note that the MS–MS analysis of some classes is described elsewhere, for example, benzodiazepines in Section 4.2.4 and barbiturates in Section 4.2.6.

Mass spectrometry, both GC–MS and LC–MS, is frequently applied in the analysis of drugs of abuse (Van Bocxlaer et al., 2000; Castiglioni et al., 2008). The goals of such studies are the identification and/or confirmation of identity, and/or quantitative analysis of these illicit drugs and/or their metabolites in samples from patients or users (clinical and forensic toxicology). Important sample matrices in such studies are urine (e.g., de Jager & Bailey, 2011), saliva (e.g., Badawi et al., 2009), hair (e.g., Kim et al., 2014; Montesano et al., 2014), breast milk (e.g., Marchei et al., 2011), and plasma (e.g., Kraemer & Paul, 2007). Identification of metabolites of drugs of abuse is of considerable importance as well. In addition, detection, identification, and metabolism of designer drugs are important. Designer drugs (or synthetic recreational drugs) are structural or functional analogs of an illicit or controlled substance. These so-called new psychoactive substances (NPSs) are designed to mimic the pharmacological effects of the original drug, while having some different structure features (and thus *m/z*). Hence, they are not readily detected in targeted drug screening tests and in fact may not yet be classified as illegal. For instance, a number of analogs of the opioid analgesic drug fentanyl (which has also found recreational use) has been reported, for example, 3-methyl-, acetyl-, acryl-, butyryl-, 4-fluoro-butyryl-, 4-fluorofenatanyl. Information of the fragmentation of various classes of designer drugs or NPSs is included in this chapter, partly based on an available data compilation (Ambach & Franz, 2012).

In the past few years, considerable attention has been paid to the analysis of illicit drugs and/or their metabolites in wastewater. It has been demonstrated that apart from monitoring environmental contamination, quantitative analysis of illicit drugs in wastewater enables a quite accurate estimation of the use of illicit drugs in the population (Thomas et al., 2012; Castiglioni et al., 2013). To this end, multiresidue analytical methods have been developed (Bijlsma et al., 2013; Borova et al., 2014). Most classes of illicit drugs are best analyzed in positive-ion mode.

4.7.2 AMPHETAMINE AND RELATED COMPOUNDS

Amphetamine, discovered in 1887, is a potent CNS stimulant of the phenethylamine class. Amphetamine and its derivatives are used as illicit stimulant drugs as well. Various subclasses can be discriminated.

4.7.2.1 Amphetamines

In its product-ion mass spectrum, amphetamine (1-phenylpropan-2-amine, $[M+H]^+$ with *m/z* 136) shows the loss of NH_3 to form 1-phenylprop-2-yl carbocation with *m/z* 119 ($[C_6H_5CH_2CHCH_3]^+$), which may undergo a H-rearrangement with charge migration to form the α-(ethyl)benzyl cation ($[C_6H_5CHCH_2CH_3]^+$), as well as the formation of a tropylium ion with *m/z* 91 ($[C_7H_7]^+$) due to cleavage of the propane $C^1—C^2$ bond. At higher collision energies, the tropylium ion may show secondary fragmentation to form the cyclopentadienyl cation ($[C_5H_5]^+$) with *m/z*

65 and cyclobutadienyl cation ($[C_4H_3]^+$) with *m/z* 51. This is the characteristic fragmentation for the complete group.

Analysis of deuterated analogs, for example, $[D_6]$- or $[D_5]$-analogs with [D]-labels on the substituent chain, reveals that the tropylium ion is not formed by a straightforward propane C^1—C^2 bond cleavage. In the $[D_6]$-analog, apart from the expected ion with *m/z* 93 ($[C_7H_5D_2]^+$), isotopologue ions with *m/z* 94 ($[C_7H_4D_3]^+$) and *m/z* 95 ($[C_7H_3D_4]^+$) are observed. Secondary fragmentation leads to fragment ions with *m/z* 65, *m/z* 66, *m/z* 67, *m/z* 68, and *m/z* 69. In the $[D_5]$-analog (with one D-label on the propane C^1), the most abundant fragment ion is *m/z* 93 ($[C_7H_5D_2]^+$), and ions with *m/z* 92 and 94 are also observed. This indicates that significant scrambling of the D-labels occurs by the propane C^1—C^2 bond cleavage (Bijlsma et al., 2011).

A wide variety of amphetamine-related compounds have been synthesized, both with *N*-substitution such as in methamphetamine, dimethamphetamine, and *N*-ethylamphetamine and/or with substitution on various positions on the phenyl ring like in 2,5-dimethoxyamphetamine (2,5-DMA). The loss of the amine substituent and the propane C^1—C^2-bond cleavage are observed for all these compounds. In addition, compound-specific fragmentation may be observed. The 2,5-dimethoxy analogs are another subclass of the phenethylamines. The characteristic fragmentation can be illustrated for 2,5-dimethoxyamphetamine (DMA, $[M+H]^+$ with *m/z* 196). Next to the loss of ammonia (NH_3) to an ion with *m/z* 179 and a subsequent loss of a methyl radical ($^{\bullet}CH_3$) to an ion with *m/z* 164, the tropylium analog with *m/z* 151, that is, the 2,5-dimethoxybenzyl cation ($[(CH_3O)_2C_6H_3CH_2]^+$), is observed. The latter ion shows two subsequent losses of formaldehyde ($H_2C{=}O$) to ions with *m/z* 121 and 91. Table 4.7.1 summarizes *m/z* values of the class-specific fragments as well as of major compound-specific fragment ion for 18 amphetamine-based designer drugs.

4.7.2.2 3,4-Methylenedioxyamphetamines

The 3,4-methylenedioxy analogs, for example, 3,4-methylenedioxyamphetamine (MDA, $[M+H]^+$ with *m/z* 180), form a phenethylamine subclass. *N*-methyl-3,4-methylenedioxymethamphetamine (MDMA, "Ecstasy", $[M+H]^+$ with *m/z* 194) shows the loss of methylamine (CH_3NH_2) to an ion with *m/z* 163 and the formation of the 3,4-methylenedioxybenzyl cation ($[C_7H_5O_2CH_2]^+$) with *m/z* 135 due to propane C^1—C^2 bond cleavage. Loss of formaldehyde ($H_2C{=}O$) from the ion with *m/z* 163 results in an ion with *m/z* 133, from which further loss of carbon monoxide (CO) leads to the ion with *m/z* 105, probably the 2° α-(methyl)benzyl cation ($[C_6H_5CHCH_3]^+$). The ion with *m/z* 105 has also been erroneously interpreted as benzoyl cation $[C_6H_5{-}C{=}O]^+$, but accurate-mass determination shows that its molecular formula is $[C_8H_9]^+$ (Bijlsma et al., 2011). Loss of the amine substituent and the formation of the ions with *m/z* 135, *m/z* 133, and *m/z* 105 are also observed for MDA and for its analogs *N*-ethyl-3,4-methylenedioxyethamphetamine (MDEA, $[M+H]^+$ with *m/z* 208), the isomeric (*N,N*-dimethyl)-3,4-methylenedioxyamphetamine (MDDMA, $[M+H]^+$ with *m/z* 208), and *N*-propyl-3,4-methyledioxyamphetamine (MDPA, $[M+H]^+$ with *m/z* 222). In the product-ion mass spectra, minor fragment ions are also observed that involve a cleavage of the propane C^1—C^2 bond with charge retention on the amine N-atom, that is, protonated *N*-ethylmethanimine ($[CH_3CH_2N{=}CH_2+H]^+$) with *m/z* 58 for MDMA, protonated *N*-dimethylethanimine ($[C_4H_9N+H]^+$) with *m/z* 72 for MDEA, protonated *N,N*-dimethylethanimine ($[C_4H_9N+H]^+$) with *m/z* 72 for MDDMA, and protonated *N*-propylethanimine ($[C_5H_{11}N+H]^+$) with *m/z* 86 for MDPA. The class-specific and major compound-specific fragments of the 3,4-methylenedioxyamphetamines are summarized in Table 4.7.2.

4.7.2.3 2-Aminopropylbenzofurans

A third class of related compounds are the *x*-(2-aminopropyl)benzofurans and *x*-(2-aminopropyl)dihydrobenzofurans, where *x* is 4, 5, 6, or 7 (Table 4.7.3). These compounds show the same class-specific fragmentation. The *m/z* values of class-specific fragments and major compound-specific fragment ions are summarized in Table 4.7.3.

4.7.2.4 Cathinones

Cathinone (2-amino-1-phenylpropan-1-one, $[M+H]^+$ with *m/z* 150) shows a fragmentation pattern involving loss of water, which probably involves a skeletal rearrangement of a methyl (CH_3) group (Table 4.7.4) to produce the α-methyl-α-(methylideneamino)benzyl cation with *m/z* 132, followed by the loss of either a methyl radical ($^{\bullet}CH_3$) to an ion with *m/z* 117 or hydrogen cyanide (HCN) to the α-(methyl)benzyl cation ($[C_6H_5CHCH_3]^+$) with *m/z* 105, thus not $[C_6H_5{-}C{=}O]^+$, as also discussed for MDA and its analogs (Section 4.7.2.2). A wide variety of related compounds (designer drugs) based on the cathinone structure have been described (Ammann et al, 2012b; Meyer et al., 2012, Shanks et al., 2012). Two subgroups can be distinguished: cathinones based on the 2-amino-1-phenylpropan-1-one skeleton and cathinones based on other 2-amino-1-phenylalkan-1-one skeletons. For the most part, they show similar fragmentation characteristics (Jankovics et al., 2011; Fornal, 2013; Strano Rossi et al., 2014; Frison et al., 2015). The *m/z* values of class-specific fragments and major compound-specific fragment ions for compounds from both subgroups are summarized in Table 4.7.4. It should be noted that accurate-mass data were not available for all compounds. The interpretation

TABLE 4.7.1 Class-specific and major compound-specific fragment ions of amphetamine and related compounds (designer drugs or NPSs).

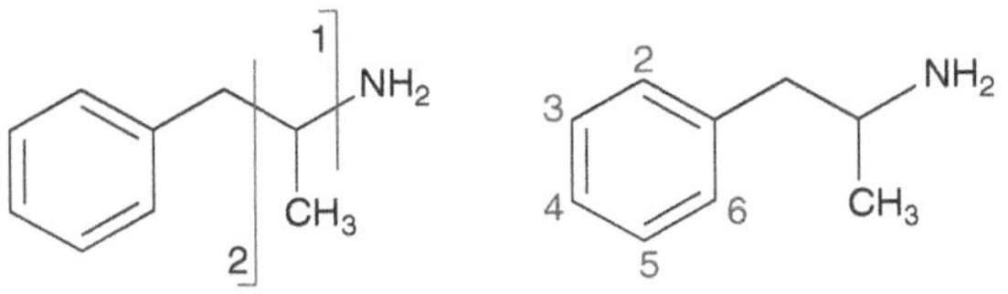

Compound	$[M+H]^+$	Features	1	2	Other Major Fragments
Amphetamine	136		119	91	65: $[C_5H_5]^+$
Methamphetamine	150	*N*-CH_3	119	91	
3-FA	154	3-F	137	109	89: 109 – HF 83: $[C_5H_4F]^+$
Dimethamphetamine	164	*N,N*-di-CH_3	119	91	
Ethylamphetamine	164	*N*-CH_2CH_3	119	91	
PMA	166	4-CH_3O	149	121	134: 149 – $^\bullet CH_3$
2-FMA	168	2-F, *N*-CH_3	137	109	89: 109 – HF 83: $[C_5H_4F]^+$
PMMA	180	4-CH_3O, *N*-CH_3	149	121	134: 149 – $^\bullet CH_3$
4-MTA	182	4-CH_3S	165	137	150: 165 – $^\bullet CH_3$ 117: $[C_9H_9]^+ \equiv$ 165 – CH_3SH
2,5-DMA	196	2,5-di-CH_3O	179	151	164: 179 – $^\bullet CH_3$ 121: 151 – $H_2C{=}O$
3,4-DMA	196	3,4-di-CH_3O	179	151	164: 179 – $^\bullet CH_3$ 136: 151 – $^\bullet CH_3$
DOM	210	2,5-di-CH_3O-4-CH_3	193	165	178: 193 – $^\bullet CH_3$ 135: 165 – $H_2C{=}O$
DOET	224	2,5-di-CH_3O-4-C_2H_5	207	179	192: 207 – $^\bullet CH_3$ 149: 179 – $H_2C{=}O$
3,4,5-TMA	226	3,4,5-tri-CH_3O	209	181	194: 209 – $^\bullet CH_3$ 178: 194 – CH_4 151: 181 – $H_2C{=}O$ 121: 151 – $H_2C{=}O$
TMA-2	226	2,4,5-tri-CH_3O	209	181	194: 209 – $^\bullet CH_3$ 178: 194 – CH_4 162: 194 – CH_3OH 151: 181 – $H_2C{=}O$ 121: 151 – $H_2C{=}O$
TMA-6	226	2,4,6-tri-CH_3O	209	181	194: 209 – $^\bullet CH_3$ 179: 209 – $H_2C{=}O$ 162: 194 – CH_3OH 151: 181 – $H_2C{=}O$ 121: 151 – $H_2C{=}O$
4-Chloro-2,5-DMA	230	2,5-di-CH_3O-4-Cl	213	185	198: 213 – $^\bullet CH_3$ 178: 213 – $Cl^\bullet$ 155: 185 – $H_2C{=}O$
DOB	274	4-Br-2,5-di-CH_3O	257	229	242: 257 – $^\bullet CH_3$ 199: 229 – $H_2C{=}O$ 178: 257 – $Br^\bullet$ 163: 178 – $^\bullet CH_3$ 150: 229 – $Br^\bullet$
DOI	322	2,5-di-CH_3O-4-I	305	277	290: 305 – $^\bullet CH_3$ 247: 277 – $H_2C{=}O$ 178: 305 – $I^\bullet$ 163: 178 – $^\bullet CH_3$ 150: 277 – $I^\bullet$

TABLE 4.7.2 Class-specific and major compound-specific fragment ions of 3,4-methylenedioxyamphetamine (MDA) and related compounds (designer drugs or NPSs).

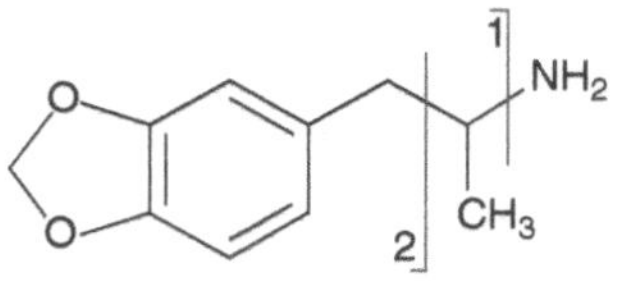

Compound	$[M+H]^+$	Features	1	2	Other Fragments
MDA	180		163	135	105: $[C_6H_5CHCH_3]^+$
MDMA	194	*N*-CH_3	163	135	105: $[C_6H_5CHCH_3]^+$ 58: $[C_3H_7N{+}H]^+$ (imine)
MDEA	208	*N*-C_2H_5	163	135	105: $[C_6H_5CHCH_3]^+$ 72: $[C_4H_9N{+}H]^+$ (imine)
MDDMA	208	*N*,*N*-di-CH_3	163	135	105: $[C_6H_5CHCH_3]^+$ 72: $[C_4H_9N{+}H]^+$ (imine)
MDPA	222	*N*-C_3H_7	163	135	105: $[C_6H_5CHCH_3]^+$ 86: $[C_5H_{11}N{+}H]^+$ (imine)

TABLE 4.7.3 Class-specific and major compound-specific fragment ions of *x*-(2-aminopropyl)benzofurans and related compounds (designer drugs or NPSs), with *x* = 4, 5, 6, or 7.

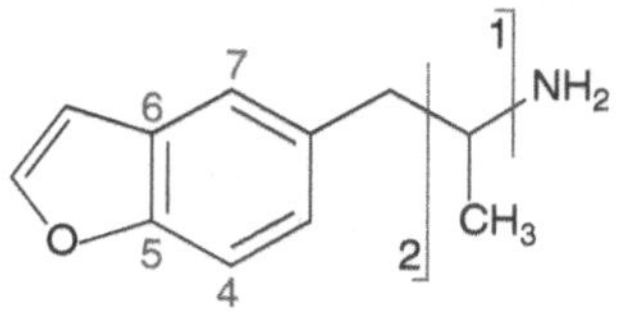

5-APB

Compound	$[M+H]^+$	Features	1	2	Other Fragments
5-APB ($C_{11}H_{13}NO$)	176	5-Benzofuran	159	131	117: $[C_9H_9]^+ \equiv 159 - CH_3OH$ 91: $[C_7H_7]^+$
6-APB	176	6-Benzofuran	159	131	116: $131 - {}^\bullet CH_3$ 91: $[C_7H_7]^+$
7-ABP	176	7-Benzofuran	159	131	116: $131 - {}^\bullet CH_3$ 91: $[C_7H_7]^+$
5-MAPB	190	5-Benzofuran, *N*-CH_3	159	131	117: $[C_9H_9]^+ \equiv 159 - CH_3OH$ 116: $131 - {}^\bullet CH_3$
5-EAPB	204	5-Benzofuran, *N*-C_2H_5	159	131	72: $[C_4H_{10}N]^+$ (imine)
4-APBD	178	4-Dihydrobenzofuran	161	133	119: $[C_9H_{11}]^+$ 105: $[C_8H_9]^+ \equiv 133 - CO$
5-APBD	178	5-Dihydrobenzofuran	161	133	146: $161 - {}^\bullet CH_3$ 105: $[C_8H_9]^+ \equiv 133 - CO$
6-APBD	178	6-Dihydrobenzofuran	161	133	119: $[C_9H_{11}]^+ \equiv 161 - CH_3OH$ 105: $[C_8H_9]^+ \equiv 133 - CO$
7-APBD	178	7-Dihydrobenzofuran	161	133	119: $[C_9H_{11}]^+ \equiv 161 - CH_3OH$ 105: $[C_8H_9]^+ \equiv 133 - CO$
5-MAPDB	192	5-Dihydrobenzofuran, *N*-CH_3	161	133	146: $161 - {}^\bullet CH_3$ 105: $[C_8H_9]^+ \equiv 133 - CO$
5-API or 5-IT	175	5-Indole	158	130	143: $138 - {}^\bullet CH_3$ 117: $[C_9H_9]^+ \equiv 158 - CH_3CN$

TABLE 4.7.4 Class-specific and major compound-specific fragment ions of cathinones (designer drugs or NPSs), with (a) cathinones based on the 2-amino-1-phenylpropan-1-one skeleton, and (b) cathinones based on other 2-amino-1-phenylalkan-1-one skeletons.

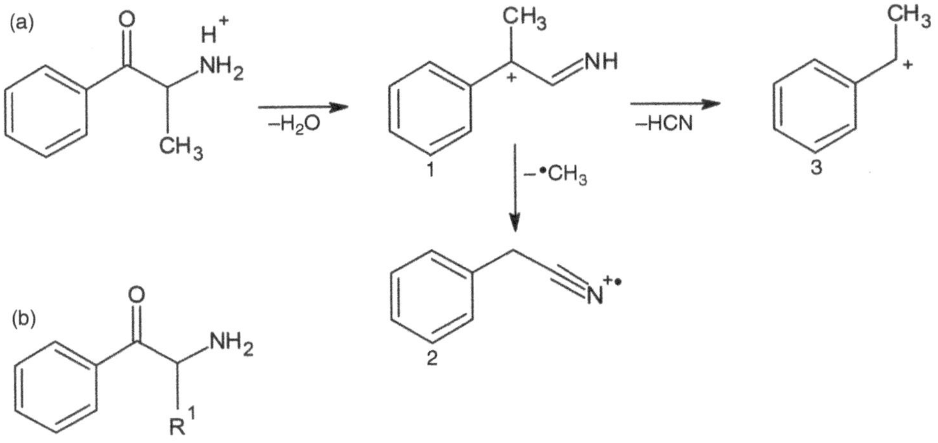

Compound	$[M+H]^+$	Features	1	2	3	Other Fragments
(a) Cathinone based on the 2-amino-1-phenylpropan-1-one skeleton						
Cathinone*	150		132	117	105	
Methcathinone	164	N-CH_3	146	131	105	58: $[C_3H_8N]^+$ (imine)
Ethcathinone	178	N-C_2H_5	160	145	105	131: $[C_9H_9N]^{+\bullet}$ 132: 160 − C_2H_4 117: 132 − $^\bullet CH_3$
3-MMC*	178	3-CH_3, N-CH_3	160	145	119	58: $[C_3H_8N]^+$ (imine)
Mephedrone*	178	4-CH_3, N-CH_3	160	145	119	58: $[C_3H_8N]^+$ (imine)
2-FMC	182	2-F, N-CH_3	164	149	123	144: 164 − HF 103: 123 − HF
3-FMC	182	3-F, N-CH_3	164	149	123	103: 123 − HF
4-FMC, flephedrone	182	4-F, N-CH_3	164	149	123	151: 182 − CH_3NH_2 103: 123 − HF
3,4-DMMC	192	3,4-di-CH_3, N-CH_3	174	159		133: $[C_{10}H_{13}]^+$ 58: $[C_3H_8N]^+$ (imine)
4-MEC*	192	4-CH_3, N-C_2H_5	174	159	119	146: 174 − C_2H_4 145: 174 − $^\bullet C_2H_5 \equiv [C_{10}H_{11}N]^{+\bullet}$ 144: 159 − $^\bullet CH_3$ 131: $[C_9H_9N]^{+\bullet}$ 105: $[C_6H_5$—C=$O]^+$ 72: $[C_4H_{10}N]^+$ (imine)

(continued)

TABLE 4.7.4 (*Continued*)

Compound	$[M+H]^+$	Features	1	2	3	Other Fragments
4-EMC*	192	4-C_2H_5, *N*-CH_3	174	159		145: 174 − $^•C_2H_5 \equiv [C_{10}H_{11}N]^{+•}$ 144: 159 − $^•CH_3$ 131: $[C_9H_9N]^{+•}$ 105: $[C_6H_5CHCH_3]^+$ 58: $[C_3H_8N]^+$ (imine)
Methedrone*	194	4-CH_3O, *N*-CH_3	176	161	135	146: 176 − $H_2C{=}O$ 145: 176 − $^•OCH_3 \equiv [C_{10}H_{11}N]^{+•}$ 105: $[C_6H_5CHCH_3]^+$ 58: $[C_3H_8N]^+$ (imine)
Methylone*	208	3,4-(—OCH_2O—), *N*-CH_3	190	175	149	177: 208 − CH_3NH_2 160: 190 − $H_2C{=}O$ 147: 177 − $H_2C{=}O$ 132: $[C_9H_{10}N]^+$ 58: $[C_3H_8N]^+$ (imine)
Ethylone*	222	3,4-(—OCH_2O—), *N*-C_2H_5	204	189		174: 204 − $H_2C{=}O$ 147: $[C_9H_7O_2]^+$ 146: $[C_9H_6O_2]^{+•}$ 131: $[C_9H_9N]^{+•}$ 72: $[C_4H_{10}N]^+$ (imine)
(b) Cathinone based on other 2-amino-1-phenylalkan-1-one skeleton						
Buphedrone, MABP*	178	$R^1 = C_2H_5$, *N*-CH_3	160	145		131: $[C_9H_9N]^{+•}$ 105: $[C_6H_5{-}C{=}O]^+$
3-Me-MABP	192	$R^1 = C_2H_5$, 3-CH_3, *N*-CH_3	174	159		161: 192 − CH_3NH_2 146: 161 − $^•CH_3$ 145: 174 − $^•C_2H_5 \equiv [C_{10}H_{11}N]^{+•}$ 105: $[C_6H_5CHCH_3]^+$
4-Me-MABP	192	$R^1 = C_2H_5$, 4-CH_3, *N*-CH_3	174	159		161: 192 − CH_3NH_2 146: 161 − $^•CH_3$ 145: 174 − $^•C_2H_5 \equiv [C_{10}H_{11}N]^{+•}$ 105: $[C_6H_5CHCH_3]^+$
Pentedrone*	192	$R^1 = C_3H_7$, *N*-CH_3	174	159	105	161: 192 − CH_3NH_2 148: 192 − C_3H_8 145: 174 − $^•C_2H_5 \equiv [C_{10}H_{11}N]^{+•}$ 132: 173 − C_3H_6 131: $[C_9H_9N]^{+•}$ 105: $[C_6H_5{-}C{=}O]^+$ 86: $[C_5H_{12}N]^+$ (imine)

(*continued*)

TABLE 4.7.4 (*Continued*)

Compound	$[M+H]^+$	Features	1	2	3	Other Fragments
Butylone*	222	3,4-(—OCH_2O—), $R^1 = C_2H_5$, *N*-CH_3	204			191: 222 – CH_3NH_2 175: 204 – $^{\bullet}C_2H_5$ 174: 204 – $H_2C{=}O$ 161: 191 – $H_2C{=}O$ 149: $[C_7H_5O_2C{=}O]^+$ 146: $[C_9H_6O_2]^{+\bullet}$ 131: $[C_9H_9N]^{+\bullet}$ 72: $[C_4H_{10}N]^+$ (imine)
Pentylone*	236	3,4-(—OCH_2O—), $R^1 = C_3H_7$, *N*-CH_3	218	203	149	205: 236 – CH_3NH_2 188: 218 – C_2H_6 175: 218 – $^{\bullet}C_3H_7$ 159: $[C_{10}H_9NO]^{+\bullet}$ 131: $[C_9H_9N]^{+\bullet}$ 86: $[C_5H_{12}N]^+$ (imine)
Eutylone	236	3,4-(—OCH_2O—), $R^1 = C_2H_5$, *N*-C_2H_5	218			191: 236 – $C_2H_5NH_2$ 189: 218 – $^{\bullet}C_2H_5$ 188: 218 – $H_2C{=}O$ 161: 191 – $H_2C{=}O$ 86: $[C_5H_{12}N]^+$ (imine)
NRG-3	242	2-Naphthyl, $R^1 = C_3H_7$, *N*-CH_3	224	209		211: 242 – CH_3NH_2 182: 224 – C_3H_6 181: 224 – $^{\bullet}C_3H_7$ 167: 182 – $^{\bullet}CH_3$ 155: $[C_{11}H_7O]^+$ 141: $[C_{11}H_9]^+$ 86: $[C_5H_{12}N]^+$ (imine)
N-Ethyl-pentylone	250	3,4-(—OCH_2O—), $R^1 = C_3H_7$, *N*-C_2H_5	232			205: 250 – $C_2H_5NH_2$ 202: 232 – $H_2C{=}O$ 190: 205 – $^{\bullet}CH_3$ 175: 205 – $H_2C{=}O$ 160: 190 – $H_2C{=}O$ 135: $[C_8H_7O_2]^+$ 100: $[C_6H_{14}N]^+$ (imine)
bk-2C-B*	274	4-Br, 2,5-di-CH_3O, $R^1 = H$	256			257: 274 – NH_3 229: $[C_9H_{10}BrO_2]^+$ 199: 229 – $H_2C{=}O$ 178: 257 – $Br^{\bullet}$ 177: 274 – H_2O – $Br^{\bullet}$ 162: 177 – $^{\bullet}CH_3$

*Interpretation confirmed by accurate-mass data.

of the product-ion mass spectra of compounds for which accurate-mass data lacked was done by analogy. Multiresidue screening of 32 cathinone analogs in serum using SRM on a TQ instrument has been reported (Swortwood et al., 2013).

4.7.2.5 Pyrrolidinophenones

Pyrrolidinophenones are structurally related to the cathinone. However, their fragmentation characteristics are different (Matsuta et al., 2015; Ibáñez et al., 2016). Thus, 3,4-methylenedioxypyrovalerone (MDPV; $[M+H]^+$ with *m/z* 276) shows class-specific fragmentation (Table 4.7.5), leading to an ion with *m/z* 205 due to the loss of pyrrolidine (C_4H_9N) and the 1-(pyrrolidin-1-yl)butyl carbocation with *m/z* 126 ($[C_8H_{16}N]^+$) due to the loss of 3,4-methylenedioxybenzaldehyde ($C_8H_6O_3$). Cleavage of the C—N bond to the 1-pyrrolidinyl group with charge retention on the N-atom can lead to either an ion with *m/z* 72 due to the protonated pyrrolidine ($[C_4H_9N+H]^+$) or an ion with *m/z* 70 due to protonated 3,4-dihydro-2*H*-pyrrole ($[C_4H_7N+H]^+$) (Table 4.7.5) (Ibáñez et al., 2016). In addition, several compound-specific fragment ions are observed. The *m/z* values for ions due to class-specific fragmentation as well as for major compound-specific fragment ions for 13 pyrrolidinophenone NPSs are summarized in Table 4.7.5.

4.7.2.6 Ephedrines

Ephedrine ($[M+H]^+$ with *m/z* 166) shows fragment ions with *m/z* 134 due to water loss, followed by the loss of a methyl radical ($^\bullet CH_3$) to an ion with *m/z* 133 or the loss of CH_3NH_2 (and H_2) to the ions with *m/z* 117 ($[C_9H_9]^+$) and *m/z* 115 ($[C_9H_7]^+$). [D]-labeling indicates that the loss of the $^\bullet CH_3$ occurs from both the N—CH_3 and the C^2—CH_3 (Bijlsma et al., 2011). Interestingly, no loss of a $^\bullet CH_3$ occurs in norephedrine ($[M+H]^+$ with *m/z* 152). In methylephedrine ($[M+H]^+$ with *m/z* 180), the ions due to loss of water and subsequent $^\bullet CH_3$ are also observed in addition to the ions with *m/z* 117 and 115. The additional fragment ion with *m/z* 135 is due to the loss of dimethylamine ($(CH_3)_2NH$) from $[M+H]^+$.

4.7.3 CANNABINOIDS

Cannabinoids are compounds that interact with the cannabinoid receptors, which are the membrane-bound G-protein-coupled receptors (GPCR) CB_1, which are primarily found in the brain, and GPCR CB_2, which are primarily found in immune-derived cells in the spleen. Compounds such as cannabinol, cannabidiol, and tetrahydrocannabinol are among the many different cannabinoids that have been isolated from the cannabis plant. Structurally related synthetic products such as JWH-018 (Section 4.7.7.1) are also available. Δ^9-Tetrahydrocannabinol (THC), the major psychoactive component in cannabis, and its related substances and metabolites can be analyzed in both positive-ion and negative-ion modes. Cannabinoid analysis is frequently performed in positive-ion mode, although methods based on negative-ion detection have been reported as well (e.g., Stephanson et al., 2008; Chebbah et al., 2010; Schwope et al., 2011).

In positive-ion mode, cannabinol (CBN, $[M+H]^+$ with *m/z* 311) shows fragment ions with *m/z* 293 due to water loss and with *m/z* 241 due to the loss of pentene (C_5H_{10}), followed by subsequent losses of water to an ion with *m/z* 223 and of carbon monoxide (CO) to *m/z* 195. In negative-ion mode ($[M-H]^-$ with *m/z* 309), little fragmentation is observed at low collision energy. The structures of the fragments with *m/z* 279 due to the loss of ethane (C_2H_6), *m/z* 222 due to the loss of pentane (C_5H_{12}) and a methyl radical ($^\bullet CH_3$), *m/z* 171 ($[C_{12}H_{11}O]^-$), and *m/z* 143 ($[C_{11}H_{11}]^-$) are not yet clear.

The fragmentation of Δ^9-tetrahydrocannabinol (THC; $[M+H]^+$ with *m/z* 315) in the positive-ion product-ion mass spectrum results in ions with *m/z* 297 due to loss of water, *m/z* 259 due to the loss of 2-methylprop-1-ene (C_4H_8), and *m/z* 193 due to a C_9H_{14} loss. Two possible structures can be proposed for the ion with *m/z* 259 (Figure 4.7.1). Based on data for the $[D_3]$-labeled analog, with labeling at C^5 of the pentyl substituent, the pentyl substituent is retained in this fragment ion. A similar controversy exists in the negative-ion mode (see later). In addition, fragment ions with *m/z* 135 ($[C_{10}H_{15}]^+$) and 93 ($[C_7H_9]^+$) are observed, for which structure proposals are also shown in Figure 4.7.1.

The fragmentation of THC ($[M-H]^-$ with *m/z* 313) in the negative-ion product-ion mass spectrum is outlined in Figure 4.7.2. In the interpretation, the data for the $[D_3]$-THC analog, with the D-labeling at C^5 of the pentyl chain, can be of help. The ion with *m/z* 245 is due to the loss of C_5H_8 from $[M-H]^-$ of THC. The $[D_3]$-THC analog shows an ion with *m/z* 248. Based on the assumption of a retro-Diels–Alder fragmentation mechanism and a subsequent C—O bond cleavage, the structure annotated as [A] in Figure 4.7.2 was proposed, which was consistent with the fact that a +3 *m/z* shift is observed in the product-ion mass spectrum of $[D_3]$-THC. However, literature data for 11-nor-9-carboxy-delta-9-tetrahydrocannabinol (THC-9-COOH) and $[D_9]$-THC-9-COOH suggest the structure annotated [B] in Figure 4.7.2 to be the correct one (Stephanson et al., 2008; Chebbah et al., 2010). For $[D_9]$-THC-9-COOH, with labels onto the two 6-methyl substituents and at C^5 of the pentyl chain, a corresponding fragment with *m/z* 254 is observed, indicating that all nine D-labels were kept (Stephanson et al., 2008; Chebbah et al., 2010). Subsequent fragment ions for THC are the ions with *m/z* 191 (consistent with C_4H_6 loss from *m/z* 245), *m/z* 179 (consistent with the loss of C_5H_6 from *m/z* 245), and

TABLE 4.7.5 Class-specific and major compound-specific fragment ions of pyrrolidinophenones (designer drugs or NPSs).

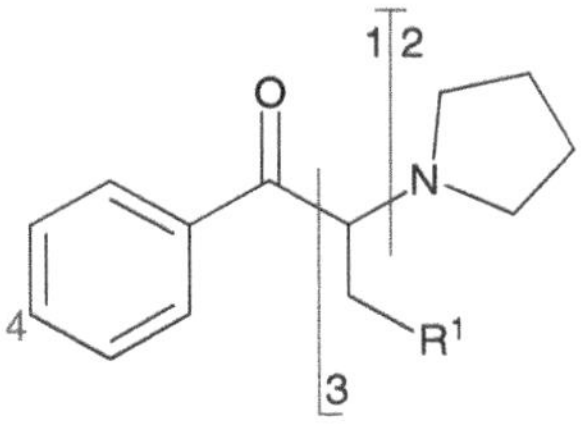

Compound	$[M+H]^+$	Features	1	2	3	Other Fragments
MPPP	218	4-CH_3, $R^1=H$	147	72	98	200: 218 – H_2O 185: 200 – $\bullet CH_3$ 119: $[C_8H_7O]^+$
α-PBP	218	$R^1=CH_3$	147	72	112	105: $[C_7H_5O]^+$
α-PVP	232	$R^1=C_2H_5$	161	70	126	214: 232 – H_2O 189: 232 – $\bullet C_3H_7$ 126: $[C_8H_{16}N]^+$ 119: $[C_9H_{11}]^+$ 105: $[C_7H_5O]^+$
4F-α-PBP	236	4-F, $R^1=CH_3$	165	70	112	137: 165 – C_2H_4/CO 109: $[C_7H_6F]^+$
Pyrovalerone	246	4-CH_3, $R^1=C_2H_5$	175	72, 70	126	203: 246 – $\bullet C_3H_7$ 157: 175 – H_2O 119: $[C_9H_{11}]^+$ 105: $[C_7H_5O]^+$
MDPPP	248	3,4-OCH_2O–	177	70	98	147: 177 – $H_2C{=}O$
PV8	260	$R^1=C_4H_9$	189	70	154	242: 260 – H_2O 119: $[C_9H_{11}]^+$
PV9	274	$R^1=C_5H_{11}$	203	70	168	256: 274 – H_2O 189: 274 – $\bullet C_6H_{13}$
MDPV	276	3,4-OCH_2O–, $R^1=C_2H_5$	205	72	126	233: 276 – $\bullet C_3H_7$ 175: 205 – $H_2C{=}O$ 149: $[C_8H_5O_3]^+$ 135: $[C_8H_7O_2]^+$
4F-PV8	278	4-F, $R^1=C_4H_9$	207	70	154	109: $[C_7H_6F]^+$
Naphyrone*	282	2-Naphthyl, $R^1=C_2H_5$	211	70	126	239: 282 – $\bullet C_3H_7$ 155: $[C_{11}H_7O]^+$ 141: $[C_{11}H_9]^+$
4-MeO-PV9	304	4-CH_3O, $R^1=C_5H_{11}$	233	72	168	135: $[C_8H_7O_2]^+$

*Naphyrone has a naphthyl instead of a phenyl group (1-(naphthalen-2-yl)-2-(pyrrolidin-1-yl)pentan-1-one).

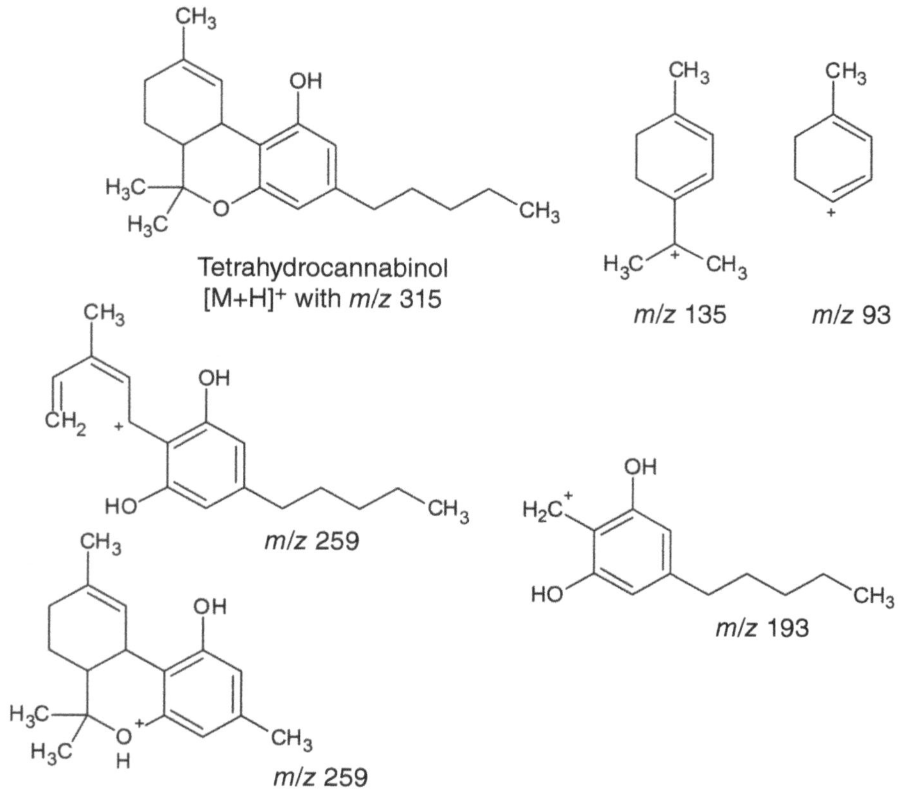

FIGURE 4.7.1 Proposed structures for the fragment ions of protonated tetrahydrocannabinol.

m/z 173 (consistent with the loss of C_4H_8O from *m/z* 245). For the ions with *m/z* 191 and 179, the structures proposed in Figure 4.7.2 are consistent with the observed +3 *m/z* shifts in the [D_3]-THC and [D_9]-THC-9-COOH spectra. The proposed structure for the ion with *m/z* 173 is consistent with no shift in the [D_3]-THC spectrum, but in contrast the ion with *m/z* 173 is not observed for [D_9]-THC-9-COOH, which supports the suggested presence of C^{11} in the fragment ion with *m/z* 173. In addition, fragment ions with *m/z* 107 ([C_7H_7O]$^-$) and *m/z* 79 ([C_6H_7]$^-$), due to the loss of CO from the ion with *m/z* 107, are observed (Figure 4.7.2).

Cannabidiol (CBD, $[M+H]^+$ with *m/z* 315) shows fragment ions with *m/z* 259 due to the loss of C_4H_8 and *m/z* 193 ([$C_{12}H_{17}O_2$]$^+$). Based on the fragmentation of THC (see earlier), the structures of these two fragments are most likely similar (Figure 4.7.1). In negative-ion mode, cannabidiol ($[M-H]^-$ with *m/z* 313) shows fragment ions with *m/z* 245, 179, and 107, again similar to THC (Figure 4.7.2).

Two major metabolites of THC are 11-hydroxy-THC (11-OH-THC) and THC-9-COOH. In positive-ion mode, the major fragments of THC-9-COOH ($[M+H]^+$ with *m/z* 345) are the ions with *m/z* 327 due to water loss, *m/z* 299 due to formic acid (HCOOH) loss, and *m/z* 193 ([$C_{12}H_{17}O_2$]$^+$) (Figure 4.7.1), *m/z* 257 due to propene ($CH_3CH{=}CH_2$) loss from the ion with *m/z* 299 (Bijlsma et al., 2011). In negative-ion mode, the mass spectrum of THC-9-COOH ($[M-H]^-$ with *m/z* 343) is very similar to that of THC. At the high-*m/z* end, losses of water and CO_2 result in fragment ions with *m/z* 325 and 299, thus involving the loss of the C^{11}-group. In addition, the fragment ions with *m/z* 245, *m/z* 191, *m/z* 179, *m/z* 107, and *m/z* 79, already discussed for THC itself, are observed for THC-9-COOH as well. The THC fragment ion with *m/z* 173 is not observed for THC-9-COOH (see earlier). The fragmentation of 11-OH-THC ($[M-H]^-$ with *m/z* 329) is different. It shows fragment ions with *m/z* 311 due to the loss of water, and *m/z* 268 due to loss of water and $C_3H_7^\bullet$ (and C_3H_8 at higher collision energy, *m/z* 267). The structure of the ion with *m/z* 173 ([$C_{12}H_{13}O$]$^-$, the same molecular formula as for THC) is not fully clear; it does not contain the intact pentyl chain anymore.

4.7.4 COCAINE AND RELATED SUBSTANCES

Cocaine is a tropane alkaloid obtained from the leaves of the coca plant (*Erythroxylum* family). It is a stimulant, an appetite suppressant, and a nonspecific sodium channel blocker. It acts as a serotonin–norepinephrine–dopamine reuptake inhibitor. It has a long history of use and abuse. Cocaine and its metabolites are analyzed in positive-ion mode.

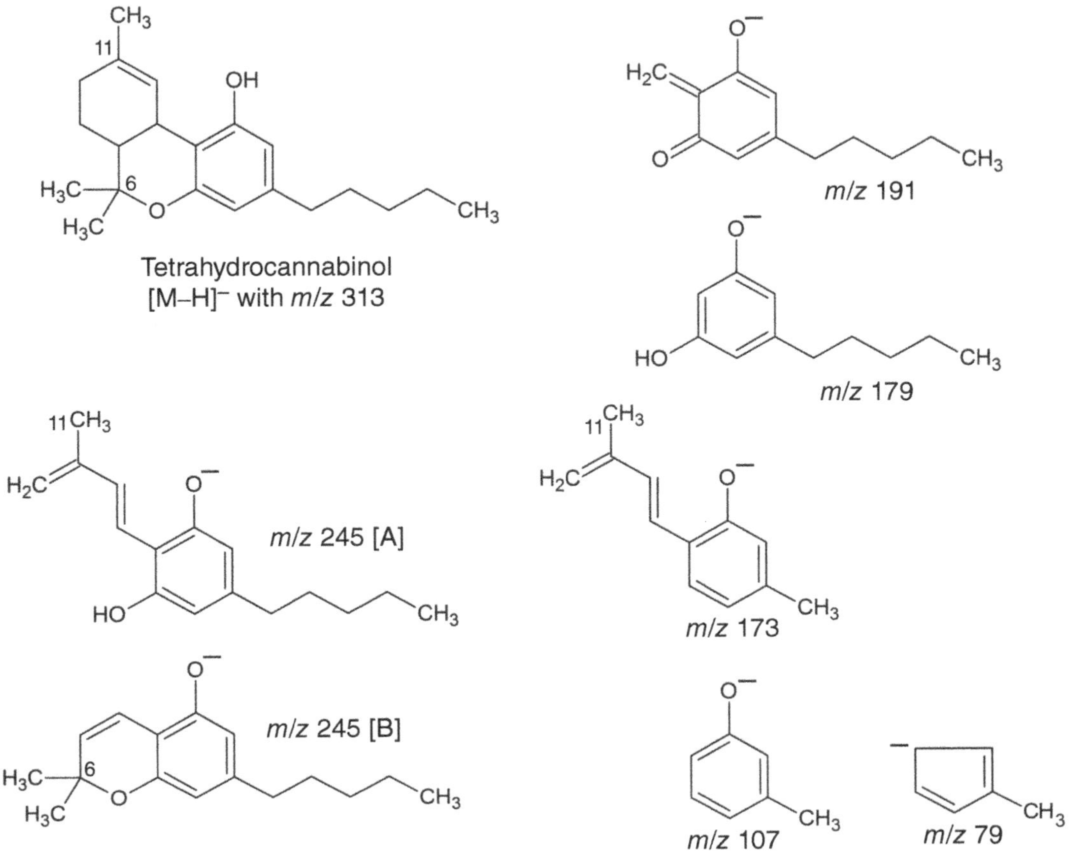

FIGURE 4.7.2 Proposed structures for the fragment ions of deprotonated tetrahydrocannabinol.

The fragmentation of cocaine ($[M+H]^+$ with *m/z* 304) is straightforward (Figure 4.7.3), showing major fragment ions with *m/z* 272 due to the loss of methanol (CH_3OH), *m/z* 182 due to the loss of benzoic acid (C_6H_5—COOH), *m/z* 150 due to CH_3OH loss from the ion with *m/z* 182, and the benzoyl cation ($[C_6H_5—C{=}O]^+$) with *m/z* 105 (Wang & Bartlett, 1998). Subsequent fragmentation of the ion with *m/z* 150 results in minor fragments ions with *m/z* 122 due to carbon monoxide (CO) loss, *m/z* 108 due to the loss of ethenone ($H_2C{=}C{=}O$), *m/z* 119 due to the loss of methylamine (CH_3NH_2), and *m/z* 91 due to a subsequent loss of CO (Figure 4.7.3) (Wang & Bartlett, 1998).

From a clinical and forensic point of view, the analysis of cocaine metabolites is of interest as well (Wang & Bartlett, 1998; Xia et al., 2000; Johansen & Bhatia, 2007). Cocaine and its principal metabolites have also been analyzed in waste and surface water, among other matrices, to monitor cocaine use (Gheorghe et al., 2008; Castiglioni et al., 2008; Bijlsma et al., 2011). The major metabolites of cocaine are benzoylecgonine (BE), ecgonine methyl ester (EME), and ecgonine. Further minor metabolites of cocaine include norcocaine, *p*-hydroxycocaine, *m*-hydroxycocaine, *p*-hydroxybenzoylecgonine (pOHBE), and *m*-hydroxybenzoylecgonine (mOHBE). If consumed with alcohol, they combine in the liver to form cocaethylene. Relevant fragment ions of these cocaine metabolites are summarized in Table 4.7.6 (Wang & Bartlett, 1998; Castiglioni et al., 2008).

4.7.5 OPIATES

Opiates are opioid analgesic alkaloids, structurally related to morphine. They are found in the opium poppy plant, *Papaver somniferum*. Important opiates are morphine, 6-monoacetylmorphine, heroin (3,6-diacetylmorphine), and codeine (3-methylmorphine). Two important morphine metabolites are morphine-3-β-glucuronide and morphine-6-β-glucuronide. Other metabolites and related compounds are the *N*-desmethyl analogs normorphine and norcodeine, dihydromorphine, and acetylcodeine. Related compounds with the morphinan substructure are dextrorphan and its isomer levorphanol.

6-Monoacetylmorphine ($[M+H]^+$ with *m/z* 328) shows fragment ions with *m/z* 286 due to the loss of ethenone ($H_2C{=}C{=}O$), *m/z* 268 due to the loss of

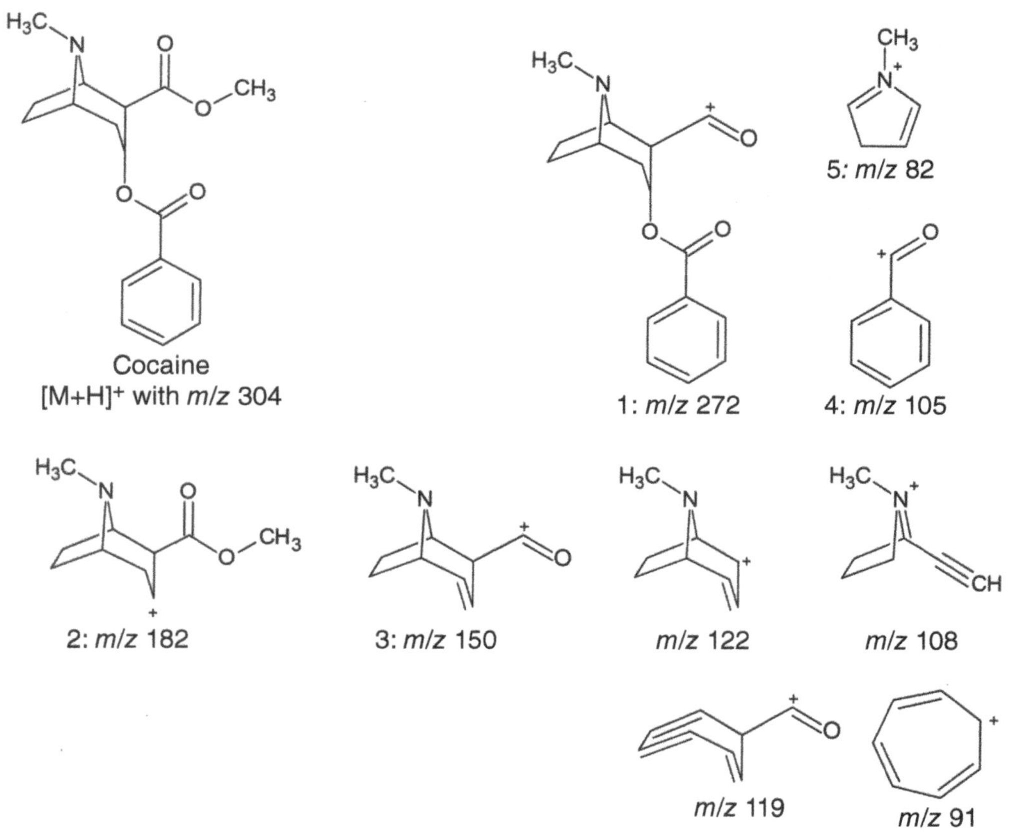

FIGURE 4.7.3 Proposed structures for the fragment ions of protonated cocaine.

acetic acid (CH_3COOH), and *m/z* 211 due to the subsequent loss of *N*-methylethanimine ($CH_3N{=}CHCH_3$) or *N*-ethylmethanimine ($CH_3CH_2N{=}CH_2$). Many fragment ions are observed between *m/z* 100 and *m/z* 200 (Castiglioni et al., 2008; Bijlsma et al., 2011). Some of these ions are easily interpreted as secondary fragmentation of the ion with *m/z* 211, such as the ions with *m/z* 193 due to the loss of water, *m/z* 183 due to the loss of carbon monoxide (CO), and *m/z* 165 due to the loss of both water and CO. Other fragment ions were elucidated with considerable care, using labeled substances and PIA (Bijlsma et al., 2011). Somewhat similar fragmentation is observed for heroin and morphine. The main fragmentation pathway for the glucuronide conjugates of morphine is the loss of dehydroglucuronic acid (176 Da, $C_6H_8O_6$), which is the characteristic fragmentation of glucuronic acid conjugates (Section 3.6.2).

4.7.6 MISCELLANEOUS DRUGS OF ABUSE

A wide variety of other compounds and compound classes are involved as illicit drugs or drugs of abuse. Some compound classes have been discussed in other sections of the book. The topic has been reviewed (Kraemer & Paul, 2007; Meyer & Maurer, 2012).

4.7.6.1 Ketamine

In its positive-ion product-ion mass spectrum, ketamine ($[M{+}H]^+$ with *m/z* 238) shows fragment ions with *m/z* 220 due to the loss of water, with *m/z* 207 due to the loss of methylamine (CH_3NH_2), which in turn shows loss of water to an ion with *m/z* 189, the loss of CO to an ion with *m/z* 179, or the loss of ethenone ($H_2C{=}C{=}O$) to an ion with *m/z* 165, and the 2-chlorobenzyl cation ($[ClC_6H_4CH_2]^+$) with *m/z* 125 (Wang et al., 2005). At higher collision energy, fragment ions are observed with *m/z* 115 and 116, consistent with the 2*H*-inden-2-ylium ion ($[C_9H_7]^+$) and the indene molecular ion ($[C_9H_8]^{+\bullet}$), respectively (Bijlsma et al., 2011).

4.7.6.2 Dimethocaine

Dimethocaine is a synthetic derivative of cocaine, currently distributed as a NPS. The fragmentation of dimethocaine ($[M{+}H]^+$ with *m/z* 279) in LIT MS^2 and MS^3 is straightforward (Figure 4.7.4) (Meyer et al., 2014a). The

TABLE 4.7.6 Fragmentation of cocaine and its metabolites (Figure 4.7.3).

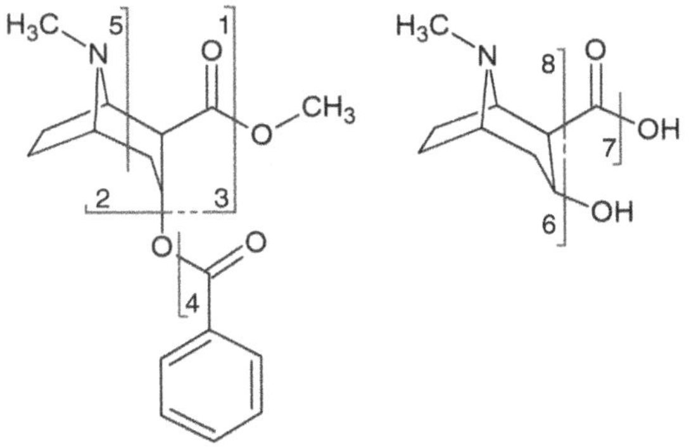

	$[M+H]^+$	1	2	3	4	5
Cocaine	304	272	182	150	105	82
Benzoylecgonine	290	272	168	150	105	82
Norcocaine	290	258	168	136	105	68
Norbenzoylecgonine	276	258	154	136	105	68
Cocaethylene	318	272	196	150	105	82
Hydroxycocaine	320	288	182	150	121	82
Hydroxybenzoylecgonine	306	288	168	150	121	82
Cocaine, *N*-oxide	320	288	198, 182	166, 150	105	82
	$[M+H]^+$	6	7	8		5
Ecgonine	186	168				82
Ecgonine methyl ester	200	182	168			82
Anhydroecgonine	168		150	122		82
Anhydroecgonine methyl ester	182		150	122		82

loss of diethylamine ($(C_2H_5)_2NH$) leads to a fragment ion with *m/z* 206. The loss of 4-aminobenzoic acid results in a fragment ion with *m/z* 142 ($[C_9H_{20}N]^+$). Cleavage of the ester group C—O bond results in two complementary ions with *m/z* 120 and 160. The *N,N*-diethylmethaniminium ion ($[(C_2H_5)_2N{=}CH_2]^+$) with *m/z* 86 is observed as well. The *in vitro* and *in vivo* metabolism of dimethocaine has been reported (Meyer et al., 2014a,b).

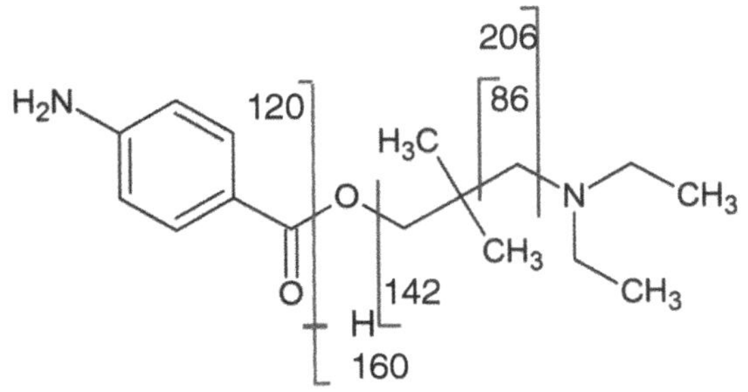

Dimethocaine
$[M+H]^+$ with *m/z* 279

FIGURE 4.7.4 Fragmentation scheme for protonated dimethocaine.

4.7.6.3 Methadone

Methadone is used as a substitution treatment for opioids. The loss of dimethylamine ($(CH_3)_2NH$) from methadone ($[M+H]^+$ with *m/z* 310), resulting in a fragment ion with *m/z* 265, is generally used in SRM for screening and quantitative analysis of methadone. Despite of accurate-mass data, some of the other fragment ions are less straightforward to identify. Secondary fragmentation of the ion with *m/z* 265 results in ions with *m/z* 247 and 223, due to the loss of water and of propene ($CH_3CH{=}CH_2$), respectively. The ion with *m/z* 223 is the 2-oxo-1,1-diphenylbut-1-yl carbocation. The ion with *m/z* 219 may result from ethene (C_2H_4) loss from the ion with *m/z* 247. Based on its molecular formula, the ion with *m/z* 195 might be due to carbon monoxide (CO) loss from the ion with *m/z* 223 which would require significant skeletal rearrangement. The ion with *m/z* 105 seems to be the benzoyl cation ($[C_6H_5{-}C{=}O]^+$), which would also require a skeletal rearrangement.

The major methadone metabolite, EDDP (2-ethylidene-1,5-dimethyl-3,3-diphenylpyrrolidine; $[M+H]^+$ with *m/z*

278), shows fragment ion with *m/z* 249 due to the loss of an ethyl radical ($^{\bullet}C_2H_5$), involving the ethylidene substituent, with *m/z* 234 due to the loss of propane (C_3H_8), most likely resulting from a methyl radical ($^{\bullet}CH_3$) loss from the ion with *m/z* 249. Another $^{\bullet}CH_3$ loss results in the fragment ion with *m/z* 219. The combined loss of a phenyl radical ($^{\bullet}C_6H_5$) and $^{\bullet}CH_3$ results in the fragment ion with *m/z* 186. This interpretation is supported by product-ion mass spectral data for a [D_3]-labeled analog.

4.7.6.4 Lysergic Acid Diethylamide (LSD)

The major fragment ion of the psychedelic compound lysergic acid diethylamide (LSD, $[M+H]^+$ with *m/z* 324) is the ion with *m/z* 223 due to the loss of diethylamine ($(C_2H_5)_2NH$) and carbon monoxide (CO) (Figure 4.7.5). Other fragment ions are the ions with *m/z* 281 due the loss of *N*-methylmethanimine ($CH_3N{=}CH_2$), *m/z* 251 due to the loss of $(C_2H_5)_2NH$, the 2-(diethylcarbamoyl)ethyl carbocation ($[C_7H_{14}NO]^+$) with *m/z* 128, and protonated diethylamine ($(C_2H_5)_2NH+H]^+$) with *m/z* 74 (Figure 4.7.5) (Cai & Henion, 1996).

4.7.6.5 Mescaline

Mescaline (2-(3,4,5-trimethoxyphenyl)ethanamine; $[M+H]^+$ with *m/z* 212) is a hallucinogenic psychoactive substance present in several cacti species, for example, the Peyote (*Lophophora williamsii*). The fragmentation is characterized by methyl radical ($^{\bullet}CH_3$) losses, resulting from the trimethoxyphenyl structure. Thus, fragment ions are observed with *m/z* 196 due to the loss of methane (CH_4), *m/z* 180 due to a combined loss of ammonia (NH_3) and $^{\bullet}CH_3$, *m/z* 165 due to a subsequent loss of $^{\bullet}CH_3$. The fragment ion with *m/z* 150 is also secondary fragmentation of the ion with *m/z* 180 due to the loss of formaldehyde ($H_2C{=}O$). The ion with *m/z* 105 is the α-(methyl)benzyl cation ($[C_6H_5CHCH_3]^+$), apparently formed by the loss of NH_3 and three times $H_2C{=}O$.

4.7.6.6 GHB (γ-Hydroxybutyric Acid)

GHB (γ-hydroxybutyric acid) is nowadays considered as a drug of abuse. It is readily analyzed in negative-ion ESI. In its product-ion mass spectrum, GHB ($[M-H]^-$ with *m/z* 103) shows the loss of water and HCOOH to fragment ions with *m/z* 85 and 57, respectively. The fragmentation is mechanistically interesting as demonstrated by the fact that in [D_6]-GHB ($[M-H]^-$ with *m/z* 109), the corresponding fragment ions are observed with *m/z* 90 (due to loss of HDO) and with *m/z* 61, consistent with the loss of DCOOD.

4.7.6.7 Ethanol

Two ethanol (C_2H_5OH) metabolites, ethyl sulfate and ethyl glucuronide, should also be discussed. Ethyl sulfate ($[M-H]^-$ with *m/z* 125) shows the loss of ethene (C_2H_4) to the hydrogen sulfate anion with *m/z* 97 (HSO_4^-) and the loss of an ethoxy radical ($C_2H_5O^{\bullet}$) to an ion with *m/z* 80 ($SO_3^{-\bullet}$). Ethyl glucuronide ($[M-H]^-$ with *m/z* 221) shows the loss of water to a fragment ion with *m/z* 203, and the subsequent loss of C_2H_5OH (46.0417 Da) rather than formic acid (HCOOH, 46.0055 Da) to an ion with *m/z* 157. The data for two [D_5]-labeled ethyl glucuronide isotopomers is

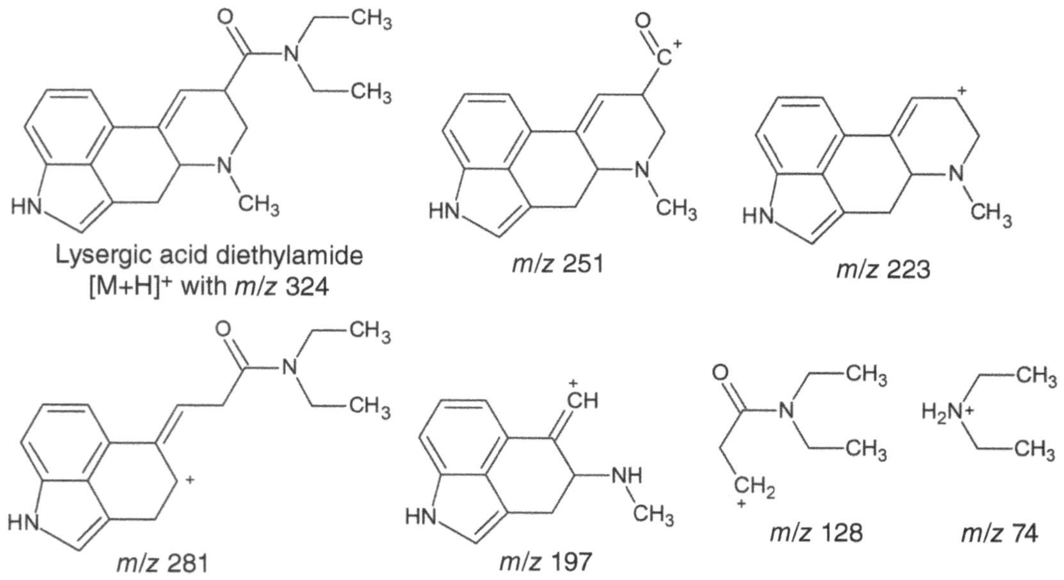

FIGURE 4.7.5 Proposed structures for the fragment ions of protonated lysergic acid diethylamide.

more confusing, suggesting a heavy-atom effect. For the isomer with [D_5]-ethyl glucuronide in the Q–LIT instrument (Dresen et al., 2009), a fragment ion with *m/z* 157 (loss of [D_5]-ethanol, C_2D_5OH) is observed. For the isomer with ethyl-[D_5]-glucuronide in a Q–TOF instrument (Agilent Technologies, 2010; Broecker et al., 2011), the loss of CO_2 rather than HCOOH or C_2H_5OH is observed, leading to a fragment ion with *m/z* 164. Secondary fragmentation of the ion with *m/z* 157 ($[C_6H_5O_5]^-$) is observed, resulting in fragment ions with *m/z* 129 ($[C_5H_5O_4]^-$, loss of CO), *m/z* 113 ($[C_5H_5O_3]^-$, loss of CO_2), *m/z* 95 ($[C_5H_3O_2]^-$, loss of CO_2 and water), *m/z* 85 ($[C_4H_5O_2]^-$, loss of CO_2 and CO), *m/z* 75 ($[C_2H_3O_3]^-$), and *m/z* 59 ($[C_2H_3O_2]^-$) (Kuuranne et al., 2000).

4.7.7 DESIGNER DRUGS

A wide diversity of synthetic variants of illicit drugs have been developed; some have already been discussed earlier. There is considerable interest in the analysis and characterization of these NPSs, especially because they are the cause of several deaths by intoxication cases over the past years (Kyriakou et al., 2015). These NPSs, also indicated as "bath salts" or "legal highs," include synthetic cannabinoids, substituted cathinones such as mephedrone, methylone, and methylenedioxypyrovalerone (Section 4.7.2.4), and synthetic hallucinogens like NBOMe. The MS–MS fragmentation of some classes of designer drugs is discussed in this section.

4.7.7.1 Synthetic Cannabinoids

There are various classes of synthetic cannabinoids, some of which structurally closely resemble THC and related compounds (Section 4.7.3), for example, HU-210, whereas other do not, for example, the aminoalkylindoles and indazoles (Znaleziona et al., 2015). The fragmentation of some of these compounds is discussed in this section.

HU-210 is a synthetic cannabinoid, structurally related to THC. The Phase I metabolism of HU-210 has been investigated (Kim et al., 2012). The fragmentation of HU-210 ($[M+H]^+$ with *m/z* 387) is rather straightforward. It shows the loss of water to an ion with *m/z* 369 or the loss of the branched alkyl chain as 2-methyloctene (C_9H_{18}) to an ion with *m/z* 261, which shows the loss of water to an ion with *m/z* 243, which in turn shows the loss of propene ($CH_3CH{=}CH_2$) to an ion with *m/z* 201 (Kim et al., 2012).

In the course of metabolite identification studies, the fragmentation of several indazole-type synthetic cannabinoids was studied (Takayama et al., 2014). Examples of this compound class include AB-FUBINACA ($[M+H]^+$ with *m/z* 369), ADB-FUBINACA ($[M+H]^+$ with *m/z* 383), and AB-PINACA ($[M+H]^+$ with *m/z* 331). The major fragments of AB-FUBINACA are the ions with *m/z* 324 due to the loss of formamide (H_2NCHO), *m/z* 253 due to the loss of 2-amino-3-methylbutanamide ($C_5H_{12}N_2O$), and the 2-fluorobenzyl cation ($[FC_6H_4CH_2]^+$) with *m/z* 109 (Figure 4.7.6). Similarly, ADB-FUBINACA shows fragment ions with *m/z* 338 due to the loss of H_2NCHO, *m/z* 253 due to the loss of 2-amino-3,3-dimethylbutanamide, and an ion with *m/z* 109. AB-PINACA shows fragment ions with *m/z* 286 due to the loss of H_2NCHO, *m/z* 215 due to the loss of 2-amino-3-methylbutanamide ($C_5H_{12}N_2O$), and (the (1*H*-indazol-3-yl)acylium ion ($[C_7H_5N_2{-}C{=}O]^+$) with *m/z* 145 due to the subsequent loss of pentene (C_5H_{10}).

Several classes of NPSs based on the aminoalkylindole structure have been synthesized and described (Wiley et al., 2014; Znaleziona et al., 2015). The LC–MS analysis of a wide range of related compounds and some of their metabolites has been reported (Ammann et al., 2012a; de Jager et al., 2012; Hutter et al., 2012; Shanks et al., 2012; Wohlfarth et al., 2013).

JWH-018 is an example of the naphthoylindole class. The fragment ions observed for JWH-018 ($[M+H]^+$ with *m/z* 342) are due to cleavages of either C—C bond of the carbonyl functional group (Figure 4.7.6) (Teske et al., 2010). A direct cleavage (Section 3.5.4) of the C—C bond between the 1-naphthyl group and the carbonyl results in a naphthalen-1-ylium ion with *m/z* 127 ($[C_{10}H_7]^+$) and (1-pentylindol-3-yl)acylium cation with *m/z* 214 ($[C_{13}H_{16}N{-}C{=}O]^+$). The cleavage of the C—C bond between the *N*-pentylindole and the carbonyl moiety results in the naphthoyl cation ($[C_{10}H_7{-}C{=}O]^+$) with *m/z* 155. The fragmentation of 20 naphthoylindole compounds, is summarized in Table 4.7.7.

Aminoalkylindoles of the phenylacetylindole class, like JWH-251 ($[M+H]^+$ with *m/z* 320), show two fragment ions due to direct cleavage of the C—C bond between the carbonyl and the benzyl group and a fragment ion due to cleavage of the C—C bond between the carbonyl group and the indole. In the latter case, H-rearrangement to the indole ring occurs. For JWH-251, this leads to an acylium ion with *m/z* 214 ($[C_{13}H_{16}N{-}C{=}O]^+$), a 2-methylbenzyl cation ($[CH_3{-}C_6H_4CH_2]^+$) with *m/z* 105, and protonated 1-pentyl-1*H*-indole with *m/z* 188 ($[C_{13}H_{17}N{+}H]^+$). Identical fragmentation is observed for two analogs, although instead of the ion with *m/z* 105 different ions are observed, that is, 2-methoxybenzyl cation with *m/z* 121 ($[CH_3O{-}C_6H_4CH_2]^+$) for JWH-250 ($[M+H]^+$ with *m/z* 336; with a 2-methoxyphenylacetyl substituent) and 2-chlorobenzyl cation with *m/z* 125 ($[ClC_6H_4CH_2]^+$) for JWH-203 ($[M+H]^+$ with *m/z* 340; with a 2-chlorophenylacetyl substituent).

4.7.7.2 2C Family of Compounds (2,5-Dimethoxyphenethylamines)

2C (2C-X) is a general name for the family of psychedelic phenethylamines containing a 2,5-dimethoxy benzene ring,

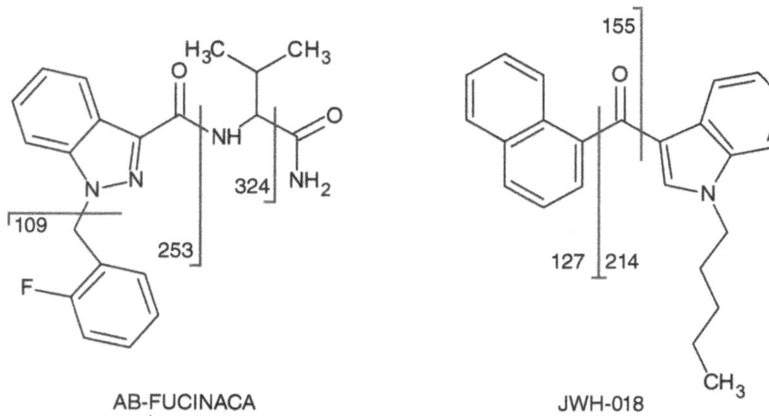

AB-FUCINACA
[M+H]+ with *m/z* 369

JWH-018
[M+H]+ with *m/z* 342

FIGURE 4.7.6 Fragmentation schemes for two protonated synthetic cannabinoids, AB-FUCINACA and JWH-018.

TABLE 4.7.7 Structure and major fragmentation of synthetic cannabinoids of the naphthoylindole class.

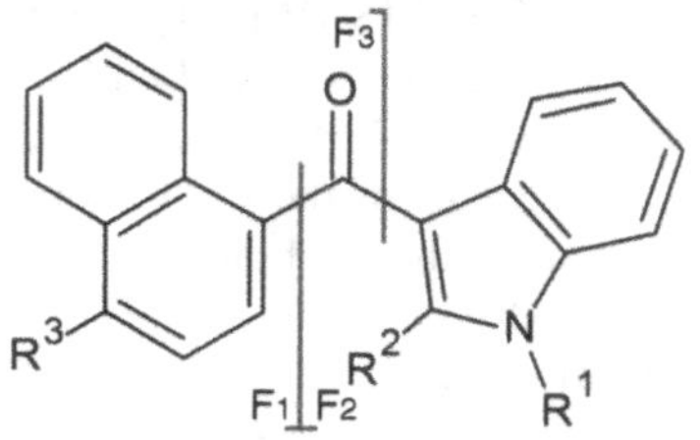

Compound	[M+H]+	R¹	R²	R³	F₁	F₂	F₃	Other Fragments
JWH-018	342	$—C_5H_{11}$	—H	—H	127	214	155	
JWH-007	356	$—C_5H_{11}$	$—CH_3$	—H	127	228	155	
JWH-015	328	$—C_3H_7$	$—CH_3$	—H	127	200	155	
JWH-019	356	$—C_6H_{13}$	—H	—H	127	228	155	
JWH-020	370	$—C_7H_{15}$	—H	—H	127	242	155	
JWH-022	340	$—C_5H_9$	—H	—H	127	212	155	
JWH-072	314	$—C_3H_7$	—H	—H	127	186	155	
JWH-073	328	$—C_4H_9$	—H	—H	127	200	155	
JWH-080	358	$—C_4H_9$	—H	$—OCH_3$	157	200	185	127: $F_1 - H_2C{=}O$
JWH-081	372	$—C_5H_{11}$	—H	$—OCH_3$	157	214	185	127: $F_1 - H_2C{=}O$
JWH-098	386	$—C_5H_{11}$	$—CH_3$	$—OCH_3$	157	228	185	127: $F_1 - H_2C{=}O$
JWH-122	356	$—C_5H_{11}$	—H	$—CH_3$	141	214	169	
JWH-180	356	$—C_3H_7$	—H	$—C_3H_7$	169	186	197	154: $F_3 - C_3H_7^{\bullet}$
JWH-182	384	$—C_5H_{11}$	—H	$—C_3H_7$	169	214	197	154: $F_3 - C_3H_7^{\bullet}$
JWH-200	385	$—(C_6H_{12}NO)$	—H	—H	127		155	114: $[C_6H_{12}NO]^+$
JWH-210	370	$—C_5H_{11}$	—H	$—C_2H_5$	155	214	183	
JWH-213	384	$—C_5H_{11}$	$—CH_3$	$—C_2H_5$	155	228	183	
JWH-387	420	$—C_5H_{11}$	—H	—Br	205	214	233	126: $F_1 - Br^{\bullet}$
JWH-398	376	$—C_5H_{11}$	—H	—Cl	161	214	189	126: $F_1 - Cl^{\bullet}$
JWH-412	360	$—C_5H_{11}$	—H	—F	145	214	173	

often with a lipophilic substituent present at the 4-position, which increases their potency, metabolic stability, and duration of action. The term "2C" refers to the two C-atoms (ethylene unit) between the benzene ring and the amine group. The 2C family includes compounds, substituted at the 4-position with a methyl (2C-D; $[M+H]^+$ with *m/z* 196), ethyl (2C-E; $[M+H]^+$ with *m/z* 210), propyl (2C-P; $[M+H]^+$ with *m/z* 224), chloro (2C-C; $[M+H]^+$ with *m/z* 216), bromo (2C-B; $[M+H]^+$ with *m/z* 260), iodo (2C-I; $[M+H]^+$ with *m/z* 308), nitro (2C-N; $[M+H]^+$ with *m/z* 227), ethylthio (2C-T2;

$[M+H]^+$ with *m/z* 242), *iso*-propylthio (2C-T4; $[M+H]^+$ with *m/z* 256), or propylthio (2C-T7; $[M+H]^+$ with *m/z* 256) group. These compounds show characteristic fragmentation. As an example, 2C-C shows the loss of a methyl radical ($^\bullet CH_3$) to an ion with *m/z* 201 or the loss of ammonia (NH_3) to an ion with *m/z* 199. The latter ion shows a loss of $^\bullet CH_3$ to an ion with *m/z* 184, a loss of formaldehyde ($H_2C{=}O$) to an ion with *m/z* 169, or a loss of a chlorine radical ($Cl^\bullet$) to an ion with 164. The latter ion in turn shows a loss of $^\bullet CH_3$ to an ion with *m/z* 149 or a loss of $H_2C{=}O$ to an ion with *m/z* 134 (Wink et al., 2015). In 2C-T2, the loss of NH_3 is followed by the loss of either the ethyl radical ($^\bullet CH_2CH_3$) or the ethylthio radical ($CH_3CH_2S^\bullet$) to ions with *m/z* 196 and 164, respectively. In 2C-T7 and 2C-T4, the loss of NH_3 is followed by either the loss of propene ($CH_3CH{=}CH_2$) to an ion with 197 or by the loss of the *n*-propylthio (2C-T7) or *iso*-propylthio (2C-T4) radical ($C_3H_7S^\bullet$) to an ion with *m/z* 164. The fragmentation of 2C-N has been studied in more detail (Zuba et al., 2012). After an initial loss of NH_3 to an ion with *m/z* 210, the loss of nitroxyl (HNO) results in an ion with *m/z* 179, which is proposed to be the α-(methyl)-2-methoxy-4,5-methylenedioxy-benzyl cation ($[C_9H_9O_3CH_2]^+$). The ion with *m/z* 195 is due to subsequent losses of NH_3 and $^\bullet CH_3$. The ion with *m/z* 151 is the 2,5-dimethoxybenzyl cation ($[(CH_3O)_2C_6H_3CH_2]^+$).

4.7.7.3 NBOMe Compounds

NBOMe compounds are *N*-2′-methoxybenzyl substituted 2C class of hallucinogens (known in the market as N-bomb, Smiles, Solaris, and Cimbi) (Kyriakou et al., 2015). The NBOMes are 2,5-dimethoxyphenylethylamines with an additional 2′-methoxybenzyl substituent on the amine functional group. NBOMe variants are available having methyl (25D-NBOMe; $[M+H]^+$ with *m/z* 316), ethyl (25E-NBOMe; $[M+H]^+$ with *m/z* 330), propyl (25P-NBOMe; $[M+H]^+$ with *m/z* 344), chloro (25C-NBOMe; $[M+H]^+$ with *m/z* 336), bromo (25B-NBOMe; $[M+H]^+$ with *m/z* 380), iodo

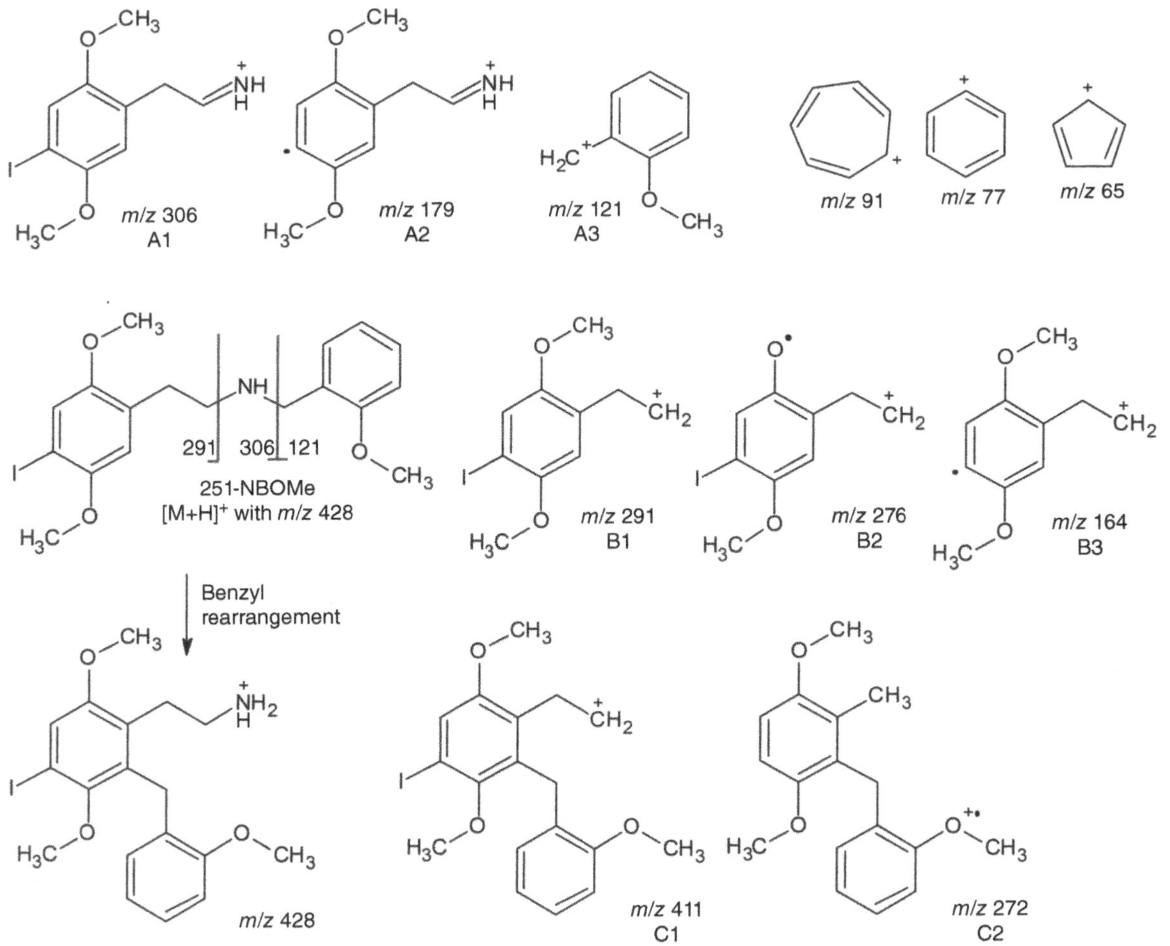

FIGURE 4.7.7 Proposed structures for the fragment ions of protonated 25I-NBOMe, that is, 2-(4-iodo-2,5-dimethoxyphenyl)-*N*-(2-methoxybenzyl)ethanamine.

TABLE 4.7.8 ***m/z*** **values for characteristic fragments of 25X-NBOMe compounds, with 2′-methoxy unless otherwise indicated.**

	Substitution	$[M+H]^+$	A1	A2	A3	B1	B2	B3	C1	C2
25I-NBOMe	4-I	428	306	179	121	291	276	164	411	272
25H-NBOMe	4-H	302	180		121	165			285	
25B-NBOMe	4-Br	380	258	179	121	243	228	164	363	
25C-NBOMe	4-Cl	336	214		121	199	184	164	319	
25D-NBOMe	4-CH_3	316	194	179	121	179		164	299	
25E-NBOMe	4-C_2H_5	330	208	179	121	193	178	164	313	
25G-NBOMe	3,4-di-CH_3	330	208	179	121	193	178		313	
25N-NBOMe	4-NO_2	347	225		121					272
25P-NBOMe	4-C_3H_7	344	222		121	207	192		327	
25T2-NBOMe*	4-C_2H_5S	362	240	179	121	225	210	164	345	
25T4-NBOMe	4-*i*-C_3H_7S	376	254	179	121	239	224	164	359	272
25T7-NBOMe	4-*n*-C_3H_7S	376	254	179	121	239	224	164	359	
25I-NBF	4-I, 2′-F	416		179	109	291	276	164		260
25I-NBD	4-I, 2′,3′-OCH_2O	442		179	135	291		164	425	
25I-NBOH	4-I, 2′-OH	414	†		107			164		

Refer to Figure 4.7.7, where structure proposals for the fragment ions and fragment identification for 25I-NBOMe are shown.
*Confusingly, in some cases, 25T2-NBOMe is indicated as the 4-methylthio (4-CH_3S—) instead of the 4-ethylthio (4-CH_2CH_2S—) analogue (Johnson et al., 2014).
†For 25I-NBOH, this fragment is the protonated 2-(4-iodo-2,5-dimethoxyphenyl)ethanamine with *m/z* 308 rather than protonated 2-(4-iodo-2,5-dimethoxyphenyl)ethanimine with *m/z* 306, since a H-rearrangement from the hydroxybenzyl group to the N-atom is possible.

(25I-NBOMe; $[M+H]^+$ with *m/z* 428), nitro (25N-NBOMe; $[M+H]^+$ with *m/z* 347), ethylthio (25T2-NBOMe; $[M+H]^+$ with *m/z* 362), *iso*-propylthio (25T4-NBOMe; $[M+H]^+$ with *m/z* 376), or *n*-propylthio (25T7-NBOMe; $[M+H]^+$ with *m/z* 376) substituents at the 4-position on the 2,5-dimethoxyphenyl ring. Analogs to 25I-NBOMe with different substituents on the benzyl ring, that is, instead of 2′-methoxy, have been described as well, for example, 25I-NBF (with 2′-fluoro; $[M+H]^+$ with *m/z* 416), 25I-NBOH (with 2′-hydroxy; $[M+H]^+$ with *m/z* 414), and 25I-NBD (with 2′,3′-methylenedioxy; $[M+H]^+$ with *m/z* 442). Multiresidue methods for the analysis of NBOMes in whole blood, plasma, and urine have been reported (Poklis et al., 2014; Johnson et al., 2014). Their metabolism has been studied as well (Brandt et al., 2015; Caspar et al., 2015; Boumrah et al., 2015).

The fragmentation of this compound class has been recently studied (Zuba et al., 2013; Brandt et al., 2015; Caspar et al., 2015; Boumrah et al., 2015). A wide variety of fragment ions has been observed, as demonstrated for 25I-NBOMe in Figure 4.7.7. The fragmentation of 25I-NBOMe (and its related compounds) can be described along three lines. (1) The loss of ammonia (NH_3) to an ion with *m/z* 411 (fragment C1 in Figure 4.7.7 and Table 4.7.8) requires a skeletal rearrangement of the 2′-methoxybenzyl group from the amino-N-atom to the 2,5-dimethoxyphenyl ring. This skeletal rearrangement was also proposed to explain a radical fragment ion with *m/z* 272 ($[C_{17}H_{20}O_3]^{+\bullet}$; C2), which based on elemental composition would involve the formal loss of CH_3NI, that is, methanimine ($H_2C{=}NH$) and a iodine radical ($I^\bullet$) (Brandt et al., 2015; Caspar et al., 2015). It is observed only for some NBOMes. Similar skeletal rearrangements involving benzyl migration have been observed in other compounds (Section 3.5.8). (2) The direct cleavage of the N—C bond between the ethanamine and the 2′-methoxybenzyl group results in fragment ions with *m/z* 306 (A1) and *m/z* 121 (A3), that is, the sum of *m/z* values of the fragment ions equals the *m/z* ($[M+H]-1$) (Section 3.5.4). Both fragment ions show secondary fragmentation. The ion with *m/z* 306 shows the loss of $I^\bullet$ to an ion with *m/z* 179 (A2). The ion with *m/z* 121 shows the loss of formaldehyde ($H_2C{=}O$) to the tropylium ion ($[C_7H_7]^+$) with *m/z* 91. In addition, the phenyl ion ($[C_6H_5]^+$) with *m/z* 77 and the cyclopentadienyl ion ($[C_5H_5]^+$) with *m/z* 65 are observed. (3) The cleavage of the N—C^1 bond in ethanamine results in the ion with *m/z* 291 (B1), which shows secondary fragmentation involving the loss of either a methyl radical ($^\bullet CH_3$) to an ion with *m/z* 276 (B2) or $I^\bullet$ to an ion with *m/z* 164 (B3). The sequence of fragmentation events in Figure 4.7.7 is rather different from what is described in literature (Caspar et al., 2015), as for instance the loss of NH from the EE^+ ion with *m/z* 306 to yield the EE^+ ion with *m/z* 291 seems highly unlikely. The *m/z* values for the fragment ions observed for 25I-NBOMe, its analogs, and other NBOMes are summarized in Table 4.7.8.

REFERENCES

Ambach L, Franz F. 2012. QTRAP® MS–MS library of designer drugs, http://www.chemicalsoft.de/qtrap_111_designer_drugs.html (accessed, 29 October 2015).

Ammann J, McLaren JM, Gerostamoulos D, Beyer J. 2012a. Detection and quantification of new designer drugs in human blood: Part 1 - Synthetic cannabinoids. J Anal Toxicol 36: 372–380.

Ammann D, McLaren JM, Gerostamoulos D, Beyer J. 2012b. Detection and quantification of new designer drugs in human blood: Part 2 - Designer cathinones. J Anal Toxicol 36: 381–389.

Badawi N, Simonsen KW, Steentoft A, Bernhoft IM, Linnet K. 2009. Simultaneous screening and quantification of 29 drugs of abuse in oral fluid by solid-phase extraction and ultraperformance LC–MS/MS. Clin Chem 55: 2004–2018.

Bijlsma L, Sancho JV, Hernández F, Niessen WMA. 2011. Fragmentation pathways of drugs of abuse and their metabolites based on QTOF MS/MS and MS(E) accurate-mass spectra. J Mass Spectrom 46: 865–875.

Bijlsma L, Emke E, Hernández F, de Voogt P. 2013. Performance of the linear ion trap Orbitrap mass analyzer for qualitative and quantitative analysis of drugs of abuse and relevant metabolites in sewage water. Anal Chim Acta 768: 102–110.

Boumrah Y, Humbert L, Phanithavong M, Khimeche K, Dahmani A, Allorge D. 2015. In vitro characterization of potential CYP- and UGT-derived metabolites of the psychoactive drug 25B-NBOMe using LC-high resolution MS. Drug Test Anal in press 26382567, doi: 10.1002/dta.1865.

Borova VL, Maragou NC, Gago-Ferrero P, Pistos C, Thomaidis NS. 2014. Highly sensitive determination of 68 psychoactive pharmaceuticals, illicit drugs, and related human metabolites in wastewater by liquid chromatography–tandem mass spectrometry. Anal Bioanal Chem 406: 4273–4285.

Brandt SD, Elliott SP, Kavanagh PV, Dempster NM, Meyer MR, Maurer HH, Nichols DE. 2015. Analytical characterization of bioactive *N*-benzyl-substituted phenethylamines and 5-methoxytryptamines. Rapid Commun Mass Spectrom 29: 573–584.

Broecker S, Herre S, Wüst B, Zweigenbaum J, Pragst F. 2011. Development and practical application of a library of CID accurate mass spectra of more than 2,500 toxic compounds for systematic toxicological analysis by LC-QTOF-MS with data-dependent acquisition. Anal Bioanal Chem 400: 101–117.

Cai J, Henion JD. 1996. Elucidation of LSD in vitro metabolism by liquid chromatography and capillary electrophoresis coupled with tandem mass spectrometry. J Anal Toxicol 20: 27–37.

Caspar AT, Helfer AG, Michely JA, Auwärter V, Brandt SD, Meyer MR, Maurer HH. 2015. Studies on the metabolism and toxicological detection of the new psychoactive designer drug 2-(4-iodo-2,5-dimethoxyphenyl)-*N*-[(2-methoxyphenyl)methyl]-ethanamine (25I-NBOMe) in human and rat urine using GC–MS, LC–MS(*n*), and LC-HR-MS/MS. Anal Bioanal Chem 407: 6697–6719.

Castiglioni S, Zuccato E, Chiabrando C, Fanelli R, Bagnati R. 2008. Mass spectrometric analysis of illicit drugs in wastewater and surface water. Mass Spectrom Rev 27: 378–394.

Castiglioni S, Bijlsma L, Covaci A, Emke E, Hernández F, Reid M, Ort C, Thomas KV, van Nuijs AL, de Voogt P, Zuccato E. 2013. Evaluation of uncertainties associated with the determination of community drug use through the measurement of sewage drug biomarkers. Environ Sci Technol 47: 1452–1460.

Chebbah C, Pozo ÓJ, Deventer K, Van Eenoo P, Delbeke FT. 2010. Direct quantification of 11-nor-Delta(9)-tetrahydrocannabinol-9-carboxylic acid in urine by liquid chromatography/tandem mass spectrometry in relation to doping control analysis. Rapid Commun Mass Spectrom 24: 1133–1141.

de Jager AD, Bailey NL. 2011. Online extraction LC–MS/MS method for the simultaneous quantitative confirmation of urine drugs of abuse and metabolites: Amphetamines, opiates, cocaine, cannabis, benzodiazepines and methadone. J Chromatogr B 879: 2642–2652

de Jager AD, Warner JV, Henman M, Ferguson W, Hall A. 2012. LC–MS/MS method for the quantitation of metabolites of eight commonly-used synthetic cannabinoids in human urine--an Australian perspective. J Chromatogr B 897: 22–31.

Dresen S, Gergoc M, Politi, L, Halter C, Weinmann W. 2009. ESI-MS–MS library of 1,253 compounds for application in forensic and clinical toxicology. Anal Bioanal Chem 395: 2521–2526.

Fornal E. 2013. Formation of odd-electron product ions in collision-induced fragmentation of electrospray-generated protonated cathinone derivatives: Aryl α-primary amino ketones. Rapid Commun Mass Spectrom 27: 1858–1866.

Frison G, Odoardi S, Frasson S, Sciarrone R, Ortar G, Romolo FS, Strano Rossi S. 2015. Characterization of the designer drug bk-2C-B (2-amino-1-(bromo-dimethoxyphenyl)ethan-1-one) by gas chromatography/mass spectrometry without and with derivatization with 2,2,2-trichloroethyl chloroformate, liquid chromatography/high-resolution mass spectrometry, and nuclear magnetic resonance. Rapid Commun Mass Spectrom 29: 1196–1204.

Gheorghe A, van Nuijs A, Pecceu B, Bervoets L, Jorens PG, Blust R, Neels H, Covaci A. 2008. Analysis of cocaine and its principal metabolites in waste and surface water using solid-phase extraction and liquid chromatography–ion trap tandem mass spectrometry. Anal Bioanal Chem 391: 1309–1319.

Hutter M, Kneisel S, Auwärter V, Neukamm MA. 2012. Determination of 22 synthetic cannabinoids in human hair by liquid chromatography–tandem mass spectrometry. J Chromatogr B 903: 95–101.

Ibáñez M, Pozo ÓJ, Sancho JV, Orengo T, Haro G, Hernández F. 2016. Analytical strategy to investigate 3,4-methylenedioxypyrovalerone (MDPV) metabolites in consumers' urine by high-resolution mass spectrometry. Anal Bioanal Chem 408: 151–164

Jankovics P, Váradi A, Tölgyesi L, Lohner S, Németh-Palotás J, Koszegi-Szalai H. 2011. Identification and characterization of the new designer drug 4'-methylethcathinone (4-MEC) and elaboration of a novel liquid chromatography–tandem mass spectrometry (LC–MS/MS) screening method for seven different methcathinone analogs. Forensic Sci Int 210: 213–220.

Johansen SS, Bhatia HM. 2007. Quantitative analysis of cocaine and its metabolites in whole blood and urine by high-performance liquid chromatography coupled with tandem mass spectrometry. J Chromatogr B 852: 338–344.

Johnson RD, Botch-Jones SR, Flowers T, Lewis CA. 2014. An evaluation of 25B-, 25C-, 25D-, 25H-, 25I- and 25T2-NBOMe via LC–MS–MS: Method validation and analyte stability. J Anal Toxicol 38: 479–484

Kraemer T, Paul LD. 2007. Bioanalytical procedures for determination of drugs of abuse in blood. Anal Bioanal Chem 388: 1415–1435.

Kim U, Jin MJ, Lee J, Han SB, In MK, Yoo HH. 2012. Tentative identification of phase I metabolites of HU-210, a classical synthetic cannabinoid, by LC–MS/MS. J Pharm Biomed Anal 64–65: 26–34.

Kim J, Ji D, Kang S, Park M, Yang W, Kim E, Choi H, Lee S. 2014. Simultaneous determination of 18 abused opioids and metabolites in human hair using LC–MS/MS and illegal opioids abuse proven by hair analysis. J Pharm Biomed Anal 89: 99–105.

Kuuranne T, Vahermo M, Leinonen A, Kostianen R. 2000. Electrospray and atmospheric pressure chemical ionization tandem mass spectrometric behavior of eight anabolic steroid glucuronides. J Am Soc Mass Spectrom 11: 722–730.

Kyriakou C, Marinelli E, Frati P, Santurro A, Afxentiou M, Zaami S, Busardo FP. 2015. NBOMe: New potent hallucinogens - pharmacology, analytical methods, toxicities, fatalities: A review. Eur Rev Med Pharmacol Sci 19: 3270–3281.

Marchei E, Escuder D, Pallas CR, Garcia-Algar O, Gómez A, Friguls B, Pellegrini M, Pichini S. 2011. Simultaneous analysis of frequently used licit and illicit psychoactive drugs in breast milk by liquid chromatography tandem mass spectrometry. J Pharm Biomed Anal 55: 309–316.

Matsuta S, Shima N, Kamata H, Kakehashi H, Nakano S, Sasaki K, Kamata T, Nishioka H, Miki A, Katagi M, Zaitsu K, Sato T, Tsuchihashi H, Suzuki K. 2015. Metabolism of the designer drug α-pyrrolidinobutiophenone (α-PBP) in humans: Identification and quantification of the phase I metabolites in urine. Forensic Sci Int 249: 181–188.

Meyer MR, Maurer HH. 2012. Current status of hyphenated mass spectrometry in studies of the metabolism of drugs of abuse, including doping agents. Anal Bioanal Chem 402: 195–208.

Meyer MR, Vollmar C, Schwaninger AE, Wolf E, Maurer HH. 2012. New cathinone-derived designer drugs 3-bromomethcathinone and 3-fluoromethcathinone: Studies on their metabolism in rat urine and human liver microsomes using GC-MS and LC-high-resolution MS and their detectability in urine. J Mass Spectrom 47: 253–262.

Meyer MR, Lindauer C, Maurer HH. 2014a. Dimethocaine, a synthetic cocaine derivative: Studies on its in vitro metabolism catalyzed by P450s and NAT2. Toxicol Lett 225: 139–146.

Meyer MR, Lindauer C, Welter J, Maurer HH. 2014b. Dimethocaine, a synthetic cocaine analogue: Studies on its in vivo metabolism and its detectability in urine by means of a rat model and liquid chromatography-linear ion-trap (high-resolution) mass spectrometry. Anal Bioanal Chem 406: 1845–1854.

Montesano C, Johansen SS, Nielsen MK. 2014. Validation of a method for the targeted analysis of 96 drugs in hair by UPLC-MS/MS. J Pharm Biomed Anal 88: 295–306.

Poklis JL, Clay DJ, Poklis A. 2014. High-performance liquid chromatography with tandem mass spectrometry for the determination of nine hallucinogenic 25-NBOMe designer drugs in urine specimens. J Anal Toxicol 38: 113–121.

Schwope DM, Scheidweiler KB, Huestis MA. 2011. Direct quantification of cannabinoids and cannabinoid glucuronides in whole blood by liquid chromatography–tandem mass spectrometry. Anal Bioanal Chem 401: 1273–1283.

Shanks KG, Dahn T, Behonick G, Terrell A. 2012. Analysis of first and second generation legal highs for synthetic cannabinoids and synthetic stimulants by ultra-performance liquid chromatography and time of flight mass spectrometry. J Anal Toxicol 36: 360–371.

Stephanson N, Josefsson M, Kronstrand R, Beck O. 2008. Accurate identification and quantification of 11-nor-delta(9)-tetrahydrocannabinol-9-carboxylic acid in urine drug testing: Evaluation of a direct high efficiency liquid chromatographic–mass spectrometric method. J Chromatogr B 871: 101–108.

Strano Rossi S, Odoardi S, Gregori A, Peluso G, Ripani L, Ortar G, Serpelloni G, Romolo FS. 2014. An analytical approach to the forensic identification of different classes of new psychoactive substances (NPSs) in seized materials. Rapid Commun Mass Spectrom 28: 1904–1916.

Swortwood MJ, Boland DM, DeCaprio AP. 2013. Determination of 32 cathinone derivatives and other designer drugs in serum by comprehensive LC-QQQ-MS/MS analysis. Anal Bioanal Chem 405: 1383–1397.

Takayama T, Suzuki M, Todoroki K, Inoue K, Min JZ, Kikura-Hanajiri R, Goda Y, Toyo'oka T. 2014. UPLC/ESI-MS/MS-based determination of metabolism of several new illicit drugs, ADB-FUBINACA, AB-FUBINACA, AB-PINACA, QUPIC, 5F-QUPIC and α-PVT, by human liver microsome. Biomed Chromatogr 28: 831–838.

Teske J, Weller JP, Fieguth A, Rothämel T, Schulz Y, Tröger HD. 2010. Sensitive and rapid quantification of the cannabinoid receptor agonist naphthalen-1-yl-(1-pentylindol-3-yl)methanone (JWH-018) in human serum by liquid chromatography–tandem mass spectrometry. J Chromatogr B 878: 2659–2663.

Thomas KV, Bijlsma L, Castiglioni S, Covaci A, Emke E, Grabic R, Hernández F, Karolak S, Kasprzyk-Hordern B, Lindberg RH, Lopez de Alda M, Meierjohann A, Ort C, Picó Y, Quintana JB, Reid M, Rieckermann J, Terzic S, van Nuijs AL, de Voogt P. 2012. Comparing illicit drug use in 19 European cities through sewage analysis. Sci Total Environ 432: 432–439.

Van Bocxlaer JF, Clauwaert KM, Lambert WE, Deforce DL, Van den Eeckhout EG, De Leenheer AP. 2000. Liquid chromatography–mass spectrometry in forensic toxicology. Mass Spectrom Rev 19: 165–214.

Wang PP, Bartlett MG. 1998. Collision-induced dissociation mass spectra of cocaine and its metabolites and pyrolysis products. J Mass Spectrom 33: 961–967.

Wang K-C, Shi T-S, Cheng S-G. 2005. Use of SPE and LC/TIS/MS/MS for rapid detection and quantitation of ketamine and its metabolite, norketamine, in urine. Forensic Sci Int 147: 81–88.

Wiley JL, Marusich JA, Huffman JW. 2014. Moving around the molecule: Relationship between chemical structure and in vivo activity of synthetic cannabinoids. Life Sci 97: 55–63.

Wink CS, Meyer MR, Braun T, Turcant A, Maurer HH. 2015. Biotransformation and detectability of the designer drug 2,5-dimethoxy-4-propylphenethylamine (2C-P) studied in urine by GC-MS, LC-MS^n, and LC-high-resolution-MS^n. Anal Bioanal Chem 407: 831–843.

Wohlfarth A, Scheidweiler KB, Chen X, Liu HF, Huestis MA. 2013. Qualitative confirmation of 9 synthetic cannabinoids and 20 metabolites in human urine using LC–MS/MS and library search. Anal Chem 85: 3730–3738.

Xia Y, Wang P, Bartlett MG, Solomon HM, Busch KL. 2000. An LC–MS–MS method for the comprehensive analysis of cocaine and cocaine metabolites in meconium. Anal Chem 72: 764–771.

Znaleziona J, Ginterová P, Petr J, Ondra P, Válka I, Ševčík J, Chrastina J, Maier V. 2015. Determination and identification of synthetic cannabinoids and their metabolites in different matrices by modern analytical techniques - a review. Anal Chim Acta 874: 11–25.

Zuba D, Sekuła K, Buczek A. 2012. Identification and characterization of 2,5-dimethoxy-4-nitro-β-phenethylamine (2C-N)--a new member of 2C-series of designer drug. Forensic Sci Int 222: 298–305.

Zuba D, Sekuła K, Buczek A. 2013. 25C-NBOMe – new potent hallucinogenic substance identified on the drug market. Forensic Sci Int 227: 7–14.

4.8

FRAGMENTATION OF ANTIMICROBIAL COMPOUNDS

An antimicrobial drug either kills or prevents or inhibits the growth of microorganisms such as bacteria, virus, fungi, or protozoa. Disinfectants are antimicrobial substances used on nonliving objects or outside the body. In this chapter, the MS–MS fragmentation of various classes of antibacterial drugs is discussed. The MS–MS fragmentation of other classes of antimicrobial agents is discussed in separate sections, that is, antifungal drugs in Section 4.9, and other classes including antiprotozoal agents, especially coccidiostatics, antiviral drugs, and disinfectants in Section 4.10.

An antibacterial or antibiotic is a compound that kills or slows down the growth of bacteria. The term "antibiotic" was introduced to describe any substance produced by a microorganism that is antagonistic to the growth of other microorganisms. Today, also synthetic antibacterial compounds such as the sulfonamides are included. In fact, most antibiotics used today are chemical modifications of various natural compounds.

Most antibiotic compounds are preferably analyzed in positive-ion mode (Niessen, 1998; Niessen, 2005), some others only in negative-ion mode, but in most cases both positive-ion and negative-ion modes can be applied. This is certainly true for the sulfonamides, penicillins, and cephalosporins discussed in the following. Numerous multiresidue analytical methods based on liquid chromatography–tandem mass spectrometry (LC–MS–MS) have been reported, especially in relation to veterinary medicine and food safety applications. In most cases, these methods rely on positive-ion ESI-MS.

4.8.1 SULFONAMIDES

Sulfonamide antibacterials are synthetic broad-spectrum antimicrobial agents. They are *N*-derivatives of 4-aminobenzene sulfonamide ($H_2N—C_6H_4—SO_2NH_2$). They act as competitive inhibitors of dihydropteroate synthetase, an enzyme involved in folate synthesis. As a result, the microorganism will be deprived of folate, which is essential for DNA synthesis, and dies. Sulfonamide antibacterials are not only widely used as veterinary drugs for prophylactic and therapeutic purposes but also because they show growth-promoting activity in animals. Typical maximum residue limits are 100 μg/kg for meat and 10 μg/kg for milk. Sulfonamide residues in food are of concern because of the possible development of antibiotic resistance and their potential carcinogenic character. In human, they are used for urinary tract infections.

Sulfonamide antibiotics have been analyzed by LC–ESI-MS from the early days onwards (Pleasance et al., 1991). Residue analysis of sulfonamide antibiotics is mostly performed in positive-ion mode, mainly in food of animal origin such as meat (Van Eeckhout et al., 2000; Yu et al., 2011), in milk (Turnipseed et al., 2011), in eggs (Heller et al., 2006; Forti & Scortichini, 2009), and also in honey (Bedendo et al., 2010), and in environmental compartments, especially surface and waste water (Lindsey et al., 2001; García-Galán et al., 2010). The identity of three characteristic product ions used in SRM analysis of 17 sulfonamides in veterinary residue analysis has been elucidated (Geis-Asteggiante et al., 2014). Some examples of the use of negative-ion mode have been reported, for example, in honey (Sheridan et al., 2008) and in milk (Segura et al., 2010).

Fragmentation of sulfonamide antibiotics in positive-ion MS–MS has been studied for some 30 compounds of our library collection, running from sulfanilamide ($[M+H]^+$ with *m/z* 173) up to sulfaphenazole ($[M+H]^+$ with *m/z* 315) (Table 4.8.1). The fragmentation follows the general fragmentation pattern for sulfonamides (Section 3.6.6). All sulfonamide antibiotics produce a number of class-specific fragment ions, illustrated for sulfadimidine ($[M+H]^+$ with *m/z* 279, sulfamethazine) with ions F_1–F_6 in Figure 4.8.1. These are the result of cleavages in the sulfonamide moiety with charge retention on the 4-aminobenzene group, that is, the 4-aminobenzenesulfonyl

TABLE 4.8.1 Characteristic fragment ions for protonated sulfonamide antibiotics.

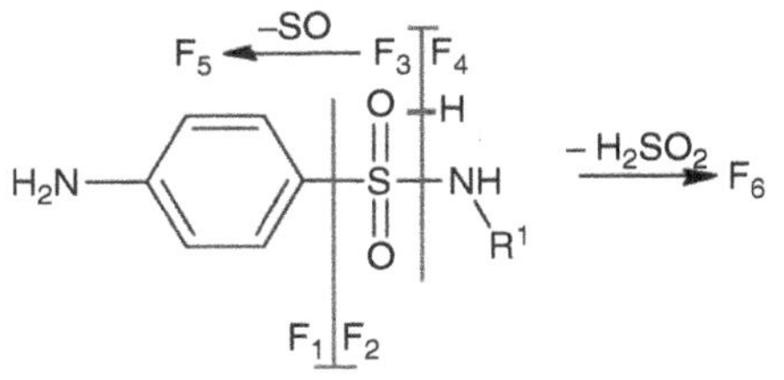

Compound	Formula	*m/z* of $[M+H]^+$	Fragment Ions (*m/z*)					
			F_1	F_2	F_3	F_4	F_5	F_6
Sulfanilamide	$C_6H_8N_2O_2S$	173.038	92		156		108	
Sulfacetamide	$C_8H_{10}N_2O_3S$	215.049	92		156		108	
Sulfaguanidine	$C_7H_{10}N_4O_2S$	215.060	92	122	156	60	108	
Sulfacarbamide	$C_7H_9N_3O_3S$	216.044	92		156		108	
Sulfathiourea	$C_7H_9N_3O_2S_2$	232.021	92		156		108	
Sulfapyridine	$C_{11}H_{11}N_3O_2S$	250.065	92		156	95	108	184
Sulfadiazine	$C_{10}H_{10}N_4O_2S$	251.060	92		156	96	108	185
Sulfamethoxazole	$C_{10}H_{11}N_3O_3S$	254.059	92		156	99	108	188
Sulfadicramide	$C_{11}H_{14}N_2O_3S$	255.080	92		156	100	108	
Sulfathiazole	$C_9H_9N_3O_2S_2$	256.021	92	163	156	101	108	190
Sulfamerazine	$C_{11}H_{12}N_4O_2S$	265.075	92	172	156	110	108	199
Sulfaperin	$C_{11}H_{12}N_4O_2S$	265.075	92	172	156	110	108	199
Sulfamoxole	$C_{11}H_{13}N_3O_3S$	268.075	92		156	113	108	
Sulfamethylthiazole	$C_{10}H_{11}N_3O_2S_2$	270.037	92	177	156	115	108	
Sulfamethizole	$C_9H_{10}N_4O_2S_2$	271.032	92	178	156	116	108	
Sulfabenzamide	$C_{13}H_{12}N_2O_3S$	277.064	92		156		108	
Sulfadimidine	$C_{12}H_{14}N_4O_2S$	279.091	92	186	156	124	108	213
Sulfamethin	$C_{12}H_{14}N_4O_2S$	279.091	92	186	156	124	108	
Sulfalene	$C_{11}H_{12}N_4O_3S$	281.070	92		156	126	108	215
Sulfamethoxypyridazine	$C_{11}H_{12}N_4O_3S$	281.070	92	188	156	126	108	215
Sulfamethoxydiazine	$C_{11}H_{12}N_4O_3S$	281.070	92	188	156	126	108	215
Sulfachloropyridazine	$C_{10}H_9ClN_4O_2S$	285.021	92		156	130	108	
Sulfaethidole	$C_{10}H_{12}N_4O_2S_2$	285.047	92	192	156	130	108	219
Sulfametrole	$C_9H_{10}N_4O_3S_2$	287.027	92		156	132	108	221
Sulfaquinoxaline	$C_{14}H_{12}N_4O_2S$	301.075	92	208	156	146	108	235
Sulfaguanole	$C_{12}H_{15}N_5O_3S$	310.097	92		156		108	
Sulfadimethoxin	$C_{12}H_{14}N_4O_4S$	311.081	92	218	156	156	108	245
Sulfadoxine	$C_{12}H_{14}N_4O_4S$	311.081	92	218	156	156	108	245
Sulfaclomid	$C_{12}H_{13}ClN_4O_2S$	313.052	92	220	156	158	108	
Sulfaphenazole	$C_{15}H_{14}N_4O_2S$	315.091	92	222	156	160	108	

ion ($[H_2N—C_6H_4—SO_2]^+$) with *m/z* 156 (F_3), protonated 4-iminocyclohexa-2,5-dienone with *m/z* 108 ($[HN—C_6H_4O+H]^+$) (F_5), and the 4-aminophenylium ion ($[H_2N—C_6H_4]^+$) with *m/z* 92 (F_3). The ion with *m/z* 108 is due to secondary fragmentation of the ion with *m/z* 156 involving a skeletal rearrangement and the loss of sulfur monoxide (SO) (Niessen, 1998; Klagkou et al., 2003; Wang et al., 2005). Two minor fragment ions due to secondary fragmentation are also observed: the protonated cyclopenta-2,4-dieneimine ($[C_5H_5N+H]^+$) with *m/z* 80 due the loss of carbon monoxide (CO) from the ion with *m/z* 108 (F_5), and the cyclopentadienyl cation ($[C_5H_5]^+$) with *m/z* 65 due to the loss of hydrogen cyanide (HCN) from the ion with *m/z* 92 (F_1). Many sulfonamides also show compound-specific fragments resulting from cleavages in the sulfonamide and charge retention on the *N*-substituted sulfonamide (Figure 4.8.1). These fragments are the ions with *m/z* ($[M+H]-93$) (F_2 in Figure 4.8.1) and *m/z* ($[M+H]-155$) (F_4). For sulfadimidine, these are the *N*-4,6-dimethyl(pyrimidin-2-yl)sulfamoyl ion ($[C_6H_7N_2—NHSO_2]^+$) with *m/z* 186, and the protonated 4,6-dimethyl(pyrimidin-2-yl)amine ($[C_6H_7N_2—NH_2+H]^+$) with *m/z* 124. The loss of ammonia (NH_3) from the ion with *m/z* ($[M+H]-155$) is sometimes observed as well.

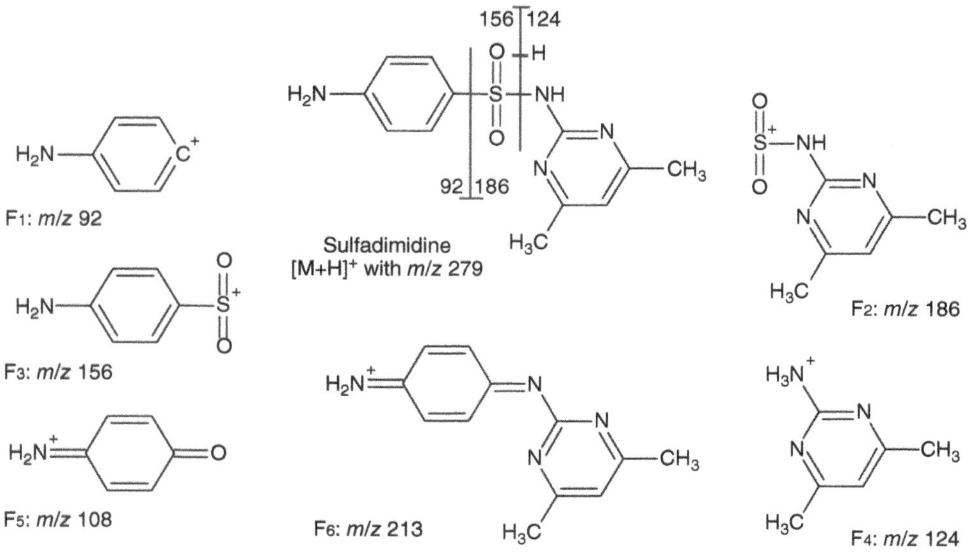

FIGURE 4.8.1 Fragmentation scheme and proposed structures for the fragment ions of protonated sulfadimidine (sulfamethazine).

Another frequently observed fragment is the ion with *m/z* ([M+H]−66) (F_6 in Figure 4.8.1) involving a skeletal rearrangement and the loss of (effectively) H_2SO_2 (Niessen, 1998; Klagkou et al., 2003). What is actually happening in this 66 Da loss has not been further investigated. MS^3 studies have demonstrated that the benzeneamine and the pirimidine (in sulfadimidine) rings are interconnected via an N-atom (Klagkou et al., 2003). The occurrence of these six characteristic positive-ion fragments for the 30 sulfonamide antibiotics studied is summarized in Table 4.8.1.

Whereas this characteristic fragmentation clearly confirms the sulfonamide character of the compound, it does not provide information on the identity of compound-specific substituent group. The latter would be especially important with isomeric sulfonamides, such as sulfamerazine and sulfaperin, or sulfalene, sulfamethoxypyridazine, and sulfamethoxydiazine. Although not demonstrated for these compounds, a pseudo-MS^3 strategy involving in-source generation of the fragment ion with *m/z* ([M+H]−155) followed by MS–MS analysis in a TQ instrument has been proposed for this purpose (Bateman et al., 1997). The resulting fragmentation pattern is not easy to interpret, but establishing a small spectral library would facilitate the identification of the isomeric species.

Early MS^2 data of sulfonamides using an ion-trap instrument show significantly different product-ion mass spectra (Heller et al., 2002). Whereas the major fragment ions for sulfadimidine from a TQ instrument are the ions with *m/z* 213, *m/z* 186, *m/z* 156, *m/z* 124, *m/z* 108, and *m/z* 92 (Figure 4.8.1), the major fragments observed in the ion-trap product-ion mass spectrum are the ions with *m/z* 218, *m/z* 204, *m/z* 186, *m/z* 174, *m/z* 156, and *m/z* 124. The ions with low *m/z*, that is, the ions with *m/z* 92 and 108, may be absent due to the low-*m/z* cut-off limitation of the ion-trap instrument (Section 1.3.3), but the ions with *m/z* 174, *m/z* 204, and *m/z* 218 are at first sight more difficult to explain. It has been demonstrated that these ions are actually due to solvent adducts of fragment ions with water or methanol (CH_3OH), generated in the ion trap as a result of the relatively long ion residence time and the ease of solvation of a sulfonyl ion ($[RSO_2]^+$) to a protonated sulfonic acid ($[RSO_3H+H]^+$) ion (Guan & Liesch, 2001). Thus, the ion with *m/z* 174 is protonated 4-aminobenzenesulfonic acid ($[H_2N—C_6H_4—SO_3H+H]^+$), the ion with *m/z* 204 protonated *N*-(4,6-dimethylpyrimidin-2-yl)sulfamic acid ($[C_6H_7N_2—NHSO_3H+H]^+$), and the ion with *m/z* 218 the protonated methyl sulfimidate of the latter ($[C_6H_7N_2—NHSO_3CH_3+H]^+$).

Phthalylsulfathiazole ($[M+H]^+$ with *m/z* 404) shows two fragment ions involving the phthalyl group, that is, the protonated phthalic anhydride ($[C_8H_4O_3+H]^+$) with *m/z* 149 and the protonated sulfathiazole with *m/z* 256 ($[C_9H_9N_3O_2S_2+H]^+$) due to the loss of phthalyl anhydride. In addition, the common fragment ions of a sulfonamide antibiotic are observed, for example, an ion with *m/z* 304, which corresponds to the class-specific fragment ion with *m/z* 156 (F_3), the ion with *m/z* 156 due to the loss of phthalic anhydride from the fragment ion with *m/z* 304, as well as the ions with *m/z* 108, *m/z* 92, and *m/z* 65 (see earlier).

Negative-ion MS–MS data for a number of sulfonamide antibiotics are available in our library collection. The general fragmentation pattern is similar to that in positive-ion mode. Characteristic fragment ions formed are the 4-aminobenzenesulfonyl anion ($[H_2N—C_6H_4—SO_2]^-$)

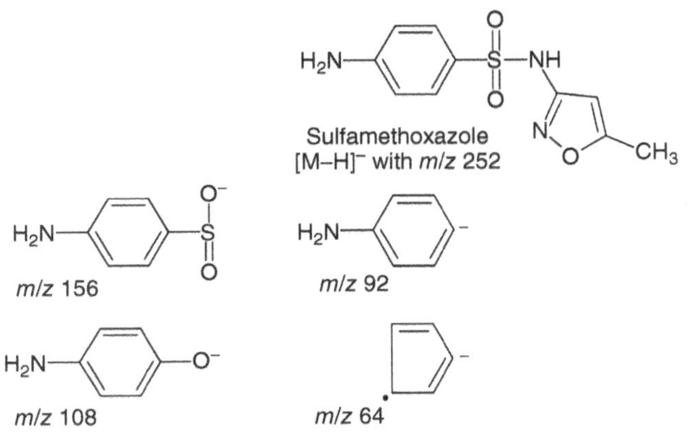

FIGURE 4.8.2 Proposed structures for the fragment ions of deprotonated sulfamethoxazole.

with *m/z* 156, 4-aminophenoxide anion ([H_2N—C_6H_4O]$^-$) with *m/z* 108, the anilinide anion ([C_6H_5NH]$^-$) or the 4-aminophenylide anion ([H_2N—C_6H_4]$^-$) with *m/z* 92, and a 2,4-cyclopentadiene radical anion with *m/z* 64 ($C_5H_4^{-\bullet}$). This type of fragmentation was observed for sulfacetamide ([M−H]$^-$ with *m/z* 213), sulfabenzamide ([M−H]$^-$ with *m/z* 275), sulfamethoxazole ([M−H]$^-$ with *m/z* 252, Figure 4.8.2), and sulfathiazole ([M−H]$^-$ with *m/z* 254). In the product-ion mass spectrum of sulfamethizole ([M−H]$^-$ with *m/z* 269), this characteristic fragmentation is obscured by an intense fragment with *m/z* 196 due to a cleavage in the thiadiazole ring involving the loss of 73 Da (C_2H_3NS). Interestingly, a loss of sulfur dioxide (SO_2) from [M−H]$^-$, as reported for sulfonamides such as sulfamerazine, sulfadiazine, was not readily observed in our data set (Wang et al., 2003; Hu et al., 2008).

Somewhat different fragmentation behavior is observed for phthalylsulfathiazole ([M−H]$^-$ with *m/z* 402) and phthalylsulfacetamide ([M−H]$^-$ with *m/z* 361). They both show losses of carbon dioxide (CO_2) and of phthalic anhydride ($C_8H_4O_3$). Another common fragment is the deprotonated *N*-phenylbenzamide ion ([$C_6H_5CONHC_6H_5$−H]$^-$) with *m/z* 196 due to the loss of both CO_2 and *N*-sulfonylideneacetamide ($CH_3CON{=}SO_2$) (for phthalylsulfacetamide) or *N*-sulfonylidene-*N*-1,3-thiazol-2-yl amine (for phthalylsulfathiazole). In phthalylsulfacetamide, an ion with *m/z* 121 ([$CH_3CON{=}SO_2$−H]$^-$) is observed. In addition, an unexpected skeletal rearrangement appears to occur as both compounds show the loss of sulfur trioxide (SO_3) and a radical anion with *m/z* 80 ($SO_3^{-\bullet}$).

4.8.2 CHLORAMPHENICOL AND RELATED COMPOUNDS

Chloramphenicol is a bacteriostatic antimicrobial. Chloramphenicol is effective against a wide variety of Gram-positive and Gram-negative bacteria, thus as a typical broad-spectrum antibiotic. Due to bacterial resistance and safety concerns, it is no longer indicated as a first-line agent, although problems regarding bacterial resistance to newer antibiotics have renewed interest in its use.

Chloramphenicol ([M−H]$^-$ with *m/z* 321) is generally analyzed in negative-ion mode. In MS–MS, it provides a series of fragment ions, of which the ions with *m/z* 257, *m/z* 194, and *m/z* 152 are the most abundant; these ions are also generally applied in screening for chloramphenicol using SRM (Section 1.5.2) in TQ instruments (Rønning et al., 2006). The ion with *m/z* 257 ([$C_{10}H_{10}ClN_2O_4$]$^-$) involves the loss of hydrogen chloride (HCl) and carbon monoxide (CO), the ion with *m/z* 194 ([$C_9H_8NO_4$]$^-$) involves the loss of 2,2-dichloroacetamide ($H_2NCOCHCl_2$), and ion with *m/z* 152 ([$C_7H_6NO_3$]$^-$) involves the loss of 2,2-dichloro-*N*-(2-hydroxyvinyl)acetamide ($HOCH{=}CHNHCOCHCl_2$) (Figure 4.8.3). In addition, other fragments observed are the ion with *m/z* 249 due to the loss of two times HCl, the ion with *m/z* 219 due to a subsequent loss of formaldehyde ($H_2C{=}O$), with *m/z* 176 due to the loss of water from the ion with *m/z* 194, with *m/z* 151 due to the loss of the $HOCH_2C^{\bullet}HNHCOCHCl_2$ radical, and the deprotonated 4-hydroxybenzaldehyde ([$C_7H_6O_2$−H]$^-$) with *m/z* 121 due to the loss of nitroxyl (HNO) from the ion with *m/z* 152 or the loss of nitrosyl radical ($^{\bullet}NO$) from the ion with *m/z* 151.

The fragmentation of eight chloramphenicol stereoisomers has recently been investigated: chloramphenicol occurs as a *para*- or a *meta*-isomer and contains two chiral centers (Berendsen et al., 2011). The most common form of chloramphenicol is the RR-*p*-configuration. It is demonstrated that the *para*-and the *meta*-isomers show different fragment ions; there are also some differences between RR-*m*- and SS-*m*-configurations and between the RS-*m*- and SR-*m*-configurations. Chiral LC is required to fully discriminate between the stereoisomers (Berendsen et al., 2011).

In azidamphenicol ([M−H]$^-$ with *m/z* 294), the —$CHCl_2$ group in chloramphenicol is replaced by —CH_2—N=N=N (an azide). The compound shows the same fragment ions as chloramphenicol, that is, ions with *m/z* 121, *m/z* 151, *m/z* 152, *m/z* 176, and *m/z* 194.

In thiamphenicol ([M−H]$^-$ with *m/z* 354), the nitro (—NO_2) group in chloramphenicol is replaced by a methylsulfonyl (CH_3SO_2–) group. To some extent, this compound shows similar fragmentation, especially in the formation of the ion with *m/z* 184 (structure analog of the chloramphenicol fragment ion with *m/z* 151). Both [M−H]$^-$ and the fragment ion with *m/z* 184 show a loss of sulfur dioxide (SO_2) due to a skeletal rearrangement. The methylsulfonyl anion ([CH_3SO_2]$^-$) with *m/z* 79 is observed as well.

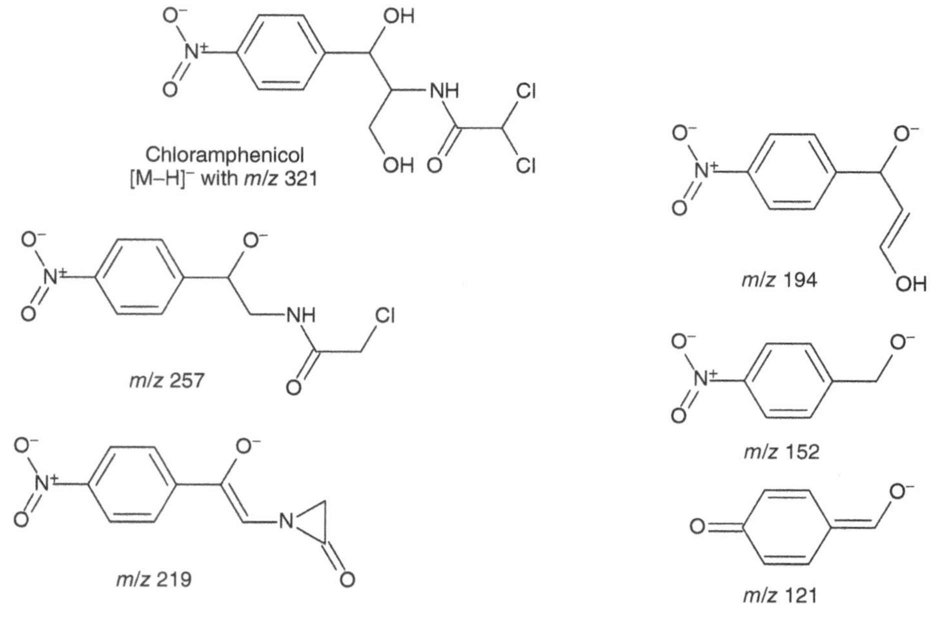

FIGURE 4.8.3 Proposed structures for the fragment ions of deprotonated chloramphenicol.

4.8.3 β-LACTAMS

β-Lactam antibiotics comprise several classes of compounds, among which the penicillins and the cephalosporins are the most important ones. The general structures and examples relevant to the present discussion of these two classes, that is, on 6-aminopenicillanic acid and 7-aminocephalosporanic acid, are given in Tables 4.8.2, and 4.8.4 and 4.8.5, respectively. The β-lactams are widely used for their antimicrobial activity against both Gram-positive and Gram-negative organisms. Because of their use in veterinary medicine for the treatment of bacterial infections, residue analysis in animal food products and milk is an important topic. Positive-ion LC–MS is mainly applied for this purpose (Niessen, 1998; Niessen, 2005). Positive-ion product-ion mass spectra of β-lactam antibiotics have been extensively studied ever since the early days of LC–MS (Suwanrumpha & Freas, 1989; Scandola et al., 1989; Voyksner et al., 1991; Ohki et al., 1992; Heller et al., 2000b; Geis-Asteggiante et al., 2014). Negative-ion product-ion mass spectra of β-lactam antibiotics, especially penicillins, have also been reported (Straub & Voyksner, 1993; Rabbolini et al., 1998).

Many examples of multiresidue methods for β-lactam antibiotics in food of animal origin or in milk have been reported, mainly based on positive-ion ESI-MS–MS in SRM mode (Bruno et al., 2001; Holstege et al., 2002; Fagerquist et al., 2005; Mastovska & Lightfield, 2008; Rezende et al., 2012). In a recently reported multiresidue LC–MS method for the analysis of veterinary drug residues in feed stuff, negative-ion ESI was used for the analysis of some penicillins, whereas other compounds were analyzed in positive-ion mode (Boscher et al., 2010).

4.8.3.1 Penicillins

In positive-ion mode, penicillins show characteristic class-specific fragmentation (Niessen, 1998; Niessen, 2005) (Table 4.8.2). This can be readily illustrated for penicillin G ($[M+H]^+$ with *m/z* 335), which forms two complementary fragments (Section 3.5.2), that is, an acylium ion with *m/z* 176 ($[C_{10}H_{10}NO_2]^+$; F_1 in Figure 4.8.4) and the protonated 5,5-dimethyl-4,5-dihydro-1,3-thiazole-4-carboxylic acid ($[C_6H_9NO_2S+H]^+$; F_2) with *m/z* 160 due to cleavages of the C^5—C^6 and C^7—N^1 bonds of the β-lactam ring (Figure 4.8.4). The ion with *m/z* 160 may be found with other *m/z* values if the carboxylic acid function is esterified, such as in pivampicillin ($[M+H]^+$ with *m/z* 464) and bacampicillin ($[M+H]^+$ with *m/z* 466) (Table 4.8.2). A third fragment ion with *m/z* 114 ($[C_5H_8NS]^+$; F_3) involves secondary loss of formic acid (HCOOH) from the fragment ion with *m/z* 160. Depending on the nature of the substituent group, two compound-specific fragment ions are observed, for example, the fragment ions with *m/z* 176 (F_1) and the tropylium ion ($[C_7H_7]^+$; F_4) with *m/z* 91 for penicillin G (Figure 4.8.4), *m/z* 183 ($[C_9H_{15}N_2O_2]^+$) and the protonated cyclohexylideneamine ion ($[C_6H_{11}N+H]^+$) with *m/z* 98 for ciclacillin ($[M+H]^+$ with *m/z* 342), *m/z* 193 ($[C_{10}H_{13}N_2O_2]^+$) and the protonated (cyclohexa-1,4-dien-1-yl)methanimine ($[C_6H_7N{=}CH_2+H]^+$) with *m/z* 108 for epicillin ($[M+H]^+$ with *m/z* 352), and *m/z* 206 ($[C_{11}H_{12}NO_3]^+$) and the

TABLE 4.8.2 Class-specific fragment ions of protonated penicillin β-lactam antibiotics.

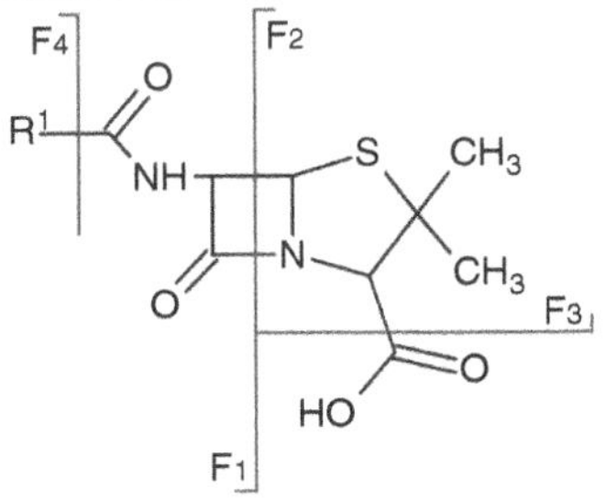

Compound	R^1	$[M+H]^+$	F_1	F_2	F_3	F_4	Other Fragments
Amoxicillin	$HOC_6H_4CHNH_2$—	366		160	114		208: $[C_{10}H_{10}NO_2S]^+$
Ampicillin	$C_6H_5CHNH_2$—	350		160	114	106	192: $[C_{10}H_{10}NOS]^+$
Azidocillin	$C_7H_6N_3$—	376		160	114		
Azlocillin	$C_{11}H_{12}N_3O_2$—	462	303	160	114	218	246: $[F_4\text{—}C{\equiv}O]^+$
							130: $[C_4H_7N_3O_2+H]^+$
Bacampicillin (ester)	$C_6H_5CHNH_2$—	466		276	114	106	232: $F_2 - CH_3CHO$
							160: $F_2 - C_5H_8O_3$
Carindacillin	$C_{17}H_{15}O_2$—	495	336	160	114		174: $[C_{10}H_8NO_2]^+$
Ciclacillin	$C_6H_{11}NH$—	342	183	160	114	98	166: $F_1 - NH_3$
							81: $F_4 - NH_3$
Dicloxacillin	$C_{10}H_6Cl_2NO$—	470	311	160	114		254: $[F_4\text{—}C{\equiv}O]^+$
Epicillin	$C_7H_{10}N$—	352	193	160	114	108	189: $[C_7H_{12}N_2O_2S+H]^+$
							176: $F_1 - NH_3$
							136: $[F_4\text{—}C{\equiv}O]^+$
							91: $F_4 - NH_3$
Flucloxacillin	$C_{10}H_6ClFNO$—	454	295	160	114		238: $[F_4\text{—}C{\equiv}O]^+$
Mezlocillin	$C_{12}H_{14}N_3O_4S$—	540	381	160	114	296	324: $[F_4\text{—}C{\equiv}O]^+$
							208: $[C_5H_9N_3O_4S+H]^+$
Penicillin G	$C_6H_5CH_2$—	335	176	160	114	91	189: $[C_7H_{12}N_2O_2S+H]^+$
							142: $F_2 - H_2O$
Pheneticillin	$C_6H_5OCHCH_3$—	365	206	160	114	121	189: $[C_7H_{12}N_2O_2S+H]^+$
Pivampicillin (ester)	$C_6H_5CHNH_2$—	464		274	114	106	244: $F_2 - H_2C{=}O$
							160: $F_2 - C_6H_{10}O_2$
							85: $[C_4H_9C{\equiv}O]^+$
							57: $[C_4H_9]^+$
Propicillin	$C_9H_{11}O$—	379	220	160	114	135	126: $F_1 - C_6H_5OH$
Temocillin	$C_6H_5O_2S$—	415	256	160	114		383: $[M+H]^+ - CH_3OH$
							365: 383 − H_2O
							339: 383 − CO_2

1-phenoxyeth-1-yl carbocation ($[C_8H_9O]^+$) with *m/z* 121 for pheneticillin ($[M+H]^+$ with *m/z* 365) (Table 4.8.2). In addition, other compound-specific fragments may be observed, for example, an ion with *m/z* 208 ($[C_{10}H_{10}NO_2S]^+$) for amoxicillin ($[M+H]^+$ with *m/z* 366), which is similar to frequently observed fragment ions for cephalosporins (cf., the ions with *m/z* 192 and 208 for cefalexin and cefadroxil, respectively; F_4 in Figure 4.8.6). For penicillin G, epicillin, and pheneticillin, a fragment ion with *m/z* 189 ($[C_7H_{12}N_2O_2S+H]^+$; F_5 in Figure 4.8.4) is observed as well.

Interpretation of negative-ion product-ion mass spectra for penicillin G and amoxicillin has been reported, proposing that the main fragment is due to cleavages of the C^5—C^6 and C^7—N^1 bonds of the β-lactam ring (Straub & Voyksner, 1993), similar to what is observed in positive-ion mode (Niessen, 1998; Niessen, 2005). An alternative interpretation has been suggested based on ion-trap MS^n data (Rabbolini et al., 1998). The latter interpretation is actually confirmed to be correct by accurate-mass data from the Q–TOF product-ion mass spectra available in our library collection.

In ion-trap MS^n, a stepwise fragmentation of dicloxacillin ($[M-H]^-$ with *m/z* 468) is achieved, involving the loss of carbon dioxide (CO_2) to an ion with *m/z* 424 in MS^2, the loss of *N*-(2,2-dimethylvinyl)isocyanate (97 Da;

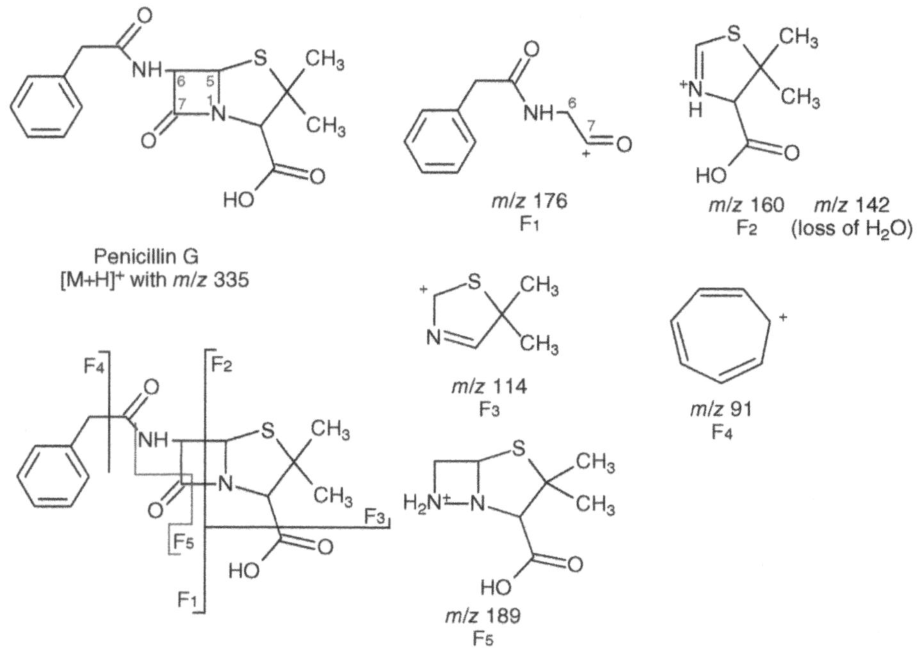

FIGURE 4.8.4 Fragmentation scheme and proposed structures for the fragment ions of protonated penicillin G.

$(CH_3)_2C{=}CH{-}N{=}C{=}O$, C_5H_7NO) to *m/z* 327 in MS^3, the loss of hydrogen chloride (HCl) to *m/z* 291 in MS^4, and finally the loss of an acetyl radical ($CH_3C{\equiv}O^\bullet$) to *m/z* 248 in MS^5 (Rabbolini et al., 1998). The loss of 97 Da (C_5H_7NO) involves cleavages of the $C^6{-}C^7$ and $C^5{-}N^1$ bonds of the β-lactam ring, as well as the $C^3{-}S^4$ bond of the thiazolidine ring, resulting in a sulfide anion (Figure 4.8.5). This sequence of events is almost completely confirmed by the Q–TOF accurate-mass data for dicloxacillin, available in our library collection, except for the last step which should be attributed to the loss of hydrogen isocyanate ($HN{=}C{=}O$; 43.0058 Da, Δ −2.2 mDa) rather than the loss of the $CH_3C{\equiv}O^\bullet$ (43.0184 Da, Δ +10.4 mDa). Based on this, a ring closure to a 2-azepinone-like structure may be suggested for the fragment ion with *m/z* 291 (Figure 4.8.5). A similar loss is observed for related structures oxacillin ($[M-H]^-$ with *m/z* 400), cloxacillin ($[M-H]^-$ with *m/z* 434), and flucloxacillin ($[M-H]^-$ with *m/z* 452).

Careful inspection of the negative-ion MS–MS data for 14 penicillin β-lactam antibiotics reveals, for most compounds, a class-specific fragmentation involving the loss of CO_2 and the combined loss of CO_2 and the group with 97 Da (C_5H_7NO) (Table 4.8.3). In addition, some other fragment ions observed are indicated in the table. If a halogen is present in the structure, the loss of $C_6H_7NO_3$ is followed by the loss of hydrogen chloride (HCl) as in cloxacillin and dicloxacillin, or the loss of hydrogen fluoride (HF) as in flucloxacillin. Otherwise, either the loss of hydrogen sulfide (H_2S), such as for ampicillin ($[M-H]^-$ with *m/z* 348) and ciclacillin ($[M-H]^-$ with *m/z* 340), or fragmentation of the aminobenzamide moiety is observed, such as the loss of NH_3 for amoxicillin ($[M-H]^-$ with *m/z* 364) (Table 4.8.3). This can be considered to be compound-specific fragmentation.

If the substituent chain is more complex, such as for apalcillin, azidocillin, carindacillin, mezlocillin, and piperacillin, the fragmentation pattern can look more complex as well. The mixing of compound-specific (or substituent-group-specific) and class-specific fragmentation can be nicely illustrated with the negative-ion product-ion mass spectrum of phenoxymethylpenicillin ($[M-H]^-$ with *m/z* 349). After the initial loss of carbon dioxide (CO_2) to an ion with *m/z* 305, it shows both the class-specific loss of 97 Da (C_5H_7NO) to an ion with *m/z* 208 and a substituent-group-specific loss of phenol (C_6H_5OH) to an ion with *m/z* 211. The latter ion shows a subsequent class-specific loss of 97 Da (C_5H_7NO) to an ion with *m/z* 114. The class-specific fragmentation route outlined may be obscured by substituent-group losses, as is for instance the case for carindacillin ($[M-H]^-$ with *m/z* 493). The major fragment is an ion with *m/z* 218, which is due to the loss of CO_2, 2,3-dihydro-1*H*-inden-5-ol ($C_9H_{10}O$) from the β-ketopropanoate moiety, and the loss of 97 Da (C_5H_7NO). Similarly, piperacillin ($[M-H]^-$ with *m/z* 516) shows an intense fragment ion with *m/z* 330 due to the loss of CO_2 and the N^1-ethyl substituted piperazine-2,3-dione ($C_6H_{10}N_2O_2$) of the benzamide substituent. Subsequent loss of 97 Da (C_5H_7NO) leads to the base peak of the spectrum, the

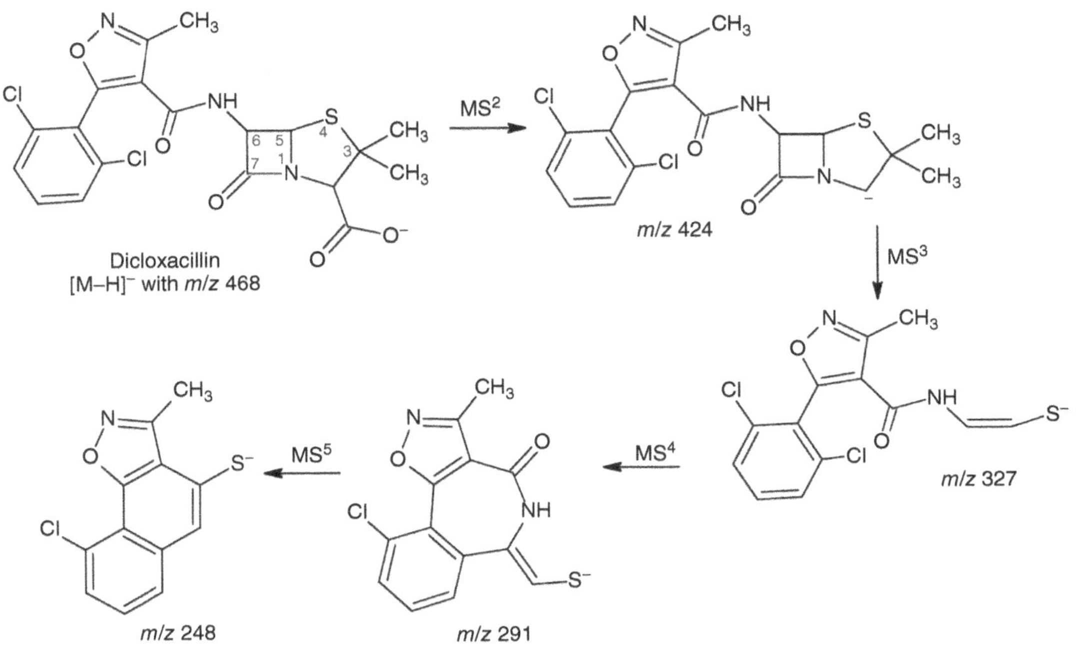

FIGURE 4.8.5 Proposed structures for the fragment ions of deprotonated dicloxacillin.

ion with *m/z* 233. Similar behavior is observed for mezlocillin ($[M-H]^-$ with *m/z* 538) (Table 4.8.3). For apalcillin ($[M-H]^-$ with *m/z* 520), the most intense fragment ion with *m/z* 188 ($[C_8H_4N_2OCONH_2]^-$) is due to the loss of the benzylpenicillin group of the molecule with charge retention on the 4-hydroxy-1,5-naphthyridin-3-yl formamide group of R^1. The subsequent loss of HN=C=O results in the fragment ion with *m/z* 145 ($C_8H_4N_2OH$). Formation of the fragment ion with *m/z* 361 is not understood yet.

The related compound sulbactam ($[M-H]^-$ with *m/z* 232) shows a different fragmentation, that is, an ion with *m/z* 188 due to the loss of CO_2, an ion with *m/z* 140 consistent with the loss of CO and the sulfur dioxide (SO_2), an ion with *m/z* 96 consistent with the loss of CO_2 from the ion with *m/z* 140, and a sulfur dioxide radical anion ($[SO_2]^{-\bullet}$) with *m/z* 64.

4.8.3.2 Cephalosporins

Penicillins contain a 5,5-dimethyl-1,3-thiazolidine-4-carboxylic acid group (Section 4.8.3.1), whereas the cephalosporins contain a 5-methyl-3,6-dihydro-2*H*-1,3-thiazine-4-carboxylic acid group. Some relevant structures of cephalosporins as well as positive fragmentation data are given in Tables 4.8.4 and 4.8.5. Regarding the fragmentation of cephalosporins in positive-ion MS–MS, three subclasses can be discriminated: (1) compounds with a 2-amino-2-phenylacetyl substituent at the amide C^7 of the β-lactam ring (cefalexin as an example; Table 4.8.4 and Figure 4.8.6), (2) compounds with a 2-(2-amino-1,3-thiazol-4-yl)-2-(methoxyimino)acetyl substituent chain at the amide C^7 of the β-lactam ring and a substituent (ester, (thio)ether, or quaternary ammonium) at the C^3 methylene (CH_2) group of the β-lactam ring (cefotaxime as an example; Table 4.8.5 and Figure 4.8.7), and (3) compounds with other structural features, such as ceftibuten, cefapirin, cefotiam, and cefotetan, which are not discussed here any further.

The characteristic fragmentation of Subclass 1 cephalosporins is outlined for cefalexin ($[M+H]^+$ with *m/z* 348) in Figure 4.8.6. Similar to the penicillin G and other penicillins, two complementary fragments (Section 3.5.2) occur due to cleavages of the C^6—C^7 and C^8—N^1 bonds of the β-lactam ring. With cefalexin, these ions are the acylium ion with *m/z* 191 ($[C_{10}H_{11}N_2O_2]^+$; F_1 in Table 4.8.4), containing the 2-amino-2-phenylacetyl substituent chain, and the protonated 5-methyl-6*H*-1,3-thiazine-4-carboxylic acid ($[C_6H_7NO_2S+H]^+$) with *m/z* 158 (F_2). Both fragments may be observed with different *m/z* values for other members of this subclass as a result of differences in substitution (Table 4.8.4), for example, for cefadrine ($[M+H]^+$ with *m/z* 350) with *m/z* 193 ($[C_{10}H_{13}N_2O_2]^+$; F_1) and *m/z* 158 (F_2) and for cefaclor ($[M+H]^+$ with *m/z* 368) with *m/z* 191 ($[C_{10}H_{11}N_2O_2]^+$; F_1) and *m/z* 178 ($[C_5H_4ClNO_2S+H]^+$; F_2). For cefalexin, secondary fragmentation of the ion with *m/z* 191 (F_1) leads to an ion with *m/z* 174 due to the loss of ammonia (NH_3). Secondary fragmentation of the ion with *m/z* 158 (F_2) leads to fragment ions with *m/z* 140 due to the loss of water and *m/z* 114 ($[C_5H_7NS+H]^+$; F_3) due to the loss of carbon dioxide (CO_2). Two other characteristic fragments are observed. One of these fragments results from the loss of ammonia and cleavages of the C^7—C^8 and C^6—N^1

TABLE 4.8.3 Structures and fragmentation of deprotonated penicillin β-lactam antibiotics.

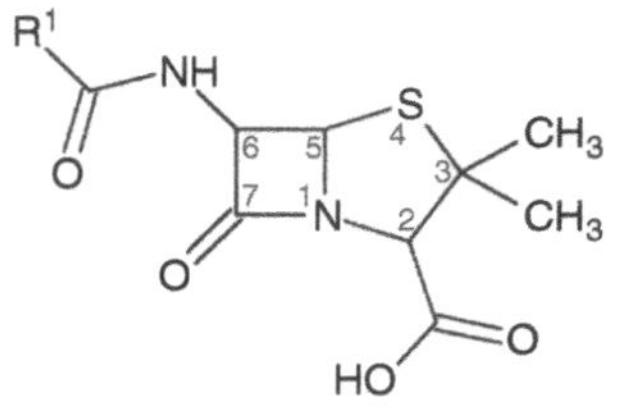

Compound	[M−H]⁻	R¹	−CO_2	−CO_2−97	Other Fragments
Amoxicillin	364	$HOC_6H_4CHNH_2$—	320	223	206: 223 − NH_3
Ampicillin	348	$C_6H_5CHNH_2$—	304	207	173: 207 − H_2S
Apalcillin	520	$C_{16}H_{12}N_3O_2$—	476	379	361: *not*: 379 − H_2O 188: $[C_9H_6N_3O_2]^-$ 145: $[C_8H_5N_2O]^-$
Carindacillin	493	$C_{17}H_{15}O_2$—			315: $[M-H]^-$ − CO_2 − $C_9H_{10}O$ 218: 315 − 97
Ciclacillin	340	$C_6H_{11}NH$—	296	199	165: 199 − H_2S
Cloxacillin	434	ClC_6H_4—C_4H_3NO—	390	293	257: 293 − HCl 214: 257 − HNCO
Dicloxacillin	468	$Cl_2C_6H_3$—C_4H_3NO—	424	327	388: 424 − HCl 291: 327 − HCl 248: 291 − HNCO
Flucloxacillin	452	$ClFC_6H_3$—C_4H_3NO—	408	311	388: 408 − HF 291: 311 − HF (major) 275: 311 − HCl (minor) 248: 291 − HNCO
Mezlocillin	538	$C_{12}H_{14}N_3O_4S$—	494	397	330: 494 − $C_4H_8N_2O_3S$ 287: 330 − HNCO 233: 330 − 97
Oxacillin	400	C_6H_5—C_4H_3NO—	356	259	216: 259 − HNCO
Penicillin G	333	$C_6H_5CH_2$—	289	192	
Pheneticillin	363	$C_6H_5OCHCH_3$—	319	222	128: 222 − C_6H_5OH 93: $[C_6H_5O]^-$
Phenoxymethylpenicillin	349	$C_6H_5OCH_2$—	305	208	211: 305 − C_6H_5OH 114: 211 − 97 93: $[C_6H_5O]^-$
Piperacillin	516	$C_{14}H_{16}N_3O_3$—	472	375	330: 472 − $C_6H_{10}N_2O_2$ 233: 330 − 97

bonds of the β-lactam ring, as well as the C^4—S^5 bond of the thiazinane ring, resulting in a fragment ion with *m/z* 192 ($[C_{10}H_{10}NOS]^+$; F_4) for cefalexin. The other fragment involves the cleavage of the C—C bond of the acetyl group of the β-lactam C^7 amide, with charge retention on the substituent chain resulting in the benzylamine cation with *m/z* 106 ($[C_6H_5CHNH_2]^+$; F_5) for cefalexin (Figure 4.8.6). Data for other compounds in this subclass are given in Table 4.8.4.

The characteristic fragmentation of Subclass 2 cephalosporins is outlined for cefotaxime ($[M+H]^+$ with *m/z* 456) and shown in Figure 4.8.7. Interestingly, most compounds in this subclass show fragment ions with identical *m/z* (Table 4.8.5). Three major fragmentation routes are outlined. (I) Similar to penicillins and Subclass 1 cephalosporins, cleavages of the C^6—C^7 and C^8—N^1 bonds of the β-lactam ring occur, although in this case only one of the two fragments is observed, often with low relative abundance, that is, the acylium ion with *m/z* 241 for most members of this group (Table 4.8.5). (II) A series of abundant ions is observed, initially involving the loss of the substituent to the C^3 methylene (CH_2) group of the thiazinane ring, leading to a carbocation with *m/z* 396 ($[C_{14}H_{14}N_5O_5S_2]^+$) (Table 4.8.5). Secondary fragmentation of this ion involves the loss of carbon monoxide (CO) from the oxime group, the loss of CO_2 from the thiazinane ring,

TABLE 4.8.4 Structures and class-specific fragment ions of cephalosporin β-lactam antibiotics with a 2-amino-2-phenylacetyl substituent chain at the amide C^7 of the β-lactam ring.

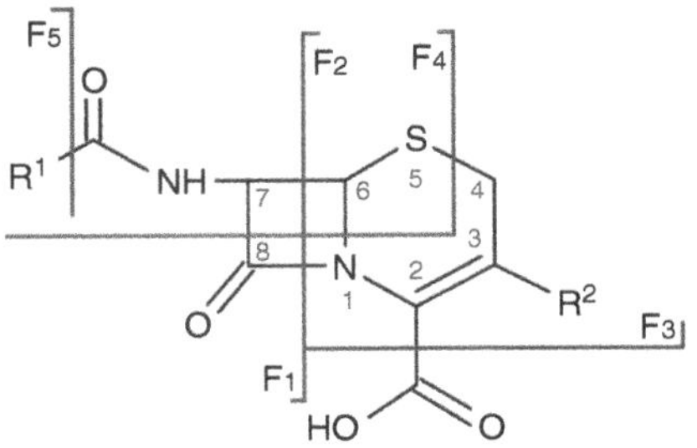

Compound	[M+H]$^+$	[M−H]$^-$	R^1	R^2
Cefalexin	348	346	$C_6H_5CHNH_2$—	—CH_3
Cefadrine	350		$C_6H_7CHNH_2$—	—CH_3
Cefadroxil	364	362	$HOC_6H_4CHNH_2$—	—CH_3
Cefachlor	368		$C_6H_5CHNH_2$—	—Cl
Cefetamet	398		$C_5H_6N_3OS$—	—CH_3
Cefoxitin*		426	$C_4H_3SCH_2$—	—CH_2OOCNH_2

Compound	[M+H]$^+$	F_1	F_2	F_3	F_4	F_5	Other Fragments
Cefalexin	348	191	158	114	192	106	174: $F_1 - NH_3$
Cefadrine	350	193	158	114	194	108	176: $F_1 - NH_3$
Cefadroxil	364	207	158	114	208	122	190: $F_1 - NH_3$
Cefachlor	368	191	178	134	192	106	332: [M+H]$^+$ − HCl 174: $F_1 - NH_3$
Cefetamet	398	241	158	114		156	351: [M+H]$^+$ − $H_2C{=}O$ − NH_2 285: $[C_9H_9N_4O_3S_2]^+$† 210: F_2 − $^\bullet OCH_3$ 126: $F_5 - H_2C{=}O$

*CH_3O— substituent at C^7 of β-lactam ring. Analyzed in negative-ion only.
†The structure of this ion is identical to the ion with *m/z* 285 observed for cefotaxime (Figure 4.8.7).

and the loss of both CO and CO_2. Thus, apart from the ion with *m/z* 396, three related ions are observed, that is, with *m/z* 368, *m/z* 352, and *m/z* 324. (III) Cleavage of the C—C bond between the oxime and the carbonyl group of the C^7 amide of the β-lactam ring results in a fragment ion with *m/z* 156 ($[C_5H_6N_3OS]^+$), which shows the subsequent loss of formaldehyde ($H_2C{=}O$) to *m/z* 126 (Table 4.8.5). Another set of class-specific ions may be considered as secondary fragments of the above-mentioned ion with *m/z* 241, which include the ion with *m/z* 211 due to the loss of $H_2C{=}O$ from the methyloxime group, and *m/z* 210 due to the loss of a methoxy radical ($CH_3O^\bullet$) from the same moiety. Finally, three other fragment ions are frequently observed, that is, the ions with *m/z* 285 ($[C_9H_9N_4O_3S_2]^+$), *m/z* 277 ($[C_{11}H_9N_4O_3S]^+$), and *m/z* 167 ($[C_7H_7N_2OS]^+$) (Figure 4.8.7). The identity of the ions with *m/z* 285 and 167 could be readily concluded from the elemental composition derived from the accurate mass. The structure proposed for the ion with *m/z* 277 is the most plausible one that could be formulated for all the compounds studied. In most spectra of cephalosporins, several other ions are observed with minor abundance; no attempt was made to explain those fragments. Also note that cefepime, cefpirome, and cefquinome contain a quaternary ammonium group in the substituent chain at the C^3-position of the thiazinane ring. Thus, they are observed as M^+ rather than $[M+H]^+$. A fragment ion related to this quaternary ammonium group actually is the base peak in the product-ion mass spectra of these compounds, that is, protonated 1-methylpyrrolidine ($[C_5H_{11}N+H]^+$) with *m/z* 86 for cefepime, protonated 2-pyrindene ($[C_8H_9N+H]^+$) with *m/z* 120 for cefpirome, and protonated 5,6,7,8-tetrahydroquinoline ($[C_9H_{11}N+H]^+$) with *m/z* 134 for cefquinome.

Negative-ion product-ion mass spectra are only available in our library collection for a limited number of cephalosporins. The Subclass 1 cephalosporin (Table 4.8.4)

TABLE 4.8.5 Structures and class-specific positive-ion fragment ions of cephalosporin β-lactam antibiotics with a 2-(2-amino-1,3-thiazol-4-yl)-2-(methoxyimino)acetyl substituent chain at the C^7amide of the β-lactam ring and a substituent (ester, (thio)ether, or quaternary ammonium (4° N)) on the methylene (CH_2) group at C^3 of the β-lactam ring.

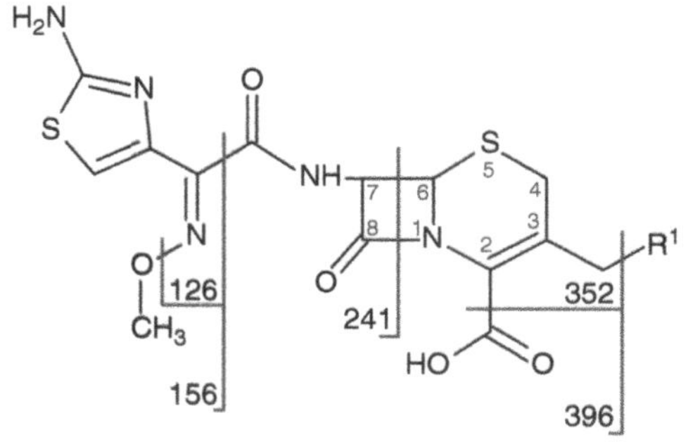

Compound	$[M+H]^+$	M^+	$[M-H]^-$	R^1	Other Fragments
Cefotaxime	456		454	$—OOCCH_3$	
Cefepime		481		$—C_5H_{11}N$ (4° *N*)	86: $[C_5H_{11}N+H]^+$
Cefmenoxime	512			$—S—C_2H_3N_4$	
Cefpirome		515		$—C_8H_9N$ (4° *N*)	120: $[C_8H_9N+H]^+$
Cefquinome		529		$—C_9H_{11}N$ (4° *N*)	134: $[C_9H_{11}N+H]^+$
Ceftriaxone	555		553	$—S—C_4H_4N_3O_2$	
Cefodizime	585			$—S—C_6H_6NO_2S$	
Ceftiofur	524			$—S—C_5H_3O_2$	
Ceftazidime*		547		$—C_5H_5N$ (4° *N*)	80: $[C_5H_5N +H]^+$
Cefpodoxime proxetil†	558			$—O—CH_3$	Not observed: *m/z* 296, 368, 352, 324 due to ester group. 410: $[M+H]^+ - C_6H_{12}O_4$ 382: 410 − CO

*2-Carboxypropan-2-yl ($C_4H_7O_2$) instead of methyl (CH_3) on oxime.
†1-({[(propan-2-yl)oxy]carbonyl}oxy)ethyl ($CH_3CHOCOOCH(CH_3)_2$; $C_6H_{11}O_3$) ester substituent at the C^2 carboxylic acid functional group.

cefalexin ($[M-H]^-$ with *m/z* 346) shows a series of sequential small-molecule neutral losses, starting with the loss of hydrogen sulfide (H_2S) to an ion with *m/z* 312, followed by the loss of CO_2 to an ion with *m/z* 268. In addition, the loss of 113 Da ($C_5H_7NO_2$) to an ion with *m/z* 233 is observed, involving cleavages of $S^5—C^4$, $N^1—C^6$, and $N^1—C^8$ bonds, followed by a loss of carbon monosulfide (CS) to an ion with *m/z* 189 ($[C_{10}H_9N_2O_2]^-$). The deprotonated 2-imino-3-methylbut-3-enoic acid with *m/z* 112 ($[C_5H_7NO_2-H]^-$) is also observed (Figure 4.8.8). The ions with *m/z* 233 and 112 result from a direct cleavage: the sum of their *m/z* values equals *m/z* ($[M-H]-1$) (Section 3.7.2). Note that this fragmentation pattern is significantly different from that of the penicillins. Apart from the class-specific fragment ions, just mentioned, cefadroxil ($[M-H]^-$ with *m/z* 362) shows two additional substituent-group-related fragment ions, that is, the phenoxide ion ($[C_6H_5O]^-$) with *m/z* 93 and 4-(iminomethyl)phenoxide ion with *m/z* 120 ($HN{=}CH—C_6H_4O]^-$).

Subclass 2 cephalosporin (Table 4.8.5) ceftriaxone ($[M-H]^-$ with *m/z* 553) shows the loss of CO_2 followed by the loss of H_2S, that is, losses in a reversed order compared to Subclass 1 cephalosporins. A fragment ion involving cleavages of the $S^5—C^4$, $N^1—C^6$, and $N^1—C^8$ bonds is observed with *m/z* 283 ($[C_9H_7N_4O_3S_2]^-$). Subsequent loss of CS is observed as well. Thus, in this respect, the fragmentation of Subclass 1 and Subclass 2 cephalosporins in negative-ion mode is more alike than in the positive-ion mode. However, the most intense fragment ion of ceftriaxone is due to the cleavage of the thioether C—S bond to the C^3 substituent chain of the β-lactam ring, that is, the ion with *m/z* 158 ($[C_4H_4N_3O_2S]^-$).

The presence of a methoxy ($CH_3O—$) group as a substituent at C^7 of the β-lactam ring in cefoxitin ($[M-H]^-$ with *m/z* 426; Table 4.8.4) apparently blocks the fragmentation of the thiazinane ring described earlier. Loss of carbamic acid (H_2NCOOH) from the C^3 substituent chain and cleavages of the $C^6—C^7$ and $C^8—N^1$ bonds of the β-lactam ring, that is, similar to what is observed in positive-ion mode (Table 4.8.4), results in deprotonated 5-methyl-2*H*-1,3-thiazine-4-carboxylic acid ($C_6H_7NO_2S-H]^-$) with *m/z* 156 (Partani et al., 2010), which upon CO_2 loss yields a fragment ion with *m/z* 112 ($[C_5H_6NS]^-$ rather than $[C_5H_7NO_2-H]^-$ as observed for

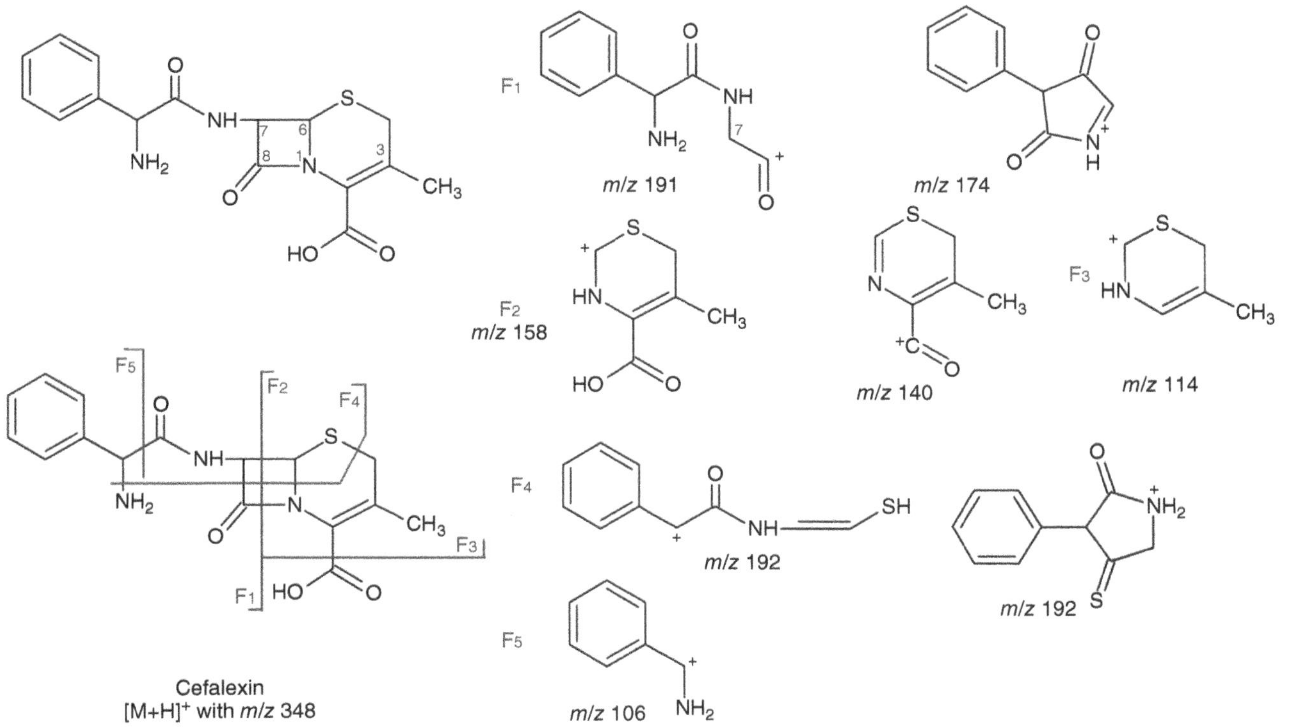

FIGURE 4.8.6 Fragmentation scheme and proposed structures for the fragment ions of protonated cefalexin.

cefalexin and cefadroxil). Subsequent loss of a methyl radical ($^{\bullet}CH_3$) results in the radical anion with *m/z* 97 ($[C_4H_3NS]^{-\bullet}$) (Figure 4.8.9). Directly from $[M-H]^-$, losses of CO_2, hydrogen isocyanate (HN=C=O), and methanol (CH_3OH) are also observed. With a [(2-furyl)-2-(methoxyimino)]acetyl amide and a (carbamoyloxy)methyl ($H_2NCOOCH_2$–) substituent at the C^3, cefuroxime ($[M-H]^-$ with *m/z* 423) resembles a Subclass 2 cephalosporin. It shows fragmentation ions due to losses of CO_2 and HN=C=O. In addition, the cleavages of the C^6—C^7 and C^8—N^1 bonds of the β-lactam ring lead, in this case, to a fragment ion with *m/z* 207 ($[C_9H_7N_2O_4]^-$), with charge retention on the substituent containing the 2-(2-furyl)-2-(methoxyimino)acetyl substituent rather than on the thiazinane ring (Partani et al., 2010). Unfortunately, the number of available reference spectra is too small to draw more elaborate conclusions.

4.8.4 (FLUORO)QUINOLONES

The quinolones are synthetic broad-spectrum antibiotics in the treatment against bacterial infections. Nalidixic acid was the first quinolone introduced in 1962 for the treatment of urinary tract infections in humans. Quinolones bind to and inhibit the bacterial DNA gyrase enzyme, thus preventing the bacterial DNA from unwinding and replicating. Quinolone-resistant bacteria frequently harbor mutated topoisomerases that resist quinolone binding.

The general structure of quinolones is based on 4-oxo-quinoline-3-carboxylic acid, with substituents at C^6 and C^7 (Figure 4.8.10). The substituent at C^7 is often piperazine or *N*-methyl-piperazine. The majority of the clinically used quinolones are fluoroquinolones, which have a fluorine atom at C^6. Despite the fact that quinolones are acidic compounds, they are exclusively analyzed in positive-ion ESI-MS. Multiresidue methods based on positive-ion LC–MS in SIM mode or LC–MS–MS in SRM mode have been reported frequently (Bogialli et al., 2007b; Mottier et al., 2008; Hermo et al., 2006; Li et al., 2009b; Biselli et al., 2013).

The fragmentation of first-generation quinolones can be described as consecutive neutral losses (Figure 4.8.11). As an example, nalidixic acid ($[M+H]^+$ with *m/z* 233) shows a water loss to an ion with *m/z* 215, followed by the loss of ethene (C_2H_4) to an ion with *m/z* 187, followed by two carbon monoxide (CO) losses leading to fragment ions with *m/z* 159 and 131, and finally the loss of hydrogen cyanide (HCN) to an ion with *m/z* 104 ($[C_7H_6N]^+$) (Volmer et al., 1997). Similar fragmentation behavior is observed for oxolinic acid ($[M+H]^+$ with *m/z* 262), piromedic acid ($[M+H]^+$ with *m/z* 289), and pipemidic acid ($[M+H]^+$ with *m/z* 304) (Figure 4.8.11). In fact, it is interesting to note the sequence

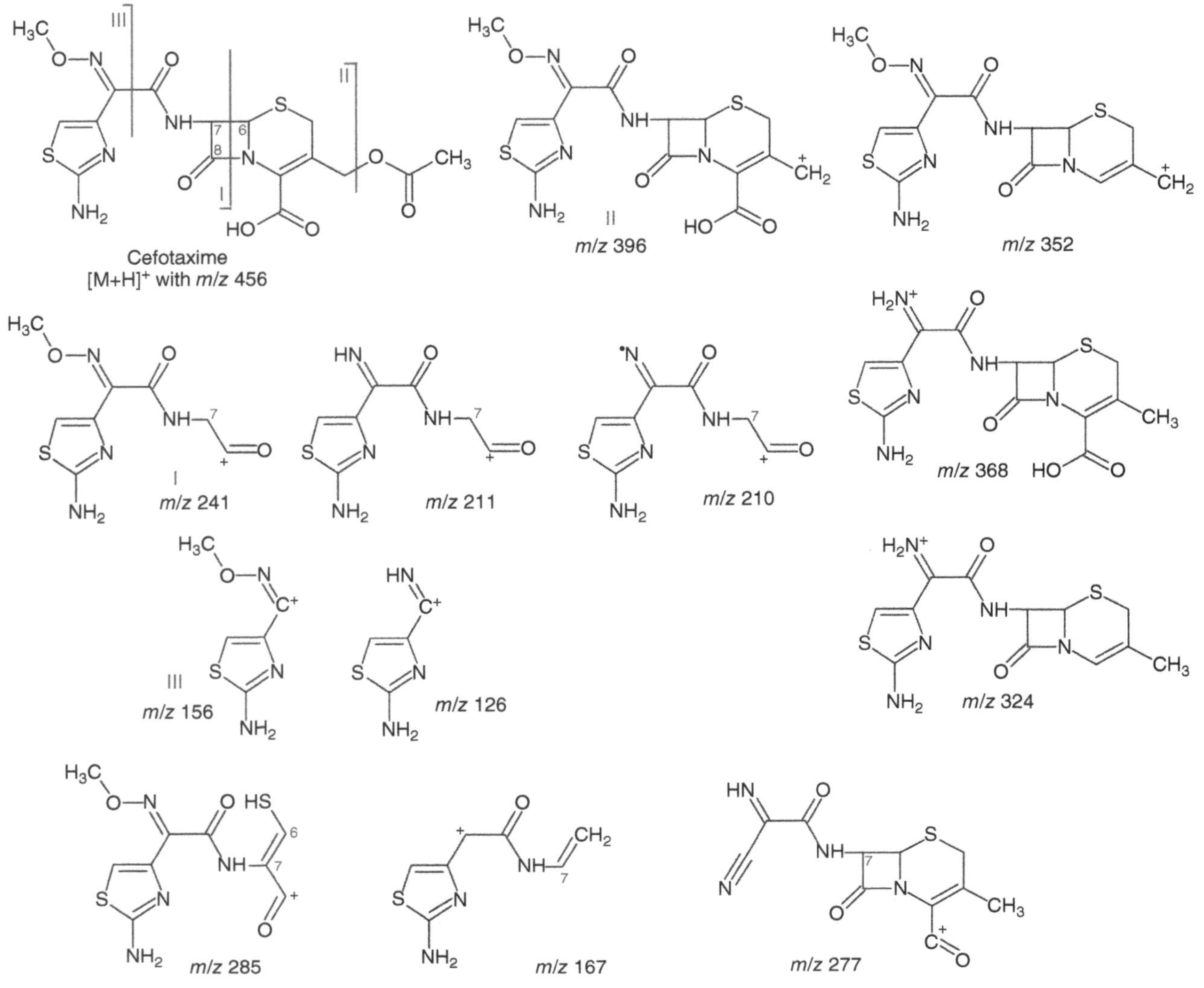

FIGURE 4.8.7 Fragmentation scheme and proposed structures for the fragment ions of protonated cefotaxime.

of water and C_2H_4 losses (and no CO loss); this sequence is confirmed by accurate-mass data. Flumequine ($[M+H]^+$ with *m/z* 262) shows subsequent losses of water, propene ($CH_3CH{=}CH_2$), two times CO, and hydrogen fluoride (HF) (Volmer et al., 1997).

Initial fragmentation of fluoroquinolones involves competition between the loss of water and the loss of carbon dioxide (CO_2). Despite the fact that many of the fluoroquinolones have an ethyl substituent at the N^1 position (similar to most first-generation quinolones discussed earlier), the loss of this group seems more difficult in the fluoroquinolones and loss of CO_2 from the C^3 position is observed instead (Volmer et al., 1997; Bogialli et al., 2007b). Subsequently, fragmentation in the piperazine ring occurs, involving a loss of ethanimine ($CH_3CH{=}NH$) for fluoroquinolones without an *N*-substituted piperazine ring as in ciprofloxacin, grepafloxacin, lomefloxacin, norfloxacin, sarafloxacin; or a loss of *N*-methylethanimine ($CH_3CH{=}NCH_3$) for compounds with an *N*-methyl-substituted piperazine ring as in difloxacin, fleroxacin, levofloxacin, ofloxacin, and pefloxacin; or 3,5-dimethyl-substituted piperazine rings as in orbifloxacin and sparfloxacin; or a loss of *N*-ethylethanimine ($CH_3CH{=}NCH_2CH_3$) for enrofloxacin with an *N*-ethyl-substituted piperazine ring.

Detailed fragmentation schemes have been reported for a number of compounds, for example, ciprofloxacin ($[M+H]^+$ with *m/z* 332) (Volmer et al., 1997), sparfloxacin ($[M+H]^+$ with *m/z* 393) (Engler et al., 1998), norfloxacin ($[M+H]^+$ with *m/z* 320) (Wang et al., 2010), prulifloxacin ($[M+H]^+$ with *m/z* 462) (Raju et al., 2011), danofloxacin ($[M+H]^+$ with *m/z* 358) (Liu et al., 2011), difloxacin ($[M+H]^+$ with *m/z* 400), and enrofloxacin ($[M+H]^+$ with *m/z* 360) (Junza et al.,

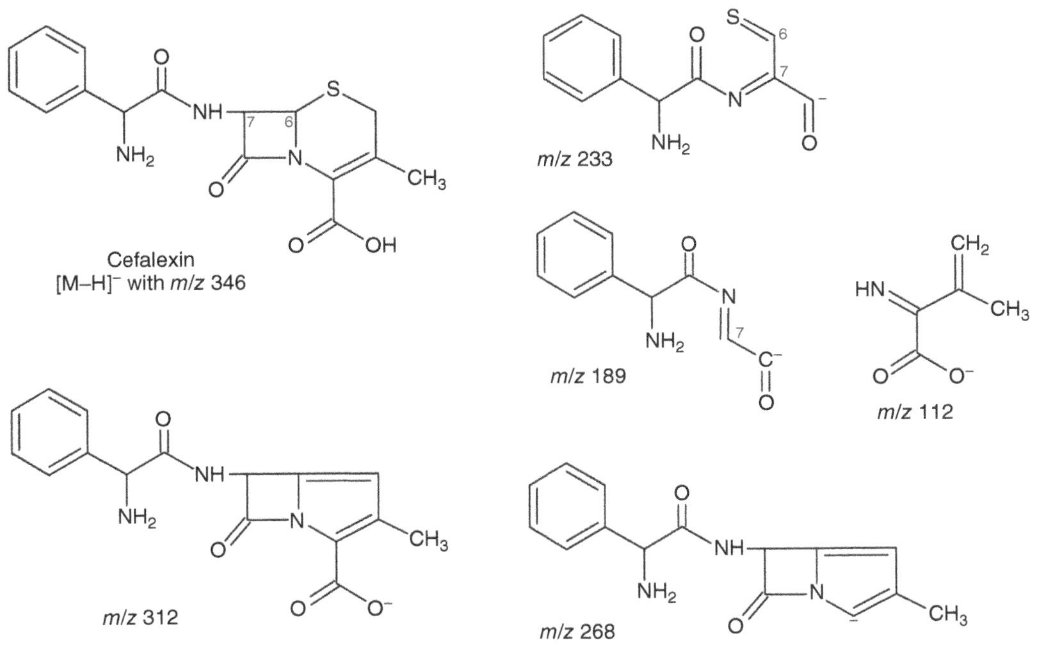

FIGURE 4.8.8 Proposed structures for the fragment ions of deprotonated cefalexin.

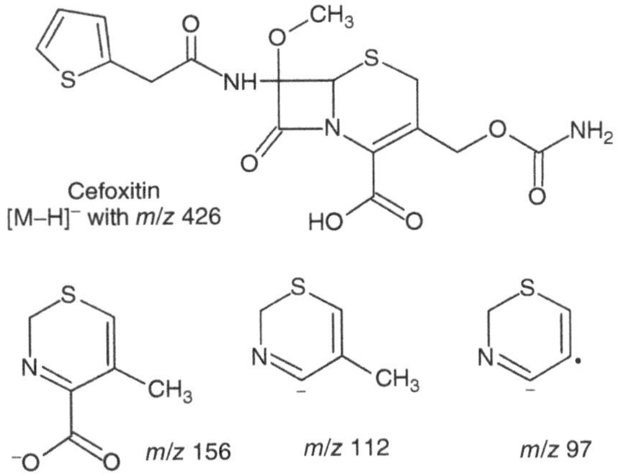

FIGURE 4.8.9 Proposed structures for the fragment ions of deprotonated cefoxitin.

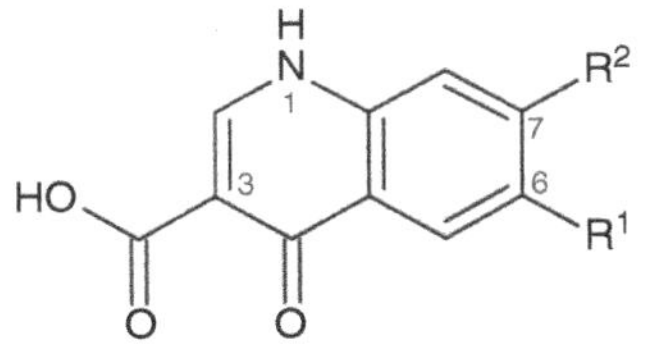

FIGURE 4.8.10 General structure of (fluoro)quinolone antibiotics.

2014). The identity of three product ions, frequently used in SRM during veterinary residue analysis, has been elucidated for seven fluoroquinolones (Geis-Asteggiante et al., 2014). The fragmentation of ciprofloxacin ($[M+H]^+$ with *m/z* 332) may serve as an example (Figure 4.8.12). The product-ion mass spectrum in our library collection contains significantly less fragments than reported elsewhere. Ciprofloxacin shows the initial loss of either water or CO_2 leading to fragment ions with *m/z* 314 and 288, respectively. Further fragmentation can be considered as secondary fragmentation, involving small-molecule losses from either of these two ions. The ion with *m/z* 314 shows losses of HF and of cyclopropene (C_3H_4) plus $CH_3CH{=}NH$ leading to fragment ions with *m/z* 294 and 231, respectively, whereas the ion with *m/z* 288 shows losses of HF and of $CH_3CH{=}NH$ leading to fragment ions with *m/z* 268 and 245, respectively (Volmer et al., 1997). Accurate-mass data for fluoroquinolones have also been reported (Hermo et al., 2006; Raju et al., 2011; Liu et al., 2011). Based on accurate-mass data, an ion with *m/z* 360 is reported for prulifloxacin (Raju et al., 2011) attributed to the loss of 3-sulfanylpropynoic acid (HS—C≡C—COOH).

In ion-trap MS^n, similar fragmentation is observed for fluoroquinolones, that is, losses of water or CO_2 followed by cleavages in the piperazine ring (Li et al., 2006b). In multiresidue analysis of fluoroquinolones, mostly SRM transitions based on losses of water, CO_2, and the cleavage of the piperazine ring are applied (Mottier et al., 2008; Li et al., 2009b; Yu et al., 2012; Biselli et al., 2013; Dorival-García et al., 2013).

There is considerable attention for the identification of (stressed) degradation products of fluoroquinolones. Studies have been reported for, among others, clinafloxacin

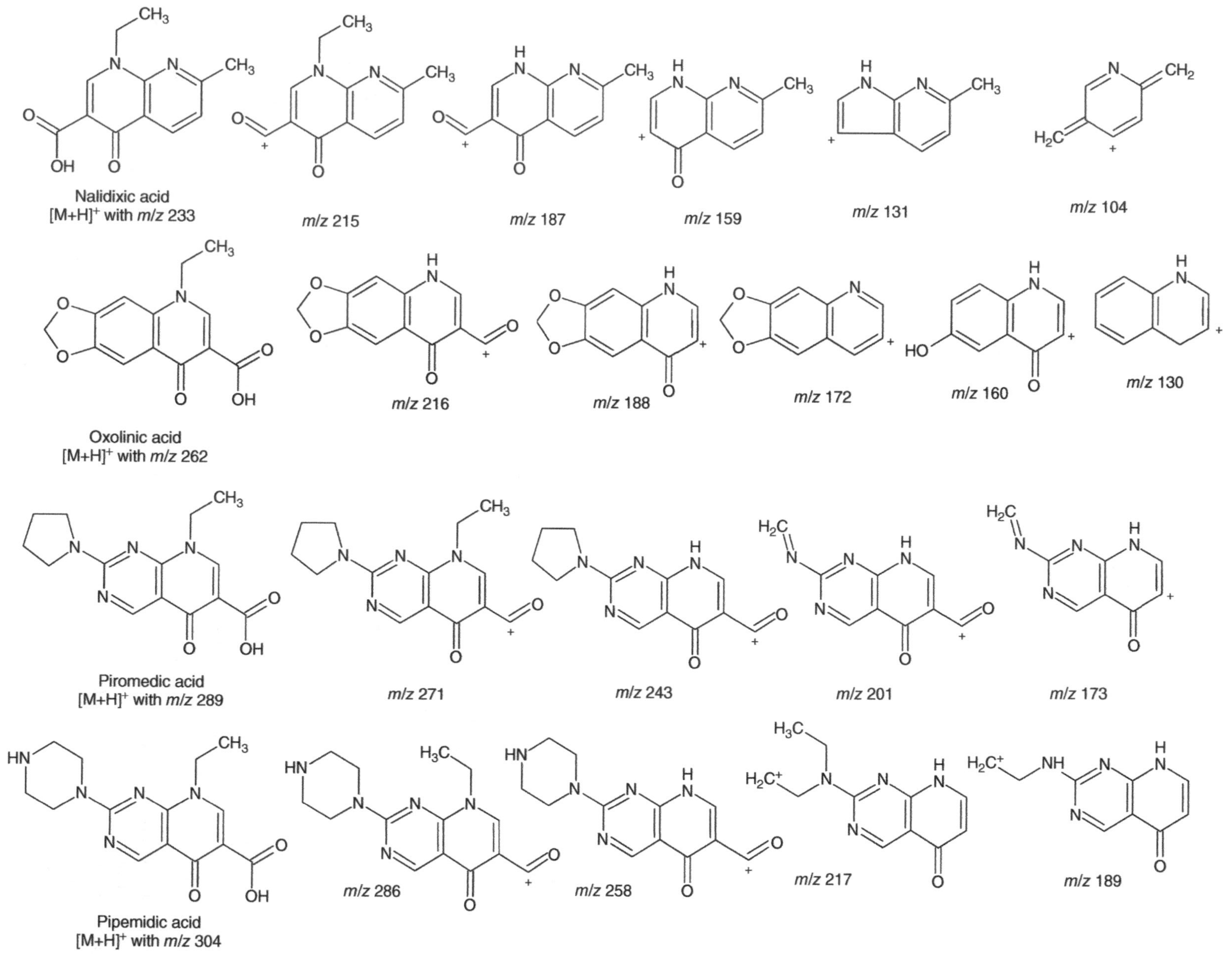

FIGURE 4.8.11 Proposed structures for the fragment ions of the protonated quinolone antibiotics nalidixic acid, oxolinic acid, piromedic acid, and pipemidic acid.

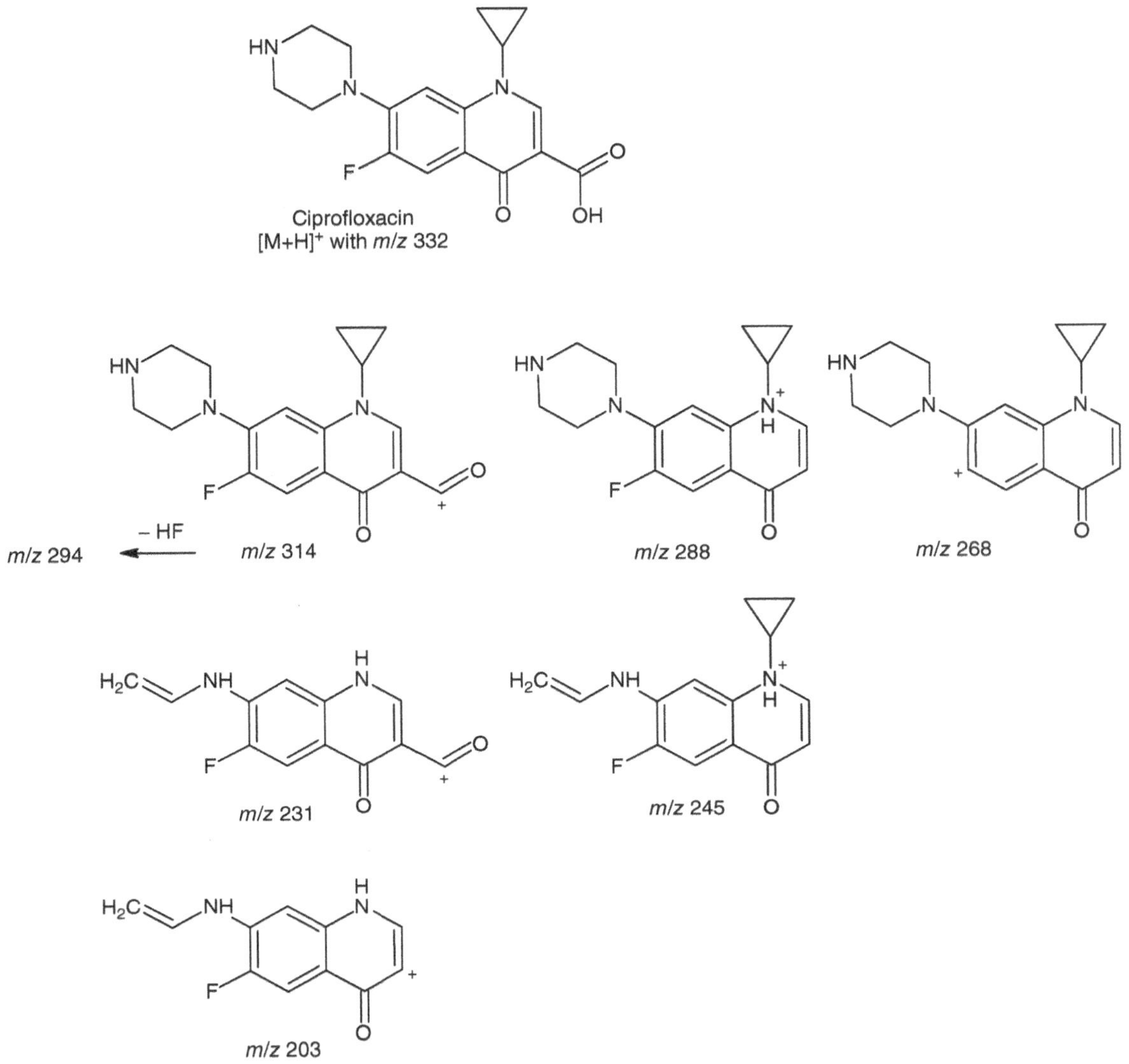

FIGURE 4.8.12 Proposed structures for the fragment ions of the protonated fluoroquinolone antibiotic ciprofloxacin.

(Lovdahl & Priebe, 2000); pruliﬂoxacin (Raju et al., 2011); danofloxacin (Liu et al., 2011); and for ofloxacin, norfloxacin, ciprofloxacin, and moxifloxacin (Maia et al., 2014).

4.8.5 AMINOGLYCOSIDES

Aminoglycosides are Gram-negative antibacterial agents via protein synthesis inhibition. They are derived from bacteria of either the *Streptomyces* genus (suffix: mycin) or the *Micromonospora* genus (suffix: micin). There are three subclasses: (1) 1,3-disubstituted deoxystreptamines such as kanamycin A, tobramycin, and gentamicin, (2) 1,2-disubstituted deoxystreptamines such as neomycin, and (3) non-deoxystreptamine aminoglycosides such as the streptamine derivative streptomycin. The LC–MS analysis of aminoglycosides is hampered by their high polarity and protolytic properties (Farouk et al., 2015). The LC separation requires either derivatization or the use of ion-pair LC, cation-exchange LC (McGlinchey et al., 2008), or zwitterionic HILIC (Chiaochan et al., 2010). Multiresidue analysis of aminoglycosides in food of animal origin has been reported (Zhu et al., 2008; van Holthoon et al., 2009; Tao et al., 2012; Almeida et al., 2012; Lehotay et al., 2013).

The fragmentation of gentamicins after LC–TSI-MS has been discussed (Getek et al., 1991). The fragmentation scheme for gentamicin C1 is drawn in Figure 4.8.13. Three fragment ions were observed, that is, a carbocation with *m/z* 160 from the garosamine unit ($[C_7H_{14}NO_3]^+$; ring A in Figure 4.8.13), protonated 2-deoxystreptamine

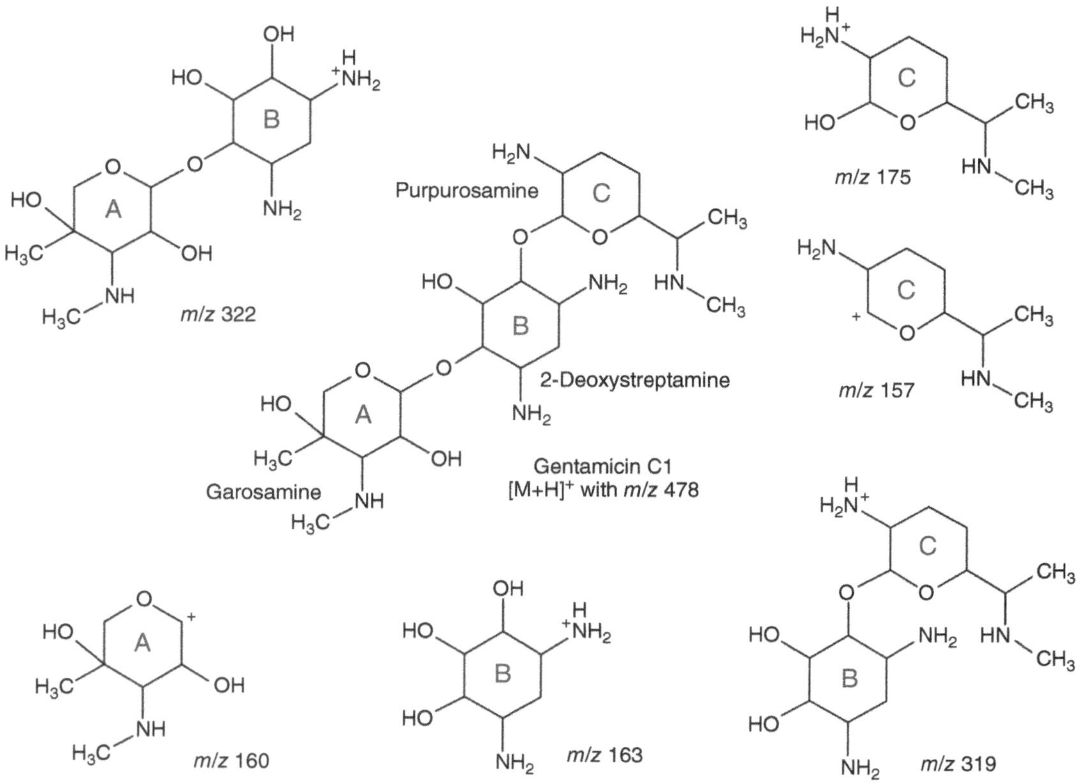

FIGURE 4.8.13 Proposed structures for the fragment ions of protonated gentamicin C1. (Source: Farouk et al., 2015. Reproduced with permission of Elsevier.)

($[C_6H_{13}N_2O_3+H]^+$; ring B) with *m/z* 163, and a carbocation due to the purpurosamine unit (ring C), being *m/z* 157 ($[C_8H_{17}N_2O]^+$) for gentamicin C1, *m/z* 129 ($[C_6H_{13}N_2O]^+$) for gentamicin C1a, and *m/z* 143 ($[C_7H_{15}N_2O]^+$) for the gentamicins C2, C2a, and C2b. In later studies, an additional ion with *m/z* 322 ($[C_{13}H_{27}N_3O_6+H]^+$) due to the loss of the purpurosamine unit was also described (Figure 4.8.13). Subsequent losses of water, ammonia (NH_3), or methylamine (CH_3NH_2) may result in secondary fragment ions.

Under ESI-MS conditions, most aminoglycosides may show doubly-charged ions, $[M+2H]^{2+}$ (McLaughlin et al., 1994). In the absence of ion-pairing agent, the doubly-charged ion is most abundant, whereas the singly-charged ion is most abundant in the presence of ion-pairing agent. The doubly-charged ion upon CID produces mostly singly-charged fragment ions.

Basically, similar fragmentation as described for the gentamicins is observed for other aminoglycosides (Table 4.8.6). This table is based on literature data (McLaughlin et al., 1994; Heller et al., 2000a; Hu et al., 2000; Li et al., 2009a; Lehotay et al., 2013). The ions that are frequently applied in SRM transitions are underlined in Table 4.8.6. Again, subsequent losses of water, NH_3, or CH_3NH_2 may occur (Hu et al., 2000). Singly-charged precursor ions are used in most of the SRM transitions. The relative abundance of fragment ions may differ depending on whether singly-charged or doubly-charged precursor ions are fragmented, for example, the most abundant fragment ions from singly-charged dihydrostreptomycin and streptomycin are the ions with *m/z* 263 and 245, whereas from doubly-charged ions the most abundant ions are *m/z* 176 and 409 for dihydrostreptomycin and *m/z* 176 and 263 for streptomycin (Bogialli et al., 2005).

In order to identify impurities in kanamycin B and tobramycin, the fragmentation of these two compounds was studied in some detail. The fragmentation pattern is greatly in agreement with Figure 4.8.13 and Table 4.8.6, although two additional fragments are discussed. The ions with *m/z* 365 (kanamycin B) and *m/z* 349 (tobramycin) are due to cross-ring cleavages in the C-ring. Subsequent loss of the A-ring results in the fragment ion with *m/z* 205 (Li et al., 2009a). Similarly, related substances of dihydrostreptomycin (Pendela et al., 2009) and gentamicins (Li et al., 2011; Grahek & Zupancic-Kralj, 2009) were identified, and detailed fragmentation schemes were proposed.

The structure elucidation of six *N*-octyl derivatives of neomycin B using LC–MS^n in an ion-trap–time-of-flight hybrid instrument (Section 1.3.6) was reported. MS^4 was

TABLE 4.8.6 Fragmentation of aminoglycosides in MS–MS (for annotation, Figure 4.8.13).

Aminoglycoside	$[M+H]^+$ (*m/z*)	$[M+2H]^{2+}$ (*m/z*)	A (*m/z*)	B (*m/z*)	C (*m/z*)	AB (*m/z*)	BC (*m/z*)	Other (*m/z*)
Amikacin	586	293.7	162	264	162, 180	425	425	102, 163, 247
Apramycin	540	270.6	162	235	145, 163	396, 378	379	199, 217
Arbekacin	553	277.2	129	264	162, 180	392	425	102, 163
Dihydrostreptomycin	584	292.6	322	245, 263	n.a.	$[M+H]^+$	n.a.	176, 246
Streptomycin	582	291.6	320	245, 263	n.a.	$[M+H]^+$	n.a.	176, 246
Gentamicin C1a	450	225.7	160	163	129, 147	322	291	112
Gentamicin C2, C2a, C2b	464	232.7	160	163	143, 161	322	305	
Gentamicin C1	478	239.7	160	163	157, 175	322	319	139
Hygromycin B	528	264.6	352	159, 177	n.a.	$[M+H]^+$	n.a.	257
Kanamycin A	484	242.6	161	163	162, 180	323	324	205, 102
Kanamycin B	485	243.1	162	163	162, 180	324	324	
Neomycin B	615	308.2	293	163	161, 179	455	323	
Paromomycin	616	308.7	293	163	162, 180	455	324	203, 161
Netilmicin	476	238.2	127	191	160, 178	317	350	458, 281
Ribostamycin	455	228.2	133	163	161, 179	295	323	
Sisomycin	448	224.6	162	163	127, 145	322	289	271, 254
Tobramycin	468	234.6	145	163	162, 180	307	324	

Source: Zhu et al., 2008; Tao et al., 2012; van Holthoon et al., 2009; Lehotay et al., 2013; Farouk et al., 2015.
The underlined *m/z* values are the ions frequently used as product ions in SRM. n.a.: not applicable.

necessary to discriminate between the various isomeric species (Giera et al., 2010).

4.8.6 TETRACYCLINES

Tetracyclines are broad-spectrum antibacterial compounds, which are indicated against organisms such as *Mycoplasma*, *Chlamydia*, and a number of other Gram-positive and Gram-negative bacteria. With the onset of bacterial resistance, their usefulness has been reduced. However, they are still widely used in veterinary medicine. Tetracyclines are derivatives of polycyclic octahydrotetracene-2-carboxamide. Residue analysis of tetracyclines using LC–MS has been frequently reported, often in conjunction with other classes of antibiotics, for example, sulfonamides, β-lactams, and fluoroquinolones (Heller et al., 2006; Ortelli et al., 2009; Bittencourt et al., 2012).

Tetracyclines are analyzed in the positive-ion mode. The most abundant fragment ions are due to the loss of water, ammonia (NH_3), and the combined loss of water and NH_3. The latter, involving a loss of 35 Da, is frequently used as quantifier ion in multiresidue analysis using SRM (Blanchflower et al., 1997; Oka et al., 1997; Halling-Sørensen et al., 2003; Khong et al., 2005; Jadhav et al., 2013). The identity of three product ions, frequently used in SRM during veterinary residue analysis, has been elucidated for three tetracyclines (Geis-Asteggiante et al., 2014).

Closer inspection of the product-ion mass spectra shows additional less abundant fragments, which are more informative and selective than ions due to water and NH_3 losses (Oka et al., 1998; Kamel et al., 2002; Khong et al., 2005). Mechanistic issues related to the water and NH_3 losses were studied in detail using H/D-exchange experiments (Kamel et al., 2002). From this, it has been established that the first water loss involves the hydroxy (HO—) group at the C^6 position via a charge-remote fragmentation, that is, a 1,2-elimination with the H-atom from the C^{5a} position (Figure 4.8.14). The fragmentation of various tetracyclines was also described in some detail in a tabular format (Kamel et al., 2002). Based on this and with the help of confirmation by accurate-mass data in our mass spectral library, a fragmentation scheme for tetracycline was drawn (Figure 4.8.14). Similar fragmentation is observed for the other widely studied tetracyclines such as chlortetracycline, oxytetracyline, or methacycline and meclocycline.

4.8.7 NITROFURANS

Nitrofurans are effective antimicrobials with a broad spectrum of activity against various Gram-positive and Gram-negative bacteria as well as protozoa. Because of their rapid metabolism, the nitrofurans are seldom analyzed as intact compounds, but either as their metabolites or as their protein-bound metabolites, from which the 3-amino-1,3-oxazolidin-2-one unit (AOZ) can be released under mildly acidic conditions (Horne et al., 1996). As their metabolites, they are mostly analyzed after derivatization with 2-nitrobenzaldehyde (NB) (Horne et al., 1996) (Figure 4.8.15). The 2-nitrophenyl derivatives can be analyzed by positive-ion APCI-MS. In MS mode, in addition to $[M+H]^+$, they produce an ion with *m/z* ($[M+H]-30$), which most likely results from a thermally induced reduction of

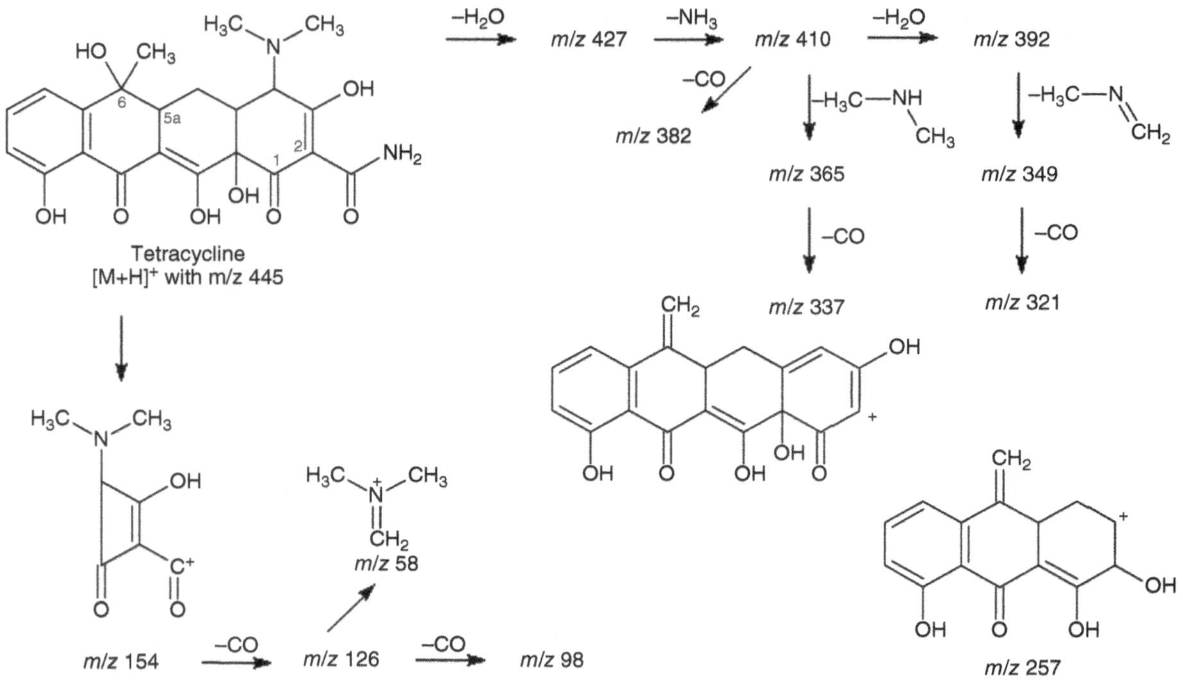

FIGURE 4.8.14 Fragmentation scheme and proposed structures for some fragment ions of protonated tetracycline.

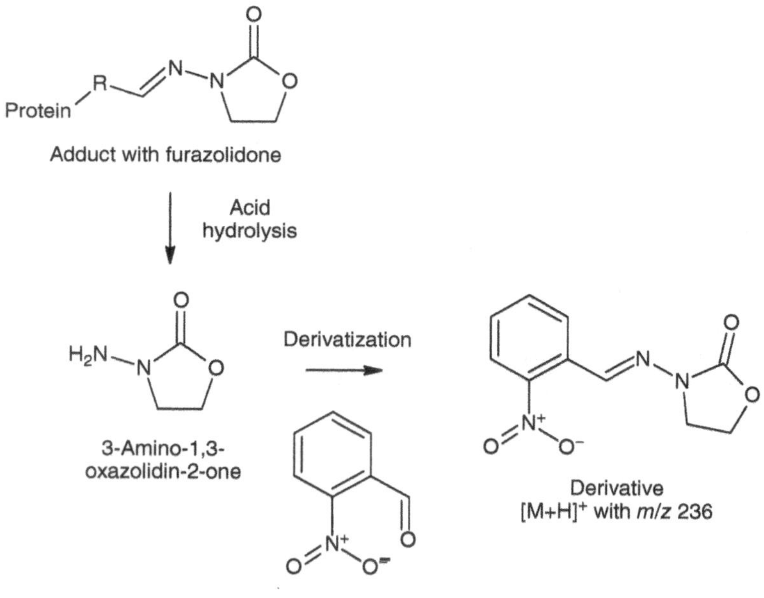

FIGURE 4.8.15 Analysis of furazolidone antibiotics: release of the 3-amino-1,3-oxazolidin-2-one unit (AOZ) from protein adducts and subsequent derivatization using nitrobenzaldehyde.

the nitro group into an amino group (Karancsi & Slegel, 1999) (Section 3.5.7). Based on the SRM transitions applied to analyze the nitrobenzyl derivatives (Leitner et al., 2001; Mottier et al., 2005; Cooper & Kennedy, 2005), structure proposals can be made for some fragment ions (Figure 4.8.16). However, these structures require further confirmation by accurate-mass determination.

When intact nitrofurans are analyzed, some of them are analyzed in positive-ion mode, and some of them in negative-ion mode (Barbosa et al., 2007; Ardsoongnearn

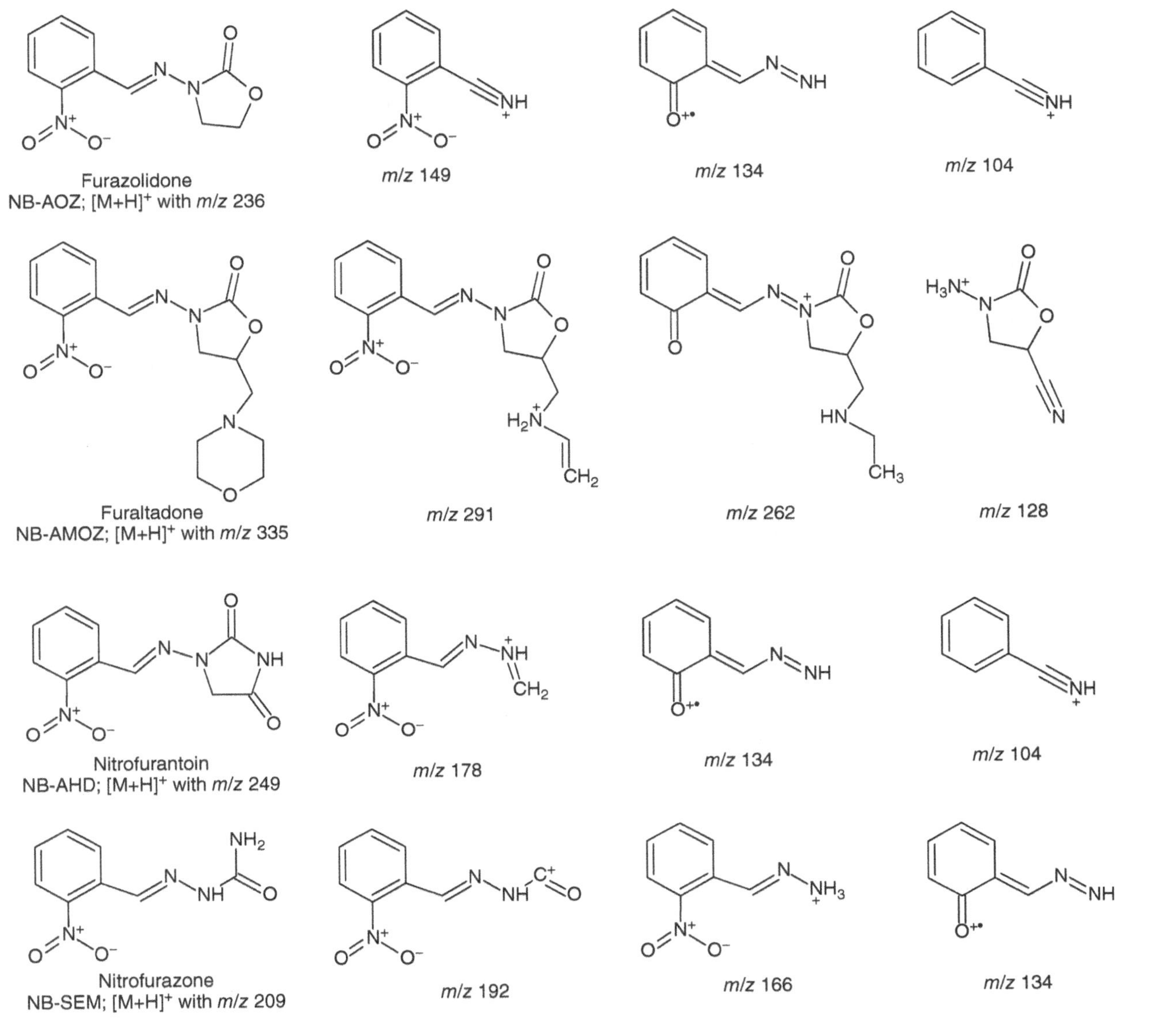

FIGURE 4.8.16 Proposed structures for the fragment ions of protonated *o*-nitrobenzene derivatives of protein-released firazolidone, furaltadone, nitrofurantoin, and nitrofurazone.

et al., 2014). SRM transitions for furaltadone ($[M+H]^+$ with *m/z* 325) are based on detecting the fragment ions with *m/z* 281 due to the loss of carbon dioxide (CO_2) and *m/z* 252 due to the formal loss of $C_2H_3NO_2$ (confirmed by accurate mass). This most likely involves the combined loss of CO_2 and methanimine ($H_2C{=}NH$), requiring a skeletal rearrangement. Furazolidone ($[M+H]^+$ with *m/z* 226) is analyzed using the ions with *m/z* 139 ($[C_5H_2N_2O_3]^+$) resulting from the loss of the oxazolidinone ring ($C_3H_5NO_2$), and *m/z* 122 due to the loss of hydroxyl radical ($^{\bullet}OH$) from the ion with *m/z* 139 (Barbosa et al., 2007), or the ion with *m/z* 113 due to the loss of nitrofuran ($C_4H_3NO_3$) (Ardsoongnearn et al., 2014). Nitrofurantoin ($[M-H]^-$ with *m/z* 237) and nitrofurazone ($[M-H]^-$ with *m/z* 197) are analyzed in negative-ion mode using the fragment ion with *m/z* 152 due to the loss of $C_3H_3NO_2$ from the imidazolidinedione ring or due to the loss of carbon monoxide (CO) and ammonia (NH_3), respectively (Barbosa et al., 2007; Ardsoongnearn et al., 2014).

4.8.8 MACROLIDES

Macrolide antibiotics consist of a 12-, 14-, or 16-membered macrocyclic lactone to which amino and deoxy sugars are attached, for example, desosamine and cladinose to erythromycin A. They are produced by various *Streptomyces* strains and are used in veterinary practice against Gram-positive bacteria, as well as in humans against various infectious diseases, where they are administered as acid-resistant esters. Abundant literature is available on LC–MS analysis of macrolide antibiotics, for example, with respect to bioanalysis in human plasma (Chen et al., 2006; Li et al., 2006a), multiresidue analysis in environmental samples (Abuin et al., 2006) or in food of animal origin (Pleasance et al., 1992; Bogialli et al., 2007a; Wang, 2009), and in impurity profiling (see later).

Macrolide antibiotics are analyzed in positive-ion ESI, mostly as $[M+H]^+$, although the use of $[M+2H]^{2+}$ as precursor ion has also been described, for example, for azithromycin (Chen et al., 2006; Shen et al., 2010). The fragmentation of macrolides has been studied in detail using various MS–MS and MS^n platforms and different dissociation techniques (Pleasance et al., 1992; Kearney et al., 1999; Gates et al., 1999; Crowe et al., 2002). In studies involving impurity profiling and/or identification of related substances, detailed fragmentation schemes have also been reported for azithromycin (Debremaeker et al., 2003), clarithromycin (Leonard et al., 2006), dirithromycin (Diana et al., 2006), erythromycin (Govaerts et al., 2000; Kumar Chitneni et al., 2004; Deubel et al., 2006; Pendela et al., 2012), josamycin (Govaerts et al., 2004; Van den Bossche et al., 2013), and tylosin (Chopra et al., 2013).

The characteristic fragmentation of macrolide antibiotics is illustrated for erythromycin A ($[M+H]^+$ with *m/z* 734) in Figure 4.8.17. The ion with *m/z* 576 is due to the loss of the dehydrocladinose ($C_8H_{14}O_3$), the ions with *m/z* 558, 540, and 522 to three subsequent water losses, and the ion with *m/z* 158 ($[C_8H_{16}NO_2]^+$) due to the cleavage of the glycosidic bond at C^1 of desosamine, with charge retention on the resulting dehydrodesosamine, eventually with a subsequent loss of ethenone ($H_2C{=}C{=}O$) to an ion with *m/z* 116. Similar fragmentation is observed for troleandromycin ($[M+H]^+$ with *m/z* 814), roxithromycin ($[M+H]^+$ with *m/z* 837), clarithromycin ($[M+H]^+$ with *m/z* 748), and azithromycin ($[M+H]^+$ with *m/z* 749). Spiramycin I ($[M+H]^+$ with *m/z* 843), josamycin ($[M+H]^+$ with *m/z* 828), and tylosin ($[M+H]^+$ with *m/z* 916) have substitution at C^4 of the desosamine pyranoside, for example, a 3,4-dihydroxy-3,5-dimethyltetrahydropyranoside group for spiramycin and tylosin. Additional fragments may be observed due to the loss of this extra sugar, as well as a fragment due to the cleavage of the glycosidic bond at C^1 of the C^4-substituted desosamine group, with charge retention on the resulting disaccharide, for example, resulting in a fragment ion with *m/z* 318 ($[C_{15}H_{28}NO_6]^+$) for spiramycin I and tylosin. The fragments discussed here are frequently used as product ions in SRM transitions for quantitative or residue analysis.

4.8.9 MISCELLANEOUS ANTIBIOTICS

Trimethoprim is a bacteriostatic chemotherapeutic agent, used among others in the treatment of urinary tract infections. In MS–MS, trimethoprim ($[M+H]^+$ with *m/z* 291) generates two fragment ions due to cleavages of either C—C bond of the methylene (CH_2) group connecting the two

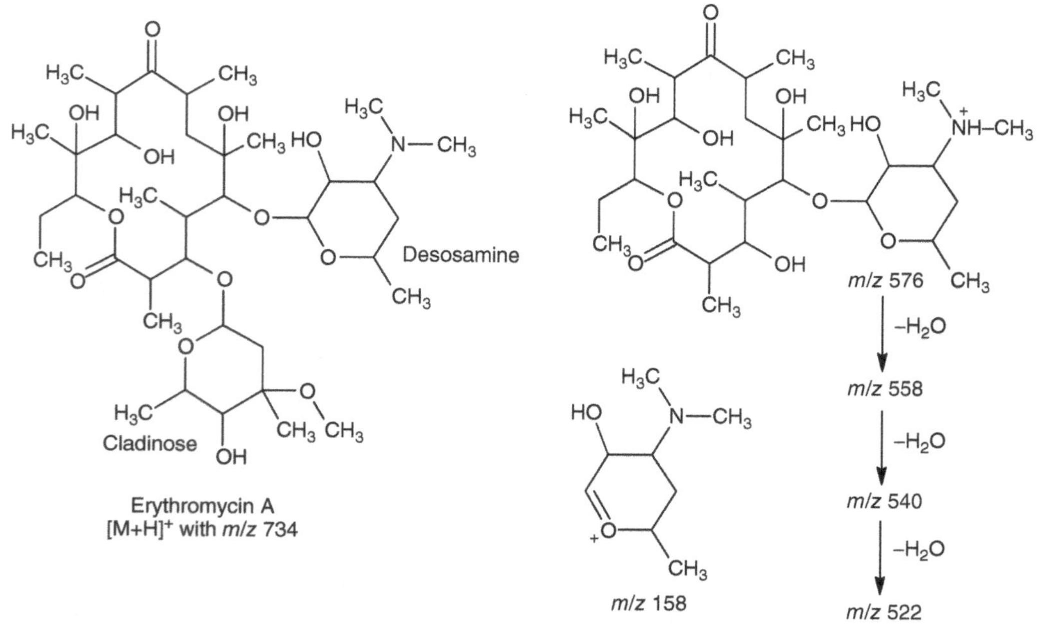

FIGURE 4.8.17 Fragmentation scheme and proposed structures for some fragment ions of protonated erythromycin A.

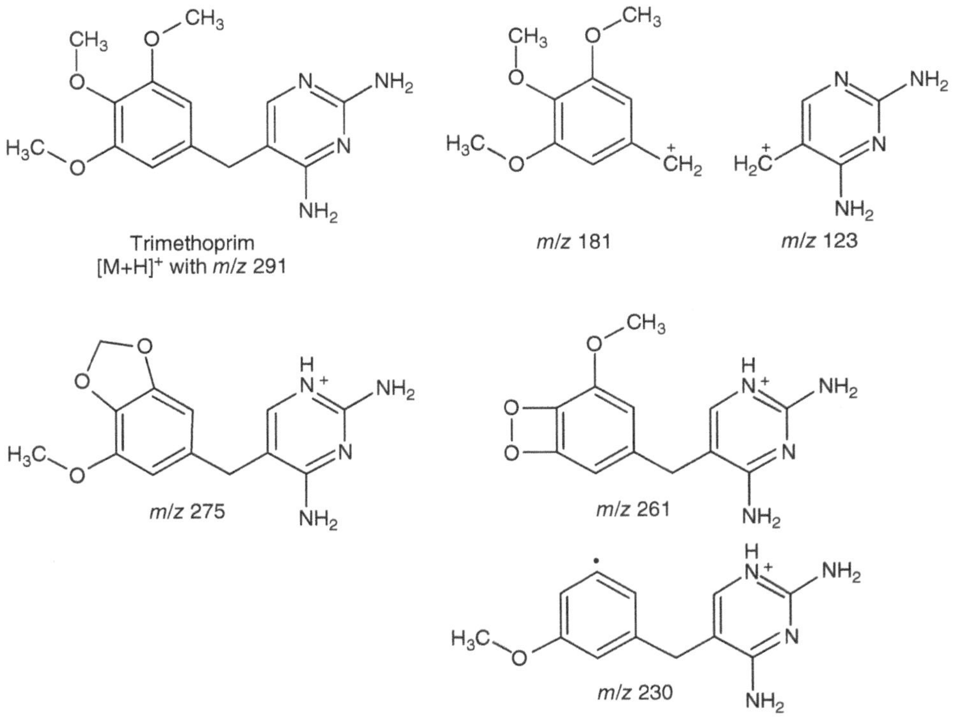

FIGURE 4.8.18 Proposed structures for some fragment ions of protonated trimethoprim.

rings, that is, the 2,4-diaminopyrimidin-5-yl-methylium ion ($[C_5H_7N_4]^+$) with *m/z* 123 and the 3,4,5-trimethoxybenzyl cation ($[C_{10}H_{13}O_3]^+$) with *m/z* 181 (Figure 4.8.18). Quite interesting small-molecule losses from $[M+H]^+$ occur. Fragment ions are observed with *m/z* 276 due to the loss of a methyl radical ($^{\bullet}CH_3$), *m/z* 275 due to the loss of methane (CH_4), thereby forming a 1,3-dioxole ring, and *m/z* 261 and 230. The last two ions have often been attributed to losses of formaldehyde ($H_2C{=}O$), and losses of $H_2C{=}O$ and a methoxy radical ($^{\bullet}OCH_3$), respectively (Lehr et al., 1999; Barbarin et al., 2002; Eichhorn et al., 2005). However, based on accurate-mass data, it has been shown that the ion with *m/z* 261 is actually due to the loss of ethane (C_2H_6) rather than of $H_2C{=}O$ (Eckers et al., 2005; Liu et al., 2012). This leads to an interesting mechanistic question about the formation of the ion with *m/z* 230, which still is due to the loss of $H_2C{=}O$ and $^{\bullet}OCH_3$ (Figure 4.8.18). This ion cannot be formed in one step, as it involves losses from two different C-atoms. Similar fragmentation is observed for tetroxoprim and diaveridine (Section 4.10.2.2).

Spectinomycin is an aminocyclitol antimicrobial, closely related to aminoglycosides, produced by the bacterium *Streptomyces spectabilis*. In LC–MS, spectinomycin forms strong solvent adducts with water or methanol (CH_3OH) (Peru et al., 2006; Zhu et al., 2008). Fragmentation schemes for spectinomycin ($[M+H]^+$ with *m/z* 333) have been proposed (Carson & Heller, 1998; Hornish & Wiest, 1998). Fragment ions are observed with *m/z* 315 due to the loss of water, *m/z* 305 due to the loss of carbon monoxide (CO), *m/z* 302 due to the loss of methylamine (CH_3NH_2), and *m/z* 289 due to the loss of carbon dioxide (CO_2). Subsequent fragmentation of the ion with *m/z* 289 results in fragment ions with *m/z* 271 due to water loss and *m/z* 245 due to CO_2 loss. The loss of CO and CO_2 from the ion with *m/z* 271 results in fragment ions with *m/z* 243 and 227, respectively. Another fragmentation route involves cleavage of the 1,4-dioxane ring to a fragment ion with *m/z* 207 ($[C_8H_{18}N_2O_4+H]^+$), which in turn shows subsequent losses of water to an ion with *m/z* 189, of CH_3NH_2 to an ion with *m/z* 158, and of water to an ion with *m/z* 140 ($[C_7H_9NO_2+H]^+$). The ion with *m/z* 140 shows the loss of water to an ion with *m/z* 122, of CO to an ion with *m/z* 112, or of ethenone ($H_2C{=}C{=}O$) to an ion with *m/z* 98 ($[C_5H_8NO]^+$). Since the fragment ions with *m/z* 207 and 189 are frequently used in SRM transitions for the analysis of spectinomycin, their proposed structures are shown in Figure 4.8.19.

Lincomycin ($[M+H]^+$ with *m/z* 407) shows fragment ions with *m/z* 389 due to the loss of water, with *m/z* 359 due to the loss of methanethiol (CH_3SH), and the 1-methyl-4-propylpyrrolidin-2-ylium ion ($[C_8H_{16}N]^+$) with *m/z* 126. The related clindamycin ($[M+H]^+$ with *m/z* 425) also shows the loss of CH_3SH and the formation of the ion with *m/z* 126.

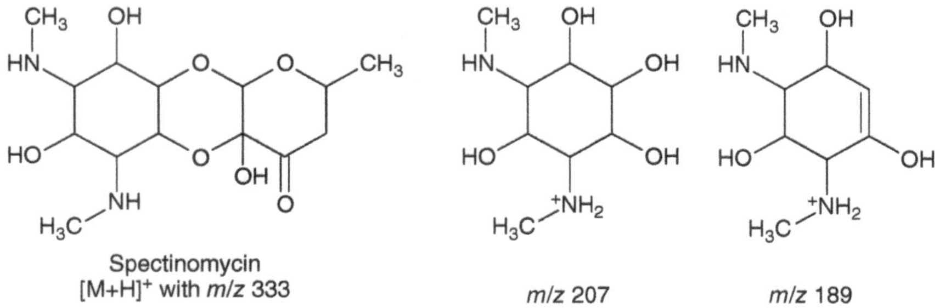

FIGURE 4.8.19 Proposed structures for the fragment ions of protonated spectinomycin. (Source: Farouk et al., 2015. Reproduced with permission of Elsevier.)

REFERENCES

Abuin S, Codony R, Compañó R, Granados M, Prat MD. 2006 Analysis of macrolide antibiotics in river water by solid-phase extraction and liquid chromatography–mass spectrometry. J Chromatogr A 1114: 73–81.

Almeida MP, Rezende CP, Souza LF, Brito RB. 2012. Validation of a quantitative and confirmatory method for residue analysis of aminoglycoside antibiotics in poultry, bovine, equine and swine kidney through liquid chromatography–tandem mass spectrometry. Food Addit Contam A 29: 517–525.

Ardsoongnearn C, Boonbanlu O, Kittijaruwattana S, Suntornsuk L. 2014. Liquid chromatography and ion trap mass spectrometry for simultaneous and multiclass analysis of antimicrobial residues in feed water. J Chromatogr B 945–946: 31–38.

Barbarin N, Henion JD, Wu Y. 2002. Comparison between liquid chromatography-UV detection and liquid chromatography–mass spectrometry for the characterization of impurities and/or degradants present in trimethoprim tablets. J Chromatogr A. 970(1–2):141–154.

Barbosa J, Moura S, Barbosa R, Ramos F, da Silveira MI. 2007. Determination of nitrofurans in animal feeds by liquid chromatography-UV photodiode array detection and liquid chromatography–ionspray tandem mass spectrometry. Anal Chim Acta 586: 359–365.

Bateman KP, Locke SJ, Volmer DA. 1997. Characterization of isomeric sulfonamides using capillary zone electrophoresis coupled with nano-electrospray quasi-MS/MS/MS. J Mass Spectrom 32: 297–304.

Bedendo GC, Jardim IC, Carasek E. 2010. A simple hollow fiber renewal liquid membrane extraction method for analysis of sulfonamides in honey samples with determination by liquid chromatography–tandem mass spectrometry. J Chromatogr A 1217: 6449–6454.

Berendsen BJ, Zuidema T, de Jong J, Stolker LA, Nielen MW. 2011. Discrimination of eight chloramphenicol isomers by liquid chromatography tandem mass spectrometry in order to investigate the natural occurrence of chloramphenicol. Anal Chim Acta 700: 78–85.

Biselli S, Schwalb U, Meyer A, Hartig L. 2013. A multi-class, multi-analyte method for routine analysis of 84 veterinary drugs in chicken muscle using simple extraction and LC–MS/MS. Food Addit Contam A 30: 921–939.

Bittencourt MS, Martins MT, de Albuquerque FG, Barreto F, Hoff R. 2012. High-throughput multiclass screening method for antibiotic residue analysis in meat using liquid chromatography–tandem mass spectrometry: A novel minimum sample preparation procedure. Food Addit Contam A 29: 508–516.

Blanchflower WJ, McCracken RJ, Haggan AS, Kennedy DG. 1997 Confirmatory assay for the determination of tetracycline, oxytetracycline, chlortetracycline and its isomers in muscle and kidney using liquid chromatography–mass spectrometry. J Chromatogr B 692: 351–360.

Bogialli S, Curini R, Di Corcia A, Laganà A, Mele M, Nazzari M. 2005. Simple confirmatory assay for analyzing residues of aminoglycoside antibiotics in bovine milk: Hot water extraction followed by liquid chromatography–tandem mass spectrometry. J Chromatogr A 1067: 93–100.

Bogialli S, Di Corcia A, Laganà A, Mastrantoni V, Sergi M. 2007a. A simple and rapid confirmatory assay for analyzing antibiotic residues of the macrolide class and lincomycin in bovine milk and yoghurt: Hot water extraction followed by liquid chromatography/tandem mass spectrometry. Rapid Commun Mass Spectrom 21: 237–246.

Bogialli S, D'Ascenzo G, Di Corcia A, Innocenti G, Laganà A, Pacchiarotta T. 2007b. Monitoring quinolone antibacterial residues in bovine tissues: Extraction with hot water and liquid chromatography coupled to a single- or triple-quadrupole mass spectrometer. Rapid Commun Mass Spectrom 21: 2833–2842.

Boscher A, Guignard C, Pellet T, Hoffmann L, Bohn T. 2010. Development of a multi-class method for the quantification of veterinary drug residues in feeding stuffs by liquid chromatography–tandem mass spectrometry. J Chromatogr A 1217: 6394–6404.

Bruno F, Curini R, Di Corcia A, Nazzari M, Samperi R. 2001. Solid-phase extraction followed by liquid chromatography–mass spectrometry for trace determination of beta-lactam antibiotics in bovine milk. J Agric Food Chem 49: 3463–3470.

Carson MC, Heller DN. 1998. Confirmation of spectinomycin in milk using ion-pair solid-phase extraction and liquid chromatography–electrospray ion trap mass spectrometry. J Chromatogr B 718: 95–102.

Chen BM, Liang YZ, Chen X, Liu SG, Deng FL, Zhou P. 2006. Quantitative determination of azithromycin in human plasma by liquid chromatography–mass spectrometry and its application in a bioequivalence study. J Pharm Biomed Anal 42: 480–487.

Chiaochan C, Koesukwiwat U, Yudthavorasit S, Leepipatpiboon N. 2010. Efficient hydrophilic interaction liquid chromatography–tandem mass spectrometry for the multiclass analysis of veterinary drugs in chicken muscle. Anal Chim Acta 682: 117–129.

Chopra S, Van Schepdael A, Hoogmartens J, Adams E. 2013. Characterization of impurities in tylosin using dual liquid chromatography combined with ion trap mass spectrometry. Talanta 106: 29–38.

Cooper KM, Kennedy DG. 2005. Nitrofuran antibiotic metabolites detected at parts per million concentrations in retina of pigs--a new matrix for enhanced monitoring of nitrofuran abuse. Analyst 130: 466–468.

Crowe MC, Brodbelt JS, Goolsby BJ, Hergenrother P. 2002. Characterization of erythromycin analogs by collisional activated dissociation and infrared multiphoton dissociation in a quadrupole ion trap. J Am Soc Mass Spectrom 13: 630–649.

Debremaeker D, Visky D, Chepkwony HK, Van Schepdael A, Roets E, Hoogmartens J. 2003. Analysis of unknown compounds in azithromycin bulk samples with liquid chromatography coupled to ion trap mass spectrometry. Rapid Commun Mass Spectrom 17: 342–350.

Deubel A, Fandiño AS, Sörgel F, Holzgrabe U. 2006. Determination of erythromycin and related substances in commercial samples using liquid chromatography/ion trap mass spectrometry. J Chromatogr A 1136: 39–47.

Diana J, Govaerts C, Hoogmartens J, Van Schepdael A, Adams E. 2006. Characterization of impurities in dirithromycin by liquid chromatography/ion trap mass spectrometry. J Chromatogr A 1125: 52–66.

Dorival-García N, Zafra-Gómez A, Cantarero S, Navalón A, Vílchez JL. 2013. Simultaneous determination of 13 quinolone antibiotic derivatives in wastewater samples using solid-phase extraction and ultra-performance liquid chromatography–tandem mass spectrometry. Microchem J 106: 323–333.

Eckers C, Monaghan JJ, Wolff JC. 2005. Fragmentation of trimethoprim and other compounds containing alkoxy-phenyl groups in electrospray ionisation tandem mass spectrometry. Eur J Mass Spectrom 11: 73–82.

Eichhorn P, Ferguson PL, Pérez S, Aga DS. 2005. Application of ion trap-MS with H/D exchange and QqTOF-MS in the identification of microbial degradates of trimethoprim in nitrifying activated sludge. Anal Chem 77: 4176–4184.

Engler M, Rüsing G, Sörgel F, Holzgrabe U. 1998. Defluorinated sparfloxacin as a new photoproduct identified by liquid chromatography coupled with UV detection and tandem mass spectrometry. Antimicrob Agents Chemother 42: 1151–1159.

Fagerquist CK, Lightfield AR, Lehotay SJ. 2005. Confirmatory and quantitative analysis of beta-lactam antibiotics in bovine kidney tissue by dispersive solid-phase extraction and liquid chromatography–tandem mass spectrometry. Anal Chem 77: 1473–1482.

Farouk F, Azzazy H, Niessen WMA. 2015. Challenges in the determination of the UV silent aminoglycoside antibiotics. Anal Chim Acta 890: 21–43.

Forti AF, Scortichini G. 2009. Determination of ten sulphonamides in egg by liquid chromatography–tandem mass spectrometry. Anal Chim Acta 637: 214–219.

García-Galán MJ, Díaz-Cruz MS, Barceló D. 2010. Determination of 19 sulfonamides in environmental water samples by automated on-line solid-phase extraction-liquid chromatography–tandem mass spectrometry (SPE-LC–MS/MS). Talanta 81: 355–366.

Gates PJ, Kearney GC, Jones R, Leadlay PF, Staunton J. 1999. Structural elucidation studies of erythromycins by electrospray tandem mass spectrometry. Rapid Commun Mass Spectrom 13: 242–246.

Getek TA, Vestal ML, Alexander TG. 1991. Analysis of gentamicin sulfate by high-performance liquid chromatography combined with thermospray mass spectrometry. J Chromatogr 554: 191–203.

Giera M, de Vlieger JS, Lingeman H, Irth H, Niessen WMA. 2010. Structural elucidation of biologically active neomycin *N*-octyl derivatives in a regioisomeric mixture by means of liquid chromatography/ion trap time-of-flight mass spectrometry. Rapid Commun Mass Spectrom 24: 1439–1446.

Geis-Asteggiante L, Nuñez A, Lehotay SJ, Lightfield AR. 2014. Structural characterization of product ions by electrospray ionization and quadrupole time-of-flight mass spectrometry to support regulatory analysis of veterinary drug residues in foods. Rapid Commun Mass Spectrom 28: 1061–1081.

Govaerts C, Chepkwony HK, Van Schepdael A, Roets E, Hoogmartens J. 2000. Investigation of unknown related substances in commercial erythromycin samples with liquid chromatography/mass spectrometry. Rapid Commun Mass Spectrom 14: 878–884.

Govaerts C, Chepkwony HK, Van Schepdael A, Adams E, Roets E, Hoogmartens J. 2004. Application of liquid chromatography–ion trap mass spectrometry to the characterization of the 16-membered ring macrolide josamycin propionate. J Mass Spectrom 39: 437–446.

Grahek R, Zupancic-Kralj L. 2009. Identification of gentamicin impurities by liquid chromatography tandem mass spectrometry. J Pharm Biomed Anal 50: 1037–1043.

Guan Z, Liesch JM. 2001. Solvation of acylium fragment ions in electrospray ionization quadrupole ion trap and Fourier transform ion cyclotron resonance mass spectrometry. J Mass Spectrom 36: 264–276.

Halling-Sørensen B, Lykkeberg A, Ingerslev F, Blackwell P, Tjørnelund J. 2003 Characterisation of the abiotic degradation pathways of oxytetracyclines in soil interstitial water using LC–MS–MS. Chemosphere 50: 1331–1342.

Holstege DM, Puschner B, Whitehead G, Galey FD. 2002. Screening and mass spectral confirmation of beta-lactam antibiotic residues in milk using LC–MS/MS. J Agric Food Chem 50: 406–411.

Heller DN, Clark SB, Righter HF. 2000a. Confirmation of gentamicin and neomycin in milk by weak cation-exchange extraction and electrospray ionization/ion trap tandem mass spectrometry. J Mass Spectrom 35: 39–49.

Heller DN, Kaplan DA, Rummel NG, von Bredow J. 2000b. Identification of cephapirin metabolites and degradants in bovine milk by electrospray ionization--ion trap tandem mass spectrometry. J Agric Food Chem 48: 6030–6035.

Heller DN, Ngoh MA, Donoghue D, Podhorniak L, Righter H, Thomas MH. 2002. Identification of incurred sulfonamide residues in eggs: Methods for confirmation by liquid chromatography–tandem mass spectrometry and quantitation by liquid chromatography with ultraviolet detection. J Chromatogr B 774: 39–52.

Heller DN, Nochetto CB, Rummel NG, Thomas MH. 2006. Development of multiclass methods for drug residues in eggs: Hydrophilic solid-phase extraction cleanup and liquid chromatography/tandem mass spectrometry analysis of tetracycline, fluoroquinolone, sulfonamide, and beta-lactam residues. J Agric Food Chem 54: 5267–5278.

Hermo MP, Barrón D, Barbosa J. 2006. Development of analytical methods for multiresidue determination of quinolones in pig muscle samples by liquid chromatography with ultraviolet detection, liquid chromatography–mass spectrometry and liquid chromatography–tandem mass spectrometry. J Chromatogr A 1104: 132–139.

Horne E, Cadogan A, O'Keeffe M, Hoogenboom LA. 1996. Analysis of protein-bound metabolites of furazolidone and furaltadone in pig liver by high-performance liquid chromatography and liquid chromatography–mass spectrometry. Analyst 121: 1463–1468.

Hornish RE, Wiest JR. 1998. Quantitation of spectinomycin residues in bovine tissues by ion-exchange high-performance liquid chromatography with post-column derivatization and confirmation by reversed-phase high-performance liquid chromatography-atmospheric pressure chemical ionization tandem mass spectrometry. J Chromatogr A 812: 123–133.

Hu P, Chess EK, Brynjelsen S, Jakubowski G, Melchert J, Hammond RB, Wilson TD. 2000. Collisionally activated dissociations of aminocyclitol-aminoglycoside antibiotics and their application in the identification of a new compound in tobramycin samples. J Am Soc Mass Spectrom 11: 200–209.

Hu N, Liu P, Jiang K, Zhou Y, Pan Y. 2008. Mechanism study of SO_2 elimination from sulfonamides by negative electrospray ionization mass spectrometry. Rapid Commun Mass Spectrom 22: 2715–2722.

Jadhav MR, Utture SC, Banerjee K, Oulkar DP, Sabale R, Shabeer TP. 2013. Validation of a residue analysis method for streptomycin and tetracycline and their food safety evaluation in pomegranate (Punica granatum L.). J Agric Food Chem 61: 8491–8498.

Junza A, Barbosa S, Codony MR, Jubert A, Barbosa J, Barrón D. 2014. Identification of metabolites and thermal transformation products of quinolones in raw cow's milk by liquid chromatography coupled to high-resolution mass spectrometry. J Agric Food Chem. PMID: 24499328, DOI: 10.1021/jf405554z

Kamel AM, Fouda HG, Brown PR, Munson B. 2002. Mass spectral characterization of tetracyclines by electrospray ionization, H/D exchange, and multiple stage mass spectrometry. J Am Soc Mass Spectrom 13: 543–557.

Karancsi T, Slegel P. 1999. Reliable molecular mass determination of aromatic nitro compounds: Elimination of gas-phase reduction occurring during atmospheric pressure chemical ionization. J Mass Spectrom 34: 975–977.

Kearney GC, Gates PJ, Leadlay PF, Staunton J, Jones R. 1999. Structural elucidation studies of erythromycins by electrospray tandem mass spectrometry II. Rapid Commun Mass Spectrom 13: 1650–1656.

Khong SP, Hammel YA, Guy PA. 2005. Analysis of tetracyclines in honey by high-performance liquid chromatography/tandem mass spectrometry. Rapid Commun Mass Spectrom 19: 493–502.

Klagkou K, Pullen F, Harrison M, Organ A, Firth A, Langley GJ. 2003. Fragmentation pathways of sulphonamides under electrospray tandem mass spectrometric conditions. Rapid Commun Mass Spectrom 17: 2373–2379.

Kumar Chitneni S, Govaerts C, Adams E, Van Schepdael A, Hoogmartens J. 2004. Identification of impurities in erythromycin by liquid chromatography–mass spectrometric detection. J Chromatogr A 1056: 111–120.

Lehotay SJ, Mastovska K, Lightfield AR, Nuñez A, Dutko T, Ng C, Bluhm L. 2013. Rapid analysis of aminoglycoside antibiotics in bovine tissues using disposable pipette extraction and ultrahigh performance liquid chromatography–tandem mass spectrometry. J Chromatogr A 1313: 103–112.

Leitner A, Zöllner P, Lindner W. 2001. Determination of the metabolites of nitrofuran antibiotics in animal tissue by high-performance liquid chromatography–tandem mass spectrometry. J Chromatogr A 939: 49–58.

Leonard S, Ferraro M, Adams E, Hoogmartens J, Van Schepdael A. 2006. Application of liquid chromatography/ion trap mass spectrometry to the characterization of the related substances of clarithromycin. Rapid Commun Mass Spectrom 20: 3101–3110.

Lehr GJ, Barry TL, Petzinger G, Hanna GM, Zito SW. 1999. Isolation and identification of process impurities in trimethoprim drug substance by high-performance liquid chromatography, atmospheric pressure chemical ionization liquid chromatography–mass spectrometry and nuclear magnetic resonance spectroscopy. J Pharm Biomed Anal 19: 373–389.

Li B, Van Schepdael A, Hoogmartens J, Adams E. 2009a. Characterization of impurities in tobramycin by liquid chromatography–mass spectrometry. J Chromatogr A 1216: 3941–3945.

Li B, Van Schepdael A, Hoogmartens J, Adams E. 2011. Mass spectrometric characterization of gentamicin components separated by the new European Pharmacopoeia method. J Pharm Biomed Anal 55: 78–84.

Li H, Kijak PJ, Turnipseed SB, Cui W. 2006a. Analysis of veterinary drug residues in shrimp: A multi-class method by liquid chromatography–quadrupole ion trap mass spectrometry. J Chromatogr B 836: 22–38.

Li W, Rettig J, Jiang X, Francisco DT, Naidong W. 2006b. Liquid chromatographic–electrospray tandem mass spectrometric determination of clarithromycin in human plasma. Biomed Chromatogr 20: 1242–1251.

Li YL, Hao XL, Ji BQ, Xu CL, Chen W, Shen CY, Ding T. 2009b. Rapid determination of 19 quinolone residues in spiked fish and pig muscle by high-performance liquid chromatography (HPLC) tandem mass spectrometry. Food Addit Contam A 26: 306–313.

Lindsey ME, Meyer TM, Thurman EM. 2001. Analysis of trace levels of sulfonamide and tetracycline antimicrobials in groundwater and surface water using solid-phase extraction and liquid chromatography/mass spectrometry. Anal Chem 73: 4640–4646.

Liu ZY, Zhou XN, Zhang HH, Wan L, Sun ZL. 2011. An integrated method for degradation products detection and characterization using hybrid ion trap/time-of-flight mass spectrometry and data processing techniques: Application to study of the degradation products of danofloxacin under stressed conditions. Anal Bioanal Chem 399: 2475–2486.

Liu ZY, Wu Y, Sun ZL, Wan L. 2012. Characterization of in vitro metabolites of trimethoprim and diaveridine in pig liver microsomes by liquid chromatography combined with hybrid ion trap/time-of-flight mass spectrometry. Biomed Chromatogr 26: 1101–1108.

Lovdahl MJ, Priebe SR. 2000. Characterization of clinafloxacin photodegradation products by LC–MS/MS and NMR. J Pharm Biomed Anal 23: 521–534.

Maia AS, Ribeiro AR, Amorim CL, Barreiro JC, Cass QB, Castro PM, Tiritan ME. 2014. Degradation of fluoroquinolone antibiotics and identification of metabolites/transformation products by liquid chromatography–tandem mass spectrometry. J Chromatogr A 1333: 87–98.

Mastovska K, Lightfield AR. 2008. Streamlining methodology for the multiresidue analysis of beta-lactam antibiotics in bovine kidney using liquid chromatography–tandem mass spectrometry. J Chromatogr A 1202: 118–123.

McGlinchey TA, Rafter PA, Regan F, McMahon GP. 2008. A review of analytical methods for the determination of aminoglycoside and macrolide residues in food matrices. Anal Chim Acta 624: 1–15.

McLaughlin LG, Henion JD, Kijak PJ. 1994. Multi-residue confirmation of aminoglycoside antibiotics and bovine kidney by ion spray high-performance liquid chromatography/tandem mass spectrometry. Biol Mass Spectrom 23: 417–429.

Mottier P, Khong SP, Gremaud E, Richoz J, Delatour T, Goldmann T, Guy PA. 2005. Quantitative determination of four nitrofuran metabolites in meat by isotope dilution liquid chromatography-electrospray ionisation-tandem mass spectrometry. J Chromatogr A 1067: 85–91.

Mottier P, Hammel YA, Gremaud E, Guy PA. 2008. Quantitative high-throughput analysis of 16 (fluoro)quinolones in honey using automated extraction by turbulent flow chromatography coupled to liquid chromatography–tandem mass spectrometry. J Agric Food Chem 56: 35–43.

Niessen WMA. 1998. Analysis of antibiotics by liquid chromatography–mass spectrometry. J Chromatogr A 812: 53–76.

Niessen WMA. 2005. Mass spectrometry of antibiotics. In: Nibbering NMM, (ed.), Encyclopedia of Mass Spectrometry. Volume 4: Fundamentals of and applications of organic (and organometallic) compounds. Elsevier Ltd, Oxford. pp. 822–837.

Ohki Y, Nakamura T, Nagaki H, Kinoshita T. 1992. Structural analysis of non-volatile compounds by liquid chromatography/mass spectrometry and liquid chromatography/tandem mass spectrometry: Thermal isomerization of benzylpenicillin in a Plasmaspray interface. Biol Mass Spectrom 21: 133–140.

Oka H, Ikai Y, Ito Y, Hayakawa J, Harada K, Suzuki M, Odani H, Maeda K. 1997. Improvement of chemical analysis of antibiotics. XXIII. Identification of residual tetracyclines in bovine tissues by electrospray high-performance liquid chromatography–tandem mass spectrometry. J Chromatogr B 693: 337–344.

Oka H, Ito Y, Ikai Y, Kagami T, Harada K. 1998. Mass spectrometric analysis of tetracycline antibiotics in foods. J Chromatogr A 812: 309–319.

Ortelli D, Cognard E, Jan P, Edder P. 2009. Comprehensive fast multiresidue screening of 150 veterinary drugs in milk by ultra-performance liquid chromatography coupled to time of flight mass spectrometry. J Chromatogr B 877: 2363–2374.

Partani P, Gurule S, Khuroo A, Monif T, Bhardwaj S. 2010. Liquid chromatography/electrospray tandem mass spectrometry method for the determination of cefuroxime in human plasma: Application to a pharmacokinetic study. J Chromatogr B 878: 428–434.

Pendela M, Hoogmartens J, Van Schepdael A, Adams E. 2009. Characterization of dihydrostreptomycin-related substances by liquid chromatography coupled to ion trap mass spectrometry. Rapid Commun Mass Spectrom 23: 1856–1862.

Pendela M, Béni S, Haghedooren E, Van den Bossche L, Noszál B, Van Schepdael A, Hoogmartens J, Adams E. 2012. Combined use of liquid chromatography with mass spectrometry and nuclear magnetic resonance for the identification of degradation compounds in an erythromycin formulation. Anal Bioanal Chem 402: 781–790.

Peru KM, Kuchta SL, Headley JV, Cessna AJ. 2006. Development of a hydrophilic interaction chromatography–mass spectrometry assay for spectinomycin and lincomycin in liquid hog manure supernatant and run-off from cropland. J Chromatogr A 1107: 152–158.

Pleasance S, Blay P, Quilliam MA, O'Hara G. 1991. Determination of sulfonamides by liquid chromatography, ultraviolet diode array detection and ion-spray tandem mass spectrometry with application to cultured salmon flesh. J Chromatogr 558: 155–173.

Pleasance S, Kelly J, LeBlanc MD, Quilliam MA, Boyd RK, Kitts DD, McErlane K, Bailey MR, North DH. 1992. Determination of erythromycin A in salmon tissue by liquid chromatography with ion-spray mass spectrometry. Biol Mass Spectrom 21: 675–687.

Rabbolini S, Verardo E, Da Col M, Gioacchini AM, Traldi P. 1998.Negative ion electrospray ionization tandem mass spectrometry in the structural characterization of penicillins. Rapid Commun Mass Spectrom 12: 1820–1826.

Raju B, Ramesh M, Srinivas R, Raju SS, Venkateswarlu Y. 2011. Identification and characterization of stressed degradation products of prulifloxacin using LC-ESI-MS/Q-TOF, MSn experiments: Development of a validated specific stability-indicating LC–MS method. J Pharm Biomed Anal 56: 560–568.

Rezende CP, Almeida MP, Brito RB, Nonaka CK, Leite MO. 2012. Optimisation and validation of a quantitative and confirmatory LC–MS method for multi-residue analyses of β-lactam and tetracycline antibiotics in bovine muscle. Food Addit Contam 29: 541–549.

Rønning HT, Einarsen K, Asp TN. 2006. Determination of chloramphenicol residues in meat, seafood, egg, honey, milk, plasma and urine with liquid chromatography–tandem mass spectrometry, and the validation of the method based on 2002/657/EC. J Chromatogr A 1118: 226–233.

Scandola M, Tarzin G, Gaviraghi G, Chiarello D, Traldi P. 1989. Mass spectrometric approaches in structural characterization of cephalosporins, Biomed Environ Mass Spectrom 18: 851–854.

Segura PA, Tremblay P, Picard P, Gagnon C, Sauvé S. 2010. High-throughput quantitation of seven sulfonamide residues in dairy milk using laser diode thermal desorption-negative mode atmospheric pressure chemical ionization tandem mass spectrometry. J Agric Food Chem 58: 1442–1446.

Shen Y, Yin C, Su M, Tu J. 2010, Rapid, sensitive and selective liquid chromatography–tandem mass spectrometry (LC–MS/MS) method for the quantitation of topically applied azithromycin in rabbit conjunctiva tissue. J Pharm Biomed Anal 52: 99–104.

Sheridan R, Policastro B, Thomas S, Rice D. 2008. Analysis and occurrence of 14 sulfonamide antibacterials and chloramphenicol in honey by solid-phase extraction followed by LC/MS/MS analysis. J Agric Food Chem 56: 3509–3516.

Straub RF, Voyksner RD. 1993. Determination of penicillin G, ampicillin, amoxicillin, cloxacillin and cephapirin by high-performance liquid chromatography-electrospray mass spectrometry. J Chromatogr 647(1): 167–181.

Suwanrumpha S, Freas RB. 1989. Identification of metabolites of ampicillin using liquid chromatography/thermospray mass spectrometry and fast atom bombardment tandem mass spectrometry. Biomed Environ Mass Spectrom 18: 983–994.

Tao Y, Chen D, Yu H, Huang L, Liu Z, Cao X, Yan C, Pan Y, Liu Z, Yuan Z. 2012. Simultaneous determination of 15 aminoglycoside(s) residues in animal derived foods by automated solid-phase extraction and liquid chromatography–tandem mass spectrometry. Food Chem 135: 676–683.

Turnipseed SB, Storey JM, Clark SB, Miller KE. 2011. Analysis of veterinary drugs and metabolites in milk using quadrupole time-of-flight liquid chromatography–mass spectrometry. J Agric Food Chem 59: 7569–7581.

Van den Bossche L, Daidone F, Van Schepdael A, Hoogmartens J, Adams E. 2013 Characterization of impurities in josamycin using dual liquid chromatography combined with mass spectrometry. J Pharm Biomed Anal 73: 66–76.

Van Eeckhout N, Perez JC, Van Peteghem C. 2000. Determination of eight sulfonamides in bovine kidney by liquid chromatography/tandem mass spectrometry with on-line extraction and sample clean-up. Rapid Commun Mass Spectrom 14: 2331–2338.

van Holthoon FL, Essers ML, Mulder PJ, Stead SL, Caldow M, Ashwin HM, Sharman M. 2009. A generic method for the quantitative analysis of aminoglycosides (and spectinomycin) in animal tissue using methylated internal standards and liquid chromatography tandem mass spectrometry. Anal Chim Acta 637: 135–143.

Volmer DA, Mansoori B, Locke SJ. 1997. Study of 4-quinolone antibiotics in biological samples by short-column liquid chromatography coupled with electrospray ionization tandem mass spectrometry. Anal Chem 69: 4143–4155.

Voyksner RD, Tyczkowska KL, Aronson AL. 1991. Development of analytical methods for some penicillins in bovine milk by ion-paired chromatography and confirmation by thermospray mass spectrometry. J Chromatogr 567: 389–404.

Wang Z, Hop CE, Kim MS, Huskey SE, Baillie TA, Guan Z. 2003. The unanticipated loss of SO_2 from sulfonamides in collision-induced dissociation. Rapid Commun Mass Spectrom 17: 81–86.

Wang H-Y, Zhang X, Guo Y-L, Dong X-C, Tang Q-H, Lu L. 2005. Sulfonamide bond cleavage in benzenesulfonamides and rearrangement of the resulting p-aminophenylsulfonyl cations: Application to a 2-pyrimidinyloxybenzylaminobenzenesulfonamide herbicide, Rapid Commun Mass Spectrom 19: 1696–1702.

Wang J. 2009. Analysis of macrolide antibiotics, using liquid chromatography–mass spectrometry, in food, biological and environmental matrices. Mass Spectrom Rev 28: 50–92.

Wang J, Aubry A, Bolgar MS, Gu H, Olah TV, Arnold M, Jemal M. 2010. Effect of mobile phase pH, aqueous-organic ratio, and buffer concentration on electrospray ionization tandem mass spectrometric fragmentation patterns: Implications in liquid chromatography/tandem mass spectrometric bioanalysis. Rapid Commun Mass Spectrom 24: 3221–3229.

Yu H, Tao Y, Chen D, Wang Y, Huang L, Peng D, Dai M, Liu Z, Wang X, Yuan Z. 2011. Development of a high performance liquid chromatography method and a liquid chromatography–tandem mass spectrometry method with the pressurized liquid extraction for the quantification and confirmation of sulfonamides in the foods of animal origin. J Chromatogr B 879: 2653–2662.

Yu H, Tao Y, Chen D, Pan Y, Liu Z, Wang Y, Huang L, Dai M, Peng D, Wang X, Yuan Z. 2012. Simultaneous determination of fluoroquinolones in foods of animal origin by a high performance liquid chromatography and a liquid chromatography tandem mass spectrometry with accelerated solvent extraction. J Chromatogr B 885–886: 150–159.

Zhu WX, Yang JZ, Wei W, Liu YF, Zhang SS. 2008. Simultaneous determination of 13 aminoglycoside residues in foods of animal origin by liquid chromatography-electrospray ionization tandem mass spectrometry with two consecutive solid-phase extraction steps. J Chromatogr A 1207: 29–37.

4.9

FRAGMENTATION OF ANTIMYCOTIC AND ANTIFUNGAL COMPOUNDS

An antimycotic or antifungal compound is used to treat fungal infections, such as tinea pedis (athlete's foot), tinea corporis (ringworm), and candidiasis (thrush). Antifungal compounds (fungicides) are also used in agriculture to prevent fungal growth in fruits, vegetables, and similar products. Therefore, multiresidue analysis of fungicides in such products is relevant. From a chemical point of view, antimycotic and antifungal agents form a diffuse group of compounds with several different structure classes. In LC–MS, antimycotic and antifungal compounds are mostly analyzed in positive-ion mode, although some data from negative-ion mode is available as well.

4.9.1 IMIDAZOLYL ANTIMYCOTIC COMPOUNDS

The conazoles featuring an imidazolyl (—$C_3H_3N_2$) or triazolyl (—$C_2H_2N_3$) moiety form part of a large group of antimycotic compounds: the azoles. These antimycotics inhibit the enzyme CYP51 14α-demethylase, which converts lanosterol into ergosterol. Depletion of ergosterol disrupts the structure as well as many functions of the fungal membranes leading to fungal growth inhibition. The LC–MS analysis of various azole antifungal drugs has been described for therapeutic drug monitoring, for example, as a multiresidue LC–MS method for conazoles (Alffenaar et al., 2010). Multiresidue environmental monitoring of conazoles has also been reported (Jeannot et al., 2000).

The imidazole antimycotics, such as bifonazole ($[M+H]^+$ with *m/z* 311), climbazole ($[M+H]^+$ with *m/z* 293), clotrimazole ($[M+H]^+$ with *m/z* 345), croconazole ($[M+H]^+$ with *m/z* 311), econazole ($[M+H]^+$ with *m/z* 381), enilconazole (imazalil; $[M+H]^+$ with *m/z* 297), fenticonazole ($[M+H]^+$ with *m/z* 455), isoconazole ($[M+H]^+$ with *m/z* 415) and its structural isomer miconazole ($[M+H]^+$ with *m/z* 415), ketoconazole ($[M+H]^+$ with *m/z* 531), omoconazole ($[M+H]^+$ with *m/z* 423), prochloraz ($[M+H]^+$ with *m/z* 376), sertaconazole ($[M+H]^+$ with *m/z* 437), and tioconazole ($[M+H]^+$ with *m/z* 387) show a characteristic loss of imidazole (68 Da, $C_3H_4N_2$) and/or a characteristic fragment ion with *m/z* 69 ($[C_3H_4N_2+H]^+$, protonated imidazole). In MS–MS, compounds with mono- or dichlorobenzyl moieties often show the production of mono- or dichlorobenzyl cations with *m/z* 125 ($[ClC_6H_4CH_2]^+$) or *m/z* 159 ($[Cl_2C_6H_3CH_2]^+$) ions, respectively, that is, substituted tropylium ions. These features are illustrated for econazole, tioconazole, and isocanazole (Figure 4.9.1). Similar fragment ions may be observed for other compounds, for example, the (2-chlorothiophen-3-yl) methylium ion ($[C_5H_4ClS]^+$) with *m/z* 131 for tioconazole (Figure 4.9.1), the [4-(phenylsulfanyl)benzyl cation ($[C_6H_5S—C_6H_4CH_2]^+$) with *m/z* 199 for fenticonazole, and the (7-chloro-1-benzothiophen-3-yl)methylium ion ($[C_9H_6ClS]^+$) with *m/z* 181 for sertaconazole. Apart from this class-specific fragmentation, further compound-specific fragment ions may be observed, sometime due to skeletal rearrangements of whole substituent groups. Econazole, tioconazole, isocanazole, and related compounds show a fragment ion due to a skeletal rearrangement with migration of the ether alkyl (often a benzyl) group to the imidazolyl ring. Thus, protonated 1-(4-chlorobenzyl) imidazole with *m/z* 193 is observed for econazole, protonated 1-[(2-chlorothiophen-3-yl)methyl]imidazole with *m/z* 199 for tioconazole, and protonated 1-(2,6-dichlorobenzyl) imidazole with *m/z* 227 for isoconazole and its 2,4-dichlorobenzyl isomer micoconazole (Figure 4.9.1). Oxiconazole ($[M+H]^+$ with *m/z* 428, an *O*-benzyloxime) and ketoconazole (a benzyl ethylene glycol ketal) do not show the fragment ion with *m/z* 69 and/or the loss of imidazole (68 Da); the molecular ion of 1-methylimidazole with *m/z* 82 ($[C_4H_6N_2]^{+\bullet}$) is observed instead.

The fragmentation of enilconazole ($[M+H]^+$ with *m/z* 297; imazalil) has also been investigated using ion-trap MS^n. After the loss of propene ($CH_3CH{=}CH_2$) to an

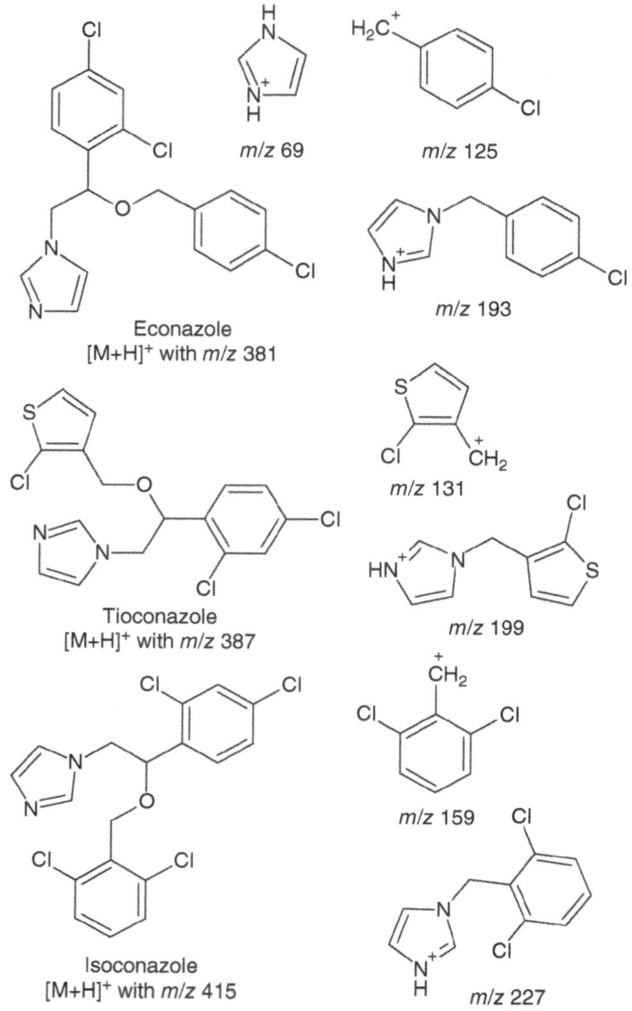

FIGURE 4.9.1 Proposed structures for the fragment ions of some protonated imidazole fungicides.

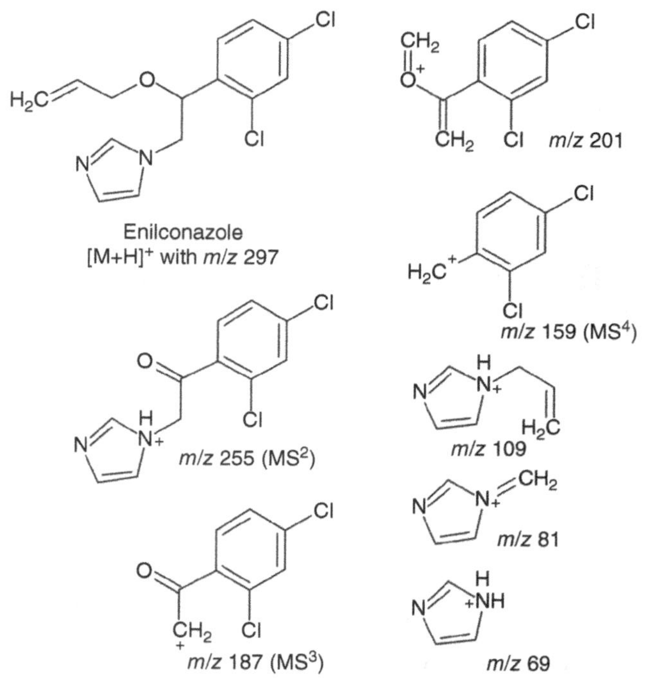

FIGURE 4.9.2 Proposed structures for the fragment ions of protonated enilconazole. The fragment ions observed in ion-trap MS^n are indicated. Other ions are observed under collision-cell CID conditions.

ion with *m/z* 255 in MS^2, further loss of imidazole ($C_3H_4N_2$) leads to the (2,4-dichlorobenzoyl)methylium ion with *m/z* 187 in MS^3, and a subsequent loss of CO with a rearrangement to the 2,4-dichlorobenzyl cation with *m/z* 159 ($[Cl_2C_6H_3CH_2]^+$) in MS^4 (Soler et al., 2007) (Figure 4.9.2). Using a Q–TOF instrument, the fragmentation was somewhat different (Picó et al., 2007). In addition to the ions with *m/z* 255 and 159 already discussed, a 2,4-dichlorostyren-α-yl(methylidene)oxonium ion ($[C_9H_7Cl_2O]^+$) with *m/z* 201 is observed. This ion effectively corresponds to the loss of 2-ethyl-1*H*-imidazole ($C_5H_8N_2$). Furthermore, protonated imidazole with *m/z* 69 ($[C_3H_4N_2+H]^+$), the 1-(methylidene)imidazol-1-ium ion ($[C_4H_4N_2]^+$) with *m/z* 81, and protonated 1-(prop-2-en-1-yl)-1*H*-imidazole with *m/z* 109 ($[C_6H_9N_2]^+$) are observed. The ion with *m/z* 109 results from a skeletal rearrangement of the allyl group to the imidazolyl N-atom (Figure 4.9.2).

4.9.2 TRIAZOLYL ANTIFUNGAL COMPOUNDS

Similarly, the fragmentation of 20 triazolyl antifungals was studied. They show a characteristic loss of 1*H*-1,2,4-triazole (69 Da, $C_2H_3N_3$) and/or a characteristic fragment ion with *m/z* 70 ($[C_2H_4N_3]^+$, protonated triazole). Compounds with a monochlorobenzyl group such as cyproconazole ($[M+H]^+$ with *m/z* 292), fenbuconazole ($[M+H]^+$ with *m/z* 337), metconazole ($[M+H]^+$ with *m/z* 320), myclobutanil ($[M+H]^+$ with *m/z* 289), and terbuconazole ($[M+H]^+$ with *m/z* 308) show the chlorobenzyl cation with *m/z* 125 ($[ClC_6H_4CH_2]^+$). Compounds with a dichlorobenzyl moiety such as bromuconazole ($[M+H]^+$ with *m/z* 376), dichlobutazole ($[M+H]^+$ with *m/z* 328), diniconazole ($[M+H]^+$ with *m/z* 326), etaconazole ($[M+H]^+$ with *m/z* 328), hexaconazole ($[M+H]^+$ with *m/z* 314), penconazole ($[M+H]^+$ with *m/z* 284), and propiconazole ($[M+H]^+$ with *m/z* 342) show the dichlorobenzyl cation with *m/z* 159 ($[Cl_2C_6H_3CH_2]^+$). The general fragmentation is shown in Figure 4.9.3 for fenbuconazole and hexaconazole. If no monochloro- or dichlorobenzyl groups are present, similar fragmentation may still be observed. Thus, the 2- or 4-fluorobenzyl cation ($[FC_6H_4CH_2]^+$) with *m/z* 109 is observed for flutriafol ($[M+H]^+$ with *m/z* 302) and the 2-chloro-4-(4-chlorophenoxy)benzyl cation ($[ClC_6H_4O—ClC_6H_3CH_2]^+$) with *m/z* 251 for difenoconazole ($[M+H]^+$ with *m/z* 406).

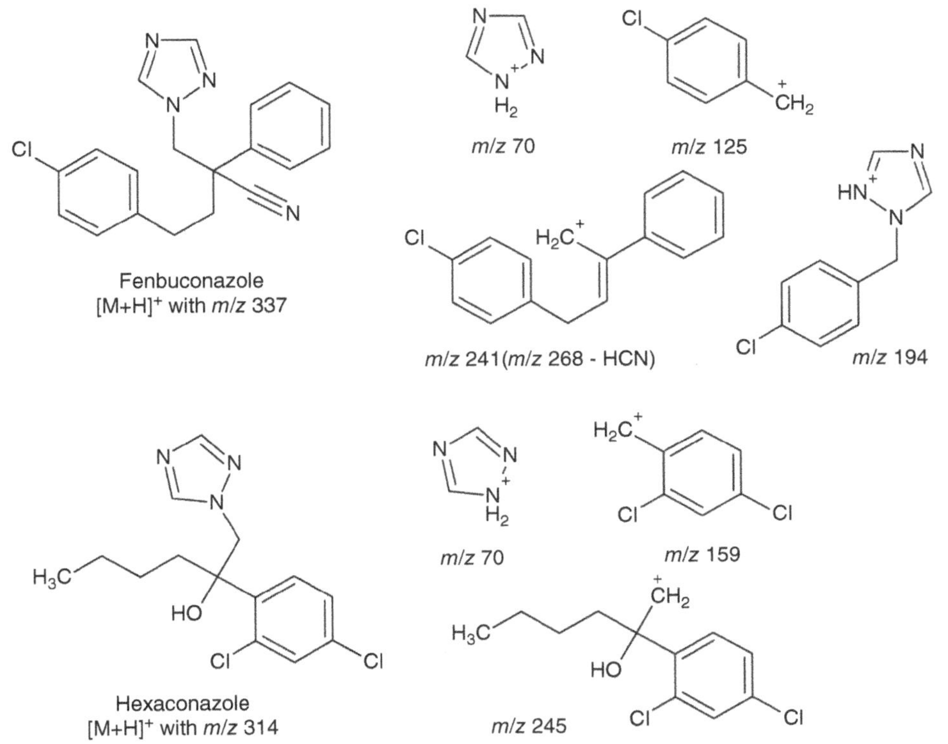

FIGURE 4.9.3 Proposed structures for the fragment ions of some protonated triazole fungicides.

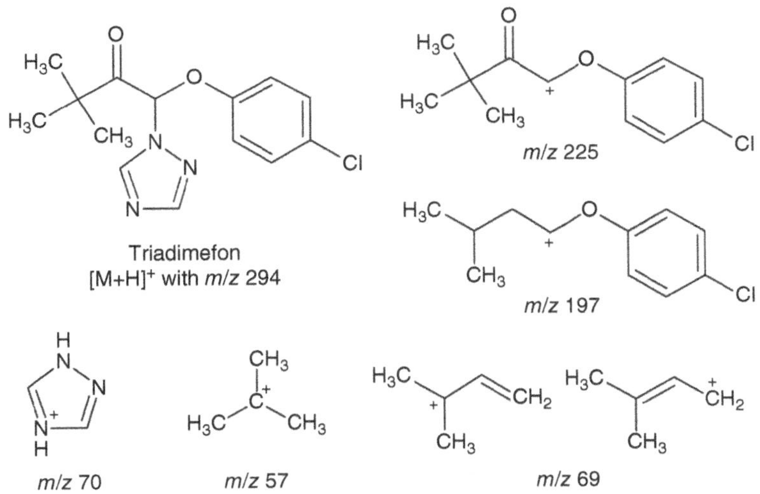

FIGURE 4.9.4 Proposed structures for the fragment ions of protonated triadimefon.

Apart from this class-specific fragmentation, which is similar to the imidazole antifungals, compound-specific fragmentation and/or other fragmentation reactions may be observed for some compounds. A skeletal rearrangement of the 4-chlorobenzyl group to the triazole ring is observed for fenbuconazole, resulting in the protonated 1-(4-chlorobenzyl)-2*H*-1,2,4-triazole with *m/z* 194 (Figure 4.9.3).

Triadimofon ($[M+H]^+$ with *m/z* 294) shows the loss of 1*H*-triazole to form an ion with *m/z* 225, an ion with *m/z* 197 due to the subsequent loss of carbon monoxide (CO) via a rearrangement of the *tert*-butyl group to the carbonyl α-C-atom, with a subsequent loss of 4-chlorophenol to an ion with *m/z* 69 ($[C_5H_9]^+$; e.g., the 3-methylbut-1-ene-3-yl carbocation), and the *tert*-butyl carbocation with 57 ($[C_4H_9]^+$) (Figure 4.9.4). The structure proposal for the

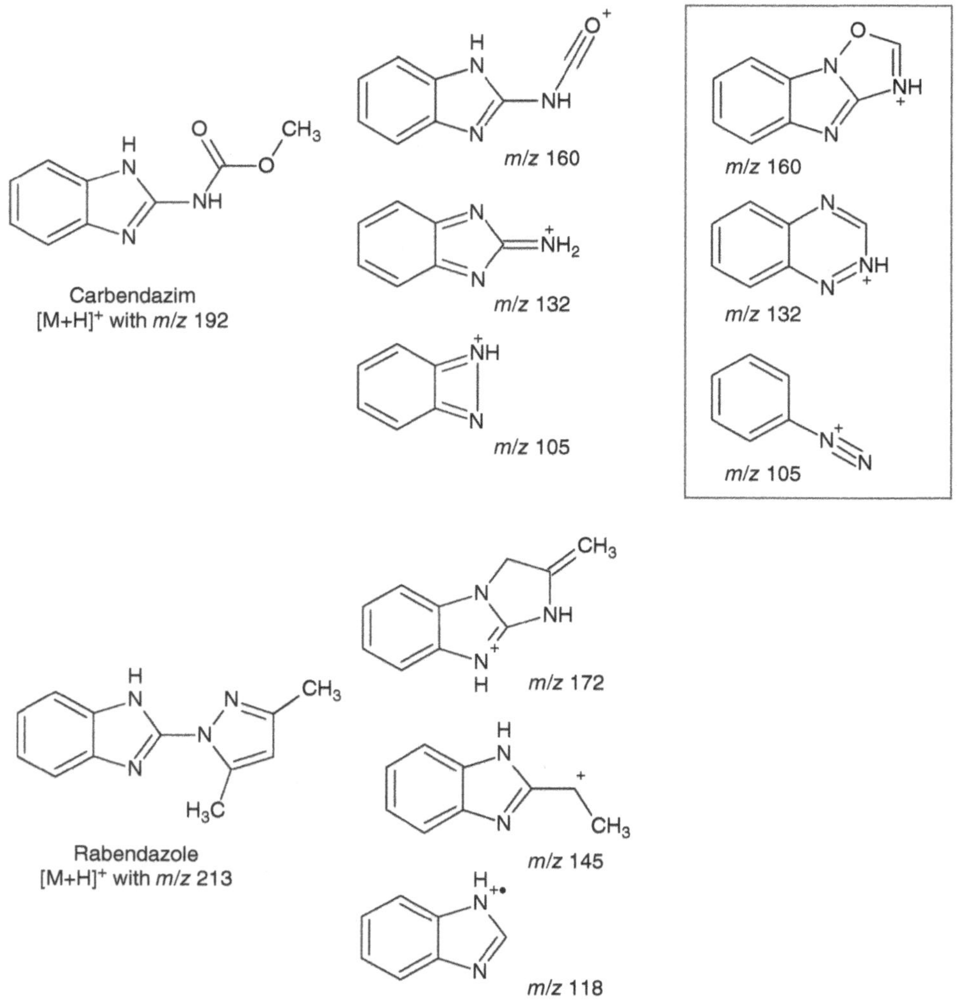

FIGURE 4.9.5 Proposed structures for the fragment ions of the protonated benzamidazole fungicides carbendazim and rabendazole. In the frame, structures proposed elsewhere are given for the fragment ions of carbendazim. (Source: Partly based and redrawn from (Picó et al., 2007) with permission of Elsevier.)

ion with *m/z* 197, involves a 3-methylbutoxy rather than a 2,2-dimethylpropoxy group, anticipating the formation of the ion with *m/z* 69. For propiconazole, apart from the ion with *m/z* 70 ($C_2H_4N_3^+$), a more abundant ion with *m/z* 69 ($[C_5H_9]^+$, e.g., the pent-1-ene-3-ylium ion) is observed, which is due to a cleavage in the 4-propyl-1,3-dioxolane group (an ethylene glycol ketal).

4.9.3 BENZAMIDAZOLE FUNGICIDES

Benzamidazole fungicides are widely used in agriculture for the conservation of fruits. Carbendazim ($[M+H]^+$ with *m/z* 192) shows subsequent losses of methanol (CH_3OH), carbon monoxide (CO), and hydrogen cyanide (HCN) to yield fragment ions with *m/z* 160, 132, and 105, respectively. The identity of the fragments was confirmed by accurate-mass determination using a Q–TOF instrument (Picó et al., 2007). The proposed structures for these fragment ions given here differ from the ones proposed elsewhere (Figure 4.9.5) (Picó et al., 2007). The formation of the proposed aromatic structure for the ion with *m/z* 132 would require (unlikely) skeletal and H-atom rearrangements. In contrast, the structure proposed here is also an aromatic ion formed by a single H-rearrangement.

Benomyl ($[M+H]^+$ with *m/z* 291) shows the loss of butylisocyanate (C_4H_9—N═C═O) to a fragment ion with *m/z* 192, which is structurally identical to carbendazim, and similar secondary fragment ions are observed, that is, ions with *m/z* 160, 132, and 105. Thiabendazole ($[M+H]^+$ with *m/z* 202), also used as anthelmintic (Section 4.10.1.1) shows major fragment ions with *m/z* 175 and 131, consistent with the subsequent losses of HCN and carbon monosulfide (CS). Some minor fragment ions, for example, the cyclopentadienyl cation with *m/z* 65 ($[C_5H_5]^+$) and the 2-aminophenylium ion with *m/z* 92 ($[H_2N—C_6H_4]^+$), are observed as well. Rabendazole ($[M+H]^+$ with *m/z* 213) shows the loss of acetonitrile ($CH_3C{\equiv}N$) to an ion with *m/z* 172 and the subsequent loss of HCN to *m/z* 145, which requires a rearrangement. The radical cation with *m/z* 118 is the molecular ion of benzimidazole ($[C_7H_6N_2]^{+\bullet}$) (Figure 4.9.5).

4.9.4 OTHER CLASSES OF FUNGICIDES

Allylamine antifungals, such as naftifine and terbinafine, show fragment ions due to cleavages of both C—N-bonds of the methylamine group, as illustrated for naftifine ($[M+H]^+$ with *m/z* 288) in Figure 4.9.6. The cleavage of the C—N bond to the 3-phenylprop-2-ene is a direct cleavage (Section 3.5.2), probably due to inability to perform a H-rearrangement. Thus, the protonated *N*-[(1-naphthyl)methyl]methanimine with *m/z* 170 ($[C_{12}H_{11}N+H]^+$), the 3-phenylprop-2-en-1-yl carbocation with *m/z* 117 ($[C_9H_9]^+$), and the (1-naphthyl)methylium ion with *m/z* 141 ($[C_{11}H_9]^+$) are observed. The same behavior is observed for terbinafine ($[M+H]^+$ with *m/z* 292), where two pairs of ions from direct cleavages of the C—N bonds on both sides of the methylamino group are observed. This results in the ($[C_{11}H_9]^+$) ion with *m/z* 141 and the protonated *N*-[2,2-dimethylhept-5-ene-3-yne-7-yl]methanimine with *m/z* 150 ($[C_{10}H_{15}N+H]^+$), and the protonated *N*-[(1-naphthyl)methyl]methanimine with *m/z* 170 ($[C_{12}H_{11}N+H]^+$) and the 6,6-dimethylhept-4-yne-2-enylium ion with *m/z* 121 ($[C_9H_{13}]^+$) (Figure 4.9.6).

The fragmentation of thiophanate-methyl ($[M+H]^+$ with *m/z* 343) involves the loss of methanol (CH_3OH) to an ion with *m/z* 311, followed by the loss of hydrogen isocyanate (HN=C=O) to *m/z* 268. Furthermore, the loss of methoxycarbonyl isothiocyanate ($CH_3OCON{=}C{=}S$) leads to an ion with *m/z* 226, which in turn shows subsequent losses

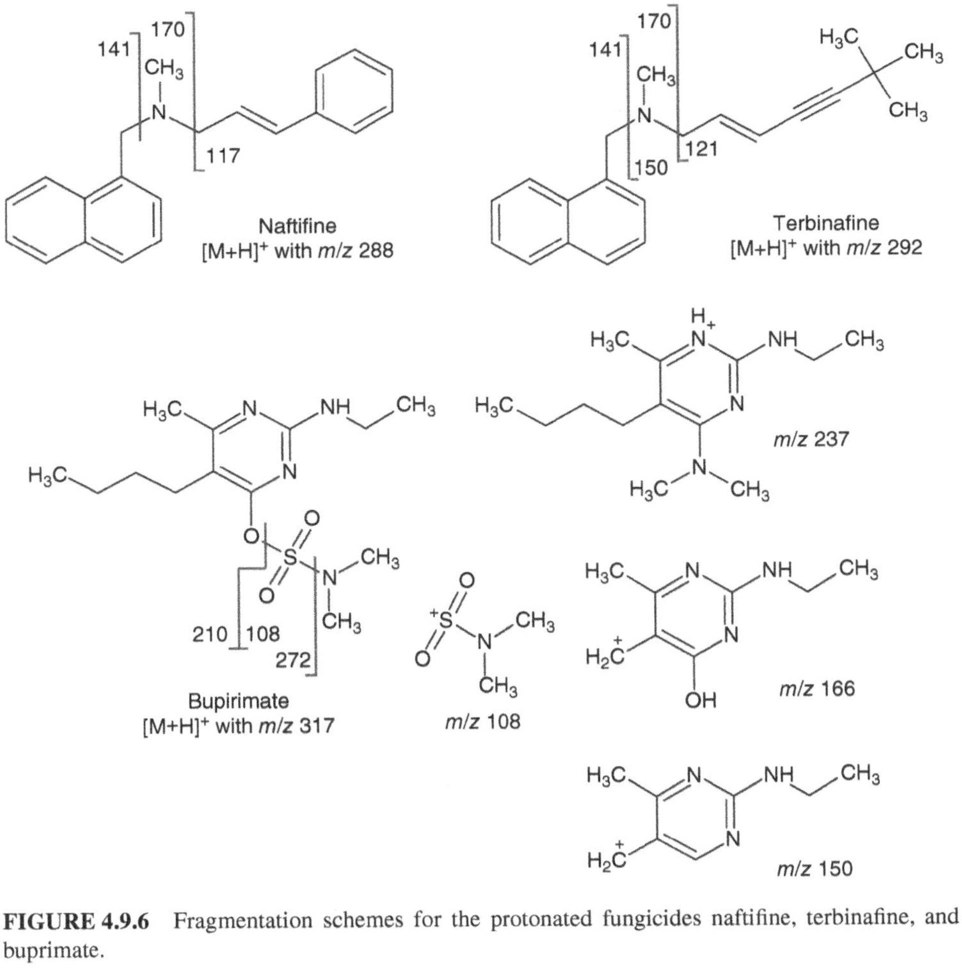

FIGURE 4.9.6 Fragmentation schemes for the protonated fungicides naftifine, terbinafine, and buprimate.

of CH_3OH to *m/z* 194 and HN=C=O to the protonated 2-isothiocyanatoaniline ($[H_2N—C_6H_4—N=C=S+H]^+$) with *m/z* 151 (Thurman et al., 2007).

The fragmentation of the related pyrimidine fungicides bupirimate and ethirimol was studied using ion-trap and Q–TOF instruments (Soler et al., 2007). The major fragment ions of bupirimate ($[M+H]^+$ with *m/z* 317, Figure 4.9.6) result from the loss of dimethyl amine ($(CH_3)_2NH$) to the ion with *m/z* 272, from a rearrangement of the dimethyl amino group to the pyrimidine ring associated with the loss of sulfur trioxide (SO_3) to an ion with *m/z* 237, and from the combined loss of *N*-methylmethanimine ($CH_3N=CH_2$) and sulfur dioxide (SO_2) to an ion with *m/z* 210 (Soler et al., 2007). The fragment ion with *m/z* 210 is the protonated molecule of ethirimol. The product-ion spectrum of ethirimol ($[M+H]^+$ with *m/z* 210) shows unusual fragmentation. However, given the errors in the calculations elsewhere (Soler et al., 2007), data from our library collection with different product ions are used here. The loss of ammonia (NH_3, which requires a rearrangement) results in the ion with *m/z* 193. The loss of ethene (C_2H_4) leads to an ion with *m/z* 182, which shows subsequent loss of a methyl radical ($^•CH_3$) to an ion with *m/z* 167. The loss of ethylamine ($C_2H_5NH_2$) leads to an ion with *m/z* 165. The ion with *m/z* 150 seems to be the result of a concerted loss of propanol (C_3H_7OH). The ion with *m/z* 140 ($[C_8H_{14}NO]^+$), involving the loss of ethylcyanamide ($C_2H_5NH—C{\equiv}N$), requires the rupture of the pyrimidine ring.

Furanilide fungicides, such as methfuroxam ($[M+H]^+$ with *m/z* 230), furalaxyl ($[M+H]^+$ with *m/z* 302), and fenfuram ($[M+H]^+$ with *m/z* 202), yield an intense acylium ion, containing the furan moiety, that is, the (2,3,5-trimethylfuran-3-yl)acylium ion ($[C_7H_9O—C=O]^+$) with *m/z* 137 for methfuroxam, the (furan-2-yl)acylium ion ($[C_4H_3O—C=O]^+$) with *m/z* 95 for furalaxyl, and the (2-methylfuran-3-yl)acylium ion with *m/z* 109 ($[C_5H_5O—C=O]^+$) for fenfuram. In addition, fragments due to the loss of the furan ring are observed for methfuroxam and fenfuram, resulting in the protonated isocyanatobenzene ($[C_6H_5—N=C=O+H]^+$) with *m/z* 120 for both compounds. Furalaxyl shows subsequent losses of methanol (CH_3OH) and carbon monoxide (CO) from the *N*-methylpropionate ester substituent to ions with *m/z* 270 and 242.

Quinomethionate ($[M+H]^+$ with *m/z* 235) shows fragment ions with *m/z* 207 due to the loss of CO, *m/z* 175 due to the loss of sulfanylidenemethanone (S=C=O), and *m/z* 163 due to the loss of CO and carbon monosulfide (CS). Structure proposals for some other fragment ions (Figure 4.9.7) are based on molecular formulae derived from accurate-mass data.

The fungicides tolnaftate ($[M+H]^+$ with *m/z* 308) and tolciclate ($[M+H]^+$ with *m/z* 324) show the same fragments. The *N*-(hydroxythioformyl)-*N*-(3-methylphenyl) methaniminium ion with *m/z* 180 results from the loss of naphthalene in tolnaftate and of benzonorbornane ($C_{11}H_{12}$) in tolciclate. A thiono–thiolo rearrangement, discussed in more detail in Section 4.11.4, is observed in the formation of the fragment ions with *m/z* 164 and 148, that is, the *N*-thioformyl- and *N*-formyl-*N*-(3-methylphenyl)methaniminium ions, respectively (Figure 4.9.8).

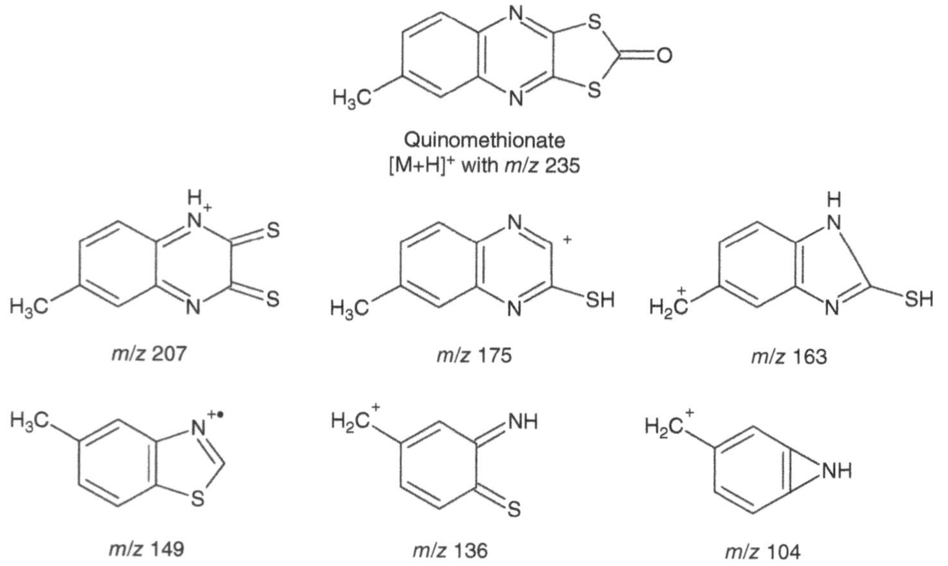

FIGURE 4.9.7 Proposed structures for the fragment ions of the protonated fungicide quinomethionate.

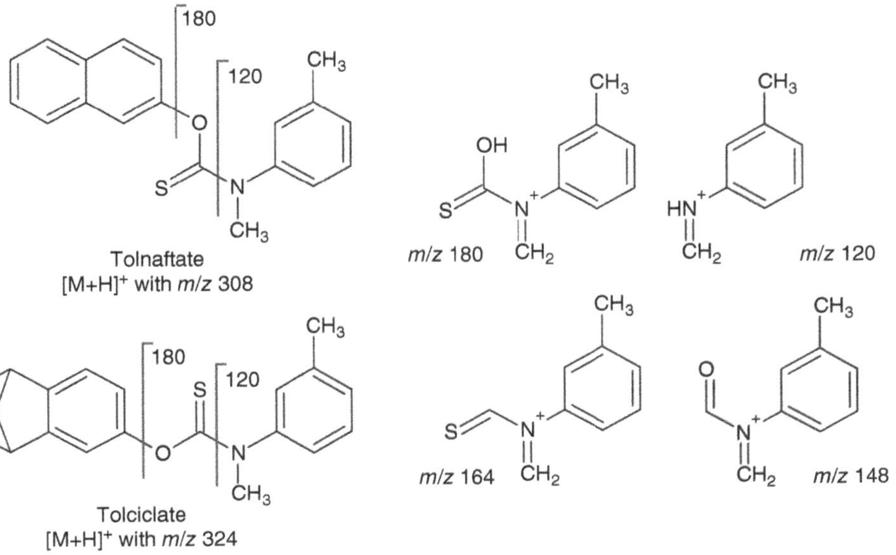

FIGURE 4.9.8 Fragmentation schemes and proposed structures for some of the fragment ions of protonated tolnaftate and tolciclate.

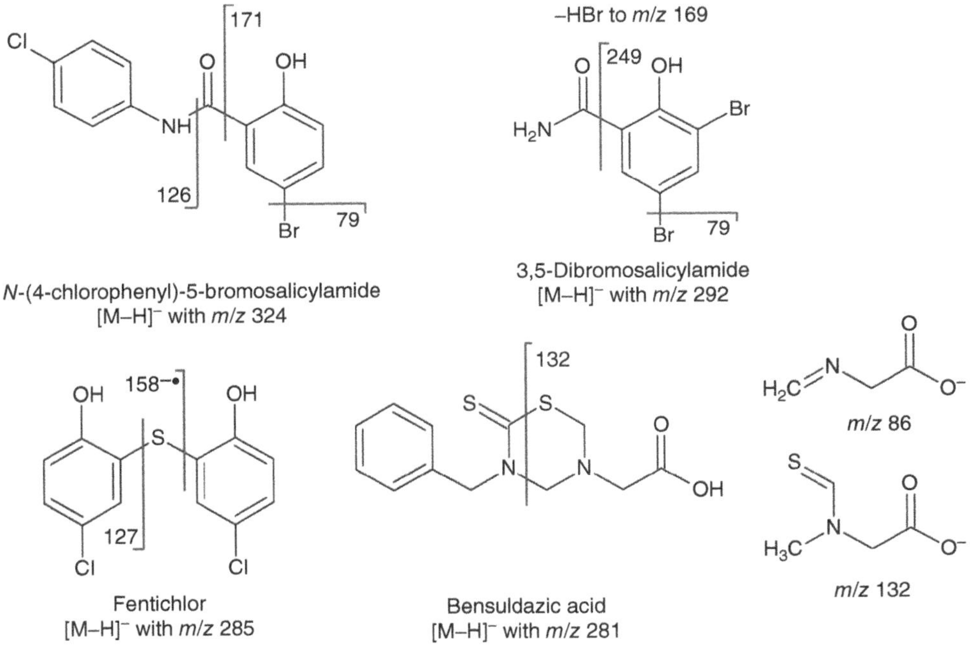

FIGURE 4.9.9 Fragmentation schemes for some deprotonated fungicides.

The negative-ion MS–MS spectra of fungicides present in our library collection are related to a wide variety of structures. Proposed fragmentation patterns for some of these compounds, that is, *N*-(4-chlorophenyl)-5-bromosalicylamide ($[M-H]^-$ with *m/z* 324), 3,5-dibromosalicylamide ($[M-H]^-$ with *m/z* 292), fentichlor ($[M-H]^-$ with *m/z* 285), and benzuldazic acid ($[M-H]^-$ with *m/z* 281), are shown in Figure 4.9.9. Clodantoin ($[M-H]^-$ with *m/z* 345) shows a major fragment ion with *m/z* 195 consistent with the loss of trichloromethanethiol ($HSCCl_3$). Fludioxonil ($[M-H]^-$ with *m/z* 247) shows a fragment ion with *m/z* 180 due to the loss of the difluoromethoxy radical ($CHF_2O^\bullet$) and with *m/z* 126 due to subsequent loss of CO and the cyano radical ($^\bullet CN$). Fenaminosulf ($[M-H]^-$ with *m/z* 228) only shows a radical anion with *m/z* 80 ($SO_3^{-\bullet}$).

Dichloran ($[M-H]^-$ with *m/z* 205) shows fragment ions with *m/z* 175 due to the loss of a nitrosyl radical ($^{\bullet}NO$), with *m/z* 159 due to the loss of a nitryl radical ($^{\bullet}NO_2$), and with *m/z* 123 due to the subsequent loss of hydrogen chloride (HCl).

REFERENCES

Alffenaar JW, Wessels AM, van Hateren K, Greijdanus B, Kosterink JG, Uges DR. 2010. Method for therapeutic drug monitoring of azole antifungal drugs in human serum using LC/MS/MS. J Chromatogr B 878: 39–44.

Fernández M, Rodríguez R, Picó Y, Mañes J. 2001. Liquid chromatographic–mass spectrometric determination of post-harvest fungicides in citrus fruits. J Chromatogr A 912: 301–310.

Jeannot R, Sabik H, Sauvard E, Genin E. 2000. Application of liquid chromatography with mass spectrometry combined with photodiode array detection and tandem mass spectrometry for monitoring pesticides in surface waters. J Chromatogr A 879: 51–71.

Picó Y, la Farré M, Soler C, Barceló D. 2007. Identification of unknown pesticides in fruits using ultra-performance liquid chromatography–quadrupole–time-of-flight mass spectrometry. Imazalil as a case study of quantification. J Chromatogr A 1176: 123–134.

Soler C, James KJ, Picó Y. 2007. Capabilities of different liquid chromatography tandem mass spectrometry systems in determining pesticide residues in food. Application to estimate their daily intake. J Chromatogr A 1157: 73–84.

4.10

FRAGMENTATION OF OTHER ANTIBIOTIC COMPOUNDS

4.10.1 ANTHELMINTIC DRUGS

Anthelmintics are compounds that are used to treat various types of intestinal infections with parasitic worms such as helminths: flatworms (platyhelminthes), tapeworms (Cestoda), and flukes (Trematoda), as well as round worms: nemathelminths; ectoparasites like leeches (Annelida) are also treated with anthelmintics. Their mode of action is either by stunning the helmiths, that is, creating an environment that forces them to leave the host (vermifuges) or by targeting key metabolic pathways that are essential to their survival (vermicides). These compounds are used mainly in veterinary medicine.

4.10.1.1 Benzimidazole Anthelmintics

Benzimidazole anthelmintics are one of the relevant compound classes, which are also used as fungicides (Section 4.9.3). Analytical methodologies for benzimidazoles have been reviewed (Danaher et al., 2007). In LC–MS, the benzimidazoles are analyzed in positive-ion mode. Multiresidue analysis based on LC–MS has been reported in edible tissues (Balizs, 1999; Kinsella et al., 2010; Chen et al., 2011) and in milk (De Ruyck et al., 2002; Kinsella et al., 2009; Xia et al., 2010, Whelan et al., 2010). Metabolite studies have also been reported (Cvilink et al., 2008). The identity of three product ions, frequently used in SRM during veterinary residue analysis, has been elucidated for several benzimidazole anthelmintics (Nuñez et al., 2015).

Benzimidazole anthelmintics with a methyl carbamate group (—$NHCOOCH_3$) show fragmentation similar to the fungicide carbendazim (Section 4.9.3). Mebendazole ($[M+H]^+$ with *m/z* 296) shows the loss of methanol (CH_3OH) from the methylcarbamate group to an ion with *m/z* 264 and the benzoyl cation ($[C_6H_5—C{=}O]^+$) with *m/z* 105. The same fragmentation is observed for flubendazole ($[M+H]^+$ with *m/z* 314), that is, an ion with *m/z* 282 due to the loss of CH_3OH and the 4-fluorobenzoyl cation ($[FC_6H_4—C{=}O]^+$) with *m/z* 123. Albendazole ($[M+H]^+$ with *m/z* 266) shows fragment ions with *m/z* 234 due to the loss of CH_3OH and with *m/z* 191 due to subsequent loss of a propyl radical ($^{\bullet}C_3H_7$) from the propylsulfanyl substituent. Fenbendazole ($[M+H]^+$ with *m/z* 300) shows fragment ions with *m/z* 268 due to the loss of CH_3OH and with *m/z* 159, that is, the molecular ion of 1*H*-(benzimidazol-2-yl)isocyanate ($[C_7H_5N_2—N{=}C{=}O]^{+\bullet}$) due to the subsequent loss of the phenylsulfanyl radical ($C_6H_5S^{\bullet}$). The ion with *m/z* 159 is also observed for albendazole sulfoxide ($[M+H]^+$ with *m/z* 282) and sulfone ($[M+H]^+$ with *m/z* 298), fenbendazole sulfoxide (oxfendazole, $[M+H]^+$ with *m/z* 316), and sulfone ($[M+H]^+$ with *m/z* 332). The fragment ion due to the loss of CH_3OH and the ion with *m/z* 159 are frequently used in SRM (Whelan et al., 2010; Chen et al., 2011).

4.10.1.2 Avermectin Anthelmintics

Avermectins are 16-membered macrocyclic lactone derivatives, generated as fermentation products by the soil-living bacteria *Streptomyces avermitilis*. These compounds have potent anthelmintic and insecticidal properties. They occur in pairs of homologue compounds (a and b), with a major component (a-component) being present in 80–90%. The a-component differs by one methylene (—CH_2) unit ($\Delta m = +14.016$ Da) from the b-component. Targeted analysis is usually directed at the major component.

Various methods for the analysis of avermectins have been reviewed (Danaher et al., 2006). Avermectins can be analyzed in both positive-ion and negative-ion modes (Durden, 2007). Analysis in ESI-MS is preferred over positive-ion APCI-MS. In negative-ion ESI-MS, $[M-H]^-$ is used as precursor ion, whereas in positive-ion ESI-MS, the precursor ion is often $[M+NH_4]^+$. If an N-atom is present in the structure as in selamectin (oxime), emamectin (amino), moxidectin (*O*-methyloxime), and eprinomectin (amide), $[M+H]^+$ can be used as precursor ion. Multiresidue LC–MS methods for avermectins, often in combination with other anthelmintics, have been reported (Durden, 2007; Kinsella et al., 2010; Whelan et al., 2010). The identity of three product ions, frequently used in SRM during veterinary residue analysis, has been elucidated for seven avermectin anthelmintics (Nuñez et al., 2015).

In APCI, thermally induced dissociation is observed, as may be concluded from the (apparent) fragment ions monitored in positive-ion mode for some avermectins using SIM (Turnipseed et al., 1999). The use of combined APCI and APPI for negative-ion LC–MS of some avermectins was also reported (Turnipseed, et al., 2005). Negative-ion APCI has been used by others (Howells & Sauer, 2001).

Fragmentation schemes of avermectins have been reported (Gianelli et al., 2000; Howells & Sauer, 2001; Beasley et al., 2006; Lehner et al., 2009). The fragmentation of ESI-MS generating $[M+Na]^+$ was studied using ion-trap MS^n (Gianelli et al., 2000) and both ion-trap MS^n and Q–TOF instruments (Lehner et al., 2009). The fragmentation observed under these conditions differs from the fragmentation of $[M+H]^+$ or $[M+NH_4]^+$ in TQ instruments. Most of the fragment ions of $[M+Na]^+$ can still be considered as sodiated molecules. Although cleavages of the glycosidic bonds with charge retention on either fragment does occur, extensive cleavages of the 16-membered macrocyclic lactone ring also occur under these conditions.

The characteristic fragmentation of $[M+H]^+$ or $[M+NH_4]^+$ of avermectins is outlined for doramectin ($[M+H]^+$ with *m/z* 899.5) in Figure 4.10.1. The interpretation is based on the fragmentation scheme reported for ivermectin B_{1a} ($H2B_{1a}$; $[M+NH_4]^+$ with *m/z* 892.5), obtained via positive-ion APCI in an ion-trap instrument (Beasley et al., 2006). The major fragments reported for ivermectin B_{1a} are the carbocations with *m/z* 713 ($[C_{41}H_{61}O_{10}]^+$) due to the loss of one L-oleandrose ($C_7H_{14}O_4$) residue and ammonia (NH_3), *m/z* 569 ($[C_{34}H_{49}O_7]^+$) due to the loss of L-oleandrosyl-L-oleandroside disaccharide, and the spiroketal fragment ion with *m/z* 307 due to a McLafferty rearrangement and subsequent rearrangement of the lactone ring ($[C_{19}H_{31}O_3]^+$, *m/z* 307.227). For doramectin, the same fragmentation leads to the ions with *m/z* 737 (loss of one sugar, $[C_{43}H_{61}O_{10}]^+$), *m/z* 593 (loss of both sugars, $[C_{36}H_{49}O_7]^+$), and the spiroketal fragment with *m/z* 331 ($[C_{21}H_{30}O_3+H]^+$). In TQ MS–MS spectra, fragment ions not only due to the same cleavages, but also with charge retention on the sugar units are observed. Most of the fragment ions of various avermectins, used in SRM transitions for residue analysis, can be explained along the same lines (Table 4.10.1). For three compounds, the fragmentation is different from the fragmentation scheme outlined in Figure 4.10.1. This is the case for moxidectin ($[M+H]^+$ with *m/z* 640.4), which lacks sugar groups. The fragmentation of emamectin ($[M+H]^+$ with *m/z* 886.5) and eprinomectin ($[M+H]^+$ with *m/z* 914.5) involves the cleavage of the glycosidic bonds, resulting in fragment ions with charge retention on the amido saccharide units.

The fragment ions generally used in SRM transitions in negative-ion mode are also reported in Table 4.10.1. The ions with *m/z* 565 for abamectin B_{1a} ($[M-H]^-$ with *m/z* 871.5), emamectin B_{1a} ($[M-H]^-$ with *m/z* 884.5), and eprinomectin B_{1a} ($[M-H]^-$ with *m/z* 912.5), with *m/z* 567 for ivermectin B_{1a} ($[M-H]^-$ with *m/z* 873.5), and with 591 for doramectin ($[M-H]^-$ with *m/z* 897.5) are due the loss of the two sugar units. For the other fragment ions, no interpretation is given, mainly because no accurate-mass data are available.

4.10.1.3 Other Anthelmintic Compounds

Tetramisole is a synthetic imidazothiazole derivative with, among other, anthelmintic properties. In MS–MS, tetramisole ($[M+H]^+$ with *m/z* 205) shows fragment ions with *m/z* 178 due to the loss of hydrogen cyanide (HCN), with *m/z* 146 due to the loss of hydrogen thiocyanate (HS=C=N), and with *m/z* 123, the (sulfanyl)benzyl cation ($[C_6H_5CHSH]^+$) resulting from a skeletal rearrangement (effective loss of $C_4H_6N_2$).

Pyrantel ($[M+H]^+$ with *m/z* 207) and morantel ($[M+H]^+$ with *m/z* 221) show very similar fragmentation because morantel only differs from pyrantel by an additional 3-methyl substituent on the thiophene ring. Since upon fragmentation the charge is retained on the thiophene rather than the pyrimidine ring, the fragment ions of pyrantel and morantel show an *m/z* difference of +14 (CH_2). The radical cation with *m/z* 174 (for pyrantel) results from the loss of the sulfanyl radical ($^{\bullet}SH$). The loss of *N*-ethylmethanimine ($C_2H_5N{=}CH_2$) results in an ion with *m/z* 150. The thenyl cation ($[C_4H_3SCH_2]^+$) with *m/z* 97 and the *N*-(propyl-3-ylium)methanimine ion with *m/z* 70 ($[C_4H_8N]^+$) due to cleavage of the pyrimidine ring are also observed. The structures proposed for the fragment ions of pyrantel are shown in Figure 4.10.2.

Praziquantel ($[M+H]^+$ with *m/z* 313), a pyrazinoisoquinolinone anthelmintic compound, shows a major fragment ion with *m/z* 203, which is due to the loss of cyclohexene (C_6H_{10}) and carbon monoxide (CO), and minor fragment ions with *m/z* 285 due to the loss of CO and with *m/z* 83 due to the cyclohexylium ion ($[C_6H_{11}]^+$).

The furocoumarin psoralene ($[M+H]^+$ with *m/z* 187) shows fragment ions with *m/z* 159 due to CO loss, *m/z* 143 due to carbon dioxide (CO_2) loss, and *m/z* 131 due to the loss of ethenedione (O=C=C=O).

Niclosamide ($[M-H]^-$ with *m/z* 325) is an anthelmintic specifically effective against tapeworms. Analyzed in negative-ion mode, its major fragment ion is due to the cleavage of the N—C carbonyl bond of the amide group, resulting in the loss of $C_7H_3ClO_2$ to the 2-chloro-4-nitroanilinide anion ($[C_6H_4ClN_2O_2]^-$) with *m/z* 171. Both $[M-H]^-$ and the fragment ion with *m/z* 171 show subsequent loss of

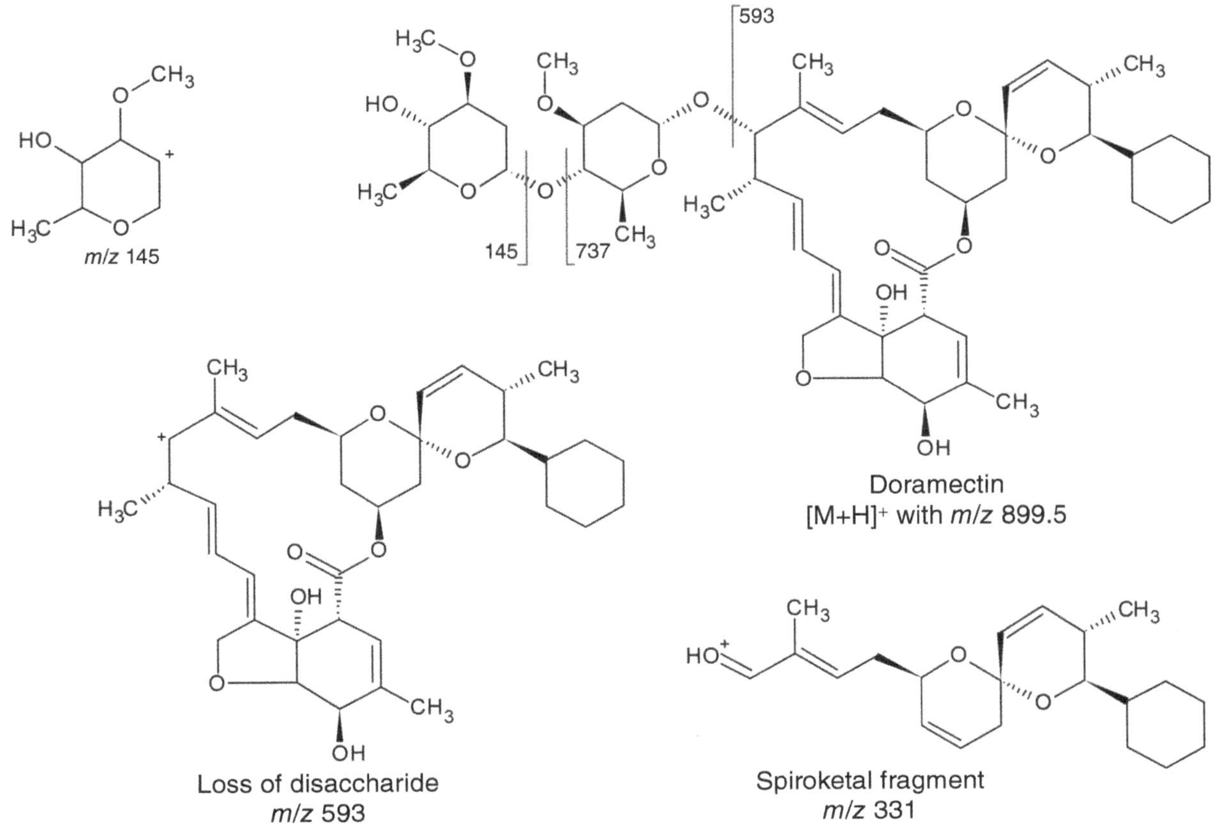

FIGURE 4.10.1 Fragmentation scheme and proposed structures for the fragment ions of protonated anthelmintic drug doramectin.

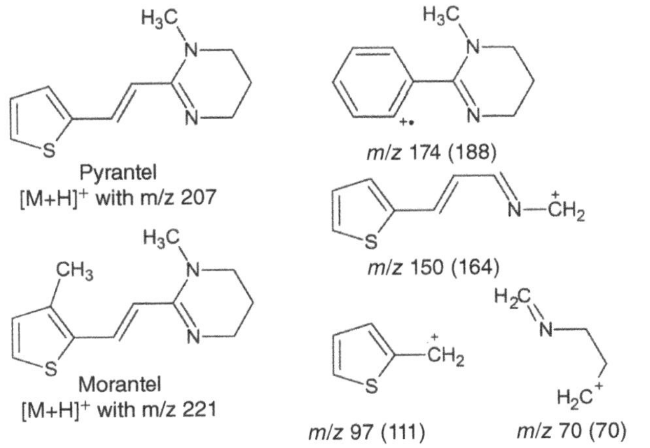

FIGURE 4.10.2 Proposed structures for the fragment ions of the protonated anthelmintics pyrantel and morantel (*m/z* of its fragment ions in parenthesis).

hydrogen chloride (HCl) to fragment ions with *m/z* 289 and 135, respectively. The ions with *m/z* 171 and 289 have been used in SRM (Caldow et al., 2009). Bromothymol ($[M-H]^-$ with *m/z* 227) yields the bromide anion (Br^-) with *m/z* 79. Nitroscanate ($[M-H]^-$ with *m/z* 271) shows fragment ions with *m/z* 239 due to the loss of atomic sulfur (S), with *m/z* 193 due to the loss of S and the nitryl radical ($^\bullet NO_2$), and the 4-isothiocyanatophen-2-ide-oxyl radical anion ($[S{=}C{=}N{-}C_6H_3O]^{-\bullet}$) with *m/z* 149 due to the loss of a 4-nitrophenyl radical ($O_2N{-}C_6H_4^\bullet$). Rafoxanide ($[M-H]^-$ with *m/z* 624) shows the iodide anion (I^-) with *m/z* 127 and the 2,4-diiodophenoxide ion with *m/z* 345 ($[I_2C_6H_3O]^-$).

4.10.2 ANTIPROTOZOAL, COCCIDIOSTATIC, AND ANTIMALARIAL AGENTS

Antiprotozoal agents are a class of drugs used for the treatment of protozoan infection. As the pathogens can be from very different origins, the drugs involved also have little common structural features. A coccidiostat is an antiprotozoal agent that acts upon *Coccidia* parasites. Antimalarial agents are used to prevent and treat infections with protozoans of the plasmodium type.

TABLE 4.10.1 Molecular formulae and characteristic fragment ions frequently used in SRM transitions for avermectin anthelmintics.

Compound	Formula	Mass (Da)	$[M+H]^+$	$[M+NH_4]^+$	Fragment +ESI (*m/z*)	Interpretation	Fragment −ESI (*m/z*)
Abamectin B_{1a}	$C_{48}H_{72}O_{14}$	872.492		890.5	567	Loss of disaccharide	565, 229
					305	Spiroketal fragment	
					145	Monosaccharide	
Avermectin B_{1a}	$C_{48}H_{72}O_{14}$	872.492		890.5	567	Loss of disaccharide	
					307	Spiroketal fragment	
Doramectin	$C_{50}H_{74}O_{14}$	898.508		916.5	593	Loss of disaccharide	591, 229
					331	Spiroketal fragment	
					145	Monosaccharide	
Ivermectin B_{1a} ($H2B_{1a}$)	$C_{48}H_{74}O_{14}$	874.508		892.5	569	Loss of disaccharide	567, 229
					307	Spiroketal fragment	
					145	Monosaccharide	
Selamectin	$C_{43}H_{63}NO_{11}$	769.440	770.4		626	Loss of saccharide	750, 722
					158	?	
Moxidectin	$C_{37}H_{53}NO_8$	639.377	640.4		528	($C_{30}H_{42}NO_7^+$); loss of $C_7H_{12}O$ from spiroketal	528, 247
					498	($C_{29}H_{40}NO_6^+$); loss of $H_2C{=}O$ from *m/z* 528	
Emamectin B_{1a}	$C_{49}H_{75}NO_{13}$	885.534	886.5		302	Disaccharide	565, 242
					158	Monosaccharide	
Eprinomectin B_{1a}	$C_{50}H_{75}NO_{14}$	913.519	914.5		330	Disaccharide	565, 270
					186	Monosaccharide	

4.10.2.1 Nitroimidazoles

Nitroimidazoles such as metronidazole can be used against anaerobic bacteria and protozoa. Nitroheterocycles may be reductively activated in hypoxic cells, and then undergo redox recycling or decompose to toxic products. Nitroimidazoles possess mutagenic and carcinogenic properties and their use as veterinary drugs is prohibited in the European Union. The nitroimidazoles are frequently analyzed together with their hydroxy metabolites in multiresidue analysis (Zeleny et al., 2009; Cronly et al., 2009; Xia et al., 2009). The identity of three product ions, frequently used in SRM during veterinary residue analysis, has been elucidated for several nitroimidazoles (Nuñez et al., 2015).

The major fragments of dimetridazole ($[M+H]^+$ with *m/z* 142) are the ions with *m/z* 112 due to the loss of a nitrosyl radical ($^\bullet NO$), *m/z* 96 due to the loss of a nitryl radical ($^\bullet NO_2$), and *m/z* 81 due to the loss of $^\bullet NO_2$ and of a methyl radical ($^\bullet CH_3$). The ion with *m/z* 54 is the 1-(methaniminyl)vinyl cation ($[H_2C{=}N{-}C{=}CH_2]^+$) (Figure 4.10.3). Its hydroxy metabolite ($[M+H]^+$ with *m/z* 158) shows fragments with *m/z* 140 due to the loss of water, *m/z* 94 due to subsequent loss of $^\bullet NO_2$, and the ion with *m/z* 55, the molecular ion of *N*-vinylmethanimine ($[H_2C{=}N{-}CH{=}CH_2]^{+\bullet}$). Metronidazole ($[M+H]^+$ with *m/z* 172) shows an ion with *m/z* 128 due the loss of ethanal (CH_3CHO) from the N^1-substituent of the imidazole ring, and subsequent loss of $^\bullet NO_2$ to an ion with *m/z* 82, the molecular ion of 2-methyl-1*H*-imidazole ($[C_4H_6N_2]^{+\bullet}$). Its hydroxy metabolite ($[M+H]^+$ with *m/z* 188) shows fragments with *m/z* 144 due to the loss of CH_3CHO, *m/z* 126 due to the subsequent loss of water, and *m/z* 123 due to the loss of nitrous acid (HNO_2) and water. Ronidazole ($[M+H]^+$ with *m/z* 201) shows the same metabolite as dimetridazole. It shows an ion with *m/z* 140 due to the loss of carbamic acid (H_2NCOOH) followed by the loss of $^\bullet NO$ to an ion with *m/z* 110, and the ion with *m/z* 55 ($[H_2C{=}N{-}CH{=}CH_2]^{+\bullet}$). Ipronidazole ($[M+H]^+$ with *m/z* 170) shows an ion with *m/z* 124 due to the loss of $^\bullet NO_2$ and an ion with *m/z* 109 due to the subsequent loss of $^\bullet CH_3$. Its hydroxy metabolite ($[M+H]^+$ with *m/z* 186) shows an ion with *m/z* 168 due to the loss of water and an ion *m/z* 122 due to the subsequent loss of $^\bullet NO_2$.

4.10.2.2 Chemical Coccidiostats

There are two classes of coccidiostats, that is, chemical coccidiostats, discussed in this section, and ionophores, discussed in the next section. The various chemical coccidiostatics show no clear structure similarities. The analysis of both types of coccidiostats in meat and eggs has been recently reviewed (Clarke et al., 2014). Multiresidue analysis of chemical coccidiostats has been reported (Mortier et al., 2005a; Dubreil-Chéneau et al., 2009; Olejnik et al., 2009; Moloney et al., 2012). Some compounds are analyzed in the positive-ion mode, some others in the negative-ion mode.

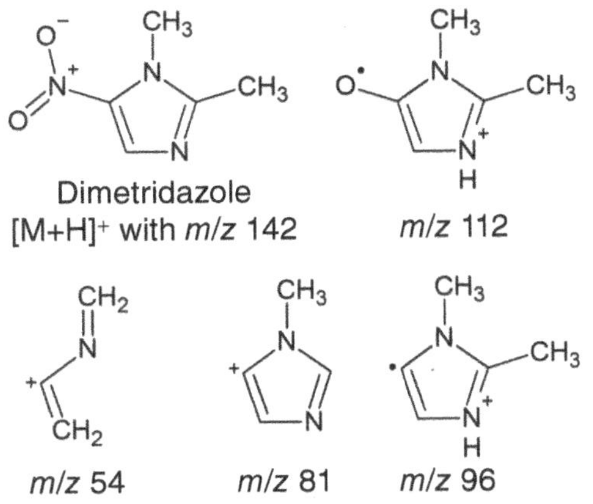

FIGURE 4.10.3 Proposed structures for the fragment ions of protonated dimetridazole.

Amprolium (M^+ with *m/z* 243) shows two fragment ions due to cleavage of the pyridinium N—C bond, that is, the protonated 2-methyl pyridine ($[CH_3—C_5H_4N+H]^+$) with *m/z* 94 and the (4-amino-2-propyl pyrimidin-5-yl)methylium ion ($[C_8H_{12}N_3]^+$) with *m/z* 150; the ion with *m/z* 122 is due to the loss of ethene (C_2H_4) from the ion with *m/z* 150 (Figure 4.10.4).

Cyromazine ($[M+H]^+$ with *m/z* 167) shows an ion with *m/z* 125 due to the loss of methanediimine (HN=C=NH) from the 1,3,5-triazine ring, an ion with *m/z* 108 ($[C_5H_6N_3]^+$) due to the subsequent loss of ammonia (NH_3), an ion with *m/z* 85 ($[C_2H_5N_4]^+$) due to the loss of cyclopropene (C_3H_4) from the ion with *m/z* 125, as well as protonated guanidine ($[H_2NCNHNH_2+H]^+$) with *m/z* 60.

Diaveridine ($[M+H]^+$ with *m/z* 261) is a structure analog of trimethoprim (Section 4.8.9) and shows similar fragmentation behavior. It shows fragment ions with *m/z* 246 due to the loss of a methyl radical ($^\bullet CH_3$) from a methoxy substituent, with *m/z* 245 due to the loss of methane (CH_4), with *m/z* 217 due to the subsequent loss of carbon monoxide (CO), with *m/z* 123 due to a cleavage of the dimethoxyphenyl C—C benzyl bond with the formation of the (2,4-diaminopyrimidin-5-yl)methylium ion ($[C_5H_7N_4]^+$), followed by the loss of HN=C=NH to an ion with *m/z* 81 ($[C_4H_5N_2]^+$.

Ethopabate ($[M+H]^+$ with *m/z* 238) shows an abundant fragment ion with *m/z* 206 due to the loss of methanol (CH_3OH), forming a substituted benzoyl cation. The ion with *m/z* 206 shows secondary fragmentation to an ion with *m/z* 178 due to the subsequent loss of C_2H_4, and to an ion with *m/z* 164 due to the subsequent loss of ethenone ($H_2C=C=O$). The (4-amino-2-hydroxy)benzoyl cation ($[C_7H_6NO_2]^+$) with *m/z* 136 is the result of all three losses indicated. Robenidine ($[M+H]^+$ with *m/z* 334) shows fragment ions with *m/z* 155 and 178 due to cleavages of C—N bonds in the guanidine group and two complementary fragment ions with *m/z* 195 and 138 due to cleavage of the N—N bond in the hydrazine moiety (Figure 4.10.4).

Halofuginone can be analyzed in both positive-ion and negative-ion modes. In positive-ion mode, halofuginone ($[M+H]^+$ with *m/z* 414) shows an ion with *m/z* 396 due to the loss of water from the 3-hydroxypiperidine substituent. The 3-hydroxypiperidin-2-ylium ion ($[C_5H_{10}NO]^+$) with *m/z* 100 and two secondary fragments of the ion with *m/z* 396 involve the loss of the 1*H*-quinazolin-4-one group, that is, the ion with *m/z* 138 ($[C_8H_{12}NO]^+$) and the ion with *m/z* 120 due to a subsequent loss of water thereof (Figure 4.10.4). Interestingly, in negative-ion mode, the fragment ions of halofuginone ($[M-H]^-$ with *m/z* 412) are based on the loss of the piperidin-3-ol ring, resulting in a fragment ion with *m/z* 313. The subsequent loss of $H_2C=C=O$ yields the ion with *m/z* 271. Deprotonated 7-bromo-6-chloro-(1*H*)-quinazolin-4-one ($[C_8H_4BrClN_2O-H]^-$) with *m/z* 257 and the bromide anion (Br^-) with *m/z* 79 are also observed (Figure 4.10.4). In this respect, complementary structure information is obtained.

Some other compounds can only be analyzed in negative-ion mode. Diclazuril ($[M-H]^-$ with *m/z* 405) shows losses of CO and hydrogen isocyanate (HN=C=O) to an ion with *m/z* 334, followed by subsequent losses of a chlorine radical ($Cl^\bullet$) to an ion with *m/z* 299 and of hydrogen chloride (HCl) to an ion with *m/z* 263. For the urea derivative dinitrocarbanilide ($[M-H]^-$ with *m/z* 301, 1,3-bis(4-nitrophenyl)urea), which is one of the two components in nicarbazin (the other being 4,6-dimethyl-1*H*-pyrimidin-2-one) shows the 4-nitroanilinide ion ($[O_2NC_6H_4NH]^-$) with *m/z* 137 and an ion with *m/z* 107 due to the subsequent loss of the nitrosyl radical ($^\bullet NO$) (Yakkundi et al., 2001). Nitromide ($[M-H]^-$ with *m/z* 210) shows subsequent losses of HN=C=O to an ion with *m/z* 167, of $^\bullet NO$ to *m/z* 137, of the nitryl radical ($^\bullet NO_2$) to *m/z* 91, and finally of CO to *m/z* 63 ($[C_5H_3]^-$).

4.10.2.3 Ionophore Coccidiostats

Polyether ionophores are fermentation-derived antibiotic compounds characterized by the presence of a carboxylic acid group and several cyclic ether units. They are widely used as coccidiostats in, among others, broiler chickens. The analysis of ionophore coccidiostats has been recently reviewed (Clarke et al., 2014). Polyether ionophores are related to crown ethers, which are well-known chelating agents, and can form macrocyclic complexes with monovalent or divalent cations. Therefore, they are mostly analyzed as $[M+Na]^+$ in positive-ion mode. Adduct ion formation with other cations (ammonium or other alkali cations) was also studied (Volmer & Lock, 1998). Multiresidue analysis has also been performed (Mortier et al., 2005b; Olejnik et al., 2009; Spisso et al., 2010; Moloney et al., 2012).

The fragmentation of polyether ionophores in positive-ion mode has been studied by various groups (Siegel et al.,

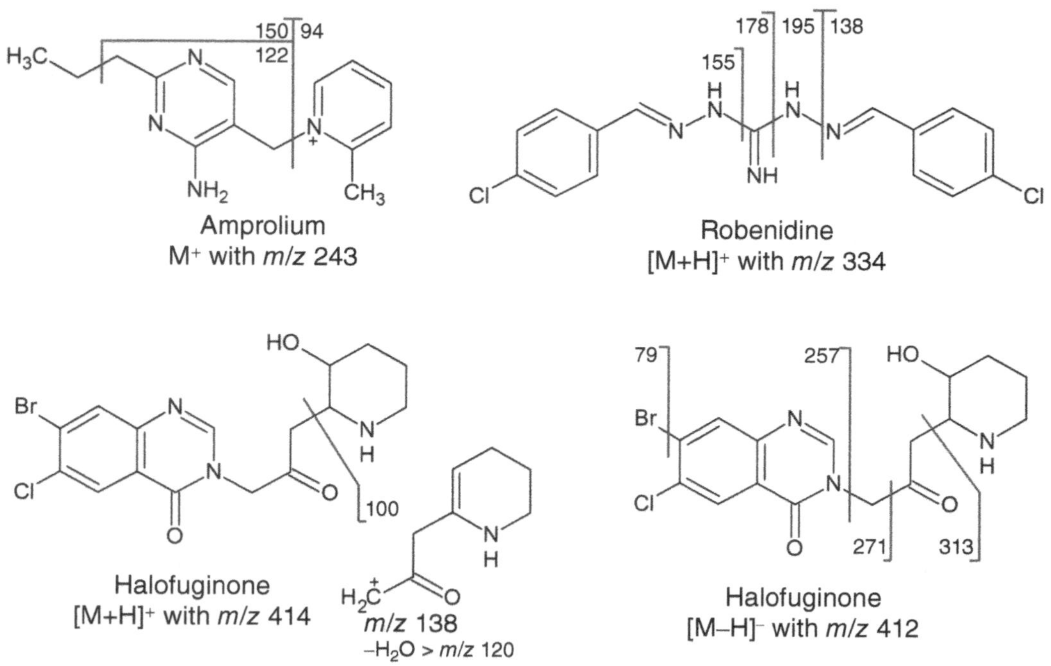

FIGURE 4.10.4 Fragmentation schemes for the protonated chemical coccidiostats amprolium, robenidine, and halofuginone and for deprotonated halofuginone.

1987; Volmer & Lock, 1998; Lopes et al., 2002a,b,c; Martínez-Villalba et al., 2009). Most of the fragment ions interpreted in the following are also used in SRM transitions for residue analysis. The sodiated molecules $[M+Na]^+$ show losses of water and/or carbon dioxide (CO_2). Maduramicin ($[M+Na]^+$ with *m/z* 939) shows a characteristic ion with *m/z* 877 due to the loss of both water and CO_2, also seen by in-source CID. The ion with *m/z* 397 ($[C_{18}H_{30}O_8+Na]^+$) is due to cleavage of both the C—O and C—C bonds on the secondary carbon of the dioxaspirodecane with charge retention on the fragment with the hydroxytetrahydropyran ring (Figure 4.10.5). If a keto group is present in the molecule, as in lasalocid A ($[M+Na]^+$ with *m/z* 613), salinomycin ($[M+Na]^+$ with *m/z* 773), and narasin A ($[M+Na]^+$ with *m/z* 787), cleavage of either of the α,β-C—C bonds relative to the carbonyl functional group occurs, resulting in the ions with *m/z* 531 ($[C_{29}H_{48}O_7+Na]^+$) and *m/z* 431 ($[C_{23}H_{36}O_6+Na]^+$) for salinomycin and narasin A and with *m/z* 377 ($[C_{21}H_{38}O_4+Na]^+$) for lasalocid A. This is illustrated for salinomycin in Figure 4.10.6.

The fragmentation of monensin A ($[M+Na]^+$ with *m/z* 693) in ion-trap MS^3 was studied. An interesting fragment of monensin A is the ion with *m/z* 479 (Figure 4.10.7). This fragment was initially interpreted to come from the cleavage of one of the oxolane rings and was assumed to contain the carboxylic acid group ($[C_{24}H_{40}O_8+Na]^+$ with *m/z* 479.262) (Volmer & Lock, 1998). Using accurate-mass data from FT-ICR-MS, the identity of the fragment turns out to be the result of cleavages in the dioxaspiro group and contains the hydroxytetrahydropyran ring ($[C_{25}H_{44}O_7+Na]^+$ with *m/z* 479.298) (Lopes et al., 2002b). Interestingly, the fragmentation of $[M+H]^+$ of monensin A with *m/z* 671 also results in an ion with *m/z* 479, also containing the hydroxytetrahydropyran ring, but with a different structure ($[C_{28}H_{47}O_6]^+$ with *m/z* 479.337) (Lopes et al., 2002c). The three structures proposed for the ion with *m/z* 479 are shown in Figure 4.10.7.

Being carboxylic acids, analysis of ionophores as $[M-H]^-$ in negative-ion mode should also be possible. The major fragment ion of lasalocid A ($[M-H]^-$ with *m/z* 589) is due to the cleavage of the α,β-C—C bond relative to the keto group, forming the 2-hydroxy-3-methyl-6-(3-methyl-4-oxobutyl)benzoate ion with *m/z* 235 ($[C_{13}H_{15}O_4]^-$). Loss of CO_2 from the ion with *m/z* 235 results in an ion with *m/z* 191. The 2,5-dimethylphenoxide ion ($[C_8H_9O]^-$) with *m/z* 121 is also found (Figure 4.10.8).

4.10.2.4 Antimalarial Drugs

Malaria is a human infectious mosquito-borne disease caused by the *Plasmodium* protozoan, which is widespread in tropical and subtropical regions. Quinine was the first

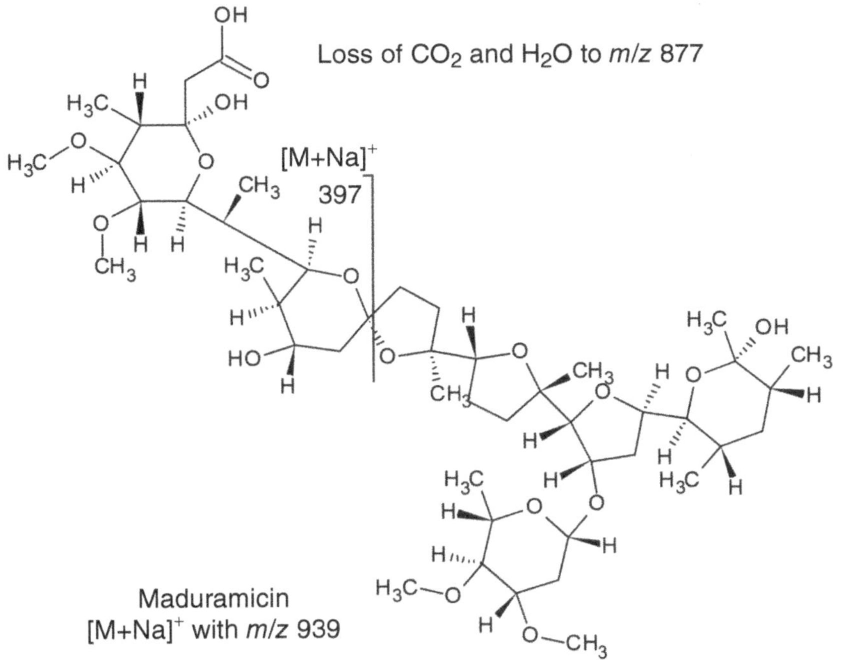

FIGURE 4.10.5 Fragmentation scheme for sodiated maduramicin, $[M+Na]^+$.

effective treatment for malaria. After the 1940s, more effective alternatives such as quinacrine, chloroquine, and primaquine have been developed. Antimalarial drugs are taken both as a preventive measure and for treatment against malaria. The LC–MS analysis of 14 antimalarial drugs and metabolites has been reported (Hodel et al., 2009).

The structures of many antimalarial drugs are based on the quinoline ring. They are analyzed mostly in positive-ion mode. Quinine ($[M+H]^+$ with *m/z* 325) shows an abundant ion with *m/z* 307 due to the loss of water. Less abundant are the (6-methoxyquinolin-4-yl)methylium ion ($[C_{11}H_{10}NO]^+$) with *m/z* 172 and the protonated 6-methoxyquinoline ($[C_{10}H_9NO+H]^+$) with *m/z* 160, both involving the loss of the 2-vinylquinuclidine group. The closely related di-(trifluoromethyl)quinoline derivative mefloquine ($[M+H]^+$ with *m/z* 379) also shows the loss of water, three subsequent hydrogen fluoride (HF) losses, as well as a protonated tetrahydropyridine ($[C_5H_9N+H]^+$) with *m/z* 84. In negative-ion mode, mefloquine ($[M-H]^-$ with *m/z* 377) shows a major fragment ion with *m/z* 294 due to the loss of tetrahydropyridine (C_5H_9N). Subsequent losses of HF, formaldehyde ($H_2C{=}O$), or trifluoromethane (CHF_3) result in fragment ions with *m/z* 274, 264, and 224, respectively. A trifluoromethanide ion (CF_3^-) with *m/z* 69 is also observed.

Primaquine and quinocide (both with $[M+H]^+$ with *m/z* 260) are isomeric (*N*-substituted-8-amino)-6-methoxyquinolines, while chloroquine ($[M+H]^+$ with *m/z* 320) and hydroxychloroquine ($[M+H]^+$ with *m/z* 336) are 4-(*N*-substituted-4-amino)-7-chloroquinolines. For all compounds, fragment ions result from cleavage of the quinoline C^4—N^1 bond to the diamino substituent group, resulting in complementary ions. For chloroquine, this leads to the protonated 7-chloroquinolin-4-amine ($[C_9H_7ClN_2+H]^+$) with *m/z* 179 and its complementary 5-(diethylamino)pent-2-ylium ion ($[C_9H_{20}N]^+$) with *m/z* 142 (Figure 4.10.9). Other fragment ions observed are characteristic of the quinoline alkylamino substituents, for example, loss of ammonia (NH_3) to an ion with *m/z* 243 for primaquine, loss of diethylamine ($(C_2H_5)_2NH$) to an ion with *m/z* 247, and the formation of the 5-(diethylamino)pent-2-ylium ion ($[C_9H_{20}N]^+$) with *m/z* 142 and the *N,N*-diethylmethaniminium ion ($[(C_2H_5)_2N{=}CH_2]^+$) with *m/z* 86 for chloroquine, and similar fragments for hydroxychloroquine (with an *m/z* difference of +16), that is, the ions with *m/z* 158 ($[C_9H_{20}NO]^+$) and *m/z* 102 (Dongre et al., 2009). The fragmentation of piperaquine ($[M+H]^+$ with *m/z* 535) is illustrated in Figure 4.10.9 (Tarning et al., 2006). It shows characteristic fragmentation of the piperazine group (Section 3.6.5), for instance, also observed for phenothiazines with a 3-(4-methylpiperazin-1-yl)propyl substituent group (Section 4.2.1 and Table 4.2.1a). Cleavage of the C—N bond between the 7-chloroquinoline and the piperazine results in the protonated 7-chloroquinoline ($[C_9H_6ClN+H]^+$) with *m/z* 164. Loss of 7-chloro-4-(piperazin-1-yl)quinoline results

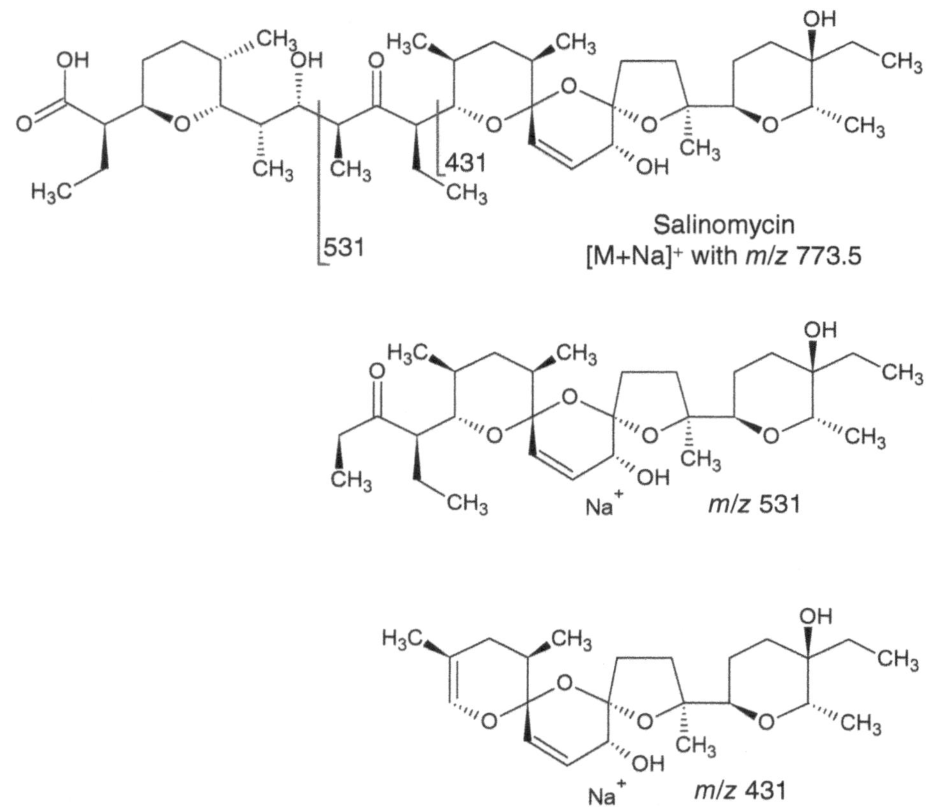

FIGURE 4.10.6 Fragmentation scheme and fragment ions for sodiated salinomycin, $[M+Na]^+$.

in a carbocation with *m/z* 288 ($[C_{16}H_{19}ClN_3]^+$). Cleavage of the β-C—C bond relative to the piperazine group yields the 4-(7-chloroquinolin-4-yl)-1-methylidenepiperazinium ion ($[C_{14}H_{15}ClN_3]^+$) with *m/z* 260. Cleavage of the C—N^1 bond of the piperazine results in a radical cation with *m/z* 245 ($[C_{13}H_{12}ClN_3]^{+\bullet}$) (Section 3.5.7). Cross-ring cleavages in the piperazine ring (Section 3.6.5) result in the [(7-chloroquinolin-4-yl)(vinyl)amino]methylium ion ($[C_{12}H_{10}ClN_2]^+$) with *m/z* 217 and the protonated 7-chloro-*N*-vinylquinolin-4-amine ($[C_{11}H_9ClN_2+H]^+$) with *m/z* 205. For primaquine, both impurity profiling (Dongre et al., 2007) and metabolite identification (Tarning et al., 2006) have been reported.

Proguanil or chloroguanide ($[M+H]^+$ with *m/z* 254; $[M-H]^-$ with *m/z* 252) can be analyzed in both positive-ion and negative-ion modes. In both modes, a series of fragment ions is observed due to cleavages in the guanidine groups, as shown in Figure 4.10.9.

Phlorizin ($[M-H]^-$ with *m/z* 435) shows a major fragment ion with *m/z* 273 due to the loss of the dehydroglucose ($C_6H_{10}O_5$), as well as fragment ions due to subsequent loss of 4-methylidenecyclohexa-2,4-dienone ($H_2C{=}C_6H_4{=}O$) to the 2-acetoyl-3,5-dihydroxyphenoxide ion ($[C_8H_7O_4]^-$) with *m/z* 167, and subsequent loss of ethenone ($H_2C{=}C{=}O$) to the 3,5-dihydroxyphenoxide ion ($[(HO)_2C_6H_3O]^-$) ion with *m/z* 125.

4.10.3 ANTIVIRAL DRUGS

Antiviral drugs inhibit the development of viruses in different ways, with specific antivirals used for clearly defined viruses. In the past years, extensive studies have been performed on the development of drugs for the treatment of the human immunodeficiency virus (HIV) infection. The two most important compound classes of HIV-related antiretroviral drugs are reverse transcriptase inhibitors (RTIs) and HIV-selective protease inhibitors. There are two classes of RTIs, that is, compounds that are similar to nucleosides, called nucleoside reverse transcriptase inhibitors (NRTIs), and compounds that are not similar to nucleosides, called non-nucleoside reverse transcriptase inhibitors (nNRTIs). They inhibit reverse transcription either by being incorporated into the newly synthesized viral DNA strand and thereby causing DNA chain termination, or by binding to the enzyme itself which inhibits reverse transcriptase, thereby preventing DNA synthesis in viral RNA.

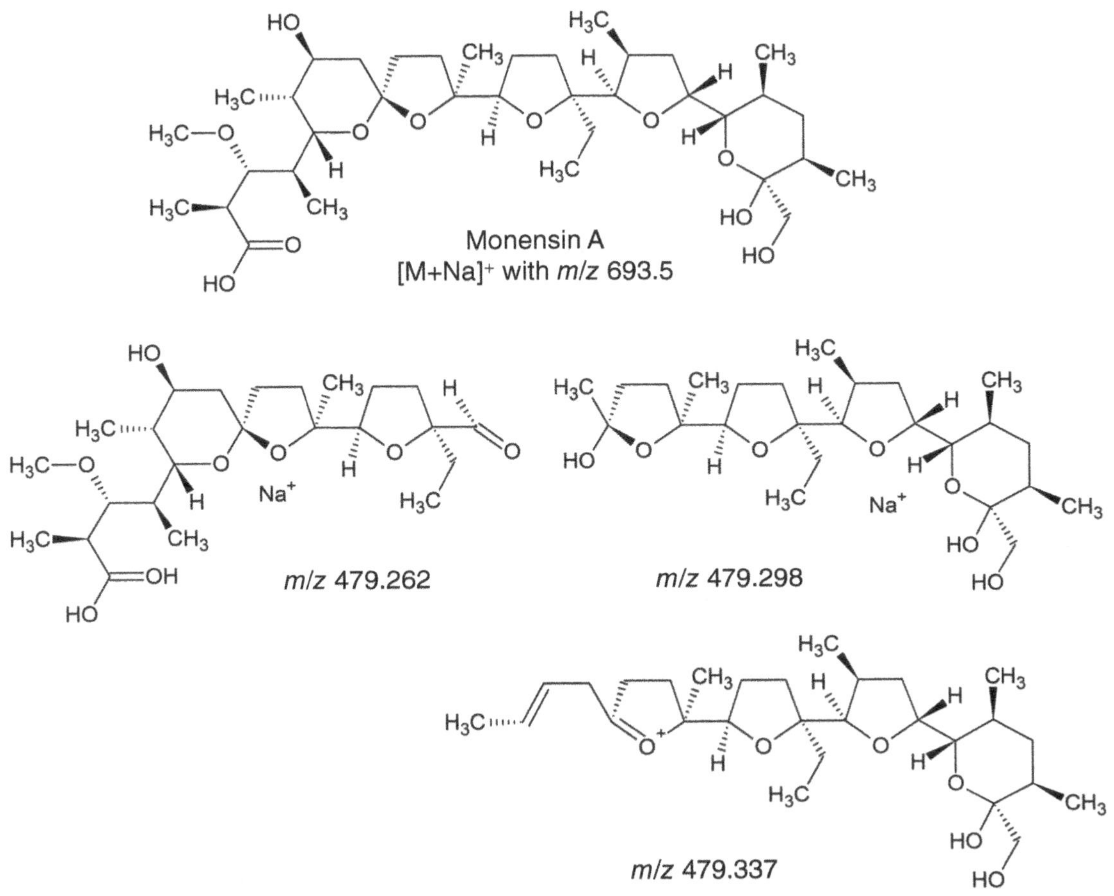

FIGURE 4.10.7 Structures for the fragment ion with *m/z* 479 from monensin A. The structure with *m/z* 479.262 was proposed as a fragment ion from $[M+Na]^+$ (Volmer & Lock, 1998). Accurate-mass data from FT-ICR-MS indicates an alternative fragment ion with *m/z* 479.298 (Lopes et al., 2002b). The fragment ion with *m/z* 479.337 was observed from $[M+H]^+$ (Lopes et al., 2002c).

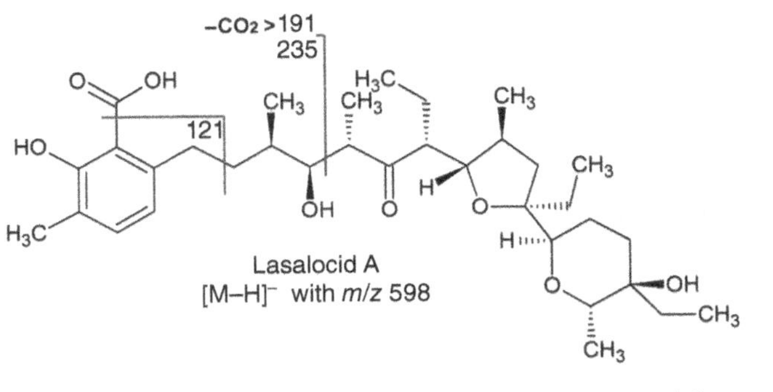

FIGURE 4.10.8 Fragmentation scheme for deprotonated lasalocid A.

Product-ion mass spectral data are also available for some antiviral drugs against the *Herpes simplex* virus or the *Influenza* virus.

4.10.3.1 Nucleoside Reverse Transcriptase Inhibitors (HIV Related)

Important examples of this compound class are zidovudine (AZT, ZDV, azidothymidine), didanosine (ddI), zalcitabine (ddC), stavudine (d4T), lamivudine (3TC), abacavir (ABC), emtricitabine (FTC), and entecavir (ETV). The analysis of NRTIs has been reviewed (Lai et al., 2008). Most NRTIs can be analyzed in both positive-ion and negative-ion modes. In positive-ion mode, the loss of the nucleoside *N*-substituent with charge retention on the nucleobase is generally the most important fragmentation route. This results in protonated 5-methyl-(1*H*,3*H*)-pyrimidine-2,4-dione ($[C_5H_6N_2O_2+H]^+$) with *m/z* 127 for zidovudine ($[M+H]^+$ with *m/z* 268) and stavudine ($[M+H]^+$ with *m/z* 225), in protonated 9*H*-purin-6-ol ($[C_5H_4N_4O+H]^+$) with *m/z* 137 for didanosine ($[M+H]^+$ with *m/z* 237), in protonated 4-amino-1*H*-pyrimidin-2-one ($[C_4H_5N_3O+H]^+$) with *m/z*

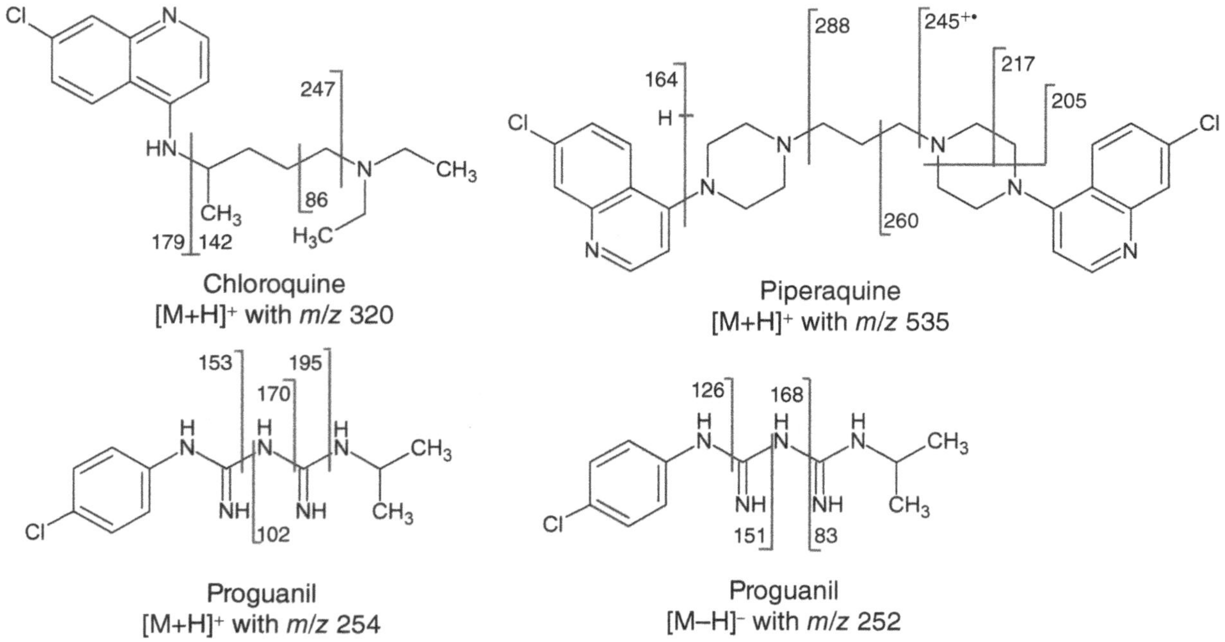

FIGURE 4.10.9 Fragmentation schemes for protonated chloroquine, piperaquine, and proguanil and for deprotonated proguanil.

112 for zalcitabine ([M+H]$^+$ with *m/z* 212) and lamivudine ([M+H]$^+$ with *m/z* 230), in protonated 2-amino-6-(*N*-cyclopropylamino)-9*H*-purine ([$C_8H_{10}N_6$+H]$^+$) with *m/z* 191 for abacavir ([M+H]$^+$ with *m/z* 287), in protonated 4-amino-5-fluoro-1*H*-pyrimidin-2-one ([$C_4H_4FN_3O$+H]$^+$) with *m/z* 130 for emtricitabine ([M+H]$^+$ with *m/z* 248), and in protonated 2-amino-9*H*-purin-6-one ([$C_5H_5N_5O$+H]$^+$) with *m/z* 152 for entecavir ([M+H]$^+$ with *m/z* 278) (Bedse et al., 2009; Li et al., 2010; Rao et al., 2011, Ramesh et al., 2014). This fragmentation behavior is similar to what is observed for nucleotides (Yang et al., 2010). For some compounds, additional fragmentation has been reported, for example, for lamivudine (Bedse et al., 2009), abacavir (Rao et al., 2011), and entecavir (Ramesh et al., 2014), mostly in the context of a study on degradation products.

In negative-ion MS–MS, the NRTIs do not show similar behavior. Zidovudine ([M−H]$^-$ with *m/z* 266) shows an abundant fragment with *m/z* 223 due to the loss of hydrogen azide (HN_3) (Font et al., 1998). Didanosine ([M−H]$^-$ with *m/z* 235) shows deprotonated 9*H*-purin-6-ol ([$C_5H_4N_4O$—H]$^-$) with *m/z* 135 due to the loss of the N^9-dideoxyribose group. Minor fragments result from the loss of hydrogen isocyanate (HN=C=O) from [M−H]$^-$ to a fragment ion with *m/z* 192 and from the ion with *m/z* 135 to a fragment ion with *m/z* 92. Zalcitabine ([M−H]$^-$ with *m/z* 210) shows a fragment ion with *m/z* 167 due to the loss of HN=C=O via a retro-Diels–Alder fragmentation (Font et al., 1998) (Section 3.5.6). Stavudine ([M−H]$^-$ with *m/z* 223) shows loss of formaldehyde ($H_2C{=}O$) to an ion with *m/z* 193, subsequent loss of HN=C=O to *m/z* 150 followed by the loss of carbon monoxide (CO) to *m/z* 122. Finally, lamivudine ([M−H]$^-$ with *m/z* 228) shows fragmentation in the N^1-(2-(hydroxymethyl)-1,3-oxathiolan-5-yl substituent, that is, the loss of ethene-1,2-diol ($C_2H_4O_2$) to an ion with *m/z* 168 and subsequent loss of the sulfanyl radical (•SH) to an ion with *m/z* 135 ([$C_6H_5N_3O$]$^{+\bullet}$).

The biologically active forms of these NRTIs are the intracellular phosphorylated anabolites. Their analysis in, for instance, peripheral blood mononuclear cells (PBMCs) poses challenges in LC–MS because ion-pair LC is required in most cases (Becher et al., 2002). The di- and triphosphorylated anabolites are analyzed in negative-ion mode using SRM with transitions involving fragmentation of [M−H]$^-$ to a common fragment ion with *m/z* 159 ([HO_6P_2]$^-$) (Becher et al., 2002; Pruvost et al., 2001).

Tenofovir ([M+H]$^+$ with *m/z* 288) is a monophosphorylated antiviral drug used in HIV treatment (often in combination with emtricitabine). In positive-ion product-ion mass spectra, tenofovir shows fragment ions with *m/z* 270 due to the loss of water, *m/z* 206 due to the loss of phosphonic acid (H_3PO_3), *m/z* 176 due to a subsequent loss of $H_2C{=}O$, and *m/z* 136 due to protonated 6-amino-9*H*-purine ([$C_5H_5N_5$+H]$^+$). The ion with *m/z* 176 is mostly used as product ion in SRM.

4.10.3.2 Non-nucleoside Reverse Transcriptase Inhibitors (HIV Related)

In positive-ion mode, the nNRTI efavirenz ($[M+H]^+$ with *m/z* 316) yields extensive fragmentation. The loss of carbon dioxide (CO_2) results in a fragment ion with *m/z* 272, which shows a subsequent loss of a chlorine radical ($Cl^•$) to an ion with *m/z* 237 and of ethene (C_2H_4) to an ion with *m/z* 244. The latter is frequently used as product ion in SRM. Additionally, the subsequent loss of water and hydrogen fluoride (HF) results in fragment ions with *m/z* 298 and 278. In negative-ion mode, efavirenz ($[M-H]^-$ with *m/z* 314) shows the loss of trifluoromethane (CHF_3) to a fragment ion with *m/z* 244; the trifluoromethanide ion ($[CF_3]^-$) with *m/z* 69 is also observed. The ion with *m/z* 250 is due to the loss of vinylidenecyclopropene (C_5H_4) from the C^4-substituent of the benzoxazinone cyclic structure (Fan et al., 2002).

In positive-ion mode, nevirapine ($[M+H]^+$ with *m/z* 267) shows the loss of a cyclopropyl radical ($^•C_3H_5$) to a fragment ion with *m/z* 226, which shows subsequent loss of carbon monoxide (CO) to an ion with *m/z* 198 ($[C_{11}H_{10}N_4]^{+•}$). At higher collision energy, cleavage of the 1,4-diazepinone ring results in protonated pyridine ($[C_5H_5N+H]^+$) with *m/z* 80.

4.10.3.3 Protease Inhibitors (HIV Related)

In the HIV virus, protease inhibitors block the activity of the protease responsible for cleaving nascent proteins for the final assembly of new virons. Because of the high inter- and intra-individual variability, therapeutic drug monitoring is required for these drugs. As a result, multiresidue LC–MS methods have been reported (Rentsch, 2003; Crommentuyn et al., 2003; Coulier et al., 2011) and reviewed (Müller & Rentsch, 2010). The protease inhibitors are generally analyzed in positive-ion mode (Müller & Rentsch, 2010). Their fragmentation involves mostly cleavages of C—N bonds in the amine and amide moieties. The position of the cleavage and the charge retention on the fragments is not always easy to predict.

The fragmentation of indinavir ($[M+H]^+$ with *m/z* 614) has been described, also in the context of metabolite studies (Lopez et al., 1998; Yu et al., 1999; Gangl et al., 2002; Anari et al., 2004) and forced degradation studies (Rao et al., 2013). The fragmentation of indinavir is outlined in Figure 4.10.10. The loss of *N-tert*-butylformamide ($(CH_3)_3CNHCHO$) yields a fragment ion with *m/z* 513. The loss of 2-amino-indan-1-ol ($C_9H_{11}NO$) results in a fragment ion with *m/z* 465. Secondary fragmentation of either of these two ions by the loss of the other substituent group, that is, the loss of $C_9H_{11}NO$ from the ion with *m/z* 513 or the loss of $(CH_3)_3CNHCHO$ from the ion with *m/z* 465, leads to the ion with *m/z* 364. Other secondary fragmentation of the ion with *m/z* 513 involving the cleavage of the piperazine N^4—C bond to the (pyridin-3-yl)methyl substituent results in two fragment ions, that is, the (pyridin-3-yl)methylium ion ($[C_5H_4NCH_2]^+$) with *m/z* 92 and the radical cation with *m/z* 421 ($[C_{25}H_{31}N_3O_3]^{+•}$) (Section 3.5.7).

Another series of fragment ions results from the cleavage of the piperazine N^1—C bond to the *iso*-propyl group. This yields two complementary fragment ions, that is, the protonated {4-[(pyridin-3-yl)methyl]piperazin-2-yl}-*N-tert*-butylcarboxamide ($[C_{15}H_{24}N_4O+H]^+$) with *m/z* 277 and the complementary carbocation with *m/z* 338 ($[C_{21}H_{24}NO_3]^+$). Finally, the 2,3,6-trihydro-1-methylidenepyrazonium ion with *m/z* 97 ($[C_5H_9N_2]^+$) and the 4-carbamoyl-5-phenylpent-1-en-3-ylium ion (or protonated 3-benzyl-3,6-dihydro-1*H*-pyridine-2-one, after ring closure) ($[C_{12}H_{14}NO]^+$) with *m/z* 188 are also observed. Water losses are observed for $[M+H]^+$ and the ions with *m/z* 421 and 338 (Figure 4.10.10).

A fragmentation scheme for nelfinavir is also provided in Figure 4.10.10. Similarly, fragmentation schemes have been reported for ritonavir ($[M+H]^+$ with *m/z* 721) (Gangl et al., 2002), atazanavir ($[M+H]^+$ with *m/z* 705), tipranavir ($[M+H]^+$ with *m/z* 603) (Crommentuyn et al., 2004), and darunavir ($[M+H]^+$ with *m/z* 548) (Rao et al., 2014).

Some negative-ion MS–MS data for protease inhibitors are available in our library collection. Proposed negative-ion fragmentation schemes for indinavir ($[M-H]^-$ with *m/z* 612) and saquinavir ($[M-H]^-$ with *m/z* 669) are given in Figure 4.10.11.

4.10.3.4 Other Antiviral Drugs

A number of (mostly guanine-related) compounds are used in the treatment of *Herpes simplex* and the herpes-related cytomegalovirus. Aciclovir ($[M+H]^+$ with *m/z* 226), ganciclovir ($[M+H]^+$ with *m/z* 256), penciclovir ($[M+H]^+$ with *m/z* 254), valaciclovir ($[M+H]^+$ with *m/z* 325), and valganciclovir ($[M+H]^+$ with *m/z* 355) all show protonated guanine ($[C_5H_5N_5O+H]^+$) as a common fragment ion with *m/z* 152 due to the loss of guanine N^9-substituents. Complementary ions due to the loss of the guanine base are observed, that is, the *O*-substituted methylene oxonium ions with *m/z* 174 ($[C_8H_{16}NO_3]^+$) and with *m/z* 204 ($[C_9H_{18}NO_4]^+$) for valaciclovir and valganciclovir, respectively. Cleavage of the carbonyl C—C bond to the aminoalkyl group results in protonated 2-methylpropan-1-imine ($[C_4H_9N+H]^+$) with *m/z* 72 for both valaciclovir and valganciclovir. The fragmentation of famciclovir ($[M+H]^+$ with *m/z* 322) involves several small-molecule losses, that is, the ion with *m/z* 280 due to the loss of ethenone ($H_2C{=}C{=}O$), *m/z* 262 due to the loss of acetic acid (CH_3COOH), which in turn leads to ions with *m/z* 220 and 202 due to subsequent loss of $H_2C{=}C{=}O$ and CH_3COOH, respectively. The protonated 2-aminopurine ($[C_5H_5N_5+H]^+$) with *m/z* 136 is due to the loss of the complete N^9-substituent.

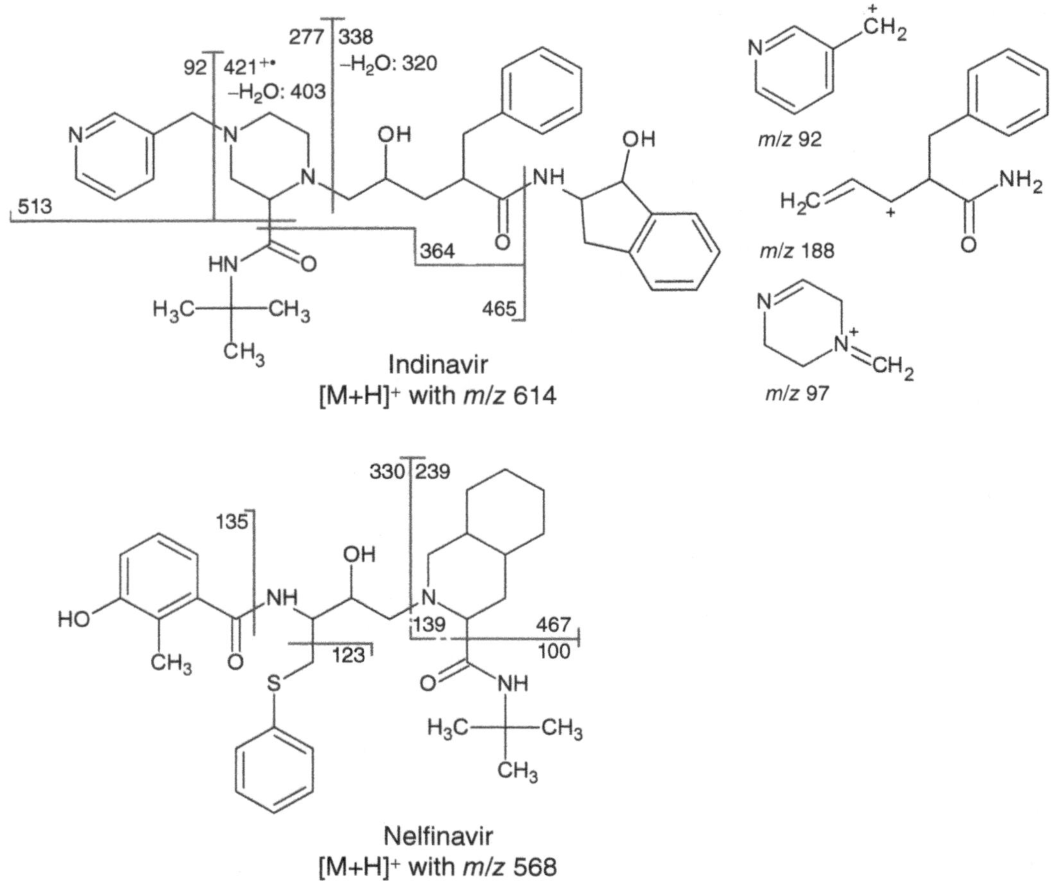

FIGURE 4.10.10 Fragmentation schemes for protonated indinavir and nelvinavir.

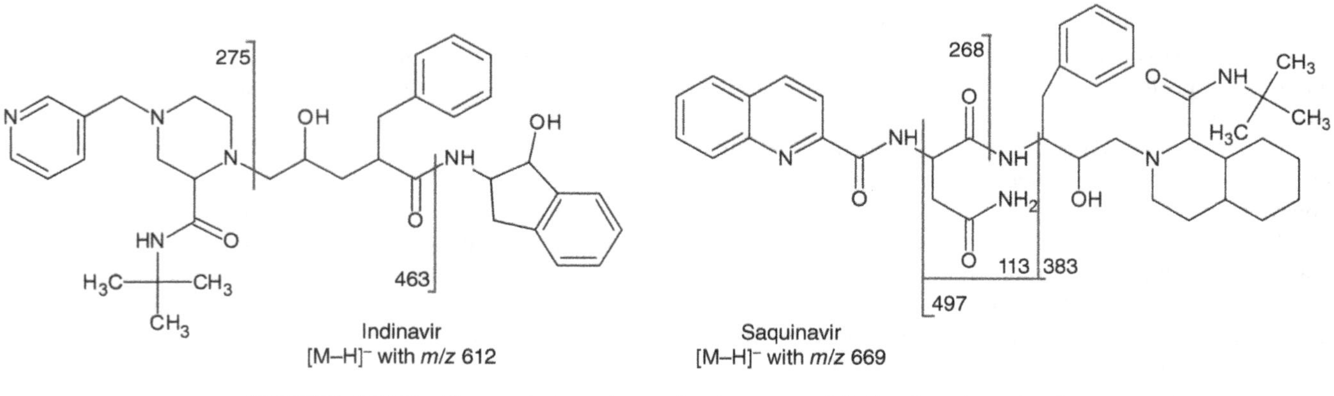

FIGURE 4.10.11 Fragmentation schemes for deprotonated indinavir and saquinavir.

Vidarabine, trifluiridine, brivudine, and idoxuridine are NRTIs, also used in the treatment of *Herpes simplex*. In positive-ion MS–MS, the loss of their N^9-substituent is observed, which for vidarabine ($[M+H]^+$ with *m/z* 268) leads to the protonated 6-aminopurine or adenine ($[C_5H_5N_5+H]^+$) with *m/z* 136, and for idoxuridine ($[M+H]^+$ with *m/z* 355) to the protonated 5-iodouracil ($[C_4H_3IN_2O_2+H]^+$) with *m/z* 239. In negative-ion MS–MS, trifluiridine ($[M-H]^-$ with *m/z* 295) primarily shows the loss of the sugar unit to the deprotonated 5-(trifluoromethyl)uracil ($[C_5H_3F_3N_2O_2-H]^+$) with *m/z* 179, whereas for brivudine ($[M-H]^-$ with *m/z* 331) and idoxuridine ($[M-H]^-$ with *m/z* 353), a minor fragment ion due to the loss of the dideoxyribose unit ($C_5H_8O_3$) is observed to the deprotonated 5-(2-bromovinyl)uracil

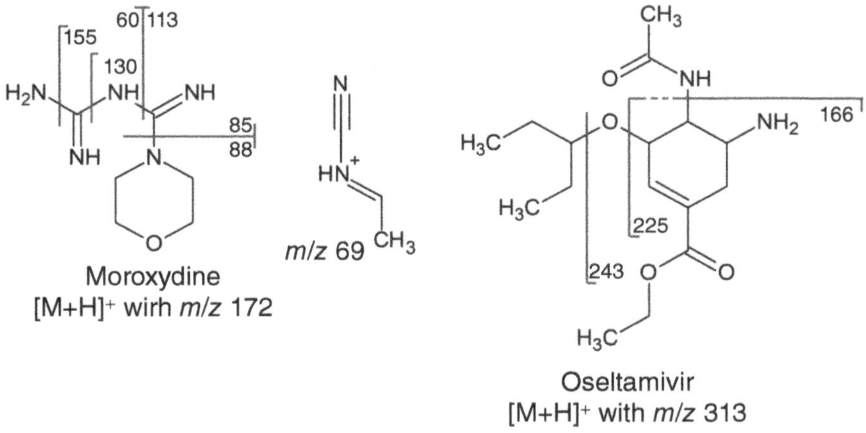

FIGURE 4.10.12 Fragmentation schemes for protonated moroxydine and oseltamivir.

($[C_6H_5BrN_2O_2-H]^-$) and 5-iodouracil ($[C_4H_3IN_2O_2-H]^-$) with *m/z* 216 and 237, respectively. However, the major fragment ions are bromide (Br^-) with *m/z* 79 and iodide (I^-) with *m/z* 127.

Ribavirin ($[M+H]^+$ with *m/z* 245) is an NRTI applied in the treatment of hepatitis. In the positive-ion product-ion mass spectrum, it shows the loss of the N^1-substituent leading to the protonated 1*H*-1,2,4-triazole-3-carboxamide ($[C_3H_4N_4O+H]^+$) with *m/z* 113, which in turn shows subsequent losses of ammonia (NH_3) or hydrogen isocyanate (HN=C=O) to the (1*H*-1,2,4-triazol-3-yl)acylium ion ($[C_2H_2N_3$—C=O$]^+$) with *m/z* 97 and the protonated 1*H*-1,2,4-triazole ($[C_2H_3N_3+H]^+$) with *m/z* 70, respectively. Adefovir ($[M+H]^+$ with *m/z* 274), also used in the treatment of hepatitis, is structurally related to tenofovir (Section 4.10.3.1) and shows similar fragmentation, that is, fragment ions with *m/z* 256 due to the loss of water, *m/z* 192 due to the loss of phosphonic acid (H_3PO_3), *m/z* 162 due to a subsequent loss of formaldehyde (H_2C=O), and with *m/z* 136 due to protonated adenine ($[C_5H_5N_5+H]^+$).

Amantadine ($[M+H]^+$ with *m/z* 152) and rimantadine ($[M+H]^+$ with *m/z* 180) are used against influenza. They are two structurally related primary amine antiviral drugs, sharing a 1-adamantyl substituent group. In their product-ion mass spectra, they both show the loss of ammonia (NH_3) as the major fragmentation route. Moroxydine ($[M+H]^+$ with *m/z* 172), also used against influenza, shows extensive fragmentation in the guanidine substituent (Figure 4.10.12). Thus, the loss of NH_3 leads to an ion with *m/z* 155 and the loss of methanediimine (HN=C=NH) to *m/z* 130. Two sets of complementary ions are observed, that is, protonated guanidine ($[H_2NCNHNH_2+H]^+$) with *m/z* 60 and protonated *N*-cyanomorpholine ($[C_5H_8N_2O+H]^+$) with *m/z* 113, and protonated N^2-cyanoguanidine ($[C_2H_4N_4+H]^+$) with *m/z* 85 and protonated morpholine ($[C_4H_9NO+H]^+$) with *m/z* 88. The ion with *m/z* 113 shows secondary fragmentation involving the loss of ethanal (CH_3CHO) to the protonated *N*-cyanoethanimine ($[CH_3CH$=N—C≡N$+H]^+$) with *m/z* 69 (Figure 4.10.12).

Oseltamivir (Tamiflu) is an antiviral against influenza, which was recently recommended as a treatment against the H1N1 influenza virus. In its product-ion mass spectrum, oseltamivir ($[M+H]^+$ with *m/z* 313) shows the loss of NH_3 to an ion with *m/z* 296 and the loss of the *O*-3-pentyl substituent as pentene (C_5H_{10}) to an ion with *m/z* 243. The latter ion shows subsequent losses of water to *m/z* 225, of NH_3 to *m/z* 208, and of ethenone (H_2C=C=O) to protonated ethyl (4-aminobenzoate) ($[C_9H_{11}NO_2+H]^+$) with *m/z* 166 (Figure 4.10.12) (Junwal et al., 2012).

4.10.4 ANTISEPTICS AND DISINFECTANTS

Antiseptics are antimicrobial substances that are applied to living tissue/skin to reduce the possibility of infection, sepsis, or putrefaction. They are distinguished from disinfectants, which are used to destroy microorganisms found on nonliving objects.

In our library collection, there are several chlorinated and brominated phenol derivatives with antiseptic and/or disinfectant use. The interpretation of the most abundant fragment ions in the negative-ion product-ion mass spectra of these compounds is summarized in Table 4.10.2. Losses of hydrogen chloride (HCl) or hydrogen bromide (HBr) is frequently observed. Dichlorophene ($[M-H]^-$ with *m/z* 267) shows the loss of HCl as well as the 4-chlorophenoxide anion with *m/z* 127. Triclosan ($[M-H]^-$ with *m/z* 287) shows two fragment ions due to cleavages of the C—O bonds of the diaryl ether linkage, that is, the 2,4-dichlorophenoxide ion ($[Cl_2C_6H_3O]^-$) with *m/z* 161 and the molecular ion of 4-chloro[1,2]benzoquinone ($[C_6H_3ClO_2]^{-\bullet}$) with *m/z* 142. Similarly, biothionol ($[M-H]^-$ with *m/z* 353) shows two fragment ions due to cleavages of C—S bonds in the diaryl thioether linkage, that is, the 2,4-dichlorophenoxide

TABLE 4.10.2 Fragmentation of chlorinated and brominated phenol derivatives used as antiseptics and disinfectants.

Compound	$[M-H]^-$ (*m/z*)	Fragment Ion (*m/z*)	Interpretation
Chlorocresol	141	105	Loss of HCl
		77	Loss of HCl and CO
Chloroxylenol	155	119	Loss of HCl
Chlorindanol	167	131	Loss of HCl
		130	Loss of HCl and $H^•$
Chlorcarvacrol	183	147	Loss of HCl
		105	Loss of HCl, C_3H_6
		77	Loss of HCl, C_3H_6, and CO
Chlorophene (2-benzyl-4-chlorophenol)	217	181	Loss of HCl
		153	Loss of HCl and CO
		101	$[C_8H_5]^-$
Chlorquinaldol	226	162	Loss of HCl and CO
Dichlorophene	267	231	Loss of HCl
		127	$[ClC_6H_4O]^-$
Triclosan	287	161	$[Cl_2C_6H_3O]^-$
		142	$[ClC_6H_3O_2]^{-•}$
Bithionol	353	192	$[Cl_2C_6H_2OS]^{-•}$
		161	$[Cl_2C_6H_3O]^-$
		125	Loss of HCl from *m/z* 161
		89	Loss of HCl from *m/z* 125
Dibromsalicil	397	199	$[BrOHC_6H_3CO]^-$
		171	Loss of CO from *m/z* 199
		79	Br^-
Hexachlorophene	403	367	Loss of HCl
		331	Loss of HCl and HCl
		303	Loss of HCl, HCl, and CO
		195	$[Cl_3C_6H_2O]^-$
		159	Loss of HCl from *m/z* 195
Tetrabromo-*o*-cresol	419	339	Loss of HBr
		79	Br^-
Bromchlorophene	423	387	Loss of HCl
		343	Loss of HBr
		205	$[BrClC_6H_3O]^-$
		125	Loss of HBr from *m/z* 205
		79	Br^-

ion ($[Cl_2C_6H_3O]^-$) with *m/z* 161 and the molecular ion of 2,4-dichloro-*o*-thioquinone ($[C_6H_2Cl_2OS]^{-•}$) with *m/z* 192.

Thymol ($[M-H]^-$ with *m/z* 149, 6-*iso*-propyl-cresol) shows fragment ions with *m/z* 134 due to the loss of a methyl radical ($^•CH_3$), *m/z* 133 due to the loss of methane (CH_4), and *m/z* 107 due to loss of propene ($CH_3CH{=}CH_2$). Gallic acid ($[M-H]^-$ with *m/z* 169) shows the loss of carbon dioxide (CO_2) to an ion with *m/z* 125 and subsequent losses of either carbon monoxide (CO) to *m/z* 97 or formic acid (HCOOH) to *m/z* 79. Nitroxoline ($[M-H]^-$ with *m/z* 189) shows the loss of a nitrosyl radical ($^•NO$) to an ion with *m/z* 159 or of a nitryl radical ($^•NO_2$) to *m/z* 143, followed by CO loss to *m/z* 115. Dibromophenolsulfonic acid ($[M-H]^-$ with *m/z* 328) and diiodophenolsulfonic acid ($[M-H]^-$ with *m/z* 425) show partly similar fragmentation. They both show loss of a halogen as radical, that is, fragment ions with *m/z* 249 and 298, respectively, and bromide (Br^-) and iodide (I^-) ions with *m/z* 79 and 127. In addition, the dibromo compound shows fragment ions with *m/z* 220 due to a combined loss of hydrogen bromide (HBr) and CO, followed by loss of sulfur dioxide (SO_2) to *m/z* 157 ($[C_5H_2BrO]^-$). The diiodo compound shows SO_2 loss from the radical anion with *m/z* 298 to an ion with *m/z* 234 and an ion with *m/z* 171 ($[C_6H_3O_4S]^-$).

A number of disinfectants and antiseptics can be analyzed in positive-ion mode. This is certainly the case for a number of quaternary ammonium compounds such as mecetronium,

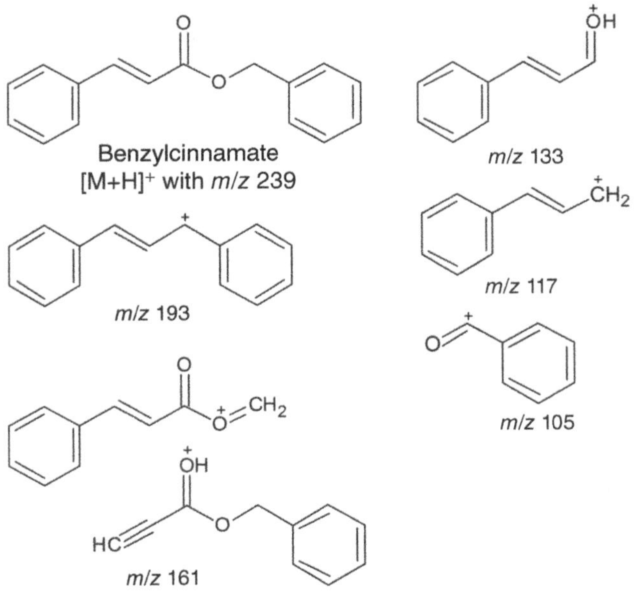

FIGURE 4.10.13 Proposed structures for the fragment ions of protonated benzylcinnamate.

benzododecinium, and cetrylpyridinium. These compounds show little fragmentation except mainly the loss of the long alkyl chains. The main fragment of mecetronium (M^+ with *m/z* 298) is protonated *N*,*N*-dimethylethanamine ($[(CH_3)_2NC_2H_5+H]^+$) with *m/z* 74 due to the loss of hexadecene. Benzododecinium (M^+ with *m/z* 304) shows the loss of toluene ($C_6H_5CH_3$) to a fragment ion with *m/z* 212. In addition, the tropylium ion ($[C_7H_7]^+$) with *m/z* 91 and the *N*,*N*-dimethylmethaniminium ion ($[(CH_3)_2N{=}CH_2]^+$) with *m/z* 58 are observed. Cetylpyridinium (M^+ with *m/z* 304) shows the loss of hexadecene to the protonated pyridine ($[C_5H_5N+H]^+$) with *m/z* 80.

Benzylcinnamate ($[M+H]^+$ with *m/z* 239) shows the loss of HCOOH to a fragment ion with *m/z* 193 ($[C_{15}H_{13}]^+$), requiring a skeletal rearrangement of the benzyl group, and the loss of 3-phenylprop-2-en-1-ol ($C_6H_5CH{=}CHCH_2OH$) to the benzoyl cation ($[C_6H_5{-}C{=}O]^+$)with *m/z* 105 (Figure 4.10.13). Other fragment ions involve the loss of benzene (C_6H_6) to an ion with *m/z* 161, of benzaldehyde (C_6H_5CHO) to *m/z* 133, and the formation of the tropylium ion with *m/z* 91 ($[C_7H_7]^+$) and the 1-phenylpropen-3-ylium ion or α-(vinyl)benzyl cation ($[C_9H_9]^+$) with *m/z* 117.

Trichlocarban ($[M+H]^+$ with *m/z* 315) shows cleavages of the N—C bonds to the carbonyl of the urea moiety, leading to the protonated 3,4-dichloroaniline ($[Cl_2C_6H_3NH_2+H]^+$) with *m/z* 162 and the protonated 4-chloroaniline ($[ClC_6H_4NH_2+H]^+$) with *m/z* 128. The latter ion shows subsequent loss of a chlorine radical ($^\bullet$Cl) to an ion with *m/z* 93. Cleavages of the N—C bonds to the carbonyl in the urea moiety also play a role in the fragmentation of aminoquinuride ($[M+H]^+$ with *m/z* 373), leading to the protonated 6-isocyanato-2-methylquinolin-4-amine ($[C_{11}H_9N_3O+H]^+$) with *m/z* 200 and the protonated 2-methylquinoline-4,6-diamine ($[C_{10}H_{11}N_3+H]^+$) with

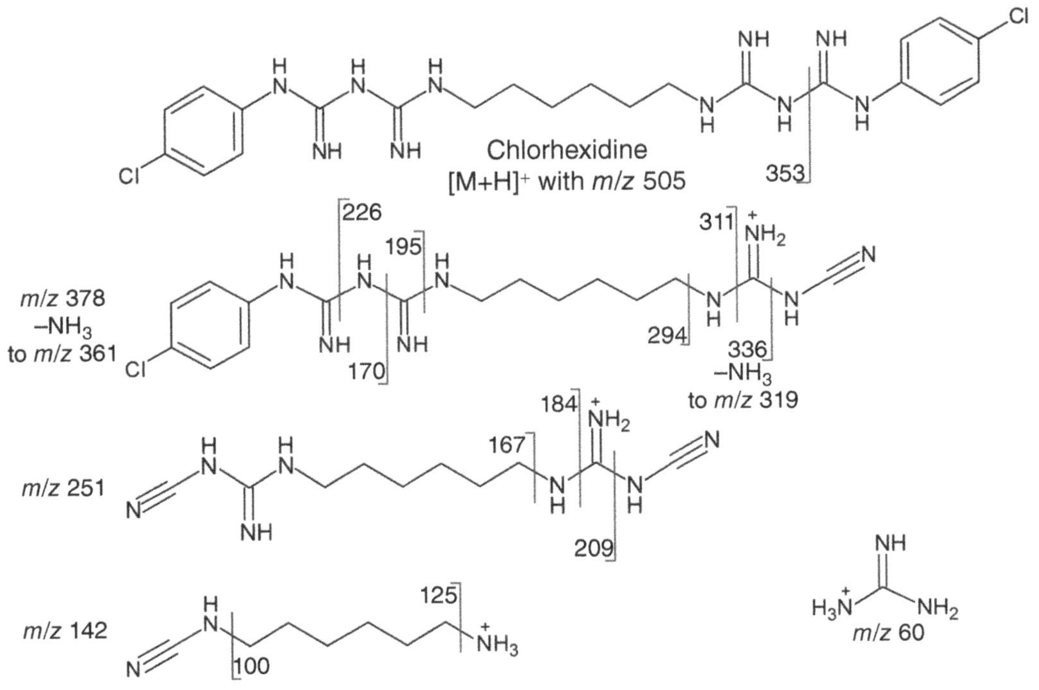

FIGURE 4.10.14 Fragmentation scheme for protonated chlorhexidine.

m/z 174. These two ions show subsequent loss of hydrogen isocyanate (HN=C=O) or ammonia (NH_3), respectively, to an ion with *m/z* 157 ($[C_{10}H_9N_2]^+$).

The extensive fragmentation of chlorhexidine ($[M+H]^+$ with *m/z* 505) is outlined in Figure 4.10.14. The fragmentation can be rationalized as secondary fragmentation by small-molecule losses from the ions formed by the initial loss of one or two of the 4-chloro-anilino groups to form the ions with *m/z* 378 and with *m/z* 251.

REFERENCES

Anari MR, Sanchez RI, Bakhtiar R, Franklin RB, Baillie TA. 2004. Integration of knowledge-based metabolic predictions with liquid chromatography data-dependent tandem mass spectrometry for drug metabolism studies: Application to studies on the biotransformation of indinavir. Anal Chem 76: 823–832.

Balizs G. 1999. Determination of benzimidazole residues using liquid chromatography and tandem mass spectrometry. J Chromatogr B 727: 167–177.

Beasley CA, Hwang TL, Fliszar K, Abend A, McCollum DG, Reed RA. 2006. Identification of impurities in ivermectin bulk material by mass spectrometry and NMR. J Pharm Biomed Anal 41: 1124–1134.

Becher F, Pruvost A, Goujard C, Guerreiro C, Delfraissy J-F, Grassi J, Benech H. 2002. Improved method for the simultaneous determination of d4T, 3TC and ddl intracellular phosphorylated anabolites in human peripheral blood mononuclear cells using high-performance liquid chromatography/tandem mass spectrometry, Rapid Commun Mass Spectrom 16: 555–565.

Bedse G, Kumar V, Singh S. 2009. Study of forced decomposition behavior of lamivudine using LC, LC–MS/TOF and MS^n. J Pharm Biomed Anal 49: 55–63.

Caldow M, Sharman M, Kelly M, Day J, Hird S, Tarbin JA. 2009. Multi-residue determination of phenolic and salicylanilide anthelmintics and related compounds in bovine kidney by liquid chromatography–tandem mass spectrometry. J Chromatogr A 1216: 8200–8205.

Chen D, Tao Y, Zhang H, Pan Y, Liu Z, Huang L, Wang Y, Peng D, Wang X, Dai M, Yuan Z. 2011. Development of a liquid chromatography–tandem mass spectrometry with pressurized liquid extraction method for the determination of benzimidazole residues in edible tissues. J Chromatogr B 879: 1659–1667.

Clarke L, Fodey TL, Crooks SR, Moloney M, O'Mahony J, Delahaut P, O'Kennedy R, Danaher M. 2014. A review of coccidiostats and the analysis of their residues in meat and other food. Meat Sci. doi: 10.1016/j.meatsci.2014.01.004.

Coulier L, Gerritsen H, van Kampen JJ, Reedijk ML, Luider TM, Osterhaus AD, Gruters RA, Brüll L. 2011. Comprehensive analysis of the intracellular metabolism of antiretroviral nucleosides and nucleotides using liquid chromatography–tandem mass spectrometry and method improvement by using ultra performance liquid chromatography. J Chromatogr B 879: 2772–2782.

Crommentuyn KM, Rosing H, Nan-Offeringa LG, Hillebrand MJ, Huitema AD, Beijnen JH. 2003. Rapid quantification of HIV protease inhibitors in human plasma by high-performance liquid chromatography coupled with electrospray ionization tandem mass spectrometry. J Mass Spectrom 38: 157–166.

Crommentuyn KM, Rosing H, Hillebrand MJ, Huitema AD, Beijnen JH. 2004. Simultaneous quantification of the new HIV protease inhibitors atazanavir and tipranavir in human plasma by high-performance liquid chromatography coupled with electrospray ionization tandem mass spectrometry. J Chromatogr B 804: 359–367.

Cronly M, Behan P, Foley B, Malone E, Regan L. 2009. Rapid confirmatory method for the determination of 11 nitroimidazoles in egg using liquid chromatography tandem mass spectrometry. J Chromatogr A 1216: 8101–8109.

Cvilink V, Skálová L, Szotáková B, Lamka J, Kostiainen R, Ketola RA. 2008. LC–MS–MS identification of albendazole and flubendazole metabolites formed ex vivo by Haemonchus contortus. Anal Bioanal Chem 391: 337–343.

Danaher M, Howells LC, Crooks SR, Cerkvenik-Flajs V, O'Keeffe M. 2006. Review of methodology for the determination of macrocyclic lactone residues in biological matrices. J Chromatogr B 844: 175–203.

Danaher M, De Ruyck H, Crooks SRH, Dowling G, O'Keeffe M. 2007. Review of methodology for the determination of benzimidazole residues in biological matrices. J Chromatogr B 845: 1–37.

Dongre VG, Karmuse PP, Ghugare PD, Gupta M, Nerurkar B, Shaha C, Kumar A. 2007. Characterization and quantitative determination of impurities in piperaquine phosphate by HPLC and LC/MS/MS. J Pharm Biomed Anal 43: 186–195.

Dongre VG, Ghugare PD, Karmuse P, Singh D, Jadhav A, Kumar A. 2009. Identification and characterization of process related impurities in chloroquine and hydroxychloroquine by LC/IT/MS, LC/TOF/MS and NMR. J Pharm Biomed Anal 49: 873–879.

Dubreil-Chéneau E, Bessiral M, Roudaut B, Verdon E, Sanders P. 2009. Validation of a multi-residue liquid chromatography–tandem mass spectrometry confirmatory method for 10 anticoccidials in eggs according to Commission Decision 2002/657/EC. J Chromatogr A 1216: 8149–8157.

Durden DA. 2007. Positive and negative electrospray LC–MS–MS methods for quantitation of the antiparasitic endectocide drugs, abamectin, doramectin, emamectin, eprinomectin, ivermectin, moxidectin and selamectin in milk. J Chromatogr B 850: 134–146.

De Ruyck H, Daeseleire E, De Ridder H, Van Renterghem R. 2002. Development and validation of a liquid chromatographic–electrospray tandem mass spectrometric multiresidue method for anthelmintics in milk. J Chromatogr A 976: 181–194.

Fan B, Bartlett MG, Stewart JT. 2002. Determination of lamivudine/stavudine/efavirenz in human serum using liquid chromatography/electrospray tandem mass spectrometry with ionization polarity switch. Biomed Chromatogr 16: 383–389.

Font E, Lasanta S, Rosario O, Rodríguez JF. 1998. Analysis of antiretroviral nucleosides by electrospray ionization mass spectrometry and collision induced dissociation. Nucleosides Nucleotides 17: 845–853.

Gangl E, Utkin I, Gerber N, Vouros P. 2002. Structural elucidation of metabolites of ritonavir and indinavir by liquid chromatography–mass spectrometry. J Chromatogr A 974: 91–101.

Gianelli L, Mellerio GG, Siviero E, Rossi A, Cabri W, Sogli L. 2000. Mass spectrometry of avermectins: Structural determination of two new derivatives of Ivermectin B(1a). Rapid Commun Mass Spectrom 14: 1260–1265.

Hodel EM, Zanolari B, Mercier T, Biollaz J, Keiser J, Olliaro P, Genton B, Decosterd LA. 2009. A single LC–tandem mass spectrometry method for the simultaneous determination of 14 antimalarial drugs and their metabolites in human plasma. J Chromatogr B 877: 867–886.

Howells L, Sauer MJ. Multi-residue analysis of avermectins and moxidectin by ion-trap LC–MS^n. 2001. Analyst 126: 155–160.

Junwal M, Sahu A, Handa T, Shah RP, Singh S. 2012. ICH guidance in practice: Degradation behaviour of oseltamivir phosphate under stress conditions. J Pharm Biomed Anal 62: 48–60.

Kinsella B, Lehotay SJ, Mastovska K, Lightfield AR, Furey A, Danaher M. 2009. New method for the analysis of flukicide and other anthelmintic residues in bovine milk and liver using liquid chromatography–tandem mass spectrometry. Anal Chim Acta 637: 196–207.

Kinsella B, Whelan M, Cantwell H, McCormack M, Furey A, Lehotay SJ, Danaher M. 2010. A dual validation approach to detect anthelmintic residues in bovine liver over an extended concentration range. Talanta 83: 14–24.

Lai J, Wang J, Cai Z. 2008. Nucleoside reverse transcriptase inhibitors and their phosphorylated metabolites in human immunodeficiency virus-infected human matrices. J Chromatogr B 868: 1–12.

Lehner AF, Petzinger E, Stewart J, Lang DG, Johnson MB, Harrison L, Seanor JW, Tobin T. 2009. ESI+ MS/MS confirmation of canine ivermectin toxicity. J Mass Spectrom 44: 111–119.

Li Z, Ding C, Ge Q, Zhou Z, Zhi X, Liu X. 2010. Simultaneous determination of lamivudine, stavudine and nevirapine in human plasma by LC–MS/MS and its application to pharmacokinetic study in clinic. Biomed Chromatogr 24: 926–934.

Lopez LL, Yu X, Cui D, Davis MR. 1998. Identification of drug metabolites in biological matrices by intelligent automated liquid chromatography/tandem mass spectrometry. Rapid Commun Mass Spectrom 12: 1756–1760.

Lopes NP, Gates PJ, Wilkins JP, Staunton J. 2002a. Fragmentation studies on lasalocid acid by accurate mass electrospray mass spectrometry. Analyst 127: 1224–1227.

Lopes NP, Stark CB, Hong H, Gates PJ, Staunton J. 2002b. Fragmentation studies on monensin A and B by accurate-mass electrospray tandem mass spectrometry. Rapid Commun Mass Spectrom 16: 414–420.

Lopes NP, Stark CB, Gates PJ, Staunton J. 2002c. Fragmentation studies on monensin A by sequential electrospray mass spectrometry. Analyst 127: 503–506.

Martínez-Villalba A, Moyano E, Galceran MT. 2009; Fast liquid chromatography/multiple-stage mass spectrometry of coccidiostats. Rapid Commun Mass Spectrom. 23(9):1255–1263.

Moloney M, Clarke L, O'Mahony J, Gadaj A, O'Kennedy R, Danaher M. 2012. Determination of 20 coccidiostats in egg and avian muscle tissue using ultra high performance liquid chromatography–tandem mass spectrometry. J Chromatogr A 1253: 94–104.

Mortier L, Daeseleire E, Van Peteghem C. 2005a. Liquid chromatographic tandem mass spectrometric determination of five coccidiostats in poultry eggs and feed. J Chromatogr B 820: 261–270.

Mortier L, Daeseleire E, Van Peteghem C. 2005b. Determination of the ionophoric coccidiostats narasin, monensin, lasalocid and salinomycin in eggs by liquid chromatography/tandem mass spectrometry. Rapid Commun Mass Spectrom 19: 533–539.

Müller DM, Rentsch KM. 2010. Therapeutic drug monitoring by LC–MS–MS with special focus on anti-infective drugs. Anal Bioanal Chem 398: 2573–2594.

Nuñez A, Lehotay SJ, Geis-Asteggiante L. 2015. Structural characterization of product ions by electrospray ionization and quadrupole time-of-flight mass spectrometry to support regulatory analysis of veterinary drug residues in foods. Part 2: Benzimidazoles, nitromidazoles, phenothiazines, and mectins. Rapid Commun Mass Spectrom 29: 719–729.

Olejnik M, Szprengier-Juszkiewicz T, Jedziniak P. 2009. Multi-residue confirmatory method for the determination of twelve coccidiostats in chicken liver using liquid chromatography tandem mass spectrometry. J Chromatogr A 1216: 8141–8148.

Pruvost A, Becher F, Bardouille P, Guerrero C, Creminon C, Delfraissy JF, Goujard C, Grassi J, Benech H. 2001. Direct determination of phosphorylated intracellular anabolites of stavudine (d4T) by liquid chromatography/tandem mass spectrometry. Rapid Commun Mass Spectrom 15: 1401–1408.

Ramesh T, Rao PN, Rao RN. 2014. LC–MS/MS method for the characterization of the forced degradation products of entecavir. J Sep Sci 37: 368–375.

Rao RN, Vali RM, Ramachandra B, Raju SS. 2011. Separation and characterization of forced degradation products of abacavir sulphate by LC–MS/MS. J Pharm Biomed Anal 54: 279–285.

Rao RN, Vali RM, Raju SS. 2013. Liquid chromatography tandem mass spectrometric studies of indinavir sulphate and its forced degradation products. J Pharm Biomed Anal 74: 101–110.

Rao RN, Ramachandra B, Sravan B, Khalid S. 2014. LC–MS/MS structural characterization of stress degradation products including the development of a stability indicating assay of darunavir: An anti-HIV drug. J Pharm Biomed Anal 89: 28–33.

Rentsch KM. 2003. Sensitive and specific determination of eight antiretroviral agents in plasma by high-performance liquid chromatography–mass spectrometry. J Chromatogr B 788: 339–350.

Siegel MM, McGahren WJ, Tomer KB, Chang TT. 1987. Applications of fast atom bombardment mass spectrometry and fast atom bombardment mass spectrometry–mass spectrometry to the maduramicins and other polyether antibiotics. Biomed Environ Mass Spectrom. 14(1):29–38.

Spisso BF, Ferreira RG, Pereira MU, Monteiro MA, Cruz TÁ, da Costa RP, Lima AM, da Nóbrega AW. 2010. Simultaneous determination of polyether ionophores, macrolides and lincosamides in hen eggs by liquid chromatography–electrospray ionization tandem mass spectrometry using a simple solvent extraction. Anal Chim Acta 682: 82–92.

Tarning J, Bergqvist Y, Day NP, Bergquist J, Arvidsson B, White NJ, Ashton M, Lindegårdh N. 2006. Characterization of human urinary metabolites of the antimalarial piperaquine. Drug Metab Dispos 34: 2011–2019.

Turnipseed SB, Roybal JE, Rupp HS, Gonzales SA, Pfenning AP, Hurlbut JA. 1999. Confirmation of avermectin residues in food matrices with negative-ion atmospheric pressure chemical ionization liquid chromatography/mass spectrometry. Rapid Commun Mass Spectrom. 13: 493–499. SIM transitions with small losses.

Turnipseed SB, Roybal JE, Andersen WC, Kuck LR. 2005. Analysis of avermectin and moxidectin residues in milk by liquid chromatography–tandem mass spectrometry using an atmospheric pressure chemical ionization/atmospheric pressure photoionization source. Anal Chim Acta 529: 159–165.

Volmer DA, Lock CM. 1998. Electrospray ionization and collision-induced dissociation of antibiotic polyether ionophores. Rapid Commun Mass Spectrom 12: 157–164.

Whelan M, Kinsella B, Furey A, Moloney M, Cantwell H, Lehotay SJ, Danaher M. 2010. Determination of anthelmintic drug residues in milk using ultra high performance liquid chromatography–tandem mass spectrometry with rapid polarity switching. J Chromatogr A 1217: 4612–4622.

Xia X, Li X, Ding S, Zhang S, Jiang H, Li J, Shen J. 2009. Determination of 5-nitroimidazoles and corresponding hydroxy metabolites in swine kidney by ultra-performance liquid chromatography coupled to electrospray tandem mass spectrometry. Anal Chim Acta 637: 79–86.

Xia X, Dong Y, Luo P, Wang X, Li X, Ding S, Shen J. 2010. Determination of benzimidazole residues in bovine milk by ultra-high performance liquid chromatography–tandem mass spectrometry. J Chromatogr B 878: 3174–3180.

Yakkundi S, Cannavan A, Elliott CT, Lovgren T, Kennedy DG. 2001. Development and validation of a method for the confirmation of nicarbazin in chicken liver and eggs using LC-electrospray MS–MS according to the revised EU criteria for veterinary drug residue analysis. Analyst 126: 1985–1989.

Yang FQ, Li DQ, Feng K, Hu DJ, Li SP. 2010. Determination of nucleotides, nucleosides and their transformation products in Cordyceps by ion-pairing reversed-phase liquid chromatography–mass spectrometry. J Chromatogr A 1217: 5501–5510.

Yu X, Cui D, Davis MR. 1999. Identification of in vitro metabolites of indinavir by "intelligent automated LC–MS/MS" (INTAMS) utilizing triple quadrupole tandem mass spectrometry. J Am Soc Mass Spectrom 10: 175–183.

Zeleny R, Harbeck S, Schimmel H. 2009. Validation of a liquid chromatography–tandem mass spectrometry method for the identification and quantification of 5-nitroimidazole drugs and their corresponding hydroxy metabolites in lyophilised pork meat. J Chromatogr A 1216: 249–256.

4.11

PESTICIDES

4.11.1 TRIAZINE HERBICIDES

The triazine herbicides belong to the 1,3,5-triazine group of compounds. Although most triazines are readily amenable to GC–MS, they are frequently studied by LC–MS. This is because sample pretreatment from aqueous samples is often more straightforward prior to LC–MS than to GC–MS. In addition, the analysis of hydroxy- and des-alkyl triazine degradation products is relevant; these are not readily amenable to GC–MS. There are three major classes of triazine herbicides: (1) chlorotriazines, (2) methylthiotriazines, and (3) methoxytriazines. In LC–MS, triazines can be analyzed using either ESI or APCI in positive-ion mode only. In most cases, protonated molecules are observed. Fragmentation due to the loss of an alkyl substituent chain may be observed under in-source CID conditions (Slobodník et al., 1996; Voyksner & Pack, 1991).

4.11.1.1 Chlorotriazines

The fragmentation of protonated chlorotriazines was investigated in TSI-MS on TQ instruments, and using H/D exchange experiments with deuterated ammonia (ND_3) (Nélieu et al., 1994; Voyksner et al., 1987). The interpretation of the product-ion mass spectra of chlorotriazines can be facilitated by (1) the acquisition of separate spectra for the ^{35}Cl- and the ^{37}Cl-containing precursor ion, (2) the selection of the complete precursor-ion isotope pattern and subsequent fragmentation, (3) a stepwise fragmentation of triazines in multistage MS^n in an ion-trap MS^n instrument (Hogenboom et al., 1998).

Atrazine ([M+H] with *m/z* 216) provides one primary fragment ion with *m/z* 174 ($[C_5H_8ClN_5+H]^+$) due to the loss of propene ($CH_3CH{=}CH_2$) from the *iso*-propyl substituent of the C^2-amine. The fragment ion with *m/z* 174 yields three secondary fragment ions, which are the protonated 6-chloro-2,4-diamino-1,3,5-triazine ion ($[C_3H_4ClN_5+H]^+$) with *m/z* 146 due to the loss of ethene (C_2H_4), the 2-amino-4-(*N*-ethylamino)-1,3,5-triazin-6-ylium ion ($[C_5H_8N_5]^+$) with *m/z* 138 due to loss of hydrogen chloride (HCl), and the N^3-ethylguanidine-N^1-chloromethylidenylium ion ($[C_4H_7ClN_3]^+$) with *m/z* 132 due to ring opening and the loss of methanediimine (HN=C=NH). Subsequent fragmentation of the ion with *m/z* 138 results in the 2,4-diamino-1,3,5-triazine-6-ylium ion with *m/z* 110 ($[C_3H_4N_5]^+$) due to C_2H_4 loss and the *N*-(1,3-diazet-4-ylium-2-yl)ethylamine ion ($C_4H_6N_3]^+$) with *m/z* 96 due to HN=C=NH loss. The ion with *m/z* 132 yields the ion with *m/z* 104 ($[C_2H_3ClN_3]^+$) due to C_2H_4 loss. The *N*-(1,3-diazet-4-ylium-2-yl)amine ion ($[C_2H_2N_3]^+$) with *m/z* 68 can be considered to be due to the loss of $CH_3CH{=}CH_2$ and subsequent losses of HN=C=NH, C_2H_4 and HCl, in either order. Other fragment ions observed for atrazine are the protonated chloroformamidine ($[H_2NCCl{=}NH_2]^+$) with *m/z* 79, the protonated ethylcyanamide ($[N{\equiv}C{-}NHCH_2CH_3+H]^+$) with *m/z* 71, and the *iso*-propyl carbocation ion ($[C_3H_7]^+$) with *m/z* 43. Proposed structures for the fragments ions are given in Figure 4.11.1.

Many of these fragment ions are class-specific fragment ions (Table 4.11.1a). Comparing product-ion mass spectra of various chlorotriazine herbicides reveals that the largest alkyl substituents is preferentially lost first, for example, in atrazine the *iso*-propyl group is lost prior to the ethyl group, that is, a loss of $CH_3CH{=}CH_2$ prior to a loss of C_2H_4. The same argument holds for the butyl and ethyl groups in sebutylazine and terbuthylazine, where the *sec*-butyl or *tert*-butyl substituent is lost as butene (C_4H_8) or 2-methylpropene ($(CH_3)_2C{=}CH_2$), respectively, prior to the loss of the ethyl substituent.

The fragmentation of cyanazine ($[M+H]^+$ with *m/z* 241) is somewhat different, featuring competitive losses of hydrogen cyanide (HCN), of HCl, and of methacrylonitrile (C_4H_5N) to fragment ions with *m/z* 214, *m/z* 205, and *m/z* 174, respectively. The ion with *m/z* 174 shows subsequent losses of HN=C=NH to *m/z* 132 and of C_2H_4 to *m/z* 146 (Table 4.11.1a).

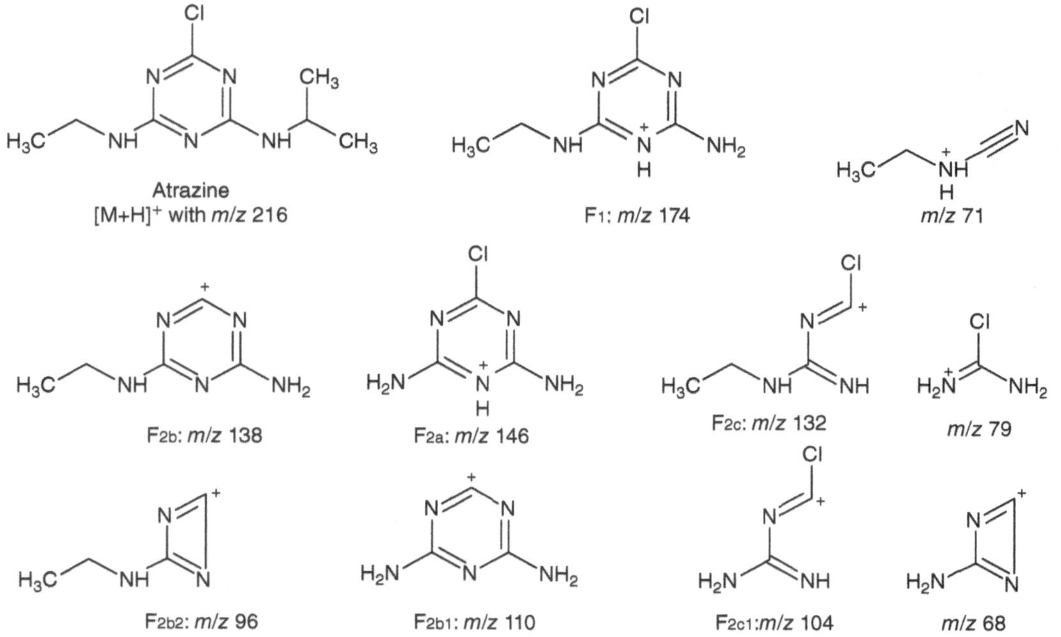

FIGURE 4.11.1 Proposed structures for the fragment ions of the protonated chlorotriazine herbicide atrazine. (Source: Niessen, 2010. Reproduced with permission of Elsevier.)

4.11.1.2 Methylthiotriazines

The methylthiotriazines follow similar fragmentation patterns. The fragmentation can be readily understood by defining three lines of fragmentation, as illustrated for ametryn ([M+H]$^+$ with *m/z* 228) (Figure 4.11.2). The first step involves the preferential loss of the largest alkyl (*iso*-propyl) group as propene ($CH_3CH{=}CH_2$) to an ion with *m/z* 186 ([$C_6H_{11}N_5S$+H]$^+$). One line involves the loss of the second alkyl (ethyl) group as ethene (C_2H_4) to an ion with *m/z* 158 ([$C_4H_7N_5S$+H]$^+$). The second line involves the loss of methanethiol (CH_3SH) to an ion with *m/z* 138 ([$C_5H_8N_5$]$^+$) and the subsequent loss of either C_2H_4 or HN=C=NH to the ions with *m/z* 110 ([$C_3H_4N_5$]$^+$) and *m/z* 96 ([$C_4H_6N_3$]$^+$), respectively. The third line involves the loss of methanediimine (HN=C=NH) and subsequent loss of C_2H_4 to the ions with *m/z* 144 ([$C_5H_9N_3S$+H]$^+$) and *m/z* 116 ([$C_3H_5N_3S$+H]$^+$), respectively. In addition, common fragment ions with *m/z* 91 due to protonated *S*-methylisothiourea ([$NH_2CNHSCH_3$+H]$^+$) and *m/z* 74 due to protonated methyl thiocyanate ([$CH_3S{=}C{=}N$+H]$^+$) are observed. Finally, the protonated ethylcyanamide ([$C_2H_5NHC{\equiv}N$+H]$^+$]) with *m/z* 71 is observed. Proposed structures for these fragments ions are given in Figure 4.11.2. Similar fragment ions are observed for desmetryn and simetryn (both [M+H]$^+$ with *m/z* 214), prometryn and terbutryn (both [M+H]$^+$ with *m/z* 242), and dimethatryn ([M+H]$^+$ with *m/z* 256) (Table 4.11.1b).

Some interesting features are observed for prometryn ([M+H]$^+$ with *m/z* 242). After the loss of the first alkyl group as $CH_3CH{=}CH_2$, two possible subsequent losses of 42 Da may occur, that is, either the loss of HN=C=NH or the loss of another $CH_3CH{=}CH_2$. Accurate-mass data (*m/z* difference of 0.025) show that only the latter one is observed. This may be explained by the fact that loss of HN=C=NH would involve aromaticity disruption and ring opening of the triazine heterocycle, whereas loss of C_3H_6 only involves the cleavage of an N—C bond with H-rearrangement. Similarly, the same two losses of 42 Da are possible for the prometryn fragment ion with *m/z* 152, which is due to subsequent losses of $CH_3CH{=}CH_2$ and CH_3SH. Surprisingly, the loss of HN=C=NH is only observed at the lower collision energy, whereas the loss of $CH_3CH{=}CH_2$ is only observed at the higher collision energy.

4.11.1.3 Methoxytriazines

Again, the methoxytriazines show similar features in their fragmentation as the other triazine classes. As an example, atraton ([M+H]$^+$ with *m/z* 212) first shows the loss of propene ($CH_3CH{=}CH_2$) to an ion with *m/z* 170 ([$C_6H_{11}N_5O$+H]$^+$). Secondary fragmentation can be considered to occur along four lines, that is, one more than for the other triazine classes. First, the loss of ethene (C_2H_4) leads to an ion with *m/z* 142 ([$C_4H_7N_5O$+H]$^+$). Second, the loss of methanol (CH_3OH) leads to an ion with *m/z* 138 ([$C_5H_8N_5$]$^+$), which shows secondary fragmentation involving the loss of either C_2H_4 to *m/z* 110 ([$C_3H_4N_5$]$^+$) or methanediimine (HN=C=NH) to *m/z* 96 ([$C_4H_6N_3$]$^+$). Third, the loss of HN=C=NH leads to an ion with *m/z* 128

TABLE 4.11.1 Fragmentation of (a) chlorotriazine herbicides (Figure 4.11.1), (b) methylthiotriazine herbicides (Figure 4.11.2), and (c) methoxytriazine herbicides (Figure 4.11.3).

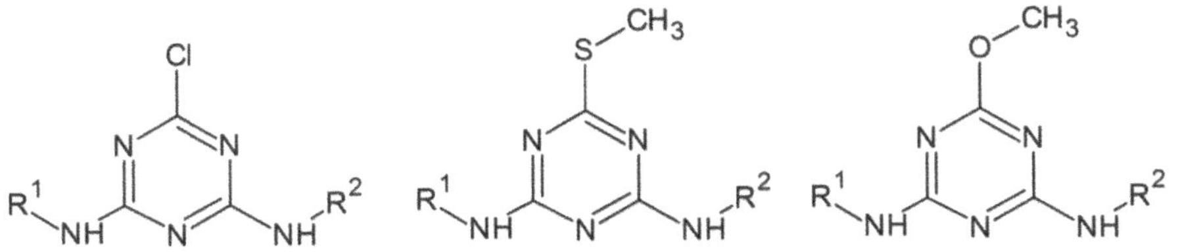

Chlorotriazine   Methylthiotriazine   Methoxytriazine

Compound	$[M+H]^+$	R^1	R^2	F_1	F_{2a}	F_{2b}	F_{2b1}	F_{2b2}	F_{2c}	F_{2c1}	Other Fragments
(a)											
Atrazine	216	$—C_3H_7$	$—C_2H_5$	174	146	138	110	96	132	104	79, 71
Simazine	202	$—C_2H_5$	$—C_2H_5$	174	146	138	110	96	132	104	79, 71
Propazine	230	$—C_3H_7$	$—C_3H_7$	188	146		110			104	79
Sebutylazine	230	$—sC_4H_9$	$—C_2H_5$	174	146	138	110	96	132	104	79, 71 57: $[C_4H_9]^+$
Terbuthylazine	230	$—tC_4H_9$	$—C_2H_5$	174	146	138	110	96	132	104	79, 71 57: $[C_4H_9]^+$
Trietazine	230	$—diC_2H_5$	$—C_2H_5$	202	174	166		124		132	146: $F_{2a}-C_2H_4$ 104: $F_{2c1}-C_2H_4$ 99: $[C_5H_{11}N_2]^+$ 71 62: $[ClCN+H]^+$
Cyanazine	241	$—C_4H_6N$	$—C_2H_5$	174	146	138	110	96	132	104	214: −HCN 205: −HCl 79, 71
(b)											
Ametryn	228	$—C_3H_7$	$—C_2H_5$	186	158	138	110	96	144	116	91, 74, 71
Desmetryn	214	$—C_3H_7$	$—CH_3$	172		124		82	130	116	91, 74, 57
Simetryn	214	$—C_2H_5$	$—C_2H_5$	186	158	138	110	96	144	116	91, 74, 71
Prometryn	242	$—C_3H_7$	$—C_3H_7$	200	158	152	110*	110†		116	91, 74, 85
Terbutryn	242	$—C_4H_9$	$—C_2H_5$	186	158	138	110	96	144	116	91, 74, 71 57: $[C_4H_9]^+$
Dimethatryn	256	$—C_5H_{11}$	$—C_2H_5$	186	158	138	110	96	144	116	91, 74, 71
(c)											
Atraton	212	$—C_3H_7$	$—C_2H_5$	170	142	138	110	96	128	100	114, 86, 75
Prometon	226	$—C_3H_7$	$—C_3H_7$	184	142	152	110*	110†	142	100	86, 75
Secbumeton	226	$—sC_4H_9$	$—C_2H_5$	170	142	138	110	96	128	100	114, 86, 75
Terbumeton	226	$—tC_4H_9$	$—C_2H_5$	170	142	138	110	96	128	100	114, 86, 75

The primary loss of (R^1-H) yields F_1, which shows subsequent losses of (R^2-H) to F_{2a}; of HCl (in chlorotriazines), CH_3SH (in methylthiotriazines), or CH_3OH (in methoxytriazines) to F_{2b}; or of HN=C=NH to F_{2c}. F_{2b} shows the loss of (R^2-H) to F_{2b1} or of HN=C=NH to F_{2b2}. F_{2c} shows loss of (R^2-H) to F_{2c1}.

*Observed at higher collision energy.

†Observed at lower collision energy.

($[C_5H_9N_3O+H]^+$), which in turn shows the loss of C_2H_4 to an ion with *m/z* 100 ($[C_3H_5N_3O+H]^+$). Lastly, the ions with *m/z* 114 ($[C_4H_7N_3O+H]^+$) and *m/z* 86 ($[C_2H_3N_3O+H]^+$) are unexpected. They seem to involve an unusual loss of methylene (:CH_2) from the methoxy moiety and the subsequent loss of C_2H_4. Proposed structures for all these fragments ions are given in Figure 4.11.3. Similar fragments are observed for the isomeric analogs prometon, secbumeton, and terbumeton ($[M+H]^+$ with *m/z* 226), again with the characteristic preferential loss of the largest alkyl substituent first (Table 4.11.1c).

4.11.2 CARBAMATES

Carbamates are a large group of compounds that are manufactured and used in large quantities as acaricides, fungicides, herbicides, insecticides, and nematicides. The general structure of a carbamate is $R^1OCONR^2R^3$. In addition, there

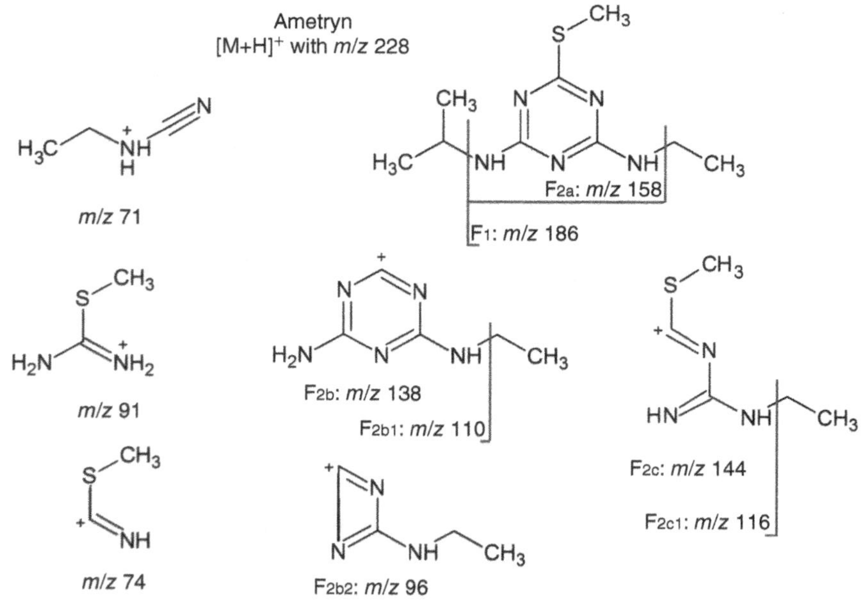

FIGURE 4.11.2 Proposed structures for the fragment ions of the protonated methylthiotriazine herbicide ametryn.

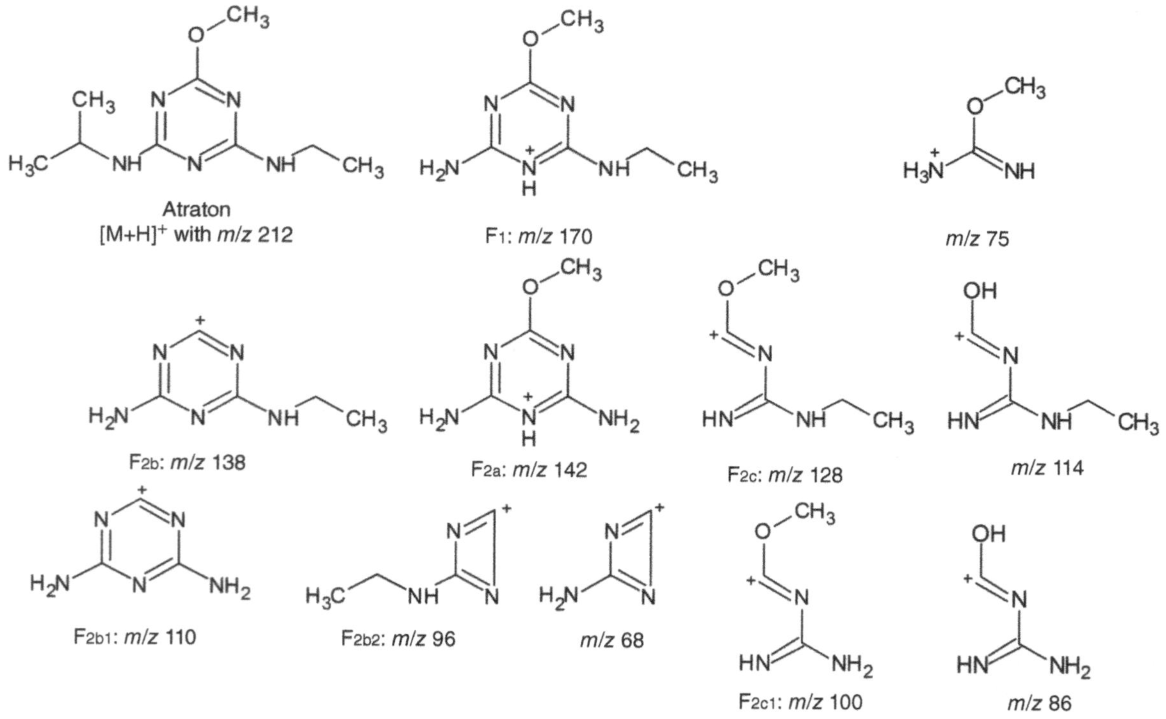

FIGURE 4.11.3 Proposed structures for the fragment ions of the protonated methoxytriazine herbicide atraton.

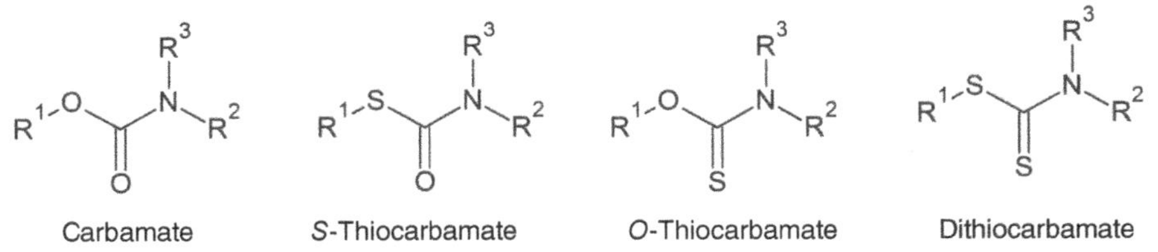

FIGURE 4.11.4 General structures for the carbamate pesticide classes.

are numerous *O*- and *S*-thiocarbamates and dithiocarbamates as well (Figure 4.11.4).

From the various subclasses of carbamates, *O*-aryl-*N*-methyl carbamates such as carbaryl, carbofuran, and propoxur, and oxime-*N*-methyl carbamates such as aldicarb, methomyl, and aldoxycarb have been most widely studied. LC–MS is the method of choice for carbamates because their thermal lability prohibits their GC–MS analysis. Carbamates are analyzed in positive-ion ESI mode, because in APCI thermal decomposition in the heated nebulizer interface and subsequent ionization of the thermal decomposition products is observed (Slobodník et al., 1996; Kawasaki et al., 1993). Due the labile character of the carbamates, fragmentation can be easily induced by in-source CID (Voyksner & Pack, 1991).

4.11.2.1 *N*-Methylcarbamates

Product-ion mass spectra of aryl *N*-methyl carbamates ($R^1OCONHCH_3$) have been investigated using TSI and either protonated or ammoniated molecules as precursor ions (Chiu et al., 1989). They show a characteristic class-specific loss of methyl isocyanate ($CH_3N{=}C{=}O$). Further fragmentation is compound specific due to losses from the resulting R^1 group; some typical examples are shown in Table 4.11.2, providing data on metolcarb ($[M+H]^+$ with *m/z* 166), carbaryl ($[M+H]^+$ with *m/z* 202), promecarb ($[M+H]^+$ with *m/z* 208), aminocarb ($[M+H]^+$ with *m/z* 209), carbofuran ($[M+H]^+$ with *m/z* 222), bendiocarb ($[M+H]^+$ with *m/z* 224), dioxacarb ($[M+H]^+$ with *m/z* 224), and methiocarb ($[M+H]^+$ with *m/z* 226). The identity of the major fragment ions of carbofuran, that is, the protonated 2,2-dimethyl-(3*H*)-benzofuran-7-ol ($[C_{10}H_{12}O_2+H]^+$) with *m/z* 165 and the 2,3-dihydroxybenzyl cation ($[(HO)_2C_6H_3CH_2]^+$) with *m/z* 123, has been confirmed and that of some minor fragments has been elucidated using LC–MS in a Q–TOF hybrid instrument (Figure 4.11.5) (Detomaso et al., 2005). The minor fragment ion with *m/z* 137 ($[C_8H_8O_2+H]^+$) is rather unexpected because it should involve losses of two substituents from the same carbon atom, which is unlikely to occur. However, the molecular formula of the ion with *m/z* 137 is in agreement with the measured accurate *m/z*.

In some cases, compound-specific fragmentation prevails over the initial characteristic class-specific loss of $CH_3N{=}C{=}O$. In fenobucarb ($[M+H]^+$ with *m/z* 208), the ion due to a primary loss of butene (C_4H_8), involving a C—C cleavage to the aromatic ring, is more abundant than the ion due to the loss of $CH_3N{=}C{=}O$, involving the cleavage of a C—O bond in the carbamate

TABLE 4.11.2 Fragmentation of protonated *N*-methyl carbamates.

Compound	$[M+H]^+$	Loss of 57 Da (*m/z*)	Secondary Fragments	
Metolcarb	166	109	*m/z* 91	109 – H_2O
			m/z 81	109 – CO
Carbaryl	202	145	*m/z* 127	145 – H_2O
			m/z 117	145 – CO
Promecarb	208	151	*m/z* 109	151 – $CH_3CH{=}CH_2$
			m/z 91	109 – H_2O
			m/z 81	108 – CO
Aminocarb	209	152	*m/z* 137	152 – $^{\bullet}CH_3$
			m/z 122	137 – $^{\bullet}CH_3$
Carbofuran	222	165	*m/z* 123	165 – $CH_3CH{=}CH_2$
Bendiocarb	224	167	*m/z* 109	167 – $(CH_3)_2C{=}O$
			m/z 81	109 – CO
Dioxacarb	224	167	*m/z* 123	167 – C_2H_4O
			m/z 95	123 – CO
Methiocarb	226	169	*m/z* 121	169 – CH_3SH

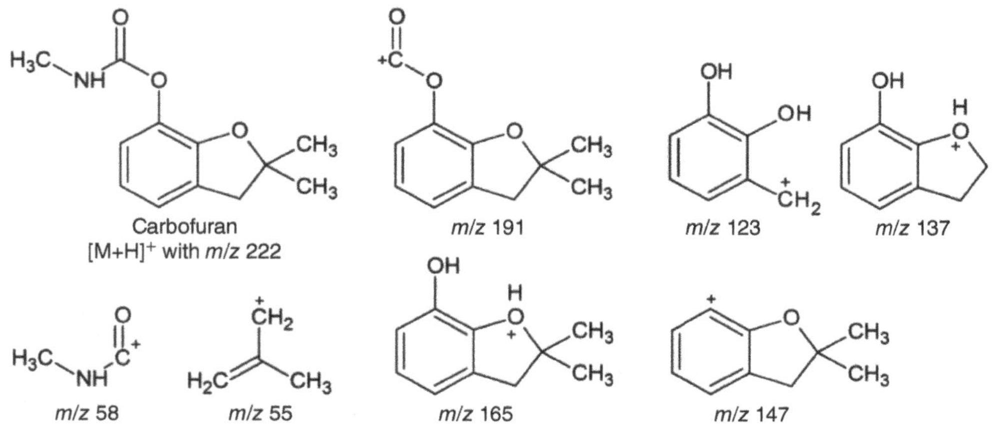

FIGURE 4.11.5 Proposed structures for the fragment ions of the protonated *N*-methyl carbamate pesticide carbofuran. (Source: Detomaso et al., 2005. Reproduced with permission of Wiley.)

group. Subsequent fragmentation results in protonated phenol with *m/z* 95 ($[C_6H_5OH+H]^+$), which is the base peak in the spectrum. Similarly, there is a competition between the loss of $CH_3N{=}C{=}O$ and of propene ($CH_3CH{=}CH_2$) in isoprocarb ($[M+H]^+$ with *m/z* 194) and in propoxur ($[M+H]^+$ with *m/z* 210). The loss of both groups leads to $[C_6H_5OH+H]^+$ with *m/z* 95 for isoprocarb and protonated catechol ($[C_6H_6O_2+H]^+$) with *m/z* 111 for propoxur. In ethiofencarb ($[M+H]^+$ with *m/z* 226), ethanethiol (CH_3CH_2SH) is lost prior to the $CH_3N{=}C{=}O$ loss, thus resulting in the 2-(*N*-methylcarbamate)benzyl cation ($[CH_3NHOCOC_6H_4{-}CH_2]^+$) with *m/z* 164 and the 2-hydroxybenzyl cation ($[HOC_6H_4CH_2]^+$) with *m/z* 107. Interestingly, in ethiofencarb sulfoxide ($[M+H]^+$ with *m/z* 242), the ion due to the primary loss of $CH_3N{=}C{=}O$ is more abundant than that due to the primary loss of CH_3CH_2SHO, whereas in ethiofencarb sulfone ($[M+H]^+$ with *m/z* 258), only the primary loss of $CH_3N{=}C{=}O$ is observed.

4.11.2.2 Oxime-*N*-Methyl Carbamates

In *N*-oxime carbamates, $R^1R^2C{=}NOCONHCH_3$, the fragmentation is mostly directed from the oxime functional group rather than from the carbamate group (Chiu et al., 1989). The major fragment ions in the product-ion mass spectrum of aldicarb ($[M+H]^+$ with *m/z* 191) are the protonated 2-methyl-2-(methylsulfanyl)propanenitrile ($[CH_3SC(CH_3)_2C{\equiv}N+H]^+$) with *m/z* 116 due to the loss of methylcarbamic acid ($CH_3NHCOOH$) and the 2-(methylsulfanyl)prop-2-yl carbocation ($[CH_3SC(CH_3)_2]^+$) with *m/z* 89. In addition, the protonated 2-methylpropanenitrilium ion ($[C_4H_8N]^+$) with *m/z* 70 and the (methylsulfanyl)methylium ion ($[CH_3SCH_2]^+$) with *m/z* 61 are observed (Figure 4.11.6). The major fragment ions of methomyl ($[M+H]^+$ with *m/z* 163) are the 2-imino-2-methylsulfanylethyl carbocation ($[CH_3SNHCCH_2]^+$) with *m/z* 88, due to the loss of $CH_3NHCOOH$, and the complementary ions with *m/z* 106 and 58, that is, the protonated methyl(methylsulfanyl)oxime ($[C_3H_8NOS]^+$), due to the

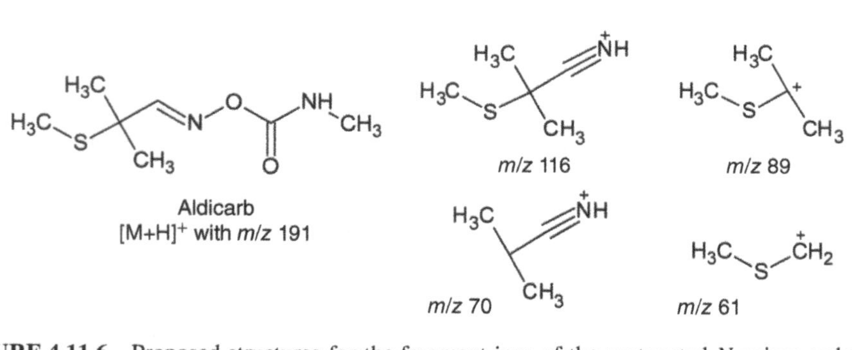

FIGURE 4.11.6 Proposed structures for the fragment ions of the protonated *N*-oxime carbamate pesticide aldicarb.

loss of $CH_3N{=}C{=}O$, and the (methylamino)acylium ion ($[CH_3NH{-}C{=}O]^+$), respectively. Oxamyl ($[M+H]^+$ with *m/z* 220) shows only two fragments, that is, the (dimethylamino)acylium ion ($[(CH_3)_2N{-}C{=}O]^+$) with *m/z* 72 and the protonated dimethylcarbamic acid with *m/z* 90 ($[(CH_3)_2NCOOH+H]^+$). The latter ion is the result of the loss of $CH_3N{=}C{=}O$ and a subsequent rearrangement involving the loss of methyl thiocyanate ($CH_3S{=}C{=}N$).

4.11.2.3 *N,N*-Dimethyl Carbamates

In the product-ion mass spectrum of the *N,N*-dimethyl carbamate insecticide pirimicarb ($[M+H]^+$ with *m/z* 239, $[C_{11}H_{19}N_4O_2]^+$), interesting rearrangements are observed. The primary fragments are the protonated 2,4-bis (dimethylamino)-5,6-dimethylpyrimidine ($[C_{10}H_{18}N_4+H]^+$) with *m/z* 195 and the protonated 2-dimethylamino-3,5,6-trimethylpyrimidin-4-one ($[C_9H_{15}N_3O+H]^+$) with *m/z* 182, which are due to a neutral loss of carbon dioxide (CO_2) and methyl isocyanate ($CH_3N{=}C{=}O$), respectively (Figure 4.11.7). The loss of CO_2 requires a rearrangement of a dimethylamino group, whereas the loss of $CH_3N{=}C{=}O$ requires a rearrangement of a methyl group to the ring. Both rearrangements were confirmed using accurate-mass determination in a Q–TOF instrument (Bobeldijk et al., 2001). The other fragments can be considered as secondary fragments of either of these two product ions (Figure 4.11.7).

4.11.2.4 Carbanilates

The related compound class of carbanilates is applied as fungicides, herbicides, and rodenticides. The general structure of a carbanilate is $R^1R^2NOCOR^3$, where R^1 is a (substituted) phenyl group and R^2 is mostly H.

Propham ($[M+H]^+$ with *m/z* 180) shows a series of fragment ions with *m/z* 138, 120, 92, consistent with subsequent losses of propene ($CH_3CH{=}CH_2$), water, and carbon monoxide (CO). Chlorpropham ($[M-H]^-$ with *m/z* 212) shows the loss of *iso*-propanol ($(CH_3)_2CHOH$) to (3-chlorophen-6-ide)isocyanate ($[ClC_6H_3{-}N{=}C{=}O]^-$) with *m/z* 152, and the complementary *iso*-propoxide anion ($[C_3H_7O]^-$) with *m/z* 59, as well as the 3-chloroanilinide ion ($[ClC_6H_4NH]^-$) with *m/z* 126 and the propen-2-oxide anion ($[C_3H_5O]^-$) with *m/z* 57. The isomeric desmedipham and phenmedipham ($[M+H]^+$ with *m/z* 301) show losses of phenylisocyanate ($C_6H_5{-}N{=}C{=}O$) or isocyanatotoluene ($CH_3C_6H_4{-}N{=}C{=}O$) to fragment ions with *m/z* 182 ($[C_9H_{11}NO_3+H]^+$) and 168 ($[C_8H_9NO_3+H]^+$), respectively, and a common (3-hydroxyanilino)acylium ion ($[HOC_6H_4NH{-}C{=}O]^+$) with *m/z* 136, which may lose CO to a fragment ion with *m/z* 108 ($[C_6H_6NO]^+$). In desmedipham, the fragment ion with *m/z* 182 shows secondary fragmentation involving the loss of either water or ethene (C_2H_4) to ions with *m/z* 164 and 154, respectively.

4.11.2.5 Thiocarbamates

The fragmentation of *S*-thiocarbamates closely resembles that of the other carbamate classes discussed here. This may be illustrated by two examples. Ethiolate ($[M+H]^+$ with *m/z* 162) shows the cleavage of the carbonyl C—N bond, resulting in the *N*-ethylethaniminium ion ($[C_4H_{10}N]^+$) with *m/z* 72 and the (ethylsulfanyl)acylium ion ($[C_2H_5S{-}C{=}O]^+$) with *m/z* 89. The cleavage of the carbonyl C—S bond results in the (diethylamino)acylium ion ($[(C_2H_5)_2N{-}C{=}O]^+$) with *m/z* 100. Loss of ethene (C_2H_4) from $[M+H]^+$ results in the ion with *m/z* 134. Orbencarb ($[M+H]^+$ with *m/z* 258)

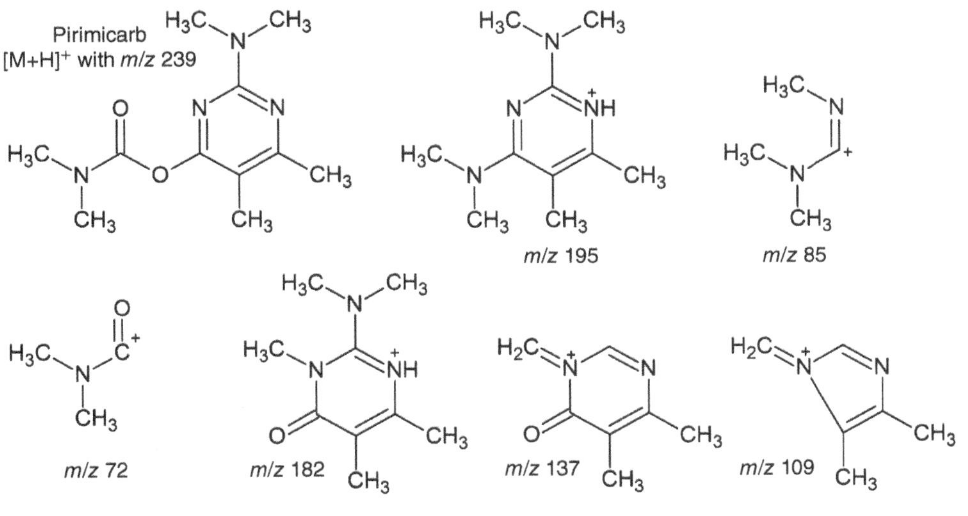

FIGURE 4.11.7 Proposed structures for the fragment ions of the protonated *N,N*-dimethyl carbamate insecticide pirimicarb.

shows the 2-chlorobenzyl cation ($[ClC_6H_4CH_2]^+$) with *m/z* 125, $[(C_2H_5)_2N—C{=}O]^+$ with *m/z* 100, and the (ethylamino)acylium ion with *m/z* 72 due to subsequent loss of C_2H_4. Similar fragmentation behavior is observed for prosulfocarb ($[M+H]^+$ with *m/z* 252) and thiobencarb ($[M+H]^+$ with *m/z* 258).

4.11.3 QUATERNARY AMMONIUM HERBICIDES

Positive-ion ESI is the obvious choice in the analysis of quaternary ammonium herbicides and plant growth regulators. Positive-ion APCI data have also been reported.

The major fragments of the plant growth regulator chlormequat ($[(CH_3)_3NCH_2CH_2Cl]^+$, M^+ with *m/z* 122) are due to the loss of the chloroethyl substituent either as chloroethane (C_2H_5Cl) or as a chloroethyl radical ($^{\bullet}CH_2CH_2Cl$), leading to the ions with *m/z* 58 ($[(CH_3)_2N{=}CH_2]^+$) and *m/z* 59 ($[(CH_3)_3N]^{+\bullet}$), respectively (Hau et al., 2000). Charge migration to form the 2-chloroethylium ion with *m/z* 63 ($[ClCH_2CH_2]^+$) is also observed (Figure 4.11.8). The product-ion mass spectrum for [D_9]-chlormequat, applied as stable isotopically labeled internal standard (SIL-IS), was reported as well. Interestingly, the relative abundance of the fragments $[CH_3)_2N{=}CH_2]^+$ and $[(CH_3)_3N]^{+\bullet}$ were reversed in the [D_9]-analogs due to a D-heavy-atom effect: the required rearrangement of a deuterium instead of a hydrogen atom in the formation of $[CH_3)_2N{=}CH_2]^+$ is more difficult (Hau et al., 2000). In the product-ion mass spectra of [D_4]-chlormequat, also applied as SIL-IS, this D-heavy-atom effect is not observed (Careri et al., 2002).

Both paraquat ($[C_{12}H_{14}N_2]^{2+}$, *m/z* 93.058) and diquat ($[C_{12}H_{12}N_2]^{2+}$, *m/z* 92.050) have two quaternary nitrogen atoms (Figure 4.11.9). The positive-ion ESI mass spectra were found to depend on the solvent composition (Song & Budde, 1996). In acetic acid or sodium acetate, the doubly-charged $[Cation]^{2+}$ with *m/z* 93 for paraquat and *m/z* 92 for diquat are the base peaks of the mass spectrum. In addition, pairs of singly-charged $[Cation-H]^+$ and $[Cation+e^-]^{+\bullet}$ were detected with *m/z* 185 and 186 for paraquat and *m/z* 183 and 184 for diquat. The radical ions with *m/z* 184 and 186 must be formed by a one-electron reduction of the EE^+ $[Cation]^{2+}$ to form an $OE^{+\bullet}$ $[Cation+e^-]^{+\bullet}$. The ions at *m/z* 183 and 185, which were the base peak in the mass spectra when ammonium is present in the solvent, were assumed to result from a proton-transfer reaction from the analyte to ammonia (NH_3) present in solution or during desolvation, that is, $[Cation-H]^+$ (Song & Budde, 1996) (Figure 4.11.9). Positive-ion ESI mass spectra of paraquat and diquat were also studied and reported by others (Marr & King, 1997; Taguchi et al., 1998; Castro et al., 1999). Under the conditions applied (Marr & King, 1997), the doubly-charged $[Cation]^{2+}$ was the most abundant ion for paraquat, whereas for diquat the singly-charged ion with *m/z* 183 due to $[Cation–H]^+$ was most abundant. In addition, singly-charged $[Cation-H]^+$ and $[Cation-CH_3]^+$ with *m/z* 185 and 171 were observed for paraquat, and the doubly-charged $[M]^{2+}$ with *m/z* 92 for diquat. Next to paraquat and diquat, tabulated mass spectra for mepiquat, chlormequat, and difenzoquat were reported, using a solvent gradient of acetonitrile in 15 mmol L^{-1} aqueous heptafluorobutyric acid, with post-column addition of acetonitrile (Castro et al., 1999). Under these conditions, $[Cation-H]^+$ was most abundant for both paraquat and diquat, while $[Cation+e^-]^+$ was most abundant for mepiquat, chlormequat, and difenzoquat.

In product-ion mass spectra using $[Cation–H]^+$ as precursor ion, paraquat (*m/z* 185) shows an ion with *m/z* 170 ($[C_{11}H_{10}N_2]^{+\bullet}$) due to the loss of a methyl radical ($^{\bullet}CH_3$), with *m/z* 158 ($[C_{11}H_{12}N]^+$) due to the loss of hydrogen cyanide (HCN), and with *m/z* 143 ($[C_{10}H_9N]^+$) due to the loss of both $^{\bullet}CH_3$ and HCN (Marr & King, 1997; Lee et al., 2004). Diquat (*m/z* 183) shows an ion with *m/z* 157 ($[C_{10}H_8N_2+H]^+$) due to the loss of acetylene (C_2H_2) and an ion with *m/z* 130 ($[C_9H_7N+H]^+$) due to the loss of C_2H_2 and HCN (Figure 4.11.9). Ion-trap product-ion mass spectra were reported for paraquat, diquat, difenzoquat, mepiquat, and chlormequat (Evans et al., 2001; Castro et al., 2001). In a subsequent study, the interpretation was confirmed using accurate-mass determination in a Q–TOF instrument (Núñez et al., 2004). For paraquat and diquat, fragment ions from both $[Cation]^{+\bullet}$ and $[Cation]^{2+}$ were tabulated (Núñez et al., 2004).

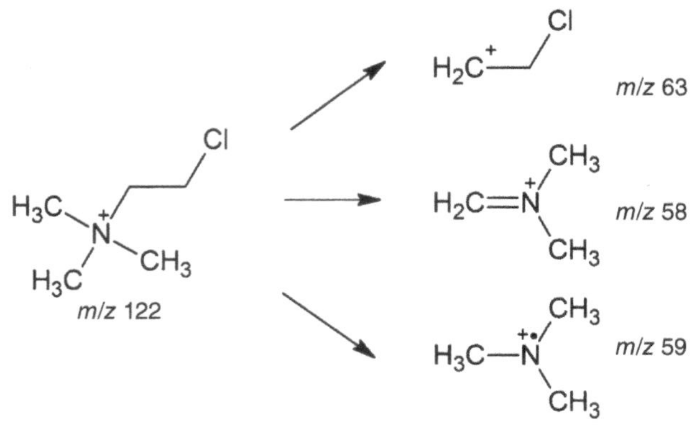

FIGURE 4.11.8 Proposed structures for the fragment ions of the quaternary ammonium herbicide chlormequat.

4.11.4 ORGANOPHOSPHORUS PESTICIDES

The MS analysis of organophosphorus compounds is relevant for both organophosphorus pesticides (OPPs) and nerve agent poisoning (John et al., 2008). The compound class of OPPs can be subdivided into various subclasses, including dimethyl and diethyl phosphate, phosphorothioate, and

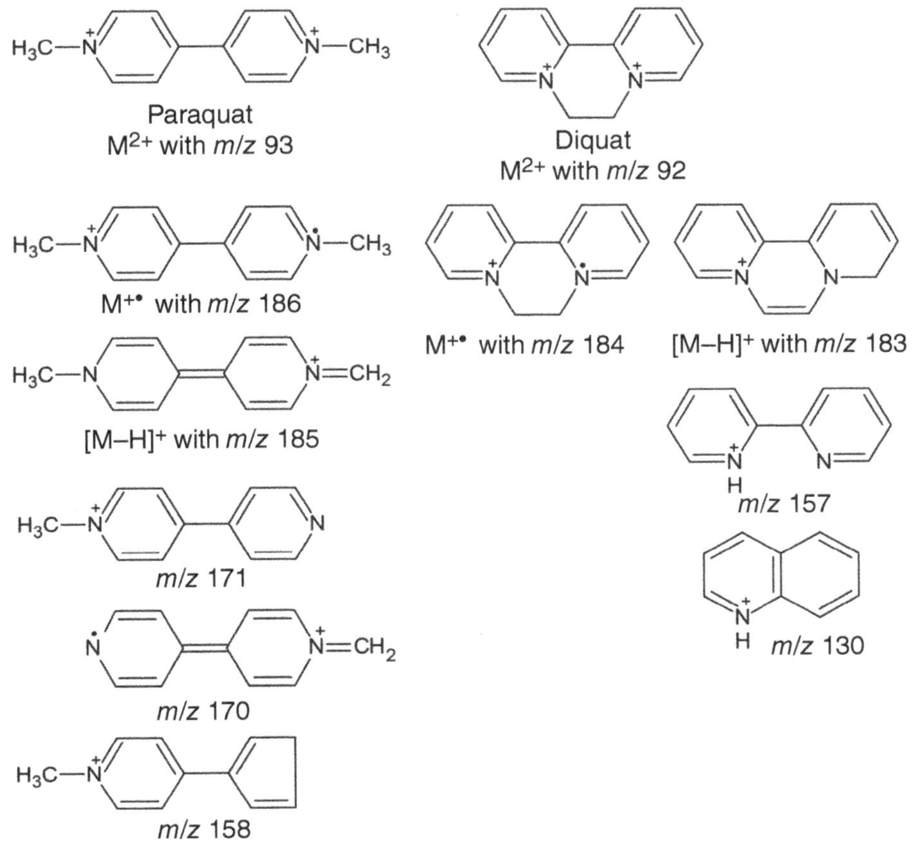

FIGURE 4.11.9 Ions observed for the quaternary ammonium herbicides paraquat and diquat, and proposed structures for the fragment ions based on $[M-H]^+$.

phosphorodithioate esters. Typical examples of these six subclasses are given in Figure 4.11.10.

Analytical methods for OPP analysis in fruits and vegetables have been reviewed (Sharma et al., 2010). Most OPPs are amenable to GC–MS, but LC–MS may be preferred because of the easier coupling to (on-line) solid-phase extraction for sample pretreatment and analyte preconcentration of aqueous samples especially. Moreover, it was shown that significantly better detection limits for OPP can be achieved by LC–MS in SRM mode than by GC–MS (Alder et al., 2006). Whether this is still true if SRM in GC–MS on a TQ instrument is applied (Fernández Moreno et al., 2008) is unclear. In LC–MS, most OPPs are analyzed in positive-ion mode, using either ESI or APCI, as demonstrated by a multiresidue method for 66 OPPs (and 32 carbamate pesticides) (Chung & Chan, 2010). OPP analysis using negative-ion mode is mainly based on the use of APCI (Aguilar et al., 1998; Fernández et al., 2001b; Blasco et al., 2004; Blasco et al., 2011).

The fragmentation of APCI-generated protonated dialkyl phosphonates and trialkyl phosphates was studied in considerable detail (Harden et al., 1993; Wensing et al., 1995). Data collections on fragmentation of protonated or deprotonated OPPs have been reported (Molina et al., 1994; Slobodník et al., 1996; Kawasaki et al., 1992; Itoh et al., 1996; Lacorte & Barceló, 1996; Lacorte et al., 1998).

Class-specific fragments that may be observed in positive-ion mode for dimethyl and diethyl phosphates, for dimethyl and diethyl phosphorothioates, and for dimethyl and diethyl phosphorodithioates are summarized in Table 4.11.3. Often, not all the fragments described are observed. Some fragments only occur at higher collision energy, which is not surprising because the fragments with lower *m/z* values result mostly from secondary fragmentation. Four types of class-specific fragment ions may be observed, as indicated by the four sections in Table 4.11.3. Proposed structures for some of these fragment ions are summarized in Figure 4.11.11.

With high *m/z* values, ions may be observed due to losses of methanol (CH_3OH) in the case of dimethyl esters or due to two subsequent losses of ethene (C_2H_4) followed by the loss of water in the case of diethyl esters. The loss of C_2H_4 from diethyl phosphate esters may involve a six-membered ring intermediate structure and a McLafferty-type *H*-rearrangement, which is similar to the

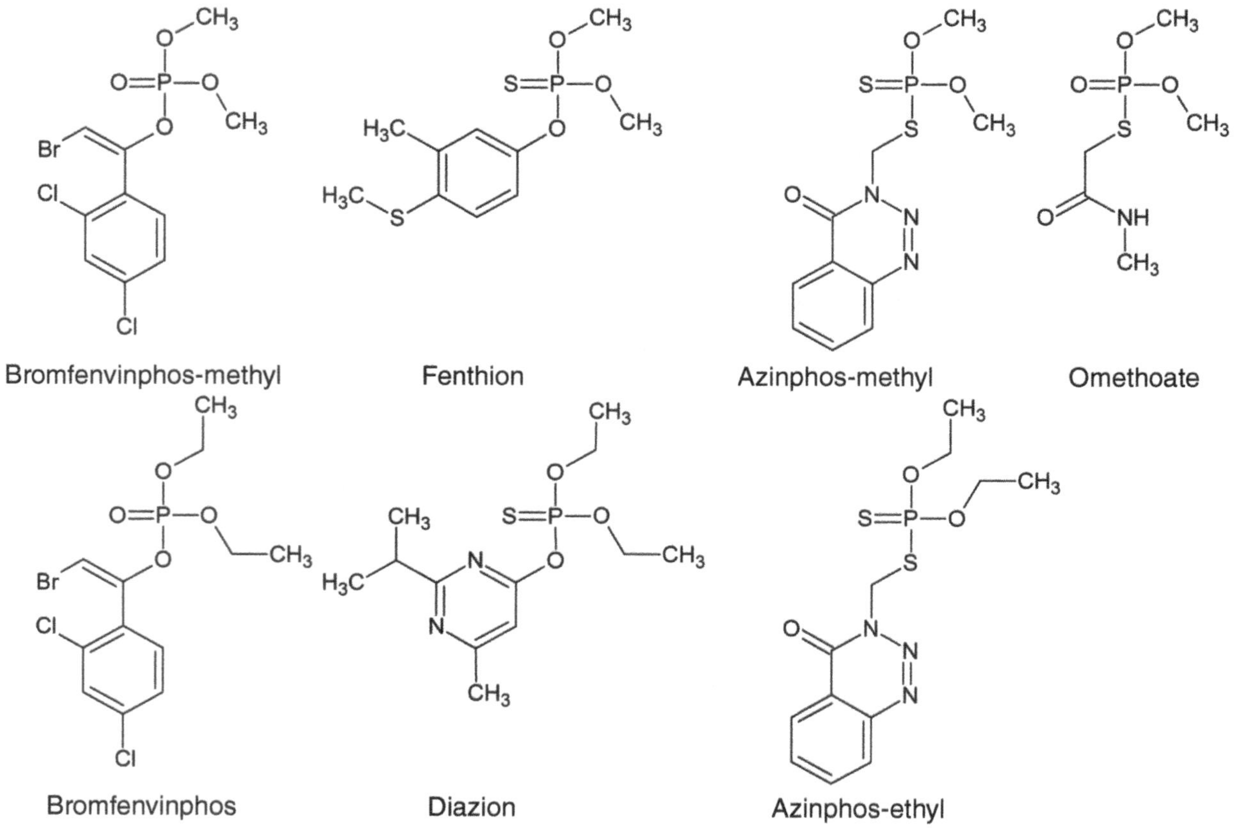

FIGURE 4.11.10 Examples of organophosphorus pesticides: bromfenvinphos-methyl is a dimethyl phosphate, fenthion a dimethyl phosphorothioate, azinphos-methyl a dimethyl phosphorodithioate, omethoate a dimethyl phosphorothioate, bromfenvinphos a diethyl phosphate, diazinon a diethyl phosphorothioate, and azinphos-ethyl a diethyl phosphorodithioate.

C_2H_4 loss from diethyl phosphates in EI-MS, although starting here from an EE^+ precursor rather than an $OE^{+\bullet}$, and thus resulting in EE^+ product ions rather than $OE^{+\bullet}$ fragment ions, as observed in EI-MS.

With low *m/z* values, a series of ions may be observed related to the loss of the (thio)phosphate group, that is, with charge retention on the (thio)phosphate group, eventually followed by subsequent secondary fragmentation of the resulting ion (Table 4.11.3). This may involve the cleavage of the C—O or C—S bond between the substituent group and the (thio)phosphate group, resulting in fragment ions with *m/z* 127 ($[(CH_3O)_2$ HO P=O+H]$^+$) or *m/z* 155 ($[(CH_3CH_2O)_2$ HO P=O+H]$^+$). Alternatively, it may involve the cleavage of the P—O or P—S bond leading to fragment ions with *m/z* 109 ($[(CH_3O)_2P{=}O]^+$), *m/z* 125 ($[(CH_3O)_2P{=}S]^+$), and *m/z* 153 ($[(CH_3CH_2O)_2P{=}S]^+$). Secondary fragmentation of these ions, for example, involving the loss of methanol (CH_3OH) for dimethyl esters and losses of ethene (C_2H_4) or ethanol (C_2H_5OH) for diethyl esters, leads to the other fragment ions observed with low *m/z* values. Some of these low-*m/z* fragment ions are also observed in the EI-MS spectra of phosphate esters.

If the (thio)phosphate ester is attached to a methylene (—CH_2—) group, as for instance with the dimethyl phosphorothioate azamethiphos, demeton-*O*-methyl, and demeton-*S*-methyl, with the dimethyl phosphorodithioate azinphos-methyl, formothion, and thiometon, and with most diethyl phosphorodithioates, cleavage of the C—C bond between the CH_2 group and the rest of the substituent group may occur, with charge retention on the (thio)phosphate group. Characteristic fragment ions resulting from this type of fragmentation are the ions with *m/z* 155 ($[(CH_3O)_2$ PO $OCH_2]^+$) for dimethyl phosphorothioates, *m/z* 171 ($[(CH_3O)_2$ PS $OCH_2]^+$) for dimethyl phosphorodithioates, and *m/z* 199 ($[(CH_3CH_2O)_2$ PS $SCH_2]^+$) for diethyl phosphorodithioates (Table 4.11.3b,c). The ion with *m/z* 199 may show two subsequent C_2H_4 losses leading to ions with *m/z* 171 and 143.

In addition, fragment ions may occur due the cleavage of the C—O or C—S bond or the P—O or P—S bond between

TABLE 4.11.3 Class-specific fragments in positive-ion product-ion mass spectra of OPP.

(a) Dimethyl and Diethyl Phosphates			
m/z	Dimethyl Phosphates	*m/z*	Diethyl Phosphates
−32.026	Loss of CH_3OH	−28.031	Loss of C_2H_4
		−56.063	Second loss of C_2H_4
		−74.073	Followed by H_2O loss
127.015	$[(CH_3O)_2\ HO\ P{=}O{+}H]^+$	155.047	$[(CH_3CH_2O)_2\ HO\ P{=}O{+}H]^+$
109.005	$[(CH_3O)_2P{=}O]^+$	127.015	$[CH_3CH_2O\ (HO)_2\ P{=}O{+}H]^+$
94.989	$[CH_3O\ HO\ P{=}O]^+$	98.984	$[(HO)_3P{=}O{+}H]^+$
80.974	$[(HO)_2P{=}O]^+$	80.974	$[(HO)_2P{=}O]^+$
78.994	$[CH_3OP{=}O{+}H]^+$		
64.979	$[HOP{=}O{+}H]^+$		
−126.008	Loss of $C_2H_7O_4P$	−154.040	Loss of $C_4H_{11}O_4P$ from alkyl.
		−136.029	Loss of $C_4H_9O_3P$ from aryl.

(b) Dimethyl and Diethyl Phosphorothioates.			
m/z	Dimethyl Phosphorothioates	*m/z*	Diethyl Phosphorothioates
−32.026	Loss of CH_3OH	−28.031	Loss of C_2H_4
		−56.063	Second loss of C_2H_4
		−74.073	Followed by H_2O loss
124.982	$[(CH_3O)_2P{=}S]^+$	153.012	$[(CH_3CH_2O)_2P{=}S]^+$
110.966	$[(CH_3O)\ HO\ P{=}S]^+$	124.982	$[(CH_3CH_2O)\ HO\ P{=}S]^+$
109.005	$[(CH_3O)_2P{=}O]^+$	96.951	$[(HO)_2P{=}S]^+$
78.994	$[CH_3O\ P{=}O{+}H]^+$	80.974	$[(HO)_2P{=}O]^+$
		64.979	$[HOP{=}O{+}H]^+$
154.993	$[(CH_3O)_2\ P{=}S\ OCH_2]^+$		
−141.985	Loss of $C_2H_7O_3PS$	−152.006	Loss of $C_4H_9O_2PS$
		−136.029	Loss of $C_4H_9O_3P$ after thiono–thiolo rearrangement

(c) Dimethyl and Diethyl Phosphorodithioates			
m/z	Dimethyl Phosphorodithioates	*m/z*	Diethyl Phosphorodithioates
−32.026	Loss of CH_3OH	−28.031	Loss of C_2H_4
		−56.063	Second loss of C_2H_4
		−74.073	Followed by H_2O loss
124.982	$[(CH_3O)_2P{=}S]^+$	158.970	$[(CH_3CH_2O)\ HO\ HS\ P{=}S{+}H]^+$
110.966	$[CH_3O\ HO\ P{=}S]^+$	153.012	$[(CH_3CH_2O)_2P{=}S]^+$
78.994	$[CH_3O\ P{=}O{+}H]^+$	124.982	$[CH_3CH_2O\ HO\ P{=}S]^+$
		96.951	$[(HO)_2P{=}S]^+$
170.970	$[(CH_3O)_2\ P{=}S\ SCH_2]^+$	199.001	$[(CH_3CH_2O)_2\ P{=}S\ SCH_2]^+$
		170.970	$[CH_3CH_2O\ HO\ P{=}S\ SCH_2]^+$
		142.938	$[(HO)_2\ P{=}S\ SCH_2]^+$
−157.963	Loss of $C_2H_7O_2PS_2$ S_2	−185.994	Loss of $C_4H_{11}O_2PS_2$ from alkyl

Source: Adapted from Niessen, 2010.

the (thio)phosphate group and the substituent group with charge retention on the substituent group. The cleavage of the P—O or P—S bond is common if the phosphate ester group is connected to an aromatic C-atom, for example, 2-ethoxy-1,3,2-dioxaphospholane ($C_4H_9O_3P$, 136 Da) is lost as a neutral. If connected to an aliphatic C-atom, $C_4H_9O_3P$ or diethyl hydrogen phosphate ($C_4H_{11}O_4P$, $(CH_3CH_2O)_2$ HO P=O, 154 Da) may be lost as a neutral. The cleavage of the C—O or C—S bond is commonly observed for dimethyl esters, thus involving the loss of dimethyl hydrogen phosphate ($C_2H_7O_4P$, 126 Da) for dimethyl phosphates, dimethoxy hydrogen phosphorothioate ($C_2H_7O_3PS$, 142 Da) for dimethyl phosphorothioates, and of dimethoxy hydrogen phosphorodithioate ($C_2H_7O_2PS_2$, 158 Da) for dimethyl phosphorodithioates (Table 4.11.3).

The occurrence of the class-specific fragments can be readily illustrated for the diethyl phosphorothioate diazinon ($[M+H]^+$ with *m/z* 305; Figure 4.11.12). With high

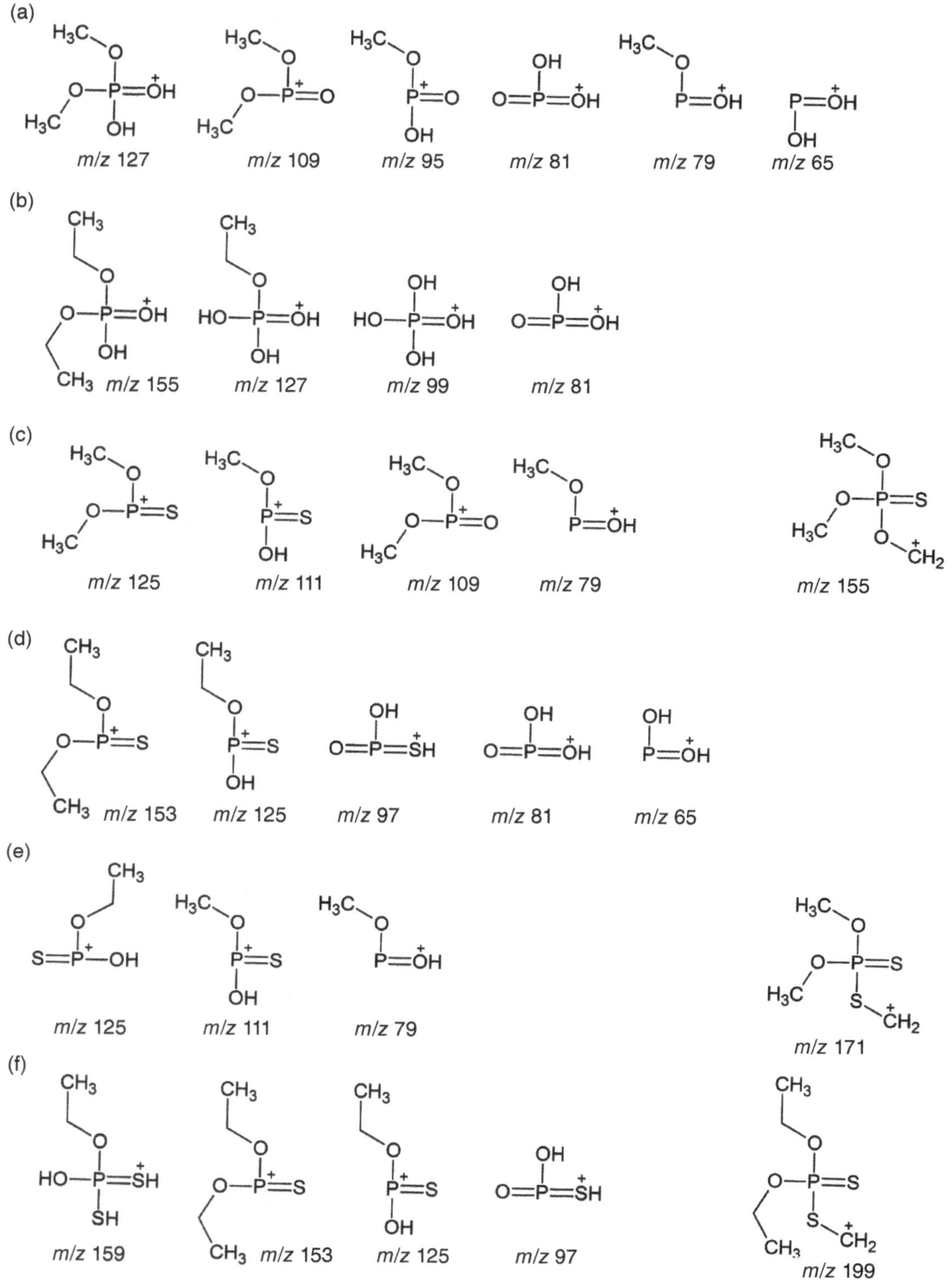

FIGURE 4.11.11 Proposed structures for the fragment ions of (a) dimethyl phosphates, (b) diethyl phosphates, (c) dimethyl phosphorothioates, (d) diethyl phosphorothioates, (e) dimethyl phosphorodithioates, and (f) diethyl phosphorodithioates, described in Table 4.11.2.

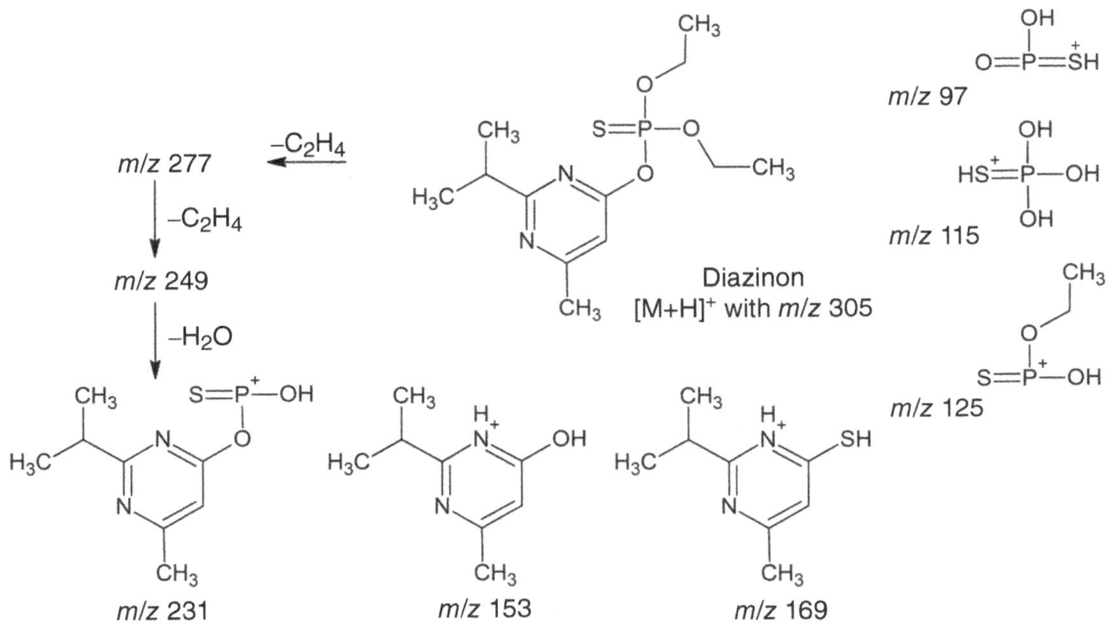

FIGURE 4.11.12 Fragmentation scheme and proposed structures for the fragment ions of the protonated phosphorothioate pesticide diazinon. (Source: Niessen, 2010. Reproduced with permission of Elsevier.)

m/z-values, ions due to subsequent losses of two C_2H_4 molecules and water are observed, resulting in the fragment ions with *m/z* 277, 249, and 231. With low *m/z*-values, the fragment ions with *m/z* 97, 115, and 125 are due to $[(HO)_2P{=}S]^+$, $[(HO)_3P{=}S{+}H]^+$, and $[CH_3CH_2O\ HO\ P{=}S]^+$, respectively. The fragment ion with *m/z* 153 is due to the cleavage of the P—O bond between the thiophosphate group and the 6-methyl-2-*iso*-propyl-pyrimidin-4-yl substituent with charge retention on the pyrimidine ring, that is, the corresponding protonated pyrimidin-4-ol ($[C_8H_{12}N_2O{+}H]^+$). The fragment ion with *m/z* 169 is due to a typical behavior observed in phosphorothioates, that is, the occurrence of a thiono–thiolo rearrangement: while the ion with *m/z* 153 has a hydroxy (—OH) substituent, the ion with *m/z* 169 has a sulfanyl (—SH) substituent, that is, ($[C_8H_{12}N_2S{+}H]^+$) (Figure 4.11.12). This behavior has been studied in detail (Barr et al., 2005). This thiono–thiolo rearrangement is observed in product-ion mass spectra of protonated diethyl phosphorothioates with an aromatic substituent group, for example, bromophos-ethyl, chlorpyriphos, chlorthiophos, coumaphos, pirimiphos-ethyl, quinalphos, triazophos, pyridafenthion, and the fungicide pyrazophos, as well as the anthelmintic dichlofenthion. In each case, two fragment ions due to losses of 2-ethoxy-1,3,2-oxathiaphospholane ($C_4H_9O_2PS$, 152 Da) and 2-ethoxy-1,3,2-dioxaphospholane ($C_4H_9O_3P$, 136 Da) are observed. It is interesting to note that the same thiono–thiolo rearrangement is observed in the EI-MS spectra of diazinon and related diethyl phosphorothioates.

In addition to these class-specific fragments, compound-specific fragment ions may be observed due to cleavages of the substituent group or cleavages in fragments still containing the substituent group. The identity of some of the fragment ions has been investigated in more detail in a study on fenthion and fenoxon and their sulfone and sulfoxide degradation products using LC–MS with ion-trap and Q–TOF instruments (Picó et al., 2007b).

In negative-ion APCI with in-source fragmentation (Fernández et al., 2001b; Blasco et al., 2004) or ion-trap MS^2 (Blasco et al., 2011), the dimethyl phosphorothioates and phosphorodithioates studied did not show $[M{-}H]^-$, but $[M{-}CH_3]^{-\bullet}$ instead. Most of the diethyl phosphorothioates and phosphorodithioates show $[M{-}H]^-$ and $[M{-}H{-}C_2H_4]^-$. The OPPs containing nonaliphatic substituent chains showed fragment ions consistent with losses of the (thio)phosphate group and charge retention on the substituent chain and/or with losses of the substituent chain and charge retention on the (thio)phosphate group. As examples, quinalphos ($[M{-}H]^-$ with *m/z* 297) yields a fragment ion with *m/z* 269 due to the loss of C_2H_4, the diethoxyphosphorothioate anion ($[(CH_3CH_2O)_2PSO]^-$) with *m/z* 169, and the quinoxalin-2-oxide anion ($[C_8H_5N_2O]^-$) with *m/z* 145. Bromophos methyl (364 Da, $[M{-}H]^-$ not observed) yields a fragment ion with *m/z* 349 ($[M{-}CH_3]^{-\bullet}$), the 4-bromo-2,5-dichlorophenoxide ion ($[BrCl_2C_6H_2O]^-$) with *m/z* 239, and the dimethoxy phosphorothioate anion ($[(CH_3O)_2PSO]^-$) with *m/z* 141 (Figure 4.11.13). The characteristic fragments and small-molecule losses

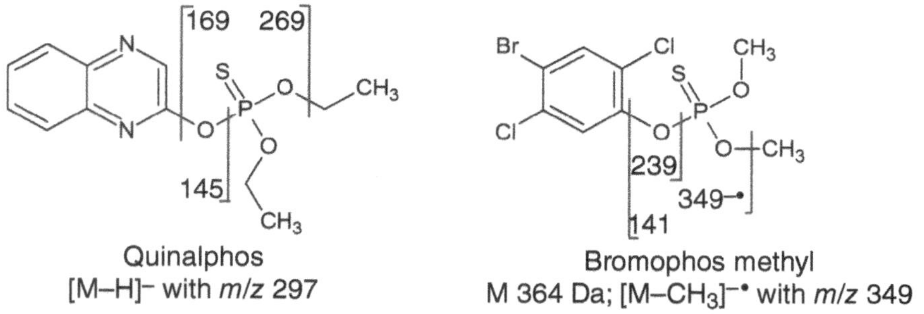

FIGURE 4.11.13 Fragmentation schemes for deprotonated organophosphorus pesticides quinalphos and bromophos methyl.

TABLE 4.11.4 Class-specific fragment ions and neutral losses in negative-ion MS–MS spectra of phosphorothioate and phosphorodithioate OPPs.

Class	Fragment Ion	*m/z*	Neutral Loss from $[M-H]^-$	Da	Other Class-Specific Fragments
$(CH_3O)_2$-Phosphorothioate	$[(CH_3O)_2POS]^-$	140.978	$C_2H_5O_2PS$	123.975	$[M-CH_3]^{-\bullet}$
$(C_2H_5O)_2$-Phosphorothioate	$[(C_2H_5O)_2POS]^-$	169.009	$C_4H_9O_2PS$	152.006	$[M-H-C_2H_4]^-$
$(CH_3O)_2$-Phosphorodithioate	$[(CH_3O)_2PSS]^-$	156.955	$C_2H_5OPS_2$	123.975	$[M-CH_3]^{-\bullet}$
$(C_2H_5O)_2$-Phosphorodithioate	$[(C_2H_5O)_2PSS]^-$	184.987	$C_4H_9OPS_2$	152.006	$[M-H-C_2H_4]^-$

of phosphorothioate and phosphorodithioate OPPs are summarized in Table 4.11.4. For some compounds, a CH_2 group is present between the (thio)phosphate group and the aromatic substituent, for example, in azinphos-ethyl, phosalone, and phosmet. For phosalone ($[M-H]^-$ with *m/z* 366) as an example, two fragment ions occur due to cleavage of either the N—C bond to the 6-chloro-1,3-benzoxazol-2-one leading to an ion with *m/z* 168 ($[C_7H_4ClNO_2-H]^-$) or the C—S bond to the diethoxy phosphorodithioate anion with *m/z* 185 ($[(CH_3CH_2O)_2PSS]^-$).

In-source fragmentation of OPPs apparently also occurs under negative-ion ESI conditions, as the dimethoxy phosphorodithioate anion ($[(CH_3O)_2PSS]^-$) with *m/z* 157 and $[(CH_3CH_2O)_2PSS]^-$ with *m/z* 185 are observed as precursor ions for azinphos-methyl ($[M-H]^-$ with *m/z* 316) and azinphos-ethyl ($[M-H]^-$ with *m/z* 344), respectively, in the product-ion mass spectra available in our library collection. This allows us to study secondary fragmentation of these phosphorodithioate esters in negative-ion mode. Fragmentation of the ion with *m/z* 157 shows the radical anion with *m/z* 142 ($[CH_3O_2PS_2]^{-\bullet}$) due to the loss of methyl radical ($^\bullet CH_3$), the oxathiaphosphiran-3-yl thiolate anion with *m/z* 111 ($[OPS_2]^-$), and ions with *m/z* 95 ($[PS_2]^-$) and *m/z* 79 ($[OPS]^-$). Terbufos ($[M-H]^-$ with *m/z* 288) and ethion ($[M-H]^-$ with *m/z* 384) also show the diethoxy phosphorodithioate anion with *m/z* 185 as precursor ion. Fragmentation of the ion with *m/z* 185 also leads to the ions with *m/z* 111, 95, and 79 along with an ion with *m/z* 157 ($[CH_3CH_2OPOS_2]^-$) due to the loss of C_2H_4.

In negative-ion mode, dimethoate ($CH_3NHCOCH_2SPS(OCH_3)_2$, M 229 Da, no $[M-H]^-$, $[M-CH_3]^{-\bullet}$ with *m/z* 214 instead) shows a (*N*-methylacetamid-2-yl)thioalkoxide anion ($[CH_3NHCOCH_2S]^-$) with *m/z* 104 and a fragment ion with *m/z* 141, attributed to $[PSH_2(OCH_3)_2CH_2]^-$ (Fernández et al., 2001a), the structure and formation of which are not readily understood yet (Figure 4.11.14). Proposed structures for the fragment ions of dimethoate ($[M+H]^+$ with *m/z* 230) in positive-ion mode are also given in Figure 4.11.14. Here, a possibly similar fragment ion is observed with *m/z* 143 (Thurman et al., 2007; Mazzotti et al., 2009). Accurate-mass determination for the ion with *m/z* 143 has resulted in a most likely molecular formula of $[C_2H_8O_3PS]^+$ (Thurman et al., 2007), that is, the protonated dimethoxy hydrogen phosphorothioate ($[(CH_3O)_2PSOH+H]^+$), thus requiring an intramolecular O-rearrangement. A similar intramolecular rearrangement in the negative-ion mode would lead to a dimethoxy phosphorothioate anion with *m/z* 141 ($[(CH_3O)_2PSO]^-$), which seems a more plausible structure for the ion with *m/z* 141 observed in negative-ion mode. Accurate-mass determination is needed to confirm this interpretation (Figure 4.11.14).

Glyphosate, glufosinate, and bialaphos (or bilanaphos) are organophosphorus broad-spectrum herbicides of frequent use. The LC–MS analysis of these compounds and their major metabolites aminomethylphosphonic acid (AMPA) and

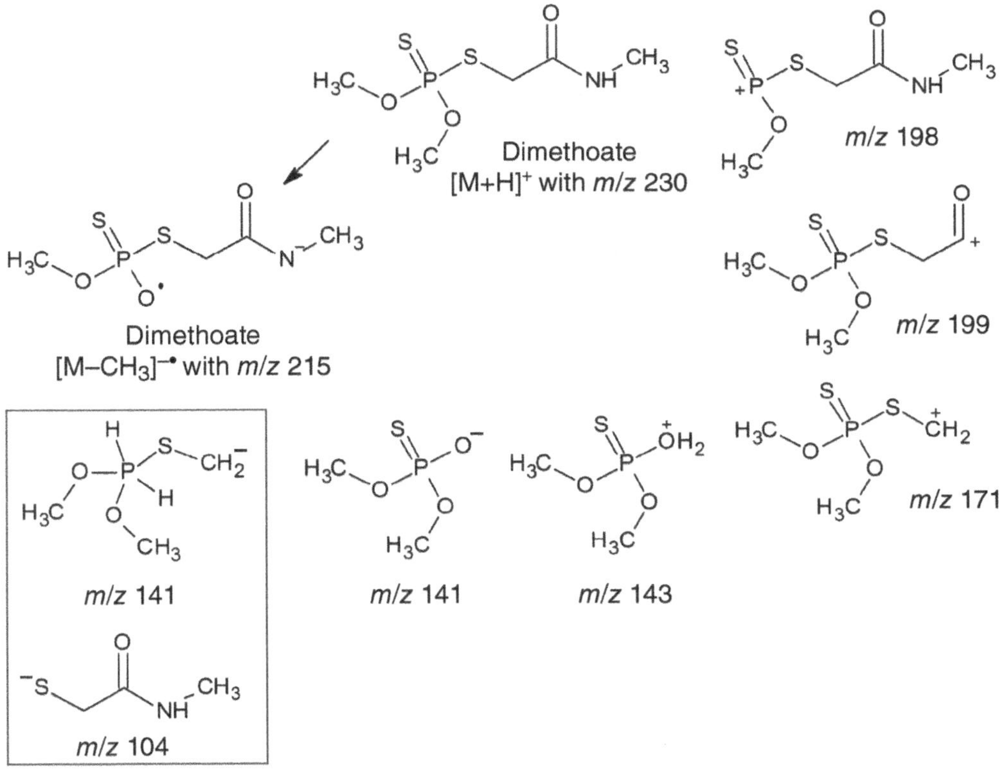

FIGURE 4.11.14 Proposed structures for the fragment ions of the protonated and deprotonated organophosphorus pesticide dimethoate. The framed structure shows the interpretation of the negative-ion product-ion mass spectrum proposed elsewhere (Fernández et al., 2001a).

3-methylphosphinicopropionic acid (MPPA) can be done either directly in negative-ion mode (Martins-Júnior et al., 2009; Yoshioka et al., 2011) or after (FMOC) derivatization in positive-ion mode (Ibáñez et al., 2005). The fragmentation of glyphosate and glufosinate and related compounds in negative-ion ion-trap MS^n has been extensively studied (Goodwin et al., 2003; Goodwin et al., 2004). For glyphosate ($[M-H]^-$ with *m/z* 168), fragment ions with *m/z* 150 due to the loss of water, with *m/z* 124 due to the loss of carbon dioxide (CO_2), with *m/z* 110 due to the loss of oxiran-2-one ($C_2H_2O_2$), and with *m/z* 81 ($[H_2PO_3]^-$) are observed in ion-trap MS^2 (Goodwin et al., 2003). Frequently used product ions in TQ SRM-based residue analysis methods are the ions with *m/z* 150, *m/z* 124, *m/z* 79 ($[PO_3]^-$), and with *m/z* 63 ($[PO_2]^-$) (Yoshioka et al., 2011). Further mechanistic studies into the fragment ions of glyphosate and AMPA observed in ion-trap MS^3 have been performed using H/D-exchange experiments (Goodwin et al., 2004). In ion-trap MS^2, glufosinate ($[M-H]^-$ with *m/z* 180) yields extensive fragmentation, resulting in fragment ions with *m/z* 163 due to the loss of ammonia (NH_3), with *m/z* 162 due to the loss of H_2O, with *m/z* 136 due to the loss of CO_2, with *m/z* 135 due to the loss of carboxyl radical ($^{\bullet}CO_2H$), with *m/z* 134 due to the loss of formic acid (HCOOH), with *m/z* 119 due to the loss of NH_3 and CO_2, and with *m/z* 95 ($[C_2H_8PO_2]^-$) due to the loss of CO_2 and acetonitrile ($CH_3C{\equiv}N$) (Goodwin et al., 2003). In SRM-based residue methods, the but-2-enoate anion with *m/z* 85 due to the loss of CH_3PO_2 and NH_3 as well as the ion with *m/z* 95 are applied (Yoshioka et al., 2011). In SRM-based residue methods, $[PO_3]^-$ with *m/z* 79 and $[H_2PO_3]^-$ with *m/z* 81 are used for AMPA ($[M-H]^-$ with *m/z* 110), while the ions with *m/z* 107 ($[C_3H_9O_2P]^-$) due to the loss of CO_2 and *m/z* 133 due to a water loss are used for MPPA ($[M-H]^-$ with *m/z* 151).

4.11.5 UREA HERBICIDES: PHENYLUREAS, BENZOYLPHENYLUREAS, AND OTHERS

The general structure of a urea herbicide is $R^1NHCON'R^2R^3$. Here, we discriminate between three classes of urea herbicides: phenylurea, benzoylphenylurea, and other urea herbicides. Because of the thermal lability of the urea moiety, urea herbicides are not amenable to GC–MS. Therefore, they are more frequently analyzed by LC–MS.

4.11.5.1 Phenylurea Herbicides

The general structure of a substituted phenylurea herbicide is $C_6H_5NHCON'CH_3R^2$. The phenyl ring is often substituted with chlorine (Cl) or bromine (Br) atoms, but methoxy (CH_3O), methyl (CH_3), trifluoromethyl (CF_3), or 2-propyl ($(CH_3)_2CH$) substitution can also be present instead. Most phenylurea herbicides are *N′*,*N′*-dimethyl substituted such as diuron, but a combination of a methyl substituent and another group also occurs like in *N′*-methyl-*N′*-butyl phenylureas such as neburon and in *N′*-methyl-*N′*-methoxy phenylureas such as linuron.

Phenylurea herbicides are mostly analyzed in positive-ion mode. In positive-ion ESI, often both protonated and sodiated molecules, that is, $[M+H]^+$ and $[M+Na]^+$, respectively, are observed. Due to the amide–iminol tautomerism, phenylurea herbicides can also be analyzed in negative-ion ESI, where, depending on the mobile-phase composition, deprotonated molecules as well as acetate or formate adducts can be observed (Thurman et al., 2001). In APCI, phenylurea herbicides can be analyzed as protonated molecules without fragmentation.

In MS–MS with the protonated molecule as precursor ion, only a few intense fragment ions are observed. For *N′*,*N′*-dimethyl phenylureas such as diuron ($[M+H]^+$ with *m/z* 233), the characteristic fragment ions are the (dimethylamino)acylium ion ($[(CH_3)_2N{-}C{=}O]^+$, F_2 in Table 4.11.5) with *m/z* 72 due the loss of dichloroaniline, protonated dimethylamine ($[(CH_3)_2NH_2]^+$, F_4) with *m/z* 46, and the corresponding fragment ions due to the loss of dimethylamine ($(CH_3)_2NH$) and *N*,*N*-dimethylformamide ($(CH_3)_2NCHO$), that is, the *N*-(3,4-dichlorophenyl) carbamoyl cation ($[Cl_2C_6H_3NH{-}C{=}O]^+$, F_3) with *m/z* 188 and the [(3,4-dichlorocyclohexa-2,5-dienylidene)-4-ylium] imine ion ($[Cl_2C_6H_4N]^+$, F_1) with *m/z* 160, respectively (Molina et al., 1995). Proposed structures for these ions are given in Figure 4.11.15. Note that the cleavage the N—C bond of the urea group leading to the fragment ions with *m/z* 160 (F_1) and *m/z* 72 (F_2) for diuron is a direct cleavage (Section 3.5.4), with the sum of the *m/z* values adding up to

TABLE 4.11.5 Structures and major fragment ions for selected phenylurea herbicides.

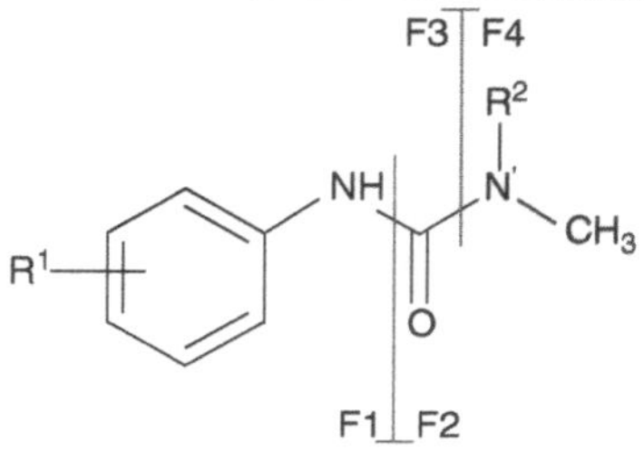

Compound	R^1	R^2	$[M+H]^+$	F_1	F_2	F_3	F_4	Other Fragments
Diuron	3,4-di-Cl	$—CH_3$	233	160	72	188	46	
Monuron	4-Cl	$—CH_3$	199	126	72	154	46	
Isoproturon	4-iC_3H_7	$—CH_3$	207	134	72	162	46	165: $- C_3H_6$
Chlortoluron	3-Cl-4-CH_3	$—CH_3$	213	140	72	168	46	
Metoxuron	3-Cl-4-OCH_3	$—CH_3$	229	156	72			149: $F_3 - Cl^•$
Fluometuron	3-CF_3	$—CH_3$	233	160	72	188	46	213: − HF 168: F_3 − HF 140: F_1 − HF
Bromuron	4-Br	$—CH_3$	243	170	72	198	46	
Monolinuron	4-Cl	$—OCH_3$	215	126	88	154	62	183: $- CH_3OH$ 148: 183 $- Cl^•$
Linuron	3,4-di-Cl	$—OCH_3$	249	160	88	188	62	217: $- CH_3OH$ 182: 217 $- Cl^•$
Metobromuron	4-Br	$—OCH_3$	259	170	88	198	62	227: $- CH_3OH$ 148: 227 $- Br^•$
Chlorbromuron	3-Cl-4-Br	$—OCH_3$	293	204	88	232	62	261: $- CH_3OH$ 226: 261 $- Cl^•$ 182: 261 $- Br^•$ 125: $F_1 - Br^•$
Buturon	4-Cl	$—C_4H_5$*	237	126	110		84	222: $- CH_3^•$ 185: $- C_4H_4$
Neburon	3,4-di-Cl	—*n*-C_4H_9	275	160	114	188	88	57: $[C_4H_9]^+$

*—C_4H_5 is butyn-3-yl.

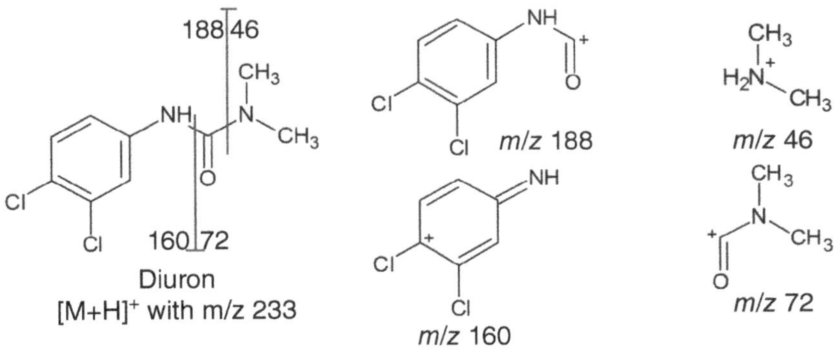

FIGURE 4.11.15 Fragmentation scheme and proposed structures for fragment ions of the protonated phenylurea herbicide diuron.

m/z ([M+H]−1), whereas the cleavage of the N′—C bond of the urea group leading to the fragment ions with *m/z* 188 (F_3) and *m/z* 46 (F_4) for diuron involves H-rearrangement and thus leads to two complementary ions, with the sum of the *m/z* values of the fragment ions equal to *m/z* ([M+H]+1) (Section 3.5.2). Similar fragmentation is observed for other *N′,N′*-dimethyl phenylureas monuron ($[M+H]^+$ with *m/z* 199), isoproturon ($[M+H]^+$ with *m/z* 207), chlortoluron ($[M+H]^+$ with *m/z* 213), fluometuron ($[M+H]^+$ with *m/z* 233), and bromuron ($[M+H]^+$ with *m/z* 243) (Table 4.11.5).

For *N′*-methyl-*N′*-methoxy phenylureas such as monolinuron ($[M+H]^+$ with *m/z* 215), linuron ($[M+H]^+$ with *m/z* 249), metobromuron ($[M+H]^+$ with *m/z* 259), and chlorbromuron ($[M+H]^+$ with *m/z* 293), similar fragment ions with *m/z* 88 (F_2), *m/z* 62 (F_4), *m/z* ([M+H]−61) (F_3), and *m/z* ([M+H]−89) (F_1) are observed (Table 4.11.5). In addition, the loss of methanol (CH_3OH) is observed as well as a subsequent loss of a chlorine radical ($Cl^•$) (for monolinuron, linuron, and chlorbromuron) or a bromine radical ($Br^•$) (for metobromuron and chlorbromuron).

Next to these class-specific fragmentation, compound-specific fragmentation is observed, for example, the loss of propene ($CH_3CH{=}CH_2$) for isoproturon ($[M+H]^+$ with *m/z* 207) (Hogenboom et al., 1997); the loss of hydrogen fluoride (HF) from the ions with *m/z* 233 ($[M+H]^+$), 188 (F_3), and 160 (F_1) for fluometuron ($[M+H]^+$ with *m/z* 233); the loss of but-1-en-3-yne (C_4H_4) for buturon ($[M+H]^+$ with *m/z* 237); and the *n*-butyl carbocation ($[C_4H_9]^+$) with *m/z* 57 for neburon ($[M+H]^+$ with *m/z* 275) (Table 4.11.5).

Negative-ion ion-trap MS^n data for the phenylurea herbicides were tabulated (Draper, 2001). For the *N′,N′*-dimethyl phenylureas diuron ($[M-H]^-$ with *m/z* 231), monuron ($[M-H]^-$ with *m/z* 197), isoproturon ($[M-H]^-$ with *m/z* 205), chlortoluron ($[M-H]^-$ with *m/z* 211), and metoxuron ($[M-H]^-$ with *m/z* 227), the two most abundant fragments are due to subsequent losses of $(CH_3)_2NH$ and hydrogen chloride (HCl), for example, resulting in fragment ions with *m/z* 186 and 150 for diuron. For the *N′*-methyl-*N′*-methoxy phenylureas monolinuron ($[M-H]^-$ with *m/z* 213), linuron ($[M-H]^-$ with *m/z* 247), metobromuron ($[M-H]^-$ with *m/z* 257), and chlorbromuron ($[M-H]^-$ with *m/z* 291), the loss of $(CH_3)_2NH$ is not observed. Minor losses of CH_3OH and formaldehyde ($H_2C{=}O$) are observed instead (Draper, 2001).

4.11.5.2 Other Urea Herbicides

The structures and fragment ions of a number of other urea herbicides are summarized in Table 4.11.6. They show a somewhat different fragmentation behavior than the phenylureas in Table 4.11.5.

The phenoxyphenylureas difenoxuron ($[M+H]^+$ with *m/z* 287, with an *N*-[4-(4-methoxyphenoxy)phenyl] and chloroxuron ($[M+H]^+$ with *m/z* 291, with an *N*-[4-(4-chlorophenoxy)phenyl] show the *N,N*-dimethylcarbamoyl cation with *m/z* 72 ($[(CH_3)_2N{-}C{=}O]^+$; F_2) and the corresponding ion due to the loss of *N,N*-dimethylformamide ($(CH_3)_2NCHO$), that is, the ions with *m/z* 214 and 218 (F_1), respectively. In addition, difenoxuron yields an oxonium ion ($[CH_3O{=}C_6H_4{=}O]^+$) with *m/z* 123. For chloroxuron, a radical cation with *m/z* 164 ($[C_9H_{12}N_2O]^{+•}$) is observed, which is due to a homolytic cleavage in the phenoxyphenyl substituent involving the loss of the 4-chlorophenoxy radical ($ClC_6H_4O^•$).

Table 4.11.6 also provides data on the fragment ions of the phenylureas thidiazuron ($[M+H]^+$ with *m/z* 221, with an *N′*-1,2,3-thiadiazol-5-yl substituent), azoluron ($[M+H]^+$ with *m/z* 231, with an *N′*-1-ethyl-pyrazol-5-yl substituent), siduron ($[M+H]^+$ with *m/z* 233, with an *N′*-2-methyl-cyclohexyl substituent). Furthermore, data are provided on the *N′,N′*-dimethyl-urea herbicide cycluron ($[M+H]^+$ with *m/z* 199, with an *N*-cyclooctyl substituent), as well as on the *N′*-methyl ureas benzthiazuron ($[M+H]^+$ with *m/z* 208) and methabenzthiazuron ($[M+H]^+$ with *m/z* 222, with *N*-methyl), both with an *N*-1,3-benzothiazol-2-yl substituent, tebuthiuron ($[M+H]^+$ with *m/z* 229, with *N*-methyl

TABLE 4.11.6 Structures and major fragment ions for selected urea herbicides.

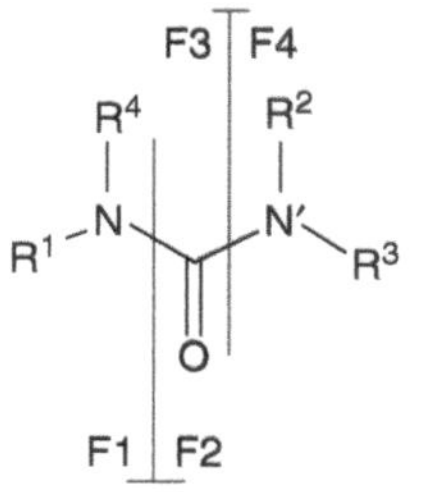

Compound	R^1	R^2	R^3	R^4	$[M+H]^+$	F_1	F_2	F_3/F_4	Other Fragments
Difenoxuron	4-(4-Methoxyphenoxy)-phenyl ($—C_{13}H_{11}O_2$)	$—CH_3$	CH_3	H	287	214	72		123: $[C_7H_7O_2]^+$
Chloroxuron	4-(4-Chlorophenoxy)-phenyl ($—C_{12}H_8ClO$)	$—CH_3$	CH_3	H	291	218	72		164: $- C_6H_4ClO^•$
Thidiazuron	Phenyl ($—C_6H_5$)	1,2,3-Thiadiazol-5-yl ($—C_2HN_2S$)	H	H	221	94		102 (F_4)	
Azoluron	Phenyl ($—C_6H_5$)	1-Ethyl-pyrazol-5-yl ($—C_5H_7N_2$)	H	H	231	94	138	112 (F_4)	110: $F_2 - C_2H_4$ 84: $F_4 - C_2H_4$
Siduron	Phenyl ($—C_6H_5$)	2-Methyl-cyclohexyl ($—C_7H_{13}$)	H	H	233	94		120 (F_3)	137: $- C_7H_{12}$ 97: $[C_7H_{13}]^+$
Cycluron	Cyclooctyl ($—C_8H_{15}$)	$—CH_3$	CH_3	H	199		72		111: $[C_8H_{15}]^+$ 89: $- C_8H_{14}$
Benzthiazuron	1,3-Benzothiazol-2-yl ($—C_7H_4NS$)	$—CH_3$	H	H	208	151		177 (F_3)	
Methabenzthiazuron	1,3-Benzothiazol-2-yl ($—C_7H_4NS$)	$—CH_3$	H	CH_3	222	165			150: $F_1 - CH_3^•$ 124: $[C_6H_6NS]^+$
Tebuthiuron	5-*t*-Butyl-1,3,4-thiadiazol-2-yl ($—C_6H_9N_2S$)	$—CH_3$	H	CH_3	229	172			157: $F_1 - CH_3^•$ 116: $F_1 - C_4H_8$
Thiazafluron	5-(Trifluoromethyl)-1,3,4-thiadiazol-2-yl ($—C_3F_3N_2S$)	$—CH_3$	H	CH_3	241	184			164: $F_1 - HF$ 111: $[C_2H_2F_3N_2]^+$ 91: $[C_2HF_2N_2]^+$ 74: $[C_2H_4NS]^+$

and *N*-5-*tert*-butyl-1,3,4-thiadiazol-2-yl substituents), and thiazafluron ($[M+H]^+$ with *m/z* 241, with *N*-methyl and *N*-5-(trifluoromethyl)-1,3,4-thiadiazol-2-yl substituents). Contrary to the phenylureas discussed in the previous section (Table 4.11.5), where one of the two N—C bond cleavages in the urea moiety involved a direct cleavage (F_1) and the other a H-rearrangement (F_4), the H-rearrangement now occurs to both N—C bond cleavages in the urea moiety (F_1 and F_4). Except for cycluron, a protonated amine ($[R^1NH_2+H]^+$; F_1) is observed for all these compounds. Compound-specific fragment ions are also provided in Table 4.11.6. Pencycuron ($[M+H]^+$ with *m/z* 329, with *N'*-cyclopentyl-*N'*-4-chlorobenzyl substituents) shows three major fragment ions, that is, an ion with *m/z* 261 due to the loss of cyclopentene (C_5H_8), the chlorobenzyl cation ($[ClC_6H_4CH_2]^+$) with *m/z* 125, and protonated *N*-(4-chlorobenzyl)aniline, ($[C_{13}H_{12}ClN+H]^+$) with *m/z* 218. The latter ion is due to the loss of cyclopentylisocyanate ($C_5H_9—N{=}C{=}O$), which is the result of a skeletal rearrangement of the benzyl group migrating to the anilino-N-atom (Section 3.5.8).

4.11.5.3 Benzoylphenylurea Herbicides

Benzoylphenylurea herbicides, $R^1CONHCON'HR^2$ with R^1 an *N*-difluorophenyl substituent and R^2 an *N'*-phenyl group with different substituents, can be analyzed in both positive-ion and negative-ion modes. Benzoylphenylurea herbicides such as chlorfluazuron, diflubenzuron, flufenoxuron, hexaflumuron, lufenuron, novaluron, triflubenzuron, and triflumuron are primarily analyzed in negative-ion mode, using either SQ instruments with detection and confirmation based on $[M-H]^-$ and $[M-H-HF]^-$ (Valenzuela et al., 2000), or SRM in TQ instruments (Sannino & Bandini, 2005). Multiple losses of hydrogen fluoride (HF) are frequently observed for most of the compounds from both $[M-H]^-$ and some fragment ions. Triflubenzuron ($[M-H]^-$ with *m/z* 379) shows abundant fragment ions due to one and two HF losses from $[M-H]^-$, flufenoxuron ($[M-H]^-$ with *m/z* 487) one and two HF losses and subsequently an HCl loss, and hexaflumuron ($[M-H]^-$ with *m/z* 460) one HF loss followed by an HCl loss. The fragmentation of benzoylphenylurea herbicides in

TABLE 4.11.7 Fragmentation of benzoylphenylurea herbicides. Cleavage sites are illustrated for hexaflumuron.

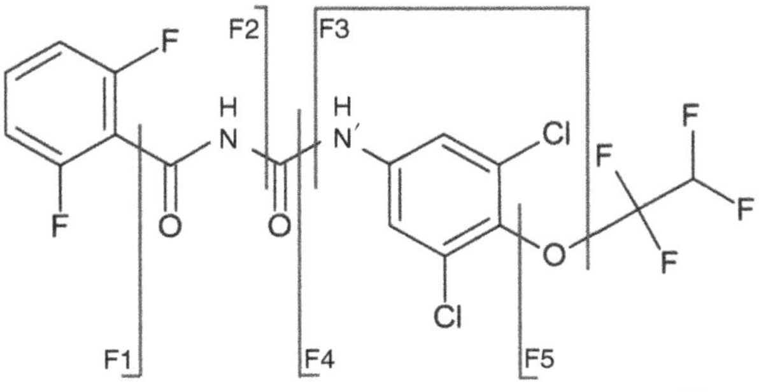

Compound	$[M{-}H]^-$	F_1 $[F_2C_6H_3]^-$	F_2 $[F_2C_6H_3CONH]^-$	F_3 $[Cl_2C_6H_3NO]^{-\bullet}$	F_4		F_5
Chlorfluazuron	538	113		175	355	$[C_{12}H_5Cl_3F_3N_2O]^-$	196 $[C_6H_2ClF_3NO]^-$
Diflubenzuron	309		156				n.a.
Flufenoxuron	487	113	156		304	$[C_{13}H_7ClF_4NO]^-$	195 $[C_7H_3ClF_3O]^-$
Hexaflumuron	459	113	156	175	276	$[C_8H_4Cl_2F_4NO]^-$	117 $[C_2HF_4O]^-$
Lufenuron*	509			175	326	$[C_9H_4Cl_2F_6NO]^-$	
Novaluron†	491	113	156	141 $[ClC_6H_4NO]^{-\bullet}$	308	$[C_9H_5ClF_6NO_2]^-$	
Triflumuron‡	357		154 $[ClC_6H_3CONH]^-$		176	$[C_7H_5F_3NO]^-$	

*N′-2,5-dichlorophenyl instead of N′-3,5-dichlorophenyl substituent.
†N′-3-dichlorophenyl instead of N′-3,5-dichlorophenyl substituent. An ion with *m/z* 85 ($[CF_3O]$) is observed instead of F_5 with *m/z* 183.
‡N-3-chlorophenyl instead of N-2,6-difluorophenyl.

negative-ion mode can be illustrated with hexaflumuron. Five fragment ions result from various cleavage sites in the molecule. Cleavage of the difluorobenzoyl C—C bond group results in the 2,6-difluorophenide anion ($[F_2C_6H_3]^-$, F_1 in Table 4.11.7) with *m/z* 113. Three fragments arise from cleavages of the N—C bonds present in the urea moiety, one with charge retention on the 2,6-difluorobenzoyl group, that is, the deprotonated 2,6-difluorobenzamide ($[F_2C_6H_3CONH]^-$, F_2) with *m/z* 156. The cleavage of the N′—C bonds of the urea moiety results in the 3,5-dichloro-4-(1,1,2,2-tetrafluoroethoxy)anilinide that is anion with *m/z* 276 ($[C_8H_4Cl_2F_4NO]^-$; F_4), which shows secondary fragmentation to the 2,6-dichloro-4-oxylanilinide radical anion ($[Cl_2C_6H_3NO]^{-\bullet}$; F_3) with *m/z* 175 due to a homolytic cleavage in the alkyl ether bond with the loss of the 1,1,2,2-tetrafluoroethyl radical ($^\bullet CF_2CHF_2$). Finally, the cleavage of the aryl C—O bond (if applicable) results in a deprotonated alcohol, that is, the 1,1,2,2-tetrafluoroethoxide anion ($[CHF_2CF_2O]^-$; F_5) with *m/z* 117. The important cleavage sites in benzoylphenylurea herbicides and the resulting fragments for several compounds, although not observed for all of them, are summarized in Table 4.11.7.

In positive-ion mode, benzoylphenylureas (except for triflumuron) yield three major fragments: Two class-specific ions, that is, the 2,6-difluorobenzoyl cation ($[F_2C_6H_3{-}C{=}O]^+$) with *m/z* 141, the protonated 2,6-difluorobenzamide ($[C_7H_5F_2NO{+}H]^+$) with *m/z* 158, and a compound-specific ion due to the loss of 2,6-difluorobenzoyl isocyanate ($C_8H_3F_2NO_2$), that is, the protonated 3,5-dichloro-2,4-difluoroaniline ($[Cl_2F_2C_6H{-}NH_2{+}H]^+$) with *m/z* 198 for teflubenzuron ($[M{+}H]^+$ with *m/z* 381), the protonated 4-(2-chloro-4-trifluoromethylphenoxy)-2-fluoroaniline ion with *m/z* 306 ($[C_{13}H_8ClF_4NO{+}H]^+$) for flufenoxuron ($[M{+}H]^+$ with *m/z* 489) (Figure 4.11.16), the protonated 3-(3-chloro-5-trifluoromethylpyridin-2-oxy)-4-chloroaniline ($[C_{12}H_7Cl_2F_3N_2O{+}H]^+$) with *m/z* 323 for fluazuron ($[M{+}H]^+$ with *m/z* 506), the protonated 2,5-dichloro-4-(2*H*-perfluoropropoxy)aniline ($[C_9H_5Cl_2F_6NO{+}H]^+$) with *m/z* 328 for lufenuron ($[M{+}H]^+$ with *m/z* 511), and the protonated 4-(3-chloro-5-trifluoromethylpyridin-2-oxy)-3,5-dichloroaniline ($[C_{12}H_6Cl_3F_3N_2O{+}H]^+$) with *m/z* 357 for chlorfluazuron ($[M{+}H]^+$ with *m/z* 540). For fluazuron and chlorfluazuron, next to the ions due to the protonated aniline derivative, that is, *m/z* 323 and 357, more abundant ring-substituted *N*-aryl carbamoyl ions with *m/z* 348 ($[C_{13}H_6Cl_2F_3N_2O_2]^+$) and *m/z* 383 ($[C_{13}H_5Cl_3F_3N_2O_2]^+$) are observed, respectively (structurally similar to F_4 in Figure 4.11.16). For triflumuron ($[M{+}H]^+$ with *m/z* 359),

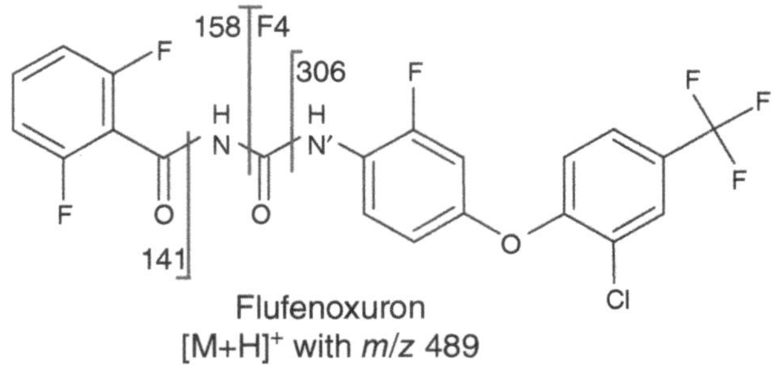

FIGURE 4.11.16 Fragmentation scheme for the protonated benzoylphenylurea herbicides illustrated for flufenoxuron. The F_4 fragment ion is only observed for fluazuron and chlorflurazone.

which has an *N*-2-chlorobenzoyl (ClC_6H_4CO—) rather than an *N*-2,6-difluorobenzoyl ($F_2C_6H_3CO$—) group, the characteristic fragment ions are the 2-chlorobenzoyl cation ($[ClC_6H_4{-}C{=}O]^+$) with *m/z* 139, the protonated 2-chlorobenzamide ($[ClC_6H_4{-}CONH_2{+}H]^+$) with *m/z* 156, and the protonated 4-(trifluoromethoxy)aniline ($[F_3COC_6H_4NH_2{+}H]^+$) with *m/z* 178.

4.11.6 SULFONYLUREA HERBICIDES

Sulfonylureas form a group of selective herbicides with R^1 $NHCON'HSO_2R^2$ as their general structure (Table 4.11.8). Depending on the identity of the R^1 group, two classes of sulfonylureas can be distinguished: pyrimidinyl sulfonylureas and triazinyl sulfonylureas. Sulfonylureas are generally analyzed using positive-ion ESI, where both protonated and sodiated molecules are observed. In addition, some fragments due to in-source CID may be observed (Reiser & Fogiel, 1994; Volmer et al., 1995). Negative-ion ESI can be applied as well (Køppen & Spliid, 1998; Bossi et al., 1998).

4.11.6.1 Pyrimidinyl Sulfonylurea Herbicides

The pyrimidinyl group is mostly 4,6-dimethoxy substituted, although 4-chloro-6-methoxy, 4,6-bis(difluoromethoxy), and 4,6-dimethyl substitution are also described (Table 4.11.8a). Product-ion mass spectra of protonated sulfonylureas were reported (Li et al., 1996). For most compounds, only three or four product ions are observed, as illustrated for foramsulfuron in Figure 4.11.17. Cleavage of the urea N—C bond to the pyrimidin-2-yl amine group generally results in two class-specific fragment ions with charge retention on the R^1 group, that is, a protonated amine ($[R^1NH_3]^+$, F_1 in Table 4.11.8a) and a protonated isocyanate ($[R^1N{=}C{=}O{+}H]^+$, F_2), for example, the protonated 2-isocyanato-4,6-dimethoxypyrimidine ($[C_7H_7N_3O_3{+}H]^+$) with *m/z* 182. Interestingly, the $[R^1NH_3]^+$ fragment, that is, the protonated 4,6-dimethoxypyrimidin-2-amine ($[C_6H_9N_3O_2{+}H]^+$) with *m/z* 156, is not always observed. Sometimes, the 2-imino-4,6-dimethoxy-2,5-dihydropyrimidin-5-ylium ion with *m/z* 154 ($[C_6H_8N_3O_2]^+$ is observed instead (Table 4.11.8a). This fragmentation characteristic is also observed for phenylureas (Section 4.11.5.1 and 2). Cleavage of the sulfonylurea S—N bond results in a fragment ion due to the corresponding sulfonyl-containing ion ($[R^2SO_2]^+$, F_4) with *m/z* ([M+H]−198, for the compounds with the 4,6-dimethoxypyrimidinyl group). For some compounds, cleavage of the urea C—N bond to the benzenesulfonyl group and charge retention on the R^2-group leads to corresponding R^2-substituted protonated benzenesulfonamide ($[R^2SO_2NH_2{+}H]^+$, F_3) with *m/z* ([M+H]−181, for the compounds with the 4,6-dimethoxy-pyrimidinyl group) (Table 4.11.8a).

However, in some cases, alternative or additional fragmentation behavior is observed. Next to the class-specific ions with *m/z* 182 and 154 as shown earlier, amidosulfuron ($[M{+}H]^+$ with *m/z* 370) yields two fragment ions involving the cleavage of the one of the sulfamoyl S—N bond with charge retention on the R^1-group rather than the R^2-group. Thus, the loss of *N*-methylmethanesulfonamide ($CH_3SO_2NHCH_3$) results in the fragment ion with *m/z* 261 ($[C_7H_9N_4O_5S]^+$), which subsequently via a rearrangement shows the loss of hydrogen isocyanate (HN═C═O) to the ion with *m/z* 218 ($[C_6H_8N_3O_4S]^+$). The same fragment ions are also observed for ethoxysulfuron ($[M{+}H]^+$ with *m/z* 399), together with a protonated cyclohexadiene-1,2-dione ($[C_6H_4O_2{+}H]^+$) with *m/z* 109. Oxasulfuron ($[M{+}H]^+$ with *m/z* 407) shows fragment ions with *m/z* 333 due to the loss of oxetan-3-ol ($C_3H_6O_2$), with *m/z* 284 due to the loss of 4,6-dimethylpyrimidin-2-amine ($C_6H_9N_3$), and with *m/z* 210 ($[C_8H_4NO_4S]^+$) due to the loss of both these groups. Pyrazosulfuron-ethyl ($[M{+}H]^+$ with *m/z* 415) shows a fragment ion with *m/z* 369 due to the loss of ethanol (C_2H_5OH) and with *m/z* 214 ($[C_6H_4N_3O_4S]^+$) due to the subsequent loss of 4,6-dimethoxypyrimidin-2-amine ($C_6H_9N_3O_2$). Trifloxysulfuron ($[M{+}H]^+$ with *m/z* 438) shows a fragment ion with *m/z* 331 ($[C_{13}H_{14}F_3N_4O_3]^+$), consistent with a skeletal rearrangement of the 3-(2,2,2-trifluoroethoxy)pyridin-2-yl group with the loss of HN═C═O and sulfur dioxide (SO_2).

In addition, some compounds show small-molecule losses from the R^2 substituent, for example, loss of methanol (CH_3OH) for the methyl esters halosulfuron-methyl and primisulfuron-methyl, loss of C_2H_5OH from the ethyl esters of chlorumuron-ethyl and pyrazosulfuron-ethyl, or loss of dimethylamine ($(CH_3)_2NH$) for foramsulfuron and nicosulfuron (Table 4.11.8a). In some cases, secondary fragmentation of the class-specific fragments ions is observed as well.

Negative-ion ion-trap MS^n data have been reported for the pyrimidinylsulfonylurea amidosulfuron and azimsulfuron (Polati et al., 2006), and for nicosulfuron including its aqueous photodegradation products (Benzi

TABLE 4.11.8 Fragmentation of sulfonylurea herbicides, illustrated for the pyrimidinyl sulfonylurea foramsulfuron and the triazinyl sulfonaylurea chlorsulfuron (Figure 4.11.16).

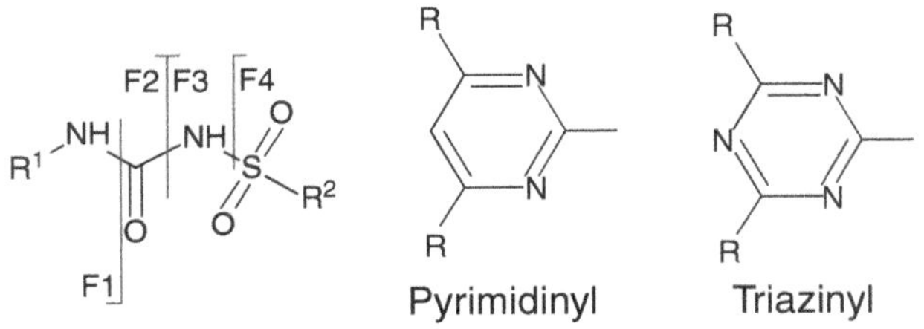

(a) Pyrimidinylsulfonylurea Herbicides.

Compound	R^1-Pyrimidine Substitution	$[M+H]^+$	F_1	F_2	F_3	F_4	Other Fragments
Amidosulfuron	4,6-Dimethoxy-	370	*154*	182			261: $- C_2H_7NO_2S$ 218: 261 − HNCO 124: $F_1 - H_2CO$
Azimsulfuron	4,6-Dimethoxy-	425		182	244	227	
Bensulfuron-methyl	4,6-Dimethoxy-	411		182		213	149: $F_4 - SO_2$
Chlorimuron-ethyl	4-Chloro-6-methoxy-	415	160	186		213	369: $- C_2H_5OH$
Cyclosulfamuron	4,6-Dimethoxy-	422				224	
Ethoxysulfuron	4,6-Dimethoxy-	399	*154*	182			261: $- C_8H_{10}O_2$ 218: 261 − HNCO 109: $[C_6H_5O_2]^+$
Foramsulfuron	4,6-Dimethoxy-	453	156	182	272	255	408: $- HN(CH_3)_2$
Halosulfuron methyl	4,6-Dimethoxy-	435		182			403: $- CH_3OH$
Imazosulfuron	4,6-Dimethoxy-	413	*154*		232		
Mesosulfuron-methyl	4,6-Dimethoxy-	504	156	182			
Nicosulfuron	4,6-Dimethoxy-	411		182		213	366: $- HN(CH_3)_2$
Oxasulfuron	4,6-Dimethyl-	407	124	150			333: $- C_3H_6O_2$ 284: $- C_6H_9N_3$ 210: $- C_3H_6O_2 - C_6H_9N_3$
Primisulfuron-methyl	4,6-Bis(difluoromethoxy)-	469		254		199	437: $- CH_3OH$
Pyrazosulfuron-ethyl	4,6-Dimethoxy-	415	156	182	234		369: $- C_2H_5OH$ 214: $- C_2H_5OH - C_6H_9N_3O_2$
Sulfometuron-methyl	4,6-Dimethyl-	365	124	150		199	
Trifloxysulfuron	4,6-Dimethoxy-	438	156	182	257		331: $- HCNO - SO_2$

(b) Triazinylsulfonylurea Herbicides.

Compound	R^1-Triazine Substitution	$[M+H]^+$	F_1	F_2	F_4	Other Fragments
Chlorsulfuron	4-Methoxy-6-methyl-	358	141	167	175	111: $[C_6H_4Cl]^+$
Cinosulfuron	4,6-Dimethoxy-	414	157	183	215	121: $[C_7H_5O_2]^+$
Iodosulfuron-methyl	4-Methoxy-6-methyl-	508	141	167	325	476: $- CH_3OH$
Metsulfuron-methyl	4-Methoxy-6-methyl-	382	141	167	199	350: $- CH_3OH$
Prosulfuron	4-Methoxy-6-methyl-	420	141	167		
Thifensulfuron-methyl	4-Methoxy-6-methyl-	388	141	167	205	
Triasulfuron	4-Methoxy-6-methyl-	402	141	167	219	359: − HNCO 121: $[C_7H_5O_2]^+$
Tribenuron-methyl	4-Methoxy-6-methyl-	396	155	181	199	364: $- CH_3OH$
Triflusulfuron-methyl	4-Dimethylamino-6-(2,2,2-trifluoroethoxy)-	493		264		461: $- CH_3OH$

Italic denotes that the fragmentation can take two different forms, leading to either an ion with *m/z* 154 or an ion with *m/z* 156 (the ions with other *m/z*-values are the result of a reaction, similar to that of the ion with *m/z* 156).

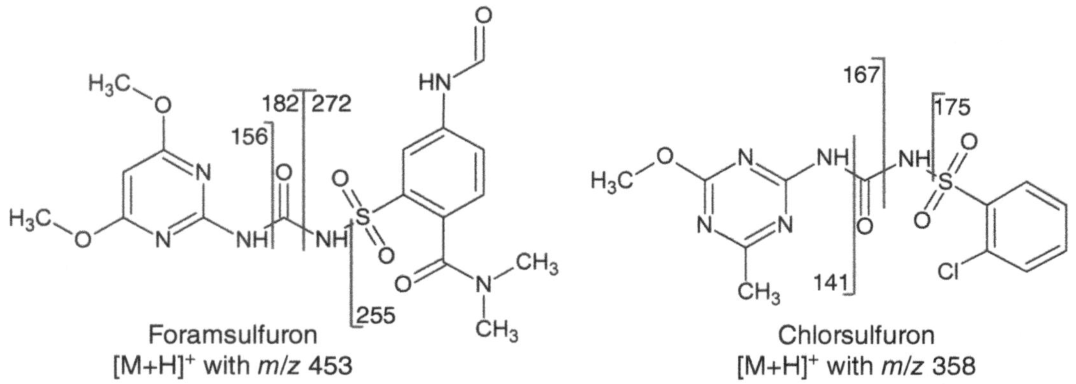

FIGURE 4.11.17 Fragmentation schemes for the protonated sulfonylurea herbicides foramsulfuron and chlorsulfuron.

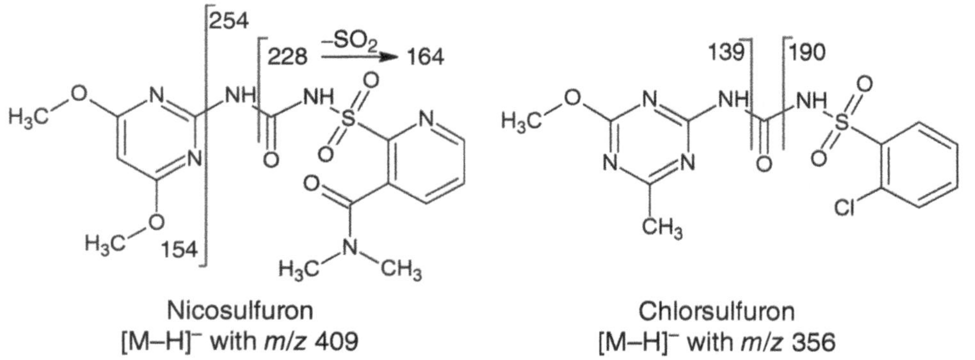

FIGURE 4.11.18 Fragmentation schemes for the deprotonated sulfonylurea herbicides nicosulfuron and chlorsulfuron.

et al., 2011). Amidosulfuron ($[M-H]^-$ with *m/z* 368) shows a fragment ion with *m/z* 259 due to the loss of *N*-(methyl)methylsulfonamide ($CH_3SO_2NHCH_3$) with charge retention on the pyrimidinyl group. The fragmentation of nicosulfuron ($[M-H]^-$ with *m/z* 409) is outlined in Figure 4.11.18.

4.11.6.2 Triazinyl Sulfonylurea Herbicides

Most triazinyl sulfonylureas contain a 4-methoxy-6-methyl-1,3,5-triazin-2-yl group (Table 4.11.8b). Some of them are methyl esters, thus showing a fragment ion due to the loss of methanol (CH_3OH). In positive-ion product-ion mass spectra, the triazinyl sulfonylureas show characteristic fragments similar to those of the pyrimidinyl sulfonylurea, as illustrated for chlorsulfuron ($[M+H]^+$ with *m/z* 358) in Figure 4.11.17. Cleavage of the urea N—C bonds generally results in two class-specific fragment ions with charge retention on the R^1 group, that is, the protonated amine $[R^1NH_3]^+$ (F_1 in Table 4.11.8b) and the protonated isocyanate $[R^1—N=C=O+H]^+$ (F_2). For chlorsulfuron, the $[R^1NH_3]^+$ fragment is the protonated 4-methoxy-6-methyl-1,3,5-triazin-2-amine ($[C_5H_8N_4O+H]^+$) with *m/z* 141 and the $[R^1—N=C=O+H]^+$ fragment is the protonated 2-isocyanato-4-methoxy-6-methyl-1,3,5-triazine ($[C_6H_6N_4O_2+H]^+$) with *m/z* 167. Cleavage of the sulfonylurea S—N bond results in a fragment ion $[R^2SO_2]^+$ (F_4) with *m/z* ($[M+H]-183$) (for the compounds with the 4-methoxy-6-methyl-1,3,5-triazinyl group). For some compounds, compound-specific fragment ions are observed such as the 2-chlorophenylium ion with *m/z* 111 ($[ClC_6H_4]^+$) for chlorsulfuron or an ion with *m/z* 121 ($[C_7H_5O_2]^+$) for cinosulfuron ($[M+H]^+$ with *m/z* 414) and triasulfuron ($[M+H]^+$ with *m/z* 402).

Negative-ion product-ion mass spectra for the triazinyl sulfonylureas chlorsulfuron, metsulfuron-methyl, thifensulfuron-methyl, and tribenuron-methyl have been tabulated (Køppen & Spliid, 1998; Bossi et al., 1998). The major class-specific fragment ion for these four compounds was the deprotonated 4-methoxy-6-methyl-1,3,5-triazin-2-amine with *m/z* 139 ($[C_5H_6N_4O{-}H]^-$), which is the deprotonated analog of the fragment with *m/z* 141 observed in positive-ion mode (Figure 4.11.18).

4.11.7 CHLORINATED PHENOXY ACID HERBICIDES

Chlorinated phenoxy acids (CPAs) are widely applied herbicides. They can be analyzed by GC–MS only after derivatization, for example, as methyl esters. Multiresidue LC–MS methods for these acid herbicides rely on negative-ion ESI (Chiron et al., 1995; Køppen & Spliid, 1998; Lindh et al., 2008; Santilio et al., 2011). Deprotonated molecules are detected, often along with a phenoxide fragment ion due to in-source fragmentation.

Negative-ion MS and MS–MS mass spectra of various CPAs including 2,4-D ((2,4-dichlorophenoxy)acetic acid), MCPA ((2-methyl-4-chlorophenoxy)acetic acid), dichlorprop, and mecoprop were tabulated (Køppen & Spliid, 1998). Negative-ion product-ion mass spectral data for seven CPA herbicides as well as MCPB (4-(4-chloro-2-methylphenoxy)butanoic acid) are given in Table 4.11.9. The interpretation of these product-ion mass spectra is straightforward, as may be illustrated using 2,4-D as example. Thus, 2,4-D ($[M{-}H]^-$ with *m/z* 219) shows an abundant 2,4-dichlorophenoxide ion ($[Cl_2C_6H_3O]^-$) with *m/z* 161 due to a cleavage of the aromatic-aliphatic ether

TABLE 4.11.9 Structures and fragmentation of chlorinated phenoxy acid herbicides (CPAs).

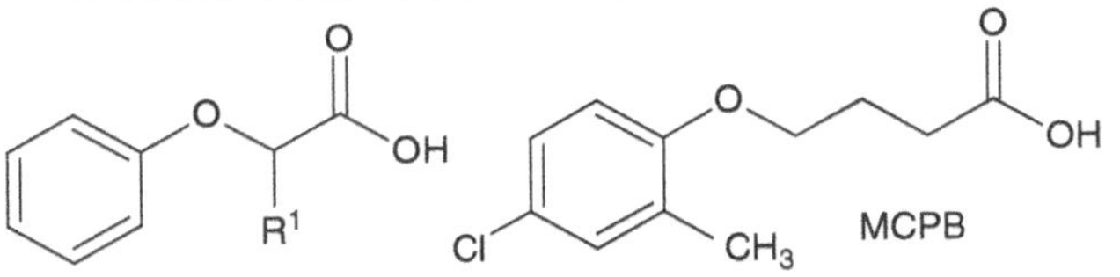

Compound	$[M{-}H]^-$	Ring Substituents	R^1	Fragmentation *m/z*	Fragmentation Interpretation
2,4,5,-T: 2,4,5-Trichlorophenoxyacetic acid	253	—Cl; —Cl; —Cl	H	195	$[Cl_3C_6H_2O]^-$
				159	195 – HCl
				123	159 – HCl
				95	123 – CO ($[C_5Cl]^-$)
2,4-D: 2,4-Dichlorophenoxyacetic acid	219	—Cl; —Cl	H	161	$[Cl_2C_6H_3O]^-$
				125	161 – HCl
				89	125 – HCl
				61	89 – CO ($[C_5H]^-$)
4-Chlorophenoxyacetic acid	185	—Cl	H	141	185 $-CO_2$
				127	$[ClC_6H_4O]^-$
				91	127 – HCl
Dichlorprop	233	—Cl; —Cl	CH_3	161	$[Cl_2C_6H_3O]^-$
				125	161 – HCl
				89	125 – HCl
				61	89 – CO ($[C_5H]^-$)
Fenoprop	267	—Cl; —Cl; —Cl	CH_3	195	$[Cl_3C_6H_2O]^-$
				159	195 – HCl
				123	159 – HCl
				95	123 – CO ($[C_5Cl]^-$)
MCPA: 2-Methyl-4-chlorophenoxyacetic acid	199	—Cl; —CH_3	H	155	199 – CO_2
				141	$[ClCH_3C_6H_3O]^-$
				105	141 – HCl
				77	105 – CO ($[C_6H_5]^-$)
MCPB: 4-(4-Chloro-2-methylphenoxy)butanoic acid	227	—Cl; —CH_3		141	$[ClCH_3C_6H_3O]^-$
				105	141 – HCl
				77	105 – CO ($[C_6H_5]^-$)
Mecoprop	213	—Cl; —CH_3	CH_3	141	$[ClCH_3C_6H_3O]^-$
				105	141 – HCl
				77	105 – CO ($[C_6H_5]^-$)

bond with charge retention on the aromatic group, consistent with the loss of oxiran-2-one ($C_2H_2O_2$). A similar cleavage is observed for all other analogs, in some cases in competition with a small carbon dioxide (CO_2) loss (Table 4.11.9). Loss of hydrogen chloride (HCl) from $[M-H]^-$ also occurs. Secondary fragmentation of the ion with *m/z* 161 results in minor fragment ions with *m/z* 125 ($[ClC_6H_2O]^-$) due to the loss of HCl, *m/z* 89 ($[C_6HO]^-$) due to another loss of HCl, and *m/z* 61 ($[C_5H]^-$) due to a subsequent loss of carbon monoxide (CO). The losses can be readily interpreted from the molecular formulae of the fragment ions, whereas proposing viable structures for these fragment ions is not an easy task. The losses and molecular formulae are confirmed by accurate-mass determination.

Similar fragmentation is observed for fluroxypyr ($[M-H]^-$ with *m/z* 253) and triclopyr ($[M-H]^-$ with *m/z* 254), which show major fragments due to the loss of $C_2H_2O_2$ to ions with *m/z* 195 and 196, respectively. The loss of hydrogen fluoride (HF) or HCl from $[M-H]^-$ is also observed.

Halogenated aryloxyphenoxypropionic acids are a relatively new class of CPA herbicides used for the selective removal of grass species. In commercial preparations, the compounds are present as alkyl esters, which in soil undergo hydrolysis to the corresponding free acids. The free acids can be analyzed in both positive-ion and negative-ion modes, while the esters are analyzed in positive-ion mode.

The positive-ion fragmentation by in-source CID of some of the esters was studied. Based on the data presented (Marchese et al., 2001), the fragmentation of fluazifop-butyl ester ($[M+H]^+$ with *m/z* 384.142) was discussed (Niessen, 2010). However, accurate-mass data show that one of the fragments was not correctly interpreted at the time: the fragment ion with *m/z* 254 does not involve the earlier presumed loss of ethene (C_2H_4) resulting in a quinone structure (*m/z* 254.042), but rather results from a methyl rearrangement and the loss of CO (*m/z* 254.079) (Figure 4.11.19). Typical positive-ion fragmentation of the halogenated aryloxyphenoxypropionic acid esters involves the following: (1) the loss of the ester alkyl group and the formation of the free acid, that is, the loss of butene (C_4H_8) to the ion with *m/z* 328 for fluazifop-butyl, (2) the subsequent loss of formic acid (HCOOH), that is, to the ion with *m/z* 282 for fluazifop-butyl, and (3) rearrangement of the methyl group to the ring and loss of CO, that is, to the ion with *m/z* 254 for fluazifop-butyl (Figure 4.11.19). The rearrangement reaction readily explains the formation of the tropylium ion ($[C_7H_7]^+$) with *m/z* 91. In addition, three other fragment ions are observed, that is, the ions with *m/z* 238, 164, and 119 for fluazifop-butyl (Figure 4.11.19). The same fragmentation, including the rearrangement of the methyl group and loss of CO, was also confirmed by accurate-mass data for clodinafop-propargyl ($[M+H]^+$ with *m/z* 350), quizalofop-ethyl ($[M+H]^+$ with *m/z* 373), haloxyfop-ethoxyethyl ($[M+H]^+$ with *m/z* 434), and propaquizafop ($[M+H]^+$ with *m/z* 444).

In positive-ion product-ion mass spectra, fluazifop ($[M+H]^+$ with *m/z* 328) shows fragments due to the loss of

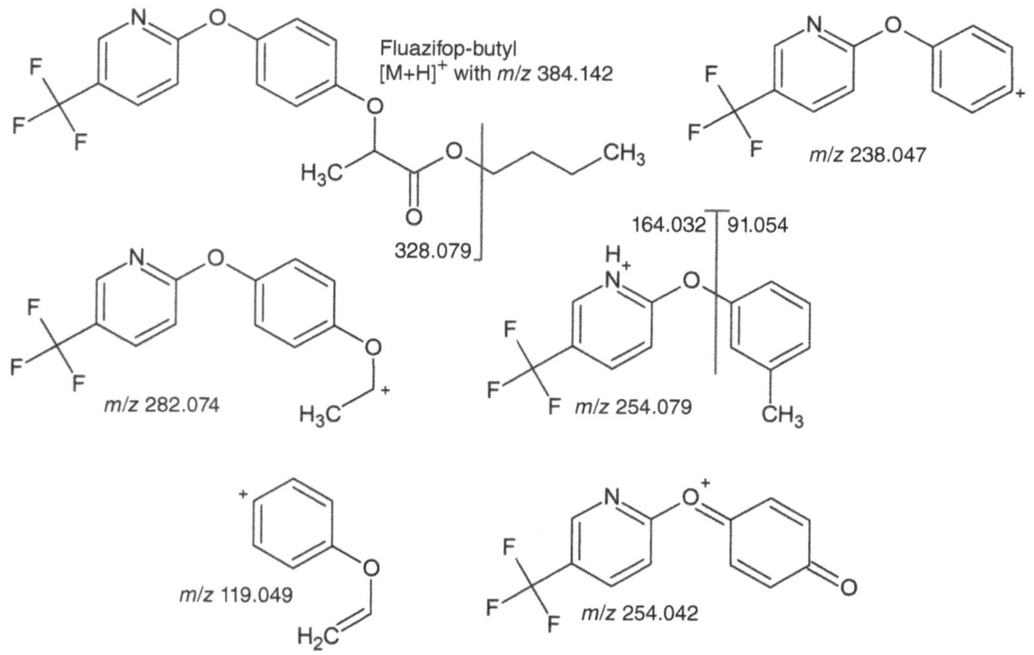

FIGURE 4.11.19 Fragmentation scheme for the protonated halogenated aryloxyphenoxypropionic acid ester herbicide fluazifop-butyl. (Source: Niessen, 2010. Reproduced with permission of Elsevier.)

HCOOH and of prop-2-enoic acid (C_2H_3COOH), that is, to ions with *m/z* 282 and 256, respectively (Bolygo & Boseley, 2000). In negative-ion mode, the free acids haloxyfop, diclofop, and fluazifop show the loss of 3-methyloxiran-2-one ($C_3H_4O_2$) (Laganà et al., 1998a; Køppen & Spliid, 1998; Rodríguez et al., 2003; Santilio et al., 2011). For fluazifop ($[M-H]^-$ with *m/z* 326), this results in a fragment ion with *m/z* 254. In addition, an anion with *m/z* 226 ($[C_{11}H_7F_3NO]^-$), consistent with the loss of CO from the ion with *m/z* 254, and the 4-oxylphenoxide radical anion with *m/z* 108 ($^{\bullet}OC_6H_4O^-$) are observed.

4.11.8 PHENOLIC COMPOUNDS

Phenolic compounds form a stable phenoxide ion and thus have acidic properties and are prone to negative-ion formation. This is true not only for some chlorinated phenols and nitrophenols, which can be considered as pesticide degradation products or as their building blocks, but also for chloroxynil, bromoxynil, and ioxynil.

Tetrachlorophenol ($[M-H]^-$ with *m/z* 229) shows fragments due to the loss of hydrogen chloride (HCl) and of HCl and carbon monoxide (CO) to anions with *m/z* 193 and 165, respectively. 4-Nitrophenol ($[M-H]^-$ with *m/z* 138) shows fragments due to the loss of a nitrosyl radical ($^{\bullet}NO$) and a nitryl radical ($^{\bullet}NO_2$) to anions with *m/z* 108 and 92, respectively. Similar behavior is observed for 2,4-dinitrophenol ($[M-H]^-$ with *m/z* 183), which shows a series of fragments due to losses of $^{\bullet}NO$ and/or $^{\bullet}NO_2$, that is, anions with *m/z* 153 (loss of $^{\bullet}NO$), *m/z* 137 (loss of $^{\bullet}NO_2$), *m/z* 123 (loss of two times $^{\bullet}NO$), *m/z* 109 (loss of $^{\bullet}NO_2$ and CO), and *m/z* 95 (loss of two times $^{\bullet}NO$ and CO).

Despite the structure resemblance, the nitrile herbicides chloroxynil, bromoxynil, and ioxynil show a somewhat different fragmentation. Chloroxynil ($[M-H]^-$ with *m/z* 186) shows subsequent losses of HCl and CO. Bromoxynil ($[M-H]^-$ with *m/z* 274) primarily shows the formation of the bromide anion (Br^-) with *m/z* 79 and a minor fragment due to the loss of $Br^{\bullet}$ and CO to an anion with *m/z* 167. Ioxynil ($[M-H]^-$ with *m/z* 370) shows the loss of $I^{\bullet}$ to an anion with *m/z* 243, the loss of both $I^{\bullet}$ and CO to an anion with *m/z* 215, and the formation of the iodide anion (I^-) with *m/z* 127. This is consistent with the tendency for the heavier halogens, for example, Br and I, to be lost as radicals (Section 3.7.8).

The phenoxy herbicide bromofenoxim ($[M-H]^-$ with *m/z* 458) shows the cleavage of the oxime-N—O bond to form two phenoxide ions, whose *m/z* values add up to *m/z* ([M−1]−1): the 2,4-dinitrophenoxide ion ($[C_6H_3N_2O_5]^-$) with *m/z* 183 and the 4-cyano-2,6-dibromophenoxide ion ($[C_7H_2Br_2NO]^-$) with *m/z* 274. In addition, the Br^- with *m/z* 79 and fragment ions consistent with one or two losses of $^{\bullet}NO$ from the anion with *m/z* 183 are observed as well.

4.11.9 MISCELLANEOUS HERBICIDES

4.11.9.1 Amide Herbicides

Isocarbamide ($[M+H]^+$ with *m/z* 186) shows the loss of 2-methylpropene ($(CH_3)_2C{=}CH_2$) to an ion with *m/z* 130 and the complimentary *iso*-butyl cation ($[C_4H_9]^+$) with *m/z* 57, as well as the protonated imidazolidin-2-one ($[C_3H_6N_2O+H]^+$) with *m/z* 87. Isoxaben ($[M+H]^+$ with *m/z* 333) shows the 2,6-dimethoxybenzoyl cation with *m/z* 165 ($[(CH_3O)_2C_6H_3{-}C{=}O]^+$) as a major fragment, and the 3-methylpentan-3-ylium ion with *m/z* 85 ($[C_6H_{13}]^+$) as well. At higher collision energy, the ion with *m/z* 165 shows fragment ions with *m/z* 150 due to the loss of a methyl radical ($^{\bullet}CH_3$), with *m/z* 122 due to subsequent loss of CO, and with *m/z* 92 due to subsequent loss of formaldehyde ($H_2C{=}O$). Tebutam ($[M+H]^+$ with *m/z* 234) shows the protonated *N*-benzyl-2,2-dimethylpropanamide ($[C_{12}H_{17}NO+H]^+$) with *m/z* 192 due to the loss of propene ($CH_3CH{=}CH_2$) from the amide *N*-*iso*-propyl substituent, the *tert*-butyl carbocation ($[(CH_3)_3C]^+$) with *m/z* 57, the tropylium ion ($[C_7H_7]^+$) with *m/z* 91, and the protonated *N*-2-allyl-2,2-dimethylpropanamide ($[C_8H_{15}NO+H]^+$) with *m/z* 142 due to the loss of toluene.

Diphenamid ($[M+H]^+$ with *m/z* 240) shows a direct cleavage of the C—C bond between the tertiary carbon and the amide carbonyl group, leading to the α-(phenyl)benzyl cation ($[(C_6H_5)_2CH]^+$) with *m/z* 167 and the (dimethylamino)acylium ion ($[(CH_3)_2N{=}C{=}O]^+$) with *m/z* 72. The elemental composition of the ion with *m/z* 134 ($[C_9H_{12}N]^+$) is consistent with the loss of CO and benzene (C_6H_6), thus requiring the rearrangement of the dimethylamino ($-N(CH_3)_2$) group to one of the phenyl rings, for example, leading to the 2-dimethylaminobenzyl cation ($[(CH_3)_2N{-}C_6H_5CH_2]^+$).

4.11.9.2 Anilide Herbicides

Chloranocryl ($[M+H]^+$ with *m/z* 230, also called dicryl) shows the 2-methylpropenacylium ion ($[C_3H_5{-}C{=}O]^+$) with *m/z* 69, as well as two other fragment ions involving skeletal rearrangements. The fragment ion with *m/z* 202 results from the extrusion of CO, leading to protonated *N*-(3,4-dichlorophenyl)dimethylmethanimine ($[Cl_2C_6H_3{-}N{=}C(CH_3)_2+H]^+$), whereas the ion with *m/z* 174 is probably due to the loss of propenaldehyde ($H_2C{=}CHCHO$) resulting in the protonated *N*-(3,4-dichlorophenyl)methanimine ($[Cl_2C_6H_3{-}N{=}CH_2+H]^+$) (Figure 4.11.20). Similarly, protonated *N*-(4-fluorophenyl) methanimine ($[FC_6H_4{-}N{=}CH_2+H]^+$) with *m/z* 124 is observed for flufenacet ($[M+H]^+$ with *m/z* 364). After the loss of 5-(trifluoromethyl)-1,3,4-thiadiazol-2-ol to an ion with *m/z* 194 ($[C_{11}H_{13}FNO]^+$) and the subsequent loss of propene ($CH_3CH{=}CH_2$) from the *N*-*iso*-propyl group, the

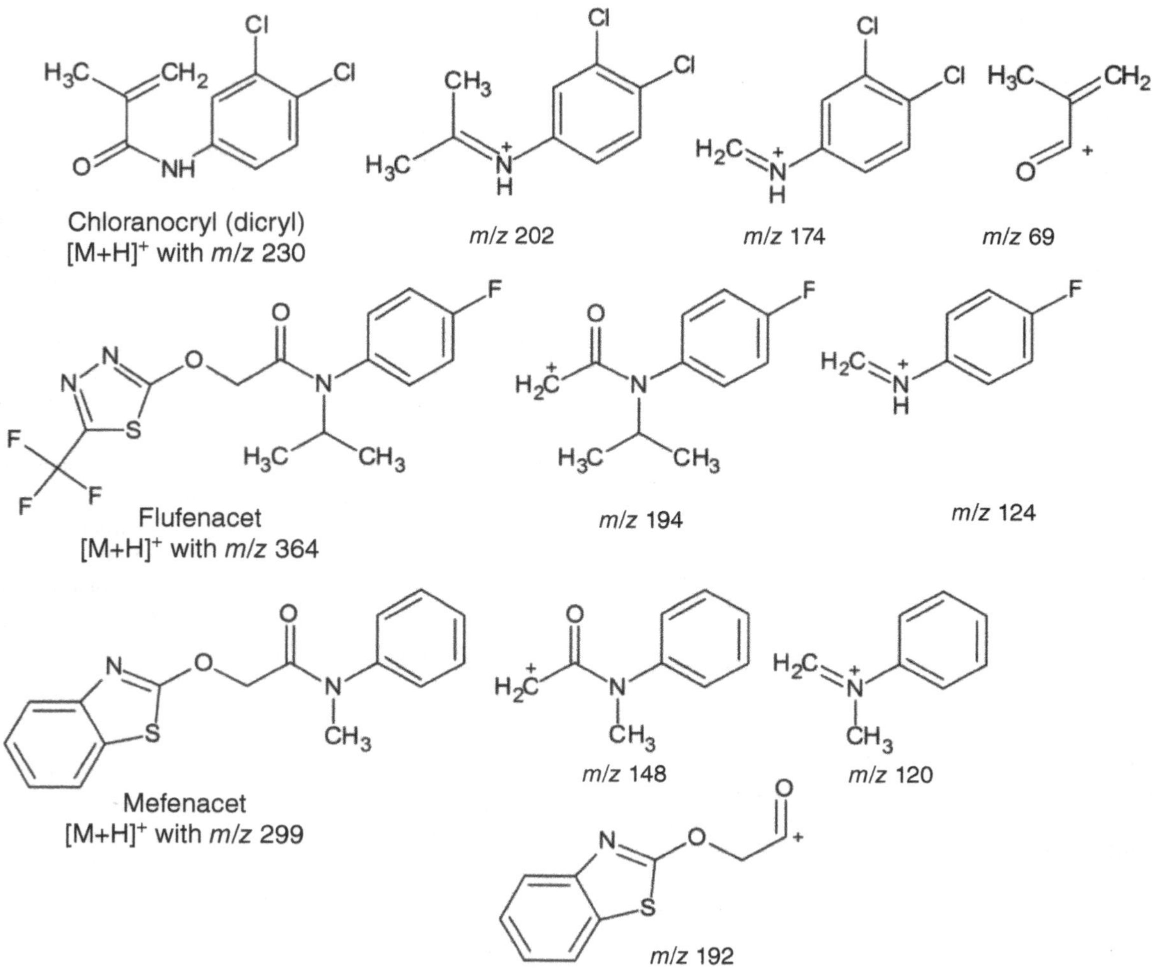

FIGURE 4.11.20 Proposed structures for the fragment ions of the protonated anilide herbicides chloranocryl, flufenacet, and mefenacet.

extrusion of CO loss seems to result in the ion with *m/z* 124 (Figure 4.11.20). Again, a similar fragmentation is observed for mefenacet ($[M+H]^+$ with *m/z* 299). The loss of 1,3-benzothiazol-2-ol results in a fragment ion with *m/z* 148 ($[C_9H_{10}NO]^+$), which after CO loss rearranges to the *N*-methyl-*N*-phenylmethaniminium ion ($[C_8H_{10}N]^+$) with *m/z* 120. The (1,3-benzothiazol-2-oxy)acetyl cation with *m/z* 192 ($[C_9H_6NO_2S]^+$) is due to the loss of *N*-methylaniline (Figure 4.11.20).

Picolinafen ($[M+H]^+$ with *m/z* 377) shows losses of water, of hydrogen fluoride (HF), and of two times HF resulting in the fragment ions with *m/z* 359, 357, and 337, respectively. The acylium ion with *m/z* 266 ($[C_{13}H_7F_3NO_2]^+$) is due to the loss of 4-fluoroaniline ($FC_6H_4NH_2$). Alternatively, the loss of 4-fluoro-phenylisocyanate (FC_6H_4—N=C=O) results in the protonated 2-[3-(trifluoromethyl)phenoxy]pyridine ($[C_{12}H_8F_3NO+H]^+$) with *m/z* 238. In addition, the 3-(trifluoromethyl)phenylium ion ($[CF_3C_6H_4]^+$) with *m/z* 145 and the 4-fluorophenylium ion ($[FC_6H_4]^+$) with *m/z* 95 are observed.

4.11.9.3 Chloracetanilide Herbicides

Chloracetanilide herbicides such as alachlor, metolachlor, and metazachlor can be analyzed in positive-ion mode, resulting in $[M+H]^+$ and $[M+Na]^+$ in ESI, and possibly a fragment due to the loss of methanol (CH_3OH) (Molina et al., 1995; Curini et al., 2000; Hogenboom et al., 2000). The fragment ion due to the loss of CH_3OH is also the most abundant ion in APCI (Hogenboom et al., 2000).

The MS–MS spectrum of alachlor ($[M+H]^+$ with *m/z* 270) was studied in detail, using MS–MS and in-source CID in an orthogonal acceleration TOF mass spectrometer (Hogenboom et al., 2000). Characteristic fragment ions are due to the loss of CH_3OH to an iminium ion with *m/z* 238, subsequent loss of water to *m/z* 220; the loss of

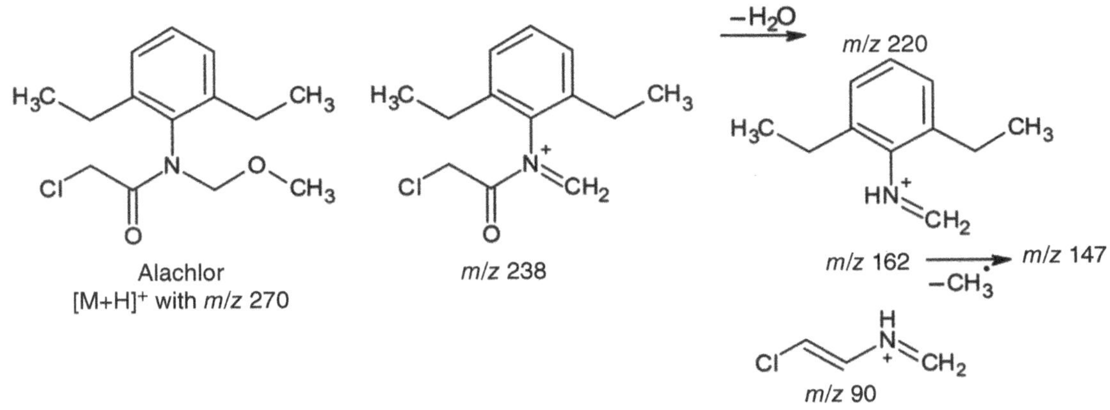

FIGURE 4.11.21 Proposed structures for the fragment ions of the protonated chloracetanilide herbicide alachlor.

TABLE 4.11.10 Structure and major fragment ions of chloracetanilide herbicides.

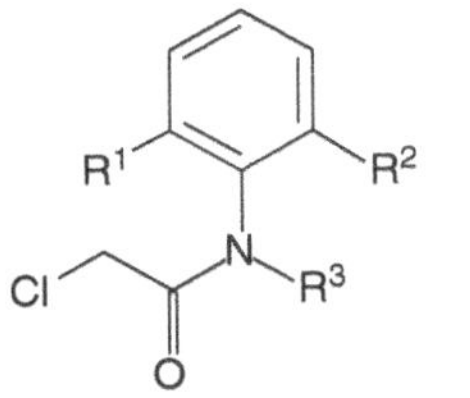

Compound	R^1	R^2	R^3	$[M+H]^+$	F_1: Loss of ROH from R^3	F_2: F_1 − ClCHCO	Other Fragments
Alachlor	$—C_2H_5$	$—C_2H_5$	$—CH_2OCH_3$	270	238	162	220: $F_1 - H_2O$ 147: $F_2 - {}^{\bullet}CH_3$ 90: $[C_3H_5ClN]^+$
Acetochlor	$—C_2H_5$	$—CH_3$	$—CH_2OCH_2CH_3$	270	224	148	133: $F_2 - {}^{\bullet}CH_3$
Propachlor	—H	—H	$—CH(CH_3)_2$	212			170: $-C_3H_6$ 152: 170 − H_2O 94: $[C_6H_5N_2+H]^+$
Dimethachlor	$—CH_3$	$—CH_3$	$—CH_2CH_2OCH_3$	256	224	148	
Metazachlor	$—CH_3$	$—CH_3$	$—CH_2—C_3H_3N_2$ (pyrazol)	278	210*	134	105: $[C_8H_9]^+$
Metolachlor	$—C_2H_5$	$—CH_3$	$—CHCH_3CH_2OCH_3$	284	252	176	212: $-C_4H_8O$ 73: $[C_4H_9O]^+$
Propisochlor	$—C_2H_5$	$—CH_3$	$—CH_2OCH(CH_3)_2$	284	224	148	
Pretilachlor	$—C_2H_5$	$—C_2H_5$	$—(CH_2)_2O(CH_2)_2CH_3$	312	252	176	147: $F_2 - {}^{\bullet}C_2H_5$ 132: 147 − ${}^{\bullet}CH_3$

*Loss of 1*H*-pyrazole instead of ROH.

2-chloroethenone (ClCH═C═O) from the ion with *m/z* 238 to the protonated *N*-(2,6-diethylphenyl)methanimine ($[C_{11}H_{15}N+H]^+$) with *m/z* 162, subsequent loss of a methyl radical (${}^{\bullet}CH_3$) to *m/z* 147 ($[C_{10}H_{13}N]^{+\bullet}$); and an ion with *m/z* 90 ($[ClCH{=}CHN{=}CH_2+H]^+$) due to consecutive losses of water, CH_3OH, and 3-ethylstryrene ($C_{10}H_{12}$) (Figure 4.11.21). Similar fragmentation is observed for the isomeric acetochlor ($[M+H]^+$ with *m/z* 270), resulting in ions with *m/z* 224 due to the loss of ethanol (CH_3CH_3OH), *m/z* 148 due to subsequent the loss of ClCH═C═O, *m/z* 133 due to subsequent loss of ${}^{\bullet}CH_3$. The loss of (part of) the alkylether chain as an alcohol followed by the loss of ClCH═C═O, thus resulting in the fragment ion with *m/z* 238 and 162 for alachlor, is also observed with other members of this class as shown in Table 4.11.10; some other fragment ions observed for these compounds are also interpreted. The related compound dimethenamid ($[M+H]^+$ with *m/z* 276, 2-chloro-*N*-(2,4-dimethylthiophen-3-yl)-*N*-(1-methoxypropan-2-yl)acetamide) shows a similar fragmentation, that is, the loss of CH_3OH to an ion

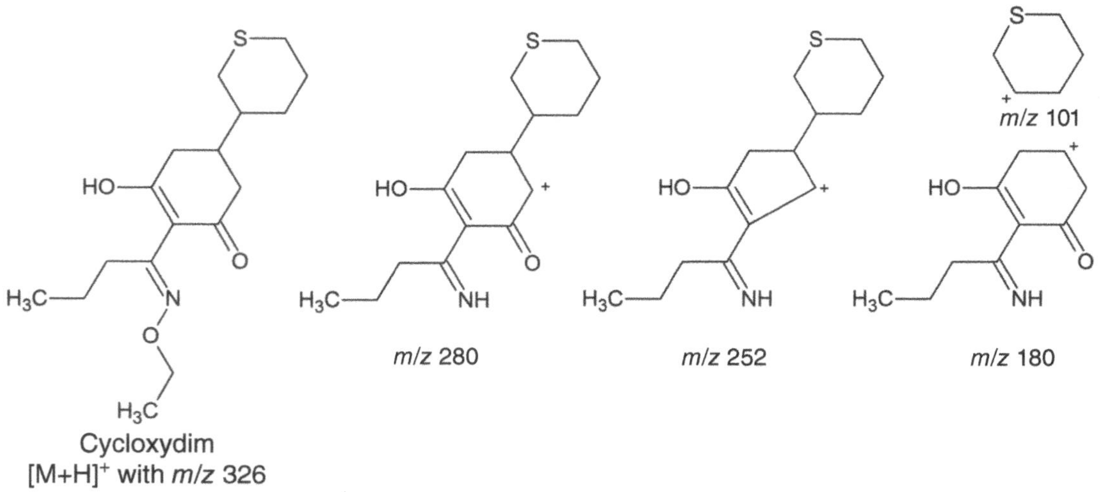

FIGURE 4.11.22 Proposed structures for the fragment ions of the protonated cyclohexanedione oxime herbicide cycloxydim.

with *m/z* 244 and the subsequent loss of ClCH=C=O to *m/z* 168. Similar to metolachlor ($[M+H]^+$ with *m/z* 284) (Table 4.11.10), the 1-methoxyprop-2-yl carbocation ($[C_4H_9O]^+$) with *m/z* 73 due to the alkylether chain is also observed.

4.11.9.4 Cyclohexanedione Oxime Herbicides

Extensive fragmentation of protonated molecules of cyclohexene oxime or cyclohexanedione oxime herbicides is observed, even at low collision energies. The characteristic fragmentation involves the loss of the oxime substituent chain as an alcohol, followed by either the loss of CO or the loss of the substituent on the cyclohexanedione ring (Marek et al., 2000). This is illustrated for cycloxydim ($[M+H]^+$ with *m/z* 326) in Figure 4.11.22. Thus, the loss of ethanol (CH_3CH_2OH) results in a fragment ion with *m/z* 280, which further fragments by the loss of either CO to *m/z* 252 or the loss of 5,6-dihydro-4*H*-thiopyran (C_5H_8S) to *m/z* 180. Charge retention on the thiopyranyl substituent results in the ion with *m/z* 101 ($[C_5H_9S]^+$). Similar fragmentation is observed for clethodim ($[M+H]^+$ with *m/z* 360), resulting in fragment ions with *m/z* 268, 240, and 164; profoxydim ($[M+H]^+$ with *m/z* 466), resulting in *m/z* 280, 238, and 180; sethoxydim ($[M+H]^+$ with *m/z* 328), resulting in *m/z* 282, 254, and 178; and tepraloxydim ($[M+H]^+$ with *m/z* 342), resulting in fragment ions with *m/z* 250, 222, and 166.

Alloxydim ($[M+H]^+$, with *m/z* 324) shows extensive fragmentation along a different fragmentation route, as is illustrated by the proposed structures of the more abundant fragment ions in Figure 4.11.23. Initial fragmentation involves the loss of methanol from the methyl ester group, followed by losses from the oxime allyl group and/or the 4-hydroxy-6,6-dimethyl-2-one-3-cyclohexene-1-carboxylate methyl ester; no loss of allyl alcohol is observed in this case.

4.11.9.5 Dinitroaniline Herbicides

Dinitroaniline herbicides such as benfluralin ($[M+H]^+$ with *m/z* 336), butralin ($[M+H]^+$ with *m/z* 296), isopropalin ($[M+H]^+$ with *m/z* 310), and pendimethalin ($[M+H]^+$ with *m/z* 282) can be analyzed in positive-ion ESI mode. Due to in-source CID or MS–MS, fragmentation due to the loss of an *N*-substituent is observed, that is, the loss of 2-pentene (C_5H_{10}) in pendimethalin; a butene (C_4H_8) in benfluralin, butralin; propene ($CH_3CH{=}CH_2$) in fluchloralin ($[M+H]^+$ with *m/z* 356); and one or two times $CH_3CH{=}CH_2$ in trifluralin ($[M+H]^+$ with *m/z* 336), isopropalin, nitralin ($[M+H]^+$ with *m/z* 346), and oryzaline ($[M+H]^+$ with *m/z* 347) (Liu et al., 2004). In addition to this common alkene loss, extensive fragmentation occurs. A characteristic sequence of fragmentation can be observed. This is illustrated in Figure 4.11.24 for benfluralin ($[M+H]^+$ with *m/z* 336). The loss of the first alkyl group (as C_4H_8 in benfluralin) is followed by the loss of either water and/or the loss of the second alkyl group. The loss of water involves an *ortho*-effect (Section 3.6.6), where the amine-H-atom of the neighboring amine alkyl group, that is, $—NHC_2H_5$ in benfluralin, is involved. The formation of the ions with *m/z* 236 and 220 for benfluralin is not readily understood. They may be formed by a concerted loss of ethyne from the amine alkyl group and water from the nitro group, thus resulting in an effective loss of ethanal (CH_3CHO) from the ions with *m/z* 280 and 262.

4.11.9.6 Imidazolinone Herbicides

Imidazolinones are a relatively new class of herbicides. Either positive-ion or negative-ion ESI-MS can be applied. In a multiresidue study, involving various herbicide classes, imazamethabenz-methyl was analyzed in positive-ion ESI, while for imazapyr, imazethapyr, imazamethabenz,

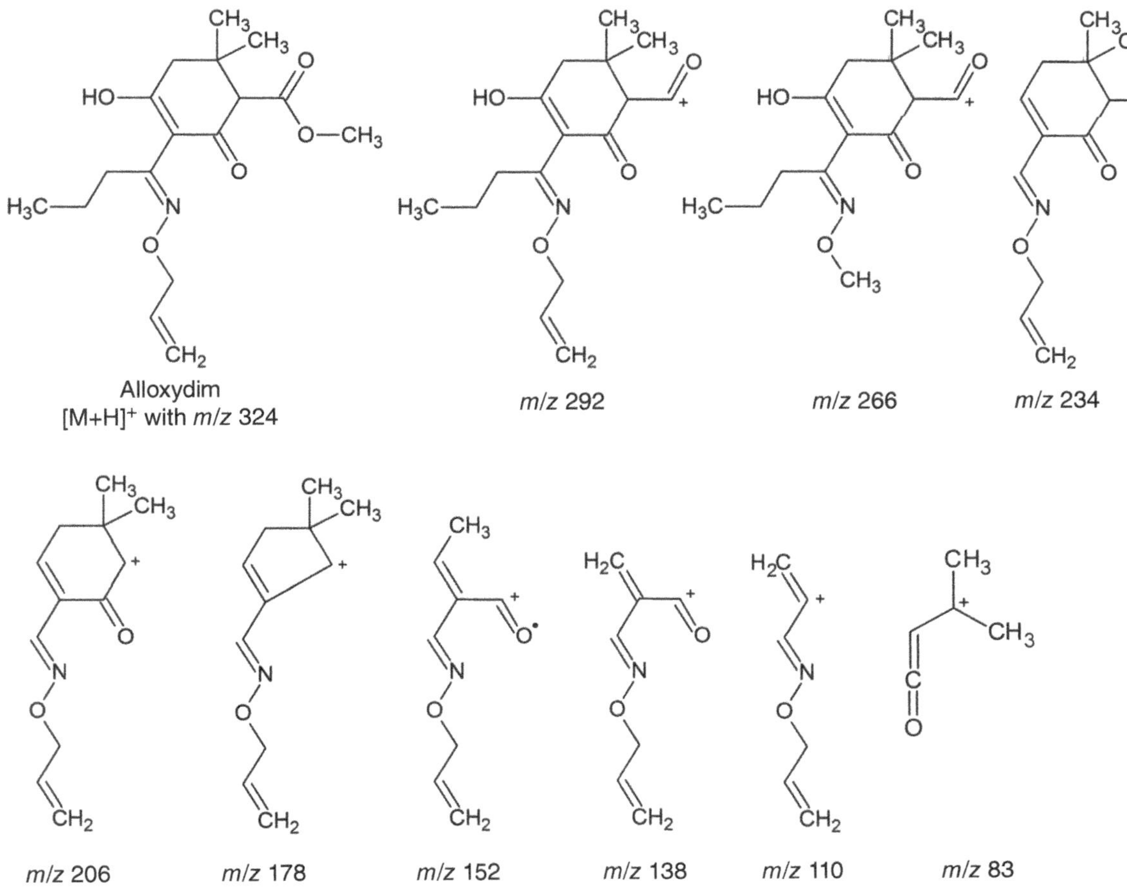

FIGURE 4.11.23 Proposed structures for the fragment ions of the protonated cyclohexanedione oxime alloxydim.

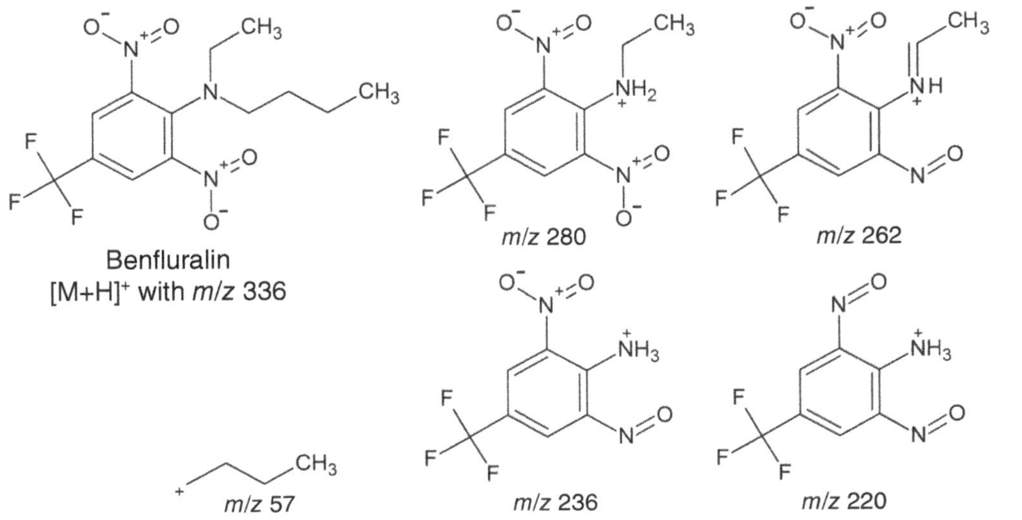

FIGURE 4.11.24 Proposed structures for the fragment ions of the protonated dinitroanaline herbicide benfluralin.

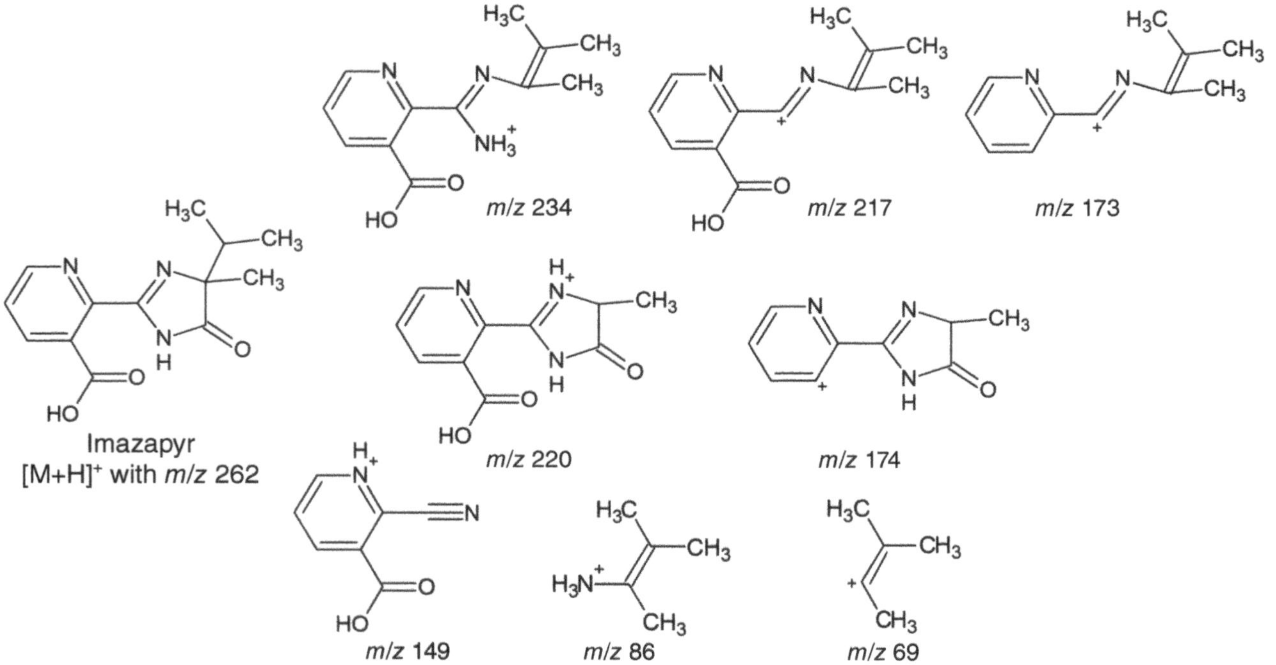

FIGURE 4.11.25 Proposed structures for the fragment ions of the protonated imidazolinone herbicide imazapyr.

and imazaquin negative-ion ESI was preferred (Curini et al., 2000).

Positive-ion fragmentation by in-source CID has been reported for imazapyr (Laganà et al., 1998b), imazamethabenz-methyl (D'Ascenzo et al., 1998a), and imazethapyr (D'Ascenzo et al., 1998b). A tentative fragmentation scheme for imazapyr has been proposed (Laganà et al., 1998a), with $OE^{+\bullet}$ fragments being proposed for several ions. However, accurate-mass data give no indication for $OE^{+\bullet}$ fragments. The major fragment ions for imazapyr ($[M+H]^+$ with *m/z* 262) can be described as consecutive small-molecule losses from the protonated molecule along two pathways (Figure 4.11.25). One pathway involves the loss of carbon monoxide (CO) to an ion with *m/z* 234, followed by the loss of ammonia (NH_3) to *m/z* 217, followed by the loss of either water or carbon dioxide (CO_2) to *m/z* 199 and 173, respectively. Another pathway involves the loss of propene ($CH_3CH{=}CH_2$) to an ion with *m/z* 220, followed by subsequent losses of water and CO to *m/z* 202 and 174, respectively. In addition, the protonated 2-cyanopyridine-3-carboxylic acid ($[C_7H_4N_2O_2{+}H]^+$) with *m/z* 149, protonated 3-methylbut-2-en-2-amine ($[C_5H_{11}N{+}H]^+$) with *m/z* 86 and the 3-methylbut-2-en-2-ylium ion ($[C_5H_9]^+$) with *m/z* 69 are observed. The same sequence of losses can be found in the product-ion mass spectrum of imazethapyr. This is in agreement with the data from in-source fragmentation (D'Ascenzo et al., 1998b), although there the ion with *m/z* 177 was erroneously indicated as the decarboxylated $[C_2H_5{-}C_5H_3N{-}C{=}NH]^+$ ($[C_8H_9N_2]^+$, calculated *m/z* 133), while in fact the COOH group is still present in the ion ($[C_9H_9N_2O_2]^+$).

4.11.9.7 Nitrophenyl Ether Herbicides

The mode of action of the nitrophenyl ether herbicides such as acifluorfen, aclonifen, bifenox, and lactofen is based on the inhibition of protoporphyrinogen oxidase. These herbicides are used for weed control in the cultivation of seeded legumes, such as soy beans. Mass spectra and product-ion mass spectra of five neutral diphenyl ether herbicides and three acid metabolites were studied (García-Reyes et al., 2007a,b; Laganà et al., 2000). Negative-ion ESI is preferred for these acidic compounds.

The negative-ion product-ion mass spectra of some nitrophenyl ether herbicides have been tabulated (Laganà et al., 2000). Data for aclonifen, acifluorfen, and fomesafen are available in our library collection. Acifluorfen ($[M-H]^-$ with *m/z* 360) shows the loss of carbon dioxide (CO_2) to an ion with *m/z* 316 followed by several subsequent small-molecule losses, for example, a loss of a nitrosyl radical ($^\bullet NO$) to a radical anion with *m/z* 286, of hydrogen chloride (HCl) to *m/z* 250, and of carbon monoxide (CO) to *m/z* 222 (Figure 4.11.26). The 2-chloro-4-trifluoromethylphenoxide ion ($[CF_3C_6H_3ClO]^-$) with *m/z* 195 is consistent with the loss of $C_7H_3NO_4$, involving the nitrobenzoic acid group. The ion with *m/z* 137

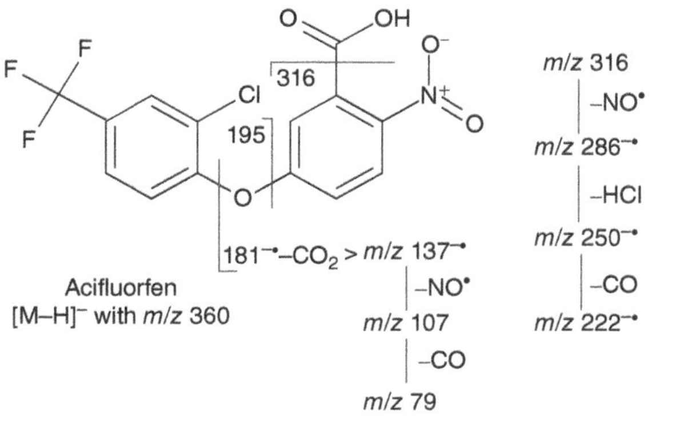

FIGURE 4.11.26 Fragmentation scheme for the deprotonated nitrophenyl ether herbicide acifluorfen.

($[C_6H_3NO_3]^{-\bullet}$) is the 2-nitrophen-5-yl-oxide radical anion; subsequent losses of $^\bullet$NO and CO yield fragment ions with *m/z* 107 and 79 (Figure 4.11.26). Aclonifen ($[M-H]^-$ with *m/z* 263) shows two radical losses from $[M-H]^-$, that is, the loss of $^\bullet$NO to a radical anion with *m/z* 233 and the loss of a phenyl radical ($^\bullet C_6H_5$) to a radical anion with *m/z* 186, which in turn shows subsequent losses of $^\bullet$NO and HCl to ions with *m/z* 156 and 120. Fomesafen ($[M-H]^-$ with *m/z* 437) shows the subsequent losses of methanesulfonyl isocyanate (CH_3SO_2—N=C=O) and $^\bullet$NO to fragment ions with *m/z* 316 and 286. The 2-chloro-4-(trifluoromethyl)phenoxide anion ($[CF_3C_6H_3ClO]^-$) with *m/z* 195 is also observed.

4.11.9.8 Bentazone and Dinitrophenol Herbicides

The herbicide bentazone ($[M-H]^-$ with *m/z* 239) shows extensive fragmentation in its negative-ion product-ion mass spectra (Køppen & Spliid, 1998), that is, loss of propene ($CH_3CH{=}CH_2$) to an ion with *m/z* 197, the loss of sulfur dioxide (SO_2) to an ion with *m/z* 175, with secondary losses of carbon monoxide (CO) to an ion with *m/z* 147 and of $CH_3CH{=}CH_2$ or the propyl radical ($^\bullet C_3H_7$) to the ions with *m/z* 133 or 132, respectively.

The isomeric dinitrophenol herbicides dinoseb and dinoterb ($[M-H]^-$ with *m/z* 239) give a range of identical fragments. However, the more abundant fragments for dinoseb are the less abundant ones for dinoterb, and vice versa. Dinoseb shows fragment ions with *m/z* 222 due to the loss of the hydroxyl radical ($^\bullet$OH), with *m/z* 194 due to the loss of both a methyl radical ($^\bullet CH_3$) and a nitrosyl radical ($^\bullet$NO), with *m/z* 193 due to the loss of an ethyl radical ($^\bullet CH_2CH_3$) from the ion with *m/z* 222, *m/z* 163 due to the loss of $^\bullet$NO from the ion with *m/z* 193. The formation of the ion with *m/z* 134 ($[C_7H_4NO_2]^-$) is not well understood yet. Our interpretation, based on accurate-mass Q–TOF data available in our library collection, differs significantly from the interpretation of the ion-trap MS^n data reported elsewhere (Juan-García et al., 2005). Dinoterb shows fragment ions with *m/z* 224 due to the loss of $^\bullet CH_3$ the subsequent loss of $^\bullet$OH to *m/z* 207, followed by the loss of $^\bullet$NO to *m/z* 177 or the loss of nitroxyl (HNO) to *m/z* 176. The loss of CO from the ion with *m/z* 177 results in the radical anion with *m/z* 149 ($[C_8H_7NO_2]^{-\bullet}$).

4.11.9.9 Uracil Herbicides

The uracil herbicides bromacil ($[M-H]^-$ with *m/z* 259) and terbacil ($[M-H]^-$ with *m/z* 215) show the same sequence of small-molecule losses, that is, losses of a butene (C_4H_8), hydrogen isocyanate (HN=C=O), and another HN=C=O leading to the 1-bromopropynide anion ($[C_3H_2Br]^-$) with *m/z* 117 and the 1-chloropropynide anion ($[C_3H_2Cl]^-$) with *m/z* 73, respectively. In addition, bromacil yields an abundant bromide anion (Br^-) with *m/z* 79.

4.11.9.10 Aromatic Acid Herbicides

The aromatic acid herbicide dicamba ($[M-H]^-$ with *m/z* 219) shows the loss of carbon dioxide (CO_2) to an ion with *m/z* 175 and a subsequent loss of formaldehyde ($H_2C{=}O$) to the 2,5-dichlorophenide anion ($[Cl_2C_6H_3]^-$) with *m/z* 145. Clopyralid ($[M-H]^-$ with *m/z* 190) shows the loss of CO_2 to an ion with *m/z* 146. Picloram ($[M-H]^-$ with *m/z* 239) shows the loss of CO_2 to an ion with *m/z* 195, followed by subsequent losses of hydrogen chloride (HCl) and hydrogen cyanide (HCN) to ions with *m/z* 159 and *m/z* 132 ($[C_4Cl_2N]^-$), for example, the 1,3-dichloro-3-cyanopropynide anion, respectively.

REFERENCES

Agilent Technologies. 2010. The accurate-mass library (Agilent Technologies, Broecker, Herre & Pragst Forensic Library) consulted is part of an LC–MS Forensic Toxicology Application Kit that is described in more detail in: http://www.chem.agilent.com/en-US/Products/Instruments/ms/Pages/forensics_qtof_ws.aspx

Aguilar CI, Ferrer I, Borrull F, Marcé RM, Barceló D. 1998. Comparison of automated on-line solid-phase extraction followed by liquid chromatography–mass spectrometry with atmospheric pressure chemical ionization and particle beam mass spectrometry for the determination of a priority group of pesticides in environmental waters, J Chromatogr A 794: 147–163.

Alder L, Greulich K, Kempe G, Vieth B. 2006. Residue analysis of 500 high priority pesticides: Better by GC–MS or LC–MS–MS? Mass Spectrom Rev 25: 838–865.

Alffenaar JW, Wessels AM, van Hateren K, Greijdanus B, Kosterink JG, Uges DR. 2010. Method for therapeutic drug monitoring of azole antifungal drugs in human serum using LC/MS/MS. J Chromatogr B 878: 39–44.

Barr JD, Bell AJ, Bird M, Mundy JL, Murrell J, Timperley CM, Watts P, Ferrante F. 2005. Fragmentations and reactions of the organophosphate insecticide diazinon and its oxygen analog diazoxon studied by electrospray ionization ion trap mass spectrometry. J Am Soc Mass Spectrom 16: 515–523.

Benzi M, Robotti E, Gianotti V. 2011. HPLC-DAD-MS^n to investigate the photodegradation pathway of nicosulfuron in aqueous solution. Anal Bioanal Chem 399: 1705–1714.

Blasco C, Fernández M, Picó Y, Font G. 2004. Comparison of solid-phase microextraction and stir bar sorptive extraction for determining six organophosphorus insecticides in honey by liquid chromatography–mass spectrometry. J Chromatogr A 1030: 77–85.

Blasco C, Vazquez-Roig P, Onghena M, Masia A, Picó Y. 2011. Analysis of insecticides in honey by liquid chromatography-ion trap-mass spectrometry: Comparison of different extraction procedures. J Chromatogr A 1218: 4892–4901.

Bobeldijk I, Vissers JPC, Kearney G, Major H, Van Leerdam JA. 2001. Screening and identification of unknown contaminants in water with liquid chromatography and quadrupole–orthogonal acceleration time-of-flight tandem mass spectrometry. J Chromatogr A 929: 63–74.

Bolygo E, Boseley A. 2000. A liquid chromatography–tandem mass spectrometry method for fluazifop residue analysis in crops. Fresenius J Anal Chem 368: 816–819.

Bossi R, Køppen B, Spliid NH, Streibig JC. 1998. Analysis of sulfonylurea herbicides in soil water at sub-part-per-billion levels by electrospray negative ionization mass spectrometry. Followed by confirmatory tandem mass spectrometry. J AOAC Int 81: 775–784.

Careri M, Elviri L, Mangia A, Zagnoni I. 2002. Rapid method for determination of chlormequat residues in tomato products by ion-exchange liquid chromatography–electrospray tandem mass spectrometry. Rapid Commun Mass Spectrom 16: 1821–1826.

Castro R, Moyano E, Galceran MT. 1999. Ion-pair liquid chromatography– atmospheric pressure ionization mass spectrometry for the determination of quaternary ammonium herbicides. J Chromatogr A 830: 145–154.

Castro R, Moyano E, Galceran MT. 2001. Determination of quaternary ammonium pesticides by liquid chromatography–electrospray tandem mass spectrometry. J Chromatogr A 914: 111–121.

Chiron S, Papilloud S, Haerdi W, Barceló D. 1995. Automated online liquid–solid extraction followed by liquid chromatography-high-flow pneumatically assisted electrospray mass spectrometry for the determination of acidic herbicides in environmental waters, Anal Chem 67: 1637–1643.

Chiu KS, Van Langenhove A, Tanaka C. 1989. High-performance liquid chromatographic–mass spectrometric and high-performance liquid chromatographic–tandem mass spectrometric analysis of carbamate pesticides. Biomed Environ Mass Spectrom 18: 200–206.

Chung SW, Chan BT. 2010. Validation and use of a fast sample preparation method and liquid chromatography–tandem mass spectrometry in analysis of ultra-trace levels of 98 organophosphorus pesticide and carbamate residues in a total diet study involving diversified food types. J Chromatogr A 1217: 4815–4824.

Curini R, Gentili A, Marchese A, Marino A, Perret D. 2000. Solid-phase extraction followed by high-performance liquid chromatography–ionspray interface–mass spectrometry for monitoring of herbicides in environmental water. J Chromatogr A 874: 187–198.

D'Ascenzo G, Gentili A, Marchese S, Marino A, Perret D. 1998a. Optimization of high performance liquid chromatography–mass spectrometry apparatus for determination of imidazolinone herbicides in soil at levels of a few ppb. Rapid Commun Mass Spectrom 12: 1359–1365.

D'Ascenzo G, Gentili A, Marchese S, Perret D. 1998b. Development of a method based on liquid chromatography–electrospray mass spectrometry for analyzing imidazolinone herbicides in environmental water at part-per-trillion levels. J Chromatogr A 800: 109–119.

Detomaso A, Mascolo G, Lopez A. 2005. Characterization of carbofuran photodegradation by-products by liquid chromatography/hybrid quadrupole time-of-flight mass spectrometry, Rapid Commun Mass Spectrom 19: 2193–2202.

Draper WM. 2001. Electrospray liquid chromatography quadrupole ion trap mass spectrometry determination of phenyl urea herbicides in water. J Agric Food Chem 49: 2746–2755.

Evans CS, Startin JR, Goodall DM, Keely BJ. 2001. Tandem mass spectrometric analysis of quaternary ammonium pesticides. Rapid Commun Mass Spectrom 15: 699–707.

Fernández M, Picó Y, Girotti S, Mañes J. 2001a. Analysis of organophosphorus pesticides in honeybee by liquid chromatography-atmospheric pressure chemical ionization-mass spectrometry. J Agric Food Chem 49: 3540–3547.

Fernández M, Rodríguez R, Picó Y, Mañes J. 2001b. Liquid chromatographic–mass spectrometric determination of post-harvest fungicides in citrus fruits. J Chromatogr A 912: 301–310.

Fernández Moreno JL, Garrido Frenich A, Plaza Bolaños O, Martínez Vidal JL. 2008. Multiresidue method for the analysis of more than 140 pesticide residues in fruits and vegetables by gas chromatography coupled to triple quadrupole mass spectrometry. J Mass Spectrom 43: 1235–1254.

García-Reyes JF, Hernando MD, Molina-Díaz A, Fernández-Alba AR. 2007a. Comprehensive screening of target, non-target and unknown pesticides in food by LC–TOF-MS. Trends Anal Chem 26: 828–841.

García-Reyes JF, Molina-Díaz A, Fernández-Alba AR. 2007b. Identification of pesticide transformation products in food by liquid chromatography–time-of-flight mass spectrometry via "fragmentation-degradation" relationships, Anal Chem 79: 307–321.

Goodwin L, Startin JR, Goodall DM, Keely BJ. 2003. Tandem mass spectrometric analysis of glyphosate, glufosinate, aminomethylphosphonic acid and methylphosphinicopropionic acid. Rapid Commun Mass Spectrom 17: 963–969

Goodwin L, Startin JR, Goodall DM, Keely BJ. 2004. Negative ion electrospray mass spectrometry of aminomethylphosphonic acid and glyphosate: Elucidation of fragmentation mechanisms by multistage mass spectrometry incorporating in-source deuterium labelling. Rapid Commun Mass Spectrom 18: 37–43.

Harden CS, Snyder AP, Eiceman GA. 1993. Determination of collision-induced dissociation mechanisms and cross-sections in organophosphorous compounds by atmospheric pressure ionization mass spectrometry. Org Mass Spectrom 28: 585–592.

Hau J, Riediker S, Varga N, Stadler RH. 2000. Determination of the plant growth regulator chlormequat in food by liquid chromatography–electrospray ionisation tandem mass spectrometry. J Chromatogr A 878: 77–86.

Hogenboom AC, Speksnijder P, Vreeken RJ, Niessen WMA, Brinkman UATh. 1997. Rapid target analysis of microcontaminants in water by on-line single-short-column liquid chromatography combined with atmospheric pressure chemical ionization tandem mass spectrometry. J Chromatogr A 777: 81–90.

Hogenboom AC, Niessen WMA, Brinkman UATh. 1998. Rapid target analysis of microcontaminants in water by on-line single-short-column liquid chromatography combined with atmospheric-pressure chemical ionization ion-trap mass spectrometry. J Chromatogr A 794: 201–210.

Hogenboom AC, Niessen WMA, Brinkman UATh. 2000. Characterization of photodegradation products of alachlor in water by on-line solid-phase extraction–liquid chromatography combined with tandem mass spectrometry and orthogonal-acceleration time-of-flight mass spectrometry, Rapid Commun Mass Spectrom 14: 1914–1924.

Ibáñez M, Pozo ÓJ, Sancho JV, López FJ, Hernández F. 2005. Residue determination of glyphosate, glufosinate and aminomethylphosphonic acid in water and soil samples by liquid chromatography coupled to electrospray tandem mass spectrometry. J Chromatogr A 1081: 145–155.

Itoh H, Kawasaki S, Tadano J. 1996. Application of liquid chromatography–atmospheric-pressure chemical-ionization mass spectrometry to pesticide analysis. J Chromatogr A 754: 61–76.

Jeannot R, Sabik H, Sauvard E, Genin E. 2000. Application of liquid chromatography with mass spectrometry combined with photodiode array detection and tandem mass spectrometry for monitoring pesticides in surface waters. J Chromatogr A 879: 51–71.

John H, Worek F, Thiermann H. 2008. LC–MS-based procedures for monitoring of toxic organophosphorus compounds and verification of pesticide and nerve agent poisoning. Anal Bioanal Chem 391: 97–116.

Juan-García A, Font G, Picó Y. 2005. Quantitative analysis of six pesticides in fruits by capillary electrophoresis-electrospray-mass spectrometry. Electrophoresis 26: 1550–1561.

Kawasaki S, Ueda H, Itoh H, Tadano J. 1992. Screening of organophosphorus pesticides using liquid chromatography–atmospheric pressure chemical ionization mass spectrometry, J Chromatogr 595 (1992) 193.

Kawasaki S, Nagumo F, Ueda H, Tajima Y, Sano M, Tadano J. 1993. Simple, rapid and simultaneous measurement of eight different types of carbamate pesticides in serum using high-performance liquid chromatography–atmospheric-pressure chemical ionization mass spectrometry. J Chromatogr 620: 61–71.

Køppen B, Spliid NH. 1998. Determination of acidic herbicides using liquid chromatography with pneumatically assisted electrospray ionization mass spectrometric and tandem mass spectrometric detection. J Chromatogr A 803: 157–168.

Lacorte S, Barceló D. 1996. Determination of parts per trillion levels of organophosphorus pesticides in groundwater by automated on-line liquid–solid extraction followed by liquid chromatography–atmospheric-pressure chemical ionization mass spectrometry using positive and negative ion modes of operation. Anal Chem 68: 2464–2470.

Lacorte S, Molina C, Barceló D. 1998. Temperature and extraction voltage effect on fragmentation of organophosphorus pesticides in liquid chromatography–atmospheric-pressure chemical ionization mass spectrometry. J Chromatogr A 795: 13–26.

Laganà A, Fago G, Marino A, Mosso M. 1998a. Soil column extraction followed by liquid chromatography and electrospray ionization mass spectrometry for the efficient determination of aryloxyphenoxypropionic herbicides in soil samples at ng/g levels. Anal Chim Acta 375: 107–116.

Laganà A, Fago G, Marino A. 1998b. Simultaneous determination of imidazolinone herbicides from soil and natural waters using soil column extraction and off-line solid-phase extraction followed by liquid chromatography with UV detection or liquid chromatography–electrospray mass spectroscopy. Anal Chem 70: 121–130.

Laganà A, Fago G, Fasciani L, Marino A, Mosso M. 2000. Determination of diphenyl-ether herbicides and metabolites in natural waters using high-performance liquid chromatography with diode array and tandem mass spectrometric detection. Anal Chim Acta 414: 79–94.

Lee XP, Kumazawa T, Fujishiro M, Hasegawa C, Arinobu T, Seno H, Ishii A, Sato K. 2004. Determination of paraquat and diquat in human body fluids by high-performance liquid chromatography/tandem mass spectrometry. J Mass Spectrom 39: 1147–1152.

Li LYT, Campbell DA, Bennett PK, Henion JD. 1996. Acceptance criteria for ultratrace HPLC-tandem mass spectrometry: quantitative and qualitative determination of sulfonylurea herbicides in soil. Anal Chem 68: 3397–3404.

Lindh CH, Littorin M, Amilon A, Jönsson BA. 2008. Analysis of phenoxyacetic acid herbicides as biomarkers in human urine using liquid chromatography/triple quadrupole mass spectrometry. Rapid Commun Mass Spectrom 22: 143–150.

Liu H, Ding C, Zhang S, Liu H, Liao XG, Qu L, Zhao Y, Wu Y. 2004. Simultaneous residue measurement of pendimethalin, isopropalin, and butralin in tobacco using high-performance liquid chromatography with ultraviolet detection and electrospray ionization mass spectrometric identification. J Agric Food Chem 52: 6912–6915.

Marchese S, Perret D, Gentili A, Curini R, Marino A. 2001. Development of a method based on accelerated solvent extraction and liquid chromatography–mass spectrometry for determination of arylphenoxyproprionic herbicides in soil. Rapid Commun Mass Spectrom 15: 393–400.

Marek LJ, Koskinen WC, Bresnahan GA. (2000) LC/MS analysis of cyclohexanedione oxime herbicides in water. J Agric Food Chem 48: 2797–2801.

Marr JC, King JB. 1997. A simple high-performance liquid chromatography–ionspray tandem mass spectrometry method for the direct determination of paraquat and diquat in water. Rapid Commun Mass Spectrom 11: 479–483.

Martins-Júnior HA, Lebre DT, Wang AY, Pires MA, Bustillos OV. 2009. An alternative and fast method for determination of glyphosate and aminomethylphosphonic acid (AMPA) residues in soybean using liquid chromatography coupled with tandem mass spectrometry. Rapid Commun Mass Spectrom 23: 1029–1034.

Mazzotti F, Di Donna L, Macchione B, Maiuolo L, Perri E, Sindona G. 2009. Screening of dimethoate in food by isotope dilution and electrospray ionization tandem mass spectrometry. Rapid Commun Mass Spectrom 23: 1515–1518.

Molina C, Honing M, Barceló D. 1994. Determination of organophosphorus pesticides in water by solid-phase extraction followed by liquid chromatography–high-flow pneumatically assisted electrospray mass spectrometry. Anal Chem 66: 4444–4449.

Molina C, Durand G, Barceló D. 1995. Trace determination of herbicides in estuarine waters by liquid chromatography–high-flow pneumatically assisted electrospray mass spectrometry. J Chromatogr A 712: 113–122.

Nélieu S, Stobiecki M, Kerhoas L, Einhorn J. 1994. Screening and characterization of atrazine metabolites by high-performance liquid chromatography–tandem mass spectrometry. Rapid Commun Mass Spectrom 8: 945–952.

Niessen WMA. 2010. Group-specific fragmentation of pesticides and related compounds in liquid chromatography–tandem mass spectrometry. J Chromatogr A 1217: 4061–4070.

Núñez O, Moyano E, Galceran MT. 2004. High mass accuracy in-source collision-induced dissociation tandem mass spectrometry and multi-step mass spectrometry as complementary tools for fragmentation studies of quaternary ammonium herbicides. J Mass Spectrom 39: 873–883.

Picó Y, la Farré M, Soler C, Barceló D. 2007a. Confirmation of fenthion metabolites in oranges by IT-MS and QqTOF-MS. Anal Chem 79: 9350–9363.

Picó Y, la Farré M, Soler C, Barceló D. 2007b. Identification of unknown pesticides in fruits using ultra-performance liquid chromatography–quadrupole–time-of-flight mass spectrometry. Imazalil as a case study of quantification. J Chromatogr A 1176: 123–134.

Polati S, Bottaro M, Frascarolo P, Gosetti F, Gianotti V, Gennaro MC. 2006. HPLC-UV and HPLC-MS*n* multiresidue determination of amidosulfuron, azimsulfuron, nicosulfuron, rimsulfuron, thifensulfuron methyl, tribenuron methyl and azoxystrobin in surface waters. Anal Chim Acta 579: 146–151.

Reiser RW, Fogiel AJ. 1994. Liquid chromatography–mass spectrometry analysis of small molecules using electrospray and fast-atom bombardment ionization. Rapid Commun Mass Spectrom 8: 252–257.

Rodríguez R, Mañes J, Picó Y. 2003. Off-line solid-phase microextraction and capillary electrophoresis mass spectrometry to determine acidic pesticides in fruits. Anal Chem 75: 452–459.

Sannino A, Bandini M. 2005. Determination of seven benzoylphenylurea insecticides in processed fruit and vegetables using high-performance liquid chromatography/tandem mass spectrometry. Rapid Commun Mass Spectrom 19: 2729–2733.

Santilio A, Stefanelli P, Girolimetti S, Dommarco R. 2011. Determination of acidic herbicides in cereals by QuEChERS extraction and LC/MS/MS. J Environ Sci Health B 46: 535–543.

Sharma D, Nagpal A, Pakade YB, Katnoria JK. 2010. Analytical methods for estimation of organophosphorus pesticide residues in fruits and vegetables: A review. Talanta 82: 1077–1089.

Slobodník J, Hogenboom AC, Vreuls JJ, Rontree JA, Van Baar BLM, Niessen WMA, Brinkman UATh. 1996. Trace-level determination of pesticide residues using on-line solid-phase extraction–column liquid chromatography with atmospheric pressure ionization mass spectrometric and tandem mass spectrometric detection. J Chromatogr A 741: 59–74.

Soler C, James KJ, Picó Y. 2007. Capabilities of different liquid chromatography tandem mass spectrometry systems in determining pesticide residues in food. Application to estimate their daily intake. J Chromatogr A 1157: 73–84.

Song X, Budde WL. 1996. Capillary electrophoresis–electrospray mass spectra of the herbicides paraquat and diquat. J Am Soc Mass Spectrom 7: 981–986.

Taguchi VY, Jenkins SWD, Crozier PW, Wang DT. 1998. Determination of diquat and paraquat in water by liquid chromatography–(electrospray ionization) mass spectrometry. J Am Soc Mass Spectrom 9: 830–839.

Thurman EM, Ferrer I, Barceló D. 2001. Choosing between atmospheric pressure chemical ionization and electrospray ionization interfaces for the HPLC/MS analysis of pesticides. Anal Chem 73: 5441–5449.

Thurman EM, Ferrer I, Fernández-Alba AR. 2005. Matching unknown empirical formulas to chemical structure using LC–MS TOF accurate mass and database searching: Example of unknown pesticides on tomato skins. J Chromatogr A 1067: 127–134.

Thurman EM, Ferrer I, Pozo ÓJ, Sancho JV, Hernández F. 2007. The even-electron rule in electrospray mass spectra of pesticides. Rapid Commun Mass Spectrom 21: 3855–3868.

Valenzuela AI, Picó Y, Font G. 2000. Liquid chromatography/atmospheric pressure chemical ionization-mass spectrometric analysis of benzoylurea insecticides in citrus fruits. Rapid Commun Mass Spectrom 14: 572–577.

Volmer D, Wilkes JG, Levsen K. 1995. Liquid chromatography–mass spectrometry multiresidue determination of sulfonylureas after on-line trace enrichment. Rapid Commun Mass Spectrom 9: 767–771.

Voyksner RD, Pack T. 1991. Investigation of collisional-activation decomposition process and spectra in the transport region of an electrospray single-quadrupole mass spectrometer, Rapid Commun Mass Spectrom 5: 263–268.

Voyksner RD, McFadden WH, Lammert SA. 1987. Application of thermospray HPLC–MS–MS for determination of triazine herbicides, in: J.D. Rosen (Ed.), Applications of new mass spectrometry techniques in pesticide chemistry, Wiley & Sons, New York, NY, pp. 247–258.

Wensing MW, Snyder AP, Harden CS. 1995. Energy resolved mass spectrometry of diethyl alkylphosphonates with an atmospheric pressure ionization tandem mass spectrometer, J Mass Spectrom 30: 1539–1545.

Yoshioka N, Asano M, Kuse A, Mitsuhashi T, Nagasaki Y, Ueno Y. 2011. Rapid determination of glyphosate, glufosinate, bialaphos, and their major metabolites in serum by liquid chromatography–tandem mass spectrometry using hydrophilic interaction chromatography. J Chromatogr A 1218: 3675–3680.

5

IDENTIFICATION STRATEGIES

5.1 INTRODUCTION

The identification of "unknown" compounds is an important challenge in many areas of analytical chemistry. Especially along with nuclear magnetic resonance spectroscopy (NMR), mass spectrometry (MS) plays a major role in this. Other spectrometric techniques that may be involved are UV–VIS and fluorescence spectroscopy, infrared and Raman spectroscopy, and X-ray diffraction. This chapter focuses on the role of MS in the identification of "unknowns". Depending on the application area, different strategies are applied for compound identification. Some of these strategies based on the use of combined liquid chromatography–mass spectrometry (LC–MS) and tandem mass spectrometry (MS–MS or MS^n; Section 1.4) are briefly discussed here.

The term "identification of unknowns" is generally a rather inaccurate term to describe the various endeavors. In a recent paper, three classes of compounds can be

Interpretation of MS–MS Mass Spectra of Drugs and Pesticides, First Edition. Wilfried M. A. Niessen and Ricardo A. Correa C.

distinguished in this context: "known knowns", "known unknowns", and "unknown unknowns" (Little et al., 2011). An identification effort directed at "known knowns" is actually best described as confirmation of identity rather than identification. For instance, a large part of the analysis in environmental and food safety application areas is directed at searching for the presence of target compounds and, when found, confirming their identity (Section 5.3). A "known unknown" is a compound that is unknown to the analytical chemist at the start of working on a particular sample, but in fact, the compound is well described in the literature and compound databases. An "unknown unknown" or "real unknown" is a compound that has not been described in literature and will not be found in compound databases (Little et al., 2011). Related substances, such as reaction by-products and drug metabolites, can generally be considered as compounds in between the "known unknowns" and the "real unknowns" (Section 5.7). Identification of "real unknowns" is the most difficult task. In principle, confirmation of identity presumes knowledge on the potential identity of the compound, for example, because it shows the same characteristics as a reference compound, but does not exclude the possibility of the presence of another compound showing the same or similar characteristics. Compound identification does not involve any *a priori* assumption about the structure of the compound, and thus requires that all other compounds be excluded (De Zeeuw, 2004; Berendsen et al., 2013). As is demonstrated in this chapter, the various classes of "unknowns" require different strategies, which may also differ between different application areas.

In this context, it should also be emphasized that in principle the structure elucidation should in many instances also address spatial descriptors related to conformation and configuration. For many biological processes, the stereoisomerism of a compound is as important a descriptor as the structure itself. In general, MS has little to offer when it comes to stereoisomerism and differentiation between enantiomers, although some interesting results have been reported. With respect to stereoisomerism, NMR is by far a more powerful tool. However, one should keep in mind that there is a factor of 100–1000 sensitivity difference between MS and NMR; the latter requiring more (pure) sample compound than MS to get an interpretable spectrum. MS analysis is readily performed within the timescale of ultra-high-performance liquid chromatography (UHPLC), whereas NMR requires longer data acquisition times (hours rather than seconds). A new dimension to MS analysis in the discrimination of stereoisomers may come from the recent introduction of hyphenated ion-mobility spectrometry–MS (IMS–MS) systems (Lapthorn et al., 2013). However, the currently available resolution of IMS systems is often not sufficient yet (Section 5.7.1).

5.2 CONFIRMATION OF IDENTITY IN FOLLOWING ORGANIC SYNTHESIS

During a multistep synthesis in an organic chemistry laboratory, it can be of interest to check the identity of the intermediate compounds as well as the end product(s) to ascertain the proper progress of the chemical synthesis. To this end, MS analysis is routinely used. Often, molecular mass determination with a soft ionization method leads to sufficient answers to the synthetic chemists. Open-access walk-up facilities based on column-bypass injection or LC–MS with a single-quadrupole (SQ) instruments have been established in organic synthesis laboratories to rapidly confirm the proper progress of the synthesis via the molecular mass determination of the intermediates and product(s) (Mallis et al., 2002). In this way, an LC–MS system is converted into a walk-up "black box" facility for synthetic chemists: Log-in to the system may be done on a remote computer, for example, via the network. One enters the sample identification code and from a menu selects the type of LC–MS experiments to be performed. As a feedback, the position(s) in the autosampler rack to be used for the sample are indicated. The sample is run automatically, for example, by column-bypass injection or a fast wide-range generic gradient LC–MS method, often in both positive-ion and negative-ion modes, and at both without and with in-source collision-induced dissociation (CID) (Section 1.4.1.1). The resulting spectra are placed onto the laboratory information management system (LIMS) network or sent to the chemist by electronic means (Tong et al., 1999).

This open-access walk-up approach was pioneered using a column-bypass thermospray MS system for the automated analysis of potential agricultural chemicals (Hayward et al., 1993). In this way, the required information was obtained in approximately 70% of the MS structural confirmations performed, thus significantly reducing the workload of specialized mass spectrometrists. A similar thermospray MS-based open-access system was also used elsewhere (Brown et al., 1994; Pullen et al., 1995). Subsequently, open-access LC–MS systems based on electrospray ionization (ESI) were introduced (Mallis et al., 2002). Dedicated open-access software modules are currently offered by most LC–MS instrument manufacturers to be used in combination with mainly SQ or orthogonal-acceleration time-of-flight (TOF) instruments. Open-access LC–MS systems are widely applied within medicinal chemistry units in the pharmaceutical industry and elsewhere by synthetic organic chemists.

The confirmation of identity of the synthetic intermediates and product(s) is thus performed based on the measurement of the *m/z* value of the protonated molecule $[M+H]^+$ in positive-ion mode only or of the *m/z* values of both $[M+H]^+$ and the deprotonated molecule $[M-H]^-$ by rapidly switching between positive-ion and negative-ion modes. This may be considered as a very rough and rather inaccurate assessment, as no differentiation is done between possible isomers. Accidental isobaric reaction products are not distinguished either, unless separation is achieved in an implemented LC step. In this respect, an accurate-mass determination by TOF-MS or other high-resolution accurate-mass (HRAM) MS approaches provides an improved confirmation of identity, but at higher costs. The measured accurate mass allows confirmation of the elemental composition. For an adequate data interpretation, the chemist should also be aware of the wide variety of adduct ions that may be generated in the column-bypass MS or LC–MS analysis (Section 2.8). Useful listings of frequently occurring adduct ions as well as artifact peaks have been reported (Tong et al., 1999).

Similar approaches have been applied for the analysis of combinatorial libraries. In addition to assessing biological activity via 96-well plate bioassays, confirmation of identity and purity assessment of such compound libraries have been performed using LC–MS approaches. Software modules for open-access operation were adapted to enable unattended data acquisition and automated data processing for large series of samples, for example, 60 samples per hour in column-bypass and up to 15 samples per hour in fast LC–MS mode using a generic solvent gradient. Sample lists can then be imported in tabular form. Dedicated data browsers have been developed to make rapid decisions on the presence and/or purity of the compounds based on color-coded representation of the 96-well plates (Hegy et al., 1996). A more recent system for purity assessment of drug discovery libraries was described, which also involves chemiluminescence nitrogen detection for quantitative purity determination (Molina-Martin et al., 2005).

5.3 CONFIRMATION OF IDENTITY IN TARGETED SCREENING BY SRM-BASED STRATEGIES

In most application areas, where confirmation of identity and/or identification is at stake, MS–MS or MS^n approaches are applied instead of SQ-MS approaches. Currently, the most widely used approach for confirmation of identity is the use of selected reaction monitoring (SRM, see Section 1.5.2) in a tandem-quadrupole (TQ) instrument. In SRM in LC–MS applications, the protonated or deprotonated molecule is selected as precursor ion in the first stage of MS. Product ions are generated by CID in the collision cell. From all fragment ions generated by CID, a preferably structure-specific product ion is selected and measured in the second stage of MS analysis. The combination of precursor ion *m/z*, product ion *m/z*, and all MS parameters to acquire their response with optimum sensitivity (and eventually the retention time of the target compound) is called an "SRM transition". Automated software procedures have been developed to search and optimize the product ions and conditions for SRM transitions for a target compound. The SRM approach is applied to achieve a (significant) gain in selectivity relative to an SQ-MS approach, in order to reach lower limits of quantification, especially in the analysis of target compounds in complex biological samples.

LC–MS on a TQ instrument operated in SRM mode is the gold standard in routine quantitative bioanalysis, as for instance performed in pharmacokinetics/pharmacodynamics (PK/PD) and absorption, distribution, metabolism, excretion (ADME) studies during drug discovery and drug development within the pharmaceutical industry and related contract-research organizations (Hopfgartner & Bourgogne, 2003; Xu et al., 2007; Jemal et al., 2010). In general, only one SRM transition per target compound is applied in these applications; a limited number of target compounds is analyzed. If a product ion with sufficient structure specificity can be selected, the use of SRM provides enhanced confirmation of identity. Therefore, one generally avoids the selection of product ions resulting from relatively common neutral losses, that is, product ions formed by the loss of 17 (ammonia, NH_3), 18 (water, H_2O), 28 (carbon monoxide, CO, or ethene, C_2H_4), 32 (methanol, CH_3OH), 35 (NH_3 and H_2O), 44 (carbon dioxide, CO_2), or 46 Da (formic acid, HCOOH), as they are generally considered not to be very structurally significant.

LC–MS (and also GC–MS) is frequently applied in residue analysis. Residue analysis is aimed at screening for residues of particular target compounds in samples in a wide variety of application areas, including environmental analysis, food safety analysis, and sports doping analysis. The analysis often aims at establishing the presence of a target compound, determining its quantity, and confirming its identity. Regulatory bodies, such as the European Commission, the Environmental Protection Agency (EPA), the Food and Drug Administration (FDA), and the World Anti-Doping Agency (WADA), have established guidelines for the analysis and especially for the way the confirmation of identity is to be performed (Stolker et al. 2000; Bogialli & Di Corcia, 2009; European Commission, 2002a,b, 2003, 2008, 2009, 2013). The current protocols that are applied on residue analysis within the European Union are mostly based on a protocol initially developed for the screening of veterinary drugs in food of animal origin (European Commission, 2002a). In brief, the protocol indicates that confirmation of identity of a particular compound can be achieved in LC–MS if a number of criteria is fulfilled: (1) the compound should elute within 2% of the retention time of the analytical standard, (2) a number of identification points should be scored based on the detection of diagnostic ions, and (3) the ion ratio of two of these diagnostic ions should be similar to that of the analytical standard within a specified tolerance criteria. A diagnostic ion can be any compound-specific ion in the (product-ion) mass spectrum with a relative abundance above 10% of a reference standard of the compound of interest, that is, the molecular ion or the (de)protonated molecule, a characteristic adduct ion, a fragment ion in MS or in MS–MS, or even an isotope peak. A diagnostic ion in a low-resolution MS spectrum delivers one identification point, whereas a diagnostic product ion in MS–MS on a TQ instrument delivers one-and-a-half identification points. For confirmation of identity of a compound with a maximum residue limit, three identification points are needed, whereas for banned substances four identification points are needed (European Commission, 2002a). The most common practice in residue analysis based on LC–MS is to use two SRM transitions, that is, a combination of a precursor ion and two product ions, to provide a quantifier transition and a qualifier transition, which can also be used for testing whether or not the ion ratio is within the tolerance criteria (European Commission, 2002a). Strengths and weaknesses of such an approach have been critically assessed (Sauvage et al., 2008; Berendsen et al., 2013). An important drawback of an SRM-based approach in many applications lies in the fact that the analysis is performed using a targeted approach and is thus limited to those analytes for which SRM transitions have been defined. In principle, any contaminant or unwanted compound present should be detected, quantified, and confirmed, whereas only a limited set of target compounds is searched for. In addition, it has been demonstrated that in the analysis of complex samples, the selectivity of many SRM transitions seems to be inadequate, especially at low target levels, as co-eluting isomeric or isobaric interferences may distort the ion-ratio determination and thereby compromise the confirmation of identity. The use of a higher number of SRM transitions has been proposed to deal with such issues (Sauvage et al., 2008; Berendsen et al., 2013).

If a large number of target compounds has to be screened for, the use of multiple SRM transitions per compound may compromise sensitivity of the analysis, unless special actions are taken. For reliable quantitative analysis, 10–20 data points per chromatographic peak should be measured for an accurate description of the peak and to achieve accurate peak-area determination. In multiresidue analysis, each data point requires multiple SRM transitions to be measured, that is, at least two SRM transitions per target compound. Given the narrow peaks in current UHPLC, a high number of SRM transitions would result in unfavorably short dwell times (<10 ms per transition). Time-scheduled SRM and dynamic SRM have been developed to cope with these problems by reducing the number of SRM transitions that has to be measured for each data point.

In time-scheduled SRM, the chromatographic analysis time is subdivided into a number of time windows. For each time window, a specific function is defined in which only the SRM transitions are monitored for the compounds that elute in that particular time window. In principle, the dwell time can be adapted for individual SRM transitions. One needs to strike a compromise between the dwell time and the number of data points. During analysis, the software steps from function to function based on the time point of the analysis. In this way, the number of SRM transitions per data point can be greatly reduced.

In dynamic SRM (also called scheduled SRM), for each individual analyte, a (narrow) time window for measuring its particular SRM transitions is defined, for example, 30–60 s, based on the known retention time of that compound. If sufficient chromatographic separation is achieved, the dynamic SRM approach further reduces the number of SRM transitions that has to be measured within one data point. As the dwell time is generally kept constant for all transitions in dynamic SRM, the number of available data points will change depending on the crowdedness of the chromatogram.

Some typical examples of confirmation of identity by SRM approaches related to various application areas are discussed in the following.

5.3.1 Environmental Analysis

In environmental analysis, residue analysis is directed at screening for a large number of target compounds such as pesticides and pharmaceuticals and their degradation products in aquatic environments, soil, and sediments. In principle, any potential hazardous and/or toxic compound may be a potential target compound. This implies that, apart from pesticides and pharmaceuticals, the analysis may be directed at targets from other compound classes such as surfactants, endocrine-disrupting compounds, and illicit drugs (Petrovic et al., 2010; Richardson & Ternes, 2011). Extensive targeted multiresidue screening methods based on LC–MS have been developed for the quantification and confirmation of identity of these classes of compounds in the environment (Petrović et al., 2005; Kim & Carlson, 2005; Gros et al., 2006; Kuster et al., 2009).

The analysis of 26 pesticides and transformation products in sediments may serve as an example (Köck-Schulmeyer et al., 2013). Pressurized liquid extraction combined with solid-phase extraction (SPE) for sample cleanup was applied as sample pretreatment. Twenty-three stable-isotope-labeled internal standards (SIL-IS) were used. Two SRM transitions for each target analyte and one SRM transition for each SIL-IS were acquired, while switching between positive-ion and negative-ion modes during the run. The limit of determination, defined as the concentration at which both quantification and confirmation of identity according to SANCO standards is still possible, ranged from 0.13 ng g^{-1} for terbuthylazine to 76.6 ng g^{-1} for MCPA. The publication also includes a short (tabular) overview of methods described in the last 10 years for pesticide analysis in sediments (Köck-Schulmeyer et al., 2013).

LC–MS analysis of inlet water of sewage treatment plants has been frequently applied for the screening of contaminants, including pharmaceuticals (Gros et al., 2006; Grabic et al., 2012) and illicit drugs (Zuccato et al., 2008; van Nuijs et al., 2011). With respect to the monitoring of illicit drugs, it has been demonstrated that in this way a reliable estimate can be achieved regarding the use of illicit drugs in a given population (Thomas et al., 2012).

5.3.2 Food Safety Analysis

Two major areas can be distinguished in food safety analysis: (1) food of animal origin, where attention is focused on the screening of veterinary drugs, especially antibiotics, and (anabolic) steroids (De Brabander et al., 2007), and (2) fruits and vegetables, where attention is paid not only to a wide variety of pesticides and fungicides (Botero-Coy et al., 2012; Núñez et al., 2012) but also to other compound classes such as mycotoxins (Malachová et al., 2014; Wang et al., 2014).

Multiresidue analysis of pesticides in fruits and vegetables requires easy and generic sample treatment approaches, which is difficult to imagine, given the high number of commodities that can be the target matrix in the analysis. In the past few years, the quick, easy, cheap, effective, rugged, and safe (QuEChERS) sample extraction and cleanup method (Anastassiades et al., 2003) has become very popular (CEN/TC, 2007). A recent example of the use of QuEChERS sample pretreatment is the multiresidue analysis of 109 pesticides in tomatoes (Golge & Kabak, 2015). Separate LC–MS runs have been established for the 102 pesticides analyzed in positive-ion mode and the 7 pesticides analyzed in negative-ion mode. Dynamic (scheduled) SRM was applied with two SRM transitions per compound (a quantifier and a qualifier transition). Validation of the method was performed in compliance with SANCO guidelines (European Commission, 2002a,b, 2003, 2008, 2009, 2013). Compound-dependent limits of quantification (LOQs) were obtained ranging from 1.3 to 30.4 μg kg^{-1}, which is comparable to the results of a GC–MS–MS method for the analysis of 186 pesticides in tomatoes (Zhao et al., 2014).

The optimization and validation of a multiresidue method for the analysis of 127 veterinary drugs in bovine muscle using UHPLC and MS–MS on a TQ instrument may serve as a typical example for veterinary drug residue analysis. The target compounds were divided into two sets (61 and 66 compounds per set), which were then analyzed by two different methods. Three SRM transitions per compound (one quantifier and two qualifiers) were used. Eight of the 127 compounds showed better performance in negative-ion mode; all others were analysis in positive-ion mode. Fast polarity switching was applied. Several sorbents were tested in dispersive SPE for cleanup of the extracts. The method turned out to be suitable for the screening of 113 compounds, out of which 87 could be quantified (with recoveries between 70% and 120% and RSD < 25%) at or below the regulatory tolerance levels (as implemented in the United States) in bovine muscle (Geis-Asteggiante et al., 2012).

5.3.3 Sports Doping Analysis

LC–MS–MS is, in addition to GC–MS and immunological detection methods, essential in modern sports doping analysis to screen, identify, and quantify prohibited compounds (Thevis et al., 2011; Pozo et al., 2012). Both targeted and untargeted methods are used. Traditionally, the fast and sensitive targeted methods directed at a preselected set of analytes are based on SRM on TQ instruments. The target compounds are measured with utmost specificity and sensitivity using diagnostic precursor–product ion pairs. The simultaneous analysis of 72 target compounds from different classes, that is, anabolic steroids, anti-estrogens, β-adrenergic drugs, diuretics, glucocorticoids, and stimulants, in human urine may serve as an example (Mazzarino et al., 2008). Time-scheduled SRM was used to reduce the number of simultaneously acquired SRM transitions. Two SRM transitions per compound were acquired; fast positive–negative polarity switching was applied. LODs in the range of 1–50 ng mL^{-1} for the steroids, and 50–500 ng mL^{-1} for other compound classes were obtained.

5.3.4 General Unknown Screening in Toxicology

In systematic toxicological analysis (STA) or general unknown screening (GUS), analytical strategies based on data-dependent acquisition (DDA, Section 1.5.4) are usually applied, thus switching between a full-spectrum MS survey mode and dependent product-ion MS–MS mode (Section 5.5). A different approach to STA is the targeted analysis of a wide variety of drugs using SRM on a TQ instrument. The procedure for the screening of 238 drugs in blood using SRM is a good example (Gergov et al., 2003). One SRM transition per compound was monitored using a compound-dependent collision energy (20, 35, or 50 eV). With 238 drugs and a dwell time of 25 ms, this adds up to a total cycle time of 6 s. Similar procedures have been reported by others (Thieme & Sachs, 2003; Di Rago et al., 2014; Remane et al., 2014). A critical review of LC–MS strategies based on SRM for drug analysis and screening indicated that the use of only one SRM transition per compound is generally insufficient. Even with two SRM transitions per compound, a significant number of false-positives may be obtained, especially if no proper attention is paid to the selection of truly selective, that is, highly structure-specific SRM transitions (Sauvage et al., 2008; Berendsen et al., 2013).

5.4 CONFIRMATION OF IDENTITY BY HIGH-RESOLUTION ACCURATE-MASS MS STRATEGIES

With the recent advent of HRAM-MS instruments, for example, TOF-MS and orbitrap MS instruments, the potential of HRAM-MS for quantitative bioanalysis (Ramanathan et al., 2011) and residue analysis (García-Reyes et al., 2007; Peters et al., 2009; Kaufmann et al., 2010; Kaufmann et al., 2011) has been investigated. Relatively recent advances in TOF-MS instrumentation have greatly removed initial limitations in quantitative analysis in terms of LOQ and linear dynamic range. It has been argued that for HRAM-MS instrument to be applicable in quantitative bioanalysis and residue analysis, it should provide a resolution exceeding 30,000, a mass accuracy better than 5 ppm, a spectrum acquisition rate in excess of 5 spectra s^{-1}, and a linear dynamic range of at least three orders of magnitude (Ramanathan et al., 2011). These instrument specifications are readily offered by current quadrupole–time-of-flight (Q–TOF) hybrid and orbitrap instruments (Van Dongen & Niessen, 2012). The major advantage of using HRAM-MS over SRM is that the analysis becomes untargeted and does not need prior optimization of (multiple) SRM transitions. This greatly speeds up the sample throughput in a bioanalytical laboratory supporting drug discovery, where one may face the task of performing high-throughput metabolic stability analysis for over 100 new chemical entities (NCEs) per week. Moreover, as the analysis is untargeted, the data acquired contain information not only on the metabolic stability of the NCE analyzed but also on the metabolites formed. As both quantitative and qualitative information is acquired, the HRAM-MS approach is sometimes indicated as a "qual/quant" strategy. The "qual/quant" strategy has already been proposed based on the application of various other MS platforms, for example, by monitoring SRM transitions of anticipated metabolites on TQ instruments (Poon et al., 1999), by application of SRM-triggered data-dependent full-spectrum MS–MS acquisition on quadrupole–linear ion-trap (Q–LIT) hybrid instruments (Li et al., 2005), and by the use of HRAM-MS systems such as a TOF-MS instrument (Zhang et al., 2000) or an orbitrap instrument (Bateman et al., 2009). Furthermore, the HRAM-MS strategy for LC–MS in quantitative bioanalysis comes with the additional advantage that other compounds of interest can be monitored either at the time of analysis or in post-acquisition data interrogation. Rapid changeover from TQ to HRAM-MS instrumentation is hindered by the financial investments required in terms of both instrument purchase and development and validation of new analytical methods. In addition, the software support for routine quantitative bioanalysis is not always suitable (Fung et al., 2013).

5.4.1 Environmental and Food Safety Analysis

Similarly, the potential of a HRAM-MS strategy has been investigated for use in residue analysis in environmental and food safety applications (García-Reyes et al., 2007; Peters et al., 2009; Kaufmann et al., 2010; Kaufmann et al., 2011). According to protocols from the regulatory bodies, adapted rules for identification points apply in HRAM-MS, delivering a higher number of identification points per ion (Ibáñez et al., 2005; Nielen et al., 2007). With HRAM-MS, the confirmation of identity of a target compound can be based on a combination of its chromatographic retention time and its accurate mass. The accurate mass can be used to determine/confirm the elemental composition of the target compound (Section 2.7). If the residue analysis is performed using an instrument that also allows MS–MS or MS^n analysis, for example, a Q–TOF hybrid instrument or an ion-trap–orbitrap or quadrupole–orbitrap hybrid instrument, further confirmation of identity can be achieved from the accurate masses of the product ions. This would require the use of DDA or data-independent acquisition (DIA; Section 1.5.5) approaches, or repeated sample analysis with targeted MS–MS or MS^n. Again, such an HRAM-MS approach provides essentially untargeted analysis, which means that unexpected and unanticipated residues may be observed as well. In fact, already acquired data can be subjected to post-acquisition data interrogation for the presence of particular compounds. An example of such post-acquisition data mining and retrospective analysis is the screening for newly identified metabolites and aquatic degradation products of omeprazole. Existing data files, previously acquired in another study on the analysis of various water compartments, were processed to screen for these metabolites and degradation products (Boix et al., 2014). Numerous examples of multiresidue analysis for food safety are available, for example, veterinary drugs in milk (Stolker et al., 2008; Ortelli et al., 2009), in meat (Kaufmann et al., 2008), or in distillers grain (Kaklamanos et al., 2013); anabolic steroids in meat (Vanhaecke et al., 2013); pesticides, veterinary drugs, and mycotoxins in bakery ingredients and food commodities (De Dominicis et al., 2012); pesticides in fruits and vegetables (Mol et al., 2012); and pesticides and veterinary drugs in baby food (Gómez-Pérez et al., 2015a). Similarly, HRAM-MS strategies have been applied in environmental analysis, for example, for the analysis of illicit drugs in urban wastewater (Bijlsma et al., 2012; Bijlsma et al., 2014; Hernández et al., 2015) and of pesticides and pharmaceuticals in environmentally relevant water compartments (Jindal et al., 2015; Masiá et al., 2014).

Comparison of traditional TQ-based and HRAM-MS-based multiresidue analysis has been reported as well

(Kaufmann et al., 2010; Kaufmann et al., 2011; Rocha et al., 2015; Kaufmann et al., 2015). Apart from the already mentioned advantages of HRAM-MS over TQ-MS for residue analysis, a reduced number of false-negative findings was reported, using criteria in which the accurate *m/z*-values of the precursor ion and just one product ion were used (Kaufmann et al., 2015). There is significant discussion and development of criteria for confirmation of identity as well as progress and improvements in sensitivity and dynamic range of the HRAM instruments. In a performance comparison between TOF and orbitrap instruments, the LOQs for most compounds were equal or somewhat better with the orbitrap than with the TOF instrument (Gómez-Pérez et al., 2015b).

5.4.2 General Unknown Screening in Toxicology

The HRAM-MS strategy, mostly based on a TOF-MS instrument enabling accurate-mass determination, has also been applied in STA (see also Section 5.5). The identification of toxic compounds in biological samples is then based on searching the molecular formula calculated from the accurate mass against a database of relevant compounds. LC retention time data may be added as an additional identification criterion to the database. Structures and thus molecular formulae of possible metabolites may be taken from the literature and also added to the database (Gergov et al., 2001; Pelander et al., 2003). This approach actually allows screening for drugs and related compounds without the availability of primary reference standards, as is often the case with drug metabolites and street drugs (Pelander et al., 2003; Laks et al., 2004). In the case of street drugs, quantification was performed using LC coupled to a chemiluminescence nitrogen detector (CLND) with caffeine as a single secondary standard. The CLND provides an equimolar response to nitrogen (Laks et al., 2004). Identification based on the elemental composition of the detected compound, derived from the measured accurate mass, was performed for STA based on combined capillary electrophoresis and ESI-HRAM-MS. A compound database with 50,500 chemical formulae was built for this purpose (Polettini et al., 2008). Similarly, UHPLC–TOF-MS for STA was applied to an alleged case of sexual assault (Birkler et al., 2012). The potential of HRAM-MS in STA has been reviewed (Ojanperä et al., 2012).

5.4.3 Sports Doping Analysis

Essentially similar strategies have been applied in sports doping analysis (Kolmonen et al., 2007). An adaptation of the method for sports doping analysis comprises a generic sample pretreatment based on mixed-mode SPE on two types of sorbents and the use of both positive-ion and negative-ion modes in LC–TOF-MS detection (Kolmonen et al., 2009). Modifications to the method extended its applicability from the analysis of 104 to 197 compounds in two runs of 8 min each (separate sample analysis in positive-ion and negative-ion modes), which is in agreement with the acceptance criteria of the WADA. A comparison has been made of the performance of TQ, TOF, and Q–TOF-based strategies for the analysis of anabolic steroids (Pozo et al., 2011). At the time, the TOF and Q–TOF approaches did not provide the sensitivity needed for some of the target analytes to reach the minimum required performance limit set by WADA (between 2 and 10 ng mL^{-1} in urine). On the other hand, untargeted screening of urine samples resulted in the detection of some metabolites revealing the use of prohibited steroids (Pozo et al., 2011). Progress in this field between 2007 and 2010 was reviewed, outlining the advantages as well as the limitations (Thevis et al., 2011).

5.5 LIBRARY SEARCHING STRATEGIES IN SYSTEMATIC TOXICOLOGICAL ANALYSIS

Given the power of computer-based mass spectral library search for the confirmation of identity of "known knowns" and identification of "known unknowns" in GC–MS using electron ionization (EI), there has been considerable interest in investigating the possibilities of building searchable mass spectral libraries for LC–MS. Obviously, other libraries are needed because the GC–MS libraries only contain mass spectra based on the fragmentation of molecular ions ($OE^{+\bullet}$), whereas in LC–MS or LC–MS–MS a library must contain spectra based on protonated molecules (EE^{+}) and/or deprotonated molecules (EE^{-}). The fragmentation of $OE^{+\bullet}$ and EE^{+} is fundamentally different (Section 3.4). One should be aware that general mass spectral libraries for GC–MS may contain mass spectra of up to 700,000 compounds (Wiley, NIST), whereas typical LC–MS libraries contain spectra for only a few thousand compounds (Oberacher, 2011; NIST; Weinmann & Dresen, 2007). For GC–MS, there are also (commercially available) library collections for particular application areas, that is, smaller libraries more dedicated to particular compound classes, for example, toxicology, steroids, flavors and fragrances.

With respect to LC–MS applications, library searching has been considered to be especially important in clinical and forensic toxicology. Clinical toxicology is among other things concerned with the diagnosis or the positive exclusion of an acute or chronic intoxication. In addition, the monitoring of addicted illegal drug users has to be performed. Forensic toxicology is mainly concerned with proof of an abuse of illegal drugs or of a murder or suicide by poisoning. Similar analytical procedures are relevant in sports doping analysis. The biological samples analyzed include especially urine, whole blood, plasma, serum, and other bodily fluids and also liver biopsy (postmortem samples) and hair.

The target analytes are typically up to a thousand toxicologically relevant compounds, including drugs of a wide variety of therapeutic classes, illicit drugs, pesticides and related compounds, antimicrobial and antibiotic compounds, as well as the metabolites of all those compound classes (Chapter 4 provides an overview of relevant target compounds). For volatile compounds, GC–MS is the gold standard for identification and quantification (Maurer, 1992; Polettini, 1999; Valli et al., 2001). For compounds not amenable to GC, identification methods based on the use of LC–UV photodiode array (LC–PDA) spectrometry have been developed (Drummer et al., 1993; Tracqui et al., 1995; Lo et al., 1997; Sadeg et al., 1997; Valli et al., 2001). With the development of reliable and robust instrumentation for LC–MS, the potential of LC–MS was evaluated for toxicological and forensic applications and for STA. This topic has been extensively reviewed (Maurer, 1998; Marquet & Lachâtre, 1999; Van Bocxlaer et al., 2000; Bogusz, 2000; Marquet, 2002; Maurer, 2007). A variety of strategies has been proposed and developed for STA using LC–MS.

Initially, methods were developed targeted at the analysis of specific classes of drugs, for example, illicit drugs, benzodiazepines, antihypertensives, neuroleptics, pesticides, using SQ or TQ instruments. Dedicated mass spectral libraries were built for that purpose. An overview of such targeted methods was given in several review papers (Maurer, 1998; Maurer, 2005; Van Bocxlaer et al., 2000; Bogusz, 2000). Examples of this approach involving an SQ instrument are targeted at, for instance, neuroleptic drugs (Kratzsch et al., 2003) or benzodiazepines (Kratzsch et al., 2004). Numerous examples are available on the use of TQ instruments for the targeted toxicological analysis of specific classes of drugs, for example, β-blockers (Umezawa et al., 2008), benzodiazepines and other drugs of abuse (Villain et al., 2005; Badawi et al., 2009), and antidepressants (de Castro et al., 2008).

Subsequently, two research groups independently developed an LC–MS approach for STA based on the use of laboratory-built mass spectral libraries (Marquet & Lachâtre, 1999; Marquet et al., 2000; Weinmann et al., 1999). The approach is based on the application of in-source CID in the atmospheric-pressure–vacuum interface of an ESI source on an SQ instrument. The libraries built contained mass spectra of ≈400 or ≈1000 drugs and other relevant compounds acquired at two or three alternating ion-sampling orifice voltages (low and high; or low, medium, and high). Mass spectra acquired under identical conditions in the analysis of patient samples were searched against the library to provide drug identification. This approach was subsequently adopted by others, also using other types of SQ instruments (Hough et al., 2000; Schreiber et al., 2000; Lips et al., 2001; Rittner et al., 2001). In order to assure the exchangeability of libraries between instruments from different manufacturers, the use of tune compounds was proposed (Weinmann et al., 2001a,b; Bristow et al., 2002). Reproducibility problems with in-source CID spectra were recognized (Bogusz et al., 1999).

Instead of using in-source CID on an SQ instrument, mass spectral libraries were built based on CID in MS–MS. In MS–MS, there are two different means to fragment the precursor ion, for example, $[M+H]^+$, that is, by collision-cell CID as is performed in TQ and Q–TOF instruments, and by ion-trap CID as is performed in three-dimensional and linear-ion-trap instruments (Section 1.4.1.1). Mass spectra obtained by the two different means share common features, but only partially, that is, the extent of fragmentation and the actual fragment ions observed may be different. In addition, the collision-energy regimes in collision-cell CID in instruments from different manufacturers may also result in distinct differences in the relative abundance and to a lesser extent in the type of fragment ions observed. Building mass spectral libraries and their application in STA have been described for both collision-cell CID in TQ instruments (Weinmann et al., 2000a,b; Gergov et al., 2000) and ion-trap CID in ion-trap instruments (Baumann et al., 2000; Liu et al., 2009). Although the CID conditions in an ion-trap instrument are easier to reproduce and with the use of normalized collision energies also easier to optimize, ion-trap instruments have not found extensive application in STA. On the other hand, STA based on LC–MS–MS on a TQ instrument has found a wide range of applications.

The library search algorithms used in both GC–MS with EI and LC–MS and LC–MS–MS have usually been the same. Advanced library searching software tools have been developed to find the best library hits (Atwater et al., 1985; Stein & Scott, 1994; Ausloos et al., 1999; Koo et al., 2013). They were initially built for the relatively reproducible mass spectra from EI, to be used with GC–MS, where both the peak position of the ions (*m/z*) and their relative abundance are taken into account. Although many researchers take the result of the library search for granted, a thorough and critical evaluation of the agreement between the experimental and the library spectrum is recommended. To successfully apply these algorithms in LC–MS, instrument manufacturer-specific mass spectral libraries should be developed, which can be considered a serious limitation for the widespread use of computer-based library searching in LC–MS–MS. The building of mass spectral MS–MS libraries and the exchangeability of these libraries between instruments from the same or from different manufacturers continues to be a topic of interest (Bristow et al., 2004; Josephs & Sanders, 2004; Milman, 2005; Jansen et al., 2005, Oberacher et al., 2009a,b; Oberacher et al., 2012). Recently, alternative library search algorithms have been proposed for specific use in LC–MS applications (Mylonas et al., 2009; Oberacher et al., 2013a,b).

Advances in instrumentation, and especially the introduction of Q–LIT hybrid (Section 1.4.4) instrument, further extended the possibilities of LC–MS in STA. In a Q–LIT instrument, the second mass analyzer can be used as either a conventional quadrupole mass analyzer or an LIT. By accumulation of ions, an LIT provides enhanced full-spectrum sensitivity compared to a conventional quadrupole (Hager, 2002). This instrument enables DDA, which is the automatic, data-dependent switching between a survey analysis mode and the (enhanced) full-spectrum MS–MS mode (Section 1.5.4). The full-spectrum MS–MS data may be searched against mass spectral libraries to assist in the identification of drugs and toxins in STA.

In the pioneering research by the group of Marquet, enhanced full-spectrum MS data were used as the survey analysis mode (Marquet et al., 2003). The full-spectrum analysis (and subsequent MS–MS analysis) was performed by alternating between positive-ion and negative-ion modes; separate libraries were generated for the positive-ion and negative-ion modes as well (Sauvage et al., 2006; Sauvage et al., 2009). Urine or blood-related samples were pretreated using SPE. The extracts were analyzed using a wide-range solvent gradient in reversed-phase LC (RPLC). Untargeted DDA was performed using three different collision energies; the cycle time was 1.36 s. The acquired enhanced product-ion (EPI) mass spectra were searched against a mass spectral library containing ≈1000 compounds analyzed in positive-ion and ≈300 compounds analyzed in negative-ion mode (Sauvage et al., 2006). Further developments involved the use of MS^2 and MS^3 libraries, as demonstrated for the analysis of pesticides in blood (Dulaurent et al., 2010).

An alternative procedure, also involving the use of a Q–LIT instrument, was proposed by the group of Weinmann, where targeted SRM with up to 1250 transitions was used in survey analysis mode, and positive-ion EPI spectra were searched against a mass spectral library (Mueller et al., 2005; Dresen et al., 2009; Dresen et al., 2010). Urine samples (at pH 9), either tenfold diluted or liquid–liquid extracted with butyl chloride, were separated on a pentafluorophenyl (PFP) column using a wide-range solvent gradient. Targeted scheduled-SRM-triggered DDA (involving ≈700 and later on even ≈1250 SRM transitions) was applied, in which EPI mass spectra were acquired at three different collision energies; the cycle time was up to 2.5 s. The spectra were searched against a mass spectral library containing ≈1250 compounds analyzed in positive-ion mode (Dresen et al., 2009; Dresen et al., 2010). A more recent version of this library also contains negative-ion EPI spectra (Weinmann & Dresen, 2007). A searchable version of this library is commercially available. This SRM-triggered DDA procedure seems to be a versatile and sensitive approach to STA, though one may argue that the inherent targeted nature of the use of SRM transitions in the survey mode excludes a direct answer to the more general clinical question of whether or not an individual had been intoxicated at all rather than intoxicated by a compound from a predefined list (Sauvage & Marquet, 2010). The use of the positive-ion mode only also narrows the detection window of the SRM-triggered DDA compared to the alternating positive-ion and negative-ion full-spectrum-triggered DDA.

As discussed earlier (Section 5.4), STA has been performed using HRAM-MS instruments in combination with a database of molecular formulae of relevant compounds (Gergov et al., 2001; Ojanperä et al., 2006; Polettini et al., 2008). Recently, STA based on a combination of accurate-mass determination on a TOF-MS system and searching of a mass spectral library generated by in-source CID was described (Lee et al., 2009). Only limited results have been reported for STA using an orbitrap HRAM-MS instrument (Virus et al., 2008). A next step in the development of LC–MS approaches in STA is the use of an MS–MS mass spectral library containing accurate-mass data, developed on a Q–TOF-MS instrument (Decaestecker et al., 2000; Broecker et al., 2011; Broecker et al., 2012). One of the procedures described (Broecker et al., 2011) involved the analysis of fivefold diluted urine samples or of blood, plasma, or serum extracts (after protein precipitation or dichloromethane liquid–liquid extraction (LLE) at pH 2 and pH 9). The samples were analyzed using a wide-range solvent gradient on a C_{18} column packed with superficially porous particles. The Q–TOF instrument was operated in untargeted DDA mode, switching between full-spectrum MS mode (0.25 s) and full-spectrum MS–MS for the three most abundant singly-charged precursor ions using an *m*/*z*-dependent ramped collision energy and a dynamic exclusion time of 0.1 min. The total cycle time was 1.1 s (Broecker et al., 2011). The library built, containing positive-ion and/or negative-ion product-ion mass spectra at three different collision energies for ≈2500 compounds and molecular formula information for another ≈7500 compounds, has been commercialized. This particular library was extensively used as a source of accurate-mass data in the preparation of Chapter 4.

Whereas this approach is based on DDA, DIA strategies have been developed as well, for example, MS^E on Waters Q–TOF instruments and MS^M on LIT–orbitrap and Q–orbitrap hybrid instruments (Plumb et al., 2006; Cho et al., 2012). In MS^E, scan-wise switching between MS and MS–MS is performed to obtain fragment ions for all precursor ions present. It has been applied in STA (Chindarkar et al., 2014). More recently, an alternative DIA strategy has been developed (Roemmelt et al., 2014; Arnhard et al., 2015). Using the AB-Sciex TripleTOF instrument, a DIA method called "sequential windowed acquisition of all theoretical fragment ion mass spectra" (SWATH) can be applied. SWATH performs fragmentation of all precursor ions in ≈20-*m*/*z*-unit wide isolation windows, going through the complete *m*/*z* range step by step. This results in a complete fragment ion map of the sample. SWATH was reported to identify more compounds and provide lower LOQs than DDA (Arnhard et al., 2015).

5.6 DEREPLICATION AND IDENTIFICATION OF NATURAL PRODUCTS AND ENDOGENOUS COMPOUNDS

Natural products from plants and other sources, including bacteria, fungi, and animal venoms, have been investigated since early human history in relation to their medicinal use for the treatment of numerous diseases. Extracts from medicinal plants and other natural sources can be screened to find potential lead compounds for drug discovery (Koehn & Carter, 2005). In such studies, both chemical and biological information should be obtained. Chemical information involves the identity of the sample constituents and, in some cases, their quantity. However, in addition to analytical strategies directed at the confirmation of identity of known compounds and the structure elucidation of unknown constituents, the pharmacological activity of the sample constituents must be established. Bioassay-guided isolation and fractionation are usually performed prior to bioactivity testing. The whole process is time consuming and laborious. Therefore, it should be directed at unknown sample constituents. Apart from NMR and various other spectroscopic techniques, LC–MS plays an important role in the rapid identification of known compounds from natural product extracts, a process often called dereplication, as well as in the identification of unknown sample constituents (Bobzin et al., 2000; Wolfender et al., 2003; Smyth et al., 2012; Wolfender et al., 2015). The term "metabolite" is often applied in this context and refers to a low-molecular-weight compound that is either essential to life via normal metabolic processes (primary metabolites such nucleic acids, amino acids, lipids, and carbohydrates) or necessary only for coping with certain conditions (secondary metabolites such as polyphenols, alkaloids, polyketides, and hormones). The secondary metabolites are also referred to as natural products in this section.

The metabolome represents the collection of all metabolites in a biological cell, tissue, organ, or organism, which are the (dynamic) end products of cellular processes. Metabolomics is directed at a nonselective and comprehensive study of the metabolites in a biological system, involving both identification and quantification. The screening of natural products is primarily directed at secondary metabolites in plants and other sources, whereas metabolomics involves both primary and secondary metabolites and may also be directed at metabolites in mammals including humans. Efforts in metabolomics may be directed at metabolite fingerprinting, metabolite profiling, or metabolite target analysis (Fiehn, 2001). In terms of natural product screening and metabolite profiling in human metabolomics, the dereplication and identification strategies are often rather similar. That is why they are treated together in this section.

Comprehensive metabolite profiling is a significant analytical challenge demanding the use of a variety of analytical approaches, both in terms of sample pretreatment and extraction and in terms of analytical platforms to be applied. NMR and MS are the most important tools in this respect. In terms of MS, this involves both GC–MS for the more volatile sample constituents and LC–MS for all other compounds. Direct infusion of complete extracts in ultra-high-resolution mass spectrometry, atmospheric-pressure desorption ionization methods, and imaging mass spectrometry can also be important additional tools. Since confirmation of identity of known unknowns and identification of real unknowns (Section 5.1) is important, HRAM-MS-based strategies are to be preferred. Given the complexity of most natural product extracts or endogenous matrices, high-resolution chromatography (both LC and GC) and even comprehensive two-dimensional chromatography ($GC \times GC$ and $LC \times LC$) must be used in combination with DDA or DIA strategies. Advanced bioinformatics tools are necessary to perform as much automated data processing as possible. Many metabolomics studies are actually directed at comprehensive metabolite profiling while comparing different states of the biological system studies, for example, different levels of development, or normal and (different) pathological or stressed states of a biological system. Therefore, comparison between (large) cohorts of samples or between samples and controls is important. Issues such as aligning complex chromatographic data sets, combining information from different analytical techniques, and application of existing information on compound identification from databases and/or mass spectra libraries are highly relevant in this context. The workflows and analytical strategies involved in such studies are best illustrated with some examples, where emphasis is put on the use of LC–MS.

Flavonoids are polyphenols that are widespread in plants. Dereplication in plant extracts is important to conduct efficient isolation procedures of unknown polyphenolic plant constituents. LC–MS is extensively applied in the analysis and structural characterization of flavonoids (Waridel et al., 2001; Wolfender et al., 2000; De Rijke et al., 2006; De Villiers et al., 2016) and flavonoid glycosides (Cuyckens & Claeys, 2004; March & Brodbelt, 2008; Vukics & Guttman, 2010). Identification of unknown and/or confirmation of identity of known flavonoid *O*- and *C*-glycosides can be achieved by a combination of multistage LC–MS^n using an ion-trap instrument, LC–HRAM-MS using a Q–TOF instrument, LC–UV using a photodiode array detector in combination with post-column added shift reagents, and LC–NMR, either in an off-line manner by fraction collection or in an on-line manner (Waridel et al., 2001). The *O*-glycosides show cleavages at the glycosidic bond, resulting in the loss of sugar units, whereas *C*-glycosides show cross-ring fragmentation in the glycosidic unit itself, resulting in characteristic neutral losses of 90, 96, 120, and 150 Da. Differentiation between 6-*C*- and 8-*C*-glycoside isomers was found to be possible based on MS–MS in the negative-ion mode, by MS^3 in the ion-trap instrument, or by a combination of in-source CID and MS–MS in either an ion-trap or a Q–TOF instrument (pseudo-MS^3). This means that for dereplication purposes complex data acquisition strategies, for example, involving switching between different in-source ion-sampling orifice voltages, are necessary (Waridel et al, 2001). In addition, data from fragmentation in MS^n and in an LIT–orbitrap instrument have been translated in fragmentation trees to help in the identification of polyphenols in complex biological samples (Van der Hooft et al., 2011).

The fragmentation of 11 steroidal alkaloids obtained from *Buxus hyrcana* in a Q–TOF instrument was investigated to identify key fragments that can be used for the dereplication of these alkaloids in plant extracts from related species such as *Buxus papillosa* (Musharraf et al., 2013). Characteristic fragment ions resulted from the presence of a cyclopropane ring in the cycloartenol skeleton and from the hydroxyl group at C^{10}. In this way, 14 compounds were identified as steroidal alkaloids in other *Buxus* species.

A comprehensive metabolomics approach involving LC–HRAM-MS was used to dereplicate microbial strains based on bioactive metabolites from marine bacteria (Forner et al., 2013). The complex data sets were processed using bucketing, presence/absence standardization, and statistical analysis tools including principal component analysis (PCA) and cluster analysis. This allowed for the grouping of *Streptomyces* bacteria isolated from geographically varied environments in addition to the identification of a number of novel chemical entities. Thus, a small bacterial library of 22 *Streptomyces* species were isolated and grown in small-scale fermentation. Several liquid extraction techniques were applied for secondary metabolite extraction. The extracts were analyzed using LC–HRAM-MS in an LIT–orbitrap hybrid instrument. After data file conversion, the data were preprocessed and then processed by a number of statistical software packages and tools. From the data, 281 buckets (*m/z* retention time) were generated (with *m/z* ranging from 191.1421 to 1483.6699, with 5 mDa mass tolerance and retention times ranging from 0.5 to 9.3 min). Subsequent normalization by presence/absence and cluster analysis revealed three main chemical groups, which can be subdivided into subgroups with differences in chemical composition. By comparison to available databases, a number of the secondary metabolites was identified as actinomycete metabolites. The developed methodology may later be used to assist when investigating the relationships between chemical composition and taxonomy (Forner et al., 2013).

This type of studies on either plant extracts or bodily fluids or tissue extracts results in large data files, which must be processed automatically. Apart from issues with the alignment of chromatograms and making the data available for processing using other software tools than what the instrument manufacturer delivers with the instrument, annotating the data with compound identities is a difficult task (Kind & Fiehn, 2010). To some extent, this can be achieved using computer searchable mass spectral libraries and data from compound databases (Section 5.8). Strategies have also been reported where compounds observed are identified based on similarities between mass spectral fragmentation trees acquired under HRAM-MS and MS^n conditions (Kind & Fiehn, 2010; Van der Hooft et al., 2011; Kasper et al., 2012; Rojas-Cherto et al., 2012). The success of such approaches depends to some extent on the accessibility of public databases of annotated fragmentation trees of a wide variety of compounds relevant to the metabolome of humans and plants. This would greatly facilitate the use of fragmentation trees for the dereplication of known constituent of natural extracts and the identification of unknown constituents, especially when substructures can be defined and used in applying fragmentation trees.

In the beginning of this section, the concept of bioactivity-guided fractionation was introduced as the means to enable bioactivity screening with bioassays of individual constituents of the plant extracts. In this context, it should be mentioned that on-line and at-line bioactivity or bioaffinity screening has been described as well (Kool et al., 2011; Malherbe et al., 2012). In these so-called high-resolution screening approaches, a flow split is applied at the end of the chromatographic column. One part of the flow is directed to the MS for chemical characterization, that is, compound identification based on accurate-mass and MS–MS or MS^n data. The other part of the flow is directed either to a system of reaction coils to perform on-line post-column continuous-flow bioassay, using often a fluorescence readout (Kool et al., 2011; Malherbe et al., 2012), or to a fractionation device that fractionates the column effluent in preferably small fractions on a 96-well of 384-well plate, which after freeze-drying can been used in a conventional bioassay using a plate reader (Mladić et al., 2015).

5.7 IDENTIFICATION OF STRUCTURE-RELATED SUBSTANCES

The structure elucidation of related substances is an important area. Within the pharmaceutical industry, it is directed either at by-products of synthesis and degradation products of active pharmaceutical ingredients or at drug metabolism. The strategies followed are more or less the same (Singh et al., 2012; Holčapek et al., 2008; Staack & Hopfgartner, 2007). The general procedure consists of a number of steps. The first step involves the development of suitable sample pretreatment and LC method for isolation and separation of the related substances. For sample pretreatment, compromises must be struck between the risk of ion source contamination and matrix effects when a very limited cleanup is applied and the risk of losing relevant compounds in the sample pretreatment procedure. Generally, wide-range generic solvent gradients are applied, mostly in RPLC. However, hydrophilic interaction chromatography (HILIC) instead of RPLC can be a useful alternative, especially for the identification of drug metabolites. The second step is the acquisition of MS, MS–MS, and/or MS^n spectra of the parent drug. The thorough interpretation of these spectra is of utmost importance for the success of the study. Then, as the third step, the relevant samples are analyzed in LC–MS mode to search for potential related substances, followed by scheduled MS–MS or MS^n analysis. Nowadays, this is mostly done by DDA or DIA MS^E or SWATH procedures, using automatic switching between survey MS and (dependent) MS–MS or MS^n experiments, preferably using HRAM-MS (Zhu et al., 2011; Xie et al., 2012; Zhu et al., 2014). The fourth step involves the interpretation of the data obtained, performing additional LC–MS or LC–MS^n experiments to confirm interpretation or to elucidate certain issues. Finally, additional experiments are done, including isolation of particular compounds, synthesis of standards, and NMR analysis, in order to confirm the identification. Without NMR, discrimination between different possible positional isomers is generally not possible.

It should be mentioned that part of the dereplication and identification of natural products is actually based on similar strategies as outlined in the following.

5.7.1 Drug Metabolites

Drug metabolism is a biotransformation of the parent drug directed at increasing its polarity to enhance its excretion via the urine, the bile, or the feces. It generally occurs in two steps. Phase I metabolism involves enhancing the water solubility of the drug by various hydrolysis, reduction, and/or oxidation reactions. In the Phase II metabolism, the drug or its Phase I metabolites are enzymatically linked or conjugated to water-soluble endogenous agents. A wide variety of enzymes are involved in these processes. For Phase I metabolites, the enzymes include carboxylesterases and peptidases for hydrolysis; cytochromes P450 and carbonyl reductases for reduction; and alcohol and aldehyde dehydrogenases, monamine oxidases, flavin monooxygenases, and again cytochromes P450 for oxidation. About 75% of the drug metabolism reactions involve various cytochrome P450 enzymes (Guengerich, 2008). For Phase II metabolites, specific cytosolic enzymes are involved in the conjugation reactions, often requiring a cofactor. Both *in vitro* and *in vivo* metabolism may be studied. For *in vitro* studies, liver microsomes, cytosols, S9 fractions, hepatocytes, or liver slices are applied. In some cases, isolated (recombinant) cytochrome P450 enzymes are employed. The *in vivo* studies may involve laboratory animals, for example, mouse, rat, dog, monkey, in the preclinical study phase and human subjects in the clinical study phases.

From an MS point of view, the biotransformation of drugs results in metabolic *m/z* shifts relative to the *m/z* of the parent drug, for example, a shift of −14.052 Da due to demethylation and +15.995 Da due to hydroxylation, N-oxide, or sulfoxide formation. Characteristic *m/z* shifts that occur in Phase I and Phase II metabolisms are collected in Tables 5.1 and 5.2, respectively. From the *m/z* shifts in a metabolite, one can often immediately conclude which biotransformation reaction has occurred. If the fragmentation of the parent drug in MS–MS has been thoroughly interpreted, the observed *m/z* shift of the metabolite can be linked to the *m/z* shift of some of the fragment ions. In this way, the *m/z* shift of the metabolite and the associated biotransformation reaction can be attributed to a change in a particular substructure of the parent drug. To this end, the parent drug can be subdivided into the so-called profile groups, to keep track of metabolic *m/z* shifts in the precursor ions and to correlate them to particular structure elements (Kerns et al., 1995; Kerns et al., 1997). Defining profile groups is easier if the fragmentation leads to a number of complementary ions (Section 3.5.2).

For routine metabolite identification (Met-ID) in drug discovery and development, the complete LC–MS workflow can be greatly supported by dedicated Met-ID application managers implemented in the MS data acquisition and data processing software of the LC–MS instrument. Data processing is based on the comparison between data files acquired for biological samples and for blank/control samples, thus searching for differences in the two data sets that can be correlated to the parent drug or its metabolites. Such application managers are widely used (Anari et al., 2004; Fandiño et al., 2006; Hakala et al., 2006; Li et al., 2007b; Ruan et al., 2008). A comparison between some of these Met-ID application managers has been reported (Hakala et al., 2006). The experimental and data interpretation procedure can further be supported by *in silico* prediction of metabolites. Numerous software tools, for example, META, Lhasa Meteor, and SyGMa, have been developed for this purpose (Anari & Baillie, 2005; Valerio, 2009; Kirchmair et al., 2012). A good example of the integration of *in silico* metabolite prediction and an automated experimental workflow is described for the identification of *in vivo* metabolites of tamoxifen using SyGMa prediction; UPHLC coupled to an LIT–orbitrap instrument operated in DDA mode to acquire MS, MS^2, and MS^3 mass spectra; and the use of an MS-vendor-independent data processing software (Jacobs et al., 2013).

The experimental and data interpretation procedures in the identification of *in vitro* Phase I drug metabolites may be illustrated for the antidepressant drug nefazodone (Niessen & Falck, 2015). In the past few years, nefazodone has frequently served as a model compound to demonstrate advances in LC–MS technology (Li et al., 2007a,b; Zhu et al., 2006). Figure 5.1 shows the structure of nefazodone and the identity of some of the fragment ions, which are used as profile groups in the following discussion. The proposed structures are checked against elemental compositions derived

TABLE 5.1 Mass shifts and mass-defect shifts due to Phase I biotransformation reactions.

Biotransformation Reaction	Formula Change	Mass Shift (Da)		Mass-Defect Shift (mDa)
		Nominal	Monoisotopic	
Nitro reduction: $RNO_2 \rightarrow RNH_2$	$-O_2 + H_2$	−30	−29.974	+25.8
2 × demethylation; de-ethylation	$-C_2H_4$	−28	−28.031	−31.3
Alcohol dehydration	$-H_2O$	−18	−18.011	−10.6
Oxidative dechlorination: $RCl \rightarrow ROH$	$-Cl + OH$	−18	−17.966	+33.9
Thiourea to urea: $(NH_2)_2C{=}S \rightarrow (NH_2)_2C{=}O$	$-S + O$	−16	−15.977	+22.8
Hydrolysis of ester: $RCOOCH_3 \rightarrow RCOOH$	$-CH_2$	−14	−14.016	−15.7
O-, N-, or S-demethylation:	$-CH_2$	−14	−14.016	−15.7
$RNHCH_3 \rightarrow RNH_2$				
$ROCH_3 \rightarrow ROH$				
$RSCH_3 \rightarrow RSH$				
Alcohol oxidation to aldehyde:	$-H_2$	−2	−2.016	−15.7
$RCH_2OH \rightarrow RCHO$				
C-hydroxylation and dehydration:	$-H_2$	−2	−2.016	−15.7
$RCH_2CH_3 \rightarrow RCHOHCH_3 \rightarrow RCH{=}CH_2$				
Desaturation (π-bond formation and cyclization)				
Oxidative defluorination: $RF \rightarrow ROH$	$-F + OH$	−2	−1.996	−4.3
Oxidative deamination to ketone: $R_2CHNH_2 \rightarrow R_2C{=}O$	$-NH_3 + O$	−1	−1.032	−31.6
Oxidative deamination to alcohol: $RNH_2 \rightarrow ROH$	$-NH + O$	+1	+0.984	−16.0
Saturation:	$+H_2$	+2	+2.016	+15.7
Ketone reduction to alcohol: $R_2C{=}O \rightarrow R_2CHOH$				
π-Bond hydrogenation				
Ring opening				
Alcohol oxidation: $RCH_2OH \rightarrow RCOOH$	$-H_2 + O$	+14	+13.979	−20.7
Mono-oxygenation, for instance hydroxylation:	$+O$	+16	+15.995	−5.1
$ArH \rightarrow ArOH$				
$RCH_3 \rightarrow RCH_2OH$				
Alkene epoxidation	$+O$	+16	+15.995	−5.1
N-oxidation: $R_3N \rightarrow R_3N{-}O$	$+O$	+16	+15.995	−5.1
N-hydroxylation: $R_2NH \rightarrow R_2NOH$				
S-oxidation:	$+O$	+16	+15.995	−5.1
Thioether to sulfoxide: $R_2S \rightarrow R_2SO$				
Sulfoxide to sulfone: $R_2SO \rightarrow R_2SO_2$				
Aldehyde oxidation: $RCHO \rightarrow RCOOH$	$+O$	+16	+15.995	−5.1
Ethyl to carboxylic acid: $RCH_2CH_3 \rightarrow RCOOH$	$-CH_4 + O_2$	+16	+15.959	−41.5
Hydration, internal hydrolysis	$+H_2O$	+18	+18.011	+10.6
Di-oxygenation, for instance dihydroxylation	$+O_2$	+32	+31.990	−10.2
Thioether to sulfone: $R_2S \rightarrow R_2SO_2$				
Alkene to dihydrodiol: $R^1CH{=}CHR^2 \rightarrow R^1CHOHCHOHR^2$	$+H_2O_2$	+34	+34.006	+5.5

from HRAM-MS, in this case involving an LIT–orbitrap instrument (Li et al., 2007b).

Initially, a Q–LIT instrument was applied to acquire EPI mass spectra in a full-spectrum DDA strategy for the metabolic profiling and identification (Li et al., 2007a). The data processing was performed using an iterative strategy, involving (1) the recognition of characteristic product ions in the MS–MS spectrum of the parent drug, for example, the ions with *m/z* 274 and 140 (Figure 5.1), (2) the generation of extracted-ion chromatogram (XICs, Section 1.3.1.1) of both samples and controls to screen for metabolites that generate these characteristic product ions in MS–MS, and (3) the careful inspection of the product-ion spectra found. The final step aims not only at the identification of the metabolites but also at the recognition of other characteristic ions that may be used in the generation of additional XICs, for example, the ions with *m/z* 290, being the oxidized form of the ion with *m/z* 274 (Li et al., 2007a). Among the 22 Phase I metabolites of nefazodone identified in this way, 7 were not previously reported, for example, metabolites involving oxidative dechlorination.

In fact, the iterative data processing strategy described earlier closely resembles an alternative experimental data acquisition strategy involving the use of precursor-ion analysis

TABLE 5.2 Mass shifts and mass-defect shifts due to Phase II biotransformation.

Biotransformation Reaction	Formula Change	Mass Shift (Da)		Mass-Defect Shift (mDa)
		Nominal	Monoisotopic	
Methylation (N, O, S)	$+CH_2$	+14	+14.0157	+15.7
Hydroxylation + methylation	$+CH_2O$	+30	+30.0105	+10.5
Acetylation	$+C_2H_2O$	+42	+42.0106	+10.6
Glycine conjugation	$+C_2H_3NO$	+57	+57.0215	+21.5
Sulfation	$+SO_3$	+80	+79.9568	−43.2
Hydroxylation + sulfation	$+SO_4$	+96	+95.9517	−48.3
Taurine conjugation	$+C_2H_5NO_2S$	+107	+107.0041	+4.1
S-Cysteine conjugation	$+C_3H_5NO_2S$	+119	+119.0041	+4.1
N-Acetylcysteine conjugation	$+C_5H_7NO_3S$	+161	+161.0147	+14.7
Glucuronidation	$+C_6H_8O_6$	+176	+176.0321	+32.1
Hydroxylation + glucuronidation	$+C_6H_8O_7$	+192	+192.0270	+27.0
S-Glutathione conjugation	$+C_{10}H_{15}N_3O_6S$	+305	+305.0682	+68.2
Desaturation + *S*-glutathione conjugation	$+C_{10}H_{17}N_3O_6S$	+307	+307.0838	+83.8

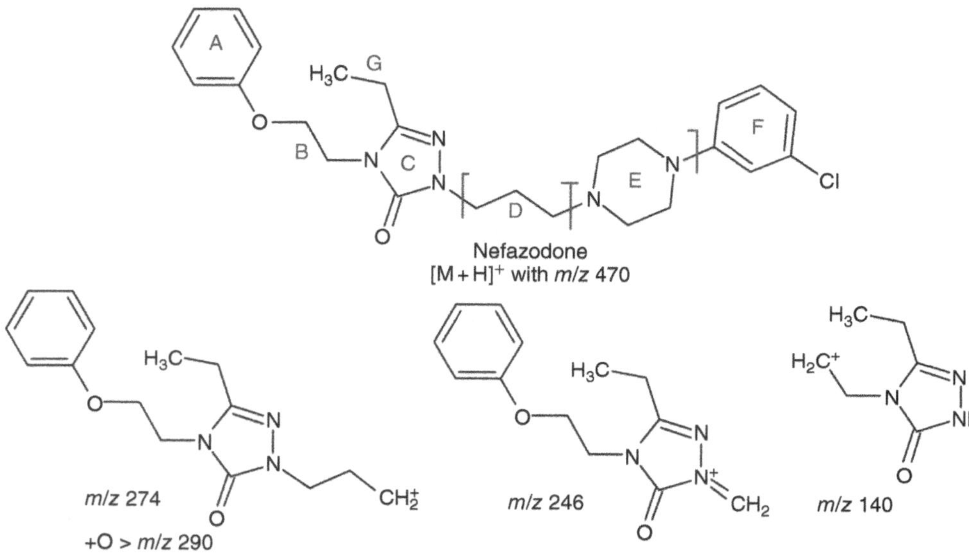

FIGURE 5.1 Structure and major fragments of nefazodone. In the top structure, both profile groups and hydrolysis positions are indicated.

(PIA) and/or neutral-loss analysis (NLA) in TQ instruments (Clarke et al., 2001; Kostiainen et al., 2003). The PIA and NLA modes allow for the screening of structurally related compounds in complex mixtures (Section 1.5.3). They are based on the detection of either a characteristic fragment ion (in the PIA mode) or a characteristic neutral loss (in the NLA mode). The NLA mode is especially powerful for the screening of Phase II drug metabolites, for example, using constant neutral losses of 80 or 176 Da for sulfate or glucuronic acid conjugates, respectively (Kostiainen et al., 2003). In fact, screening for nefazodone metabolites in PIA mode using the characteristic product ions with *m/z* 246, 274, and 290 (Figure 5.1) and in NLA mode using constant neutral losses of 196, 224, and 212 Da has also been demonstrated (Li et al., 2007b).

Four additional metabolites were found and the elemental composition of all 26 metabolites found was confirmed in a subsequent study, where the complete procedure was repeated using an LIT–orbitrap hybrid system. Of the 26 identified metabolites, 14 involved oxidation (hydroxylation or N-oxidation) of the parent drug. In many cases, the exact position could not be established but only pinpointed to a particular profile group. Other metabolites resulted from N-dealkylation or hydrolysis reactions, resulting in the loss of relatively large groups and oxygenated derivatives thereof (Li et al., 2007b).

It should be pointed out that in the current example, DDA strategies were applied using both unit-mass resolution and HRAM-MS instruments. Alternatively, with HRAM-MS instruments, DIA strategies could have been used as well. In DIA (using MS^E, MS^M, or SWATH), a continuous scan-wise switching between the MS mode at low collision energy and the MS–MS or MS^n mode at high collision energy is performed. In this way, fragment ions are obtained for all precursor ions present. Software tools have been developed to obtain separate MS–MS spectra for individual precursor ions (Plumb et al., 2006; Cho et al., 2012).

HRAM-MS instruments are frequently used in Met-ID studies (Xie et al., 2012), because the availability of accurate-mass data readily provides confirmation of the identity of the metabolic *m/z* shifts and thereby facilitates identification. Powerful additional tools have become available for the screening, profiling, and identification of drug metabolites. One of these tools is mass-defect filtering (MDF) (Zhang et al., 2003; Zhang et al., 2009). A metabolic *m/z* shift is accompanied by a shift in the mass defect of the metabolite, for example, of −0.0157 Da due to demethylation and −0.0051 Da due to hydroxylation. If the biotransformation does not lead to major structural changes in the parent drug, that is, if no N- or O-dealkylation or hydrolysis of large substituent groups occurs, the mass-defect shifts are limited to ±40 mDa for Phase I metabolites and ±70 mDa for Phase II metabolites (Tables 5.1 and 5.2). Software tools have been developed to process HRAM-MS data and to perform MDF within a ±40 or ±70 mDa wide window of the exact *m/z* of the parent drug (Zhang et al., 2003; Zhang et al., 2009). The use of MDF has been successfully demonstrated for nefazodone (Zhu et al., 2006). In the case of nefazodone, the biotransformation also involves major structural changes due to N-dealkylation (Figure 5.1). Therefore, additional mass-defect filters must be defined and used (Zhu et al., 2006; Mortishire-Smith et al., 2009).

Other tools that can assist in identification of oxygenated Phase I metabolites are hydrogen/deuterium (H/D)-exchange experiments and the use of IMS–MS instruments. Both tools aim at elucidating possible isomerism issues that may occur. In an H/D-exchange experiment, the *m/z* shift is measured between a compound and its H/D-exchange product. The *m/z* shift results from the exchange of any protons by deuteriums at —OH, —COOH, and —NH groups and the formation of $[M+D]^+$ rather than $[M+H]^+$ (Section 3.1) (Liu & Hop, 2005; Shah et al., 2013). If multiple oxygenated metabolites are observed, H/D-exchange experiments can be performed to discriminate between C-hydroxylation and N- or S-oxidation and epoxidation. According to Table 5.1, these three types of metabolites all show an *m/z* shift of +15.9949. If the metabolic mixture is re-analyzed in a mobile phase containing D_2O rather than H_2O, H/D exchange will be observed for the hydroxylated metabolites, whereas the epoxidation and N- or S-oxygenated products do not show H/D exchange (Liu & Hop, 2005; Shah et al., 2013). Various ways to perform the H/D-exchange experiments have been reviewed by Shah et al. (2013).

Differentiation between various isomeric C-hydroxylated metabolites is often difficult. However, as these isomers have different spatial conformation and thus may show different collision cross sections, IMS–MS may be of help. Unfortunately, with the currently available resolution of the IMS separation, the differences in drift time for isomeric hydroxylated metabolites are generally too small (Dear et al. 2010). However, it has been demonstrated that a selective derivatization of aromatic hydroxyl groups of drug metabolites into *N*-methyl pyridinium derivatives sufficiently increases the differences in collision cross sections and thus in drift times, allowing differentiation between isomeric forms of aromatic hydroxylated metabolites (Shimizu & Chiba, 2013). By correlating theoretically predicted collision cross sections to measured drift times of a parent drug and its fragments, a calibration plot was generated, which could subsequently be used to differentiate between chromatographically separated isomeric *N*-methyl pyridinium derivatives of hydroxylated metabolites.

In this section, an elaborate toolbox for the profiling and identification of drug metabolites has been described, involving a variety of data acquisition and data processing strategies. Depending on the properties of the parent drug, this toolbox may even be extended. The LC–MS analysis of metabolic mixtures from *in vitro* or *in vivo* experiments is performed in DDA or DIA mode and results in basically two data sets, a full-spectrum MS and a full-spectrum MS–MS data set. The full-spectrum HRAM-MS data set may be interrogated using XIC, MDF, isotope-pattern filtering (if specific isotopic features, for example, Cl or Br atoms, are present in the parent drug), and background subtraction. The DDA and/or DIA full-spectrum MS–MS and/or MS^n data sets may be interrogated using product-ion filtering or PIA and neutral-loss filtering or NLA (Zhang et al., 2009). The great potential of IMS–MS in this area must be explored further, especially for the separation and MS analysis of closely related structural compounds, for example, structural and configurational isomeric compounds.

A similar toolbox is also applied in the screening for reactive metabolites by glutathione trapping (Ma & Zhu, 2009). Based on the specific properties of glutathione in MS–MS, the NLA mode can be used to screen for characteristic losses of 129 Da (pyroglutamic acid) or 307 Da (glutathione), whereas the characteristic product ion with *m/z* 130 (protonated pyroglutamic acid) can be used for screening in the PIA mode (Kostiainen et al., 2003; Dieckhaus et al., 2005). Enhanced selectivity in NLA is achieved in a DIA MS^E strategy with a post-acquisition selection of the precursor ions that show the loss of 129.043 Da, which is the accurate mass of pyroglutamic acid (Castro-Perez et al.,

2005). These experiments, performed in positive-ion mode, are not always successful. Thus, performing PIA using the common fragment ion with *m/z* 272 in negative-ion mode may provide additional results (Dieckhaus et al., 2005). Additional analytical strategies for the screening of glutathione-conjugated reactive metabolites involve SRM-triggered DDA on a Q–LIT instrument (Zheng et al, 2007), and the use of stable-isotope-labeled or chemically modified glutathione (Yan & Caldwell, 2004; Rousu et al., 2009), and MDF (Zhu et al., 2007).

5.7.2 Impurities and Degradation Products

In the previous section, an identification strategy is outlined applicable to drug metabolites, but in fact also to other types of related compounds. After interpreting the MS–MS or MS^n spectra of the parent compound, the structure is subdivided into profile groups corresponding to particular fragment ions (or neutral losses in some cases). Subsequently, one tries to identify the *m/z* difference between the parent compound and the related compounds found and to correlate this *m/z* shift to one or more of the profile groups in order to elucidate the position of the change in the structure.

The applicability of this procedure to the identification of degradation products or (related) impurities in drugs can be demonstrated with the identification of two by-products from the synthesis of trimethoprim ($[M+H]^+$ with *m/z* 291) (Lehr et al., 1999; Barbarin et al., 2002). In MS–MS, trimethoprim generates two characteristic fragment ions with *m/z* 123 and 181 (Section 4.8.9). They are due to cleavages of either vicinal C—C bond to the methylene group connecting the pyrimidine and the phenyl rings, thus resulting in the 2,4-diaminopyrimidin-5-yl methylium ion and the 3,4,5-trimethoxybenzyl cation, respectively (Figure 4.8.18) (Lehr et al., 1999; Barbarin et al., 2002; Eckers et al., 2005). These fragment ions can be used in the data interpretation as profile groups A and B, respectively. Some other fragment ions are relevant as well, especially the fragment ions with *m/z* 261 and 230, which are due to the loss of ethane (C_2H_6) and of formaldehyde ($H_2C{=}O$) and the methoxy radical ($H_3CO^\bullet$), respectively (Eckers et al., 2005).

In this discussion, we focus on two reported reaction by-products of trimethoprim (Lehr et al., 1999), showing $[M+H]^+$ with *m/z* 305 and 339. From the mass shift (+14 Da), it may be concluded that the compound with *m/z* 305 could be consistent with the presence of an additional CH_2 group. In the product-ion mass spectrum, the ion with *m/z* 123 is still present, but the ion with *m/z* 181 is not; an ion with *m/z* 195 appears instead. This indicates that the change is in the profile group B of trimethoprim. The presence of ions with *m/z* 275 and 230 and the virtual absence of the fragment ion with *m/z* 244 indicate that one of the methoxy groups is replaced by an ethoxy group. The mass shift of +48 Da for the by-product with *m/z* 339 is not so straightforward to recognize. However, for this compound, a characteristic mono-bromine isotope pattern was observed in the MS spectrum. From this, the mass shift can be interpreted as the replacement of a methoxy group by a bromine atom. The profile group ions, observed with *m/z* 123 and 229 (the latter with a mono-bromine isotope pattern), indicate that the change is indeed in the profile group B of trimethoprim. Finally, NMR was used to determine the exact positions of the ethoxy group and the bromine atom in the ring (Lehr et al., 1999).

Another example, from the field of toxicology, involves the identification of a compound found in a designer drug (Vande Casteele et al., 2005). An inexplicable peak was found in postmortem patient samples of a user deceased from an overdose of 3,4-methylenedioxymethamphetamine (MDMA, ecstasy; Section 4.7.2). In addition to MDMA and the unknown, 3,4-methylenedioxyethylamphetamine (MDEA) and 3,4-methylenedioxyamphetamine (MDA) were observed. Using LC–MS on a Q–TOF instrument, it was found that the unknown was isomeric to MDEA ($[M+H]^+$ with *m/z* 208), but showed a different retention time. The isomeric unknown also showed the same characteristic fragment ions with *m/z* 163, 135, 133, 105, and 72 (Section 4.7.2), although with some different relative abundance. The fragment ions with *m/z* 135, 133, and 105 indicate that the 3,4-methylenedioxyphenyl skeleton is unchanged. The fragment ions with *m/z* 163 and 72 indicate that the structural differences should be in nitrogen substitution, thus 3,4-methylenedioxy-*N*,*N*-dimethylamphetamine rather than MDEA was identified as the unknown. Subsequently, this compound was synthesized, and it showed the MS–MS spectrum observed for the unknown compound as well as the same retention time in spiked biological samples (Vande Casteele et al., 2005).

Impurity profiling including identification and profiling of degradation products of drugs is a challenging task, especially in the light of the ever increasing demands from regulatory bodies, for example, also in relation to genotoxicity. As a result, significant efforts have to be made, involving advanced LC–MS technologies as well as other analytical tools (Singh et al., 2012; Jain & Basniwal, 2013).

5.8 IDENTIFICATION OF KNOWN UNKNOWNS AND REAL UNKNOWNS

The identification of known unknowns generally consists of a number of steps. Depending on the identification problem, somewhat different procedures have been described (Thurman et al., 2005; Kind & Fiehn, 2007; Little et al., 2011; Milman, 2015). Some important steps in the procedure are outlined in the following.

First, one needs to collect as much information on the compound(s) to be identified as possible. Often, information on the origin of the sample, solvent composition, and possibly underlying chemistry is highly valuable. Establishing a good information flow between the inquiring party and the analytical chemist is of utmost importance in these cases. In the first place, one needs to establish whether the sample is actually amenable to MS analysis. In principle, a wide variety of MS technologies can be tested and applied. The use of both hard and soft ionization techniques may be useful. Therefore, testing the amenability of the sample to both GC–MS in EI mode and LC–MS in either positive-ion or negative-ion ESI mode (or preferably both) is useful. Alternatively or additionally, other analyte ionization strategies may be tested, including MALDI-TOF-MS and DESI-MS or similar desorption ionization strategies.

The choice of mass spectrometer is often determined by the availability in the laboratory. For a proper identification, both accurate-mass information and MS–MS and/or MS^n fragmentation are needed. The use of HRAM-MS at an early stage is recommended, as this will enable the calculation of the elemental composition of the known unknown, especially when a soft ionization technique is applied. Software is available to perform the calculation (Section 2.7). Advances in software development have been reported and implemented in available tools (Ojanperä et al., 2006; Kind & Fiehn, 2007; Rojas-Chertó et al., 2011; Pluskal et al., 2012). More advanced software tools also take advantage of the information available in the measured isotope pattern, that is, relative abundance of M+1 and M+2 isotope peaks (Section 2.4). Ultra-high-resolution instruments provide the possibility to separate the contributions of individual elements to the isotope peak as demonstrated, for instance, with the identification of an unknown constituent in an onion bulb (Nakabayashi et al., 2013).

Based on the elemental composition found, compound databases available on the Internet may be searched for known compound structures. In searching databases, one has to keep in mind that MS gives information on the accurate monoisotopic mass and the elemental composition of the ion, whereas the compound databases generally contain information on the average molecular weight and elemental composition of the molecule. Thus, for a protonated molecule, the proton must be subtracted before searching the database. In the case of mass spectral libraries, this is not an issue. There are the more general databases such as ChemSpider, PubMed, and SciFinder, and there are also more dedicated compound databases such as the human and plant metabolome database, pesticide index, lipid maps structure database (Table 5.3).

Some of these databases contain (searchable) MS–MS and/or MS^n spectral data, which can be compared to the acquired MS–MS and/or MS^n data. Note that searching MS–MS data from collision-cell CID against an MS^n library from ion-trap CID (or vice versa) is not necessarily successful, as fragment ions may be present or absent in one data set due to differences in collision energy and fragmentation behavior. Otherwise, one has to establish whether the observed fragmentation can be explained from the available hits in the database. This means that one filters the known structures from the database search by checking the observed fragmentation behavior against predicted fragmentation of the database-retrieved structures. In some cases, substructure searches based on recognized substructures in the MS–MS and/or MS^n spectra may help reducing the number of hits. Alternatively or additionally, the use of software tools to predict fragmentation behavior of compound structures may be useful. Various tools have become available (Table 5.4) and they can be used with varying success. In favorable cases, this leads to a (number of) potential structure proposal(s) for the unknown.

If the procedure resulted in one or a few possible structure proposals, the next step would be the purchase or synthesis of standards for these compounds. These standards are then

TABLE 5.3 Selected compound databases either commercially available or publicly available on the internet.

Compound Class	Database	Website
General	ChemSpider	http://www.chemspider.com/
	PubMed	http://pubchem.ncbi.nlm.nih.gov/search/search.cgi
	SciFinder	https://www.cas.org/products/scifinder
	CHEMnetBASE	http://www.chemnetbase.com/
	Wiley Registry of Mass Spectral Data	http://eu.wiley.com/WileyCDA/WileyTitle/productCd-047052037X.html
	NIST/EPA/NIH MS/MS Mass Spectral Library	http://www.nist.gov/srd/nist1a.cfm
Metabolomics	HMDB	http://www.hmdb.ca/metabolites
	METLIN	https://metlin.scripps.edu/index.php
	Phenol Explorer	http://phenol-explorer.eu/compounds/advanced
	PMDB	http://scbt.sastra.edu/pmdb/
Pesticides	Compendium of Pesticide Common Names	http://www.alanwood.net/pesticides/index_cn_frame.html
Lipids	Lipid Maps Structure Database	http://www.lipidmaps.org/data/structure/index.html
Toxicology	ChemIDplus	http://chem.sis.nlm.nih.gov/chemidplus/chemidlite.jsp

TABLE 5.4 Selected software tools for *in silico* prediction of fragmentation of small molecules in MS–MS or MS^n.

Software Tool	Website
Mass Frontier	http://www.thermoscientific.com/en/product/mass-frontier-7-0-spectral-interpretation-software.html
ACD/MS Fragmenter	http://www.acdlabs.com/products/adh/ms/ms_frag/
Iontree	http://www.bioconductor.org/packages/release/bioc/html/iontree.html
MetFrag	http://msbi.ipb-halle.de/MetFrag/
Sirius	http://bio.informatik.uni-jena.de/software/sirius/

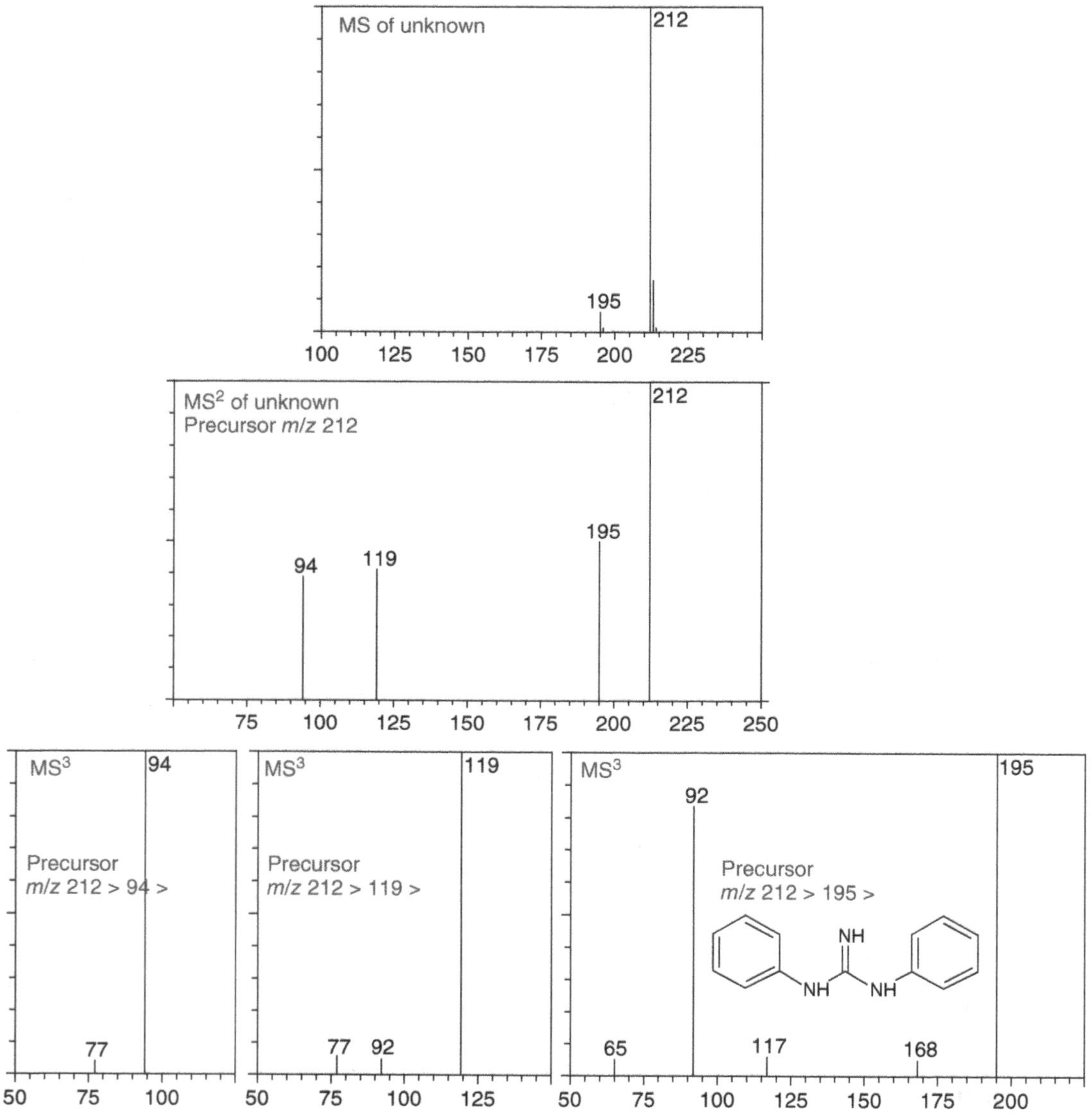

FIGURE 5.2 MS, MS^2, and MS^3 mass spectra of the "known unknown" discussed, which is identified as 1,3-diphenylguanidine. (Source: Lin et al., 2007. Reproduced with permission of Elsevier; Niessen & Honing, 2015. Reproduced with permission of Wiley.)

analyzed to check retention time, fragmentation behavior, and if possible other pertinent properties. Additional tools such NMR, IR, and other spectroscopic methods will most likely be necessary to solve the puzzle and to obtain an unambiguous identification, especially when stereochemical issues are of importance as well.

The procedure outlined for the identification of known unknowns can be illustrated with an example involving an unknown found in, but not related to, an active pharmaceutical ingredient during routine LC–UV analysis of sample batches (Lin et al., 2007; Niessen & Honing, 2015). The analyte shows UV absorbance at 236 nm, which is characteristic for a monosubstituted aromatic ring. The LC–MS analysis was performed in positive-ion ESI with two different mobile-phase compositions. A trifluoroacetic acid containing mobile phase was used in order to exclude that the major ion observed using an ammonium acetate containing mobile phase was an ammoniated molecule $[M+NH_4]^+$ with *m/z* 212 rather than a protonated molecule $[M+H]^+$. In addition to the ion with *m/z* 212, an ion with *m/z* 195 was observed, which can now be interpreted as being due to the loss of ammonia (NH_3). Thus, the unknown has a molecular mass of 211 Da (odd number of N-atoms).

As these data were acquired using an ion-trap instrument, available at the department, the logical next step is the acquisition of MS^n spectra and interpretation of these data. Apart from the expected fragment ion with *m/z* 195 owing to the loss of NH_3, the MS^2 spectrum of the precursor ion with *m/z* 212 shows two fragment ions with *m/z* 94 and 119 (Figure 5.2). As the sum of their *m/z* values is equal to *m/z* ([M+H]+1), these two ions could be complementary ions, representing two parts of the molecule (Section 3.5.2). The nitrogen rule indicates that the ion with *m/z* 94 contains an odd number of N-atoms and the ion with *m/z* 119 contains zero or an even number. Subsequently, MS^3 spectra are acquired with the ions with *m/z* 94 and 119 as precursor ions (Figure 5.2). The ion with *m/z* 94 shows the loss of NH_3 to a fragment ion with *m/z* 77, which is most likely the phenyl cation ($[C_6H_5]^+$), accounting for the UV activity at 236 nm. The ion with *m/z* 119 shows fragment ions with *m/z* 92 due to the loss of hydrogen cyanide (HCN) and with *m/z* 77 (again $[C_6H_5]^+$) due to the loss of methanediimine (HN=C=NH). From this small MS^n fragmentation tree, a number of substructures were recognized, which now have to be combined into a structure proposal. In the first place, the identification of substructures leads to a proposed elemental composition for the ion with *m/z* 212 of $[C_{13}H_{14}N_3]^+$. This elemental composition was confirmed by accurate-mass determination on a TOF-MS instrument. From the measured accurate mass of $[M+H]^+$ with *m/z* 212.120, two possible elemental compositions were calculated (the following constraints were taken into account: EE^+, tolerance 10 mDa; C, 0–20; H, 0–40; N, 0–6; O, 0–6; RDBE ≥ 3.5; odd N; no indication for the presence of other elements in MS and MS^n data), that is, $[C_{13}H_{14}N_3]^+$ (RDBE = 8.5; mass error +1.4 mDa), as already anticipated, and $[C_8H_{14}N_5O_2]^+$ (RDBE = 4.5; +5.4 mDa).

The next step in the procedure is a search of compound databases for possible structures with $C_{13}H_{13}N_3$. A search in the large ChemSpider database resulted in ≈400 isomeric structures. These structures must be inspected, especially in their ability to produce the primary fragment ions with *m/z* 195, 119, and 94. In order to illustrate this step of the procedure, a search of the much smaller Sigma-Aldrich database/catalog was done, resulting in five possible isomeric structures (Figure 5.3). Here, our prediction on the ability to produce the fragment ions with *m/z* 195, 119, and 94 are also provided. The fragmentation of $[M+H]^+$ of these five structures in MS–MS was also predicted by Mass Frontier software (Table 5.5). Some of these standards should be purchased and analyzed using LC–MS to check their retention time, UV properties, and fragmentation behavior. From the data available now, we conclude that the unknown compound is 1,3-diphenylguanidine (Lin et al., 2007). The elemental composition of the precursor ion is in agreement with this. The main fragments in the product-ion mass spectra could be assigned and are in agreement with the proposed structure. Further experiments may be needed to further confirm the structure.

The identification of real unknowns is the most challenging aspect. The general procedure is identical to the procedure outlined for the identification of known unknowns. Actually, one only becomes aware of the fact that the searched compound is a real unknown if the database searches do not lead to a result. In such cases, the use of substructures, from which the complete structures must be put together, and the application of other structure elucidation tools, especially NMR and IR, are highly important. In these cases, the puzzle can definitively not be solved by using MS technologies alone.

In the end, one comes with a structure proposal, although it would often be better to provide a number of alternative structure proposals. Depending on the experimental evidence, one can make statements on the level of confidence. In general, one can state that the elemental composition of the proposed structure is in agreement with the accurate mass measured for the intact molecule, that the main fragment ions can be assigned, that they are in agreement with the proposed structure, and that (hopefully) no anomalies are found. Whenever possible, the structure proposal must be supported by other evidence either from the analysis of standards (which often does not rule out potential stereoisomerism issues) and/or by other spectroscopic techniques.

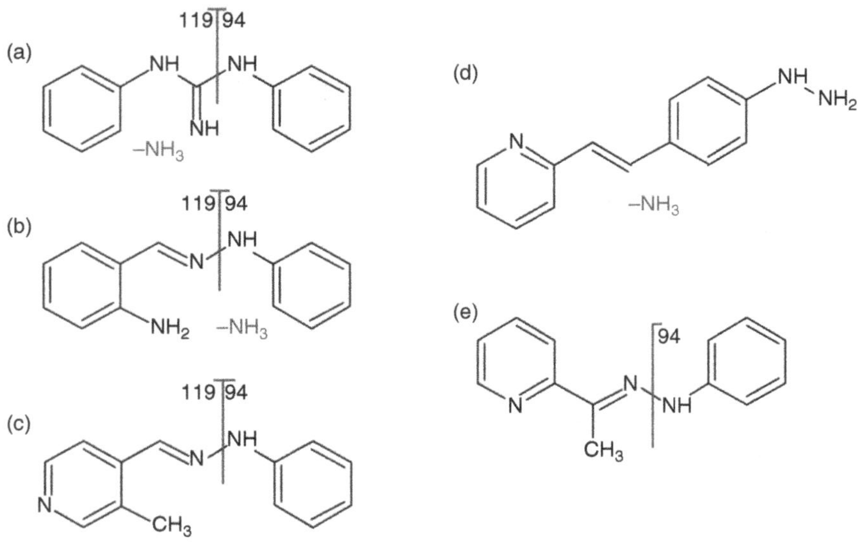

FIGURE 5.3 Hits for the search of $C_{13}H_{13}N_3$ in the Sigma-Aldrich database/catalog. Our prediction on the ability to generate the fragment ions with *m/z* 195 ($-NH_3$), 119, and 94 is indicated. (Source: Niessen & Honing, 2015. Reproduced with permission of Wiley.)

TABLE 5.5 Mass Frontier prediction of the five characteristic fragment ions of the unknown with an $[M+H]^+$ with *m/z* 212 ($[C_{13}H_{13}N_3+H]^+$) for the five structure proposals obtained from searching the Sigma-Aldrich catalog (Figure 5.3). Mass Frontier (version 5) was used.

Structure	*m/z* 195	*m/z* 119	*m/z* 94	*m/z* 92	*m/z* 77
A	×	×	×	–	×
B	×	×	–	–	×
C	–	×	–	–	×
D	–	–	–	–	–
E	–	–	–	–	×

×, predicted; –, not predicted.

REFERENCES

Anari MR, Sanchez RI, Bakhtiar R, Franklin RB, Baillie TA. 2004. Integration of knowledge-based metabolic predictions with liquid chromatography data-dependent tandem mass spectrometry for drug metabolism studies: Application to studies on the biotransformation of indinavir. Anal Chem 76: 823–832.

Anari MR, Baillie TA. 2005. Bridging cheminformatic metabolite prediction and tandem mass spectrometry. Drug Discov Today 10: 711–717.

Anastassiades M, Lehotay S J, Stajnbaher D, Schenck FJ. 2003. Fast and easy multiresidue method employing acetonitrile extraction/partitioning and "dispersive solid-phase extraction" for the determination of pesticide residues in produce. J AOAC Int 86: 412–431.

Arnhard K, Gottschall A, Pitterl F, Oberacher H. 2015. Applying 'sequential windowed acquisition of all theoretical fragment ion mass spectra' (SWATH) for systematic toxicological analysis with liquid chromatography-high-resolution tandem mass spectrometry. Anal Bioanal Chem 407: 405–414.

Atwater BL, Stauffer DB, McLafferty FW, Peterson DW. 1985. Reliability ranking and scaling improvements to the probability based matching system for unknown mass spectra. Anal Chem 57: 899–903.

Ausloos P, Clifton CL, Lias SG, Mikaya AI, Stein SE, Tchekhovskoi DV, Sparkman OD, Zaikin V, Zhu D. 1999. The critical evaluation of a comprehensive mass spectral library. J Am Soc Mass Spectrom 10: 287–299.

Badawi N, Simonsen KW, Steentoft A, Bernhoft IM, Linnet K. 2009. Simultaneous screening and quantification of 29 drugs of abuse in oral fluid by solid-phase extraction and ultra-performance LC-MS/MS, Clin Chem 55: 2004–2018.

Barbarin N, Henion JD, Wu Y. 2002. Comparison between liquid chromatography-UV detection and liquid chromatography-mass spectrometry for the characterization of impurities and/or degradants present in trimethoprim tablets. J Chromatogr A 970: 141–154.

Bateman KP, Kellmann M, Muenster H, Papp R, Taylor L. 2009. Quantitative–qualitative data acquisition using a benchtop Orbitrap mass spectrometer. J Am Soc Mass Spectrom 20: 1441–1450.

Baumann C, Cintora MA, Eichler M, Lifante E, Cooke M, Przyborowska A, Halket JM. 2000, A library of atmospheric pressure ionization daughter ion mass spectra based on wideband excitation in an ion trap mass spectrometer. Rapid Commun Mass Spectrom 14: 349–356.

Berendsen BJ, Stolker LA, Nielen MW. 2013. The (un)certainty of selectivity in liquid chromatography tandem mass spectrometry. J Am Soc Mass Spectrom 24: 154–163.

Bijlsma L, Emke E, Hernández F, de Voogt P. 2012. Investigation of drugs of abuse and relevant metabolites in Dutch sewage water by liquid chromatography coupled to high resolution mass spectrometry. Chemosphere 89: 1399–1406.

Bijlsma L, Beltrán E, Boix C, Sancho JV, Hernández F. 2014. Improvements in analytical methodology for the determination of frequently consumed illicit drugs in urban wastewater. Anal Bioanal Chem 406: 4261–4272.

Birkler RID, Telving R, Ingemann-Hansen O, Charles AV, Johannsen M, Andreasen MF. 2012. Screening analysis for medicinal drugs and drugs of abuse in whole blood using ultra-performance liquid chromatography time-of-flight mass spectrometry (UPLC-TOF-MS) – Toxicological findings in cases of alleged sexual assault. Forensic Sci Int 222: 154–161.

Bobzin SC, Yang S, Kasten TP. 2000. LC-NMR: A new tool to expedite the dereplication and identification of natural products. J Ind Microbiol Biotechnol 25: 342–345.

Bogialli S, Di Corcia A. 2009. Recent applications of liquid chromatography-mass spectrometry to residue analysis of antimicrobials in food of animal origin. Anal Bioanal Chem 395: 947–966.

Bogusz MJ, Maier R-D, Krüger KD, Webb KS, Romeril J, Miller ML. 1999. Poor reproducibility of in-source collisional atmospheric pressure ionization mass spectra of toxicologically relevant drugs. J Chromatogr A 844: 409–418.

Bogusz MJ. 2000. Liquid chromatography–mass spectrometry as a routine method in forensic sciences: A proof of maturity. J Chromatogr B 748: 3–19.

Boix C, Ibáñez M, Zamora T, Sancho JV, Niessen WMA, Hernández F. 2014. Identification of new omeprazole metabolites in wastewaters and surface waters. Sci Total Environm, 468–469: 706–714.

Botero-Coy AM, Marín JM, Ibáñez M, Sancho JV, Hernández F. 2012. Multi-residue determination of pesticides in tropical fruits using liquid chromatography/tandem mass spectrometry. Anal Bioanal Chem 402: 2287–2300.

Bristow AWT, Nichols WF, Webb KS, Conway B. 2002. Evaluation of protocols for reproducible electrospray in-source collisionally induced dissociation on various liquid chromatography–mass spectrometry instruments and the development of spectral libraries. Rapid Commun Mass Spectrom 16: 2374–2386.

Bristow AWT, Webb KS, Lubben AT, Halket J. 2004. Reproducible product-ion tandem mass spectra on various liquid chromatography–mass spectrometry instruments for the development of spectral libraries. Rapid Commun Mass Spectrom 18: 1447–1454.

Broecker S, Herre S, Wüst B, Zweigenbaum J, Pragst F. 2011. Development and practical application of a library of CID accurate mass spectra of more than 2,500 toxic compounds for systematic toxicological analysis by LC-QTOF-MS with data-dependent acquisition. Anal Bioanal Chem 400: 101–117.

Broecker S, Herre S, Pragst F. 2012. General unknown screening in hair by liquid chromatography-hybrid quadrupole time-of-flight mass spectrometry (LCQTOF-MS). Forensic Sci Int 218: 68–81

Brown DV, Dalton M, Pullen FS, Perkins GL, Richards D. 1994. An automated, open-access service to synthetic chemists: Thermospray mass spectrometry. Rapid Commun Mass Spectrom 8: 632–636.

Castro-Perez J, Plumb R, Liang L, Yang E. 2005. A high-throughput liquid chromatography/tandem mass spectrometry method for screening glutathione conjugates using exact mass neutral loss acquisition. Rapid Commun Mass Spectrom 19: 798–804.

CEN/TC 275 (European Committee for Standardization/Technical Committee). 2007. Determination of pesticides using GC–MS and/or LC–MS/MS following acetonitrile extraction/partitioning and cleanup by dispersive SPE-QuEChERS method. European Committee for Standardization, Brussels.

Chindarkar NS, Wakefield MR, Stone JA, Fitzgerald RL. 2014. Liquid chromatography high-resolution TOF analysis: Investigation of MS^E for broad-spectrum drug screening. Clin Chem 60: 1115–1125.

Cho R, Huang Y, Schwartz JC, Chen Y, Carlson TJ, Ma J. 2012. MS^M, an efficient workflow for metabolite identification using hybrid linear ion trap Orbitrap mass spectrometer. J Am Soc Mass Spectrom 23: 880–888.

Clarke NJ, Rindgen D, Korfmacher WA, Cox KA. 2001. Systematic LC/MS metabolite identification in drug discovery. Anal Chem 73: 430A–439A.

Cuyckens F, Claeys M. 2004. Mass spectrometry in the structural analysis of flavonoids. J Mass Spectrom 39: 1–15.

Dear GJ. Munoz-Muriedas J, Beaumont C, Roberts A, Kirk J, Williams JP, Campuzano I. 2010, Sites of metabolic substitution: Investigating metabolite structures utilising ion mobility and molecular modelling. Rapid Commun Mass Spectrom 24: 3157–3162.

De Brabander HF, Le Bizec B, Pinel G, Antignac JP, Verheyden K, Mortier V, Courtheyn D, Noppe H. 2007. Past, present and future of mass spectrometry in the analysis of residues of banned substances in meat-producing animals. J Mass Spectrom 42: 983–998.

Decaestecker TN, Clauwaert KM, Van Bocxlaer JF, Lambert WE, Van den Eeckhout EG, Van Peteghem CH, De Leenheer AP. 2000. Evaluation of automated single mass spectrometry to tandem mass spectrometry function switching for comprehensive drug profiling analysis using a quadrupole time-of-flight mass spectrometer. Rapid Commun Mass Spectrom 14: 1787–1792.

de Castro A, Concheiro M, Quintela O, Cruz A, López-Rivadulla M. 2008. LC–MS/MS method for the determination of nine antidepressants and some of their main metabolites in oral fluid and plasma. Study of correlation between venlafaxine concentrations in both matrices. J Pharm Biomed Anal 48: 183–193.

De Dominicis E, Commissati I, Suman M. 2012. Targeted screening of pesticides, veterinary drugs and mycotoxins in bakery ingredients and food commodities by liquid chromatography-high-resolution single-stage Orbitrap mass spectrometry. J Mass Spectrom 47: 1232–1241.

De Zeeuw RA. 2004. Substance identification: The weak link in analytical toxicology. J Chromatogr B 811: 3–12.

Dieckhaus CM, Fernández-Metzler CL, King R, Krolikowski PH, Baillie TA. 2005. Negative ion tandem mass spectrometry for the detection of glutathione conjugates. Chem Res Toxicol 18: 630–638.

Di Rago M, Saar E, Rodda LN, Turfus S, Kotsos A, Gerostamoulos D, Drummer OH. 2014. Fast targeted analysis of 132 acidic and neutral drugs and poisons in whole blood using LC-MS/MS. Forensic Sci Int 243: 35–43.

de Rijke E, Out P, Niessen WMA, Ariese F, Gooijer C, Brinkman UATh. 2006. Analytical separation and detection methods for flavonoids. J Chromatogr A 1112: 31–63.

de Villiers A, Venter P, Pasch H. 2016. Recent advances and trends in the liquid-chromatography-mass spectrometry analysis of flavonoids. J Chromatogr A 1430: 16–78.

Dresen S, Gergoc M, Politi, L, Halter C, Weinmann W. 2009. ESI-MS–MS library of 1,253 compounds for application in forensic and clinical toxicology. Anal Bioanal Chem 395: 2521–2526.

Dresen S, Ferreirós N, Gnann H, Zimmermann R, Weinmann W. 2010. Detection and identification of 700 drugs by multi-target screening with a 3200 QTRAP® LC–MS–MS system and library searching. Anal Bioanal Chem 396: 2425–2434.

Drummer OH, Kotsos A, McIntyre IM. 1993. A class-independent drug screen in forensic toxicology using a photodiode array detector. J Anal Toxicol 17: 225–229.

Dulaurent S, Moesch C, Marquet P, Gaulier JM, Lachâtre G. 2010. Screening of pesticides in blood with liquid chromatography-linear ion trap mass spectrometry. Anal Bioanal Chem 396: 2235–2249.

Eckers C, Monaghan JJ, Wolff JC. 2005. Fragmentation of trimethoprim and other compounds containing alkoxy-phenyl groups in electrospray ionisation tandem mass spectrometry. Eur J Mass Spectrom 11: 73–82.

European Commission. 2002a. Decision 2002/657/EC implementing Council Directive 96/23/EC concerning the performance of analytical methods and the interpretation of results. Off J Eur Commun L221: 8–36. Available from: http://eur-lex.europa.eu/legal-content/EN/ALL/?uri=CELEX:32002D0657 (Accessed: 9 Aug 2015).

European Commission. 2002b. Directive 2002/32/EC of the European Parliament and of the Council of 7 May 2002 on undesirable substances in animal feed, L 140/10 (consolidated version: 2002 L0032—ES— 01.01.2012—014.002—1). Available from: http://eur-lex.europa.eu/legal-content/EN/TXT/?uri=CELEX:32002L0032 (Accessed: 9 Aug 2015).

European Commission. 2003. Regulation (EC) No 1831/2003 of the European Parliament and of the Council of 22 September 2003 on additives for use in animal nutrition. Off J Eur Union L 268/29 [Internet]. Available from: http://eur-lex.europa.eu/legal-content/EN/ALL/?uri=CELEX:32003R1831 (Accessed: 9 Aug 2015).

European Commission. 2008. Regulation (EC) No 299/2008 of the European Parliament and of the Council of 11 March 2008 amending Regulation (EC) No 396/ 2005 on maximum residue levels of pesticides in or on food and feed of plant and animal origin, as regards the implementing powers conferred on the Commission. Off J Eur Union L 97/67. Available from: http://eur-lex.europa.eu/legal-content/EN/ALL/?uri=CELEX:32008R0299 (Accessed: 9 Aug 2015).

European Commission. 2009. Commission Directive 2009/8/EC of 10 February. 2009. Amending Annex I to Directive 2002/32/EC of the European Parliament and of the Council as regards maximum levels of unavoidable carryover of coccidiostats or histomonostats in nontarget feed. Off J Eur Union L 40/19. Available from: http://eur-lex.europa.eu/legal-content/EN/TXT/?uri=CELEX:02002L0032-20131227 (Accessed: 9 Aug 2015).

European Commission. 2013. SANCO/12571/2013. Guidance document on analytical quality control and validation procedures for pesticide residues analysis in food and feed. Document No. SANCO/12571/2013. Available from: http://www.eurl-pesticides.eu/docs/public/tmplt_article.asp?CntID=727 (Accessed: 9 Aug 2015).

Fandiño AS, Nägele E, Perkins PD. 2006. Automated software-guided identification of new buspirone metabolites using capillary LC coupled to ion trap and TOF mass spectrometry. J Mass Spectrom 41: 248–255.

Fiehn O. 2001. Combining genomics, metabolome analysis, and biochemical modelling to understand metabolic networks. Comp Funct Genomics 2: 155–168.

Forner D, Berrué F, Correa H, Duncan K, Kerr RG. 2013. Chemical dereplication of marine actinomycetes by liquid chromatography-high resolution mass spectrometry profiling and statistical analysis. Anal Chim Acta 805: 70–79.

Fung EN, Jemal M, Aubry AF. 2013. High-resolution MS in regulated bioanalysis: Where are we now and where do we go from here? Bioanalysis 5: 1277–1284.

García-Reyes JF, Hernando MD, Molina-Díaz A, Fernández-Alba AR. 2007. Comprehensive screening of target, non-target and unknown pesticides in food by LC-TOF-MS. Trends Anal Chem 26: 828–841.

Geis-Asteggiante L, Lehotay SJ, Lightfield AR, Dutko T, Ng C, Bluhm L. 2012. Ruggedness testing and validation of a practical analytical method for 100 veterinary drug residues in bovine muscle by ultrahigh performance liquid chromatography-tandem mass spectrometry. J Chromatogr A 1258: 43–54.

Gergov M, Robson JN, Duchoslav E, Ojanperä I. 2000. Automated liquid chromatographic/tandem mass spectrometric method for screening β-blocking drugs in urine. J Mass Spectrom 35: 912–918.

Gergov M, Boucher B, Ojanperä I, Vuori E. 2001. Toxicological screening of urine for drugs by liquid chromatography/time-of-flight mass spectrometry with automated target library search based on elemental formulas. Rapid Commun Mass Spectrom 15: 521–526.

Gergov M, Ojanperä I, Vuori E. 2003. Simultaneous screening for 238 drugs in blood by liquid chromatography–ionspray tandem mass spectrometry with multiple-reaction monitoring. J Chromatogr B 795: 41–53.

Golge O, Kabak B. 2015. Evaluation of QuEChERS sample preparation and liquid chromatography-triple-quadrupole mass spectrometry method for the determination of 109 pesticide residues in tomatoes. Food Chem 176: 319–332.

Gómez-Pérez ML, Romero-González R, Luis Martínez VJ, Garrido Frenich A. 2015a. Analysis of pesticide and veterinary drug residues in baby food by liquid chromatography coupled to Orbitrap high resolution mass spectrometry. Talanta 131: 1–7.

Gómez-Pérez ML, Romero-González R, Martínez Vidal JL, Garrido Frenich A. 2015b. Analysis of veterinary drug and pesticide residues in animal feed by high-resolution mass spectrometry: Comparison between time-of-flight and Orbitrap. Food Addit Contam A 32: in press. DOI:10.1080/19440049.2015.1061703 (PMID: 26212769).

Grabic R, Fick J, Lindberg RH, Fedorova G, Tysklind M. 2012. Multi-residue method for trace level determination of pharmaceuticals in environmental samples using liquid chromatography coupled to triple quadrupole mass spectrometry. Talanta 100: 183–195.

Gros M, Petrović M, Barceló D. 2006. Multi-residue analytical methods using LC-tandem MS for the determination of pharmaceuticals in environmental and wastewater samples: A review. Anal Bioanal Chem 386: 941–952.

Guengerich FP. 2008. Cytochrome P450 and chemical toxicology. Chem Res Toxicol 21: 70–83.

Hager JW. 2002. A new linear ion trap mass spectrometer. Rapid Commun Mass Spectrom 16:512–526.

Hakala KS, Kostiainen R, Ketola RA. 2006. Feasibility of different mass spectrometric techniques and programs for automated metabolite profiling of tramadol in human urine. Rapid Commun Mass Spectrom 20: 2081–2090.

Hayward MJ, Snodgrass JT, Thomson ML. 1993. Flow-injection thermospray ionization mass spectrometry in the automatic analysis of agricultural chemicals. Rapid Commun Mass Spectrom 7: 85–91.

Hegy G, Görlach E, Richmond R, Bitsch F. 1996. High-throughput ESI-MS of combinatorial chemistry racks with automated contamination surveillance and result reporting. Rapid Commun Mass Spectrom 10: 1894–1900.

Hernández F, Ibáñez M, Botero-Coy AM, Bade R, Bustos-López MC, Rincón J, Moncayo A, Bijlsma L. 2015. LC-QTOF MS screening of more than 1,000 licit and illicit drugs and their metabolites in wastewater and surface waters from the area of Bogotá, Colombia. Anal Bioanal Chem 407: 6405–6416.

Holčapek M, Kolářová L, Nobilis M. 2008. High-performance liquid chromatography–tandem mass spectrometry in the identification and determination of phase I and phase II drug metabolites. Anal Bioanal Chem 391: 59–78.

Hopfgartner G, Bourgogne E. 2003. Quantitative high-throughput analysis of drugs in biological matrices by mass spectrometry. Mass Spectrom Rev 22: 195–214.

Hough JM, Haney CA, Voyksner RD, Bereman RD. 2000. Evaluation of electrospray transport CID for the generation of searchable libraries. Anal Chem 72: 2265–2270.

Ibáñez M, Sancho JV, Pozo ÓJ, Niessen WMA, Hernández F. 2005. Use of quadrupole time-of-flight mass spectrometry in the elucidation of unknown compounds present in environmental water. Rapid Commun Mass Spectrom 19: 169–178.

Jacobs PL, Ridder L, Ruijken M, Roding H, Jager NGL, Beijnen JH, Bas RR, Van Dongen WD. 2013. Identification of drug metabolites in human plasma or serum integrating metabolite prediction, LC–HRMS and untargeted data processing. Bioanalysis 5: 2115–2128.

Jain D, Basniwal PK. 2013. Forced degradation and impurity profiling: Recent trends in analytical perspectives. J Pharm Biomed Anal 86: 11–35.

Jansen R, Lachâtre G, Marquet P. 2005. LC–MS–MS systematic toxicological analysis: Comparison of MS–MS spectra obtained with different instruments and settings. Clin Biochem 38: 362–372.

Jemal M, Ouyang Z, Xia YQ. 2010. Systematic LC-MS/MS bioanalytical method development that incorporates plasma phospholipids risk avoidance, usage of incurred sample and well thought-out chromatography. Biomed Chromatogr 24: 2–19.

Jindal K, Narayanam M, Singh S. 2015. A systematic strategy for the identification and determination of pharmaceuticals in environment using advanced LC-MS tools: Application to ground water samples. J Pharm Biomed Anal 108: 86–96.

Josephs JL, Sanders M. 2004. Creation and comparison of MS–MS spectral libraries using quadrupole ion trap and triple-quadrupole mass spectrometer. Rapid Commun Mass Spectrom 18: 743–759.

Kaklamanos G, Vincent U, von Holst C. 2013. Multi-residue method for the detection of veterinary drugs in distillers grains by liquid chromatography-Orbitrap high resolution mass spectrometry. J Chromatogr A 1322: 38–48.

Kasper PT, Rojas-Chertó M, Mistrik R, Reijmers T, Hankemeier T, Vreeken RJ. 2012. Fragmentation trees for the structural characterisation of metabolites. Rapid Commun Mass Spectrom 26: 2275–2286.

Kaufmann A, Butcher P, Maden K, Widmer M. 2008. Quantitative multiresidue method for about 100 veterinary drugs in different meat matrices by sub 2-microm particulate high-performance liquid chromatography coupled to time of flight mass spectrometry. J Chromatogr A 1194: 66–79.

Kaufmann A, Butcher P, Maden K, Walker S, Widmer M. 2010. Comprehensive comparison of liquid chromatography selectivity as provided by two types of liquid chromatography detectors (high resolution mass spectrometry and tandem mass spectrometry): "where is the crossover point?". Anal Chim Acta 673: 60–72.

Kaufmann A, Butcher P, Maden K, Walker S, Widmer M. 2011. Quantitative and confirmative performance of liquid chromatography coupled to high-resolution mass spectrometry compared to tandem mass spectrometry. Rapid Commun Mass Spectrom 25: 979–992.

Kaufmann A, Butcher P, Maden K, Walker S, Widmer M. 2015. Reliability of veterinary drug residue confirmation: High resolution mass spectrometry versus tandem mass spectrometry. Anal Chim Acta 856: 54–67.

Kerns EH, Volk KJ, Hill SE, Lee MS. 1995. Profiling new taxanes using LC/MS and LC/MS/MS substructural analysis techniques. Rapid Commun Mass Spectrom 9: 1539–1545.

Kerns EH, Rourick RA, Volk KJ, Lee MS. 1997. Buspirone metabolite structure profile using a standard liquid chromatographic–mass spectrometric protocol. J Chromatogr B 698: 133–145.

Kim S-C, Carlson K. 2005. LC–MS^2 for quantifying trace amounts of pharmaceutical compounds in soil and sediment matrices. Trends Anal Chem 24: 635–644.

Kind T, Fiehn O. 2007. Seven Golden Rules for heuristic filtering of molecular formulas obtained by accurate mass spectrometry. BMC Bioinf, 8: 105.

Kind T, Fiehn O. 2010. Advances in structure elucidation of small molecules using mass spectrometry. Bioanal Rev 2: 23–60.

Kirchmair J, Williamson MJ, Tyzack JD, Tan L, Bond PJ, Bender A, Glen RC. 2012. Computational prediction of metabolism: Sites, products, SAR, P450 enzyme dynamics, and mechanisms. J Chem Inf Model 52: 617–648.

Köck-Schulmeyer M, Olmos M, López de Alda M, Barceló D. 2013. Development of a multiresidue method for analysis of pesticides in sediments based on isotope dilution and liquid chromatography-electrospray-tandem mass spectrometry. J Chromatogr A 1305: 176–187.

Koehn FE, Carter GT. 2005. The evolving role of natural products in drug discovery. Nat Rev Drug Discov 4: 206–220.

Kolmonen M, Leinonen A, Pelander A, Ojanperä I. 2007. A general screening method for doping agents in human urine by solid phase extraction and liquid chromatography/time-of-flight mass spectrometry. Anal Chim Acta 585: 94–102.

Kolmonen M, Leinonen A, Kuuranne T, Pelander A, Ojanperä I. 2009. Generic sample preparation and dual polarity liquid chromatography–time-of-flight mass spectrometry for high-throughput screening in doping analysis. Drug Test Anal 1:250–266.

Koo I, Kim S, Zhang X. 2013. Comparative analysis of mass spectral matching-based compound identification in gas chromatography-mass spectrometry. J Chromatogr A 1298: 132–138.

Kool J, Giera M, Irth H, Niessen WMA. 2011. Advances in mass spectrometry-based post-column bioaffinity profiling of mixtures. Anal Bioanal Chem 399: 2655–2668.

Kostiainen R, Kotiaho T, Kuuranne T, Auriola S. 2003. Liquid chromatography/atmospheric pressure ionization–mass spectrometry in drug metabolism studies. J Mass Spectrom 38: 357–372.

Kratzsch C, Peters FT, Kraemer T, Weber AA, Maurer HH. 2003. Screening, library-assisted identification and validated quantification of fifteen neuroleptics and three of their metabolites in plasma by liquid chromatography/mass spectrometry with atmospheric pressure chemical ionization. J Mass Spectrom 38: 283–295.

Kratzsch C, Tenberken O, Peters FT, Weber AA, Kraemer T, Maurer HH. 2004. Screening, library-assisted identification and validated quantification of 23 benzodiazepines, flumazenil, zaleplone, zolpidem and zopiclone in plasma by liquid chromatography/mass spectrometry with atmospheric pressure chemical ionization. J Mass Spectrom 39: 856–872.

Kuster M, López de Alda M, Barceló D. 2009. Liquid chromatography-tandem mass spectrometric analysis and regulatory issues of polar pesticides in natural and treated waters. J Chromatogr A 1216: 520–529.

Laks S, Pelander A, Vuori E, Ali-Tolppa E, Sippola E, Ojanperä I. 2004. Analysis of street drugs in seized material without primary reference standards. Anal Chem 76: 7375–7379.

Lapthorn C, Pullen F, Chowdhry BZ. 2013. Ion mobility spectrometry–mass spectrometry (IMS–MS) of small molecules: Separating and assigning structures to ions. Mass Spectrom Rev 32: 43–71.

Lee HK, Ho CS, Iu YP, Lai PS, Shek CC, Lo Y-C, Klinke HB, Wood M. 2009. Development of a broad toxicological screening technique for urine using ultra-performance liquid chromatography and time-of-flight mass spectrometry. Anal Chim Acta 649: 80–90.

Lehr GJ, Barry TL, Petzinger G, Hanna GM, Zito SW. 1999. Isolation and identification of process impurities in trimethoprim drug substance by high-performance liquid chromatography, atmospheric pressure chemical ionization liquid chromatography–mass spectrometry and nuclear magnetic resonance spectroscopy. J Pharm Biomed Anal 19: 373–389.

Li AC, Alton D, Bryant MS, Shou WZ. 2005. Simultaneously quantifying parent drugs and screening for metabolites in plasma pharmacokinetic samples using selected reaction monitoring information-dependent acquisition on a QTrap instrument. Rapid Commun Mass Spectrom 19: 1943–1950.

Li AC, Gohdes MA, Shou WZ. 2007a. 'N-in-one' strategy for metabolite identification using a liquid chromatography/hybrid triple quadrupole linear ion trap instrument using multiple dependent product ion scans triggered with full mass scan. Rapid Commun Mass Spectrom 21: 1421–1430.

Li AC, Shou WZ, Mai TT, Jiang X-y. 2007b. Complete profiling and characterization of in vitro nefazodone metabolites using two different tandem mass spectrometric platforms. Rapid Commun Mass Spectrom 21: 4001–4008.

Lin M, Li M, Rustum A. 2007. Identification of an unknown extraneous contaminant in pharmaceutical product analysis. J Pharm Biomed Anal, 45: 747–755.

Lips AGAM, Lameijer W, Fokkens RH, Nibbering NMM. 2001. Methodology for the development of a drug library based upon collision-induced fragmentation for the identification of toxicologically relevant drugs in plasma samples. J Chromatogr B 759: 191–207.

Little JL, Cleven CD, Brown SD. 2011. Identification of "known unknowns" utilizing accurate mass data and chemical abstracts service databases. J Am Soc Mass Spectrom 22: 348–359.

Liu DQ, Hop CECA. 2005. Strategies for characterization of drug metabolites using liquid chromatography–tandem mass spectrometry in conjunction with chemical derivatization and on-line H/D exchange approaches. J Pharm Biomed Anal 37: 1–18.

Liu HC, Liu RH, Ho HO, Lin DL. 2009. Development of an information-rich LC–MS/ MS database for the analysis of drugs in postmortem specimens. Anal Chem 81: 9002–9011.

Lo DST, Chao TC, Ng-On SE, Yao YJ, Koh TH. 1997. Acidic and neutral drugs screen in blood with quantitation using microbore high-performance liquid chromatography–diode array detection and capillary gas chromatography–flame ionization detection. Forensic Sci Int 90: 205–214.

Ma S, Zhu M. 2009. Recent advances in applications of liquid chromatography–tandem mass spectrometry to the analysis of reactive drug metabolites. Chem Biol Interact 179: 25–37.

Malachová A, Sulyok M, Beltrán E, Berthiller F, Krska R. 2014. Optimization and validation of a quantitative liquid chromatography-tandem mass spectrometric method covering 295 bacterial and fungal metabolites including all regulated mycotoxins in four model food matrices. J Chromatogr A 1362: 145–156.

Malherbe CJ, de Beer D, Joubert E. 2012. Development of on-line high performance liquid chromatography (HPLC)-biochemical detection methods as tools in the identification of bioactives. Int J Mol Sci 13: 3101–3133.

Mallis LM, Sarkahian AB, Kulishoff JM, Jr, Watts WL Jr. 2002. Open-access LC–MS in drug discovery environment. J Mass Spectrom 37: 889–896.

March R, Brodbelt J. 2008. Analysis of flavonoids: Tandem mass spectrometry, computational methods, and NMR. J Mass Spectrom 43: 1581–1617.

Marquet P, Lachâtre G. 1999. Liquid chromatography–mass spectrometry: Potential in forensic and clinical toxicology. J Chromatogr B 733: 93–118.

Marquet P, Venisse N, Lacassie É, Lachâtre G. 2000. In-source CID mass spectral libraries for the "general unknown" screening of drugs and toxicants. Analusis 28: 925–934.

Marquet P. 2002. Is LC–MS suitable for a comprehensive screening of drugs and poisons in clinical toxicology? Ther Drug Monit 24: 125–133.

Marquet P, Saint-Marcoux F, Gamble TN, Leblanc JCY. 2003. Comparison of a preliminary procedure for the general unknown screening of drugs and toxic compounds using a quadrupole-linear ion-trap mass spectrometer with a liquid chromatography–mass spectrometry reference technique. J Chromatogr B 789: 9–18.

Masiá A, Campo J, Blasco C, Picó Y. 2014. Ultra-high performance liquid chromatography-quadrupole time-of-flight mass spectrometry to identify contaminants in water: An insight on environmental forensics. J Chromatogr A 1345: 86–97.

Maurer HH. 1992. Systematic toxicological analysis of drugs and their metabolites by gas chromatography–mass spectrometry. J Chromatogr B 580: 3–41.

Maurer HH. 1998. Liquid chromatography–mass spectrometry in forensic and clinical toxicology. J Chromatogr B 713: 3–25.

Maurer HH. 2005. Multi-analyte procedures for screening for and quantification of drugs in blood, plasma, or serum by liquid chromatography-single stage or tandem mass spectrometry (LC-MS or LC-MS/MS) relevant to clinical and forensic toxicology. Clin Biochem 38: 310–318.

Maurer HH. 2007. Current role of liquid chromatography-mass spectrometry in clinical and forensic toxicology. Anal Bioanal Chem 388: 1315–1325.

Mazzarino M, de la Torre X, Botrè F. 2008. A screening method for the simultaneous detection of glucocorticoids, diuretics, stimulants, anti-oestrogens, beta-adrenergic drugs and anabolic steroids in human urine by LC–ESI-MS/MS. Anal Bioanal Chem 392: 681–698.

Milman BL. 2005. Towards a full reference library of MS^n spectra. Testing a library containing 3126 MS^2 spectra of 1743 compounds. Rapid Commun Mass Spectrom 19: 2833–2839.

Milman BL. 2015. General principles of identification by mass spectrometry. Trends Anal Chem 69: 24–33.

Mladić M, Scholten DJ, Niessen WMA, Somsen GW, Smit MJ, Kool J. 2015. At-line coupling of LC–MS to bioaffinity and selectivity assessment for metabolic profiling of ligands towards chemokine receptors CXCR1 and CXCR2. J Chromatogr B 1002: 42–53.

Mol HG, Zomer P, de Koning M. 2012. Qualitative aspects and validation of a screening method for pesticides in vegetables and fruits based on liquid chromatography coupled to full scan high resolution (Orbitrap) mass spectrometry. Anal Bioanal Chem 403: 2891–2908

Molina-Martin M, Marin A, Rivera-Sagredo A, Espada A. 2005. Liquid chromatography-mass spectrometry and related techniques for purity assessment in early drug discovery. J Sep Sci 28: 1742–1750.

Mortishire-Smith RJ, Castro-Perez JM, Yu K, Shockcor JP, Goshawk J, Hartshorn MJ, Hill A. 2009. Generic dealkylation: A tool for increasing the hit-rate of metabolite rationalization, and automatic customization of mass defect filters. Rapid Commun Mass Spectrom 23: 939–948.

Mueller CA, Weinmann W, Dresen S, Schreiber A, Gergov M. 2005. Development of a multi-target screening analysis for 301 drugs using a QTrap liquid chromatography/tandem mass spectrometry system and automated library searching. Rapid Commun Mass Spectrom 19: 1332–1338.

Musharraf SG, Goher M, Shahnaz S, Choudhary MI, Atta-ur-Rahman. 2013. Structure-fragmentation relationship and rapid dereplication of Buxus steroidal alkaloids by electrospray ionization-quadrupole time-of-flight mass spectrometry. Rapid Commun Mass Spectrom 27: 169–178.

Mylonas R, Mauron Y, Masselot A, Binz PA, Budin N, Fathi M, Viette V, Hochstrasser DF, Lisacek F. 2009. X-Rank: A robust algorithm for small molecule identification using tandem mass spectrometry. Anal Chem 81: 7604–7610.

Nakabayashi R, Sawada Y, Yamada Y, Suzuki M, Hirai MY, Sakurai T, Saito K. 2013, Combination of liquid chromatography-Fourier transform ion cyclotron resonance-mass spectrometry with ^{13}C-labeling for chemical assignment of sulfur-containing metabolites in onion bulbs. Anal Chem 85: 1310–1315.

Nielen MW, van Engelen MC, Zuiderent R, Ramaker R. 2007. Screening and confirmation criteria for hormone residue analysis using liquid chromatography accurate mass time-of-flight,

Fourier transform ion cyclotron resonance and orbitrap mass spectrometry techniques. Anal Chim Acta 586: 122–129.

Niessen WMA, Falck D. 2015. Introduction to mass spectrometry, a tutorial. Ch. 1 in: Kool J & Niessen WMA (Ed.): Analyzing biomolecular interactions by mass spectrometry. Wiley-VCH Verlag. ISBN 978-3-527-33464-3.

Niessen WMA, Honing H. 2015. Mass spectrometry strategies in the assignment of molecular structure: Breaking chemical bonds before bringing the pieces of the puzzle together. Ch. 4 in: Cid MM, Bravo J (Ed.): Structure elucidation in organic chemistry: The search for the right tools. Wiley-VCH Verlag. ISBN 978-3-527-33336-3.

NIST MS Library (http://www.nist.gov/srd/nist1a.cfm)

Núñez O, Gallart-Ayala H, Ferrer I, Moyano E, Galceran MT. 2012. Strategies for the multi-residue analysis of 100 pesticides by liquid chromatography-triple quadrupole mass spectrometry. J Chromatogr A 1249: 164–180.

Oberacher H, Pavlic M, Libiseller K, Schubert B, Sulyok M, Schuhmacher R, Csaszar E, Köfeler HC. 2009a. On the inter-instrument and inter-laboratory transferability of a tandem mass spectral reference library: 1. Results of an Austrian multicenter study. J Mass Spectrom 44: 485–493.

Oberacher H, Pavlic M, Libiseller K, Schubert B, Sulyok M, Schuhmacher R, Csaszar E, Köfeler HC. 2009b. On the inter-instrument and the inter-laboratory transferability of a tandem mass spectral reference library: 2. Optimization and characterization of the search algorithm. J Mass Spectrom 44: 494–502.

Oberacher H. 2011. Wiley registry of tandem mass Spectral Data, MSforID. Hoboken, Wiley, ISBN: 978-1-118-03744-7.

Oberacher H, Pitterl F, Siapi E, Steele BR, Letzel T, Grosse S, Poschner B, Tagliaro F, Gottardo R, Chacko SA, Josephs JL. 2012. On the inter-instrument and the interlaboratory transferability of a tandem mass spectral reference library. 3. Focus on ion trap and upfront CID. J Mass Spectrom 47: 263–270.

Oberacher H, Whitley G, Berger B. 2013a. Evaluation of the sensitivity of the 'Wiley registry of tandem mass spectral data, MSforID' with MS/MS data of the 'NIST/NIH/EPA mass spectral library'. J Mass Spectrom 48: 487–496.

Oberacher H, Whitley G, Berger B, Weinmann W. 2013b. Testing an alternative search algorithm for compound identification with the 'Wiley Registry of Tandem Mass Spectral Data, MSforID'. J Mass Spectrom 48: 497–504.

Ojanperä S, Pelander A, Pelzing M, Krebs I, Vuori E, Ojanperä I. 2006. Isotopic pattern and accurate mass determination in urine drug screening by liquid chromatography/time-of-flight mass spectrometry. Rapid Commun Mass Spectrom 20: 1161–1167.

Ojanperä I, Kolmonen M, Pelander A. 2012. Current use of high-resolution mass spectrometry in drug screening relevant to clinical and forensic toxicology and doping control. Anal Bioanal Chem 403: 1203–1220.

Ortelli D, Cognard E, Jan P, Edder P. 2009. Comprehensive fast multiresidue screening of 150 veterinary drugs in milk by ultra-performance liquid chromatography coupled to time of flight mass spectrometry. J Chromatogr B 877: 2363–2374.

Pelander A, Ojanperä I, Laks S, Rasanen I, Vuori E. 2003. Toxicological screening with formula-based metabolite identification by liquid chromatography/time-of-flight mass spectrometry. Anal Chem 75: 5710–5718.

Peters RJB, Bolck YJC, Rutgers P, Stolker AAM, Nielen MWF. 2009. Multi-residue screening of veterinary drugs in egg, fish and meat using high-resolution liquid chromatography accurate mass time-of-flight mass spectrometry. J Chromatogr A 1216: 8206–8216.

Petrović M, Hernando MD, Díaz-Cruz MS, Barceló D. 2005. Liquid chromatography-tandem mass spectrometry for the analysis of pharmaceutical residues in environmental samples: A review. J Chromatogr A 1067: 1–14.

Petrovic M, Farré M, de Alda ML, Perez S, Postigo C, Köck M, Radjenovic J, Gros M, Barcelo D. 2010. Recent trends in the liquid chromatography-mass spectrometry analysis of organic contaminants in environmental samples. J Chromatogr A 1217: 4004–4017.

Plumb RS, Johnson KA, Rainville P, Smith BW, Wilson ID, Castro-Perez JM, Nicholson JK. 2006. UPLC/MS^E; a new approach for generating molecular fragment information for biomarker structure elucidation. Rapid Commun Mass Spectrom 20: 1989–1994.

Pluskal T, Uehara T, Yanagida M. 2012. Highly accurate chemical formula prediction tool utilizing high-resolution mass spectra, MS/MS fragmentation, heuristic rules, and isotope pattern matching. Anal Chem 84: 4396–4403.

Polettini A. 1999. Systematic toxicological analysis of drugs and poisons in biosamples by hyphenated chromatographic and spectroscopic techniques. J Chromatogr B 733: 47–63.

Polettini A, Gottardo R, Pascali JP, Tagliaro F. 2008. Implementation and performance evaluation of a database of chemical formulas for the screening of pharmaco/toxicologically relevant compounds in biological samples using electrospray ionization-time-of-flight mass spectrometry. Anal Chem 80: 3050–3057.

Poon GK, Kwei G, Wang R, Lyons K, Chen Q, Didolkar V, Hop CE. 1999. Integrating qualitative and quantitative liquid chromatography/tandem mass spectrometric analysis to support drug discovery. Rapid Commun Mass Spectrom 13: 1943–1950.

Pozo ÓJ, Van Eenoo P, Deventer K, Elbardissy H, Grimalt S, Sancho JV, Hernández F, Ventura R, Delbeke FT. 2011. Comparison between triple quadrupole, time of flight and hybrid quadrupole time of flight analysers coupled to liquid chromatography for the detection of anabolic steroids in doping control analysis. Anal Chim Acta 684: 98–111.

Pozo ÓJ, Marcos J, Segura J, Ventura R. 2012. Recent developments in MS for small molecules: Application to human doping control analysis. Bioanalysis 4: 197–212.

Pullen FS, Perkins GL, Burton KI, Ware RS, Taegue MS, Kiplinger JP. 1995. Putting MS in the hands of the end users, J Am Soc Mass Spectrom 6: 394–399.

Ramanathan R, Jemal M, Ramagiri S, Xia YQ, Humpreys WG, Olah T, Korfmacher WA. 2011. It is time for a paradigm shift in drug discovery bioanalysis: From SRM to HRMS. J Mass Spectrom 46: 595–601.

Remane D, Wetzel D, Peters FT. 2014. Development and validation of a liquid chromatography-tandem mass spectrometry (LC-MS/MS) procedure for screening of urine specimens for 100 analytes relevant in drug-facilitated crime (DFC). Anal Bioanal Chem 406: 4411–4424.

Richardson SD, Ternes TA. 2011. Water analysis: Emerging contaminants and current issues. Anal Chem 83: 4614–4648.

Rittner M, Pragst F, Neumann J. 2001. Screening method for seventy psychoactive drugs or drug metabolites in serum based on high-performance liquid chromatography-electrospray ionization mass spectrometry. J Anal Toxicol 25: 115–124.

Rocha DG, Santos FA, da Silva JC, Augusti R, Faria AF. 2015. Multiresidue determination of fluoroquinolones in poultry muscle and kidney according to the regulation 2002/657/EC. A systematic comparison of two different approaches: Liquid chromatography coupled to high-resolution mass spectrometry or tandem mass spectrometry. J Chromatogr A 1379: 83–91.

Roemmelt AT, Steuer AE, Poetzsch M, Kraemer T. 2014. Liquid chromatography, in combination with a quadrupole time-of-flight instrument (LC QTOF), with sequential window acquisition of all theoretical fragment-ion spectra (SWATH) acquisition: Systematic studies on its use for screenings in clinical and forensic toxicology and comparison with information-dependent acquisition (IDA). Anal Chem 86: 11742–11749.

Rojas-Chertó M, Kasper PT, Willighagen EL, Vreeken RJ, Hankemeier T, Reijmers TH. 2011. Elemental composition determination based on MS^n. Bioinformatics 27: 2376–2383.

Rojas-Chertó M, Peironcely JE, Kasper PT, van der Hooft JJ, de Vos RC, Vreeken RJ, Hankemeier T, Reijmers T. 2012. Metabolite identification using automated comparison of high-resolution multistage mass spectral trees. Anal Chem 84: 5524–5534.

Rousu T, Pelkonen O, Tolonen A. 2009. Rapid detection and characterization of reactive drug metabolites in vitro using several isotope-labeled trapping agents and ultra-performance liquid chromatography/time-of-flight mass spectrometry. Rapid Commun Mass Spectrom 23: 843–855.

Ruan Q, Peterman S, Szewc MA, Ma L, Cui D, Humphreys WG, Zhu M. 2008. An integrated method for metabolite detection and identification using a linear ion trap/Orbitrap mass spectrometer and multiple data processing techniques: Application to indinavir metabolite detection. J Mass Spectrom, 43: 251–261.

Sadeg N, François G, Petit B, Duterte-Catella H, Dumontet M. 1997. Automated liquid chromatographic analyzer used for toxicology screening in a general hospital: 12 months' experience. Clin Chem 43: 498–504.

Sauvage F-L, Saint-Marcoux F, Duretz B, Deporte D, Lachâtre G, Marquet P. 2006. Screening of drugs and toxic compounds with liquid chromatography-linear ion trap tandem mass spectrometry. Clin Chem 52: 1735–1742.

Sauvage F-L, Gaulier JM, Lachâtre G, Marquet P. 2008. Pitfalls and prevention strategies for liquid chromatography-tandem mass spectrometry in the selected reaction monitoring mode for drug analysis. Clin Chem 54: 1519–1527.

Sauvage F-L, Picard N, Saint-Marcoux F, Gaulier JM, Lachâtre G, Marquet P. 2009. General unknown screening procedure for the characterization of human drug metabolites in forensic toxicology: Applications and constraints. J Sep Sci 32: 3074–3083.

Sauvage F-L, Marquet P, 2010. Letter to the Editor: ESI-MS–MS library of 1,253 compounds for application in forensic and clinical toxicology. Anal Bioanal Chem 396: 1947.

Schreiber A, Efer J, Engewald W. 2000. Application of spectral libraries for high-performance liquid chromatography–atmospheric pressure ionisation mass spectrometry to the analysis of pesticide and explosive residues in environmental samples. J Chromatogr A 869: 411–425.

Shah RP, Garg A, Putlur SP, Wagh S, Kumar V, Rao V, Singh S, Mandlekar S, Desikan S. 2013. Practical and economical implementation of online H/D exchange in LC-MS. Anal Chem 85: 10904–10912.

Shimizu A, Chiba M. 2013. Ion mobility spectrometry-mass spectrometry analysis for the site of aromatic hydroxylation. Drug Metab Dispos 41: 1295–1299.

Singh S, Handa T, Narayanam M, Sahu A, Junwal M, Shah RP. 2012. A critical review on the use of modern sophisticated hyphenated tools in the characterization of impurities and degradation products. J Pharm Biomed Anal 69: 148–173.

Smyth WF, Smyth TJP, Ramachandran VN, O'Donnell F, Brooks P. 2012. Dereplication of phytochemicals in plants by LC-ESI-MS and ESI-MS^n. Trends Anal Chem 33: 46–54.

Staack RF, Hopfgartner G. 2007. New analytical strategies in studying drug metabolism. Anal Bioanal Chem 388: 1365–1380.

Stein SE, Scott DR. 1994. Optimization and testing of mass spectral library search algorithms for compound identification. J Am Soc Mass Spectrom 5: 859–866.

Stolker AAM, Stephany RW, van Ginkel LA. 2000. Identification of residues by LC–MS. The application of new EU guidelines. Analusis 28: 947–951.

Stolker AA, Rutgers P, Oosterink E, Lasaroms JJ, Peters RJ, van Rhijn JA, Nielen MW. 2008. Comprehensive screening and quantification of veterinary drugs in milk using UPLC-ToF-MS. Anal Bioanal Chem 391: 2309–2322.

Thevis M, Thomas A, Schänzer W. 2011. Current role of LC-MS(/MS) in doping control. Anal Bioanal Chem 401: 405–420.

Thieme D, Sachs H. 2003. Improved screening capabilities in forensic toxicology by application of liquid chromatography–tandem mass spectrometry. Anal Chim Acta 492: 171–186.

Thomas KV, Bijlsma L, Castiglioni S, Covaci A, Emke E, Grabic R, Hernández F, Karolak S, Kasprzyk-Hordern B, Lindberg RH, Lopez de Alda M, Meierjohann A, Ort C, Pico Y, Quintana JB, Reid M, Rieckermann J, Terzic S, van Nuijs AL, de Voogt P. 2012. Comparing illicit drug use in 19 European cities through sewage analysis. Sci Total Environ 432: 432–439.

Thurman EM, Ferrer I, Fernández-Alba AR. 2005 Matching unknown empirical formulas to chemical structure using LC/MS TOF accurate mass and database searching: Example of unknown pesticides on tomato skins. J Chromatogr A, 1067: 127–134.

Tong H, Bell D, Tabei K, Siegel MM. 1999. Automated data massaging, interpretation and E-mailing modules for high-throughput open-access mass spectrometry, J Am Soc Mass Spectrom 10: 1174–1187.

Tracqui A, Kintz P, Mangin P. 1995. Systematic toxicological analysis using HPLC/DAD. J Forensic Sci 40: 254–262.

Umezawa H, Lee XP, Arima Y, Hasegawa C, Izawa H, Kumazawa T, Sato K. 2008. Simultaneous determination of beta-blockers in human plasma using liquid chromatography-tandem mass spectrometry. Biomed Chromatogr 22: 702–711.

Valerio LG, Jr. 2009. *In silico* toxicology for the pharmaceutical sciences. Toxicol Appl Pharmacol 241: 356–370.

Valli A, Polletini A, Papa P, Montagna M. 2001. Comprehensive drug screening by integrated use of gas chromatography – mass spectrometry and REMEDi HS. Ther Drug Monit 23: 287–294.

Van Bocxlaer JF, Clauwaert KM, Lambert WE, Deforce DL, Van den Eeckhout EG, De Leenheer AP 2000. Liquid chromatography–mass spectrometry in forensic toxicology. Mass Spectrom Rev 19: 165–214.

Vande Casteele SR, Bouche M-PL, Van Bocxlaer JF. 2005. LC-MS/MS in the elucidation of an isomer of the recreational drug methylenedioxy ethylamphetamine: Methylenedioxy dimethylamphetamine. J Sep Sci 28: 1729–1734.

Van der Hooft JJ, Vervoort J, Bino RJ, Beekwilder J, de Vos RC. 2011. Polyphenol identification based on systematic and robust high-resolution accurate mass spectrometry fragmentation. Anal Chem 83: 409–416.

Van Dongen WD, Niessen WMA. 2012. LC-MS systems for quantitative bioanalysis. Bioanalysis 4: 2391–2399.

Vanhaecke L, Van Meulebroek L, De Clercq N, Vanden Bussche J. 2013. High resolution Orbitrap mass spectrometry in comparison with tandem mass spectrometry for confirmation of anabolic steroids in meat. Anal Chim Acta 767: 118–127.

van Nuijs ALN, Mougel J-F, Tarcomnicu I, Bervoets L, Blust R, Jorens PG, Neels H, Covaci A. 2011. Sewage epidemiology – a real-time approach to estimate the consumption of illicit drugs in Brussels, Belgium. Environ Int 37: 612–621.

Villain M, Concheiro M, Cirimele V, Kintz P. 2005. Screening method for benzodiazepines and hypnotics in hair at pg/mg level by liquid chromatography–mass spectrometry/mass spectrometry. J Chromatogr B 825: 72–78.

Virus ED, Sobolevsky TG, Rodchenkov GM. 2008. Introduction of HPLC/orbitrap mass spectrometry as screening method for doping control. J Mass Spectrom 43: 949–957.

Vukics V, Guttman A. 2010. Structural characterization of flavonoid glycosides by multi-stage mass spectrometry. Mass Spectrom Rev 29: 1–16.

Wang S, Cheng L, Ji S, Wang K. 2014. Simultaneous determination of seventeen mycotoxins residues in Puerariae lobatae radix by liquid chromatography-tandem mass spectrometry. J Pharm Biomed Anal 98: 201–209

Waridel P, Wolfender J-L, Ndjoko K, Hobby KR, Major HJ, Hostettmann K. 2001. Evaluation of quadrupole time-of-flight tandem mass spectrometry and ion-trap multiple-stage mass spectrometry for the differentiation of C-glycosidic flavonoid isomers. J Chromatogr A 926: 29–41.

Weinmann W, Lehmann N, Renz M, Wiedemann A, Svoboda M. 2000a. Screening for drugs in serum and urine by LC/ESI/CID-MS and MS/MS with library searching. Probl Forensic Sci 42: 202–208.

Weinmann W, Gergov M, Goerner M. 2000b. MS–MS libraries with triple quadrupole MS–MS for drug identification and drug screening. Analusis 28: 934–941.

Weinmann W, Wiedemann A, Eppinger B, Renz M, Svoboda M. 1999. Screening for drugs in serum by electrospray ionization/collision-induced dissociation and library searching. J Am Soc Mass Spectrom 10: 1028–1037.

Weinmann W, Stoertzel M, Vogt S, Wendt J. 2001a.Tune compounds for electrospray ionisation/in-source collision-induced dissociation with mass spectral library searching. J Chromatogr A 926: 199–209.

Weinmann W, Stoertzel M, Vogt S, Svoboda M, Schreiber A. 2001b. Tuning compounds for electrospray ionization/in-source collision-induced dissociation with mass spectral library searching. J Mass Spectrom 36: 1013–1023.

Weinmann W, Dresen S. 2007. Database of QTrap LC-MS-MS mass spectra. The 2007 version of the library (~1250 compounds) Available from: http://www.chemicalsoft.de (accessed: 9 Aug 2015).

Wiley MS Library (http://eu.wiley.com/WileyCDA/WileyTitle/productCd-1118616111.html).

Wolfender J-L, Waridel P, Ndjoko K, Hobby KR, Major HR, Hostettmann K. 2000. Evaluation of Q-TOF-MS/MS and multiple stage IT-MS^n for the dereplication of flavonoids and related compounds in crude plant extracts. Analusis 28: 895–906.

Wolfender J-L, Ndjoko K, Hostettmann K. 2003. Liquid chromatography with ultraviolet absorbance-mass spectrometric detection and with nuclear magnetic resonance spectroscopy: A powerful combination for the on-line structural investigation of plant metabolites. J Chromatogr A 1000: 437–455.

Wolfender J-L, Marti G, Thomas A, Bertrand S. 2015. Current approaches and challenges for the metabolite profiling of complex natural extracts. J Chromatogr A 1382: 136–164.

Xie C, Zhong D, Yu K, Chen X. 2012. Recent advances in metabolite identification and quantitative bioanalysis by LC-Q-TOF MS. Bioanalysis 4: 937–959.

Xu RN, Fan L, Rieser MJ, El-Shourbagy TA. 2007. Recent advances in high-throughput quantitative bioanalysis by LC-MS/MS. J Pharm Biomed Anal 44: 342–355.

Yan Z, Caldwell GW. 2004. Stable-isotope trapping and high-throughput screenings of reactive metabolites using the isotope MS signature. Anal Chem 76: 6835–6847.

Zhang H, Zhang D, Ray K. 2003. A software filter to remove interference ions from drug metabolites in accurate mass liquid chromatography/mass spectrometric analyses. J Mass Spectrom 38: 1110–1112.

Zhang H, Zhang D, Ray K, Zhu M. 2009. Mass defect filter technique and its applications to drug metabolite identification by high-resolution mass spectrometry. J Mass Spectrom 44: 999–1016.

Zhang N, Fountain ST, Bi H, Rossi DT. 2000. Quantification and rapid metabolite identification in drug discovery using API time-of-flight LC/MS. Anal Chem 72: 800–806.

Zheng J, Ma L, Xin B, Olah T, Humphreys WG, Zhu M. 2007. Screening and identification of GSH-trapped reactive metabolites using hybrid triple quadruple linear ion trap mass spectrometry. Chem Res Toxicol 20: 757–766.

Zhao P, Huang B, Li Y, Han Y, Zou N, Gu K, Li X, Pan C. 2014. Rapid multiplug filtration cleanup with multiple-walled carbon nanotubes and gas chromatography–triple-quadrupole mass spectrometry detection for 186 pesticide residues in tomato and tomato products. J Agric Food Chem 62: 3710–3725.

Zhu M, Ma L, Zhang D, Ray K, Zhao W, Humphreys WG, Skiles G, Sanders M, Zhang H. 2006. Detection and characterization of metabolites in biological matrices using mass defect filtering of liquid chromatography/high resolution mass spectrometry data. Drug Metab Dispos 34: 1722–1733.

Zhu M, Ma L, Zhang H, Humphreys WG. 2007, Detection and structural characterization of glutathione-trapped reactive metabolites using liquid chromatography-high-resolution mass spectrometry and mass defect filtering. Anal Chem 79: 8333–8341.

Zhu M, Zhang H, Humphreys WG. 2011. Drug metabolite profiling and identification by high-resolution mass spectrometry. J Biol Chem 286: 25419–25425.

Zhu X, Chen Y, Subramanian R. 2014. Comparison of information-dependent acquisition, SWATH, and MS(All) techniques in metabolite identification study employing ultrahigh-performance liquid chromatography-quadrupole time-of-flight mass spectrometry. Anal Chem 86: 1202–1209.

Zuccato E, Chiabrando C, Castiglioni S, Bagnati R, Fanelli R. 2008. Estimating community drug use by sewage analysis. Environ Health Perspect 116: 1027–1032.

COMPOUND INDEX

Interpretation of MS–MS Mass Spectra of Drugs and Pesticides, First Edition. Wilfried M. A. Niessen and Ricardo A. Correa C.

SUBJECT INDEX

Interpretation of MS–MS Mass Spectra of Drugs and Pesticides, First Edition. Wilfried M. A. Niessen and Ricardo A. Correa C.

WILEY SERIES ON MASS SPECTROMETRY

John R. de Laeter · *Applications of Inorganic Mass Spectrometry*

Michael Kinter and Nicholas E. Sherman · *Protein Sequencing and Identification Using Tandem Mass Spectrometry*

Chhabil Dass · *Principles and Practice of Biological Mass Spectrometry*

Mike S. Lee · *LC/MS Applications in Drug Development*

Jerzy Silberring and Rolf Eckman · *Mass Spectrometry and Hyphenated Techniques in Neuropeptide Research*

J. Wayne Rabalais · *Principles and Applications of Ion Scattering Spectrometry: Surface Chemical and Structural Analysis*

Mahmoud Hamdan and Pier Giorgio Righetti · *Proteomics Today: Protein Assessment and Biomarkers Using Mass Spectrometry, 2D Electrophoresis, and Microarray Technology*

Igor A. Kaltashov and Stephen J. Eyles · *Mass Spectrometry in Structural Biology and Biophysics: Architecture, Dynamics, and Interaction of Biomolecules, Second Edition*

Isabella Dalle-Donne, Andrea Scaloni, and D. Allan Butterfield · *Redox Proteomics: From Protein Modifications to Cellular Dysfunction and Diseases*

Silas G. Villas-Boas, Ute Roessner, Michael A.E. Hansen, Jorn Smedsgaard, and Jens Nielsen · *Metabolome Analysis: An Introduction*

Mahmoud H. Hamdan · *Cancer Biomarkers: Analytical Techniques for Discovery*

Chabbil Dass · *Fundamentals of Contemporary Mass Spectrometry*

Kevin M. Downard (Editor) · *Mass Spectrometry of Protein Interactions*

Nobuhiro Takahashi and Toshiaki Isobe · *Proteomic Biology Using LC-MS: Large Scale Analysis of Cellular Dynamics and Function*

Agnieszka Kraj and Jerzy Silberring (Editors) · *Proteomics: Introduction to Methods and Applications*

Ganesh Kumar Agrawal and Randeep Rakwal (Editors) · *Plant Proteomics: Technologies, Strategies, and Applications*

Rolf Ekman, Jerzy Silberring, Ann M. Westman-Brinkmalm, and Agnieszka Kraj (Editors) · *Mass Spectrometry: Instrumentation, Interpretation, and Applications*

Christoph A. Schalley and Andreas Springer · *Mass Spectrometry and Gas-Phase Chemistry of Non-Covalent Complexes*

Riccardo Flamini and Pietro Traldi · *Mass Spectrometry in Grape and Wine Chemistry*

Mario Thevis · *Mass Spectrometry in Sports Drug Testing: Characterization of Prohibited Substances and Doping Control Analytical Assays*

Sara Castiglioni, Ettore Zuccato, and Roberto Fanelli · *Illicit Drugs in the Environment: Occurrence, Analysis, and Fate Using Mass Spectrometry*

Ángel Garciá and Yotis A. Senis (Editors) · *Platelet Proteomics: Principles, Analysis, and Applications*

Luigi Mondello · *Comprehensive Chromatography in Combination with Mass Spectrometry*

Jian Wang, James MacNeil, and Jack F. Kay (Editors) · *Chemical Analysis of Antibiotic Residues in Food*

Walter A. Korfmacher (Editor) · *Mass Spectrometry for Drug Discovery and Drug Development*

Toshimitsu Niwa (Editor) Uremic Toxins

Igor A. Kaltashov, Stephen J. Eyles Mass Spectrometry in Structural Biology and Biophysics: Architecture, Dynamics, and Interaction of Biomolecules, 2nd Edition

Alejandro Cifuentes (Editor) · *Foodomics: Advanced Mass Spectrometry in Modern Food Science and Nutrition*

Christine M. Mahoney (Editor) · *Cluster Secondary Ion Mass Spectrometry: Principles and Applications*

Despina Tsipi, Helen Botitsi, and Anastasios Economou (Editors) · *Mass Spectrometry for the Analysis of Pesticide Residues and their Metabolites*

Xianlin Han · *Lipidomics: Comprehensive Mass Spectrometry of Lipids*

Jack F. Kay, James D. MacNeil, and JianWang (Editors) · *Chemical Analysis of Non-antimicrobial Veterinary Drug Residues in Food*

Wilfried M. A. Niessen and Ricardo A. Correa C. · *Interpretation of MS-MS Mass Spectra of Drugs and Pesticides*

John R. Griffiths and Richard D. Unwin (Editors) · *Analysis of Protein Post-Translational Modifications by Mass Spectrometry*

Henk Schierbeek · *Mass Spectrometry and Stable Isotopes in Nutritional and Pediatric Research*